ADVANCED BUILDING & JOINERY SKILLS

ADVANCED BUILDING & JOINERY SKILLS

Greg Cheetham

CONSTRUCTION + PLUMBING Skills Series

4e

Advanced Building and Joinery Skills
4th Edition
Greg Cheetham

Portfolio manager: Sophie Kaliniecki
Product manager: Sandy Jayadev
Senior content developer: Kylie Scott
Senior project editor: Nathan Katz
Text designer: Cengage Creative Studio
Cover designer: Cengage Creative Studio
Permissions/Photo researcher: Helen Mammides
Editor: Pete Cruttenden
Proofreader: James Anderson
Indexer: KnowledgeWorks Global Ltd.
Art direction: Danielle Maccarone
Cover: Illustration: Studio Fable/Antonia Pesenti
Typeset by KnowledgeWorks Global Ltd.

This fourth edition published in 2024

For product information and technology assistance,
in Australia call **1300 790 853**;
in New Zealand call **0800 449 725**

For permission to use material from this text or product, please email
aust.permissions@cengage.com

National Library of Australia Cataloguing-in-Publication Data
ISBN: 9780170462853
A catalogue record for this book is available from the National Library of Australia.

Cengage Learning Australia
Level 5, 80 Dorcas Street
Southbank VIC 3006 Australia

For learning solutions, visit **cengage.com.au**

Printed in China by 1010 Printing International Limited.
1 2 3 4 5 6 7 27 26 25 24 23

BRIEF CONTENTS

CONTENTS

Guide to the text

As you read this text you will find a number of features in every chapter to enhance your study of Advanced Building and Joinery Skills and help you understand how the theory is applied in the real world.

PART-OPENING FEATURES

Part openers introduce each of the chapters within the part and give an overview of how the chapters in the text relate to each other

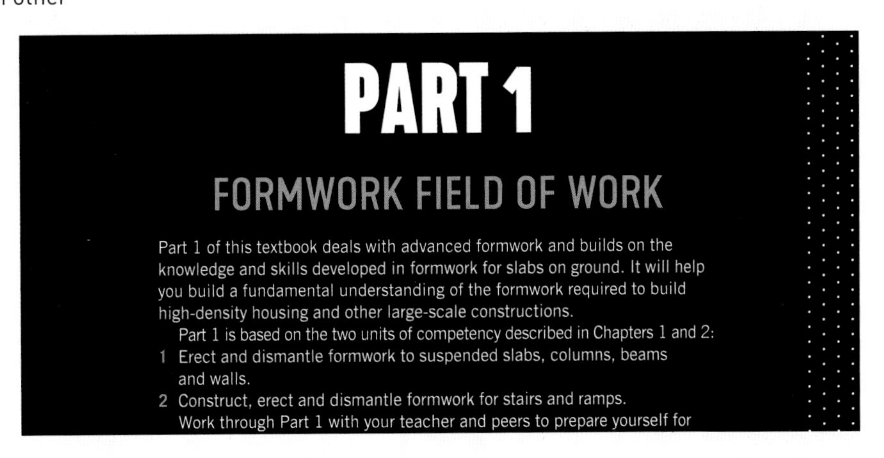

CHAPTER-OPENING FEATURES

Identify the key topics and competencies covered along with the Learning Objectives at the start of each chapter. They will indicate what you should be able to do after reading the chapter within the part and give an overview of how the chapters in the text relate to each other.

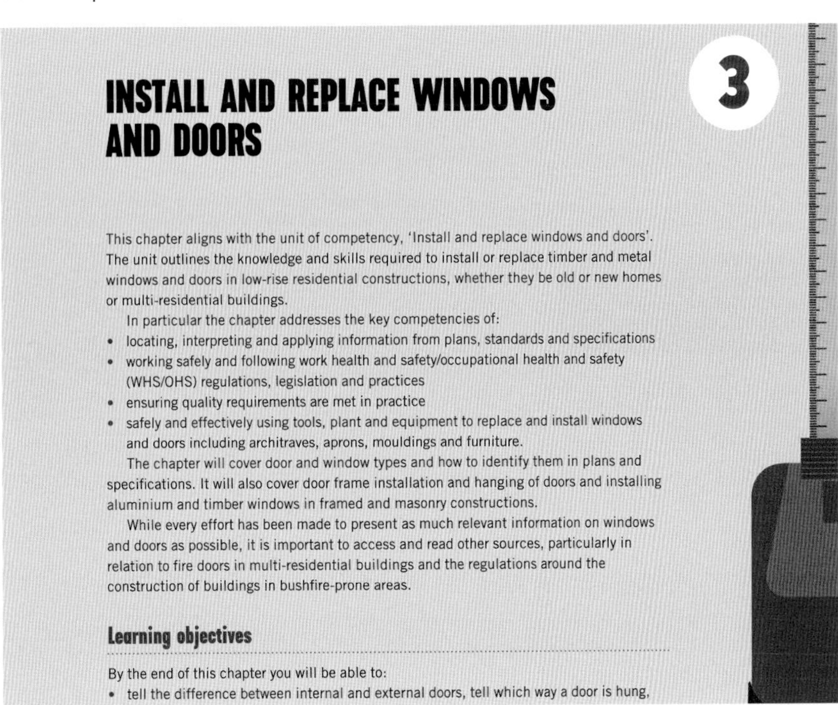

FEATURES WITHIN CHAPTERS

Learning Tasks encourage you to practically apply the knowledge and skills that you have just read about.

LEARNING TASK 1.2

FORMWORK MATERIALS COMPLIANCE ISSUES

Obtain a copy of the Engineered Wood Products Association of Australasia's 'Non-conforming building products Submission 12'. Read the executive summary and then identify and discuss the potential issues for builders using non-compliant formply. https://www.aph.gov.au/Parliamentary_Business/Committees/Senate/Economics/Non-conforming_products/Submissions

How To boxes provide step-by-step instructions on how to perform specific tasks/processes.

HOW TO 1

HOW TO

SET UP A FORMWORK FOR A SIMPLE SUSPENDED CONCRETE SLAB

Before beginning any formwork construction work, make sure all necessary formwork plans, hazard identification and risk management plans have been completed and approved by site supervisors or engineers.

Steps involved		
1	Set out the sole plates on which the formwork frames and props will sit. You can only do this based on the formwork plan or grid and the system being used. If you use long lengths the loads will be spread over greater distance and make setting out easier. See **Figures 1.26** and **1.27**. A scaffold can be erected in lieu of extending formwork to provide a working platform. **Note:** Large floor areas are rarely poured all at once.	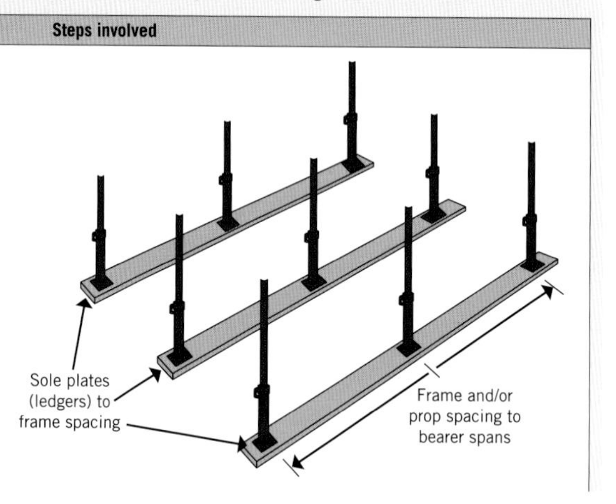

ICONS

GREEN TIP

It is always good practice to source products that are sustainable and environmentally friendly.

Green Tip boxes highlight material that relates to environmentally-sustainable workplace practices.

Stripping formwork can be dangerous, particularly soffit formwork. Make sure you wear a hard hat and other PPE to protect yourself from falling formwork sheets when working from below.

Caution boxes highlight material relating to workplace health and safety.

COMPLETE WORKSHEET 4

Worksheet icons indicate when it is appropriate to stop reading and complete a worksheet at the end of the chapter.

END-OF-CHAPTER FEATURES

At the end of each chapter you will find several tools to help you to review, practise and extend your knowledge of the key learning objectives.

Chapter summaries highlight the important concepts covered in each chapter as well as link back to the key competencies.

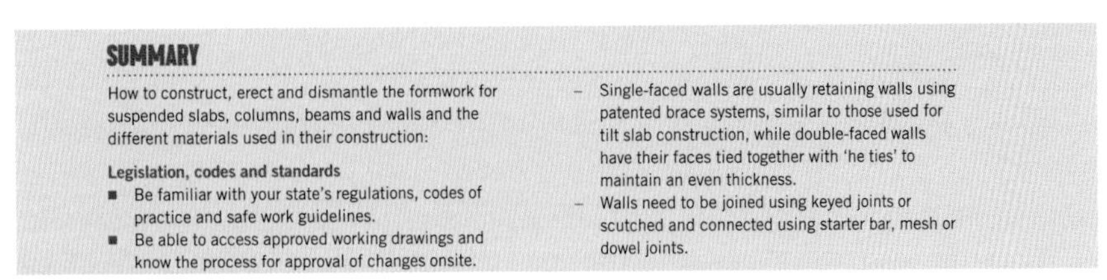

SUMMARY

How to construct, erect and dismantle the formwork for suspended slabs, columns, beams and walls and the different materials used in their construction:

Legislation, codes and standards

- Be familiar with your state's regulations, codes of practice and safe work guidelines.
- Be able to access approved working drawings and know the process for approval of changes onsite.

– Single-faced walls are usually retaining walls using patented brace systems, similar to those used for tilt slab construction, while double-faced walls have their faces tied together with 'he ties' to maintain an even thickness.

– Walls need to be joined using keyed joints or scutched and connected using starter bar, mesh or dowel joints.

The **References and Further Reading** sections provide you with a list of each chapter's references, as well as links to important text and web-based resources.

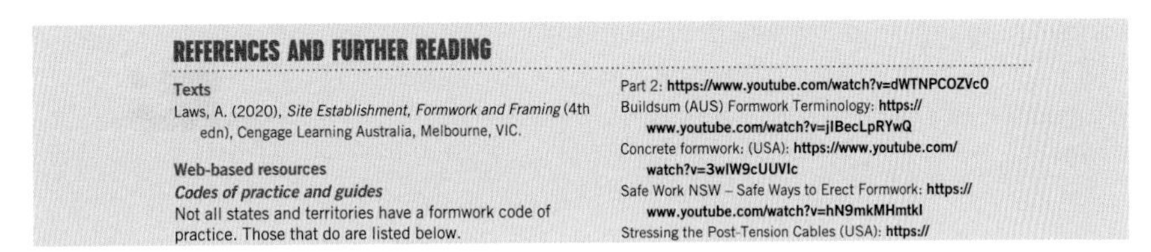

REFERENCES AND FURTHER READING

Texts

Laws, A. (2020), *Site Establishment, Formwork and Framing* (4th edn), Cengage Learning Australia, Melbourne, VIC.

Web-based resources

Codes of practice and guides

Not all states and territories have a formwork code of practice. Those that do are listed below.

Part 2: **https://www.youtube.com/watch?v=dWTNPCOZVcO**

Buildsum (AUS) Formwork Terminology: **https://www.youtube.com/watch?v=jlBecLpRYwQ**

Concrete formwork: (USA): **https://www.youtube.com/watch?v=3wlW9cUUVlc**

Safe Work NSW – Safe Ways to Erect Formwork: **https://www.youtube.com/watch?v=hN9mkMHmtkl**

Stressing the Post-Tension Cables (USA): **https://**

The **Get It Right** photo case study shows an incorrect technique or skill and encourages you to identify the correct method and provide reasoning.

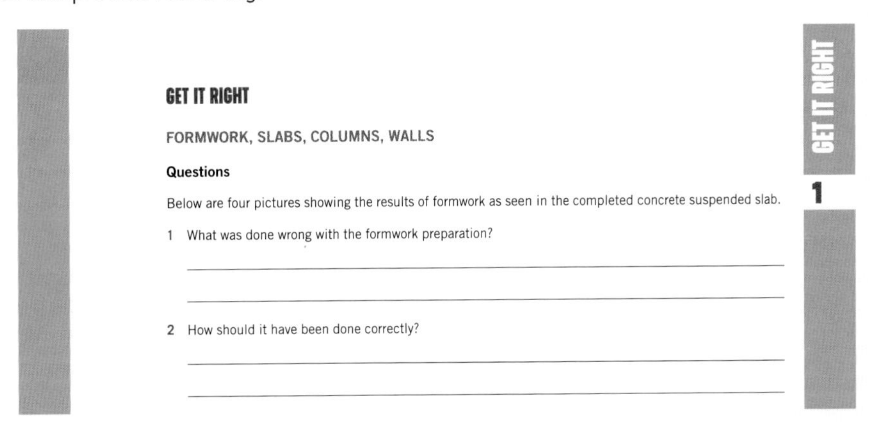

GET IT RIGHT

FORMWORK, SLABS, COLUMNS, WALLS

Questions

Below are four pictures showing the results of formwork as seen in the completed concrete suspended slab.

1 What was done wrong with the formwork preparation?

2 How should it have been done correctly?

GET IT RIGHT 1

Worksheets help assess your understanding of the theory and concepts in each chapter.

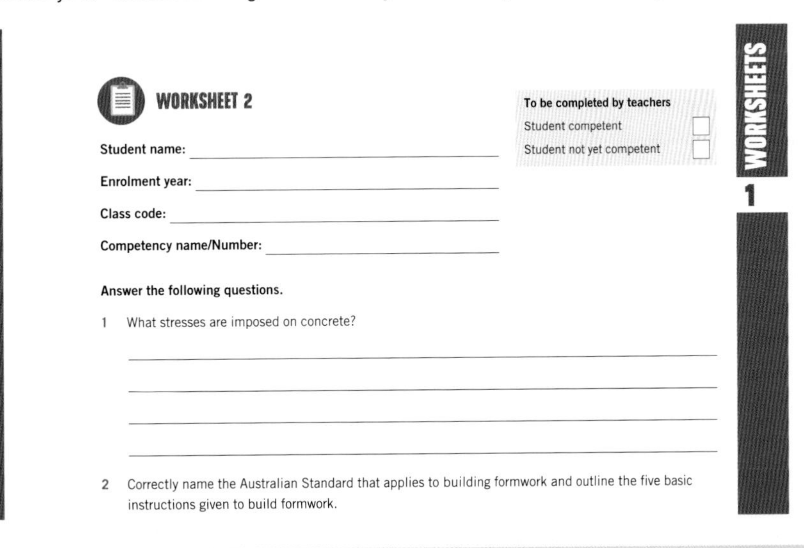

WORKSHEET 2

To be completed by teachers

Student competent ☐

Student not yet competent ☐

Student name: ______

Enrolment year: ______

Class code: ______

Competency name/Number: ______

Answer the following questions.

1 What stresses are imposed on concrete?

2 Correctly name the Australian Standard that applies to building formwork and outline the five basic instructions given to build formwork.

WORKSHEETS 1

Guide to the online resources

FOR THE INSTRUCTOR

MINDTAP

Premium online teaching and learning tools are available on the *MindTap* platform – the personalised eLearning solution.

MindTap is a flexible and easy-to-use platform that helps build student confidence and gives you a clear picture of their progress. We partner with you to ease the transition to digital – we're with you every step of the way.

MindTap for **Certificate III in Carpentry** is full of innovative resources to support critical thinking, and help your students move from memorisation to mastery! Includes:

- Basic Building and Construction Skills, 7e eBook
- Site Establishment, Footings and Framework, 5e eBook
- Advanced Building and Joinery Skills, 4e eBook
- Construction Skills, 4e eBook.
- Instructional Videos
- Worksheets
- Revision Quizzes
- And more!

MindTap is a premium purchasable eLearning tool. Contact your Cengage learning consultant to find out how *MindTap* can transform your course.

INSTRUCTOR RESOURCE PACK

Premium resources that provide additional instructor support are available for this text, including Mapping Grid, Worksheets, Testbank, PowerPoints, Solutions Manual and more. These resources save you time and are a convenient way to add more depth to your classes, covering additional content and with an exclusive selection of engaging features aligned with the text.

SOLUTIONS MANUAL

The **Solutions Manual** includes solutions to Learning Tasks, end-of-chapter Worksheets and Get It Right case studies.

POWERPOINT™ PRESENTATIONS

Use the chapter-by-chapter **PowerPoint slides** to enhance your lecture presentations and handouts by reinforcing the key principles of your subject.

ARTWORK FROM THE TEXT

Add the digital files of graphs, pictures and flow charts into your course management system, use them in student handouts, or copy them into your lecture presentations.

WORD-BASED TEST BANK

This bank of questions has been developed in conjunction with the text for creating quizzes, tests and exams for your students. Deliver these through your LMS and in your classroom.

COMPETENCY MAPPING GRID

The downloadable Competency Mapping Grid demonstrates how the text aligns to the Certificate III in Carpentry.

The Instructor Resource Pack is included for institutional adoptions of this text when certain conditions are met. The pack is available to purchase for course-level adoptions of the text or as a standalone resource. Contact your Cengage learning consultant for more information.

FOR THE STUDENT

MINDTAP

MindTap is the next-level online learning tool that helps you get better grades!

MindTap gives you the resources you need to study – all in one place and available when you need them. In the MindTap Reader, you can make notes, highlight text and even find a definition directly from the page.

If your instructor has chosen *MindTap* for your subject this semester, log in to *MindTap* to:

- Get better grades
- Save time and get organised
- Connect with your instructor and peers
- Study when and where you want, online and mobile
- Complete assessment tasks as set by your instructor.

When your instructor creates a course using *MindTap*, they will let you know your course link so you can access the content. Please purchase MindTap only when directed by your instructor. Course length is set by your instructor.

PREFACE

Advanced Building and Joinery Skills sets out to provide the basic knowledge and skills required in advanced carpentry and joinery across a range of specialist subject areas. Subjects covered include wall lining and cladding, wet area construction, stairs, doors and windows, and advanced roofing subjects. The advanced concrete construction subjects of formwork for columns, beams, suspended slabs and walls, and concrete stairs have also been covered.

In developing this book, a clear eye was set on providing a textbook that incorporated modern technologies mixed with classic skills and techniques. The book supports teachers and students with relevant worksheets containing questions and work projects associated with each chapter. This edition has been updated with advice on changes to the National Construction Code 2022 and new sections on pre- and post-tensioned concrete, revised worksheets and updated links to more video resources with an emphasis on sustainable building practices. The fourth edition of this book has been a further investment of love over the past year and it could not have been completed without the love and support of my wife, Lindy, and children, Milly and Tom.

I would like to acknowledge the help given by my father, George Cheetham. His 60-plus years as a builder and NSW Builders Licensing Board inspector was a valuable backstop when my own personal experience and knowledge were tested.

I would like to thank Chris Montgomery (Greenpoint Construction) for his time in reviewing various chapters and providing relevant photographs and graphics to support these chapters. His construction knowledge and experience has provided a valuable contribution to this work.

Pete's Rule (Peter Costin 1930–1999)

Pete's Rule (which is quoted in this book) was originally adapted and used by a renowned joiner from Geelong, Peter Costin. Peter's 'rule of thumb' provides quick, accurate solutions to calculating enough length of a stair stringer, incorporating allowances for joints and waste.

Greg Cheetham

Technical artwork colour palette

This textbook incorporates a range of artwork that has been coloured in line with the Australian Building Standards. Additionally, each coloured image includes a legend that identifies each building material colour. At the time of publication, every effort had been made to colour materials in line with the palette provided in the Australian Standards. We recommend consulting the most current edition of the Standards, and your employer or institution, prior to implementing a colour scheme for any project.

National Construction Code 2022

The National Construction Code (NCC) of Australia (2022) includes mandatory minimum accessible housing provisions for all new classes of buildings where required. A building will be required to provide, as far as is reasonably practical, safe, equitable and dignified access for people to the services and facilities of a building. Effectively, this will mean proposed minimum design requirements will influence the design of homes, including (among other things) movement into and inside the dwelling, comprising:

- a safe continuous path of travel from the street entrance and/or parking area to a dwelling entrance (i.e. stairs and ramps)
- at least one step-free entrance door
- wider internal doors and corridors (this is further discussed in Part 2 of this book)
- stairways designed to reduce likelihood of injury.

For Class 1a (and Class 10) residential houses, the performance requirements are outlined in the National Construction Code 2022 – Volume Two, Part H8 Livable housing design. All Class 1a dwellings must comply with the Australian Building Codes Board's Standard for Livable Housing Design or meet appropriate performance requirements, except in NSW where the Building Sustainability Index (BASIX) is the measure of sustainability and energy efficiency in Class 1 and Class 1a residential dwellings.

National Construction Code 2022 – Volume One, Part G7 Livable housing design outlines the requirements for all new Class 2 sole-occupancy units (Note: For Tasmania, Part G7 does not take effect until 1 October 2024). These must be designed so they are easy to enter and navigate around and/or be capable of cost-effective renovations to accommodate people with mobility issues.

The NCC also references AS 1428.1 Design for access and mobility as a guide for building practitioners, particularly for Class 2 to 9 buildings. The Standard relates directly to the accessibility of the built environment for people with a disability and specifies the design requirements for new building work, as required by the Building Code of Australia (BCA).

ABOUT THE AUTHOR

Greg Cheetham has over 40 years' experience working in the building, construction and detailed joinery industries, and in the vocational education sector (29 years), training apprentices in detailed joinery trades. Greg remains involved in Construction Training Package development, working with the BuildSkills Jobs and Skills Council. He is an elected representative on the Acceptable Standards of Construction Committee.

ACKNOWLEDGEMENTS

The publisher would like to thank the following for their constructive feedback during the development of this fourth edition:

- David Leete, SkillsTech Australia
- John Friedman, Sydney TAFE
- James Folk, Canberra Institute of Technology
- Josef Fritzer, TAFE NSW
- Carpentry Teachers Wollongong TAFE
- Jim Gott
- Ed Hawkins
- Gordon Slater.

Every effort has been made to trace and acknowledge copyright. However, if any infringement has occurred, the publishers tender their apologies and invite the copyright holders to contact them.

LIST OF FIGURES

COLOUR PALETTE FOR TECHNICAL DRAWINGS

Colour name	Colour	Material
Light Chrome Yellow		Cut end of sawn timber
Chrome Yellow		Timber (rough sawn), Timber stud
Cadmium Orange		Granite, Natural stones
Yellow Ochre		Fill sand, Brass, Particle board, Highly moisture-resistant particle board (Particle board HMR), Timber boards
Burnt Sienna		Timber – Dressed All Round (DAR), Plywood
Vermilion Red		Copper pipe
Indian Red		Silicone sealant
Light Red		Brickwork
Cadmium Red		Roof tiles
Crimson Lake		Wall and floor tiles
Very Light Mauve		Plaster, Closed cell foam
Mauve		Marble, Fibrous plasters
Very Light Violet Cake		Fibreglass
Violet Cake		Plastic
Cerulean Blue		Insulation
Cobalt Blue		Glass, Water, Liquids
Paynes Grey		Hard plaster, Plaster board
Prussian Blue		Metal, Steel, Galvanised iron, Lead flashing
Lime Green		Fibrous cement sheets
Terra Verte		Cement render, Mortar
Olive Green		Concrete block
Emerald Green		Terrazzo and artificial stones
Hookers Green Light		Grass
Hookers Green Deep		Concrete
Raw Umber		Fill
Sepia		Earth
Van Dyke Brown		Rock, Cut stone and masonry, Hardboard
Very Light Raw Umber		Medium Density Fibreboard (MDF), Veneered MDF
Very Light Van Dyke Brown		Timber mouldings
Light Shaded Grey		Aluminium
Neutral Tint		Bituminous products, Chrome plate, Alcore
Shaded Grey		Tungsten, Tool steel, High-speed steel
Black		Polyurethane, Rubber, Carpet
White		PVC pipe, Electrical wire, Vapour barrier, Waterproof membrane

PART 1

FORMWORK FIELD OF WORK

Part 1 of this textbook deals with advanced formwork and builds on the knowledge and skills developed in formwork for slabs on ground. It will help you build a fundamental understanding of the formwork required to build high-density housing and other large-scale constructions.

Part 1 is based on the two units of competency described in Chapters 1 and 2:

1 Erect and dismantle formwork to suspended slabs, columns, beams and walls.
2 Construct, erect and dismantle formwork for stairs and ramps.

Work through Part 1 with your teacher and peers to prepare yourself for the practical skills exercises you will be required to undertake.

The key elements in Part 1 are:

- Planning and preparing for formwork by selecting and quantifying appropriate tools, equipment and material; performing risk assessments; and applying the National Construction Code and Australian Standards (AS) and safe work methods in all activities including planning for stripping.
- Setting out, assembling and erecting the formwork by:
  - checking for accuracy against plans, designs, specifications and codes
  - accurately placing, fixing and bracing formwork structures.
- stripping formwork safely in correct sequence ensuring adequate back propping when and where required according to formwork plans.
- cleaning up including de-nailing, recycling practices, stacking and preparation for safe site removal or transition; and checking tools and equipment for safe use, operability, maintenance and fault rectification.

The learning outcomes for each chapter are a good indicator of what will be required to perform, know, understand and apply on completion of each chapter. Teachers and students should discuss the full practical knowledge and evidence requirements of each unit of competency before undertaking any activities.

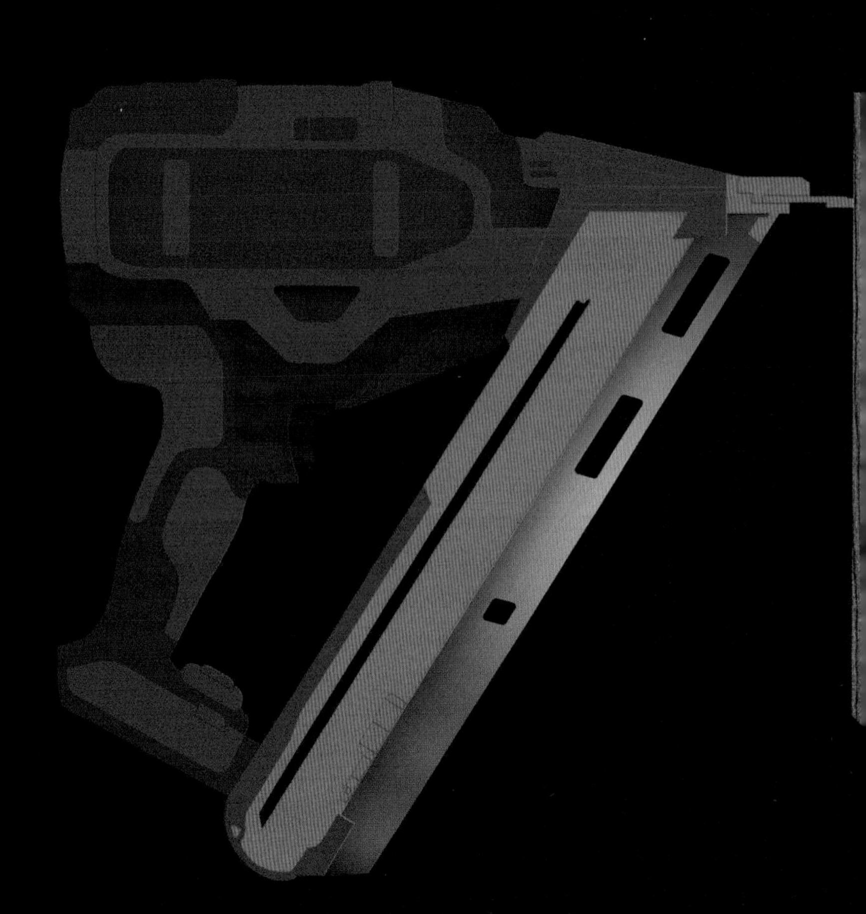

1

FORMWORK FOR SLABS, COLUMNS, BEAMS AND WALLS

This chapter aligns with the unit of competency, 'Erect and dismantle formwork to suspended slabs, columns, beams and walls'. The unit outlines the knowledge and skills required to construct, erect and dismantle formwork for suspended slabs, beams, columns and walls in accordance with Australian Standard 3610 Formwork for concrete and the relevant state or national codes of practice for formwork.

Chapter 1 builds on the knowledge and skill developed in Chapter 4 of *Site Establishment, Formwork and Framing* 5e (Laws 2023) where we looked at the excavation and formwork required for constructing an on-ground slab and strip footings.

For this chapter, you will need a basic understanding of large-scale construction systems used for multistorey buildings (office buildings and shopping centres), industrial buildings (factories) and infrastructure (roads, bridges, dams, etc.). You will also need an understanding of the various Australian Standards and codes of practice that apply to them.

Learning objectives

By the end of this chapter you will be able to:

- know the Australian Standards, regulations, codes of practice and Safe Work Australia guides and documentation related to safety issues associated with setting out and constructing formwork
- understand the materials and impacts of applied stresses on formwork for concrete slabs, columns, beams and walls
- construct and erect formwork/falsework; identify, make and use formwork/falsework components; understand bracing requirements; and establish a pouring pattern. You will also learn about calculating loads, lateral forces and the different types of column and beam construction
- strip formwork/falsework safely by applying the correct procedure, timing and sequence. This includes the need to back prop and reshore as part of the process when necessary
- use your knowledge of modular patented formwork systems to help select the appropriate formwork system for jobs.

Legislation, codes and standards

When building any structure there are certain legal requirements that need to be taken into consideration. The legislation, codes of practice and standards work together to provide a framework of advice and regulation backed by consequences and penalties in order to ensure safe and effective construction takes place. Under work health and safety legislation, everyone has a duty of care to each other. Specific responsibilities for formwork include:

1. the persons that design, certify, construct and dismantle the formwork
2. designers, manufacturers, suppliers and importers of specialist formwork systems
3. other site workers and visitors directly or indirectly involved with the formwork process (e.g. carpenters, plumbers and electricians).

See your state's code of practice for details.

Codes of practice

Most states in Australia have a comprehensive code of practice and guides for formwork. The websites for each state's code of practice are available at the end of this chapter.

The codes of practice and guides generally deal with safety-related issues concerning the installation and removal of **formwork/falsework**. They provide guidance on how to achieve appropriate health, safety and welfare standards when building formwork/falsework and they should be followed unless a better system or procedure is available. By not following the code of practice and guides for your state, you leave yourself open to prosecution or penalty.

The codes of practice and guides recommend that the prime contractor should consult with formwork/falsework contractors, employers and subcontractors to determine and plan the requirements of the job in accordance with the relevant code of practice. Consultation should consider the following points:

- the nature of the work being undertaken
- the type, height and size of formwork/falsework to be used
- the size of the deck
- the availability of equipment
- interactions with other trades
- workplace access
- public safety
- the location of intermediate working decks.

The codes of practice and guides also set out the responsibilities of the principal and subcontractors and the formwork/falsework designers. Before any work is done, a risk assessment needs to be conducted, identifying the risks and appropriate control measures for erecting the formwork/falsework, placement of reinforcement/steel, pouring of concrete, and stripping and removal of the formwork/falsework. This can be particularly important for stairs, their landings and raked concrete forms because of the eccentric (offset) loads involved.

There are a number of SafeWork NSW guidelines superseding the earlier Code of Practice for Formwork. Safe Work Australia's General Guide for Formwork and Falsework states, 'in situations where the next formwork deck is at a height that would require people to stand at heights of 2 m or more above the finished formwork deck to install bearers and joists for the next formwork deck, a continuous "false" deck, which is a full deck the same area as the floor being formed, should be provided' (Safe Work Australia, CC BY 4.0, Licence https://creativecommons.org/licenses/by/4.0/). As such, you are advised to check your state's code of practice and guides for the latest information.

Various methods are available to meet these criteria, such as providing a plank to a minimum of 450 mm wide, providing guard rails and handrails, or using fall arrest harness systems. See **Figure 1.1** and the SafeWork NSW Construction Work Code of Practice, and Formwork guide for more details.

Guard rails and handrails must finish 900 mm above the finished height of the concrete, whether the working platform is integrated in the formwork/falsework or is a separate scaffold platform.

There are many other points to know and understand in the codes of practice, so make sure you access a copy of your state's code of practice and keep it handy when building formwork/falsework.

State and Federal laws also apply to the Environmental requirements. Each site is required to have an Environmental Plan that is communicated to all on site. It can include information about waste, recycling, reuse, concerns for abatement that may impact local waterways, and the use of tools and equipment in relation to dust, noise and vibration.

Where possible, erect formwork for columns and beams and walls on the ground, and lift into position. Do not climb on formwork.

Safe Work Australia

Safe Work Australia was established with the primary responsibility to improve workplace health and safety and workers' compensation arrangements across Australia.

The WHS/OHS legislation in each state will vary but remains the primary legislation applicable to the construction industry in each state and territory, with the harmonisation rules applying to each state to ensure that a minimum uniform set of regulations is applied.

In NSW, the Work Health and Safety Regulations 2017 applies – in particular Chapters 5 and 6 of the regulations, which deal specifically with the construction industry and persons conducting a business or

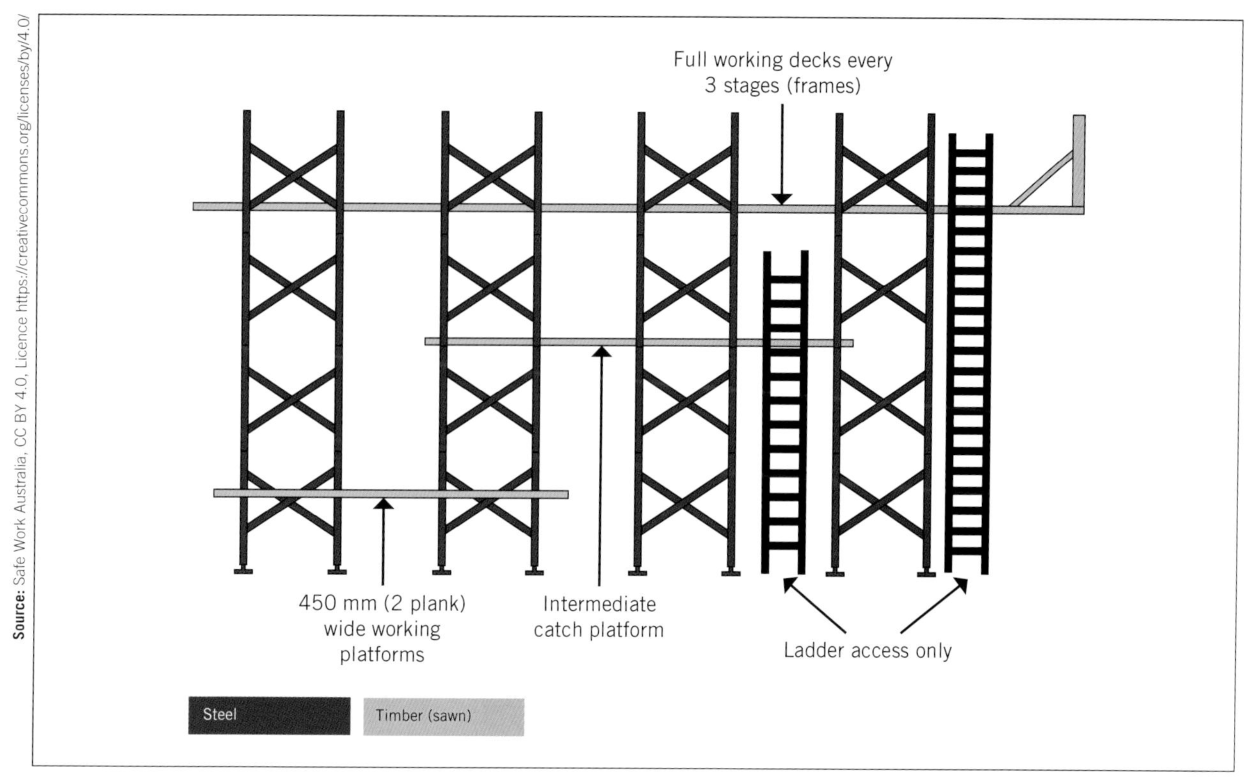

FIGURE 1.1 Formwork set out according to the SafeWork NSW *Code of Practice: Formwork*

undertaking (PCBUs). Students in other states should familiarise themselves with corresponding legislation in each state. The relevant legislation in other states is:

- Victoria – *Occupational Health and Safety Act 2004*
- South Australia – *Work Health and Safety Act 2012*
- Queensland – *Work Health and Safety Act 2011*
- Western Australia – *Work Health and Safety Act 2020*
- Tasmania – *Work Health and Safety Act 2012*
- Northern Territory and Australian Capital Territory – both come under the Federal *Work Health and Safety Act 2012*.

AS 3610.1 Formwork for concrete

AS 3610.1 Formwork for concrete is the primary standard for formwork for concrete and sets out in detail the 'requirements for the design, fabrication, erection and stripping of formwork', including testing.

In particular, formwork should be constructed and dismantled in such a way as to comply with the requirements of Section 4.6.3 of the Standard; that is, to:

- resist movement and misalignment at construction joints in the concrete
- ensure proprietary components (i.e. props, scaffold frames, tube and couplings, and steel or metal formwork systems) are used in accordance with the manufacturer's directions
- be adjustable and permit the controlled movement of formwork during stripping
- allow some movement but prevent catastrophic collapse and dislodgement, or become otherwise unstable
- be braced, built and dismantled safely by using **staging** platforms, guard rails and handrails.

There are a number of supplements and amendments to the Standard that may also impact on your work with formwork/falsework and concrete.

Documentation

According to AS 3610.1 Formwork for concrete, Part 1: Documentation and surface finish, all formwork must have documentation that includes the following basic information:

- working drawings sufficient to depict the general arrangement of the formwork, including bracing
- any important or unusual features of the design
- reference to documentation for proprietary items
- staging and stacked load areas (stacked loads are the first concrete placement points)
- method of provision for field adjustment of the forms prior to and during placement of the concrete
- location of weep holes, vibrator holes, clean-out holes and inspection points
- sequence of concrete placement
- wrecking strips and other details relating to stripping of the forms
- any testing regime required
- footing and foundation conditions and bearing capacities.

The Standard also considers live and **dead loadings** such as the placement of concrete on the forms, the weight of the formwork components, horizontal impact loads (i.e. accidental collision and impacts) and wind loadings. Calculating the loads will be discussed later in this chapter.

Stages of construction

The Standard also identifies the three main stages of formwork construction and removal as being:

- Stage 1: Prior to placement of concrete (including handling and erection of the formwork)
- Stage 2: During the placement of concrete (where continuous observation of formwork performance is monitored for safety and movement or potential catastrophic collapse)
- Stage 3: After placement of concrete, and during the curing and setting process.

Inspecting the formwork

The formwork must be inspected at the end of erection and before reinforcement and concrete placement (Stage 2) to make sure the formwork complies with the requirements of the Australian Standard, the formwork and project documentation, and safe work practices; that it is properly braced, adjustable and strippable, and adequately connected; and that it can achieve the required surface finish.

If formwork/falsework is not constructed or inspected properly it may collapse during the Stage 2 phase of placement in concrete.

Stripping formwork

The stripping of formwork has to be done in a controlled manner so the loads can be gradually transferred from formwork to structure and footings. Note that in multistorey work, stripping must be done in a sequence to ensure the new slab remains safe to work on while the concrete is curing. Back propping may be required. See the section on stripping formwork on page 26. Loads on formwork come from the weight of the components used in constructing the formwork, reinforcement and the concrete itself (wet and cured), and **live loads** such as people, tool machinery, machinery and equipment.

Quality assurance

Formwork is a moulding process and, as such, any defects present in the mould will be passed to the finished surface of the concrete. Knots and splits in plywood surface veneers will be visible. Nail heads and bent-over nails will show, as will hammer blows. This may not be a problem if the concrete is to be covered up by render or some other surface finish, but when left exposed the defects will be clearly visible.

The classes of concrete surface finish are:

- Class 1: The highest standard and is suitable for selected small elements or areas of special importance in limited quantities.
- Class 2: A high quality finish specified for most good quality architectural concrete finishes usually on external and internal façades that can be viewed in detail.
- Class 3: Specified where visual aspects are important, but have less architectural qualities. The differences between 2 and 3 are the type, number and dimensions of permitted surface defects.
- Classes 4 and 5: The visual quality is not critical, and the finish surfaces are concealed.

There are specialty surface finishes available for concrete, such as etched finishes made by **scabbling** and animal forms set in shape using rubber or foam moulds. Exposed **aggregate** and polished concrete surfaces are especially popular today and these require specialised services to achieve the appropriate finishes.

To control quality in formwork, make sure all joints and sheet edges are tight to prevent the wet concrete slurry seeping through gaps. This leakage forms fins, which will need to be cleaned off, and because of the loss of the cement product, a honeycombing effect may show up in the surface of the concrete when the formwork is removed. You can also use a low-density self-adhesive open-cell foam strip to seal edges. In all cases ensure joints are well fitted first.

Obtain copies of Cement Concrete & Aggregates Australia's *Concrete Basics* and *Guide to Off-form Concrete Finishes* for more information on concrete surface finishes and quality assurance issues (see the website link at the end of this chapter). **Note:** It is free to sign up to access documents.

Remember, formwork needs to be inspected and in specified circumstances signed off by an engineer after Stage 1 and before Stage 2 of the construction (i.e. before the concrete is placed).

Using the correct release agent is also a quality assurance issue. Using the wrong release agent can cause staining and discolouration of the finished surface, while insufficient release agent can prevent form faces separating easily from the surface.

LEARNING TASK 1.1

AS 3610 AND CODES OF PRACTICE

For this activity you will need to obtain a copy of the relevant code of practice for formwork or guidance materials provided by your state's SafeWork regulator.

Discuss the documentation required for the erection and dismantling of formwork/falsework and then answer the following questions:

1. Using an example, explain the difference between formwork and falsework (you can use pictures or photographs, properly labelled).
2. What measures can be taken to reduce slips, trips and falls during formwork construction?
3. Before stripping formwork a **competent person** should provide written confirmation that the permanent structure is self-supporting, and the formwork can be removed. Discuss the WHS Regulation 2017 definition of a competent person.

Constructing concrete slabs, columns, beams and walls

In this section we discuss the construction of formwork for columns, slabs, beams and walls, in which you need to take into account the material stresses produced on the construction.

Materials and stresses

The stresses imposed on concrete come from its own weight and the loads placed upon it. In earlier texts you learnt about in-ground slab construction where the various forces imposed upon the concrete mass were evenly spread over the entire surface of the slab. In column, beam and slab construction, you will notice that the loads are transferred to the ground (see Figure 1.2), creating hotspots of pressure where joints, slabs and beams can fail catastrophically if not constructed and reinforced properly, causing death and injury on a large scale.

Remember, concrete in its wet state is easily moulded and has a high compressive strength, but it has a low tensile and shear strength until reinforcing is added.

Pre- and post-tensioned concrete

Concrete can have its tensile strength increased by other engineered solutions. These can save on reinforcing material costs and reduced slab thicknesses. Such systems include pre- and post-tensioning of concrete.

A good understanding of post-tensioning reinforcing systems is important for form workers. A formwork plan should be used to ensure correct placements and ease of stripping. In preparing formwork for post-tensioned slabs, it is important the positioning of side form supports does not interfere with the post-tensioning pocket formers or the strands. Careful fixing and sealing of the pocket formers to prevent movement or leakage of the concrete is important. Use the correct size bit to drill the holes for the loose ends of the tendons to poke through. With multistrand tendons, make sure any holes cut in the side forms are fully supported to prevent the formwork from collapse.

Pre-tensioning

In a pre-tensioned concrete slab (and/or beam), wire strands or tendons are carefully laid out and tensioned to a predetermined plan and design load. The concrete is poured around them and allowed to harden until it achieves the required strength. The ends of steel tendons are then released, and the stress is transferred to the concrete element by the bond between the two materials.

Pre-tensioning is normally confined to off-site precast elements such as bridge decks, girders and beams.

Post-tensioning

Post-tensioning is more flexible in its application. Before the concrete member is to be poured, the steel strands or tendons are inserted into ducts or conduits previously laid into the formwork to a predetermined plan. The concrete is poured and allowed to harden in the formwork before tension is applied to the ends of the strands (see Figure 1.3). The hardened concrete provides the stop for the stressing jacks. **Note:** there are single strand and multi-strand systems.

Advantages of pre-stressed and post-tensioned concrete:

- It reduces or eliminates shrinkage cracking, so no joints (or fewer joints) are needed.
- Cracks that do form are held tightly together.
- It allows slabs and other structural members to be thinner.
- It allows us to build slabs on expansive or soft soils.

Watch the YouTube video titled 'Post-tensioned slab procedure' (https://www.youtube.com/watch?v=1MrQD2NamUE) to see how reinforcing and post-tensioning of concrete works.

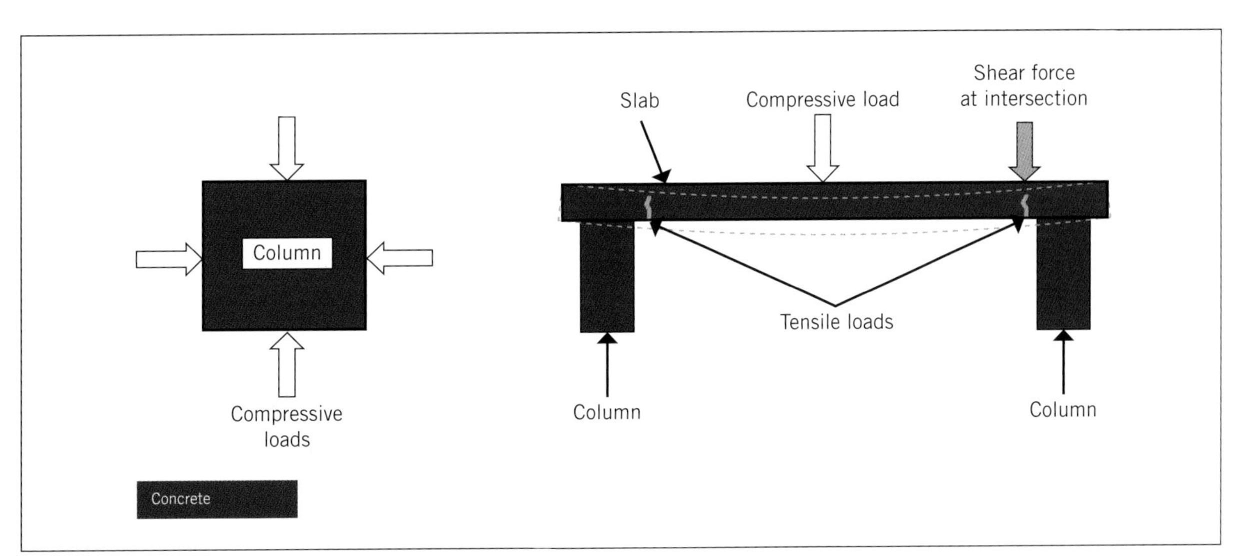

FIGURE 1.2 Concrete tension and compression

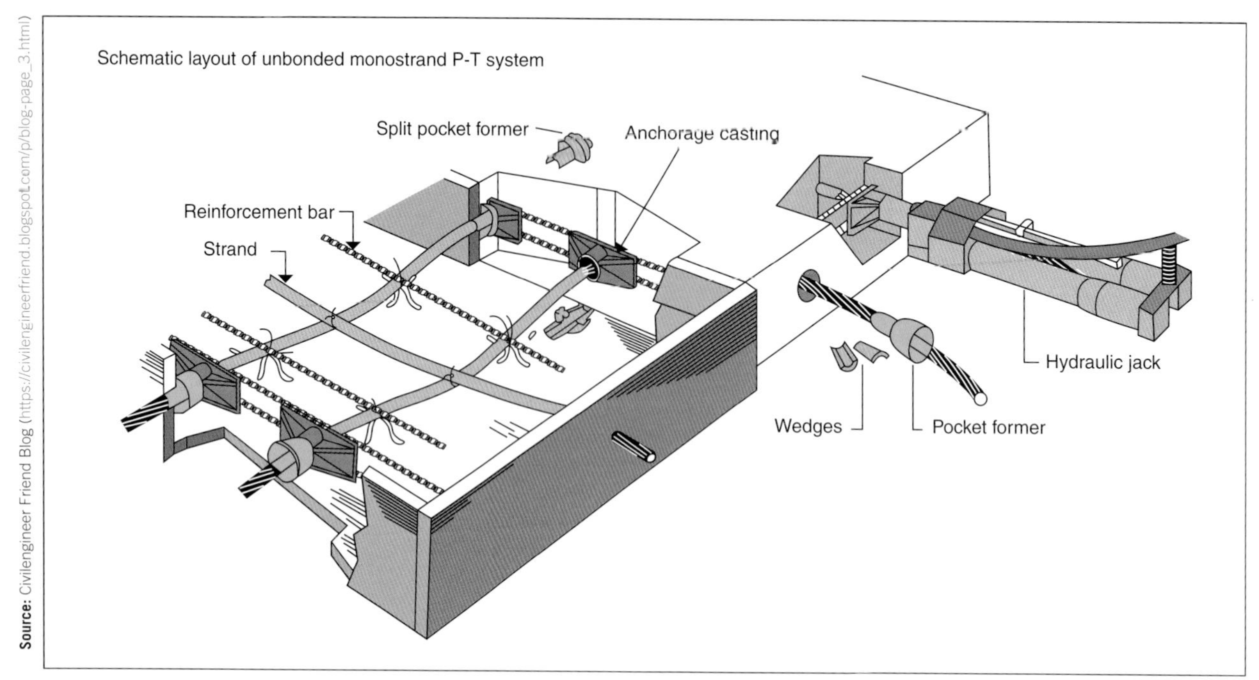

Source: Civilengineer Friend Blog (https://civilengineerfriend.blogspot.com/p/blog-page_3.html)

FIGURE 1.3 Typical mono-strand placement

Placing of steel reinforcement

Placing the steel inside the formwork has to be done carefully so that even coverage of the steel is achieved and the steel does not move during pouring and compacting. Steel too close to the edge will rust and cause spalling in concrete (this is known as concrete cancer) and may result in the potentially catastrophic failure of the entire structure.

Columns and slabs are poured at different times during the construction process so it is often crucial to the structural integrity of the building that the joints are formed correctly according to architects' plans and structural engineers' specifications.

In Chapter 4 of *Site Establishment, Formwork and Framing* we looked at concrete slabs on ground and strip footings as the means of forming the base for the structure. When concrete columns and vertical walls are attached to slabs and strip footings, the steel reinforcing is left protruding from the footings and slabs so that the column or wall forms a solid connection with the adjoining slab or footing, or column (see Figure 1.4) or beam.

More information on jointing concrete can be obtained from Cement Concrete & Aggregates Australia. Visit the website listed at the end of the chapter.

Source: Chris Montgomery

FIGURE 1.4 Steel protruding from footings and piers

Tools and Equipment

Tools and equipment used to erect and dismantle formwork for suspended slabs, columns, beams and walls can include: hammers, measuring and levelling equipment, drills, circular saw and saws. Heavy lifting equipment like cranes and forklifts are used by specialist operators.

Always conduct a visual and operational check of tools and equipment are safe to use and working to manufacturer's specifications. Look for obvious signs of damage or wear and ensure they are clean, as this could interfere with their functioning. Check for cracks in handles, buttons, sharpness of blades, hinges or other components are functioning correctly. Check power cords have been tagged and tested.

Move any tools and equipment with obvious faults away from functioning tools. Label and report them as faulty so as not to confuse other tradespersons from using them.

Always check work instructions for the type of tools, equipment and PPE that will be required throughout the job. It may vary depending on whether or not the activity requires working at height or using specialist equipment.

COMPLETE WORKSHEET 1

LEARNING TASK 1.2

FORMWORK MATERIALS COMPLIANCE ISSUES

Obtain a copy of the Engineered Wood Products Association of Australasia's 'Non-conforming building products Submission 12'. Read the executive summary and then identify and discuss the potential issues for builders using non-compliant formply. https://www.aph.gov.au/Parliamentary_Business/Committees/Senate/Economics/Non-conforming_products/Submissions

Constructing and erecting formwork/falsework

In this section we will look at the components used in formwork, the setting out of formwork and bracing the formwork.

Formwork components

There are a number of components used for forming suspended slabs. Scaffold frames and adjustable props are used with bearers and joists to support the sheets that form the floor and column/beam intersections.

The falsework frames are set and braced into position and bearers are set in **stirrup heads** (see Figure 1.5). Work begins on the set-out for any **control joints**, wall and column connections, and recessed **services** like plumbing and drainage, electrical cabling, and other block-outs and penetrations. Control joints are critical in the design as they confine any cracks to one section. Engineers' drawings will show where the control joints are to be located.

FIGURE 1.5 Stirrups, bearers and joists

Plywood/formply for formwork

Formply (sometimes called 'hardface') is available in thicknesses of 12, 17 and 18 mm and in a regular size of 2440 × 1220 mm (other sizes are available).

Regular formply has a radiata pine core bonded together with A or B type bond adhesive, with hardwood veneered faces covered in a high-density resin overlay. It is usually F11 to F14 engineering strength grade.

Formply with a hardwood core and the same surface finish, providing mechanical F14 to F22 stress grades, is also available. This material is available with a textured face as well.

For more information about formply, obtain a copy of Engineered Wood Products Association of Australasia's *Plywood in Concrete Formwork* from the website listed at the end of this chapter.

Starter bars

Starter bars are necessary to tie in columns and slabs and unite other elements of the construction. Starter bars are tied into column and beam reinforcing 'cage' work and are left to protrude from the poured concrete component ready to be tied into and bonded to the adjoining concrete component, whether it be a suspended slab, column, beam or flight of stairs and ramps.

Remember to cover the ends of exposed starter bars with the appropriate caps – see Figure 1.19.

Props

Props are needed to align and keep columns plumb in each direction and keep them aligned throughout the process of construction and pouring of the concrete. The vertical studs are fixed to the plywood form panel to provide stiffness to the structure and the column clamps are used to secure the column panels together during the pour and curing process (see Figure 1.6).

Tube and coupling components

Tube and coupling components are used to brace and support slab formwork when frame supports cannot be used (see Figure 1.7).

Wall ties

Wall ties are inserted at specified distances to keep the form panels evenly spaced. The wall panels are supported using braces, studs and **walers**. The braces and props in Figure 1.8 must be set at $> 45°$ and $< 60°$.

Types of props

Props are the principal means of supporting the formwork and are used to set heights, locate columns and brace walls, columns, beams and other concrete constructions.

Adjustable props

Adjustable props are used as additional support for slabs, columns, beams and walls, and can be used with patented steel systems as support members. Special

FIGURE 1.6 Props, column clamps and column studs

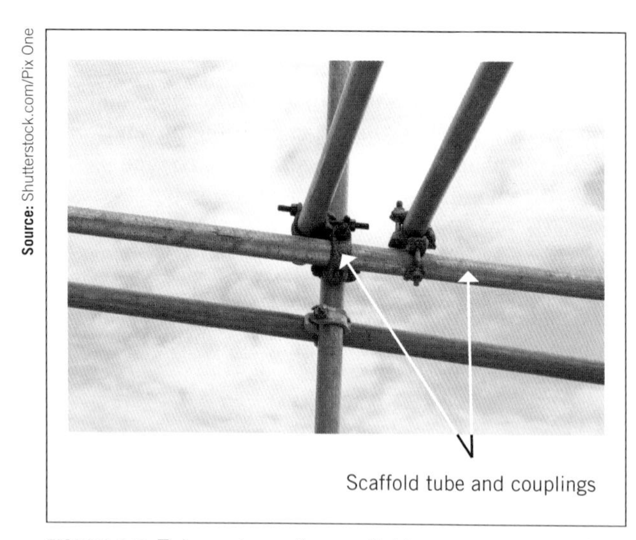

FIGURE 1.7 Tube and coupling scaffold components used in difficult areas

props with adjustable feet can be used for diagonal bracing of columns. These props are commonly used to support tilt slab wall construction systems once the panels have been lifted into place.

There is another adjustable prop system that has a second prop adaptor fitted that allows the formwork to be stripped away while the props continue to support the slab or **soffit** (see Figure 1.9). Adjustable length steel

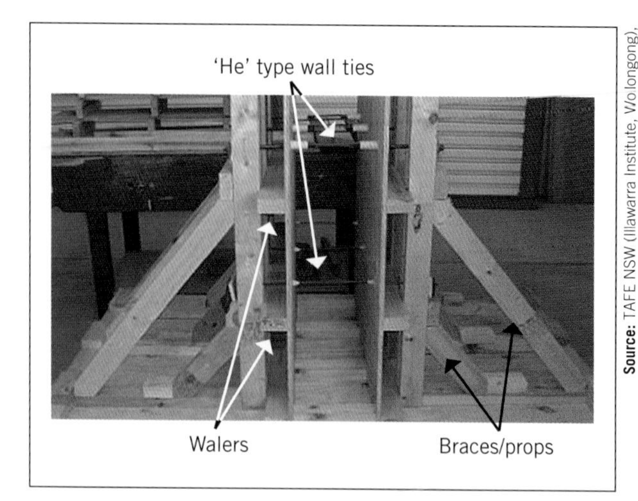

FIGURE 1.8 Wall forms – wall ties, walers and braces

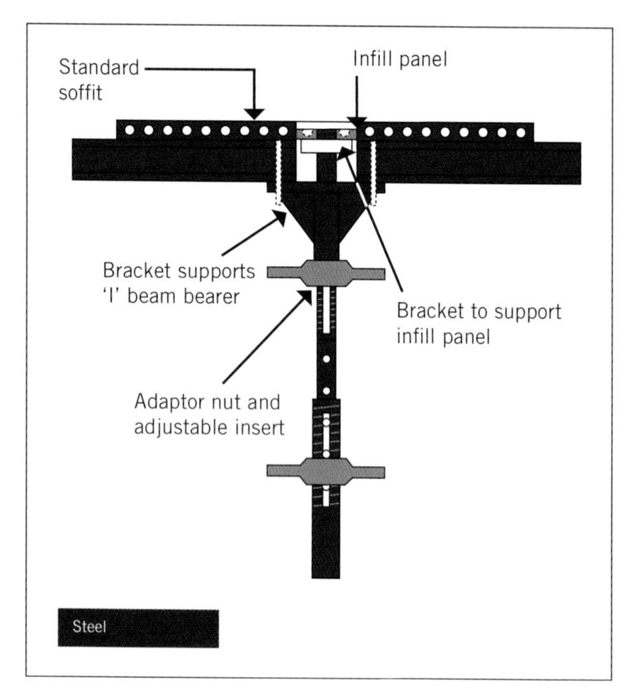

FIGURE 1.9 Prop with special adaptor

bearers and joists are also available. Some adjustable bearers do not require intermediate propping, while others do, so check with the system manufacturer and the formwork plans for further details. More information on props and scaffold frames can be found in the links given at the end of this chapter.

Timber props

Timber props, sometimes called 'toms', can be made up onsite, and are used in supporting beams, small slabs, and often stairs and ramp formwork. They have a 'T' head to support a number of bearers, but if they are used to support beam formwork the 'T' head is not needed.

All joints should be bolted for strength and safety.

As can be seen in Figure 1.10, folding wedges are used to adjust the height. Once in position, they must be braced the same as for all propping.

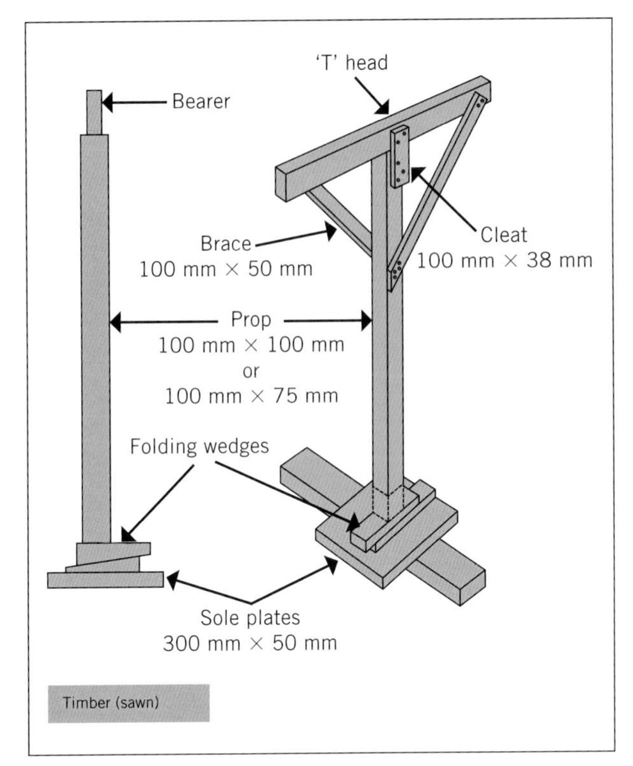

FIGURE 1.10 Timber prop

Source: iStockphoto/JoeGough

FIGURE 1.11 Adjustable bracing props with adjustable feet

Australian hardwoods are best used for timber props and must conform to the stress grading specified and be spaced according to the loads being supported. If in doubt, consult the formwork engineer for the project.

Partly seasoned timber is recommended because seasoned timber may swell when coming into contact with wet concrete, while unseasoned timber may shrink, twist, bow or warp as it dries out before the concrete is poured. Shrinkage may also cause gaps in formwork, allowing leakage of wet concrete.

Scaffold frames

Scaffold frames are the major supports for suspended slab falsework, and adjustable props are used in combination with scaffold frames and as braces for columns. When set at an angle other than perpendicular, adjustable props need to be well located, with secure blocking to prevent movement. The blocking can be fixed to the floor using masonry fasteners or an explosive power tool. In Figure 1.11, the props are bracing a wall vertically and are not genuine load-bearing props, but are propping for plumb, and in columns, placed in both directions.

Bracing props and formwork

While there are no specific regulations around the number and location of bracing required for formwork, AS 3610, Section 4.3.2 stipulates that formwork must resist overturning uplift, sliding and sideways movement and be stiff enough so it does not exceed the design loads. In constructing suspended slabs, columns and beams always consult the formwork design documentation provided. Safe Work Australia's *General Guide for Formwork and Falsework* states that, 'Adequate bracing should be provided longitudinally and transversely to ensure that the falsework is stable and that significant horizontal movements resulting from the applied loads are limited' (Safe Work Australia, CC BY 4.0 Licence https://creativecommons.org/licenses/by/4.0/).

You must consult the formwork design documentation provided for each job to ensure you meet the specifications for the job being undertaken and get signed-off approval from the formwork design engineer and site supervisor if changes need to be made.

As a general guide, individual props should be horizontally braced 300 mm from the top and bottom, with diagonal braces between 30° and 60° across the props. When propping exceeds 4 m, an intermediate row of horizontal bracing should be installed. The bracing should be installed with consideration for the direction of concrete placement.

Figure 1.12 shows typical bracing patterns.

Setting out for slabs, columns and beams

Formwork is set out using a number of different methods and combinations of methods:

1 centre lines
2 offsets
3 grids
4 datums
5 reduced levels.

To begin, centre lines are set out across a site and measurements are taken from this centre line. This is a bit different to setting out a regular building site where measurements are taken from a boundary line.

Datums, or datum points, can be given for vertical as well as horizontal positions. Vertical datums are referenced to sea level and are called Australian Height Datums (AHD). At the start of the project the AHD for a site is marked by a surveyor.

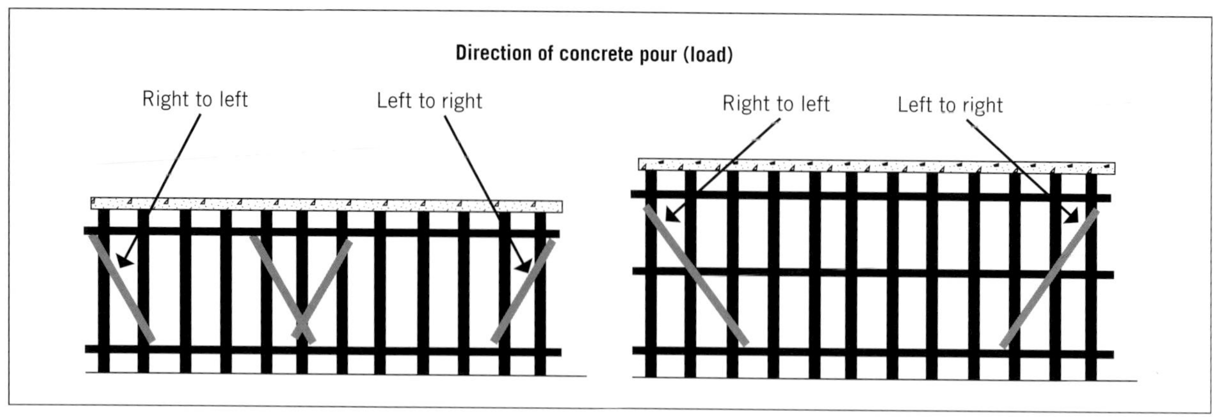

FIGURE 1.12 Bracing patterns

Reduced levels (RLs) are heights/levels above or below the site AHD.

Horizontal datums measured by a surveyor are calculated from grids or coordinates of longitude and latitude. The grids form part of the Map Grid of Australia (MGA). These grids are used on working drawings and site plans, in a similar way to a regular road map grid, where a column might be referenced at B2, while a penetration for a services duct may be referenced at G8.

More information on datums and the MGA can be obtained from Geoscience Australia – see the website listed at the end of this chapter.

Slabs

The slab is set out in roughly equal-sized sections and a pouring pattern is determined (see Figure 1.13). This may require the use of control joints (see Figure 1.14). Control joints may be formed up by formworkers or cut in by the concreter during or after construction, depending on the type specified.

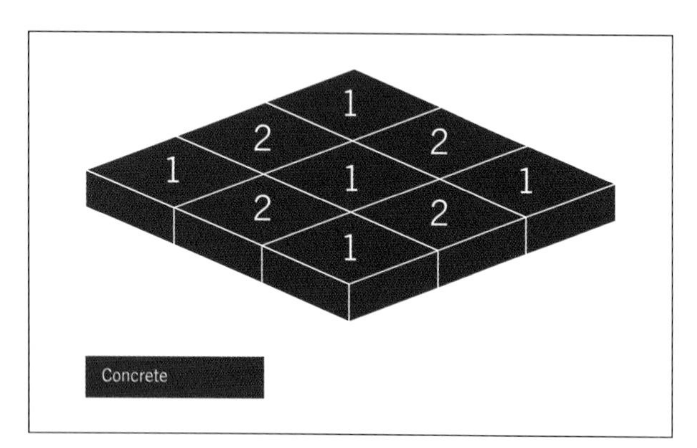

FIGURE 1.13 Simple pouring pattern

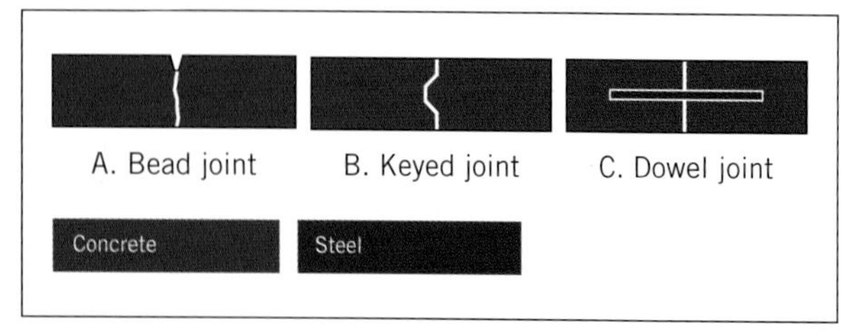

FIGURE 1.14 Control joints

Calculating loads on formwork

Calculating the loads on formwork is complex because it depends on many variables such as the type of structure; for example, multistorey (including suspended slabs) or ground-floor construction, dam or bridge construction, and many other issues. Section 4 of AS 3610 deals with the calculations in greater detail. Following are the basic principles used in calculating loads.

Use caution in making calculations. Double check and add a margin of error to calculations. Any formwork collapse could injure or even kill someone.

The total loads on formwork come from:

- the weight of the components used in constructing the formwork – such as frames, props, bearers, joists, shutters and braces and weight of reinforcement
- the concrete loads – wet concrete is a liquid, and the pressure of the liquid increases the deeper you go (e.g. the foot of a column is affected by the weight of the concrete above)
- the live loads – such as the weight of workers and their equipment and building materials (including stored materials).

Note: Avoid mounding of concrete during placement or starting the pour at the wrong point.

Lateral forces such as wind loads can cause loads to change and footings to shift and must be considered. The action and pressure from water if building a dam, bridge or wharf must also be accounted for.

Loads must be considered at all three stages of the building process, as identified earlier under 'Materials and stresses'. Some loads can be easily calculated, while others need to be estimated because they can change during the construction process.

The loads imposed on formwork from wet concrete are expressed as pressure. Kilopascals are the units of measure for pressure and are used to calculate loads (the symbol used is kPa). One pascal is about 100 g spread over 1 m^2. Consider spilling a little less than half a cup of water over a square metre of concrete. That's how much *pressure* is in one pascal – not very much really. As materials and loads are generally calculated by weight, we have to be able to convert these to kPa.

So, if concrete weighs approximately 2400–2500 kg per cubic metre (m^3), including reinforcing, the kPa value for a cubic metre of concrete can be calculated by multiplying the cubic metre rate by 0.1 kPa and then multiplying this by the thickness of the slab (see Figure 1.15). For example:

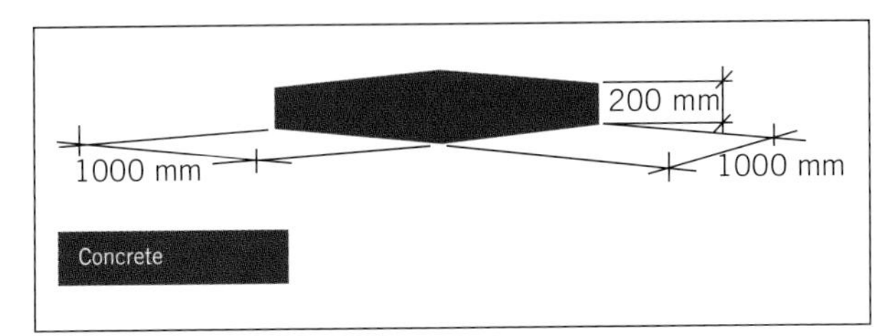

FIGURE 1.15 Calculating the kPa value of concrete

Using 2500 kg/m^3 $\times$ 0.2 m (thickness of slab) $\div$ 100 kPa

$$2500 \times 0.2 \div 100 = 5 \text{ kPa}$$

Added to the wet concrete pressures are the other predetermined kPa allowances. From AS 3610 Formwork for concrete:

Formwork components	= 0.1 kPa
Workers allowances	= 1.0 kPa (when only one floor is being supported)
Concrete mounding allowance	= 3.0 kPa (over a 1.6 × 1.6 m square at any location)
Building materials	= 4.0 kPa

Giving a total kPa/m^2 for a 200 mm slab = 8.1 kPa + 5.0 kPa for 200 mm slab = **13.1 kPa**

The eccentricity or 'offset/cantilevering' of formwork also has a bearing on any calculations.

The formwork designer usually makes the calculations and designs the formwork and falsework (support system) around the calculations to ensure a safe construction.

Floor slab and beam construction

Beams usually support floor slabs (but not always). There are primary and secondary beams. Primary beams usually run around edges and connect columns, while secondary beams are generally smaller and intersect primary beams (see Figure 1.16).

Slab and beam construction relies upon the primary and secondary beams to support a thin slab. Flat floor

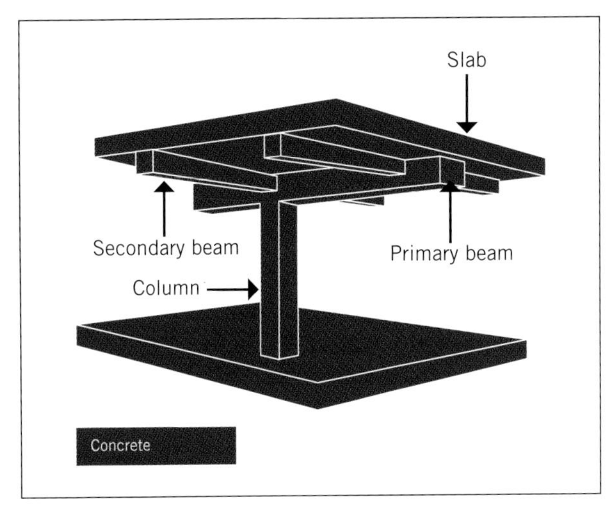

FIGURE 1.16 Primary and secondary beams

slabs without beams rely upon the thickness of the slab and the reinforcing materials for their integrity (i.e. their ability to stay up). They have much simpler formwork requirements, being a flat deck between columns, whereas primary and secondary beams are formed up and poured with the slab.

Sometimes floor slabs sit on column heads that vary in size and shape to the column.

Square and rectangular sections above columns but below the main slab are called 'drop panels'. Sometimes geometric shapes are used on top of columns (see Figure 1.17). These column heads are known as capitals or mushroom heads.

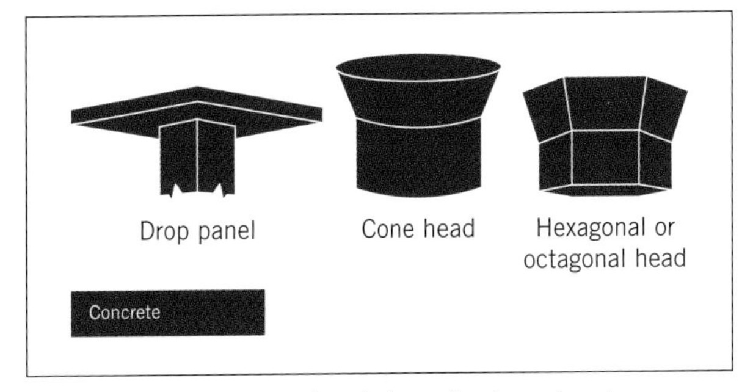

FIGURE 1.17 A drop panel and shaped column heads

Drop panels and enlarged column heads are used to reduce the chance of diagonal cracking and to resist shear forces at slab and column intersections.

Building and erecting formwork/falsework

The main points to consider when constructing formwork and falsework are accuracy, stability, safety and economy.

- *Accuracy:* Be careful in the set-out of reinforced concrete construction because once set it cannot be changed. The finished sizes of the concrete components are the starting points for measurements; the formwork material thickness is then added to the finished concrete dimensions. Allowances have to be made in slab formwork so that bracing can be applied to the formed-up sides. Use storey rods when setting up props and checking dimensions.

- *Stability:* The formwork must be stable during the concrete pour so that sides do not bow out or slabs deflect, as this can lead to complete failure in the construction.
- *Safety and economy:* Scaffolding, barricade tapes (see Figure 1.18), handrails and safety signs need to be incorporated into the design and erection of formwork. Keeping people out from under formwork during a pour is critical, as is the use of safety caps on top of starter bars (see Figure 1.19).

Source: Ed Hawkins

FIGURE 1.18 Barricade tape

Source: Ed Hawkins

FIGURE 1.19 Safety caps on top of starter bars

When erecting the falsework, never substitute scaffold or adjustable prop parts. This is illegal. Ordinary reinforcing steel is not strong enough for use as prop pins.

When setting up formwork for slabs, you will inevitably end up working near an edge that often has a large drop. A scaffold should be used to provide a means of access to the area and it should incorporate toe boards, handrails and mid-rails, with a minimum gap between the scaffold and formwork edge. Where no standard scaffold is available, it is appropriate to use a mobile scaffold for access. A personal harness attached to a fixed point should be worn when working near an unprotected edge.

Materials handling and storage

Formwork is often owned by a specialist formwork company and hired out, and as such is used and reused on multiple sites. Some basic tips are:

- Don't cut material if you don't have to. There is no problem leaving some components over-length and it saves on waste (see Figure 1.20).
- Look for smaller pieces before cutting a full length or sheet.
- Remove nails after each use.
- Careful storage and stacking will reduce waste significantly (see Figure 1.21).

Source: Ed Hawkins

FIGURE 1.20 Leave material over length

Source: Ed Hawkins

FIGURE 1.21 Proper stacking

Remember: Reuse = Profit

Remember: Formwork codes of practice should be followed throughout the construction of formwork. Links to the codes of practice are provided at the end of this chapter.

Shutters and sheaths

Shutters, or shuttering, is the actual mould shape created by the arrangement of sheaths in falsework. Solid timber can be edged, butted and cleat-fixed to form **sheathing** (see Figure 1.22) for shutters. Timber is used as the sheathing when concrete is to be finished directly off the form, highlighting the timber grain in the concrete. Dressed timber is commonly used when the grain and texture of the timber is to be revealed.

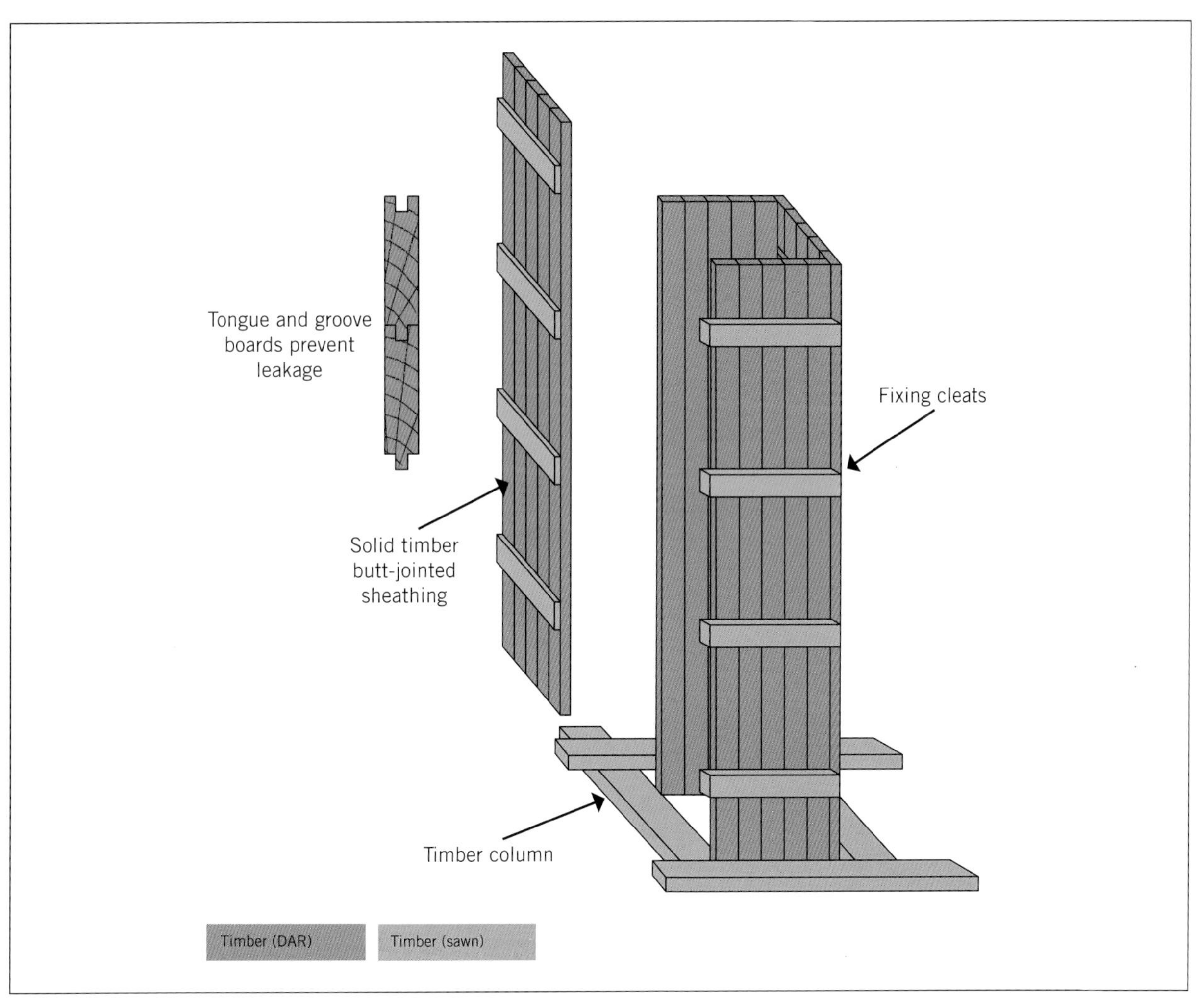

FIGURE 1.22 Column shutters

Rough-sawn and, in some cases, sand-blasted timber may be used for special effect. Spacing of the cleats or battens will be the best determinant of the thickness of the timber, but in general 19 mm solid is the minimum thickness used.

Plywood is the main sheathing material used to construct shutters and is backed up using studs and walers. For plywood sheathing, the basic sheet sizes apply, with 2400 × 1200 × 18 mm being the most common. The edges of the plywood should be sealed before assembly to prevent swelling from moisture contact. The most common sealants are silicones, mastics, rubber-based paints and waterproof adhesives. These also prevent cement sticking. The main Australian Standard that applies to plywood used in formwork is AS 6669 Plywood – Formwork. This Standard specifies the requirements for the manufacture, grading, finishing and branding of plywood used specifically in formwork.

AS 2269 Plywood – Structural – Specifications sets out the manufacturing, grading, finishes and labelling for structural plywood. The veneer quality, bond quality, standard lay-up construction, dimensional tolerances, joints, moisture content, and characteristic strength and stiffness values for the various grades (A, B, C and D) are specified. Type A, adhesive bond, is specified for structural plywood.

When installing plywood make sure the face grain runs parallel to the span (i.e. across the joists) and for columns vertically (see Figures 1.23 and 1.24).

Fillets are commonly used in column and beam joints to give a **chamfered** edge (see Figure 1.27). This reduces the amount of chipping along the edges of the column and beams.

Source: Gordon Slater

FIGURE 1.23 Ply sheathing

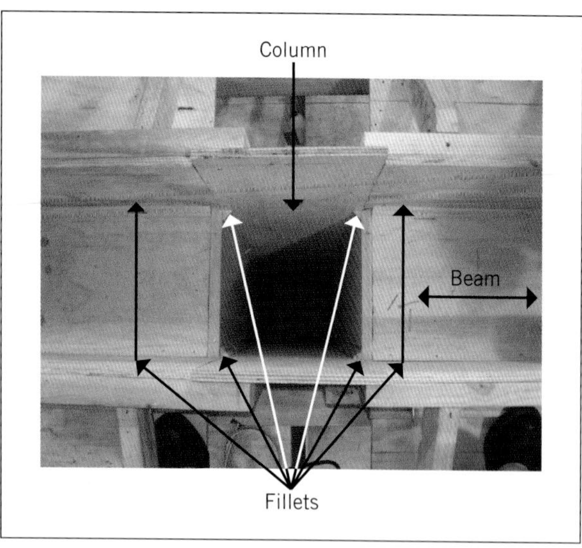

FIGURE 1.24 Fillets inserted

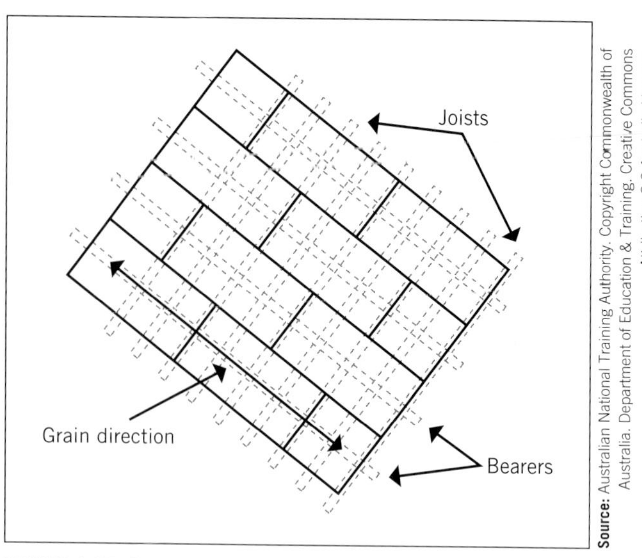

FIGURE 1.25 Grain direction in ply

If you have to lie the sheets at 90° to the span, consult the engineer's design drawing for specified materials.

Note: It may be difficult to determine the grain directions because of the plastic-coated face (see Figure 1.25), so look carefully at the edges of the sheet.

Make sure joints are watertight by making tight joints, sealing the edges and using clamps to pull the joints up tight. A self-adhesive filler tape may be used to seal the edges, but it has to be trimmed flush. You may find it difficult to stick to the edges because of previous release agents that have been used. It is better to make a tight joint than to rely on tapes.

Nail off the sheets at the corners to prevent movement. If necessary, use smaller nails for intermediate joints. Fewer fastenings will make stripping easier.

Figure 1.26 shows how to set out sheets for slabs.

Installing soffit formwork

When installing soffit formwork for slabs you will need to allow extra widths (two plank widths or 450 mm) at each edge to allow for bracing of the slab side soffit formwork and working room. Handrails are also required (see Figure 1.27).

When starting to build the falsework unit frame, a small section of soffit is constructed to act as a staging point for further tools, materials and equipment. The soffit formwork is built up using full sheets that can be reused. The soffit formwork, which is normally cantilevered (see Figure 1.27), needs to be properly supported just as the formwork for the slab is supported. Additional props can be installed for full support. Safe Work Australia's *General Guide for*

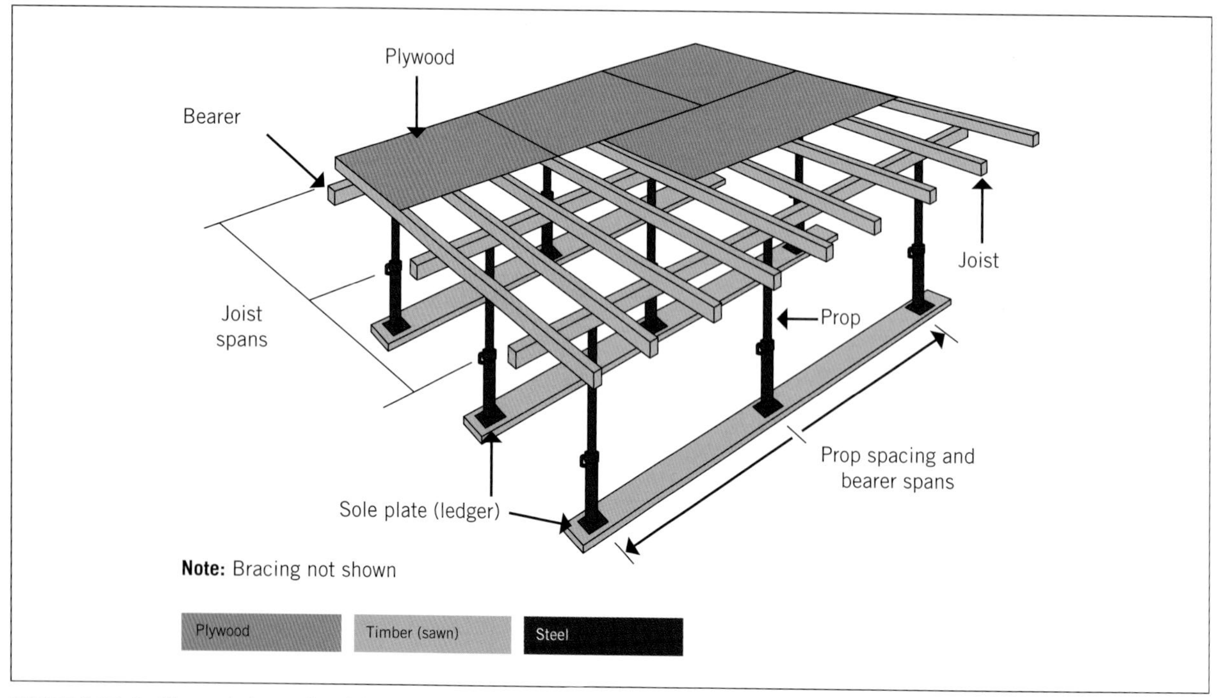

FIGURE 1.26 Setting out sheets for slabs

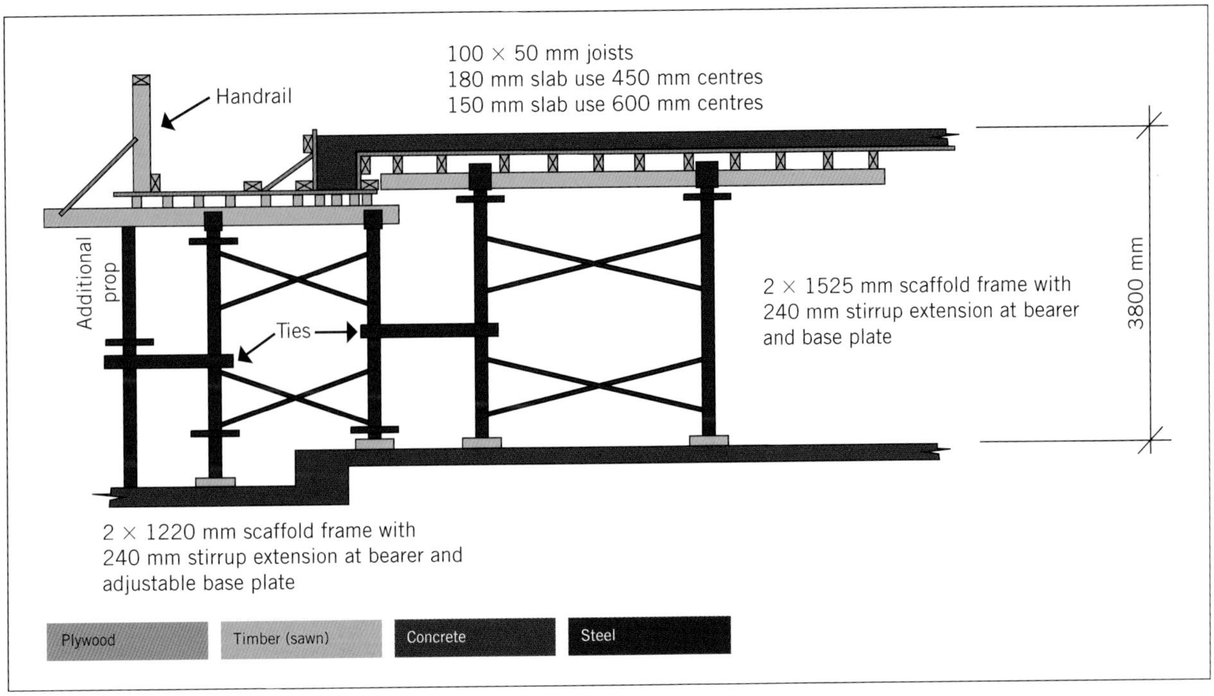

FIGURE 1.27 Typical formwork elevation

Formwork and Falsework specifies that the safe working platform outside the formwork area (including braces) must be a minimum of 450 mm or two planks wide, and if the working platform is more than 2 m (or a regular prop or scaffold frame height) above the ground, it must be constructed with toeboards and handrails.

Slab and beam falsework is constructed using unit frame scaffolds to erect the bearers and joists. A stack of joists can be marked out at the same time, making it easier to keep the right spread on the bearers and an equal overhang so that braces and ledges can be installed.

Formwork for columns

Square or rectangular columns usually have formwork made from plywood sheaths, with vertical timber studs spaced at intervals across the sheeting (see Figure 1.28). Two sides will be the same width as the column, while the adjoining sheaths will be the width of the column plus the width of two plywood sheaths so that they overlap.

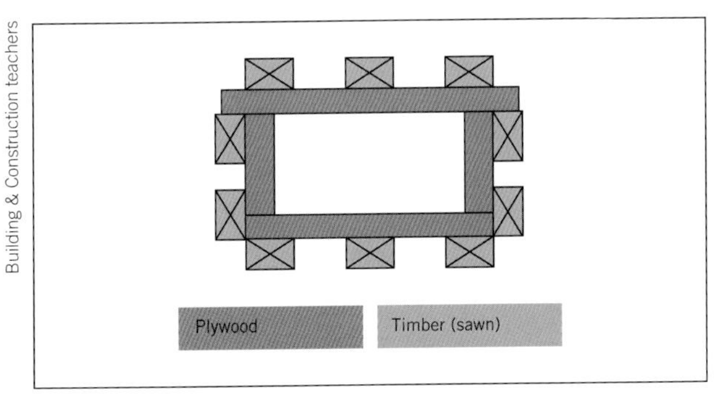

FIGURE 1.28 Column shutters

Source: TAFE NSW (Illawarra Institute, Wollongong), Building & Construction teachers

Alternatively, steel shutters can be used for regular shaped columns.

Rectangular columns are usually formed up with a clean-out access hole incorporated near the bottom of the column to allow for the removal of any rubbish, timber off-cuts or other debris that might accumulate.

One column sheath can be left out to allow for final clean-out and inspection of steel reinforcement. Once the formwork and steel work has been approved, you can insert the remaining sheath and column clamps (see Figure 1.29).

Column capitals will often be used to support drop panels.

A drop panel and a hopper head may be used together (see Figure 1.30).

Column clamps

There are many different types of column clamps available and they are often patented as part of a formwork/falsework system. Those shown here are typical of the types often used in Australia.

Column clamps (also called '**yokes**') are designed to interlock around column formwork components. One end is bent and returned to fit over the next clamp arm and a wedge is driven through slots in the bar to tighten the clamp. The spacing for clamps is determined by chart and manufacturer's specifications, but in most cases there need to be more clamps towards the bottom of the formwork to resist the significant forces generated at the base of the column from concrete mass (see Figure 1.31). A rod for spacing column clamps can be made so spacings are the same on each column.

Nails can be placed around the column to support the column clamp bars while they are being tightened.

It is crucial when setting up the column formwork that the forms are checked diagonally for square (see Figure 1.32).

Source: Chris Montgomery

FIGURE 1.29 Column left open for clean-out and inspection

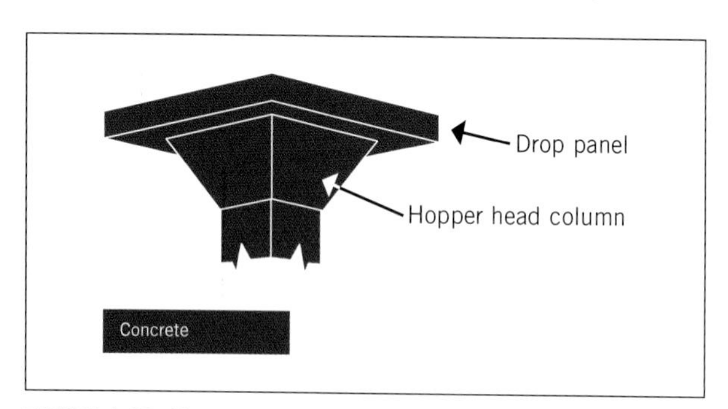

FIGURE 1.30 Hopper head column

If a row of columns is being erected, a string line or laser should be used to ensure the columns are properly aligned at the head and foot in accordance with the grid lines already marked out (see Figure 1.33).

Round columns can be formed up using cardboard or plastic tubes (see Figure 1.34) as well as patented steel forms (see Figure 1.35).

Steel formwork is available for forming up columns and tanks (see Figure 1.36).

Hollow fibre-cement columns can also be used as column formwork, filled with concrete and left in position. These do not require any stripping but are difficult to brace without causing some damage to the outer skin.

Formwork for beams

When setting out beam formwork, it is essential to consider the increasing weight of concrete as it is being placed. When placing the concrete for beams, the

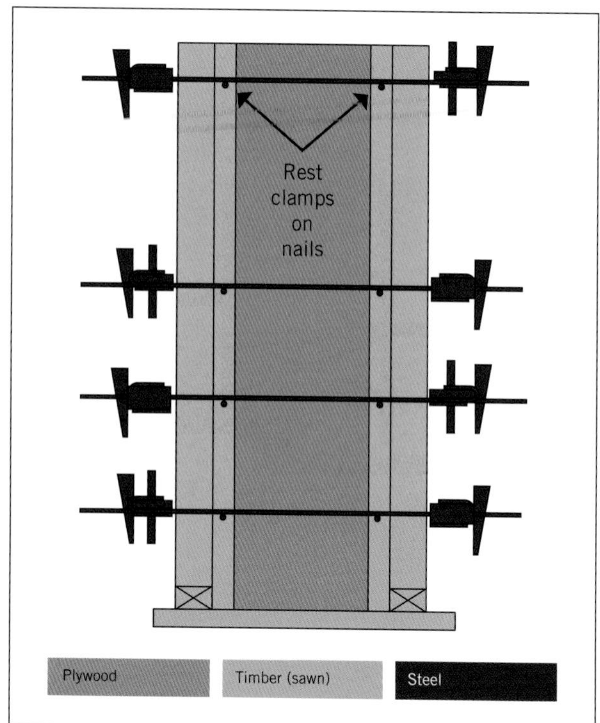

FIGURE 1.31 Column clamp spacing and installation

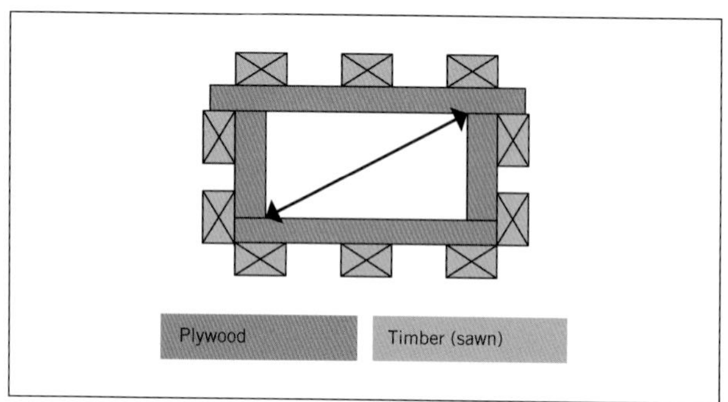

FIGURE 1.32 Check diagonals for square

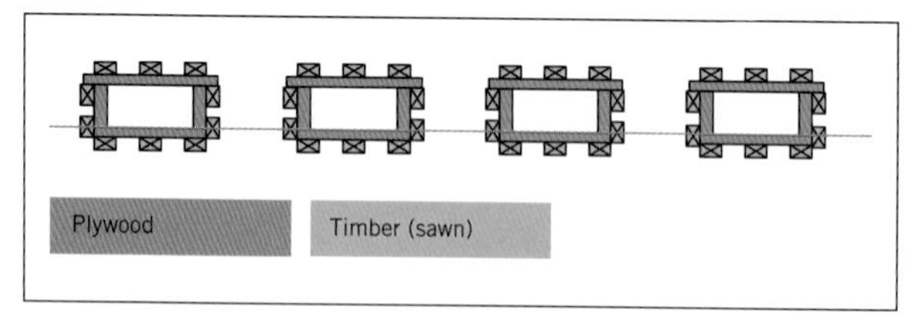

FIGURE 1.33 Align the tops to the inside of the ply shutter

Source: (left) Chris Montgomery; (right) Ed Hawkins

FIGURE 1.34 Plastic column formwork

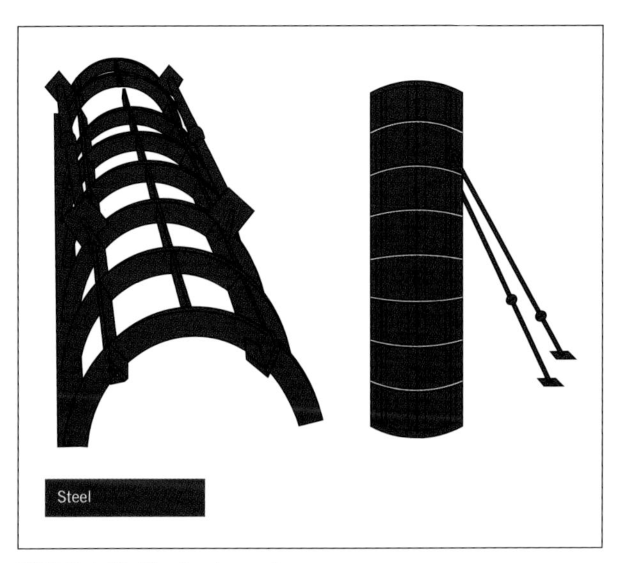

FIGURE 1.35 Steel column formers

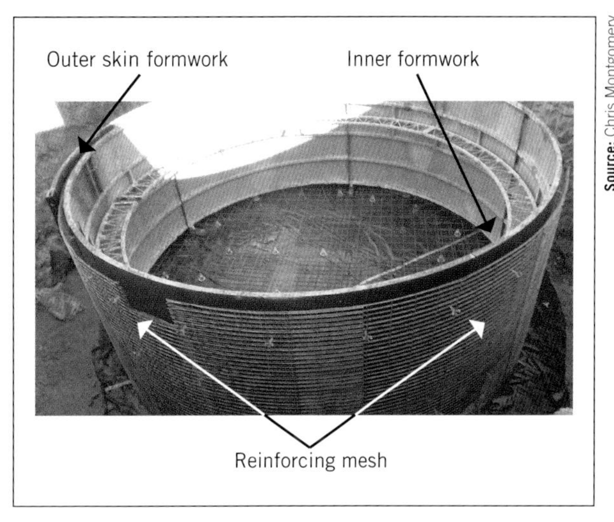

FIGURE 1.36 Tank formwork

Source: Chris Montgomery

concrete will have a tendency to push out the bottom of the sides because that is where most of the weight will settle first, so strengthening the bottom edge is necessary.

If the formwork is not supported properly by spreading the concrete being poured and keeping the load inside the scaffolding and props (see Figure 1.37), there is a risk that the unit frame scaffold and formwork could move and create a catastrophic collapse. Sole plates can be used to spread the load instead of propping directly to the concrete floor.

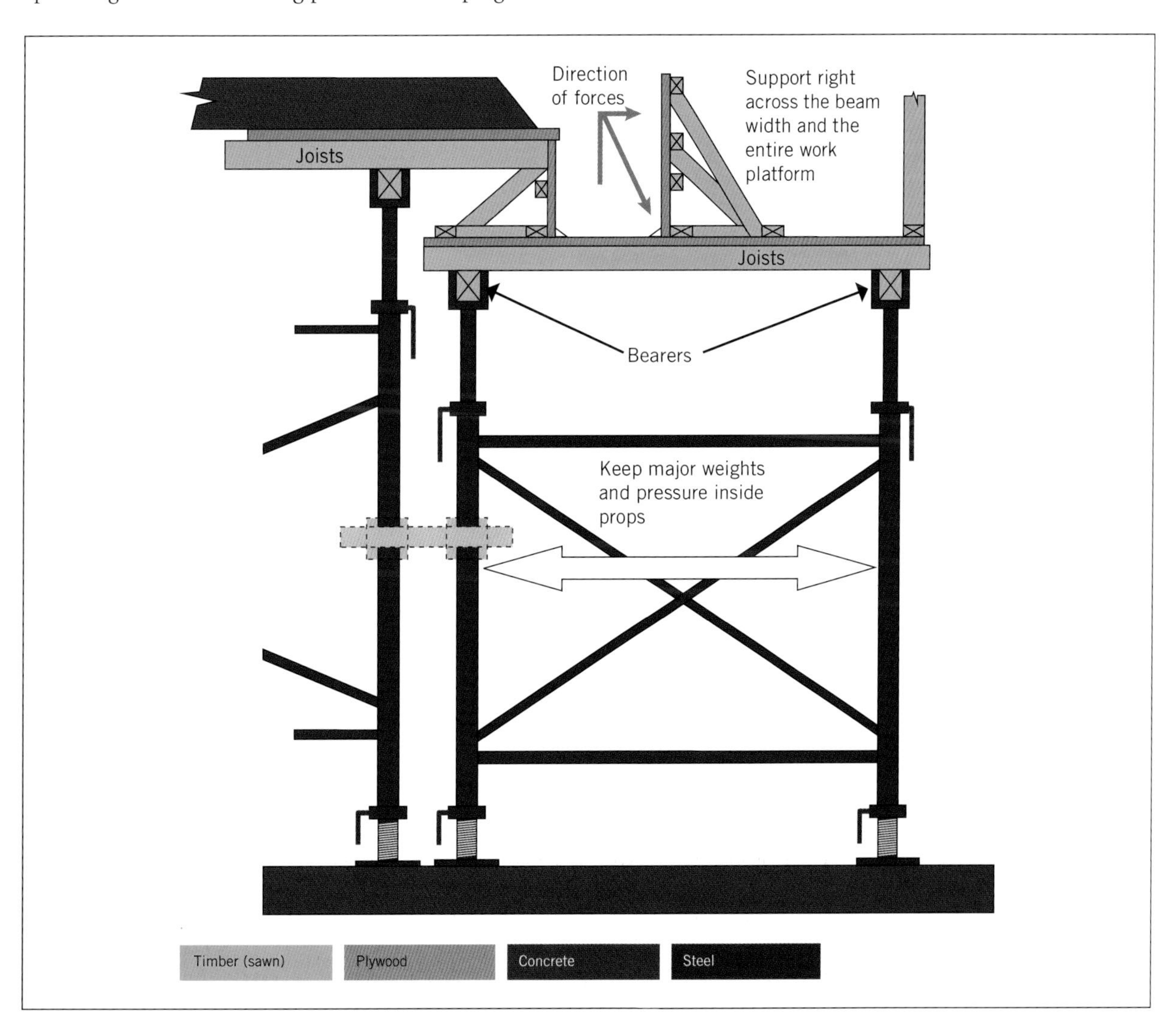

FIGURE 1.37 Correct placement of props

Section 4.4.3 of AS 3610 Formwork for concrete provides significant advice on, and formulas to be applied in, determining the amount of eccentricity that can be applied (see Figure 1.38). This will be dependent on a number of factors, including additional strutting, bracing and the age of the components being used. It is better to avoid a potential collapse wherever practical.

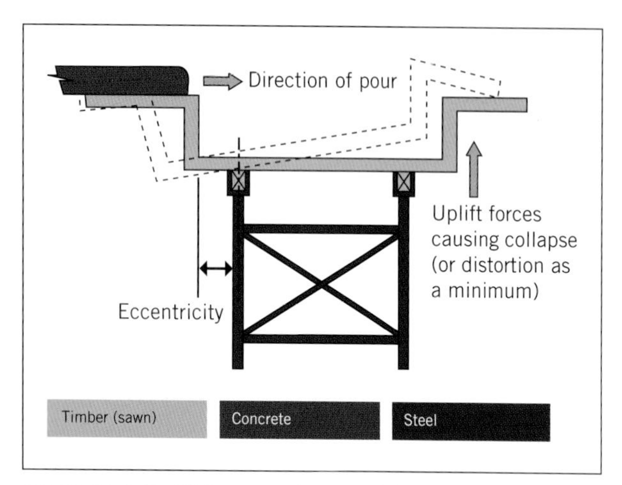

FIGURE 1.38 Uplift forces and eccentricity

The stripping of formwork for beams can be made easier by the inclusion of a stripping band around the perimeter (see Figure 1.39), which is attached to the suspended slab and prevents disturbing the beam when dropping the slab formwork.

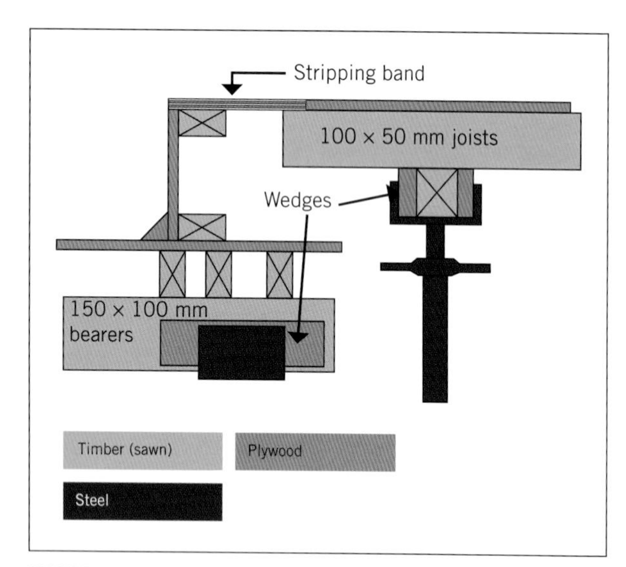

FIGURE 1.39 Stripping band

An alternative to using a stripping band is to build a shutter infill panel shorter than the height required, and then to use an aluminium angle to create the corner intersection (see Figure 1.40). This also allows for easy stripping of the beam formwork so as not to disturb the main slab formwork.

Wall ties may be needed in deep beams (see Figures 1.41 and 1.42). A rod for spacing can be made so that spacings are the same on each deep beam. The bottom edges of shutters can be braced back to the slab formwork. Setting out of joists for beams is similar to floor construction, where double joists are used to support the shutter (see Figure 1.42).

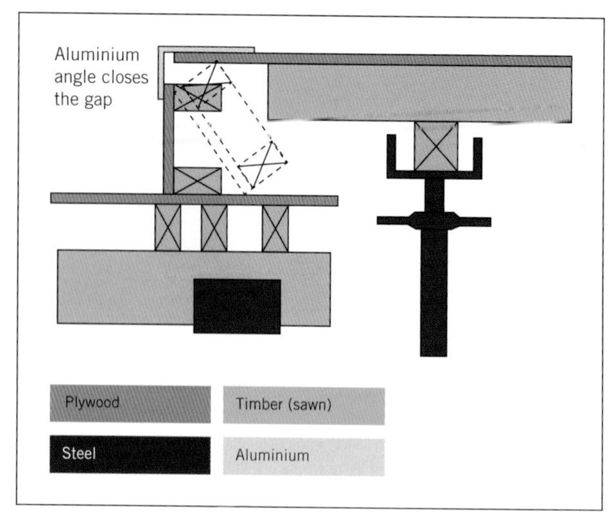

FIGURE 1.40 Alternative beam formwork (for clarity, braces are not shown)

Source: TAFE NSW (Illawarra Institute, Wollongong), Building & Construction teachers

FIGURE 1.41 Three different types of wall ties

Formwork for walls

Walls can be used to span between floors and beams and floor to floor, providing additional support to beams. This can help reduce the necessary size of beam required and the amount of steel reinforcement used in the structure.

Formwork for walls is either plywood sheaths or patent-steel forms. The sheaths are made up similarly to column sheaths using vertical and horizontal studs/walers to provide support for the ply panels. The studs and walers are aligned with the plywood, depending on the direction of the face grain of the plywood. Plywood will bend more easily in one direction than the other (see Figure 1.43).

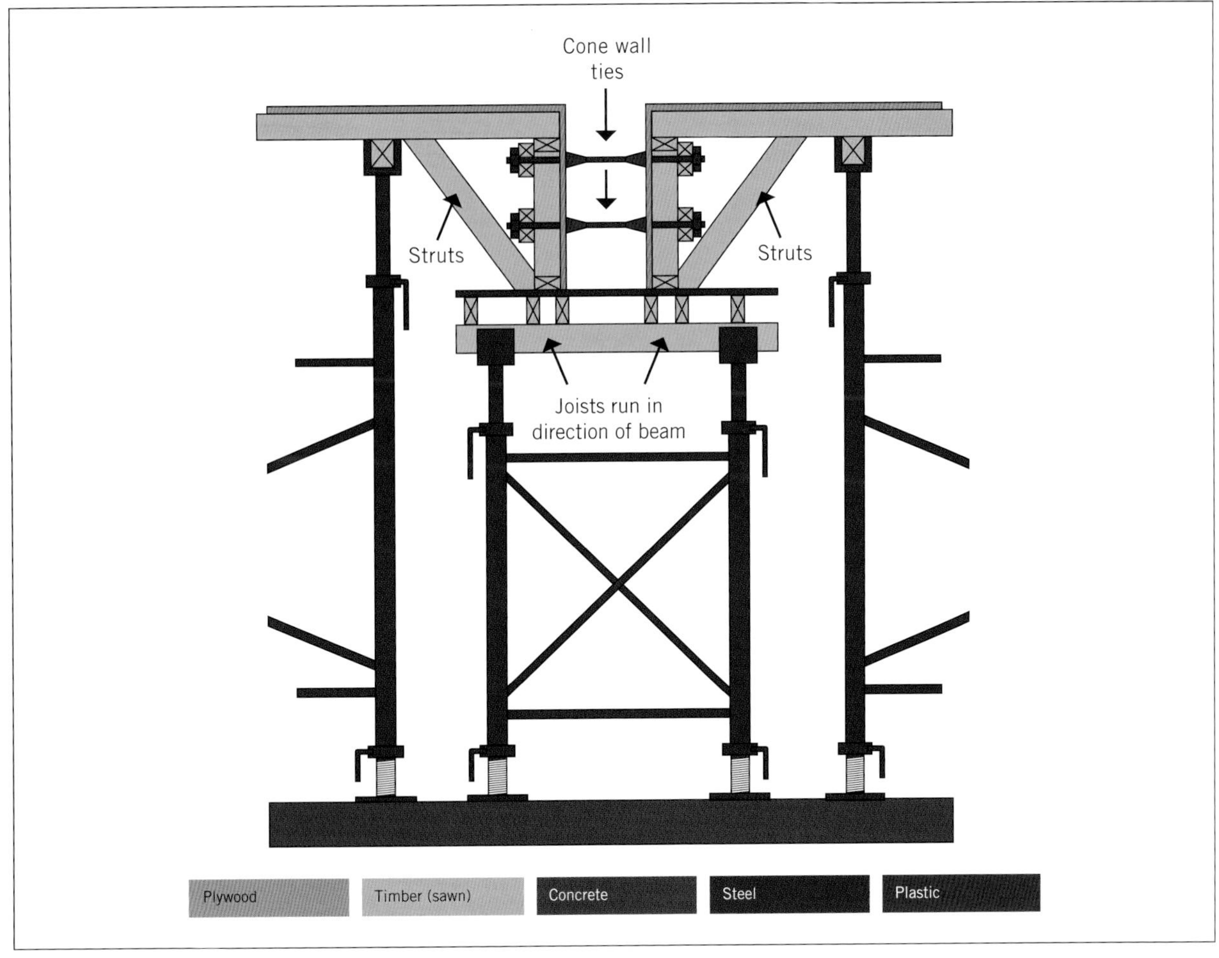

FIGURE 1.42 Deep beam formwork

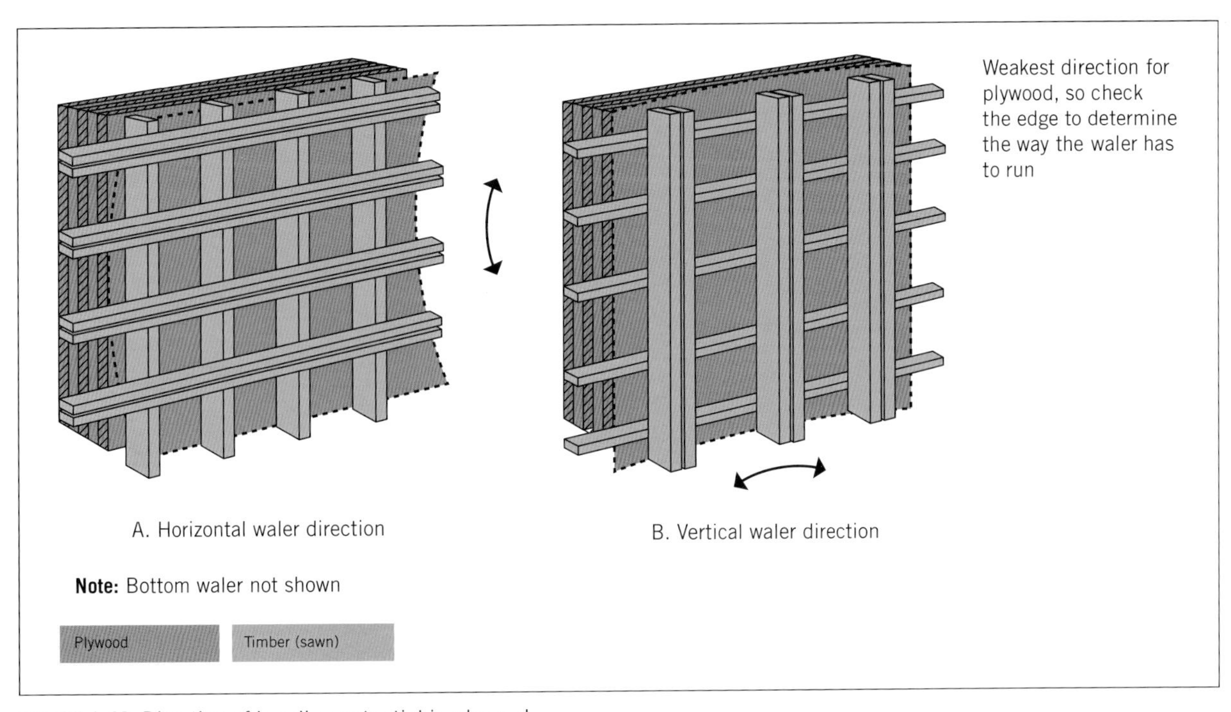

FIGURE 1.43 Direction of bending potential in plywood

The direction of the studs and walers depends on the requirements of the job. Generally, horizontal walers are used when the wall is a continuation of an existing wall in a horizontal direction, while vertical walers are used when the wall is a continuation of an existing wall in a vertical direction.

When the studs are run vertically, top and bottom plates are used to support the bottom edge of the plywood and to ensure it aligns properly with any nib wall that may already be cast in the floor slab (see Figure 1.44).

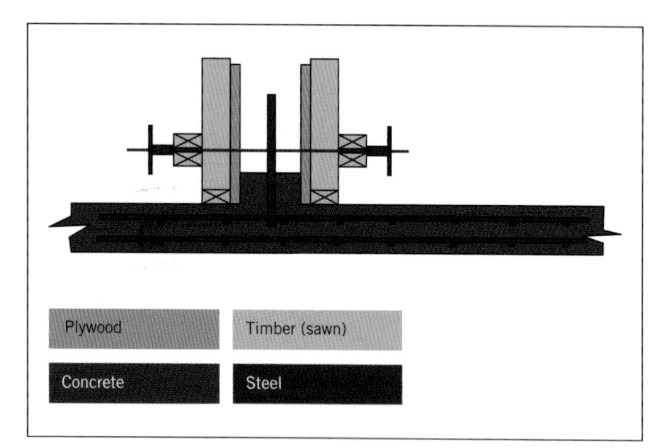

FIGURE 1.44 Nib wall intersection using dowel joint – section view

Single-faced walls

Single-faced walls are used mainly as retaining walls. This means the earth provides the formwork for the placement of concrete on one face and the form is used for the other face. The bottom of the form is fastened to previously poured piers, with cleats fastened with masonry anchors and bolting through the waler. Angled wedges are cut and bolted to the top of the wall form to provide a fixing or bracing point for the adjustable props. Metal or timber pickets are driven into the ground to a depth of about 450 to 600 mm (depending on the thickness of the wall). The thicker the wall, the deeper the picket should go. The picket is braced back against a secondary peg. The ledger is placed across sole plates and fixed/braced against the picket for maximum support (see Figure 1.45).

Steel prefabricated panels can also be used for both single- and double-skinned walls to prop and brace the wall frames (see Figure 1.46).

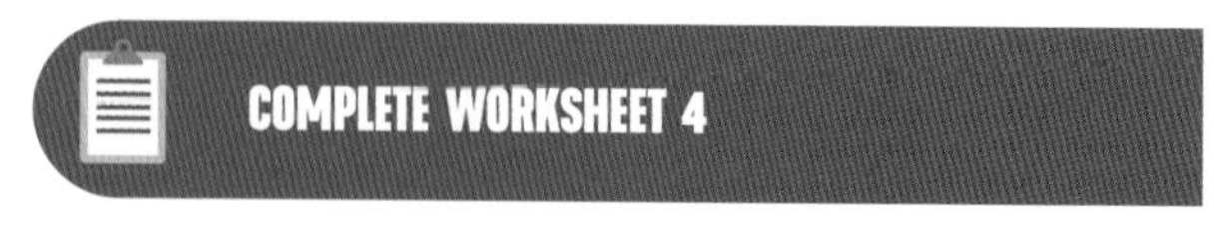

Double-faced concrete walls

Double-faced wall forms are also affected by pressures from wet concrete and wind, and these forces are counteracted by tying the two faces together using patented wall tie systems. The ties reduce the distortion of the formwork when the concrete is poured. Ties are spaced evenly and located close to joints to stop grout loss and bulging.

Large wall forms are usually lifted into place using team lifting techniques, or a crane is used to lift and support the sheaths while they are being plumbed and propped. If they are to be lifted by crane, sling points have to be incorporated into the sheaths to prevent buckling.

Wall forms are built on the concrete slabs, with the framework being laid out first and plywood being laid down on top. The panelling is turned over and walers are installed (see Figure 1.47).

If the wall frame positions have been established by the casting of nib walls into slabs, setting out is relatively easy. Otherwise, a chalk line can be laid down according to the set-out plans or the coordinates and grids, with flick lines being put down for either side of the wall to align the formwork panels. Alternatively, a laser level may be set out to cast a light beam along the wall. This has the advantage of acting as a plumb line as well.

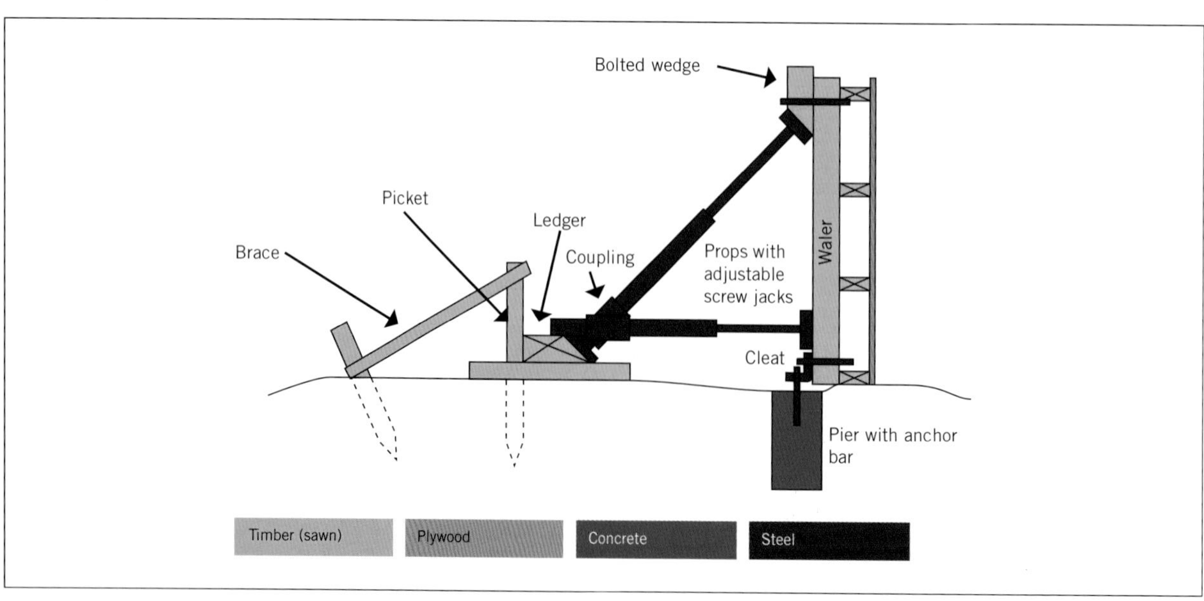

FIGURE 1.45 Single wall form with brace back to earth

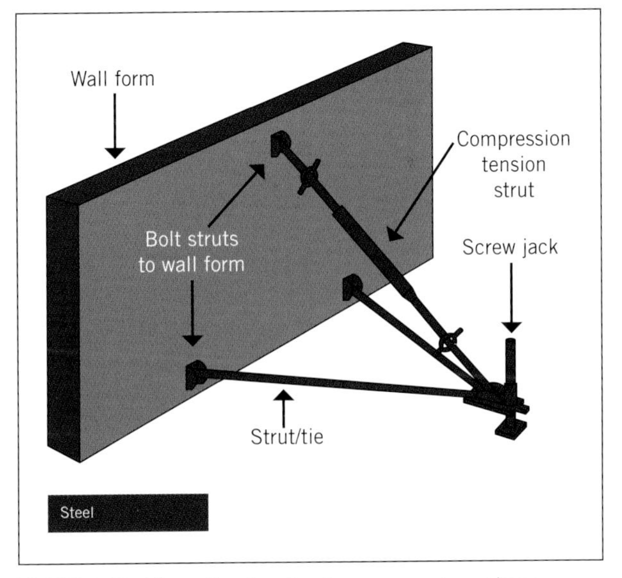

FIGURE 1.46 Alternative bracing frame propping using a proprietary system

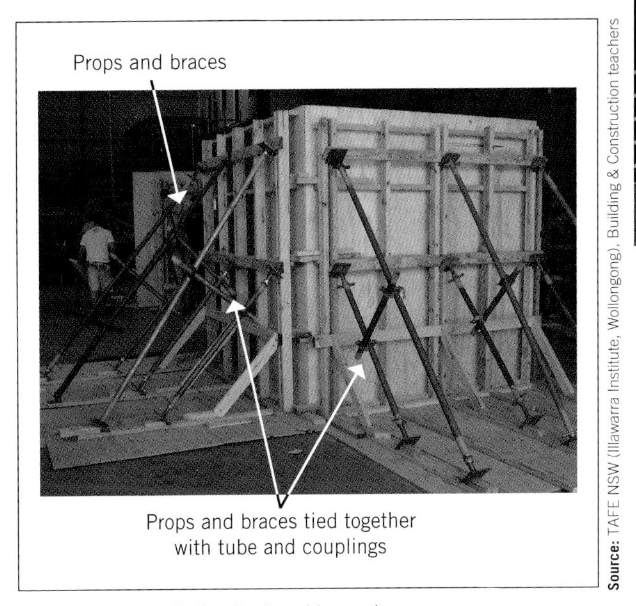

Source: TAFE NSW (Illawarra Institute, Wollongong), Building & Construction teachers

FIGURE 1.48 Wall plumbed and braced

Source: TAFE NSW (Illawarra Institute, Wollongong), Building & Construction teachers

FIGURE 1.47 Wall form laid out on floor

One side of the wall form panels is then lifted into place and plumbed, braced and joined. In Figure 1.48, note how the bottoms of the wall forms are well braced to the floor to prevent movement when the concrete is placed.

The ends of the wall forms have to be blocked off to keep the concrete within the form. This is done using stop ends that are braced (wedged) against the wall ties to prevent movement (see Figure 1.49).

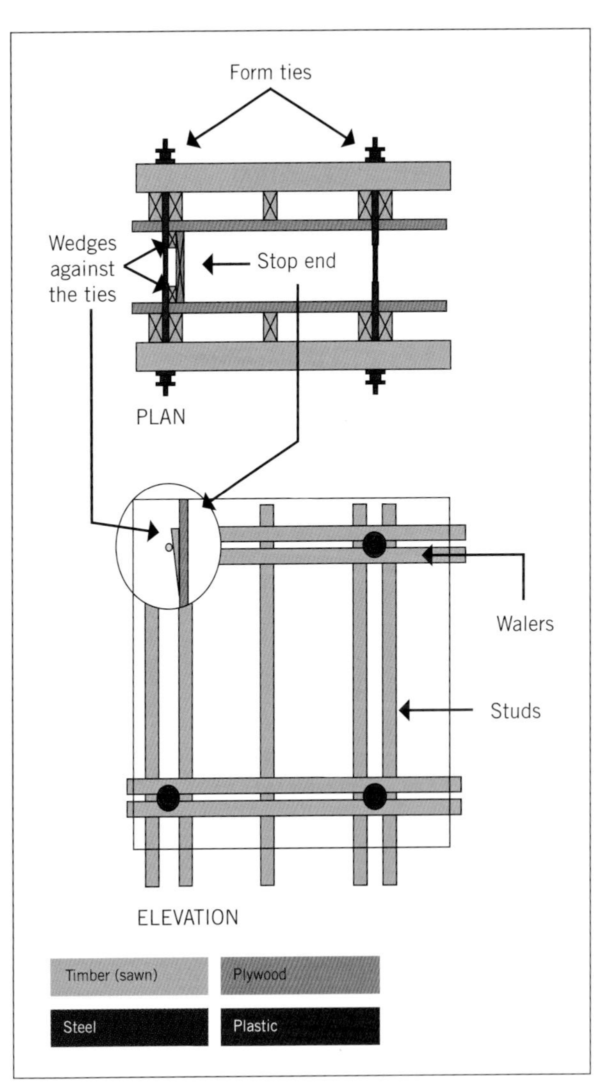

FIGURE 1.49 Stop ends

When one wall intersects another wall, joints are formed, and often one section of a wall has to be integrated so that the wall appears to be a continuous construction. This can be achieved by using steel mesh and starter bars or by scutching (roughing) the end joining surface so it keys to the next section and increases the bond strength with the concrete. You can also shape the form of the end to produce a 'key' shape for the new concrete to fit into (see Figure 1.50). The new formwork should extend over the end of the existing wall and the joint should be reinforced to prevent buckling and loss of grout.

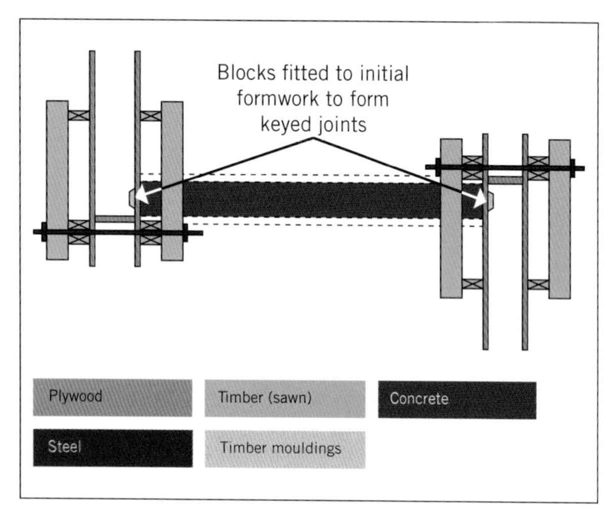

FIGURE 1.50 Keying walls together

Assembling the formwork

To assemble formwork, first locate the position of the wall on the plans. Nib walls can be used to align formwork; otherwise footings or a kicker plate should be installed.

The form face panels can be fixed to the studs in various ways, depending on the panels being used and the finish required. Panels should be joined on studs for maximum strength. If necessary, before butting panels together use a compressible foam or lay a tape over the joint to seal any gaps. If good joints have been made, no foam sealer or tape is necessary. Remember, whatever method you choose will show up in the final finish when the formwork is stripped.

Assemble panels on a flat work surface and stack them face to face for tie rod drilling and alignment.

A release agent is then broomed, rolled or painted on before lifting the first panels into position. Locate any penetrations and window and door block-outs, and then lift the second panel into position.

'He' bolts are often used in wall form construction because they hold the walls the correct distance apart and prevent bulging, and resist inward movement too (see Figure 1.41 earlier).

If 'she' bolts are used, the erection methods will be different. For more information on wall tie systems, visit the RMD Australia website (see the website at the end of this chapter).

When pouring concrete into forms, you need to consider the pressures imposed on the formwork and how that changes, depending on the location and type of form (e.g. floors, columns, walls or inclined forms). Ground floor slabs are well supported on the ground and the edge beams, and intermediate beams are usually well supported with stakes embedded into the surrounding ground (see Figure 1.51 and 1.52).

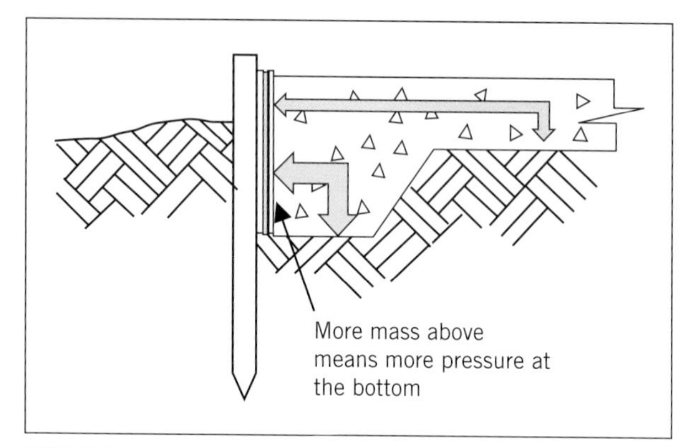

FIGURE 1.51 Slab on the ground

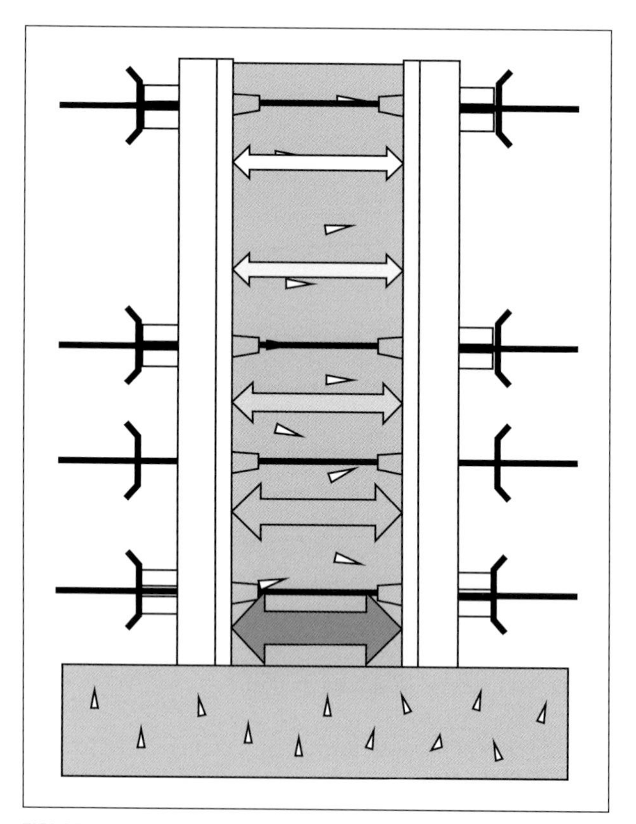

FIGURE 1.52 Concrete pressure during pour and vibrating

Block-outs and penetrations

Penetrations and **block-outs** are areas in floor slabs, columns, beams and walls where concrete is not being placed. Penetrations may be for pipes, ductwork, lifts and stairwells, as well as for doors and windows (see Figures 1.53 and 1.54).

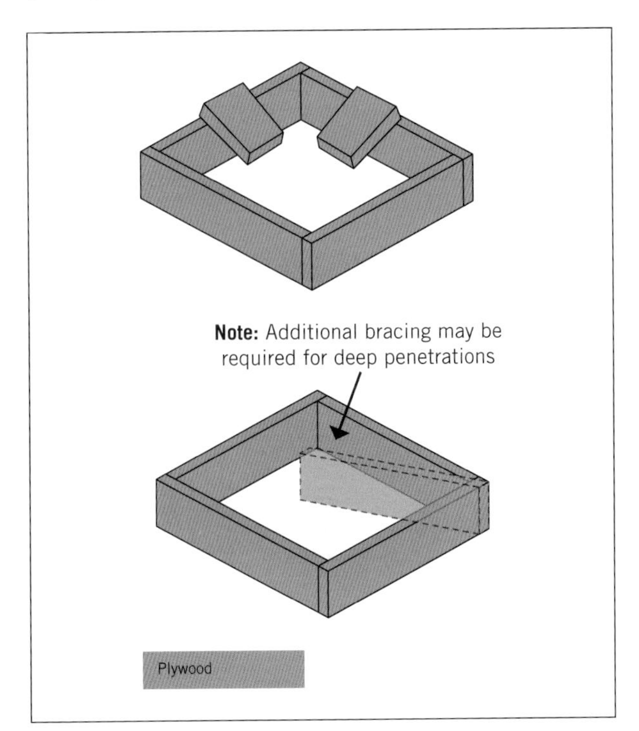

FIGURE 1.53 Square and rectangular penetrations – forming and stripping

Source: Gordon Slater

FIGURE 1.54 Cast-in services

Block-outs are recesses in the concrete surface that may be used to reduce the thickness of material or to allow for conduits or cast-in other fitments. Block-outs may also be used to form keyed joints in wall constructions.

It is important to look for block-outs on plans and carefully set them out according to the details provided, which may be grid references and coordinates or simple measurements from a given datum.

The edges of the penetration can be formed up and braced with plywood, metal or timber (see Figure 1.55). Pipework and conduit plastic sections and foam blocks (depending on the size and shape required) can be set in position before the steel is set in position.

For plumbing services, the penetrations are made larger than required so that the final plumbing works

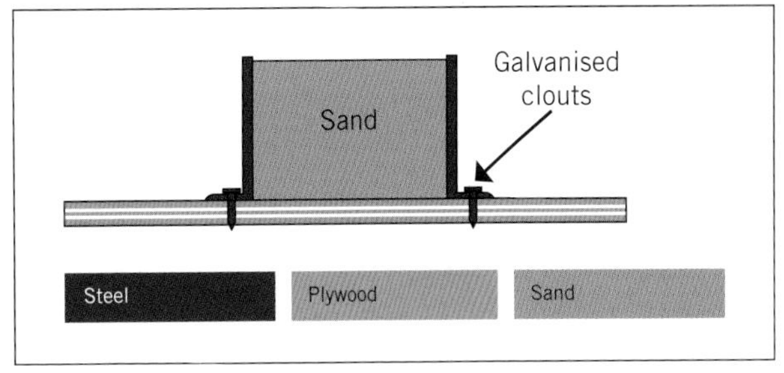

FIGURE 1.55 Sheet metal forms filled with sand

can be inserted. Any gaps are usually filled with fire-retardant foam fill or a similar product.

Formwork is constructed with a view to being able to strip out the penetration-forming material easily.

Penetrations must be located and fixed to the formwork accurately so that they do not move when the concrete is poured (see Figure 1.56). They must be recyclable wherever possible.

Source: TAFE NSW (Illawarra Institute, Wollongong), Building & Construction teachers

FIGURE 1.56 Blocking fitted for penetrations

If you are using low-density rigid plastic foam, cut the material slightly wider than the formwork gap so it is held in place once the formwork is tight.

For larger penetrations such as doors and windows, a shrinkage allowance is required.

Release agents

Release agents make it easier to remove the formwork from the concrete and easier to remove concrete from the formwork. Some release agents can affect the surface colour of the concrete so only use what has been specified. There are various types of release agents, including the following:

- Neat oil is not recommended as it can increase the number of blow holes in the surface finish. This can be reduced by using an oil diluted with a surfactant (wetting agent).

- Water-soluble emulsions are also not recommended because they can produce a dark, porous surface.
- Mould cream emulsions are an emulsion of oil and water similar to that used in the milling of aluminium and steel. They can be removed or affected by wet weather but are well regarded.
- Chemical release agents are chemicals suspended in a thin oil distillate (like diesel). They react with cement to form a soapy interface and are recommended for a high-quality finish.
- Wax emulsions dry off and are very useful in harsh climates.

Release agents can be applied with a broom, roller or brush. For larger areas, they can be sprayed on. Check the release agent manufacturer's specifications for the correct application procedure. If any personal protective equipment is recommended, check the safety data sheet (SDS; previously known as material safety data sheet, or MSDS) for the product before applying it.

Clean up

After formwork is completed and before the release agent is applied, remove debris, sawdust and other waste materials according to the work site requirements and workplace procedures. Clean the area, sort materials for landfill, recycling and reuse.

Complete a visual inspection of all tools and equipment to ensure safe operation. Regular checks and maintenance on tools and equipment should be done each time they are used. If faults are found, follow workplace procedures to record problems, tag the tools or equipment, and relocate them away from other functioning tools.

LEARNING TASK 1.3

CONSTRUCTING AND ERECTING FORMWORK/FALSEWORK

Search for the case study 'Safety and health lessons learnt: Formwork incident' on the website of the Office of the Federal Safety Commissioner (Department of Employment and Workplace Relations): https://www.fsc.gov.au/search.

Read the case study and then discuss the following questions in class or in small groups:

1. Who was responsible for the failure of the formwork and why?
2. When, why and how should back propping be done?
3. How will you know if someone doing a formwork task is competent? Who is responsible for ensuring their competence?

Provide a photographic example of poor practices in back propping. Explain why it is a poor example and how the back propping should have been done.

Before the concrete is poured and after the steelwork has been placed and tied it is a good idea to go over the formwork and remove any debris such as steel tie snips that can cause rust and stain the surface of the finished concrete. A vacuum cleaner (see Figure 1.57) is the best tool for this work as it can get down between the steel and to the bottom of deep beams and through the clean-out panels at the foot of column sheaths.

Source: Milwaukee Power Tools Australia

FIGURE 1.57 Milwaukee vacuum cleaner

Stripping formwork

Formwork needs to remain in place long enough for the concrete to become strong enough to support itself. This process is known as curing. Stripping the formwork means first removing the falsework and braces, and then finally removing the formwork that gives the concrete its profile.

Stripping formwork for suspended slabs walls, columns and beams can be very dangerous. Watch the SafeWork NSW videos listed at the end of this chapter to see examples of good stripping practices.

Stripping (or dropping or striking) the formwork is done only when the concrete is able to support its own load plus the load of workers, material and equipment that will be placed on it during ongoing construction activities.

A gradual transfer of the load is recommended, so impacts and sudden shocks must be avoided. Weather conditions should also be considered and stripping should be avoided during extreme changes in weather conditions.

Sudden shocks can be avoided by leaving some of the formwork or props in place. Beams can be extremely vulnerable to shear and compression forces that, because of their additional mass, can cause distortions. Special props with adjustable collars and beams can be used as formwork for beams. The props bear directly on the ply formwork, while the special adjustable collars support the beams against the plywood. This means that you can drop the collar and remove the beams, leaving the props in place to support the formply (see Figure 1.58).

Other types of patented steel formwork systems operate in similar ways.

Before stripping, make sure you have written permission from the site engineer.

The minimum periods for stripping or striking formwork as set out in AS 3610, Section 5.4, Table 5.4.1 are dependent on the temperature of the surrounding environment (known as the ambient temperature) and are as follows:

- vertical forms such as columns and walls: 12 to 18 hours
- slab soffits: four to six days
- beam soffits and props to slabs: 10 to 15 days
- props to beams: 14 to 21 days.

Note: The colder the weather, the longer the formwork must remain in position.

In general, slabs can be stripped before beams because beams are heavier than slabs and may cause the slab to deflect or bend if not properly supported after stripping. Stripping is generally done before the concrete achieves full strength; this allows the formwork to be cleared and reused straight away.

The process of stripping formwork depends on the formwork system used, but in all cases stripping must avoid excessive physical shocks and the transfer of the load should be done gradually.

Back propping and reshoring

Back propping and **reshoring** are processes used in multistorey construction and involve the replacement of falsework/formwork with temporary props, allowing the slab columns and beams to cure to full strength while construction work continues. This allows the formwork to be reused elsewhere onsite. Back propping can be complex and requires an engineer's design, specifications and documentation, and it can only be done once the elements have cured for the specified period in the documentation as outlined in AS 3610.1.

The back-propping process requires the installation of additional props directly to the formply only *and* is done before any other falsework materials are removed.

With the falsework around columns and beams removed, the falsework and formwork can be lowered one sheet at a time, with back propping directly to the bare concrete done immediately after and before the next sheet of formwork is dropped.

Reshoring is a process where multiple slabs and beams in multistorey constructions are supported by props until the lowest level has fully cured. The reshore props remain as an integrated structural component that supports the live load from freshly poured slabs and beams. Reshoring helps prevent cracking and deformation in slabs beams in the previously poured slabs.

Both back propping and reshoring should be conducted in a controlled manner.

When block-outs and penetrations have been used, first remove the panel not attached to them. Wall ties are then removed – before the pressure is released from the propping. Often the formwork will come away from

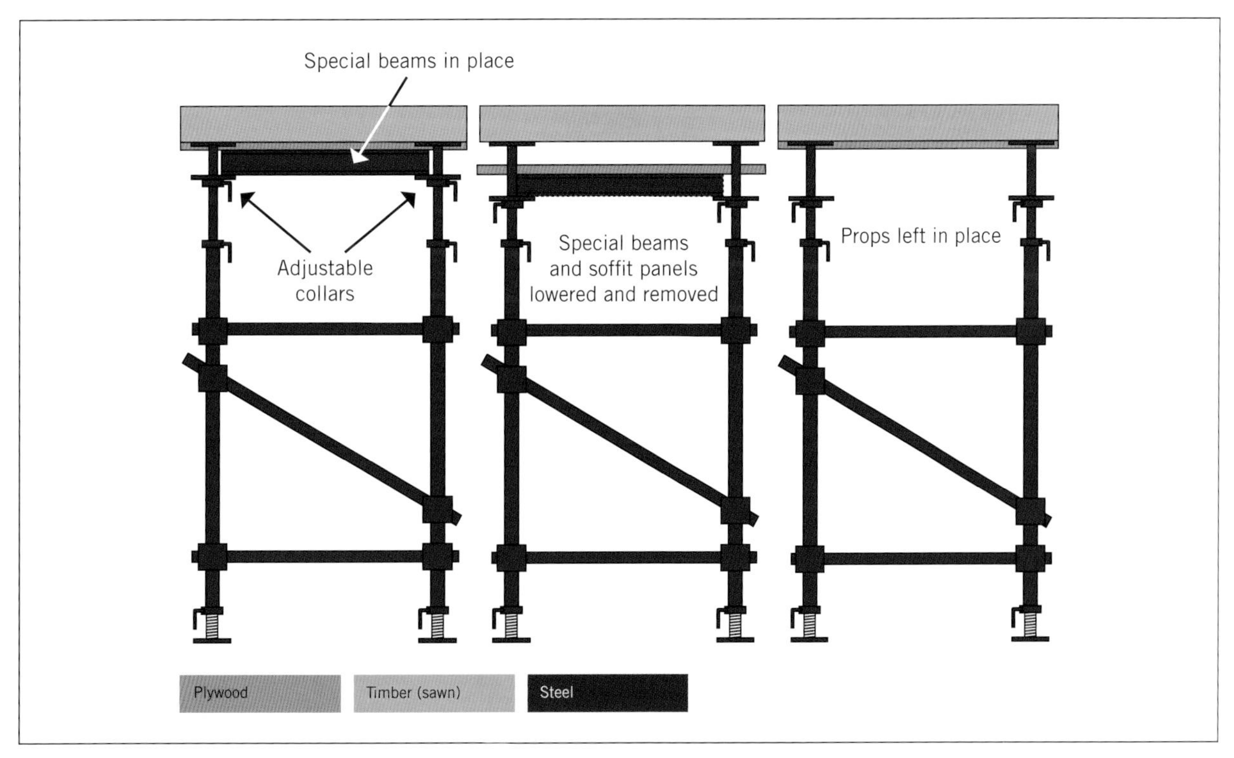

FIGURE 1.58 Formwork incorporating props with collars to enable stripping without removing the props

the face of the concrete with little or no effort. However, sometimes a little persuasion is required. In this situation, the stripping should begin at the top edges and softwood wedges driven between the form face and the concrete. Do not use steel tools as these are likely to damage the face of the concrete.

Finally, loosen the props and remove the panel, taking care not to damage the face of the concrete. The panel can be dismantled or erected elsewhere.

When stripping columns, start in the reverse order used to erect them by first removing ties or props. Remove all the clamps *except for the very top one*. Use temporary cleats to fasten all four sides together, then remove the last clamp. Finally, remove the sides one by one.

LEARNING TASK 1.4 STRIPPING FORMWORK

Research the following:

1 Working in groups of three, conduct an internet search of a work site accident using one of the search items listed below:
   a Barton Highway bridge collapse: 2010
   b 2000 Commonwealth Avenue: 5 January 1971 + formwork failure
   c Skyline Plaza: 2 March 1973 + formwork failure
   d Harbour Cay Condominium: March 1981 + formwork failure
   e The Tropicana Casino parking garage in Atlantic City, NJ: 30 October 2003
   f New York Coliseum: 9 May 1955 + formwork failure.
2 Discuss your chosen case study within your group and present your conclusions to your class by identifying, explaining and considering:
   a the cause of the accident
   b the impact on workers onsite
   c the outcomes of subsequent investigations
   d who was at fault and what should have been done to avoid the accident occurring
   e for the US incidents, what would be the likely outcome if a similar accident occurred in Australia
   f how your group would act to prevent this type of accident occurring
   g if there were a partial collapse due to insufficient propping under a bearer, whether this would be considered a reportable incident to the relevant Safe Work authority. If so, how would you make this report?

Because formply is recyclable and will be used as many times as possible, care has to be taken in cleaning the faces. Use rags and stiff brushes or brooms. Soft timber can be used on tougher patches, but *never* use a wire brush. If the shutter is to be reused, stack it vertically, out of the way but where it can be accessed easily, and give it another coat of release agent. If the shutter is to be taken apart, clean it and stack it flat under cover. Good manual handling and timber storage procedures should be used to avoid damaging the timber.

Formwork components, especially those made of wood, can absorb moisture from the concrete leading to swelling, warping, or degradation over time. Applying a layer of release agent or form oil before pouring concrete creates a barrier between the formwork and the concrete. This prevents adhesion, making it easier to strip the formwork after the concrete has set. Oiling also extends the lifespan of the formwork by protecting it from the corrosive effects of concrete.

Proper storage or stacking of formwork components prevents damage, warping, or deterioration. Components should be stored in a dry, well-ventilated area, away from direct sunlight and extreme temperatures.

COMPLETE WORKSHEET 6

Modular patented formwork systems

Today there are a number of patented formwork systems designed for faster more efficient erection and stripping.

They fall into three main categories:

- steel prefabricated formwork
- aluminium formwork systems
- polymer (plastic and foam) systems – designed to remain in situ.

In most cases the formwork design team will nominate the system to be used and provide engineered solutions for difficult circumstances.

There are many 'systems' available, suitable for columns, walls and slabs. The materials used can be a combination of polymer-based materials and steel construction or all-aluminium shutter and sheath systems. Many have a minimal number of components for walls, columns and slabs, making it easy and efficient to form up with only a minimal number of different system components.

The modules or shutter panels are usually 0.9 m to 2 m in length. The beams and formwork are typically set by hand and pinned, clipped or bolted together.

Modular systems offer these advantages:

- faster construction methods using semi-skilled labour
- modules that can be removed after concrete sets, leaving only back propping in situ during a full curing process.

Steel prefabricated formwork systems

There are many different systems available; some are full steel while others combine materials to create walls, floors, and column sheaths and shutters. Some wall formwork systems use a galvanised frame with insert ply or composite panels to create modular shutters and sheaths (see Figure 1.59). Hinged corner joints provide flexibility of design forms. Careful design specifications and selection must be made to ensure the systems used are suitable for the kPa rating of the concrete being used. In most cases, panel modules can be assembled by hand, but once joined on the ground have to be craned into position; otherwise scaffolding will be required to reach the heights if erected in situ.

Source: RMD Kwikform

FIGURE 1.59 RMD Rapid Ply steel-framed formwork panels for walls

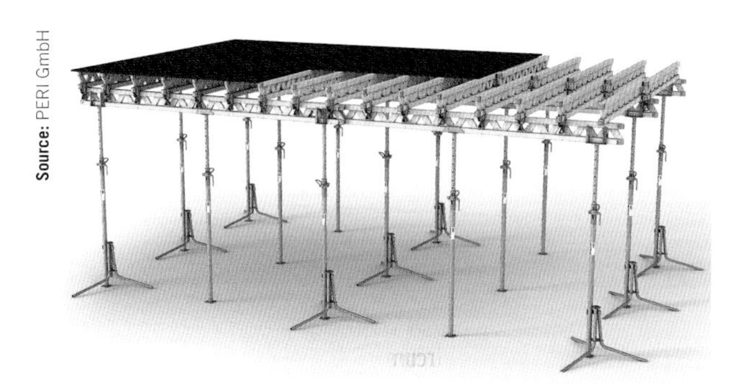

Source: PERI GmbH

FIGURE 1.60 PERI MULTIFLEX girder system

Many of the steel-based floor formwork systems are designed to remain in place and form part of the structural reinforcement system.

Some steel decking forms can reduce the need for propping entirely. Each deck or slab has to be designed precisely.

The advantages of steel prefabricated formwork systems include that they:

- span large distances without propping
- can accommodate additional reinforcing
- are pre-cambered to minimise concrete deflection
- require reduced concrete volume because of the reduced number of beams and columns
- offer quicker access for trades and services to continue work
- result in slabs that often achieve a higher fire rating than conventional column and beam construction
- can be reused many times making them more environmentally sustainable.

These systems can make them suitable for carparks, industrial buildings and open plan offices.

When using plywood as temporary formwork in conjunction with metal decking, care must be taken to minimise the risk of corrosion due to prolonged contact with wet or damp plywood or joists. Moisture from wet concrete or rain can soak into the plywood or joists and cause corrosion. The chemicals contained in some formply may also contain corrosive chemicals. By failing to prevent contact between wet plywood, lumber manufacturers' warranties can be voided. To minimise, use a protective barrier such as foam strips between the metal decking and the plywood, remove any corrosion on the surface of the decking, and treat with a zinc-rich paint.

Some brands include:

- TRUEDEK®: https://www.truedek.com.au/
- Formdeck Australia: https://www.formdeckaustralia.com.au/
- RMD Kwikform – Minima: www.rmdkwikform.com/au/products/minima-formwork/
- Form Direct: https://formdirect.com.au/products/formwork-reinforcing/steel-ply-modular-formwork.html
- Ultrafloor: https://www.ultrafloor.com.au

Aluminium formwork systems

Aluminium systems are usually patented. They allow an entire structure to be formed and poured with speed and efficiency. In most cases the extrusions are powder coated, which provides for a smooth surface finish and the use of less release agent, making cleaning easier. Being aluminium they are fully recyclable and can be repaired when minor damage to surfaces and edges occurs. (See Figure 1.61.)

FIGURE 1.61 Aluminium fence formwork

Some systems use combinations of aluminium beams with regular propping and patented bracing systems. Beam sizes range from 0.9–7 m. Aluminium beam systems can be up to 60 per cent lighter than timber beams used for soldiers and walers. Aluminium beams have greater bending and shear resistance than timber, allowing for more weight to be supported over greater spans, saving time and onsite costs (see **Figure 1.62**).

Source: PERI GmbH

FIGURE 1.62 PERI GRIDFLEX simple panel installation

The panel can be aluminium framed and fitted with a 10–12 mm formply face. The formply is usually fully enclosed in the extruded and welded frames to protect the edges, giving a longer life span.

Different systems have different specifications, but a common frame depth of 140 mm. Panel sizes range from a typical 900 × 450 mm to 1800 × 900 mm.

Polymer-based systems

Polymer-based materials are light and strong and can be handled manually, thereby reducing the need for many crane operations. Live- and dead-load capacity must be calculated by a qualified and experienced formwork engineer.

There are systems suitable for slabs, wall, and column and beam construction. The panels come in a range of sizes suited to most situations. They are lightweight, reusable and recyclable. See **Figures 1.63** and **1.64**. Some manufacturers have panels that provide decorative textured finishes such as block, stonework and timber panelling.

Most systems use wall ties, bracing and propping, and must be secured to the groundwork before pouring concrete. Some polymer-based formwork systems are used for wall constructions and remain in place, removing the need for stripping out. In below-ground areas this can reduce waterproofing requirements.

Structural load-bearing wall systems allow floor and other vertical elements to be built simultaneously. They can reduce the need for columns and beams, thereby using less concrete and less reinforcement.

In all systems, the panel layout must be planned to allow for the most efficient and safest erection, and ease of stripping.

Source: PERI GmbH

FIGURE 1.63 PERI Duo system
Note: Workers are not wearing hi-vis vests.

Source: Allcon Group

FIGURE 1.64 Geoplast lift shaft

Some polymer-based brands include:

- PERI – Duo system: https://www.peri.com.au
- Geoplast: https://www.geoplast.com.au
- Dincel Structural Walling: https://www.dincel.com.au

When considering using imported building materials, some of which may be non-compliant products or non-conforming, builders and head

LEARNING TASK 1.5

In groups of three, research a modular patented formwork system from the companies listed below. (Teacher should assign a different company to each group, and may provide other companies as alternatives.)

- PERI Australia – MULTIFLEX Girder Slab Formwork and PERI Duo system
- Acrow – ACROWALL-60
- Ancon – QwikForm Modular Formwork System
- AFS Rediwall® – permanent PVC formwork walling solution
- Ihita Formworks Pty Ltd – aluminium formwork systems
- TRUEDEK® – permanent concrete formwork system.

Each group is to explain their product advantages and disadvantages to the group and seek feedback and views in group discussion.

contractors must be aware that they may become liable for the use or misuse of those products. Builders should check on the country of origin for imported products, and ensure sufficient information and evidence is available to verify the products are 'fit for purpose' before using the product.

3D concrete printing

The common method of casting concrete in formwork is set to be revolutionised by 3D-printed concrete. This technology combines digital robotic technologies with advanced geopolymer material technologies to produce difficult-to-form construction shapes without the use of formwork.

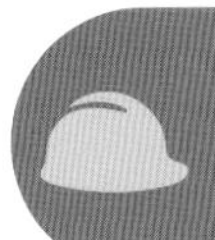

The materials used for 3D concrete printing may be geopolymer-cement-based products manufactured from industrial byproducts, creating an ecologically sustainable alternative to cement-based systems.

Unless the high costs of specially shaped formwork are met, the typical formwork/falsework used to shape concrete limits the design capabilities to basic rectangular forms in columns and beams. A further advantage of the curved form is its strength, having greater resistance to shear and tensile forces.

Reinforcement of the concrete printed structures still has to be considered, but additives such as fibreglass or metal pins can vastly increase the strength.

The advantage of 3D-printed concrete forms, especially for columns and beams, is the ability to precast unique design shapes off-site and deliver them to site, fully cured.

Advantages of 3D-printing:

- reduced cost: eliminating formwork/falsework construction saving significant time in construction
- reduced injuries: eliminating site hazards and risks
- reduced mistakes: highly accurate computer-based machine integration is used in manufacture.
- high-tech job creation: of design and manufacturing skills.

Disadvantages of 3D-printing:

- size and cost of 3D concrete printers
- set-up times and site logistics
- access to quality printable materials
- additional surface finishes and treatments.

While the technology is still in its infancy, material technologies are still being refined, and there is no doubt its many advantages will drive future acceptance in the industry.

LEARNING TASK 1.6

In groups of three, use the search term '3D concrete printing' to watch at least three YouTube videos on concrete. (Teacher is to assign a topic to each group.)

Discuss and consider:

- the potential impact on on-site labour components and site setting out
- how it might impact site setting out footings on sloping ground
- what new skills carpenters and builders will require in working with this technology
- how the work of a formworker may change into the future
- the impact on other trades such as bricklayers, plumbers, electricians, plasterers and painters.

Each group is to explain their point to the group and seek feedback views in group discussion.

SUMMARY

How to construct, erect and dismantle the formwork for suspended slabs, columns, beams and walls and the different materials used in their construction:

Legislation, codes and standards

- Be familiar with your state's regulations, codes of practice and safe work guidelines.
- Be able to access approved working drawings and know the process for approval of changes onsite.
- Understand:
  - stages of construction
  - requirements for inspecting formwork
  - the stripping of formwork requirements.

Quality assurance

- Formwork must be inspected and approved before steel fixing and concrete placement.
- The quality of concrete finishes can be affected by poor-quality formwork, especially at corners and joints.

Materials and stresses

- The stresses imposed on concrete come from its own weight and the loads placed upon it.
- Calculations are required to determine live and dead loads.
- The proper placement of steel and formwork supports is important.
- There is a difference between shear and compression forces.
- When setting out beams, it is important to consider the increasing weight of concrete as it is poured and the need to strengthen the bottom edges of beam formwork.
- Offsets or 'eccentricities' in the design of formwork need to be allowed for.
- Control joints are used to limit cracking in slabs.
- Slabs are set out in similar sized sections and a pouring plan is worked out.

Constructing and erecting formwork

- Common materials used in formwork construction include solid timber and formply. Other common components include:
  - props, walers, braces, shutters and sheaths
  - column clamps
  - wall ties (he and she).

- The key points in constructing formwork are accuracy, stability, safety, economy and quality assurance as follows:
  - When forming up walls, formply sheaths or patent-steel forms are used with studs and walers aligned, depending on the direction of the face grain of the plywood.
  - When vertical studs are used, a top and bottom plate is used to support the bottom edge of the sheath.
  - Single-faced walls are usually retaining walls using patented brace systems, similar to those used for tilt slab construction, while double-faced walls have their faces tied together with 'he ties' to maintain an even thickness.
  - Walls need to be joined using keyed joints or scutched and connected using starter bar, mesh or dowel joints.

Stripping formwork

- Plan erection of formwork with the ease of stripping in mind.
- Formwork remains in place long enough for the concrete to cure and support itself and the load of workers, material and equipment that will be placed on it, depending on the weather conditions and the type of form being supported.
- Incorporating stripping bands and shutter infills into beam formwork makes stripping easier.
- Deep beam construction requires wall ties and double joists.
- Release agents make it easier to remove formwork, but some release agents can affect the surface colour of the concrete.
- When stripping formwork, sudden shocks must be avoided. Back propping and reshoring may be required, especially in multistorey work.
- Back propping and reshoring:
  - must be engineer designed
  - is done to allow work to continue
  - allows formwork to be used elsewhere on-site while the concrete fully cures.

Modular patented formwork systems

- Patented modular systems are available in steel, aluminium and polymer (plastic).
- Steel system advantages:
  - reusable and sustainable
  - can span large distances without propping.
- Steel system disadvantage:
  - heavy to handle and prone to corrosion.
- Aluminium system advantages:
  - relatively easy to handle
  - reusable and recyclable sustainable
  - patterned faces easily achieved.
- Aluminium system disadvantage:
  - easy to dent or damage.
- Polymer-based system advantages:
  - lightweight
  - reusable and sustainable
  - can reduce the need for columns and beams.
- Polymer-based system disadvantage:
  - not suited for complex shapes and forms.

REFERENCES AND FURTHER READING

Texts

Laws, A. (2020), *Site Establishment, Formwork and Framing* (4th edn), Cengage Learning Australia, Melbourne, VIC.

Web-based resources

Codes of practice and guides

Not all states and territories have a formwork code of practice. Those that do are listed below.

Queensland – *Formwork Code of Practice 2016*: **https://www.worksafe.qld.gov.au/__data/assets/pdf_file/0019/15823/formwork-cop-2016.pdf**

Safe Work Australia – *General Guide for Formwork and Falsework*: **https://www.safeworkaustralia.gov.au/system/files/documents/1702/formwork-falsework-general-guide.pdf**

Safe Work Australia – *National Code of Practice for Precast Tilt-Up and Concrete Elements in Building Construction:* **https://www.safeworkaustralia.gov.au/system/files/documents/1702/codeofpractice_precasttiltupandconcreteelementsbuildingconstruction_2008_pdf.pdf**

SafeWork NSW – *Guide Formwork*: **https://www.safework.nsw.gov.au/hazards-a-z/formwork**

WorkSafe Victoria – *Construction safety focus: Formwork* **https://www.worksafe.vic.gov.au/resources/construction-safety-focus-formwork**

Training.gov.au **https://training.gov.au/Home/Tga**

Industry organisations

Cement Concrete & Aggregates Australia, *Guide to Off-form Concrete Finishes* (2006) and *Joints in Concrete Buildings* (2004): **https://www.ccaa.com.au/**
Note: use the full title of each in the website's search tool.

Engineered Wood Products Association of Australasia: **http://ewp.asn.au/**

National Precast Concrete Association Australia, online learning resources: **https://nationalprecastonline.com.au/**

Video resources

BM Formwork: How to build a suspended concrete slab:
Part 1: **https://www.youtube.com/channel/UCFoolRoyayXyDrqFMfWvr3w**
Part 2: **https://www.youtube.com/watch?v=dWTNPCOZVcO**

Buildsum (AUS) Formwork Terminology: **https://www.youtube.com/watch?v=jIBecLpRYwQ**

Concrete formwork: (USA): **https://www.youtube.com/watch?v=3wlW9cUUVlc**

Safe Work NSW – Safe Ways to Erect Formwork: **https://www.youtube.com/watch?v=hN9mkMHmtkI**

Stressing the Post-Tension Cables (USA): **https://www.youtube.com/watch?v=PDgfnGqPj1c**

Wall formwork hardware: (USA): **https://www.youtube.com/watch?v=bxLJcJqYFQY**

Formwork failure videos

Beam formwork failure: **https://www.youtube.com/watch?v=GfHVm_XPhvs**

Formwork blowout with explanations: **https://www.youtube.com/watch?v=hd_o730PWZk**

Illegal formwork stripping: **https://www.youtube.com/watch?v=mIS2Xk59M48**

Safe way to strip formwork: Safe Work NSW: **https://www.youtube.com/watch?v=0kcVIGxKYoE**

Shuttering/formwork failure: **https://www.youtube.com/watch?v=0VkhhOYOKRQ**

Corporate

Acrow Formwork & Scaffolding (Acrow): **http://www.acrow.com.au**

Bluescope Steel: **http://www.bluescopesteel.com.au/building-products/formwork-and-slabs**

Carter Holt Harvey – ECOPLY® FORMRITE® FORMPLY: **https://chhply.com.au/ranges/ecoply/products/plyfloor-flooring/**

Gunnersen: **https://www.woodsolutions.com.au/wood-species/wood-products/plywood**

Layher – modular scaffolding: **http://www.layher.com.au**

PERI Australia – Formwork systems: **https://www.periaus.com.au/products/formwork.html**

RMD Kwikform: **https://www.rmdkwikform.com/au/**

Datums and coordinates

Geoscience Australia: **http://www.ga.gov.au/scientific-topics/positioning-navigation/geodesy/datums-projections**

Relevant Australian Standards

AS 3610.1 Formwork for concrete. This Standard and its supplements sets out requirements for the design, fabrication, erection and stripping of formwork.
AS 6669 Plywood – Formwork. This Standard specifies requirements for the manufacture, grading, finishing and branding of plywood used specifically in formwork, with a maximum length of 3100 mm and width of 1500 mm, intended to meet off-form surface finish requirements.
AS/NZS 2098.0 Methods of test for veneer and plywood.
AS/NZS 2269.0 Plywood – Structural – Specifications.
Australian Standards can be purchased from the Standards Australia website: **https://www.standards.org.au/**, or see your teacher or librarian for assistance in accessing Australian Standards online.

GET IT RIGHT

FORMWORK, SLABS, COLUMNS, WALLS

Questions

Below are four pictures showing the results of formwork as seen in the completed concrete suspended slab.

1 What was done wrong with the formwork preparation?

______

______

2 How should it have been done correctly?

______

______

3 What should be done to rectify these problems?

______

______

4 What could happen if nothing is done?

______

______

Source: Adrian Laws

Source: Adrian Laws

Source: Adrian Laws

Source: Adrian Laws

HOW TO 1

HOW TO

SET UP A FORMWORK FOR A SIMPLE SUSPENDED CONCRETE SLAB

Before beginning any formwork construction work, make sure all necessary formwork plans, hazard identification and risk management plans have been completed and approved by site supervisors or engineers.

<table>
<tr><th colspan="3">Steps involved</th></tr>
<tr><td>1</td><td>Set out the sole plates on which the formwork frames and props will sit.
You can only do this based on the formwork plan or grid and the system being used. If you use long lengths the loads will be spread over greater distance and make setting out easier. See Figures 1.26 and 1.27.
A scaffold can be erected in lieu of extending formwork to provide a working platform.
Note: Large floor areas are rarely poured all at once.</td><td>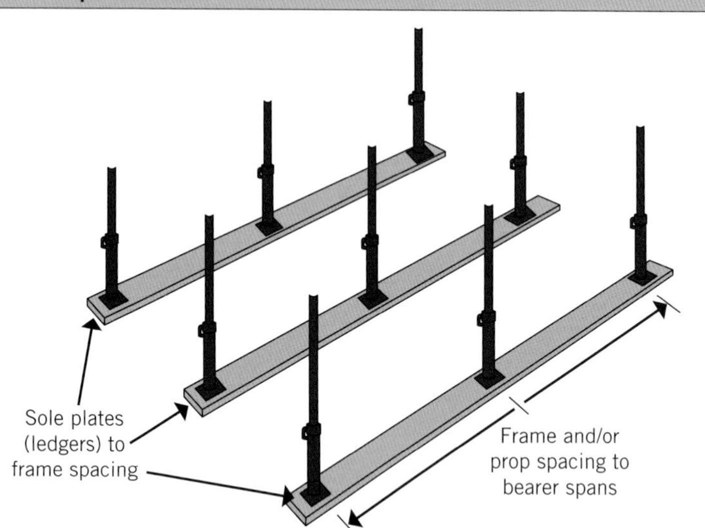</td></tr>
<tr><td>2</td><td>Set the height of the foot and head screw jacks by working back from the soffit height and making allowances for:
• slab thickness
• decking thickness
• bearer and joist widths
• stirrup heads
• formwork frame heights.
You may need to make two calculations if beams are required as these will have a different soffit height.</td><td>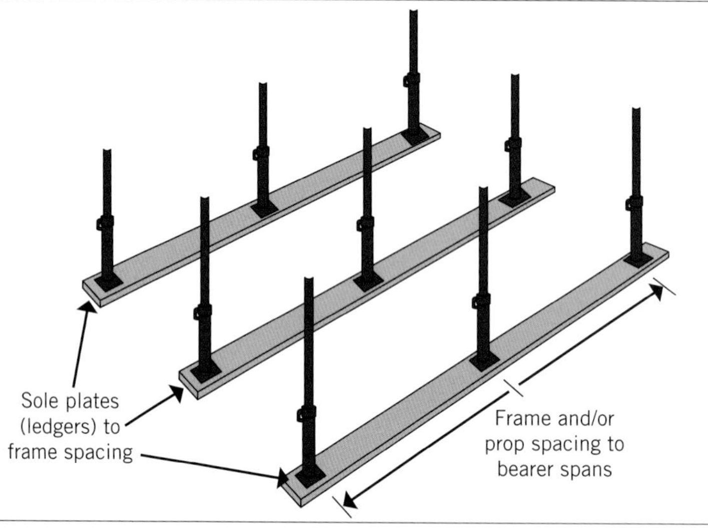</td></tr>
<tr><td>3</td><td>Set out the frames on the screw jacks. Make any adjustments and add in ties and braces and any additional props for stability and safety.</td><td rowspan="2">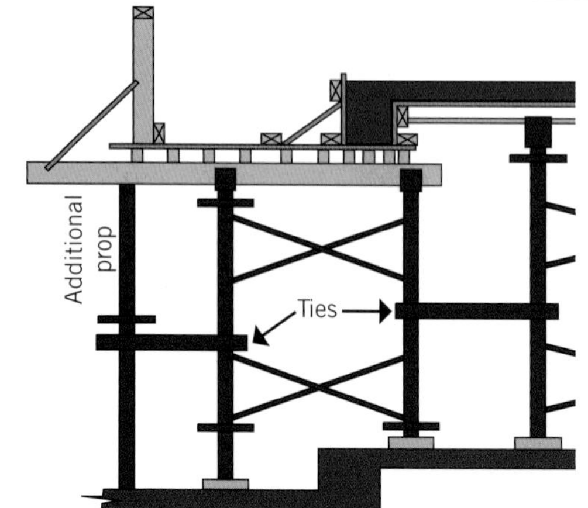</td></tr>
<tr><td>4</td><td>Install the stirrup heads and bearers (usually 1200 mm centres) making sure they are aligned to the grid shown on the formwork plan before installing the joists (usually 400–450 mm centres).
Note: Beam soffits are set at the same time.</td></tr>
</table>

>>

>>

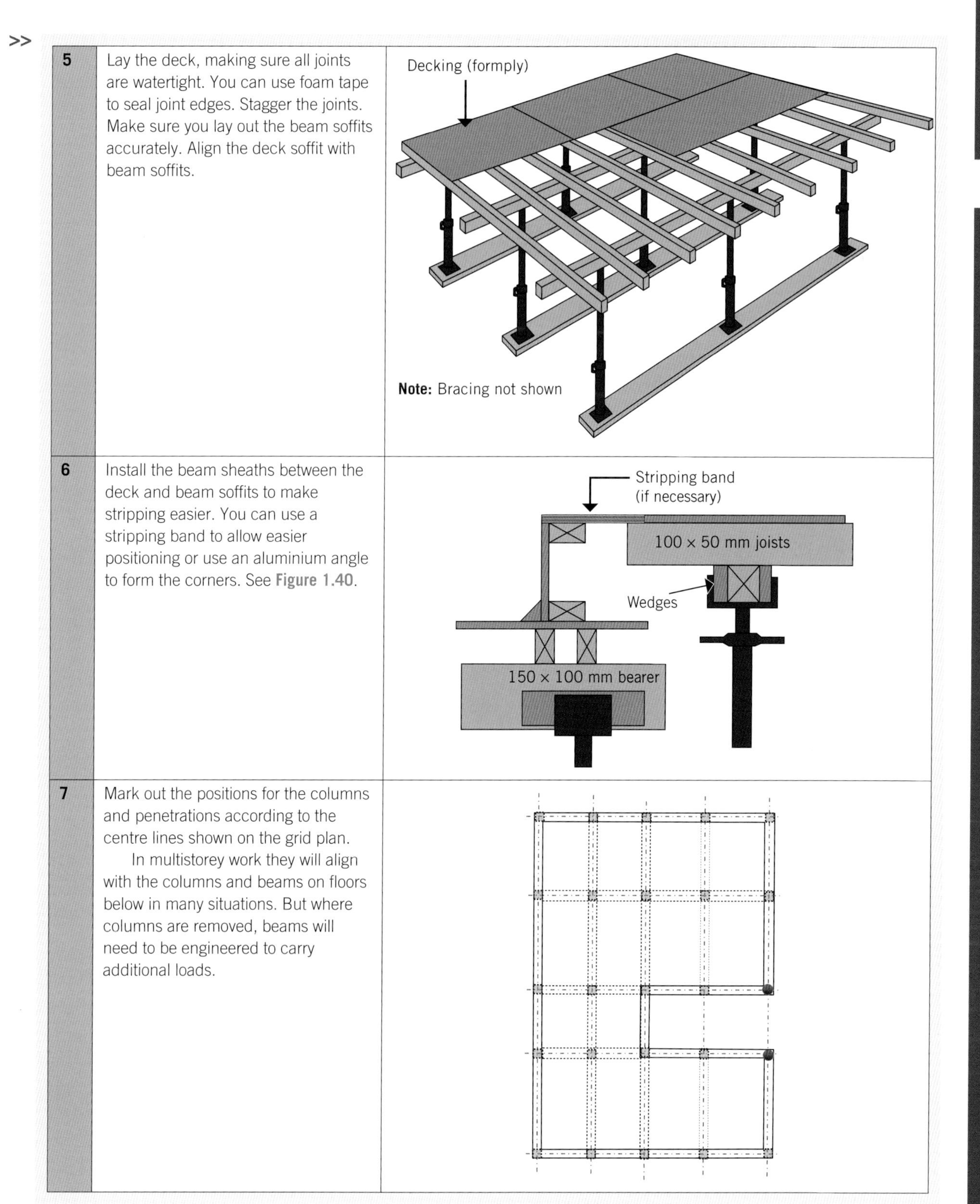

5	Lay the deck, making sure all joints are watertight. You can use foam tape to seal joint edges. Stagger the joints. Make sure you lay out the beam soffits accurately. Align the deck soffit with beam soffits.	
6	Install the beam sheaths between the deck and beam soffits to make stripping easier. You can use a stripping band to allow easier positioning or use an aluminium angle to form the corners. See **Figure 1.40**.	
7	Mark out the positions for the columns and penetrations according to the centre lines shown on the grid plan. In multistorey work they will align with the columns and beams on floors below in many situations. But where columns are removed, beams will need to be engineered to carry additional loads.	

WORKSHEET 1

To be completed by teachers	
Student competent	☐
Student not yet competent	☐

Student name: ______________________

Enrolment year: ______________________

Class code: ______________________

Competency name/Number: ______________________

Practical exercise

Set out and form up for the columns, beams and suspended slab shown in the drawing below:

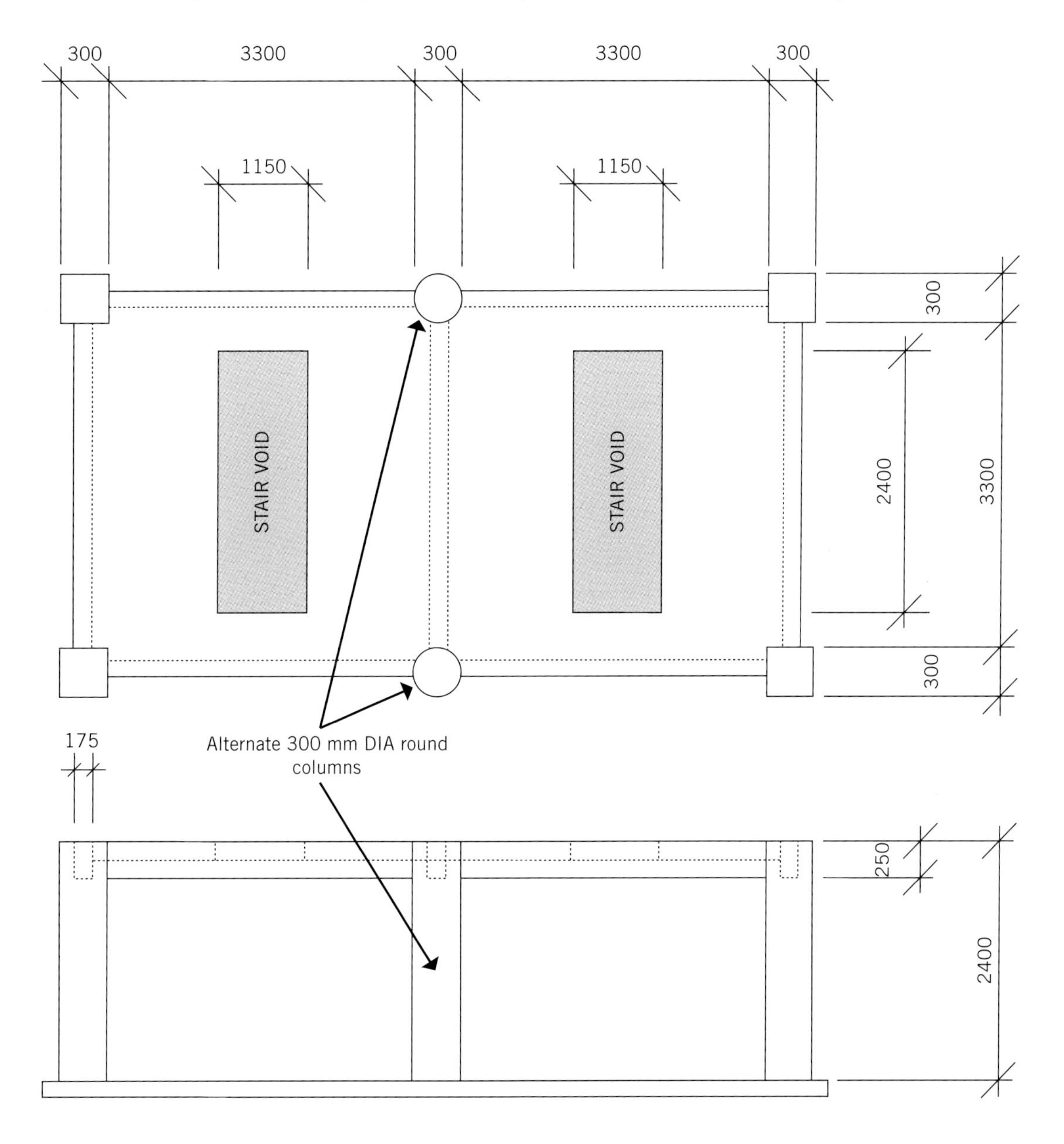

Task details

Students are to work in small teams (groups of three for columns and beams, and larger groups of eight for soffit work).

A safe work method statement must be completed by each student before commencing work. Teacher is to sight and sign.

All students are to develop and draw a formwork plan for the overall project.

Students are to develop a quality assurance checklist for the project.

Students are to develop a full materials list for the project, including sheet materials, bearers, joists, props, scaffold frames, jacks and stirrups.

Students are to calculate:

1 the total volume of concrete required for the job.

2 the loads involved for each column, slab and beam, and the entire project.

WORKSHEET 2

To be completed by teachers	
Student competent	☐
Student not yet competent	☐

Student name: ______________________

Enrolment year: ______________________

Class code: ______________________

Competency name/Number: ______________________

Answer the following questions.

1 What stresses are imposed on concrete?

____

____

____

____

2 Correctly name the Australian Standard that applies to building formwork and outline the five basic instructions given to build formwork.

Australian Standard:

____

a ______________________

b ______________________

c ______________________

d ______________________

e ______________________

3 What are three main stages of construction identified in AS 3610.1 Formwork for concrete?

a ______________________

b ______________________

c ______________________

4 When should formwork be inspected?

____

____

5 What are the three main points to consider when constructing formwork and falsework?

a ______________________________

b ______________________________

c ______________________________

6 Outline three quality assurance procedures or tasks carried out in formwork construction.

a ______________________________

b ______________________________

c ______________________________

7 Explain the term 'camber' in formwork construction in relation to building formwork for suspended slabs (see 'Definitions' in AS 3610.1).

____

____

____

8 Complete the following sentence by writing the correct words in the spaces provided.

Surface finishes are available for concrete such as ____________ made by scabbling and animal forms set in shape using _________ or _________ moulds. Exposed _________ and _________ concrete surfaces are especially popular.

9 Explain two important factors in placing steel inside formwork.

a ______________________________

b ______________________________

10 Explain how joints are formed between columns and slabs.

____

____

____

11 Explain why fillets are used in formwork.

____

____

____

12 Explain why and what types of solid timber will be used to make shutters for columns and beams.

13 How do you determine the way the face grain in plywood should run in relation to the formwork?

14 How can you make sure joints in formply are watertight?

15 When setting up for soffit formwork you need to allow extra width for working and bracing. As a minimum, how wide should the working strip be?

16 Explain the use of a clean-out panel at the base of column formwork.

17 When fastening sheets in suspended slab formwork, where should nails be placed to secure the sheets to prevent them from moving, and why are they used?

18 Explain how the spacing out of column clamps is done for individual columns and multiple columns.

19 What can be done to eliminate the risks in building formwork for concrete columns and beam construction?

WORKSHEET 3

To be completed by teachers

Student competent ☐

Student not yet competent ☐

Student name: ____________________

Enrolment year: ____________________

Class code: ____________________

Competency name/Number: ____________________

Answer the following questions.

1 According to the SafeWork NSW General Guide for Formwork and Falsework, what steps should be taken to identify hazards before constructing formwork and falsework?

2 Explain the difference between a block-out and a penetration.

3 When making joints in timber props, what type of joints should be used?

4 How should the diagonal bracing of frames be carried out?

5 What are four of the five main methods used to locate and set out concrete slabs, columns, beams and walls?

a ____________________

b ____________________

c ____________________

d ____________________

6 Investigate and answer the following question.

Why are pouring patterns for slabs important and how are they determined?

____

____

____

Suggested resource: Cement Concrete & Aggregates Australia: https://www.ccaa.com.au/.

7 What is the unit of measure used in determining the loads in concrete and formwork placement and how is it determined?

____

____

____

8 The total loads on formwork consist of what three main factors?

a ____

b ____

c ____

9 Calculate the load for the slab (only) that was used in the practical exercise given in Worksheet 1.

____

____

____

10 Calculate the load of one concrete column (not including scaffold and live loads) that was used in the practical exercise given in Worksheet 1.

____

____

____

11 Fill in the missing words below:

Primary beams usually ______________ and ______________ while beams are generally ______________ primary beams.

12 Answer True or False to the following statements by placing a circle around True or False.

	Sample statement: The moon is made from cheese.	True	(False)
1	Slab and beam construction relies upon the quantity and size of steel to support a thin slab.	True	False
2	Square and rectangular recesses in the slab are called drop panels.	True	False
3	The finished sizes of concrete components are the starting points for the measurements.	True	False
4	The use of storey rods in the set-up of props and checking of dimensions and positioning can make the task more complex.	True	False
5	Keeping people out from under formwork during a pour is critical, as is the use of safety caps on top of starter bars.	True	False
6	When installing plywood you should make sure the face grain runs parallel to the span (i.e. across the joists).	True	False
7	For slab set-out you should nail off the sheets at 450 mm intervals to prevent the sheet moving during the pour.	True	False
8	Rectangular columns are usually formed up with a clean-out access hole incorporated near the bottom of the column.	True	False
9	More column clamps are located near the bottom of column formwork.	True	False

13 Provide an example of where pre-tensioned reinforcing of concrete components may be used.

____

____

14 Describe the benefits of post-tensioning concrete slabs.

____

____

15 Provide an example of where lost or in situ formwork may be used.

____

____

WORKSHEET 4

To be completed by teachers	
Student competent	☐
Student not yet competent	☐

Student name: ______________________________

Enrolment year: ______________________________

Class code: ______________________________

Competency name/Number: ______________________________

Practical exercise

Set out and form up for the wall shown in the drawing below.

Note: The formwork is to include window penetration and light switch (of junction box) block-out in the positions indicated.

Note: The knowledge components in this task in part satisfy elements 1, 2, 3 & 4 of CPCCCA3019 and associated performance criteria.

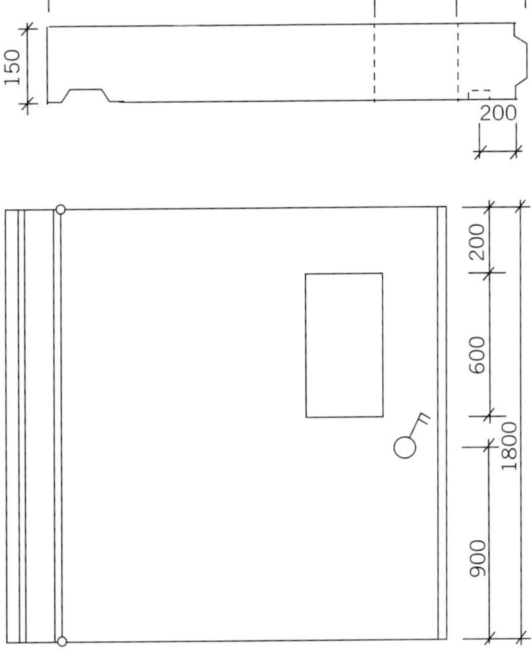

Answer the following questions.

1 Write a plan sequence of work for this job, including hazard identification and safe work method statement.

____

____

____

____

2 Which Australian Standard applies to formwork construction?

____

3 Calculate the volume of concrete required for the wall.

____

4 How many wall ties will be required?

____

5 How many walers will be required?

____

6 For a Class 2 finish, what grade of formply would you need?

____

7 Make a basic tool list for this task.

____

____

8 List the PPE required when performing this task.

____

____

Note: The practical components in this task in part satisfy elements 1, 2, 3 and 4 of CPCCCA3019 and associated performance criteria.

1 Establishing and clearing the work area, to include signage and barricades.

2 Setting out the formwork.

Note: Drawing is not to scale.

WORKSHEET 5

To be completed by teachers	
Student competent	☐
Student not yet competent	☐

Student name: ______________________

Enrolment year: ______________________

Class code: ______________________

Competency name/Number: ______________________

Answer the following questions.

1 When pouring beams, what is considered the weakest point? Describe or sketch a method to reduce the risks.

____

____

____

____

____

____

2 Write the correct response (True or False) in the box provided.

a The major weight of wet concrete in pouring beams should be kept inside the props to prevent a catastrophic collapse of the formwork. ☐

b A stripping band prevents the easy removal of formwork from the sides of columns. ☐

c A shutter infill panel is not an alternative to using a stripping band. ☐

d The studs and walers are aligned with the plywood, depending on the direction of the face grain of the plywood. ☐

3 How are wall positions marked out on a slab if new nibs are cast in?

____

____

____

4 Neat oils (motor oils) are recommended as suitable release agents for formwork. Discuss.

____

____

____

5 Describe three methods that can be used to apply release agents to wall forms.

a ______________________________

b ______________________________

c ______________________________

6 Where large wall panels are required, outline two methods used to set them in place.

a ______________________________

b ______________________________

7 How are the positions of through-bolts and tie rods drilled?

____

____

8 Write the correct response (True or False) in the box provided.

a Angled props on wall shutters are tied together with braces. [ ]

b Block-outs are recesses in the concrete surface, while penetrations are holes that pass right through the finished concrete slab or wall. [ ]

c Gaps in penetrations are often filled using fire-retardant foam fill or a similar product. [ ]

d Some modern steel decking formwork systems require no propping and remain in situ after concrete curing. It reduces the number of columns and beams required. [ ]

WORKSHEET 6

To be completed by teachers

Student competent ☐

Student not yet competent ☐

Student name: ______________________________

Enrolment year: ______________________________

Class code: ______________________________

Competency name/Number: ______________________________

Answer the following questions.

1 How long should formwork for the following elements remain in place before stripping?

a Slab soffits ______________________________

b Vertical forms such as columns and walls ______________________________

c Props to beams ______________________________

d Beam soffits and props to slabs ______________________________

2 Give one example that may increase or decrease formwork/falsework stripping times.

____

____

3 What is the difference between falsework and formwork?

Falsework:

____

Formwork:

____

4 Using AS 3610.1 Formwork for concrete, Part 1: Documentation and surface finish, define the terms 'back propping' and 'reshoring' and explain the difference between them.

Back propping:

____

____

____

Reshoring:

____

____

____

5 Formwork can sometimes 'stick' to concrete, making it hard to strip. What is the best method to release the formwork?

____

____

6 Fill in the missing words in the following statement:

In multistorey work, reshoring is done to ensure there are no _______________ to the permanent structure from the loading of _______________.

7 What are the three main types of modular formwork?

____

____

____

8 Provide three advantages of modular formwork.

a ____

b ____

c ____

9 What tool is best used to assist in stripping formwork and denailing components?

____

FORMWORK FOR STAIRS AND RAMPS

2

This chapter aligns with the unit of competency, 'Construct, erect and dismantle formwork for stairs and ramps'. The unit outlines the skills and knowledge required for the construction, erection and dismantling of basic formwork for stairs and ramps involving one or more flights for access between floors and/or landings. It includes timber, metal or prefabricated formwork.

In particular the chapter addresses the key competencies of:

- planning and preparing for erecting formwork for stairs and ramps including:
  - reading and interpreting documents, plans, specifications and drawings from a variety of sources
  - interpreting and applying quality assurance requirements methods
- setting out of formwork for stairs and ramps including:
  - applying mathematics to calculate material quantities
- assembling, erecting and stripping of formwork for stairs and ramps including:
  - understanding and interpreting construction terminology including WHS regulations and materials handling information
  - selecting formwork materials
  - materials storage and environmentally friendly waste management
  - selecting appropriate plant, tools and equipment types and characteristics, and communicating on their uses and limitations
  - applying correct processes for setting out, erecting and stripping stair and ramp formwork/falsework, including compliance and certification requirements.

In Chapter 1, we included an outline of the Australian Standard 3610.1 Formwork for concrete. This information also applies to stairs and ramps and needs to be read, together with information relating to stairs and stair construction as outlined in the National Construction Code and Chapter 6 of this book.

It is important to note that states and territories may require risk assessment to be done as part of their 'codes of practice'. Check with your state's work health and safety regulator.

Learning objectives

By the end of this chapter you will be able to:

- know and understand the additional loads placed on stair formwork, and account for the forces of concrete placement on different points of the stair formwork by ensuring the structural integrity of the formwork at the time of construction
- set out the formwork for straight flights from slab to slab, and know the advantages of prefabricated steel formwork, and the related Australian Standards

- apply knowledge and skill to construct and erect stair and landing formwork to support the top of the flight, soffit set-out in landings, and the return flight
- set out and construct raked soffits to account for the throat; set out strings, and position and reinforce riser boards on wide flights; and learn how to adapt formwork for stairs against masonry walls and construct formwork/falsework for external flights
- understand and apply design techniques for access and mobility ramps, and their special construction methods
- set out and construct other sloping forms such as vaults and sloping concrete roofs, noting the similarities to formwork/falsework for concrete walls
- select and use the right release agent for use on formwork
- apply appropriate timing sequence and steps involved in removing formwork/falsework safely.

Loads on stair formwork

Concrete in stairs is often placed at the same stage as for slabs, beams and columns. Sometimes concrete stairs are manufactured off-site and craned into position as the floors are constructed (see **Figure 2.3**).

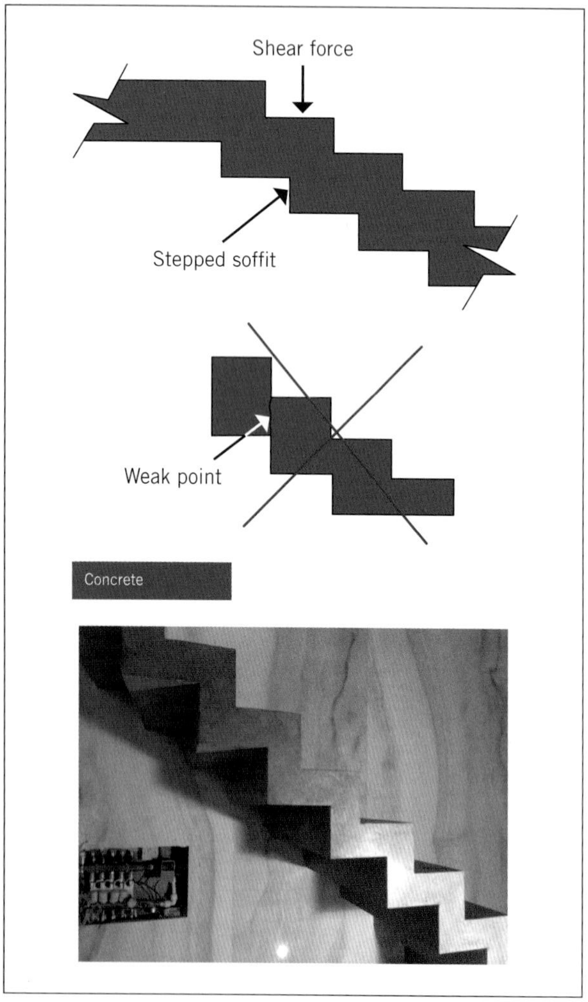

Shear force

Stepped soffit

Weak point

Concrete

Source: NS Formwork and Concreting 2023

FIGURE 2.1 Stepped soffit

Stairs need to be designed to cope with the additional loads imposed by sloping soffits and stepped soffits, especially shear forces (see **Figure 2.1**). Loads are also transferred to adjoining slabs and beams at the top end.

In stair formwork, the load is transferred to the floor at the base and through the struts. It is therefore critical that the fixing points remain firmly anchored so that the load points do not shift during concrete placement and that, wherever possible, the **sole plates** extend back to a wall or other **brace** point (see **Figure 2.22**).

Side forms must be rigid. These are supported similarly to the sides of beams. **Figure 2.2** shows a section through a typical stair soffit form, with bearers, joists, soffit, sides and braces. In some circumstances, **joists** may be forgone and only bearers used below the soffit panels. The side braces shown in **Figure 2.2** hold the stair form sides vertical. Less load is placed on the sides in the stair form than in slab and beam construction because most of the loads are

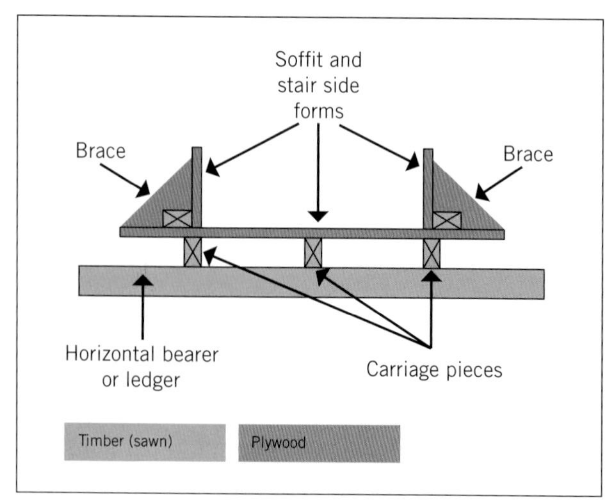

FIGURE 2.2 Section through stair form (riser boards and struts not shown)

Source: Shutterstock.com/Greg Patton

FIGURE 2.3 Precast stairs

directed towards the foot of the stairs and the floor via the struts or props. As well as the side braces, the **riser boards** (not shown) act to brace the sides together.

Note that stairs are a form of eccentric formwork, so must be well braced and supported to prevent collapse.

Setting out stair formwork/falsework

In this chapter we cover setting out the formwork for straight flights, principally from slab to slab.

In recent years there have been a number of innovations that may impact on the way concrete stairs are formed. For example, concrete stairs may be precast away from the site and later delivered to the site and craned into position. This can save construction time but may increase the costs associated with craning the precast stairs into position. This will require careful planning.

Always us a licensed crane operator to lift and place precast concrete components. Make sure the precast products have been designed and certified by an appropriately qualified structural engineer and made to the specifications.

Precast concrete elements provide many sustainable advantages in construction including:

- less formwork onsite, which provides a cleaner worksite
- less onsite labour required
- less waste onsite and better waste management at the precast site
- a faster build.

Watch the video from the National Precast Association on YouTube: https://www.youtube.com/watch?v=2Dm-KY4sRfA&t=127s

One disadvantage may be quality control. Make sure, if casting offsite, the appropriate certifications are made to ensure engineering specifications are maintained (i.e. steel reinforcement sizing and placement).

Another innovation has been the development of permanent steel prefabricated formwork systems that remain in place after construction has been completed. This saves the cost of building the formwork onsite and there is no need for stripping out the formwork once the concrete has been poured and set.

LEARNING TASK 2.1

CONCRETE STAIR FORMWORK INNOVATIONS

Use the internet to investigate a range of concrete stair formwork innovations, including formwork tools and precast staircases. YouTube videos are a good source of information, but remember that they do not always show best practice.

Some useful sites include:

- National Precast Concrete Association Australia: https://www.nationalprecast.com.au
- Fast Tread: https://www.ftigroup.com.au/
- Stairform P/L: https://www.stairform.com.au/stairs.aspx

Before commencing work, always check the plans and drawings for the location and measurements of the stairs. Engineers' drawings should be consulted to ensure load bearing on the adjourning walls will not compromise the structure.

Check the measurements from the plans and drawings at a site inspection. If stairs are made off-site, it is important they will fit into the space.

Ensure the most current version of the plans and drawings are used. Changes and variations are often made to plans and drawings, and sometimes updates are made during construction.

After checking measurements, calculate the quantities and materials required to complete the work. Plan and sequence the work activities to ensure work is completed in an efficient manner.

Complete a site risk assessment before commencing work onsite. You may need to complete a site induction, so check with onsite personnel. Refer to the SWMS and JSA to ensure compliance with all WHS measures on site. Ensure hazards are identified and apply risk-control measures, including signage and barricades where required.

Check the Environmental Plan for requirements in relation to dust, vibration and noise. Locate the waste and recycling points so as to maintain a clean workspace in order to reduce trip hazards and comply with environmental laws.

Tools and equipment

The equipment to set out will vary on the job requirements. Check plans and drawings to determine what tools and equipment are needed.

Tools and equipment can include: hammer, tape measure, square, pencil, levels, chalk line, pegs, shovel, screwdriver or drill.

Always conduct a visual and operational check of tools and equipment to ensure they are safe to use and working to the manufacturer's specifications.

Remove any tools and equipment with obvious faults away from functioning tools. Label and report them as faulty so that other tradespersons won't use them.

It is important to select and use personal protective equipment (PPE) for the task at hand. To comply with SWMS and JSA requirements ensure PPE is in good working order. Boots, hats, sunscreen, gloves and eye protection are commonly used for setting out. If there are asbestos hazards or other materials in the ground that can cause harm when disturbed, a mask may need to be worn. Adjust the required PPE to meet the safety needs of each task.

Steel reinforcement in concrete

Before setting out any formwork you will need to consider:

1 How and when the steel reinforcing will be placed (e.g. Z-bars; see Figure 2.4).
2 The type and size of steel required. A structural engineer will usually specify the quantity, size and type of reinforcing required.
3 If you need to drill the edge of the top slab to insert joining dowels or if the slab reinforcing has to be extended to allow the staircase reinforcing mesh and bars to be tied in.

A number of Australian Standards relate to the use of steel in concrete structures. The main ones are:

- AS/NZS 4671 Steel reinforcing materials
- AS 3600 Concrete structures

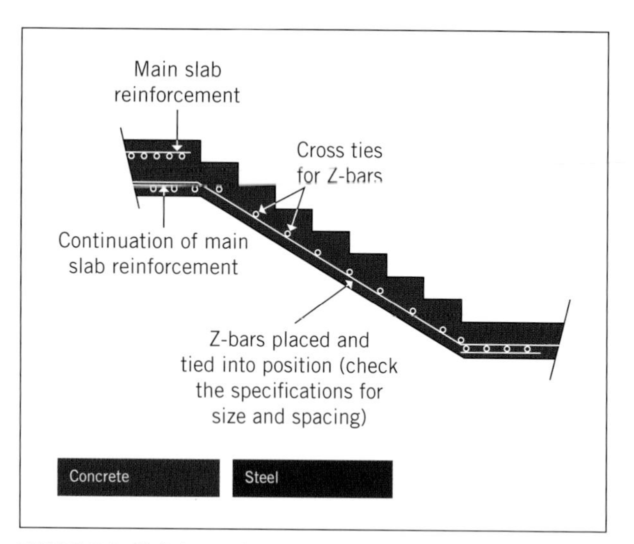

FIGURE 2.4 Slab in steel

- AS 2870 Residential slabs and footings – Construction
- AS/NZS 1170 Structural design actions
- AS 1554.3 Structural steel welding. Part 3: Welding of reinforced steels.

LEARNING TASK 2.2 REINFORCING STEEL

Investigate the types of reinforcing steel used in construction, and write a report providing the following information:

1 The different diameters used for reinforcing bars.
2 The name and description of three different types of reinforcing mesh. Provide an example with pictures of where each is used.
3 Methods of joining reinforcing bars.
4 Different accessories used in steel reinforcing. Describe where and how they are used.

To calculate the quantities of materials required for a concrete set of stairs, you need to know:

1 the overall rise, which will determine the number of rises
2 the overall run, which will determine the number and width of each step
3 the width of the stairs, which will be determined by the job requirements (e.g. if the stairs are set between walls, if one side is captive and the other side is open, or if both sides are open)
4 if the landing is already formed or will be part of the stair formwork
5 if it will need an edge beam at the landing.

The engineer's plan, rather than the architectural plans, will provide more detail, especially around steel selection and placement and other critical dimensions such as throat size.

You can use a regular cutting list to record the sizes of the components or one similar to that used in the video 'Concrete Stair Formwork Calculations' (https://www.youtube.com/watch?v=wjJ04YeV5eM).

When setting out stairs, you need to collect all the information about the stairs, including any toppings such as tiles, rubber or other surface finishes that will be used, as these materials and finishes will affect your final calculations (see Figure 2.5). If the stairs have cast-in newel posts (balusters), you will need to adjust your calculations to cater for them.

Note that most stairs in commercial or institutional buildings will usually be at least 1 m wide between any handrails and may include intermediate handrails. The requirements for stairs and ramps are generally contained in AS 1428.1 Design for access and mobility. Part 1: General requirements for access – New building work. Any fillets or filler pieces that may be used to mould the shape of the steps also need to be taken into account.

Finished floor levels must be checked off the datum and recorded either on a rod or against any walls to which the stairs may be attached if the walls have been built.

When setting out, you will be working from the bottom of the soffit when applying the measurements.

The same rules apply for concrete stairs as apply for stairs constructed of other materials; that is: (2 × Rise) + Going = 550 mm to 700 mm (see Chapter 9 for more detail on calculating the pitch of stairs), but the rules are particularly dependent on the class of building in which the stairs are located.

Note: see National Construction Code – Housing Provisions Standards, Part 11.2.2(2) on requirements for a stairway serving only non-habitable rooms.

Figure 2.6 shows the raked (carriage) bearer and horizontal joists.

It is critical to determine the exact location of the first rise and the last rise. The top of the last rise may already be formed by the edge of a slab or it may need to be determined carefully. The exact rise must be calculated from directly below the last rise, levelling with a straight edge **out to the first rise position** (see Figure 2.7). Note that the floor may be out of level, which can affect your calculations.

Allowing for headroom in a concrete flight of stairs is just as important as in a set of joinery stairs. Sometimes the height can be influenced by the depth of an edge beam, in the same way that a landing trimmer can affect the headroom of regular stairs.

To calculate the headroom when the rise and **going (tread)** are known, multiply the number of risers from the first floor down to the tread directly under the landing beam by the height of the rise and subtract the depth of the edge beam (see Figure 2.8).

For example:

15 rises × 150 mm – 300 mm edge beam
= (15 × 150) – 300 mm
= 2250 – 300 mm
= 1950 mm (does not comply)

Source: Getty Images/Mint Images

FIGURE 2.5 Combination floor finishes affect calculations

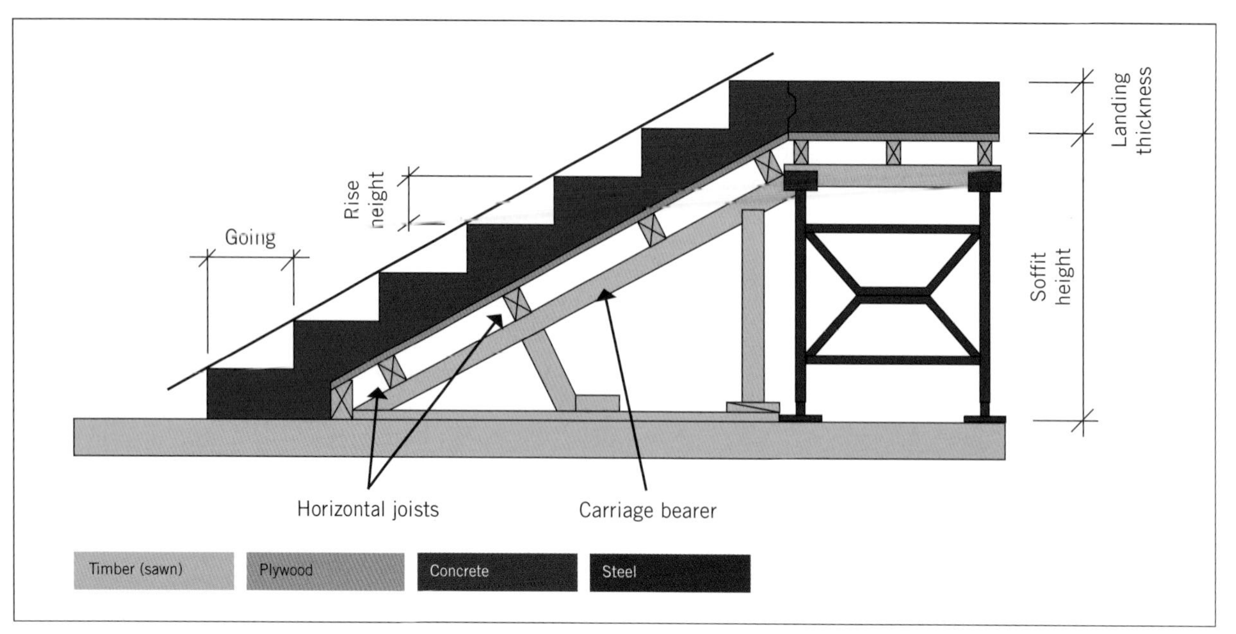

FIGURE 2.6 Raked bearer and horizontal joists

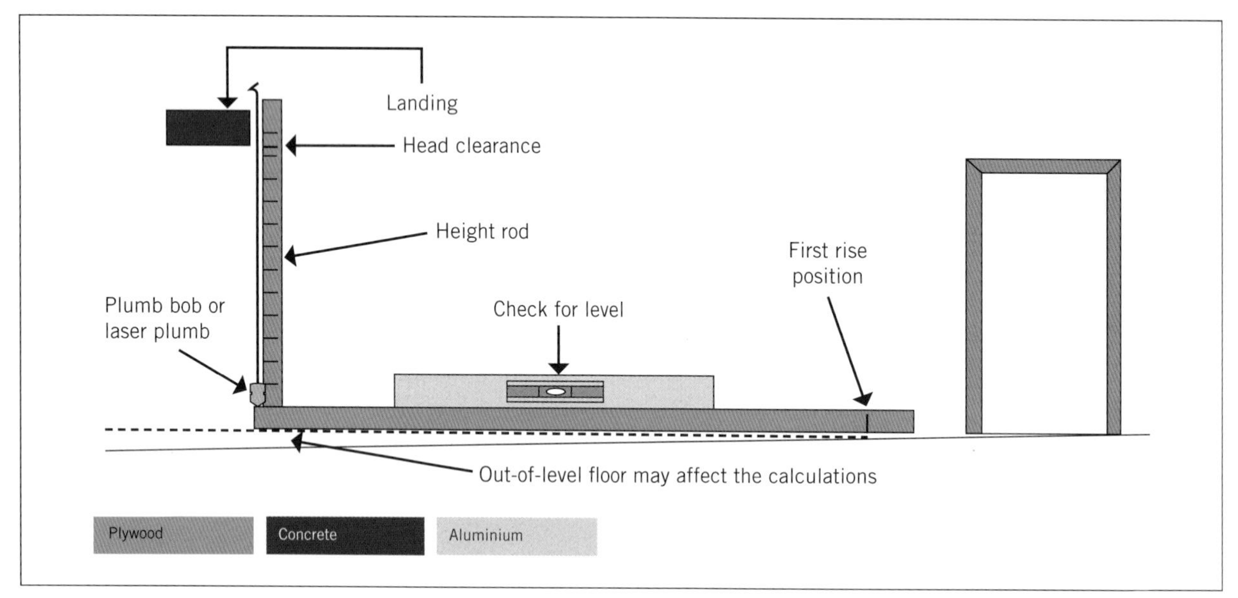

FIGURE 2.7 Determining the first rise position

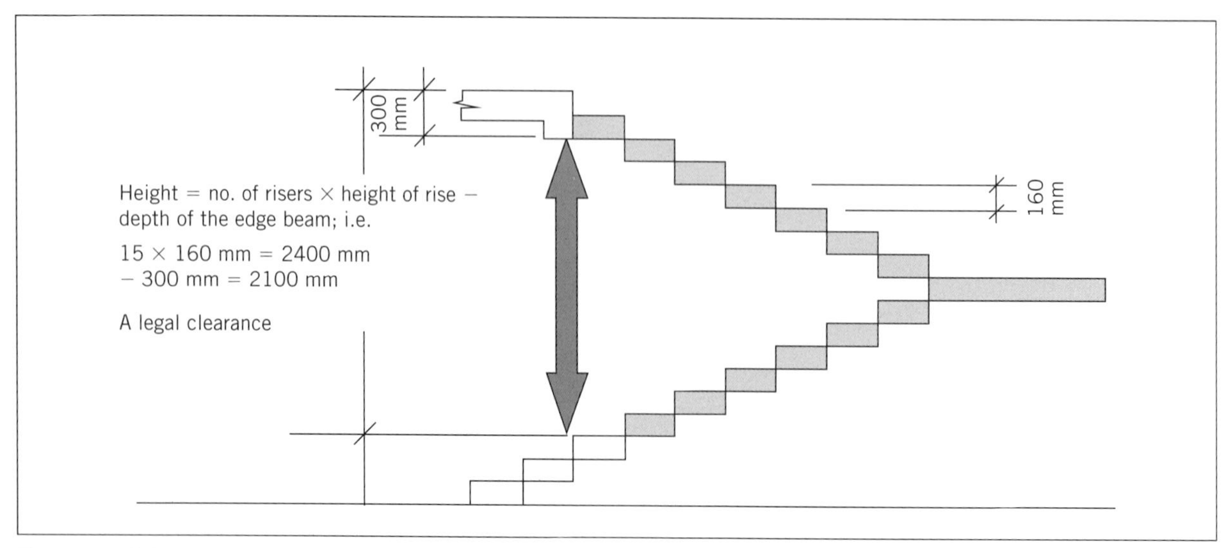

FIGURE 2.8 Calculating headroom

But

15 rises at 165 mm − 300 mm edge beam
= (15 × 165) − 300 mm
= 2475 − 300 mm
= 2175 mm (does comply)

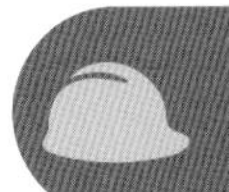

When erecting or dismantling stair formwork always wear a hard hat. You will likely bump your head on the stair formwork soffit.

Setting out the soffit

The formwork/falsework must be braced, strutted and propped so the loads are transferred to the slab below. The **decking** is commonly formply, but tongue and grooved boards with a hardboard overlay is an alternative. As seen in Figure 2.6, joists can be run horizontally across **bearers (ledgers)** or on the rake across horizontal bearers.

LEARNING TASK 2.3

Make the calculations for soffits sheets, carriage pieces and bearers for formwork for the stairs shown below. **Note:** overall rise 900 mm, overall going 1500 mm, width 1000 mm rise and going: 150 mm × 300 mm. Teacher guidance may be required.

Materials

- Formply: 1800 mm × 1200 mm; 17 mm F17 grade
- Joist and bearers: 3.6 m; 95 × 65 LVL

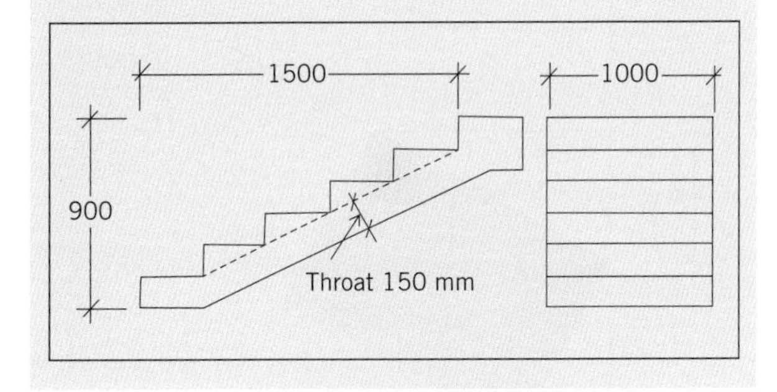

Adjoining walls, if they have been constructed, are **chased** or scabbled, or drilled to take reinforcing dowels to tie the stairs to the rest of the structure (see Figure 2.9). Most commonly, concrete stairs are not bound to walls until after they have been constructed. This makes it easier to access the formwork and place the concrete, and allows for the stairs to butt up hard against the wall.

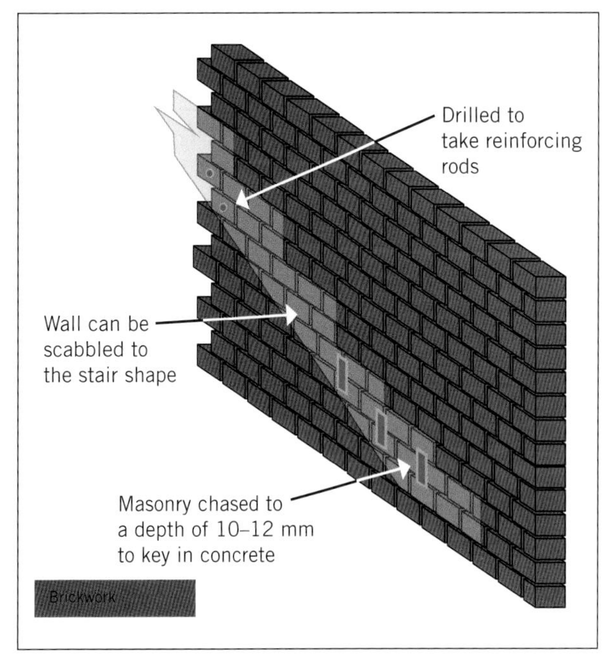

FIGURE 2.9 Chasing, scabbling or dowels tie stairs to walls

When the stairs are to be constructed against an existing wall, the position of the stairs, including rise and goings, landing and **throat** thicknesses, are marked on the wall (see Figure 2.10). The soffit lines can then be drawn in and the position and the lengths of the members, carriage bearers, joists and props can be determined.

When setting out the string/shutters for the stair sides, the margin line is aligned to the rise and going intersection at the bottom, as opposed to the top, in a set of joinery stairs. This distance is called the throat

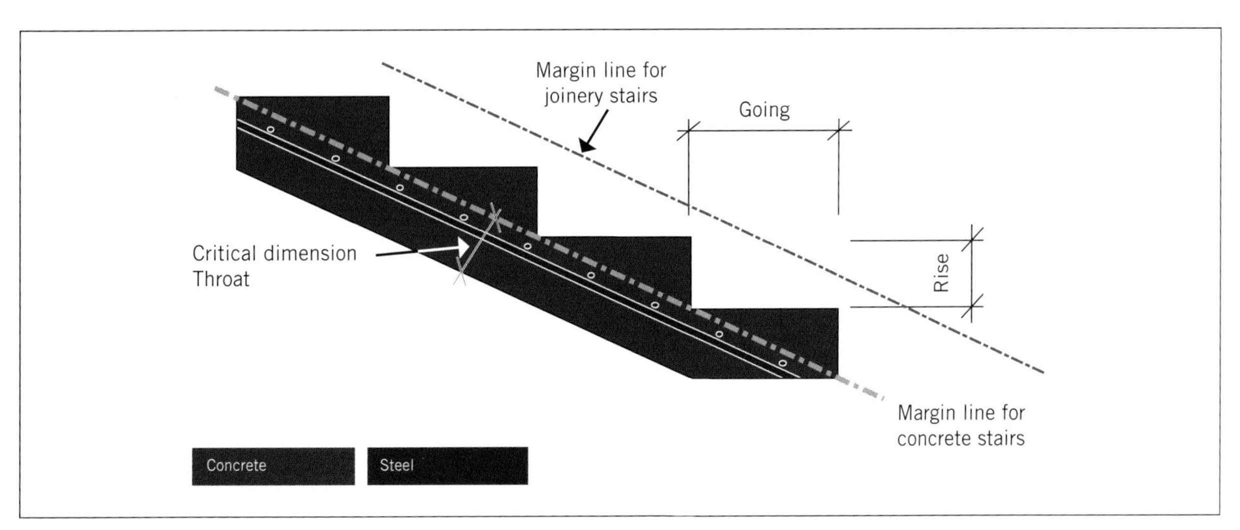

FIGURE 2.10 Margin lines

distance. By allowing a decent throat depth, sufficient cover is provided for the steel reinforcement, which helps eliminate corrosion, and also allows for the soffit lining positions to be calculated. The thickness of the raked slab will be provided in the engineer's and steel drawings (see **Figure 2.11**).

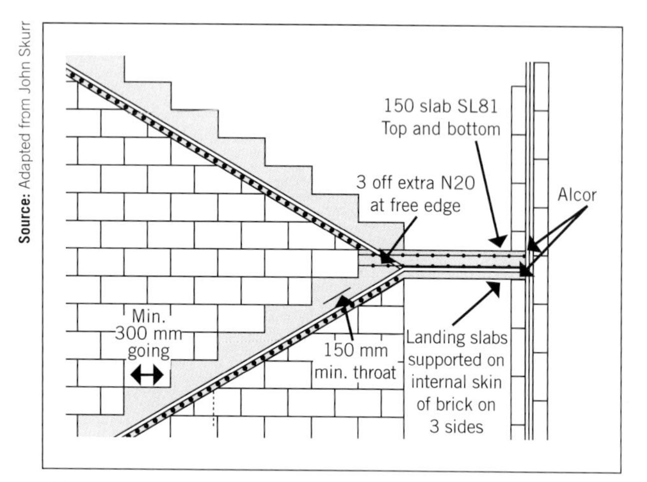

FIGURE 2.11 Engineer's drawings – throat distance

The height of the slab may be given as a height or as a relative level (see **Figure 2.12**). Otherwise, the distance can be given as a measurement or as the number of rises.

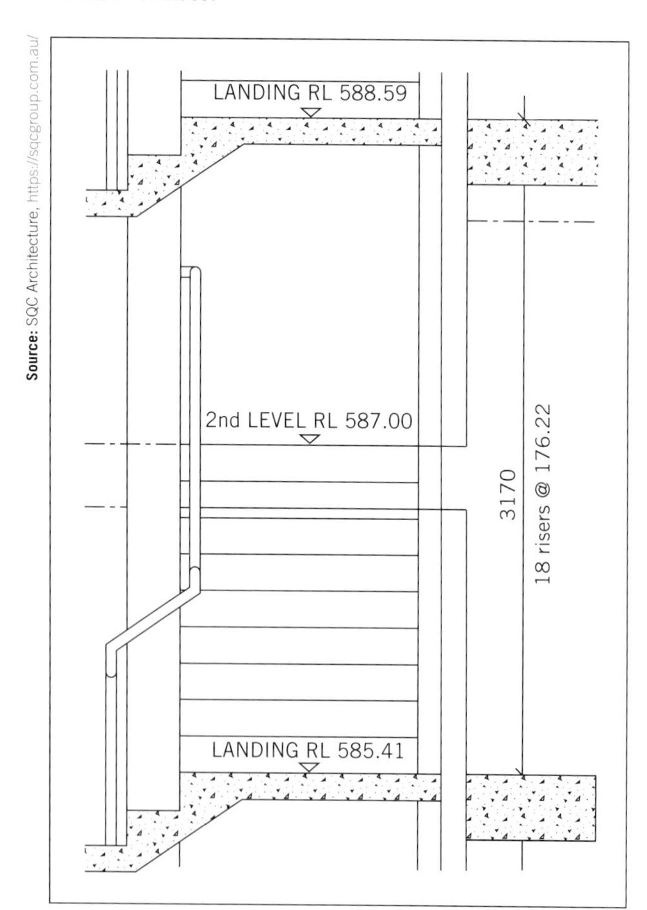

FIGURE 2.12 Relative levels and rise heights for landings

A set-out drawing is best done to establish the pitch of the stairs.

Making a pitch board similar to that shown in Chapter 6 (**Figure 6.21**) is a good starting point. (A steel square and stair buttons can be used as an alternative – see **Figure 2.13**).

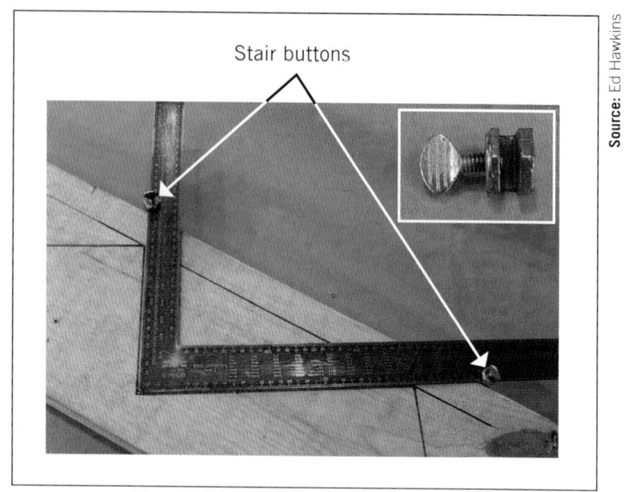

FIGURE 2.13 Steel square and stair buttons

Begin by replacing the margin line with the throat distance (see **Figure 2.14**).

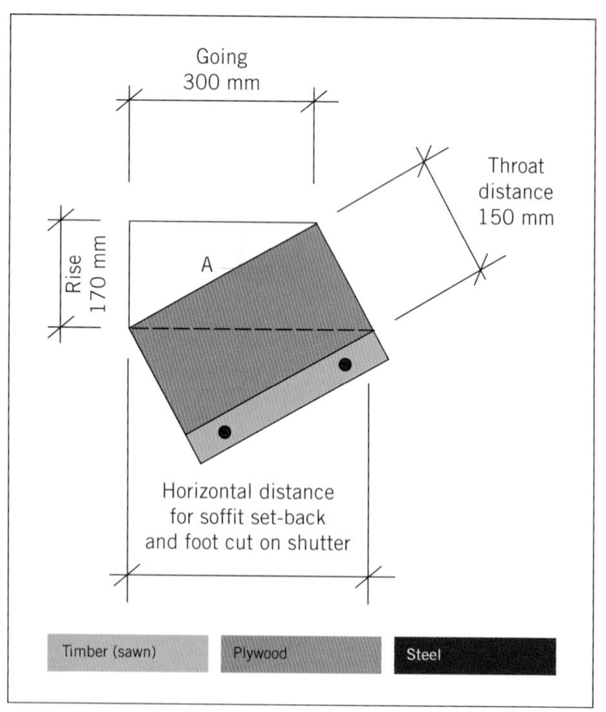

FIGURE 2.14 Pitch board

The pitch board can be used to set out the rise and goings along the face of the string/shutter boards and to develop the position of the soffit, plumb and foot cuts.

To find the starting position of the soffit, the position of the first rise is established from the rise and going calculations and plumbed marks.

Use **Pythagoras' theorem** to calculate the soffit set-back distance (see Figures 2.15 and 2.16).

Using a rise height of 170 mm and a going distance of 250 mm:

$$A = \sqrt{170^2 + 250^2} = \sqrt{91400} = 302.3 \text{ mm}$$

Soffit setback $= \sqrt{302.3^2 + 150^2}$ (throat)

(see Figures 2.15 and 2.16) $= 483.9$ (say, 484 mm)

Beams at top of flight

Where large stairwell openings (half space landings) are required in floors, a reinforced concrete beam, using steel rod or 'I' beams or similar, is created at the leading edge of the landing area to support the floor and stairs. Figure 2.17 shows the formwork/falsework for a reinforced concrete beam, consisting of a soffit and two shutter sides, supported by props. The right-hand side is supporting the floor concrete and reinforcement. The space at the end of the joist is for **wedges** to ensure that the joists do not bind tightly to the beam. The left-hand side shows the formwork for the upper

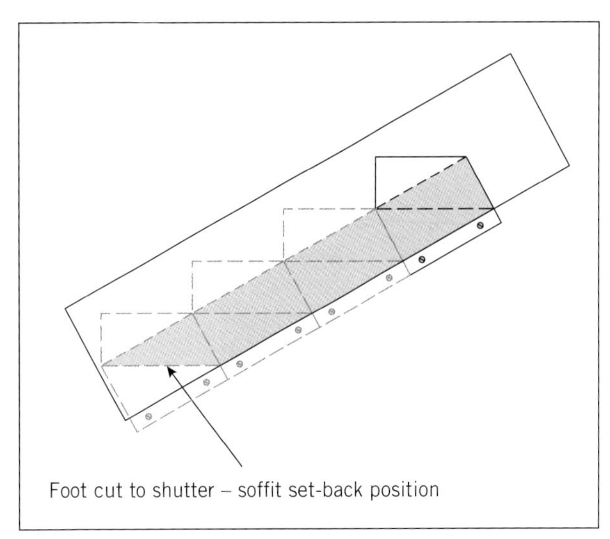

FIGURE 2.15 Soffit set-back position

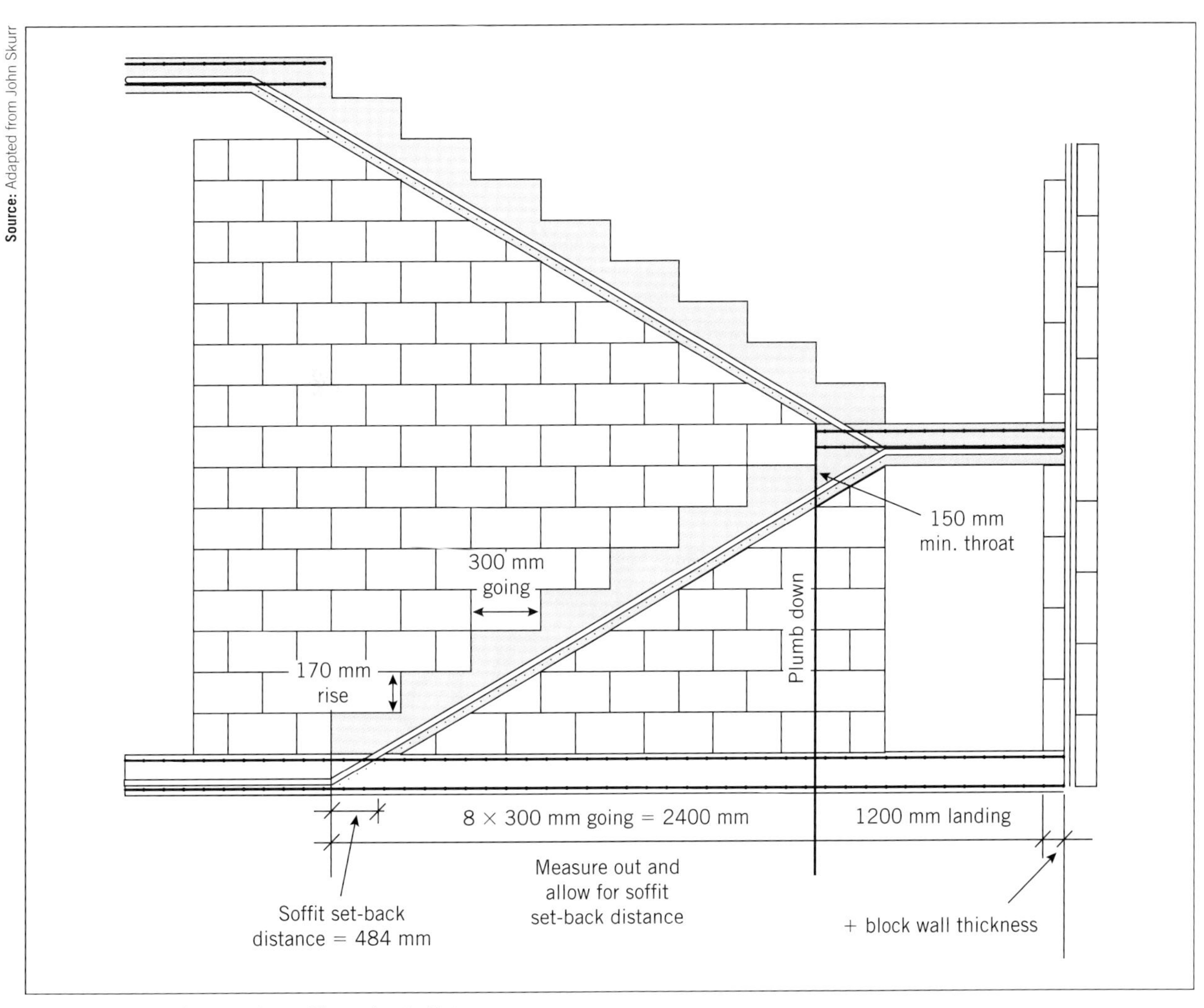

FIGURE 2.16 Setting out the soffit set-back distance

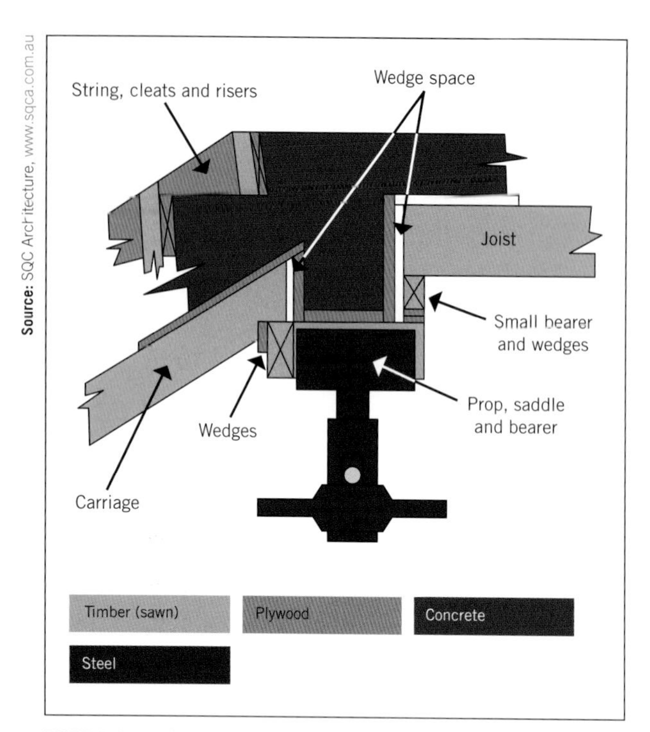

FIGURE 2.17 Soffit set-out for a landing with a beam

portion of the concrete stair and beam shutter, with the carriage joist **bird's mouth** cut over the bearer that would be supported by another prop (not shown), or may be set on another bearer. There are a number of different methods available and the engineer's details or formwork plans should show exactly how it is to be done.

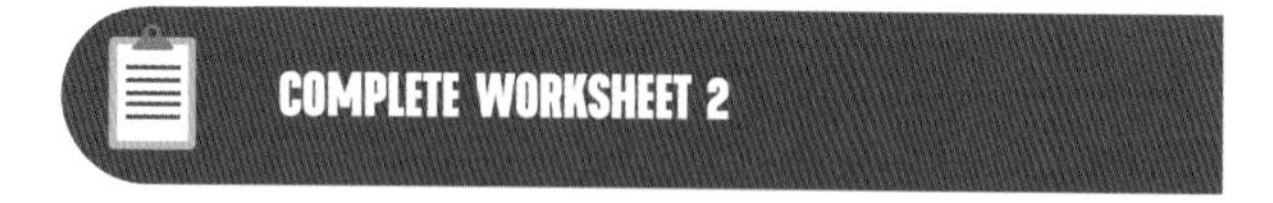

Construction of formwork for reinforced concrete stairs

When there is a change of direction in the flight of stairs at a landing, it is sensible to set up the landing soffit height first so that you can establish the position for the first rise. With the position of the soffits identified and the landing soffit set, the raked soffits are installed. **Runners** are then laid out on the soffits to locate the position of the shutters.

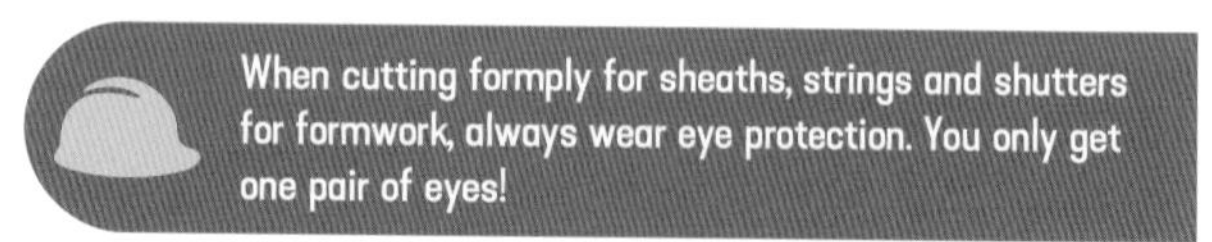

String/shutters, like columns and beams, are commonly made using formply. Tongue and grooved boards can be joined to form the shutters. Jointed boards can be used but must be well sealed to prevent leakage. **Cleats** can be used to secure the timbers together.

Note: Once the string/shutters are in place, the steel reinforcing has to be placed and secured before the riser boards are installed; otherwise you cannot position the steel reinforcement.

Bracing

The formwork/falsework must be braced, strutted and propped so the loads are transferred to the slab below.

Positioning the riser boards

A builder's square and stair buttons (see Figure 2.13) or a pitch board (see Figure 2.14) is used to set out the position of the rises and goings as with a regular stair (see Figure 2.18). The riser board is the only board

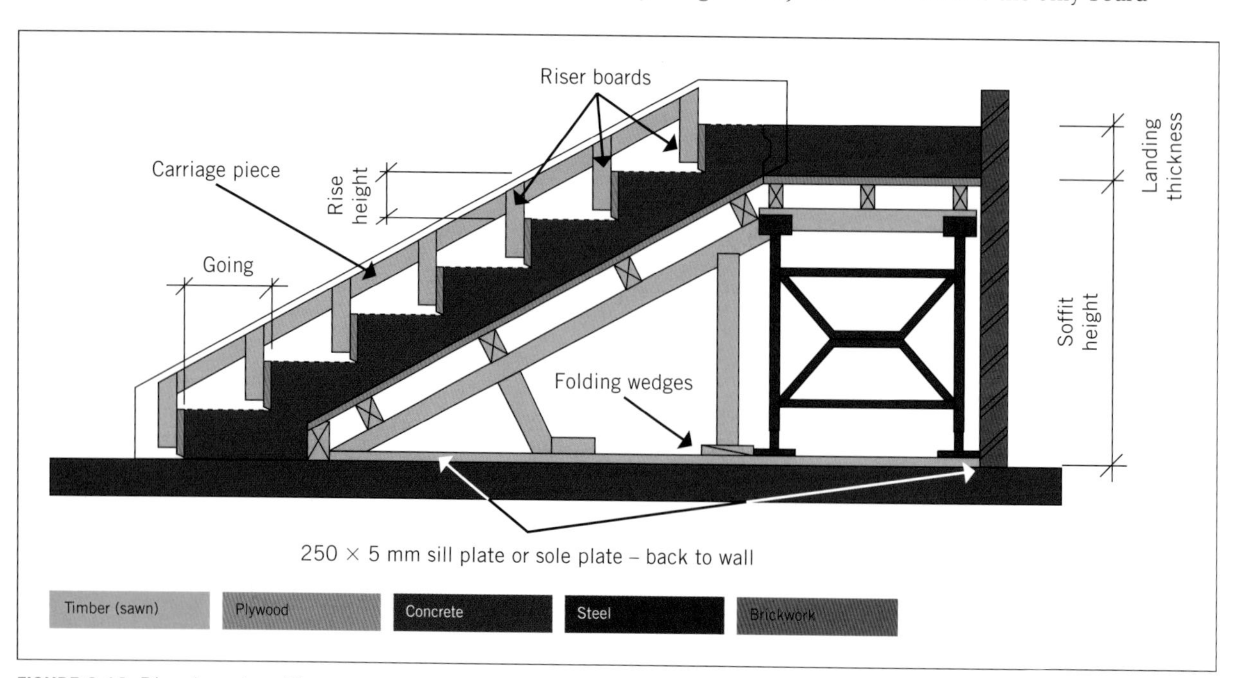

FIGURE 2.18 Riser board positions

that is installed, and it is installed level with the height of the tread position so that when the concrete is poured it will naturally flow to the correct level. Any overspill can flow over and be screeded or trowelled level.

A cut former, similar to a cut string, can be used, or a wide shutter from formply may be used (see Figure 2.22), with cleats fastened to the shutter to support the riser boards (see Figure 2.19). When pre-fixing the cleats you need to allow for the thickness of the riser boards.

Source: Ed Hawkins

FIGURE 2.19 Cleats used to support risers

Riser boards can be set out to provide a number of different finishes (see Figure 2.20).

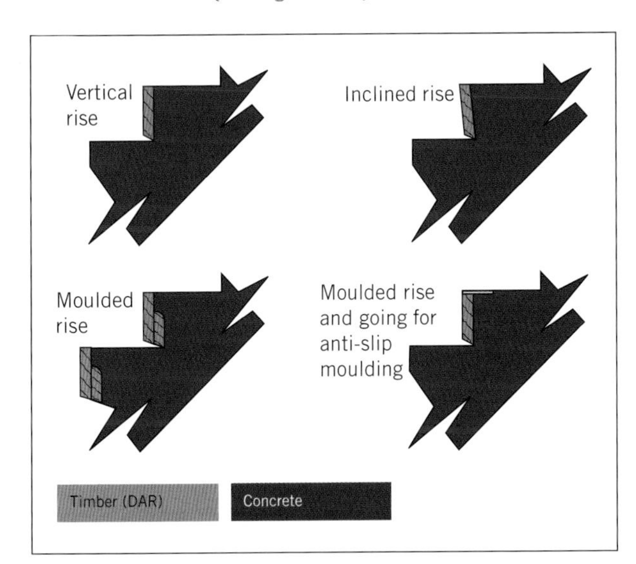

FIGURE 2.20 Riser board variations

The bottom edge of the riser board is usually bevelled to allow the concreter to finish the surface of the tread right into the corner intersection of the tread and riser (see Figure 2.21).

To prevent the riser boards from bending under the pressure of the concrete, a secondary riser cleat can be used to stiffen the riser boards. In wider stairs (over 1 m wide), a stiffener (hanging beam or **strongback**) is placed over the centre of the riser boards, and hangers or struts are fixed against the riser boards to prevent deflection (see Figures 2.22 and 2.23). The stiffener can be a piece of formply (minimum of 100 mm × 50 mm × 19 mm).

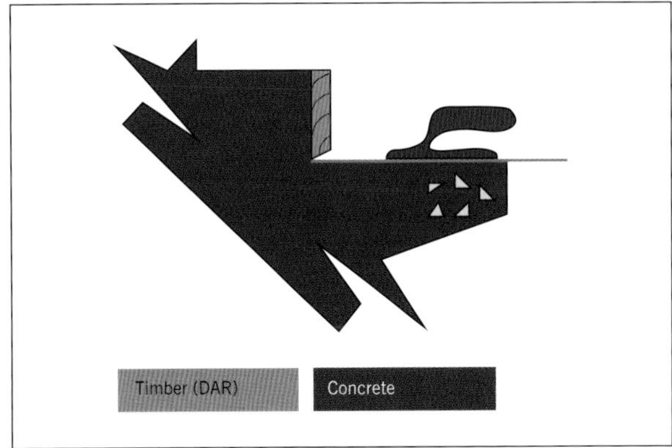

FIGURE 2.21 Chamfer the bottom edge of risers

Source: NS Formwork and Concreting 2023

FIGURE 2.22 Riser board with cleats and strongback inserted

Source: Creative Concrete & Formwork Pty Ltd (www.creativeconcreteformwork.com)

FIGURE 2.23 External stair formwork

To prevent movement, the stiffener can be strutted to a nearby wall or column or fixed to the floor via blocking.

When the stairs are formed up against an existing masonry or concrete wall, the formwork/falsework is suspended from a wall string that is firmly attached to the wall (see Figures 2.19 and 2.23). There are alternative methods for supporting the riser boards including that shown in Figure 2.24.

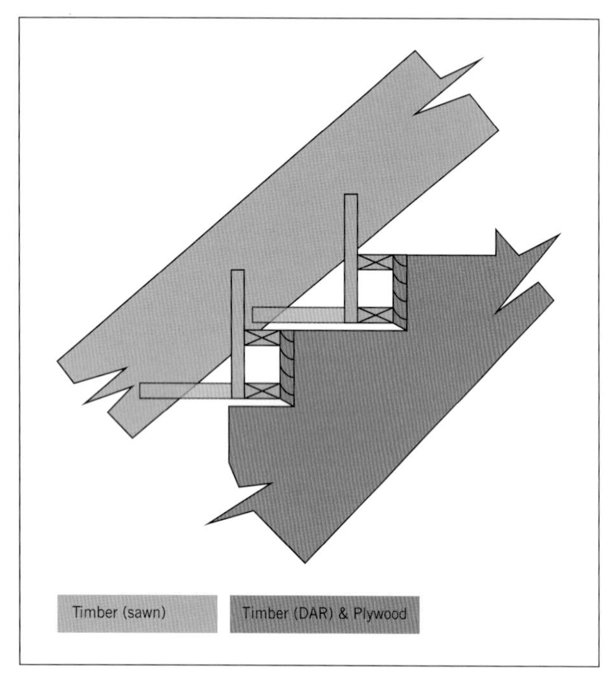

FIGURE 2.24 Alternative riser support

Constructing external stairs

Some external stairs do not require a soffit to be constructed. Earth, with sand and road base materials, is often used to form up under external stairs. How the ramped material is to be constructed is often shown in the landscape plans or specifications for the job. Special equipment such as a compactor may be needed to compact the soil and fill materials to the specified density. Ensure that the ramped material provides a relatively even throat thickness as for internal stairs. A vapour barrier can be used. Ensure reinforcement materials are set at the correct height to allow for the correct concrete cover.

Figure 2.25 shows that the stairs are likely to be formed and poured after the initial slab has been laid. Note the use of a keyed joint between the slab and the stairway.

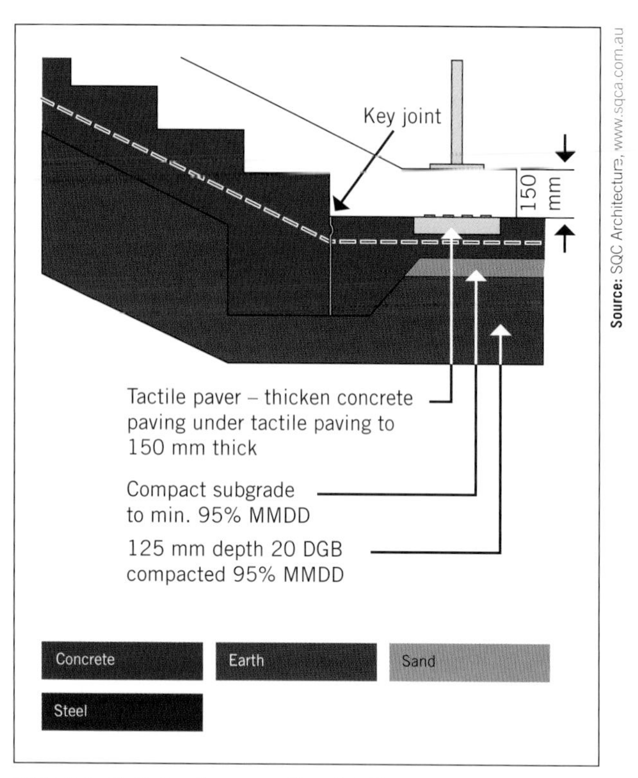

FIGURE 2.25 Subsoil compaction details

The foundation, or compacted fill, is heaped or cut away as necessary, allowing adequate space at the sides to insert shutters and bracing. Beams are used at the bottom of each flight (see Figure 2.26) to support the weight (mass) of each flight so that the full load is not transferred to the next flight through the landing area. This reduces the stress on the landing section and limits cracking or buckling at the landing.

Note: The landing has a 2° fall built in to allow for drainage to the bottom of the stairs where a drainage grate is usually installed to direct the water to the draining system. These steps will have a metal non-skid nosing fitted, but other options are available.

Figure 2.27 is a plan view of an external stair with side walls.

However, all building sites vary, so the type of formwork/falsework shown here is only one way of setting out the formwork/falsework for this type of stair work (see Figure 2.28). Depending on the style and size and the engineer's drawings, different techniques may be needed. For example, bridging beams can be used to support edge formwork/falsework and internal riser boards (see Figures 2.29, 2.30 and 2.31).

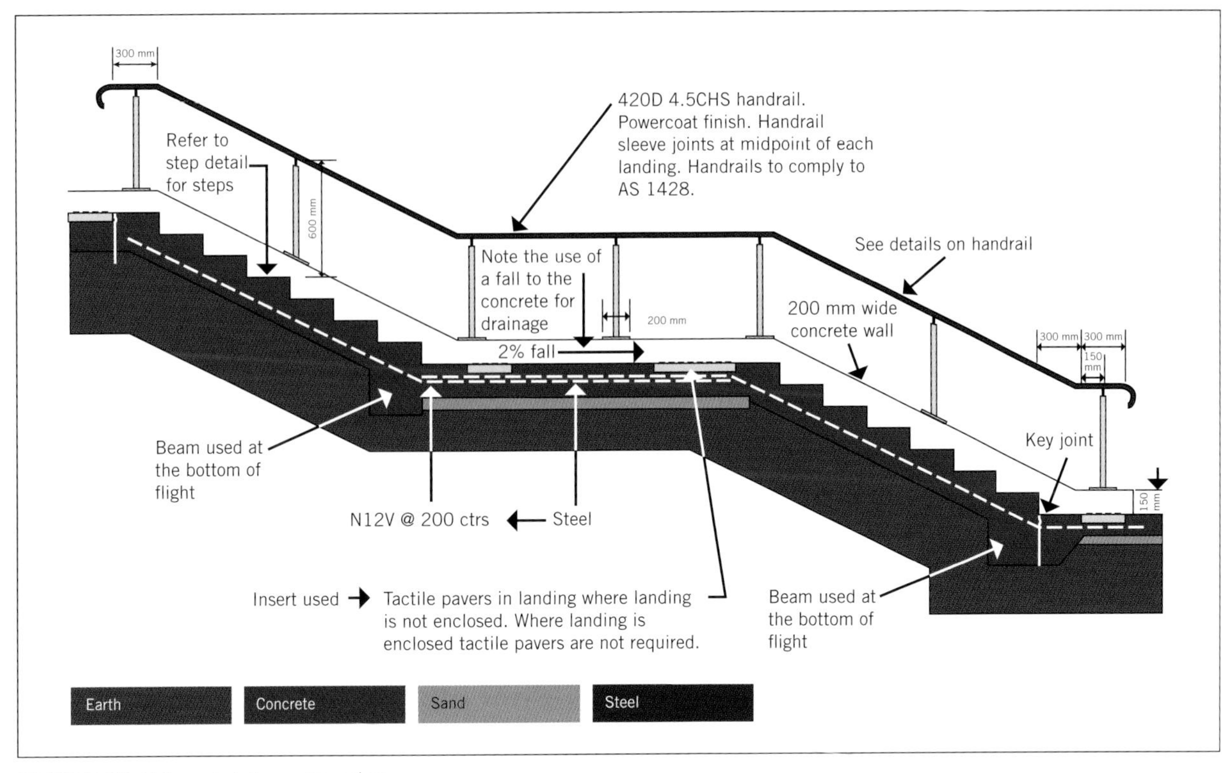

FIGURE 2.26 External stair construction

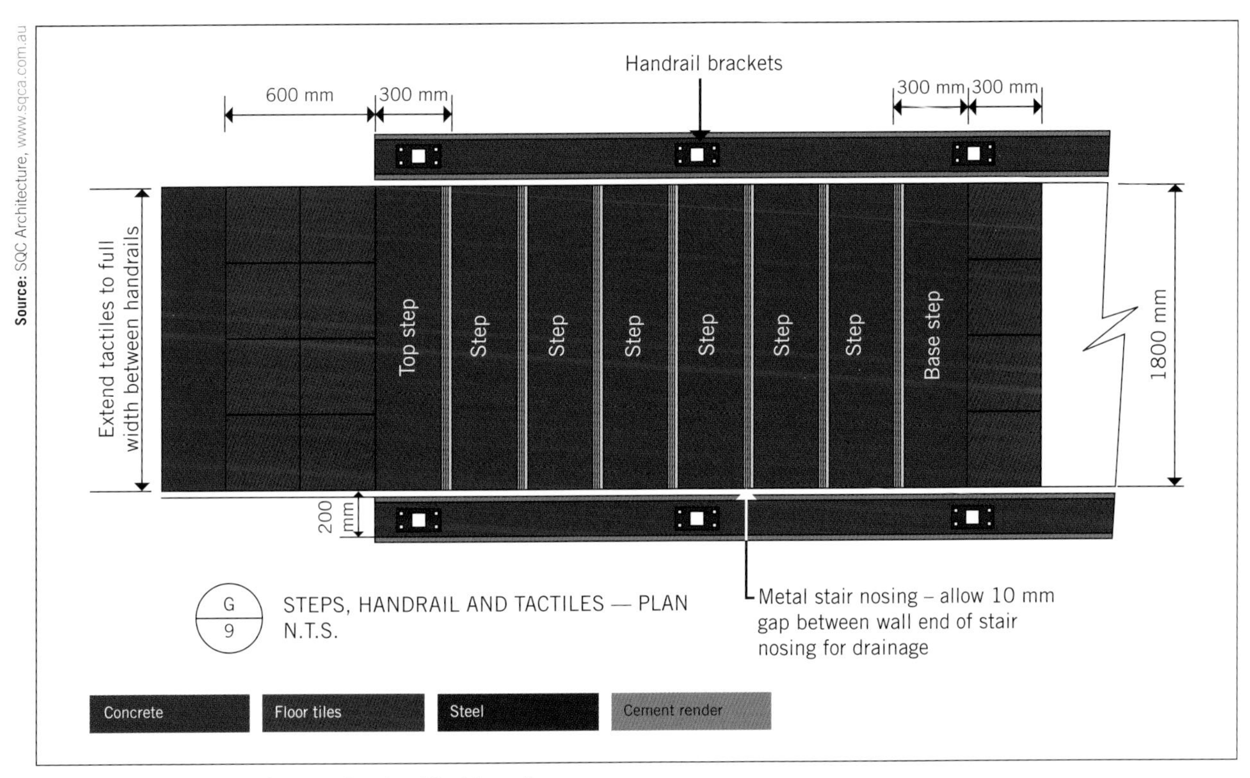

FIGURE 2.27 Plan view of external stair with side walls

Source: SQC Architecture, www.sqca.com.au

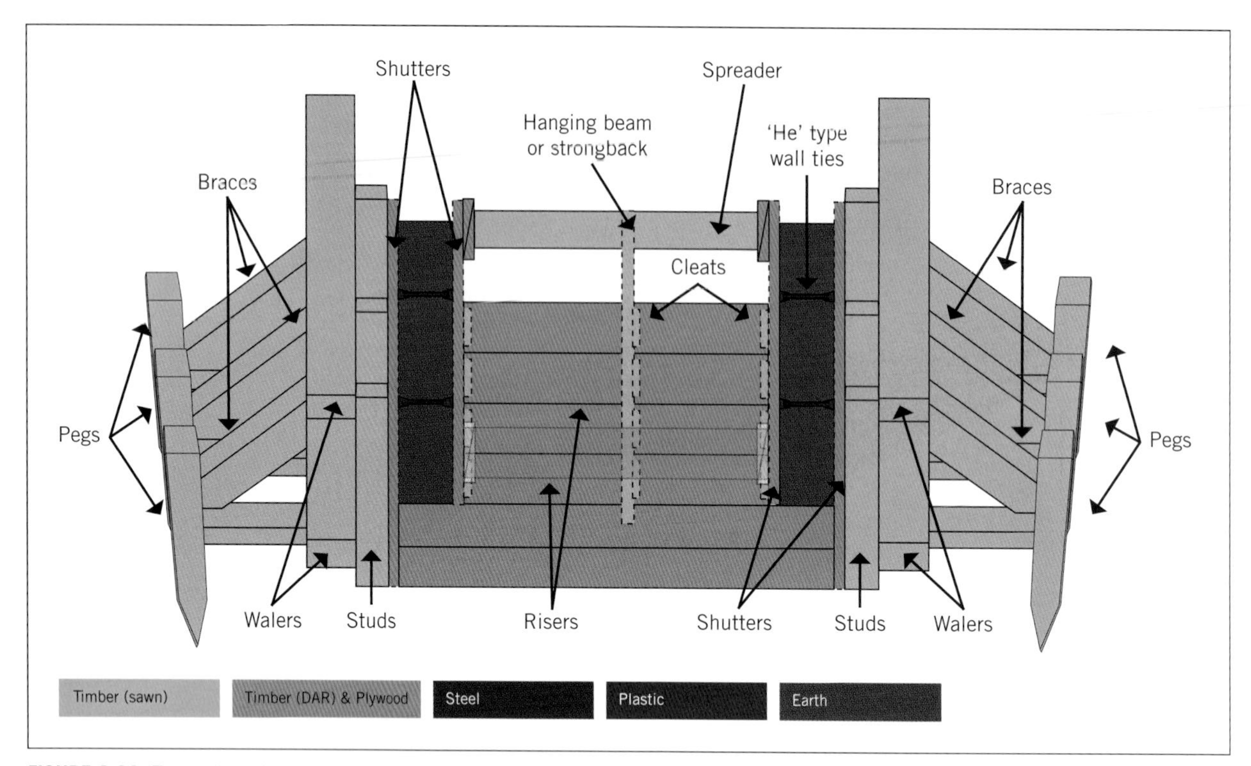

FIGURE 2.28 Front elevation of external stairs showing typical formwork

Source: Creative Concrete & Formwork Pty Ltd (www.creativeconcreteformwork.com)

FIGURE 2.29 Side forms supported by walers and studs

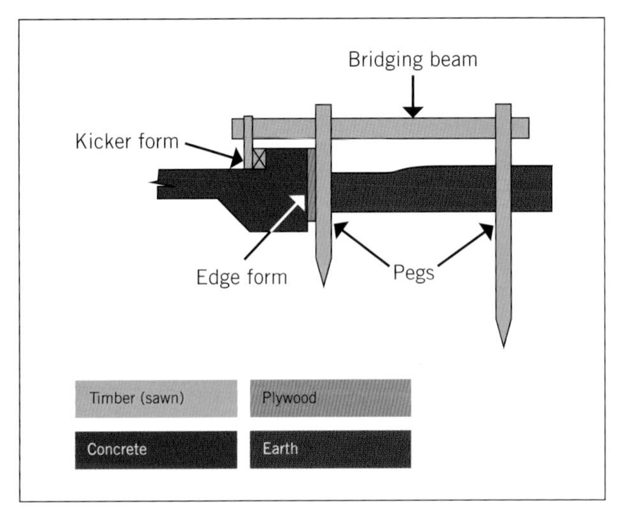

FIGURE 2.30 Bridging beam

Ramps and raked formwork/falsework

Ramps are sloping or **raked** soffits used as a means of access and egress (see Figure 2.33). Ramps are particularly useful in addressing the access and egress needs of people with disabilities.

AS 1428 (Set) Design for access and mobility sets out the requirements for the building of access ramps in commercial buildings and institutions such as hospitals, schools and nursing homes. Section 45.3 Ramps provides details on the maximum length, width and gradient of ramps.

The National Construction Code 2022 – Housing provisions, Part 11.2 provides stairway and ramp construction details and performance requirements.

Any balustrades or barriers required for the sides of ramps need to comply with the National Construction Code 2022 – Volume One, Part D Access and Egress; and Volume Two, Section H and ABCB Housing Provisions, Part 11.3.

Construction methods

Because of the slight rake on a ramp it is possible to pour concrete on the form without it slumping to the

FIGURE 2.32 Rubble screed for ramp

bottom of the form. This means it is not necessary to build in riser stops. However, it is best if the ramp is poured in sections to keep the different slab sections even in thickness. A sequence for pouring will need to be worked out. For long ramps, control joints are inserted to control and reduce cracking. Spacing of the control joints is dependent on the length and width of the slab, but as a general rule the control joints should be spaced every 1.2 to 1.5 m. A control joint should be incorporated at changes of direction, and an expansion joint wherever specified in the engineer's drawings. The joints should be designated in the engineering specifications and or formwork/falsework design.

The props and raked bearers and joists can be set out similarly to stair formwork/falsework, with scaffold frames and adjustable screw jacks for levelling, and the **canted** props supported with blocks on sole plates and adjustments made using wedges (see Figure 2.34).

Alternative method

Sometimes the formwork/falsework for ramps is built in and designed not to be removed. This is known as 'lost formwork'. This method is sometimes used for small ramps, such as those used to access low suspended slabs in external environments (e.g. verandahs and patios). Specified members such as 100 × 50 mm bearers are fastened to the masonry walls and a 100 × 50 mm joist is laid across the bearers. A sheet of formply (or patented steel former or corrugated iron) laid across

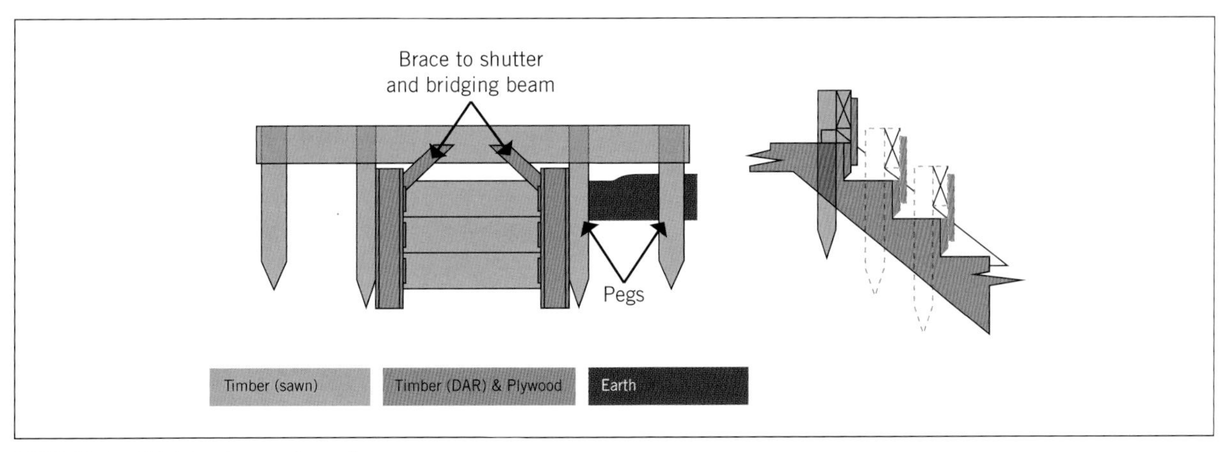

FIGURE 2.31 Bridging beam alternative

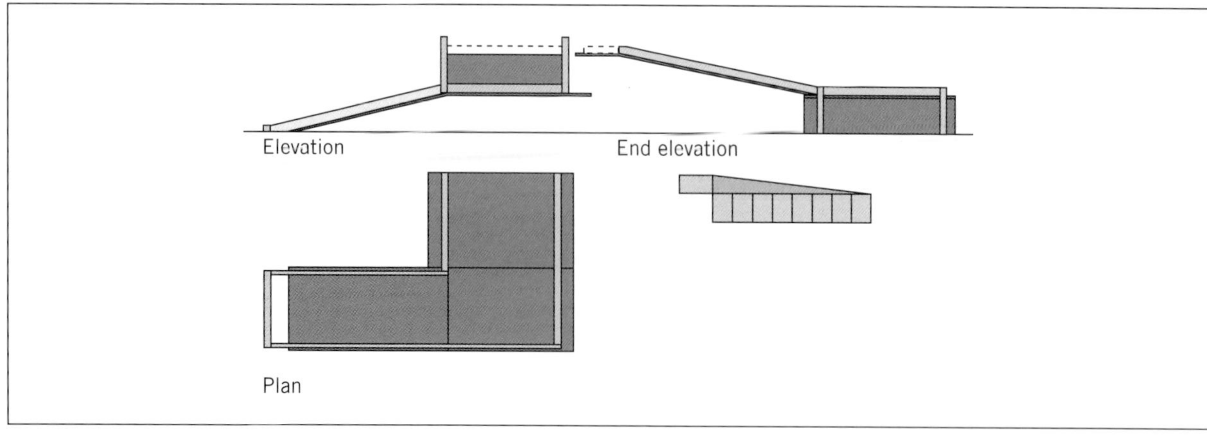

FIGURE 2.33 Ramp soffit

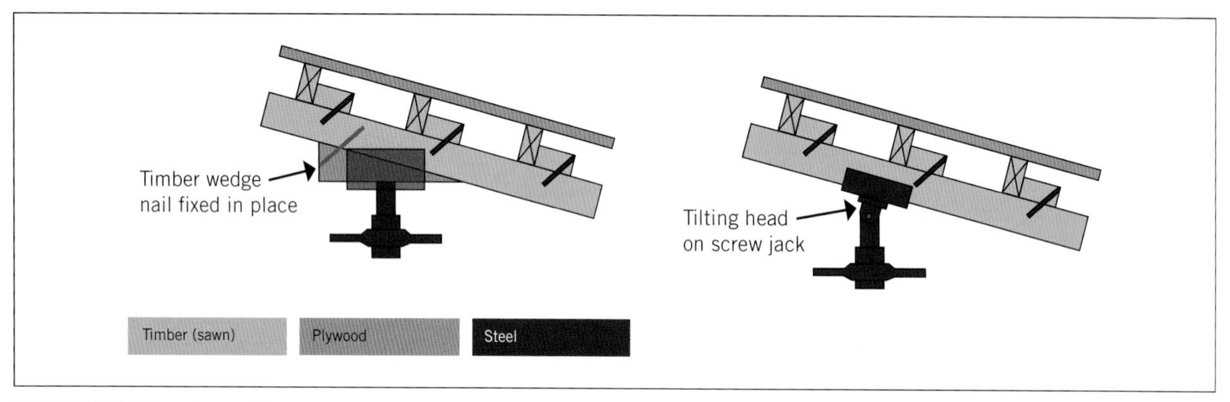

FIGURE 2.34 Ways to cant formwork

the joist can be used as a soffit lining. Walls can be chased to tie the structure together (see Figure 2.9 earlier).

Alternatively, dowels can be drilled into the masonry to tie the slab and masonry together to prevent the cured concrete mass moving down the slope. When walls are used as formwork, they need to be braced with pegs or joists and walers and braces (depending on the height) to prevent the concrete mass pushing the walls over.

Do not use timber for lost formwork as it can attract termites to undetectable sites.

Note: Whenever a concrete slab is supported by another masonry wall, a slip joint needs to be formed between the two surfaces (see Figure 2.35). This usually involves layers of galvanised iron being set between the two surfaces. This allows for differing rates of expansion and contraction between the two construction materials and reduces the risk of cracking or damage.

Formwork for other sloping concrete forms

Concrete may also be used to form sloping sides for sloping roof shapes, vaulted ceilings and other steeply sloping concrete forms. For these purposes, the required formwork/falsework is the same as for wall forms, with a top and bottom form (see Figure 2.36). This prevents the concrete bulging at the bottom and provides an even thickness of finish.

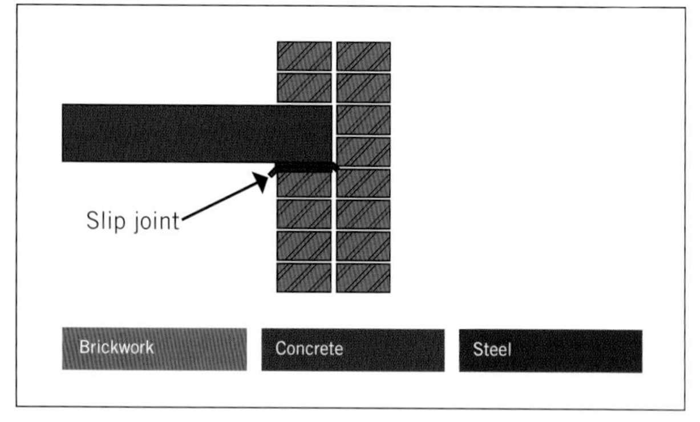

FIGURE 2.35 Slip joint

Release agents

In general, release agents make it easier to remove formwork from the concrete and to remove concrete from the formwork so it can be easily cleaned and reused. Some release agents can affect the surface colour of the concrete, so only use what has been specified. Different types of release agents include the following (see the 'Release agents' section in Chapter 1 for further details on each of these):

- neat oil
- water-soluble release agents
- mould cream emulsions

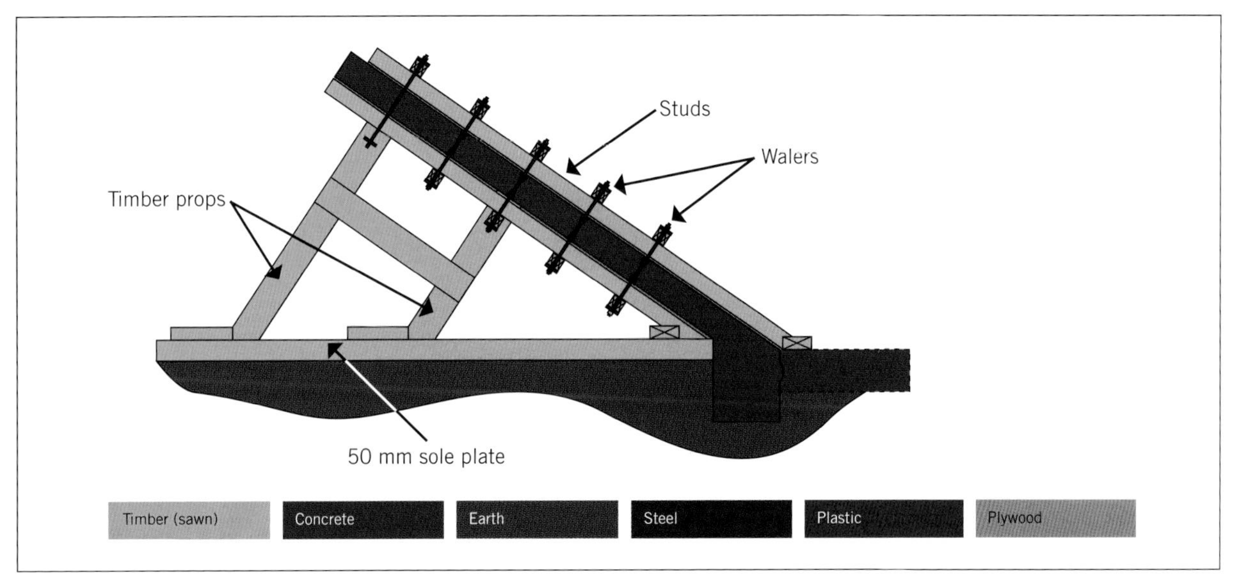

FIGURE 2.36 Steep slope slab

- chemical release agents (suspended in a thin oil distillate)
- wax emulsions.

Release agents can be applied with a broom, roller, brush or spray. It is easier to apply the release agent to the soffits and strings before installing the steel and risers.

GREEN TIP

It is always good practice to source products that are sustainable and environmentally friendly.

There are a number of manufacturers providing these types of products for use in forms of release agents. Be sure to check or ask for the safety data sheets (SDS) before deciding on which product to purchase. Check the hazards warnings provided. Wear safety glasses, goggles or face shields, and chemical impervious gloves and safety boots during application. In some instances, respirators are advised if the agent is being applied by spray. Dispose of any waste materials as directed or regulated.

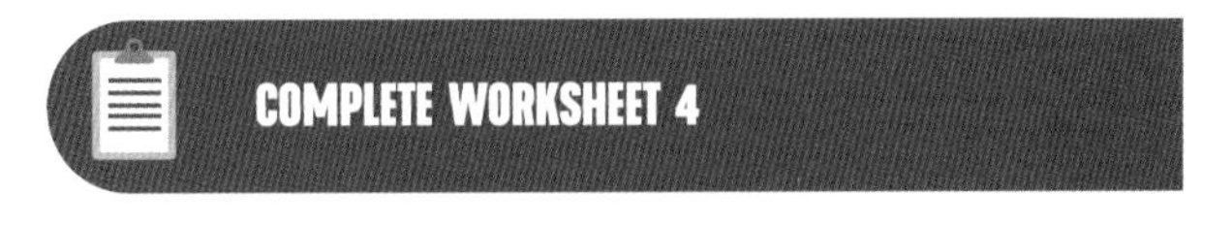

Stripping formwork

Stripping formwork/falsework from stairs and ramps involves the same process as discussed in Chapter 1. Importantly, the formwork/falsework needs to remain in place long enough for the concrete to become strong enough to support itself, which can be anywhere from 14 to 28 days. Check with your site supervisor before stripping any formwork.

When erecting or stripping formwork, always wear gloves, hearing protection and eye protection. You only get one pair of eyes and ears.

When stripping formwork, first remove the falsework and braces and then the formwork/falsework that gives the concrete its profile. This is done when the concrete is able to support its own load plus the load of workers, material and equipment. A gradual transfer of the load is recommended so that impacts and sudden shocks are avoided. Avoid sudden shocks by back propping to the engineer's specifications. Extreme changes in weather conditions should also be considered when planning any stripping operation.

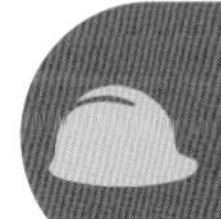

Stripping formwork can be dangerous, particularly soffit formwork. Make sure you wear a hard hat and other PPE to protect yourself from falling formwork sheets when working from below.

Straight flight stairs, once cured, are generally self-supporting, transferring loads directly between floors.

Stairs that change direction using a landing only transfer the loads between landings, so it is *critical that the landing remains supported* by props or formwork/falsework frames until the appropriate walls are formed up at the perimeter to take over the support (see Figure 2.37).

Stripping of formwork/falsework was covered in Chapter 1 under 'Stripping formwork'. The key points covered include:

1. Ensure written permission is obtained from the site engineer.
2. Follow the minimum period for striking conditions as set out in AS 3610.1, Section 5.4.
3. Avoid shocks, and transfer loads gradually.
4. Clean off and reuse formwork wherever possible.

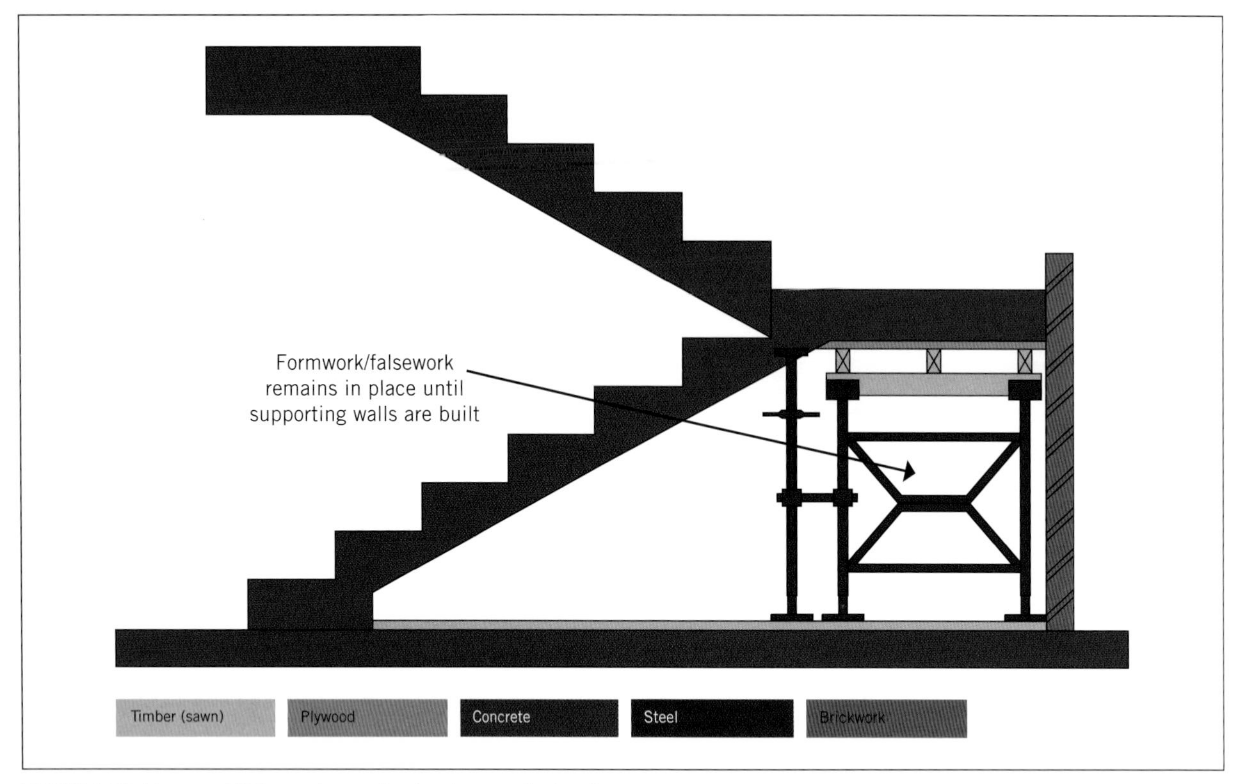

FIGURE 2.37 Formwork/falsework remains in place under landings

Many formwork components, especially those made from wood, are assembled using nails, screws, or other fasteners. After each use, these fasteners should be safely removed (de-nailed) to prevent potential hazards during the next assembly.

Formwork components, especially those made of wood, can absorb moisture from the concrete, leading to swelling, warping, or degradation over time. Applying a layer of release agent or form oil before pouring concrete creates a barrier between the formwork and the concrete. This prevents adhesion, making it easier to strip the formwork after the concrete has set. Oiling also extends the lifespan of the formwork by protecting it from the corrosive effects of concrete.

Proper storage or stacking of formwork components prevents damage, warping, or deterioration. Components should be stored in a dry, well-ventilated area, away from direct sunlight and extreme temperatures.

AS 3850.1 Prefabricated concrete elements – Part 1: General requirements provides detail on the general requirements for precast concrete elements such as stairs used in building construction. It focuses on:

- concrete, aggregate, additives, reinforcement and the need to comply with AS 3600. It includes testing and other compliance requirements such as recording and reporting of results.
- lifting inserts and brace inserts. Cast-in-place or added after components are formed and cured. Lifting inserts also have to comply through testing and reporting in manufacture and their pull-out capacity if cast-in-place.
- documentation required for drawing or pictorial representation that clearly identifies the component or system to identify each component including drawings, specifications, dimensions and weight and other specific details.

AS 3850.2 Prefabricated concrete elements – Part 2: Building construction covers the 'requirements for planning, construction, design, casting, transportation, erection and incorporation … of prefabricated concrete elements in building construction'. In particular it covers the different methods of rigging components for craning such as flat, edge or face lifting.

The Standard provides guidance on the requirements for:

- planning, including:
  - access and egress routes to and from site and in and around the site for trucks and cranes
  - public safety
  - proximity to power lines
  - ground conditions – including suspended slabs
  - excavations and underground services
  - suitable space for the crane set-up and dismantling if required
- on-site casting of elements, including setting up casting beds to cope with imposed loads of formwork and wet concrete

- compatibility of release agents with concrete compounds
- transport, including loading and unloading, and stability of loads and trucks during operations
- cranes and lifting – cranes must be able to lift the total load and place the component safely
- installation and inspection of temporary bracing and propping.

Pre-formed stair formwork

There are a number of off-site manufactured stair formwork systems available that:

- are made to site measurements
- reduce time in construction – less erection time and no stripping
- require less propping
- come to site with reinforcement in place.

The disadvantages are:

- they have to be crane handled into position
- there are fewer design options with some systems.

Figure 2.38 shows precast concrete treads bolted to steel 'C' channel carriage beams.

Source: National Precast Concrete Association Australia

FIGURE 2.38 Precast stairs on steel 'strings'

LEARNING TASK 2.4

Students to break into groups of three or four and watch a series of teacher-nominated videos on sustainable construction available on YouTube and the National Precast Concrete Association websites.

Each group is to spend 10 minutes discussing the key points of each video, then report back to the class on those key points. In your discussion consider any points or issues that may impact how your industry or job role is (or may be) impacted now and into the future.

Teacher instructions: Select one or two different videos for each group to watch.

Note: You may select alternative videos.

Suggested video titles:

- Precast concrete – building a sustainable Australia: https://www.youtube.com/watch?v=2Dm-KY4sRfA
- Concrete for a sustainable world: https://www.youtube.com/watch?v=vD7UD2jFLm8
- Concrete sustainability – Part 1: https://www.youtube.com/watch?v=2A9091CStVo [at 2:20]
- Eleven green building materials way better than concrete: https://www.youtube.com/watch?v=VsahyCrZ9sk
- Construction of concrete stairway start to finish: https://www.youtube.com/watch?v=xoNmEFEM6fE. This video goes for 25 minutes, but shows the full process of setting out, pouring and stripping the formwork for a set of external stairs and landing outdoors on solid ground. Identify and report on all the safety issues you see.
- Three disadvantages of reinforced concrete: https://www.youtube.com/watch?v=Ru724BjZOxY

COMPLETE WORKSHEET 5

SUMMARY

In this chapter we have covered the topics outlined below relating to the formwork for stairs.

Setting out for concrete-reinforced stairs

- When setting out formwork for stairs, the design must account for the additional loads, especially shear forces, imposed by sloping soffits and stepped soffits (see Figure 2.1).
- The loads are transferred to adjoining slabs and beams, with fixing points firmly anchored so that the load points don't shift, and sole plates extend to a wall or other brace point.
- Stairs can be poured at the same stage as slabs, beams and columns, or manufactured off-site and craned into position. Steel pre-fabricated formwork can be used as 'lost formwork'. It is manufactured off-site and craned into position, saving time on site.
- Side forms, known as strings or shutters, must be rigid and supported similarly to the sides of beams.
- Work from the bottom of the soffit when setting out and applying the measurements.
- The same rules apply for concrete stairs as apply for stairs constructed of other materials; that is:

  (2 × Rise) + Going = 550 mm to 700 mm

- The exact location of the first and last rise must be pinpointed, and a rod used to record and transfer measurements to walls and floors.
- The set-back distance for the soffit location can be calculated using Pythagoras' theorem.
- Strings can be set up with rise and foot cuts.
- Side shutters are constructed by allowing the shutter to be supported against a bearer and joist ends.
- At a change of direction, the landing soffit height must be set up first so that you can establish the positions for the first rise. You will also need to check for headroom clearances.
- Plumb and foot cuts are made in the same way as for regular stairs. Allow for the thickness of riser material and cleats or blocking used to support the riser boards. When cut strings are used, bevel the bottom edge of the riser boards so steel floats can be used to finish the tread surface.
- On wide stairs, a cleat or batten can be used to stiffen the riser board, or a stiffener (strong back or hanging beam) can be used with cleats to prevent them bowing out. The stiffener should be strutted to a wall or fixed point for leverage.

External stairs

- Soils and fill materials need to be compacted to specifications or plan details before setting out external stairs.
- Beams need to be incorporated at the foot of each flight.
- Different methods are used to build the formwork/falsework, depending on the type of construction required.
- Pegs, braces, joists and walers can be used to construct a side retaining wall if incorporated into the construction.

Ramps and raked soffits

- Ramps and sloping or raked soffits are useful in addressing the access and egress needs of people with disability.
- As a guide, the maximum gradient of a ramp should not exceed a pitch of 1:14 (or 7.12 degrees from the horizontal) and the surface finish must be non-slip.
- Long ramps with control joints need to be poured in sections and a sequence for pouring determined.
- Steeply sloping ramps or roof shapes (> 1:8) need formwork/falsework the same as wall forms to prevent the concrete bulging at the bottom and to provide an even thickness finish.

Stripping formwork/falsework

- Formwork/falsework must remain in place long enough for the concrete to cure and support itself and the load of workers, materials and equipment that will be placed on it.
- Release agents make it easier to remove formwork/falsework, although it is important to be aware that some release agents can affect the surface colour of the concrete.
- Written permission from the supervising engineer or site supervisor must be obtained to strip formwork/falsework before commencing work.
- Sudden shocks must be avoided.
- The minimum period of time that formwork/falsework must be left in place depends upon the weather conditions and the form that is being supported.
- Landings must remain supported by props or formwork/falsework frames until the appropriate walls are formed up at the perimeter to take over the support.
- Formply faces must not be damaged when removing or cleaning. Rags and stiff bristle brushes are used for cleaning.
- Formwork must be stacked and stored neatly to avoid damage and to make later access easy.

REFERENCES AND FURTHER READING

National Construction Code 2022

Copies of the NCC can be obtained free of charge from the Australian Building Codes Board website: **http://www.abcb.gov.au**.

National Construction Code 2022 – Volume Two, Class 1 and Class 10 Buildings and ABCB Housing Provisions.

Texts

Note: There are very few local textbooks available that deal with formwork. Cement Concrete & Aggregates Australia (CCAA), the peak body for the concrete industry in Australia, has an extensive library of documents and publications dealing with concrete in all its forms. Some documents are immediately available to download online, while for others you must register and apply via the website. To see what is available visit the website: **http://www.ccaa.com.au/iMIS_Prod**.

In this chapter we reference CCAA's *New South Wales Concrete By-product Recycling and Disposal Industry Guidelines* and the *Concrete Pump Delivery Industry Guidelines*.

Your teacher can also guide you to a range of suitable resources that may be available through your TAFE library service.

Special requirements for access for people with disability

AS 1428 (Set) Design for access and mobility. General requirements for access: New building work can be purchased from the Standards Australia website (**https://www.standards.org.au/**) or see your teacher or librarian for assistance in accessing Australian Standards online.

Web-based resources

See Chapter 1 References and further reading section for web-based resources related to formwork.

Video resources

Eleven minutes to install concrete stairs between walls: **https://www.youtube.com/watch?v=cJOJVVHYwtO**

How to mark out stairs for two-storey house suspended concrete slab: **https://www.youtube.com/watch?v=-QIOaAsPUxs**

Precast concrete: building a sustainable Australia: **https://www.youtube.com/watch?v=2Dm-KY4sRfA**

Concrete for a sustainable world: **https://www.youtube.com/watch?v=vD7UD2jFLm8**

Concrete sustainability – Part 1: **https://www.youtube.com/watch?v=2A9091CStVo [at 2:20]**

Eleven green building materials way better than concrete: **https://www.youtube.com/watch?v=VsahyCrZ9sk**

Three disadvantages of reinforced concrete: **https://www.youtube.com/watch?v=Ru724BjZOxY**

National Precast Association on YouTube: **https://www.youtube.com/@nationalprecastconcreteass282**

Relevant Australian Standards
AS 3610.1 Formwork for concrete; and supplements. This Standard and its supplements sets out requirements for the design, fabrication, erection and stripping of formwork.
AS 6669 Plywood – Formwork. This Standard specifies requirements for the manufacture, grading, finishing and branding of plywood used specifically in formwork, with a maximum length of 3100 mm and width of 1500 mm, intended to meet off-form surface finish requirements.
AS/NZS 2098.0 Methods of test for veneer and plywood
AS/NZS 2269.0 Plywood – Structural – Specifications
AS 3850.1:2015 General requirements (Part 1 of AS 3850:2015 Prefabricated Concrete Elements)
AS 3850.2:2015 Building construction (Part 2 of AS 3850:2015 Prefabricated Concrete Elements)
AS 4357 Structural laminated veneer lumber
AS 1428.1 Design for access and mobility, Part 1: General requirements for access – New building work

GET IT RIGHT

FORMWORK FOR STAIRS AND RAMPS

Questions

Below are some pictures showing formwork nearly completed and ready for pouring concrete.

1 What would need to be done just prior to pouring concrete for a complex concrete stairs job like this? (Hint: A vacuum cleaner may be needed.)

____

____

2 What special care should be taken as this formwork progresses towards pouring concrete?

____

____

3 What could happen if nothing is done?

____

____

WORKSHEET 1

To be completed by teachers

Student competent ☐

Student not yet competent ☐

Student name: ______________________

Enrolment year: ______________________

Class code: ______________________

Competency name/Number: ______________________

The practical exercise outlined below is best performed in conjunction with the formwork exercise in Chapter 1 (Worksheet 1) on formwork for slabs, columns, beams and walls.

Practical exercise

Set-out and form up for the stairs shown in the drawing below. **Note:** the drawing is not to scale and does not depict the number of treads and risers accurately. Your set-out must have a legal pitch of 2R + G = 550 to 700 mm and include a landing and quarter turn.

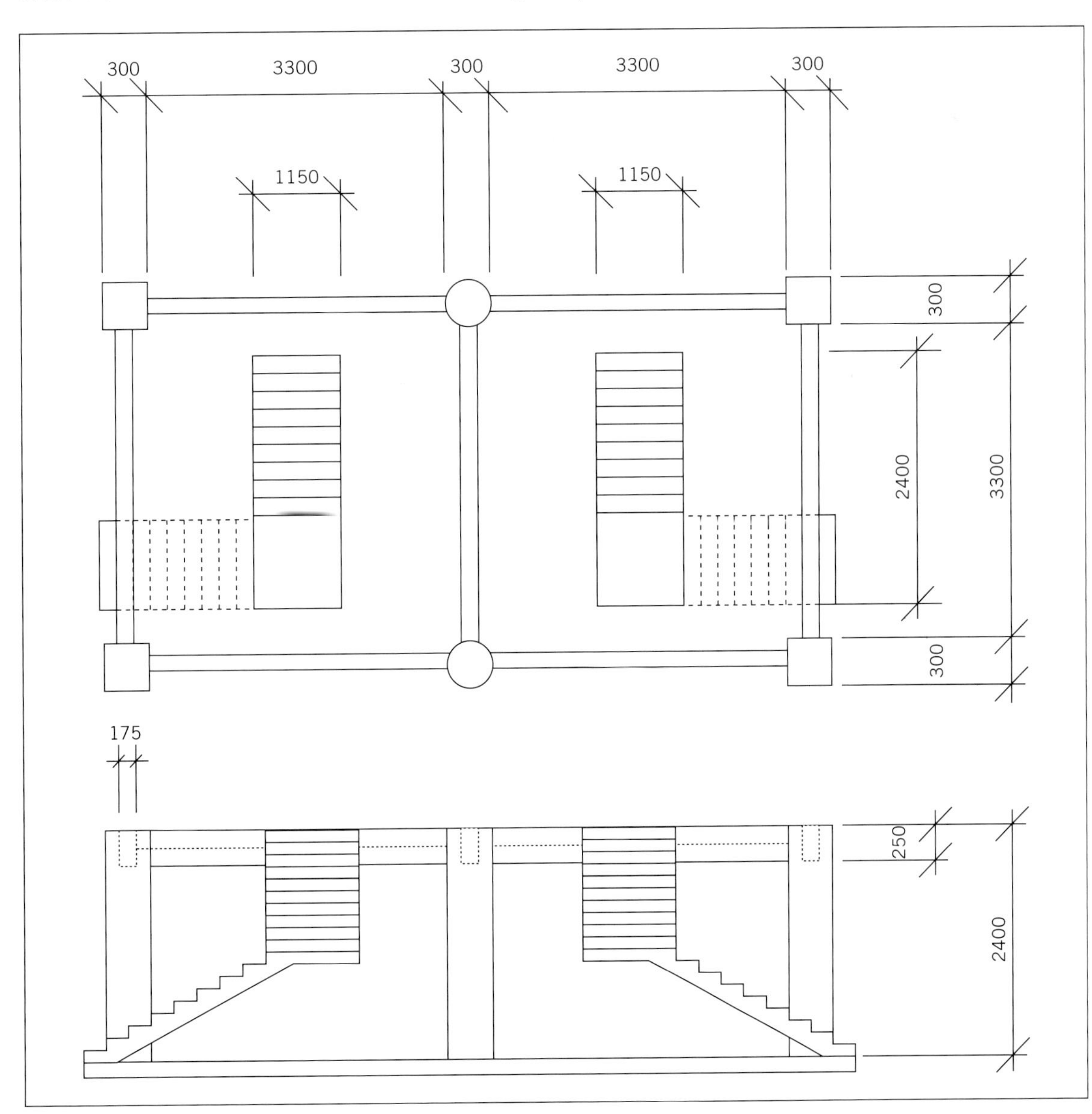

Task details

Students are to work in small teams (groups of three or four for stair formwork).

A safe work method statement (SWMS) must be completed by each student before commencing work. Teacher to sight and sign.

All students are to review the formwork/falsework plan for the overall project, noting any specific measurements and codes of practice/standards relevant to the plan/formwork set-out; for example, head height clearances, stair widths, rise and going issues.

Students are to participate in the development of a quality assurance checklist for the project.

Students are to develop a full materials list for the stair project, including sheet materials, bearers, joists, props, scaffold frames, jacks and stirrups.

Students are to calculate:

1 a suitable rise and going for the flights of stairs

2 the total volume of concrete required for one set of stairs including landing.

As part of the project, students will:

- produce a height and going rod for the staircase
- set out the landing position on the floor in the correct position.

WORKSHEET 2

To be completed by teachers

Student competent ☐

Student not yet competent ☐

Student name: ______________________________

Enrolment year: ______________________________

Class code: ______________________________

Competency name/Number: ______________________________

Answer the following questions.

1 The horizontal supports for formwork for concrete are known as:

a sole pieces
b props
c joint
d bearers.

2 Tick either True or False.

The horizontal piece of timber placed outside the bottom of the beam side to resist lateral pressure is called a runner.

True ☐ False ☐

3 Tick either True or False.

A soffit is the underneath lining of the formwork that supports the wet concrete.

True ☐ False ☐

4 Which term best describes the construction materials used to support the vertical face of a step?

a Riser board
b Rise of flight
c Nosing
d Rise

5 When setting out concrete stairs, list three things that could affect the height of the riser board in the formwork.

a ______________________________

b ______________________________

c ______________________________

6 Tick either True or False.

The horizontal distance from the front edge of the landing to the face of the bottom riser is called the 'going'.

True ☐ False ☐

7 The flat or horizontal portion upon which the foot rests while descending or ascending a stair is known as the:

a carriage piece
b going
c tread
d string.

8 Fill in the missing words.

Finished floor levels must be checked off ________ and recorded either ________ or against ________ to which the stairs may be attached.

WORKSHEET 3

To be completed by teachers	
Student competent	☐
Student not yet competent	☐

Student name: ______________________________

Enrolment year: ______________________________

Class code: ______________________________

Competency name/Number: ______________________________

Practical exercise

Task details

Students are to calculate the headroom distance for the flight of stairs shown in Worksheet 1 and determine whether it satisfies the National Construction Code.

____

____

____

Students are to design an alternative stair construction that does satisfy the National Construction Code for the project shown in Worksheet 1.

As part of this project students may:

- determine a suitable rise and going distance
- produce a height and going rod for the new design
- set out the new landing position on the floor in the correct position.

WORKSHEET 4

To be completed by teachers

Student competent ☐

Student not yet competent ☐

Student name: ______________________

Enrolment year: ______________________

Class code: ______________________

Competency name/Number: ______________________

Answer the following questions.

1 Name the parts of the stair formwork in the picture below:

2 Fill in the missing words in the following sentences.

a The AS 3610 Formwork for concrete outlines the details on the __________ and __________ of formwork and falsework.

b Battens can be used to _______ the riser board or ____________ can be used with ____________________________ of the riser boards.

c Loads are also transferred ____________ and ______ at the top end.

d It is critical to determine the exact location of __________________.

e Adjoining walls can be _________ to accept the stairs.

f When there is ______________ in the flight at a landing, it is sensible to set up the ________________________ so that you can establish the position for the _________.

g Special equipment such as ___________ may be needed to compact the soil and fill materials.

h In external stair construction __________________ can be used to support edge formwork/ falsework and internal riser boards.

i The maximum gradient of a ramp should not exceed a pitch ________.

3 List the two standards regulating the use of precast concrete stairs in building, and describe the difference in their application.

a ___

b ___

4 What are the four main advantages of using precast stair components?

a ___

b ___

c ___

d ___

WORKSHEET 5

To be completed by teachers

Student competent ☐

Student not yet competent ☐

Student name: ______________________________

Enrolment year: ______________________________

Class code: ______________________________

Competency name/Number: ______________________________

Answer the following questions.

1 What are the main benefits of ramps or raked soffits?

____

2 In what parts of the National Construction Code 2022 do you find references to ramps for access and egress?

____

3 Where in the National Construction Code 2022 do you find information on access and egress for buildings for people with disability?

____

4 To which Australian Standards does the National Construction Code 2022 refer you for specific requirements for building ramps for access for people with disability?

____

____

5 Answer True or False to the following question.

According to the National Construction Code 2022, a series of connected ramps must not have a combined vertical rise of more than 3.6 m.

True ☐ False ☐

6 What is the maximum gradient (or pitch) of a ramp designed for access and egress allowing for people with disability?

____

7 Answer True or False to the following question.

Steeply sloping ramps used in Class 1a buildings should not be less than a gradient (or pitch) of 1:7.

True ☐ False ☐

8 Fill in the missing words.

According to the National Construction Code – Volume Two, Part 3.9.1.3, an external ramp serving an external doorway or a ramp within a building must be provided with landings complying at the top and bottom of the ramp and at intervals not greater than ____ m.

9 According to AS 1428.1 – 2009, Clause 6.5.1, what is the minimum size of a landing in a series of connected access ramps in order for a wheelchair to make a 60° to 90° turn?

___

___

10 Fill in the missing words.

A building will be required to provide, as far as is reasonably practical, __________, __________ and __________ access for people to the services and facilities of a building.

PART 2

INSTALLATION FIELD OF WORK

Part 2 of this textbook deals with some of the complexities of the fit-out and finish requirements associated with residential construction, in particular the installation and replacement of windows and doors, wet area fixtures, special wall linings, panelling and mouldings, and the construction, assembly and installation of internal and external timber stairs. It will help you build a fundamental understanding of the skills and knowledge required to construct and install the materials and fixtures to National Construction Code and Australian Standards (AS).

It is based on a number of units of competency and the content described in Chapters 3, 4, 5 and 6:

3 Install and replace windows and doors
4 Install wet area fixtures
5 Install lining, panelling and moulding
6 Construct, assemble and install internal and external timber stairs.

The key elements in Part 2 are:

- Planning and preparing for installation of fitments, fixtures and materials by selecting and quantifying appropriate tools, equipment and material; performing risk assessments; and applying the National Construction Code and Australian Standards and safe work methods in all activities.
- Setting out, assembling, constructing and installing fitments, fixtures and panelling by
  - checking for accuracy against plans, designs, specifications and codes
  - accurately placing, fixing and detailing.
- Installing fitments, fixtures and materials using techniques such as scribing, mitring and developing and constructing raked mouldings using profiled moulding, architraves and skirtings.
- Cleaning up, including de-nailing, recycling practices, stacking and preparation for safe site removal or transition. Checking tools and equipment for safe use, operability, maintenance and fault rectification.

Additional information on basic waterproofing techniques when installing wet area fixtures is also included.

The learning outcomes for each chapter are a good indicator of what will be required to perform, know, understand and apply on completion of each chapter. Teachers and students should discuss the full practical and knowledge evidence requirements of each unit of competency before undertaking any activities.

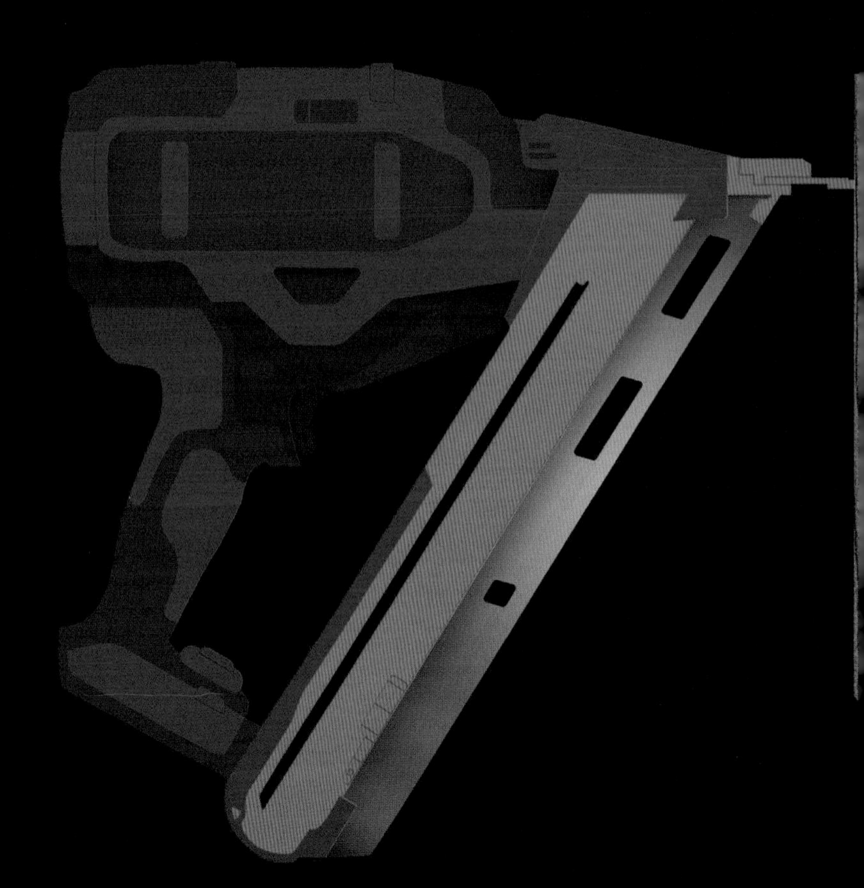

3

INSTALL AND REPLACE WINDOWS AND DOORS

This chapter aligns with the unit of competency, 'Install and replace windows and doors'. The unit outlines the knowledge and skills required to install or replace timber and metal windows and doors in low-rise residential constructions, whether they be old or new homes or multi-residential buildings.

In particular the chapter addresses the key competencies of:

- locating, interpreting and applying information from plans, standards and specifications
- working safely and following work health and safety/occupational health and safety (WHS/OHS) regulations, legislation and practices
- ensuring quality requirements are met in practice
- safely and effectively using tools, plant and equipment to replace and install windows and doors including architraves, aprons, mouldings and furniture.

The chapter will cover door and window types and how to identify them in plans and specifications. It will also cover door frame installation and hanging of doors and installing aluminium and timber windows in framed and masonry constructions.

While every effort has been made to present as much relevant information on windows and doors as possible, it is important to access and read other sources, particularly in relation to fire doors in multi-residential buildings and the regulations around the construction of buildings in bushfire-prone areas.

Learning objectives

By the end of this chapter you will be able to:

- tell the difference between internal and external doors, tell which way a door is hung, either right or left, and identify window types and glazing requirements from plans and specifications
- plan and prepare door and window schedules, and handle and store doors and windows safely onsite
- prepare doors and windows with flashings to window heads and sills, and door heads and thresholds
- install different types of door frames in external and internal openings, including split jambs in timber and metal
- select and install the right hinge type and door furniture, including locks and latches to doors, and hang the door in the door jamb

- install different types of windows including timber framed and aluminium in different types of wall constructions such as:
  - timber framed and clad
  - brick veneer
  - cavity brick
- remove with minimal damage, door and window joinery from different built forms.

National Construction Code 2022

The National Construction Code (NCC) has been updated, with NCC 2022 replacing the previous NCC 2019 version.

Changes include a new clause numbering system, which students will need to become familiar with. Guidance on the new system is available in each volume of the code – see the 'Introduction to the National Construction Code (NCC)' sections.

The NCC has also been updated to provide additional guidance on access and egress to buildings, plus fire safety and energy efficiency. This particularly relates to doors and windows in in new buildings and for renovation and retro-fittings. For example, the NCC now includes:

- specific performance requirements for each aspect of a building, along with guidance on verification methods of products and processes
- higher levels of energy efficiency through building envelope sealing – see NCC 2022 – Volume 2, H6V3 (formerly NCC 2019 – Volume 2, 2.6.2.3) – requiring the construction and fitting of doors and windows with better seals and glazing properties.

The objective of NCC 2022 – Volume 2, Part H8 Livable housing design is to 'ensure that housing is designed to meet the needs of the community, including older Australians and those with a mobility-related disability'. The aim is to ensure homes are easy to enter and move around. This will impact door sizing and threshold construction as well.

One change for Class 1 and 10 buildings has been to the structural integrity of glazing. The NCC 2022 says installations (including doors) that may be subjected to human impact must have glazing that:

(a) if broken on impact, will break in a way that has a lower than 0.13% chance of penetrating adult skin

(b) resists foreseeable human impact without breaking and

(c) is visually distinct with:

  (i) markings within the visual range of the occupants which achieve a 30% luminance contrast to both the floor and visual background, and of sizes no less than—

    (A) *in parts of a building required to be accessible, 75,000 mm^2 of glass marked per metre of width, for the full width of the installation; or*

    (B) *in parts of a building not required to be accessible, 20,000 mm^2 of glass marked per metre of width; or*

  (ii) other measures which achieve an equivalent level of visual impact.

Other changes impacting on windows and doors manufacture and installation include:

1. a minimum number and type of fixings needed for installing windows and doors, depending on wall structures (e.g. masonry, timber steel or lightweight steel frame and Hebel©).
2. entrance doors to be a minimum 870 mm with an external landing of 1.2 m × 1.2 m.

LEARNING TASK 3.1

In this teacher-led discussion, students should make a list of building elements in all classes of buildings that may be impacted by changes to the National Construction Code.

1. Compare and contrast the changes in the National Construction Code (NCC), such as:
   - NCC 2022 Volume 1 – B1P3 Glass installations at risk of human impact compared to NCC 2019 Volume 1 2019 – BP1.3
   - NCC 2022 Volume 2 – H1P1 Structural stability and resistance compared to NCC 2019 Volume 2 – P2.1.1.
2. Discuss how the manufacture of doors and other glazed building elements will be impacted.
3. Consider what forms of visual distinctions to achieve a 30% luminance contrast may be required for glazed elements in a building.

Door and window styles and specifications

Doors

Doors fall into two main categories:

- internal
- external.

Household doors are commonly made from solid timber and wood-based materials such as plywood, solid core (or block board), medium density fibreboard (MDF) and particleboard.

Exterior doors are manufactured to provide protection from the elements (wind, rain, snow, etc.) and some security against intruders. Internal doors are not manufactured to withstand the extremities of weather (i.e. wind, rain, heat and cold), and should never be used for entrances.

Being able to tell which way a door swings or identify a sliding door or cavity sliding door on plans is critical in preparing to install or replace them. Examples of what to look for on building plans for various types of doors are illustrated in Figures 3.1 to 3.8.

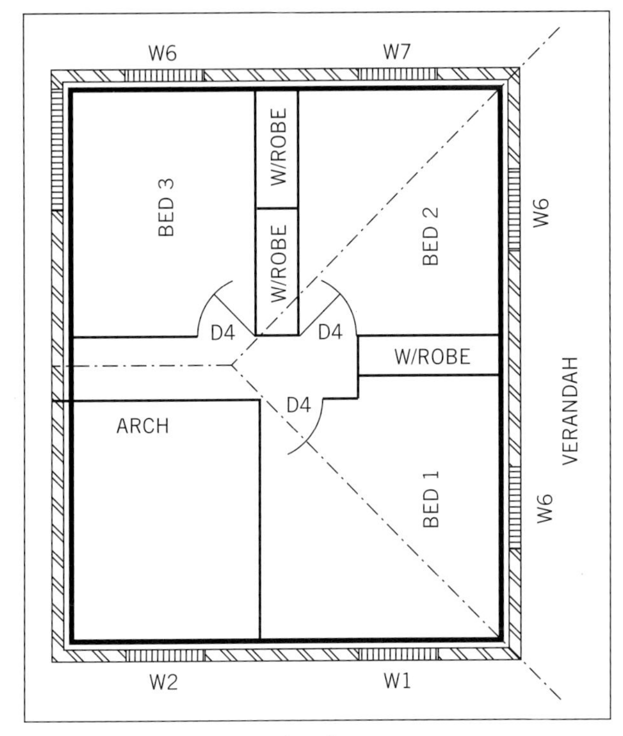

FIGURE 3.1 Swing doors – plan view

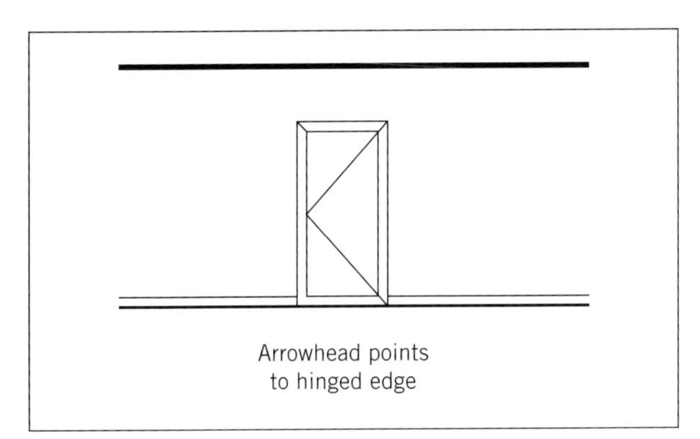

FIGURE 3.2 Swing doors – elevation view

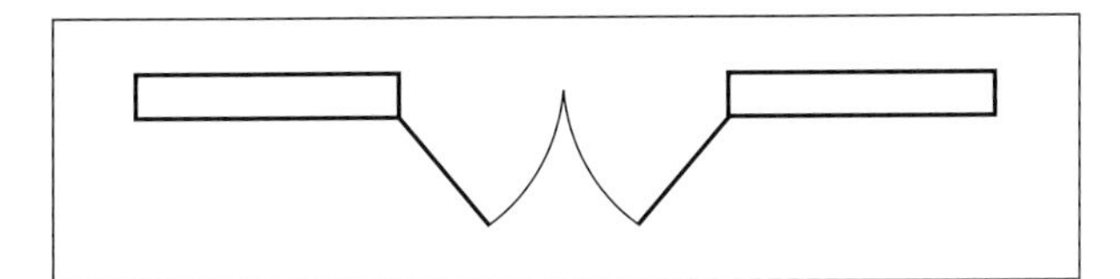

FIGURE 3.3 Double doors – plan view

To determine if a door is to be right-hand or left-hand hung, stand on the side of the door where the hinge knuckles or pivots are visible: if the knuckles or

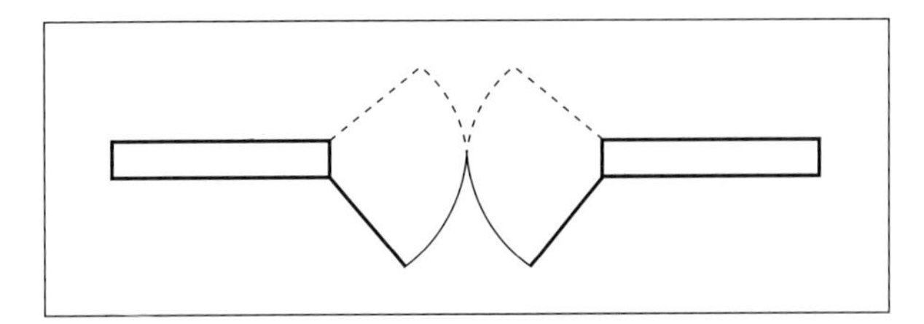

FIGURE 3.4 Double swinging (saloon) door – plan view

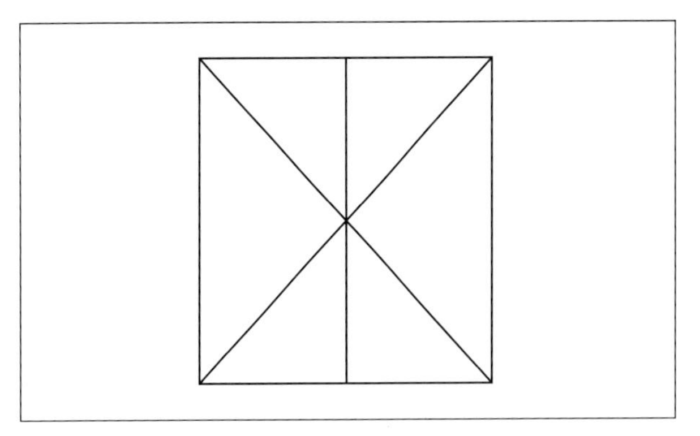

FIGURE 3.5 Double doors – elevation view

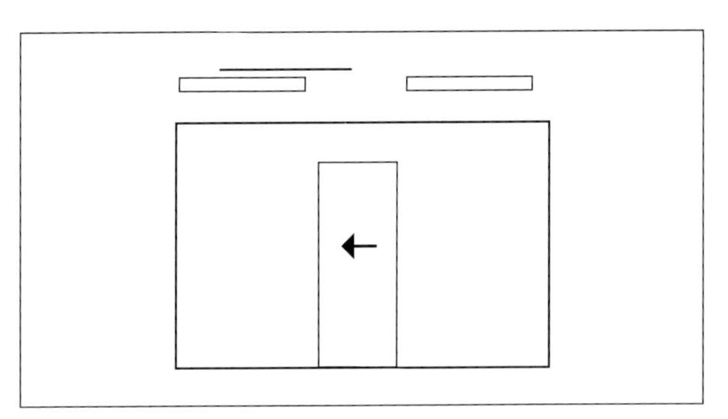

FIGURE 3.6 Sliding door – plan and elevation views

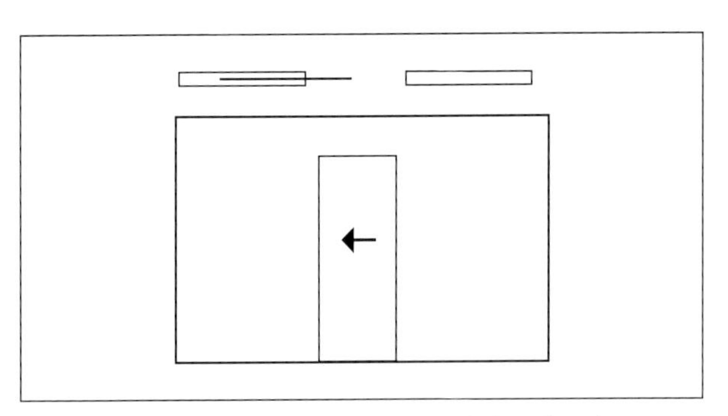

FIGURE 3.7 Cavity sliding door – plan and elevation views

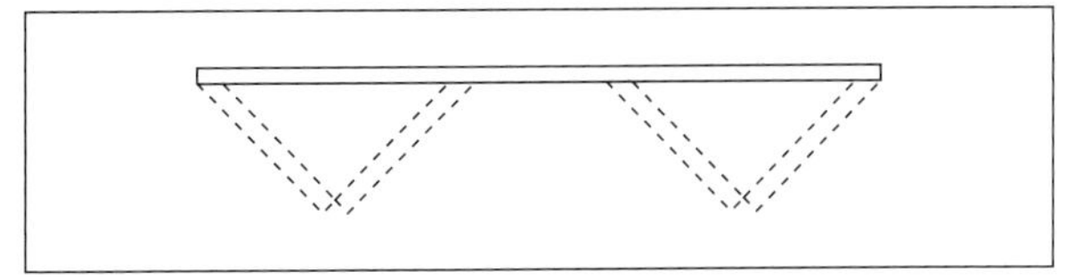

FIGURE 3.8 Bifold doors – plan view

pivots are on your left, it is a left-hand hung door; if the knuckles or pivots are on your right, it is a right-hand hung door (see Figure 3.10). Bifold doors use pivots to operate (see Figure 3.9).

Source: Greg Cheetham

FIGURE 3.9 Bifold doors

When there is a pair of doors (double or **French doors**), the hand of the first door you open, whether they are rebated or close with a weather strip, determines whether they are right-hand or left-hand hung (see Figures 3.10 and 3.11).

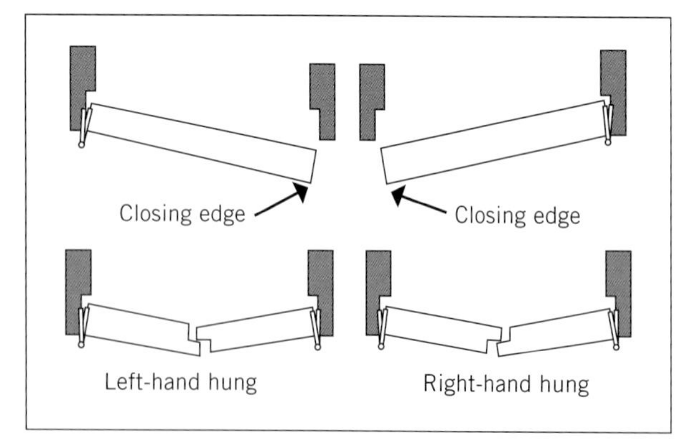

FIGURE 3.10 Determining left-hand and right-hand hung doors

Windows

Windows are any glazed opening in a building that transfers light from outside into a room or space. NCC 2022 Volume Two – BCA Class 1 and 10 buildings – Schedule 1 Glossary says a 'window includes a roof light, glass panel, glass block or brick, glass **louvre**, glazed **sash**, glazed door, or other device which transmits natural light directly from outside a building to the room concerned when in the closed position'.

Source: Greg Cheetham

FIGURE 3.11 Double doors – left-hand hung

Window sashes and frames are made from a range of materials including solid timber, aluminium and PVC (polyvinyl chloride). Because PVC windows are not common in Australia, we have not addressed their specific installation needs in this text. PVC windows are generally retro-fitted to buildings to improve the thermal and acoustic performance of the buildings. They serve this purpose because of the use of double-glazed units in their construction. Note that double-glazed units can be fitted to both aluminium and timber-framed windows as well, and the methods used are similar for both. Consult with the window manufacturer for more detailed information.

GREEN TIP

Double glazing adds to the sustainability (or liveability) of a building by reducing heating and cooling costs. Double glazing can also assist in meeting the mandated seven-star rating for new buildings.

Understanding how windows operate, while not always essential to installing them, is critical to identifying the different types from plans so that you order the right type and size. Figures 3.12 to 3.20 show a variety of different styles of window.

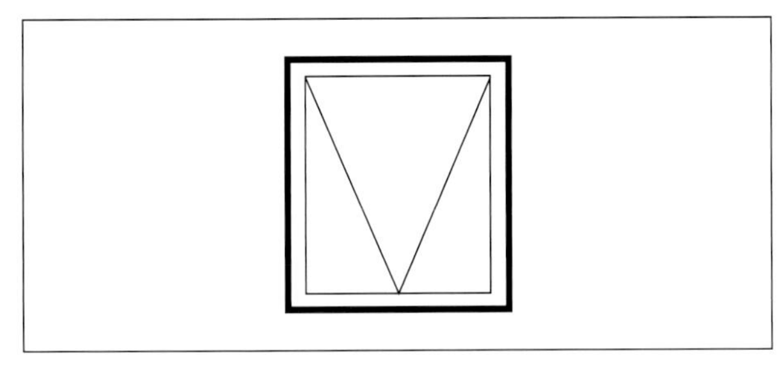

FIGURE 3.12 Awning window

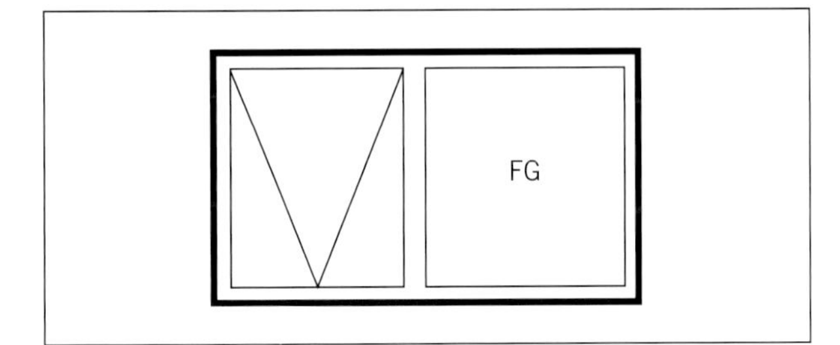

FIGURE 3.13 Awning in combination with fixed pane

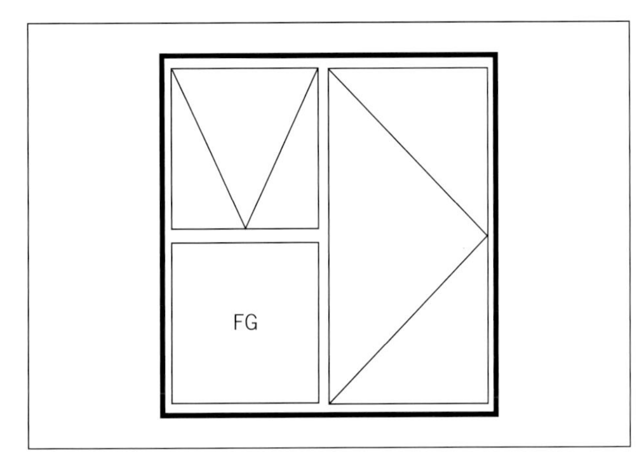

FIGURE 3.14 Awning in combination with door

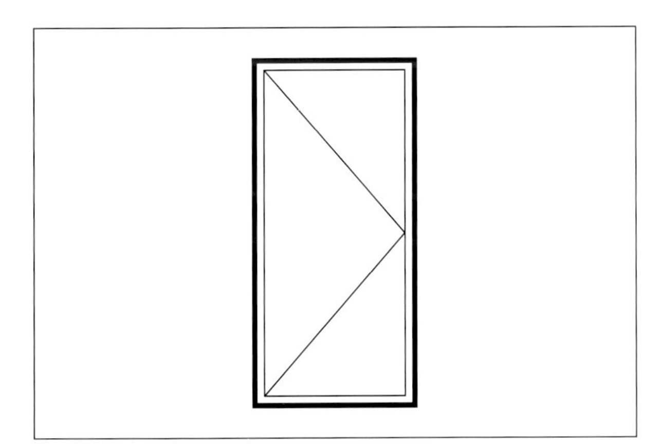

FIGURE 3.15 Casement sash

All window types can be made in combination with doors, including transom lights and fanlights (see Figure 3.21).

To view a greater range of door and window design styles, visit some of the manufacturers' websites listed at the end of this chapter.

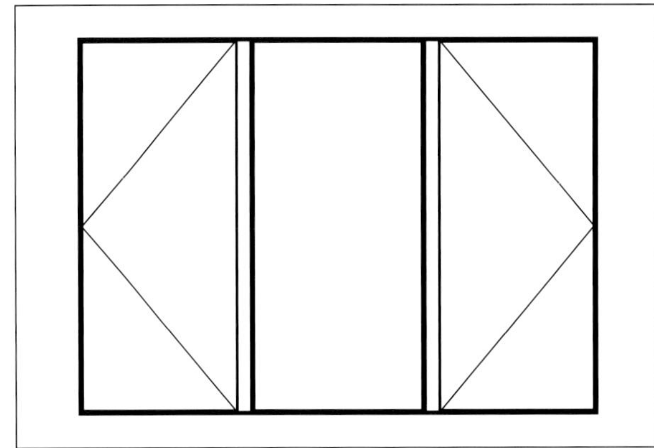

FIGURE 3.16 Casement in combination with fixed pane

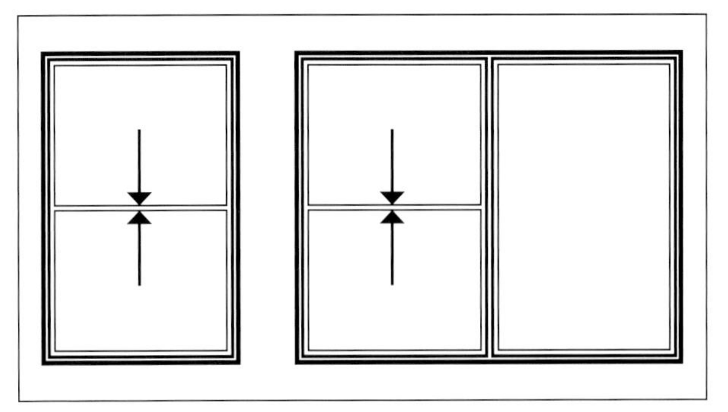

FIGURE 3.17 Double-hung sashes, and in combination with fixed pane

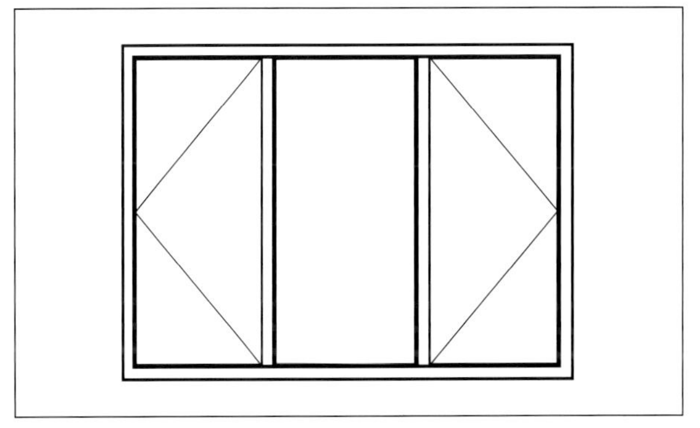

FIGURE 3.18 Casement sash with fixed pane

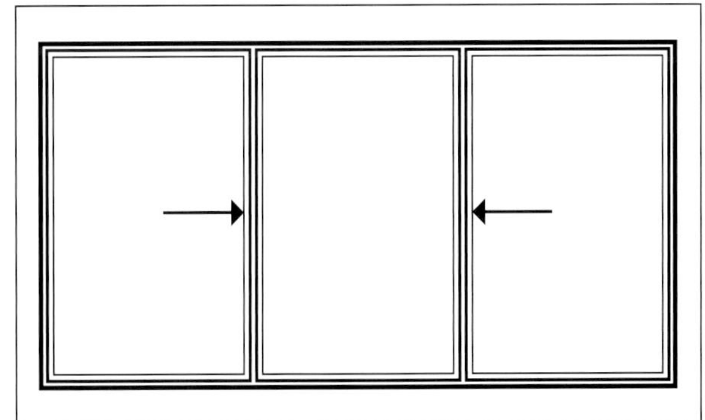

FIGURE 3.19 Double sliding sashes in combination with fixed pane

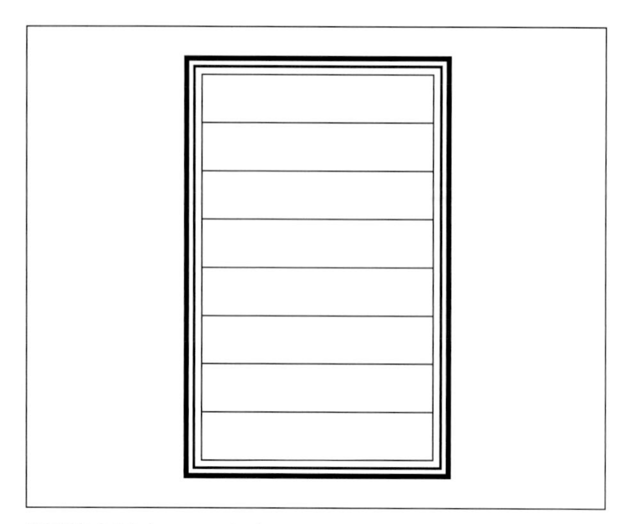

FIGURE 3.20 Louvre window

FIGURE 3.21 Door frame with sidelights and fanlight

Note: The term '**casement**' refers to any glazed hinged sash. In Australia, casements are commonly broken up into categories that reflect their opening action; for example, side hung are casements, top hung are awning sashes, and bottom hung are hopper sashes. There are also fixed sashes and both horizontal and pivoting casement sashes.

Door sizes

Doors and windows come in a large range of sizes to suit most openings and styles. When a standard size from a manufacturer is not available, custom joinery will need to be ordered or made onsite.

The standard size for an internal door is 2040 × 820 × 32–35 mm. While external doors are generally the same height and width, they are usually 40 mm thick. In newly built Class 1 residential buildings, all entrance doors need to be 870 mm wide. Thickness can vary from manufacturer to manufacturer and may depend upon the type of door. Full-height doors to 2340 mm high (approximately) are available in a range of widths to cater for wheelchair access, bifolds and double (or French) doorsets: 870 mm, 770 mm, 720 mm and 620 mm are common manufactured sizes.

For non-standard sizes, you may need to make your own door. When making your own door, you need to comply with the Australian Standards.

Window sizes

There is no standard size for windows; however, most major manufacturers have a standard range of sizes to choose from.

Windows must comply with a number of Australian Standards:

- AS 2047 Windows in buildings – Selection and installation.

  The objective of AS 2047 is to provide window designers and manufacturers with a generic window code, setting out the performance requirements and specifications in the design and manufacture of all windows, regardless of materials.

  Other standards you need to be aware of include:
  - AS 1288 Glass in buildings – Selection and installation. Specifies procedures for the design, selection and installation of glass in residential and commercial buildings.
  - AS 4420.0 Windows. A series of standards that provides a general introduction and list of methods that cover air infiltration tests, operating force tests, ultimate strength tests and water penetration resistance tests.
  - AS/NZS 2208 – Safety glazing materials in buildings. Specifies the functional properties of various safety glazing materials, including toughened glass, laminated glass, wired glass, organic-coated glass and plastic.
  - AS/NZS 4666 – Insulating glass units. Details requirements and guidelines for the testing, glazing and other associated aspects to do with insulating glass units.
  - AS/NZS 1170.2 – Structural design actions – Wind actions. Provides design values of wind actions for use in structural design in conjunction with AS/NZS 1170.0.
  - AS/NZS 4667 – Quality requirements for cut to size and processed glass. Sets out the quality requirements for cut sizes of flat, transparent, clear ordinary annealed, tinted heat-absorbing, patterned and wired glass for general glazing and/or further processing.
  - AS 3959 – Construction of buildings in bushfire-prone areas. Details requirements for the construction of buildings and a methodology for assessing categories of bushfire attack in respect of designated bushfire-prone zone.

For more detailed information about these Standards, visit the Standards Australia website: http://www.standards.org.au/.

Windows must also comply with the light and ventilation requirements of the NCC (see Part F4F6 – Light and ventilation for Class 2 to 9 buildings, and H4P4 and

H4P5 for lighting and ventilation for Class 1 and 10 buildings), which contains minimum requirements in relation to room sizes and natural light and ventilation. A habitable room must be provided with windows that provide an 'average daylight factor' of not less than 2 per cent. The building designer needs to be familiar with this. In the superseded NCC (2019), the size of the glazed area of a window was no less than 10 per cent of the overall floor space of each room. Each room should be able to be naturally vented to the outside of the building; where a window cannot be opened for security or privacy reasons, an exhaust fan should be installed.

Glass

The type and size of glass used in windows is governed not only by environmental and aesthetic factors but also by AS 1288 Glass in buildings – Selection and installation, which provides guidance on the type, thickness and installation requirements for glass in buildings. The Standard was updated in 2021 to include greater guidance on wind loads, human impact, overhead glazing (skylights), barriers and glazed balustrades, and framed, partly framed and unframed glazing. It does not cover glass in widows and doors in heritage buildings or lead lights.

The range of glass available includes:

- drawn sheet – most common in old buildings; it usually provides a distorted view and is not used today
- float glass – it provides complete distortion-free view
- laminated glass – consists of two sheets of float glass bonded together with a polyvinyl butyral (PVB) interlayer
- toughened glass – heat-treated glass designed to increase face-only impact strength (from five to seven times stronger); it breaks into fine pieces and must be shaped or drilled before toughening
- wired glass – the wire holds glass pieces together if it breaks and increases security (not common today)
- solar glass, including:
  - heat-reflecting glass – reflects heat
  - heat-absorbing glass – absorbs heat and reduces penetration
  - glare-reducing glass – has tinted film to help control glare
- cyclone-resistant glass – similar to laminated glass, but with a tougher interlayer to resist wind-blown objects.

Note: Safety glass and plastics used for doors and windows in buildings should comply with AS/NZS 2208 Safety glazing materials in buildings.

Labelling windows

All window manufacturers should label their windows with the appropriate information, as outlined in AS 2047 Windows in buildings – Selection and installation, including:

- manufacturer's identification mark
- serviceability limit – state residential rating as identified during testing in accordance with AS 4420.3
- ultimate limit – state residential rating as identified during testing in accordance with AS 4420.6
- water penetration resistance in accordance with AS 4420.5.

A window energy rating scheme accredited by the Australian Fenestration Rating Council (AFRC) provides a quick and easy guide to correctly rated windows and glazed doors that comply with NCC requirements for regions in Australia; however, it does not cover all door and window manufacturers. Windows can be selected by manufacturer, frame material, window type, glazing type, star rating and UV protection. For those manufacturers not covered, you will need to check the relevant information in other Australian Standards and your state's regulators to ensure the products comply with any new building materials compliance legislation.

For more information, check the Australian Glass and Window Association website at: https://www.agwa.com.au.

COMPLETE WORKSHEET 1

LEARNING TASK 3.2

SELECTING THE RIGHT MATERIALS FOR WINDOWS AND DOORS

In a teacher-led discussion consider the following topic: Selecting the right materials for windows and doors. In particular, discuss:

1. What are the key considerations in choosing the right window?
2. Aluminium and timber are the two most common materials used in Australia to manufacture windows. What are the advantages and disadvantages of these two materials?

Make two checklists (one for windows and one for doors) that can be used in client–builder discussions on critical points to consider when selecting doors and windows.

Consider for doors:

- how they will be hung or operated
- whether they are internal or external
- materials to be used
- security issues – locks, levers, handles, knockers, glazing, pets and technology.

Consider for windows:

- acoustic control
- thermal and energy efficiency
- ventilation, building orientation and window sizing
- compliance – Australian Standards and bushfire zones.

Planning and preparation

Read plans and specifications

It is very important to read plans and specifications carefully to identify:

- door and window types, so that you can develop a schedule (list of types and sizes required)
- materials to be used (e.g. for wall construction, door framing and thresholds)
- specifications for door and window hardware (sometimes called door and window furniture)
- environmental and quality control issues, before work commences, such as storage and security for:
  - door and window hardware
  - bulk window deliveries
  - glass types
  - climate zone and Bushfire Attack Level (BAL); that is, additional fire protection such as shutters and screens
  - many construction businesses will have quality assurance policies and standards in place to ensure work is completed to a high standard and call backs and costly repairs are reduced.
- protection required for installed doors and windows during final building works; for example, thresholds need to be protected from damage by workers during the remainder of construction activities, and anodised aluminium window and door frames need protection from damage by cement render works, plaster materials, foot traffic and wheelbarrows
- protection required for glass – it is often a good idea to arrange separate installation of glazed sashes and panels, and frames, particularly in areas where glass is likely to be damaged
- Many construction businesses will have quality assurance policies and standards in place to ensure work is completed to a high standard so callbacks and costly repairs are reduced
- Depending on the works to be carried out, work areas should be isolated with the use of barricades. Signage should be installed warning other people in the area that construction work is in progress
- the relevant safety data sheets (SDS) necessary for materials being used in the construction need to be kept onsite.

Take the time to refresh yourself on the various OHS requirements related to construction work by reviewing the chapter CPCCWHS2001 Apply WHS requirements, policies and procedures in the construction industry.

When using chemicals or cleaning agents, always check the SDS to see what personal protective equipment (PPE) or special handling instructions may be needed for inclusion in an SWMS.

Prepare door and window schedules

Door and window **schedules** are prepared for buildings to ensure that the right styles, sizes and finishes for doors and windows are ordered, and they are fitted in their correct position according to the plans and specifications. In large projects, these schedules are partly prepared by the architect, while for a smaller cottage construction, for example, it is sometimes left to the builder to determine the quantities and order the joinery.

Table 3.1 shows a typical door schedule for a small cottage construction (see also Figure 3.22).

In window schedules, a code is often used to specify timber and aluminium window requirements for a job. The code used will depend on the manufacturer's catalogue and order codes. You can see examples of the various company order codes by visiting the manufacturers' websites at the end of this chapter. A specialist joinery company can supply special sizes

TABLE 3.1 Door schedule

Supplier	Timber doors – Stegbar			Site address		
Door type	**Location**	**Size**	**Qty**	**Code**	**Hinge (l/r)**	**Finish**
Entry door	D1 – South elevation	2040 × 820 mm	1	Stegbar Balmoral design	LHH	Traditional cedar stained – clear polyurethane finish
External flush panel	D2 – North elevation	2040 × 820 mm	1	Stegbar Solid core	RHH	Traditional cedar stained – clear polyurethane finish
Internal	D7 – see floor plan	2040 × 820 mm	1	Stegbar Tudor design	LHH	Gloss paint to specifications
Internal	D8 – see floor plan	2040 × 820 mm	1	Stegbar Tudor design	RHH	Gloss paint to specifications
Internal	D10	2040 × 820 mm	1	Stegbar Tudor design	LHH	Gloss paint to specifications
Internal sliding	D3, D4, D5, D6, D9	2040 × 820 mm	5	Stegbar Tudor design	Cavity slide unit	Gloss paint to specifications

Source: Stegbar, http://www.stegbar.com.au

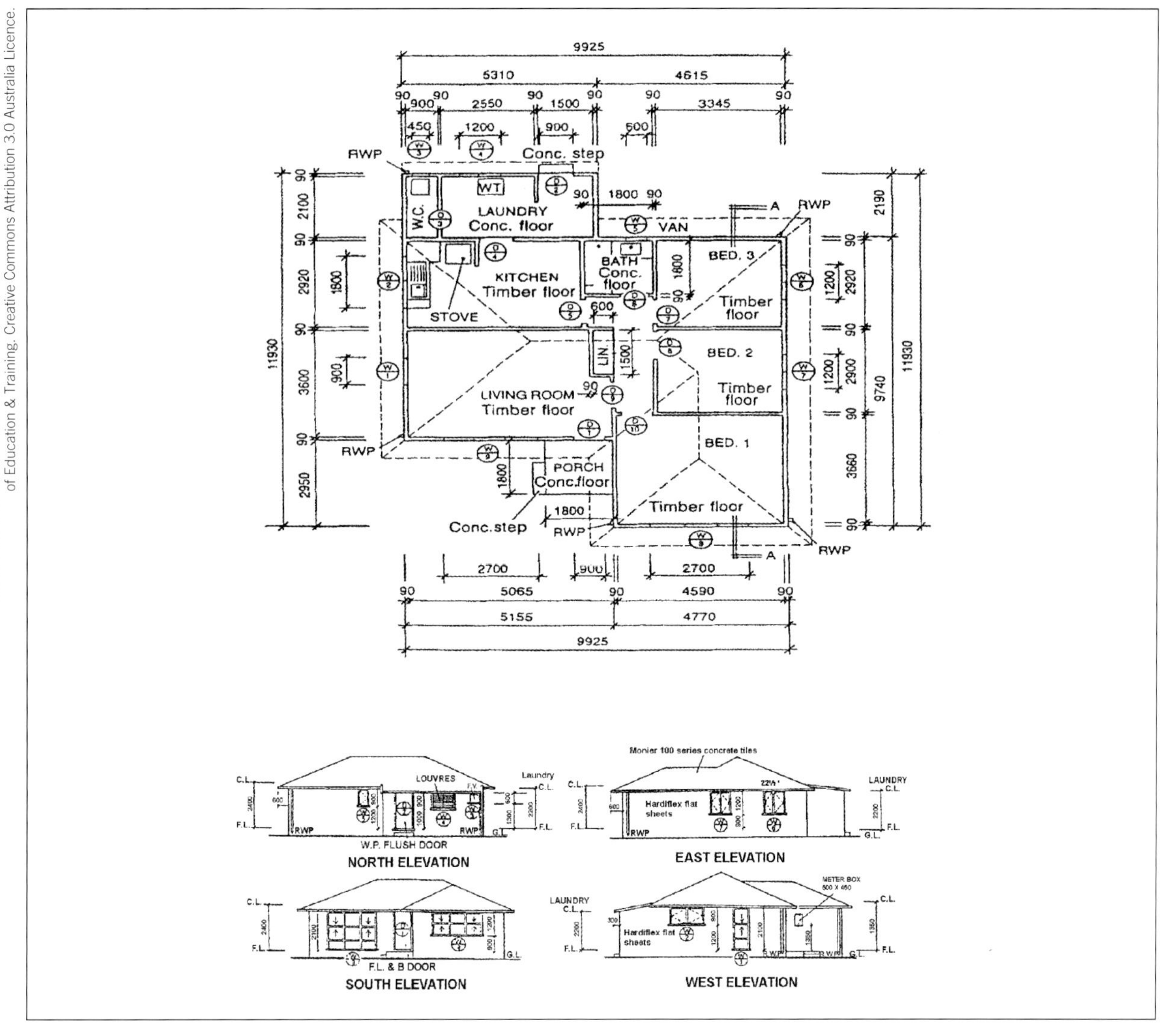

FIGURE 3.22 House plan and elevations

on request, but you would need to be able to provide accurate opening sizes or frame sizes, with the height being the first measurement followed by the width (e.g. 1320 mm high × 1720 mm wide). Lead times may be critical.

Table 3.2 provides a typical schedule for aluminium windows (see also Figure 3.22).

Before ordering doors and windows, **check the opening sizes** required. The width of door and window openings should be the width of the frame + 10 mm on either side for clearance and + 10 mm above for frame head (or lintel) clearance (see Figure 3.23).

TABLE 3.2 Window schedule

Supplier	Aluminium – Stegbar joinery			Site address	
Window type	**Location**	**Size**	**Qty**	**Glass**	**Finish**
Aluminium double hung	W1 – West elevation	2100 × 1215 mm	1	Laminated 4.32	Bronze anodised to spec
Sliding	W2 – West elevation	900 × 1800 mm	1	Laminated 4.32	Bronze anodised to spec
Fixed vented	W3 – North elevation	600 × 450 mm	1	Obscure	Bronze anodised to spec
Louvre	W4 – North elevation	900 × 1200 mm	1	Obscure	Bronze anodised to spec
Casement	W5 – North elevation	900 × 600 mm	1	Laminated 4.32	Bronze anodised to spec
Sliding	W6 & W7	1220 × 1200 mm	2	Laminated 4.32	Bronze anodised to spec
Double hung, fixed light in centre	W8 & W9	2700 × 1200 mm	2	Laminated 4.32	Bronze anodised to spec

Source: Stegbar, http://www.stegbar.com.au

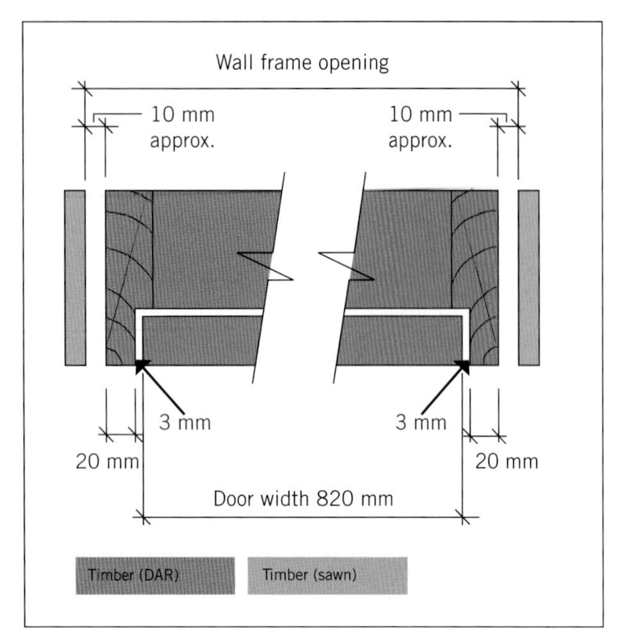

FIGURE 3.23 Door frame clearance

Allowances also need to be made for door and window sill and threshold thicknesses.

In masonry construction, the position of doors and windows should be located using a brick gauge or rod to ensure they can be fitted neatly into the building, with even brickwork on either side of the openings.

Brick gauges are pieces of timber, one for width gauging and the other for height gauging, marked out with brick heights and widths + 10 mm allowances for mortar joints (see Figure 3.24). This means brickwork can be easily checked and doors and windows easily located to ensure they work to half or full bricks.

In many cases, door and window manufacturers provide brochures that outline their standard range of sizes and a guide to appropriate opening sizes, the number of brick courses both vertically and horizontally, and the window and appropriate stud spacings.

For more information and examples, see the manufacturers' websites listed at the end of this chapter.

Onsite storage and handling

Doors need to be stored on a flat, even surface, preferably off the ground on gluts or packing to prevent the doors from twisting and cupping and soaking up moisture. Most entry doors will come with some form of protective packaging, such as corrugated paper or bubble wrap. Leave the packaging on until the door is to be installed.

Glazed panel doors and sidelights should be stored in light traffic areas to reduce the risk of damage.

It is a good idea to have the doors delivered when you are ready to install them. Door frames need to be ordered early, especially if they are to be installed in full masonry constructions, as these will have to be located and plumbed so that brickwork can progress.

The same applies to windows: install the window frames first and the sashes and glazed units only when necessary to achieve lock-up.

Bushfire-prone areas

Special consideration needs to be given at the design stage in assessing the requirements within bushfire-prone areas. This is especially true in regards to timber door and window frames and the glass used in construction. You may need to be able to assess the Bushfire Attack Level (or **BAL**) of the building or site. The BAL is based on:

1. weather conditions of the area
2. surrounding vegetation and distance from vegetation
3. land gradients (slope).

You should obtain a copy of AS 3959 Construction of buildings in bushfire-prone areas (updated in 2018) for more detail on BALs, and you can also do an internet search using the search term 'bushfire attack level'. In NSW the NCC has been varied so that the construction solutions for BAL in AS 3959 are not the minimum. The Flame Zone has been established, where radiant heat levels and flame contact are likely to threaten building integrity and result in significant risk to residents. Your Rural Fire Service may recommend other protection measures such as drenching systems and radiant heat barriers. Check the requirements for your state. See also Learning task 3.4 later in this chapter.

Tools, equipment and PPE

Tools and equipment required for the installation of doors and windows may include a:

- hammer
- pinch bar
- spirit level
- wood chisel
- plugging chisel
- drill
- fixing gun.

Before using tools and equipment it is important that they are checked for faults and reported. Tools should be used in accordance with the manufacturer's instructions and should not be used outside of those parameters set within the instructions.

Further information on tools and PPE used in this type of work can be located in CPCCCA2002 Use carpentry tools and equipment.

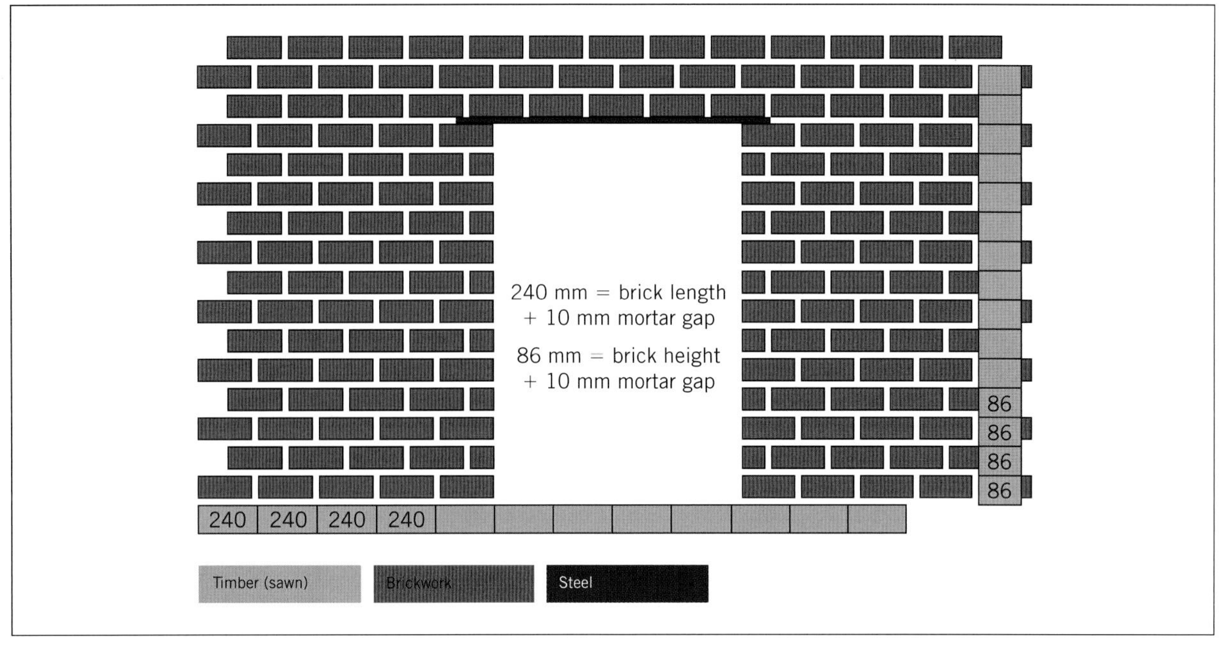

FIGURE 3.24 Brick gauges

Personal protective equipment (PPE) must be used in accordance with the workplace policy procedures and be properly fitted and fit for purpose in line with the risk assessment that is carried out for the work to be completed. Further information on PPE can be found in CPCCWHS2001 Apply WHS requirements, policies and procedures in the construction industry.

COMPLETE WORKSHEET 2

LEARNING TASK 3.3 ALTERNATIVE TASK

Students are to gather a set of house plans and specifications and then create a door and window schedule from the plans and specifications.
The schedules should include:

- supplier details and site address
- door/window types
- the location of doors/windows
- size – width first, height second
- quantity required
- supplier code if applicable
- door hinge detail; for example, **no. off** – handing (left or right), finish (painted, stained, etc.) and special detail
- window details for glass type; for example, handing (left or right if applicable) and finish (powder coated or anodised).

Students must provide copies of the drawings and copies of specific references in the specifications with their schedules. Alternatively, the teacher may provide sets for students to use.

Flashings in doors and windows

Flashings are membranes that are installed in external openings in walls above and below doors and windows to prevent water penetration into the building.

Flashings in timber-framed construction

Thresholds and sills

A flashing material is laid below the **threshold** and **sills** and folded up the back to form a tray (good practice is no less than between 12 and 20 mm). The flashing is laid in place before the door or window frame is set in place. Flashings are commonly made from bitumen-coated aluminium (e.g. Alcore™) or polyethylene with a texture finish to prevent wicking of moisture, but they can also be made from lead (no longer recommended due to health and safety concerns) or zinc. Bitumen-coated flashings may not be suitable for some fire-prone areas.

Figure 3.25 shows how flashings are formed and Figure 3.26 shows a flashing in place.

Note: When forming the tray, fold the corners up. Do not cut them. The back edge of the threshold is rebated slightly to accommodate the thickness of the tray, and to allow flooring and window nosing and reveal-lining material to butt hard against the threshold or sill. The rebate may be done by the frame manufacturer or onsite by the carpenter.

In window installation in timber-framed constructions, the flashing for the sill is cut into the window studs to locate the tray and make installing the frame easier, because the tray does not move (see Figure 3.27). The saw cuts should be approximately 50 mm up from the sill trimmer and angled towards the outside and only 6 mm

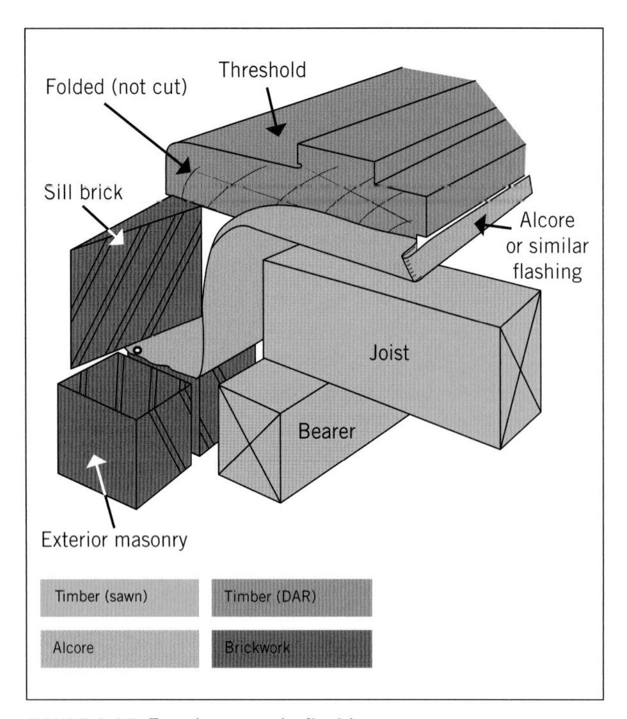

FIGURE 3.25 Forming trays in flashings

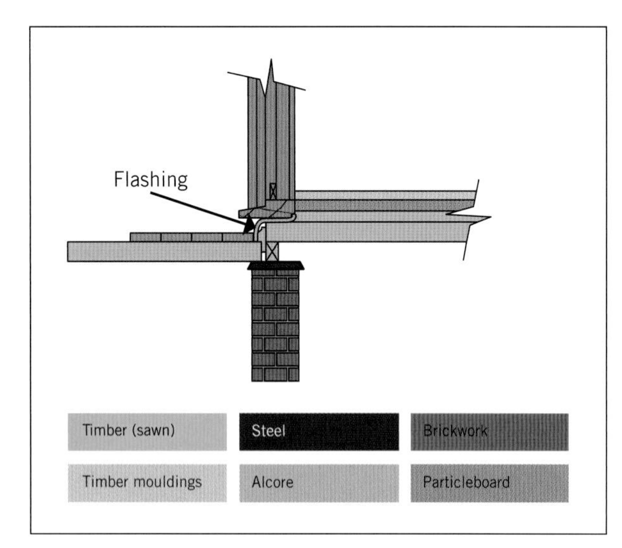

FIGURE 3.26 Flashing

deep. Without penetrating the flashing, use a couple of **clouts** to hold the flashing in place. Allow the flashing to overhang about 25 mm past the cladding so that it can be folded down over the top edge of the **architrave**, apron piece or quad.

Above windows and doors

Above windows and doors, the flashing material is fastened to the door studs and jack studs using 25 mm clouts (though it can also be stapled) and is laid so that the bottom edge of the flashing covers the top edge of the architrave, **storm mould** or door frame depending upon the construction method being used (see Figure 3.28). In Figure 3.28, note the use of a blocking piece, the same thickness as the cladding

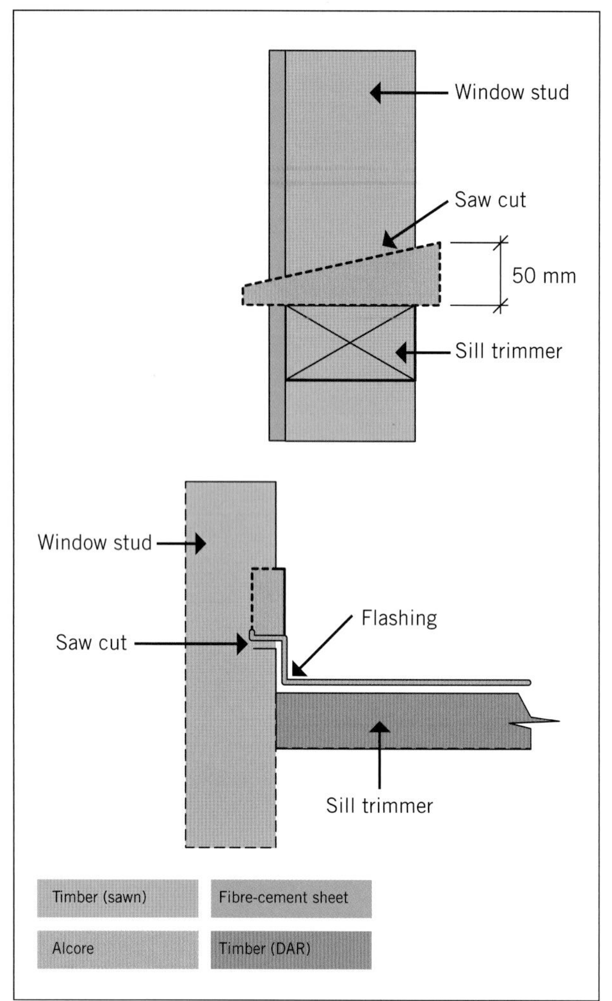

FIGURE 3.27 Cutting in sill flashing for windows

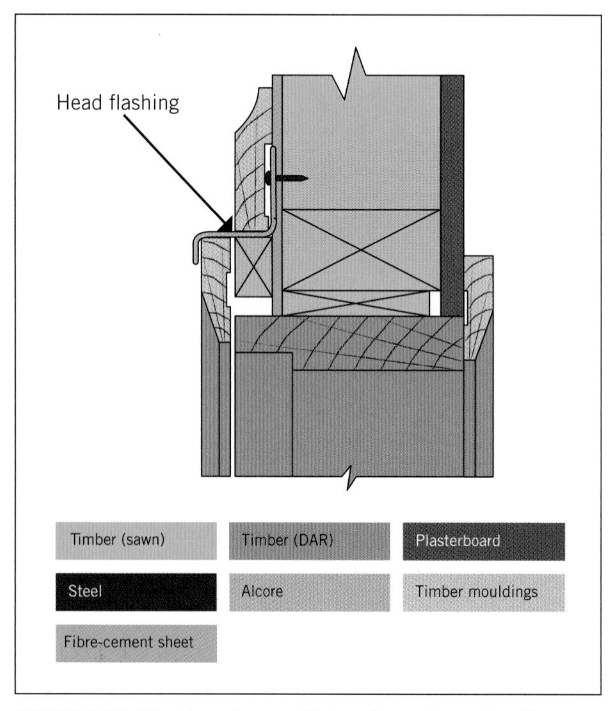

FIGURE 3.28 Flashing above a timber-framed construction

material, to allow the flashing materials to come out and over the architrave. The blocking can be a piece of the cladding material cut to suit the opening.

Flashings in brick veneer construction

In brick veneer construction, a tray is formed at the sill, as for a timber-framed construction, but you must make sure you provide sufficient width of flashing material to allow the flashing to be embedded in the masonry skin below the weep hole course (see **Figure 3.29**).

In brick veneer construction, the flashing above the window or door is fixed to the studs using clouts and is embedded in the masonry skin above the galvanised metal **lintel** so that any water can escape through the **weep holes** (see **Figures 3.30** and **3.31**).

In brick veneer construction, the flashing is embedded in the brickwork below the weep hole.

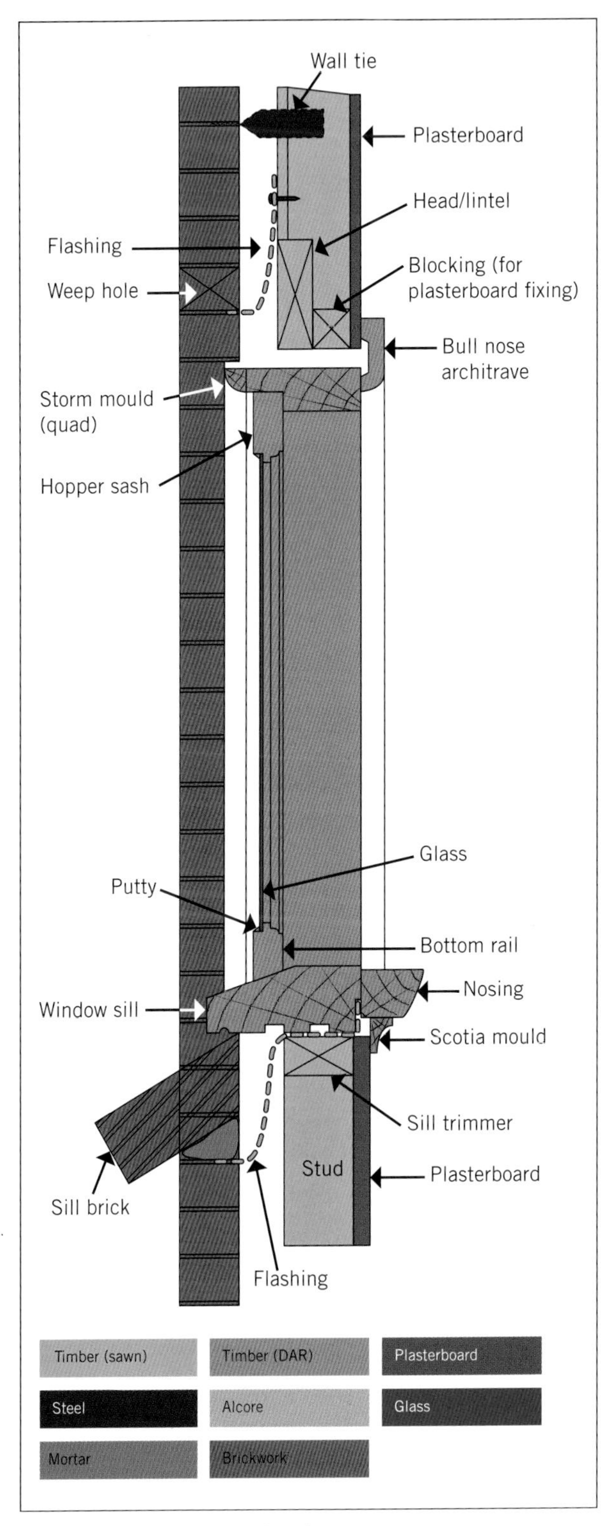

FIGURE 3.29 Brick veneer flashing

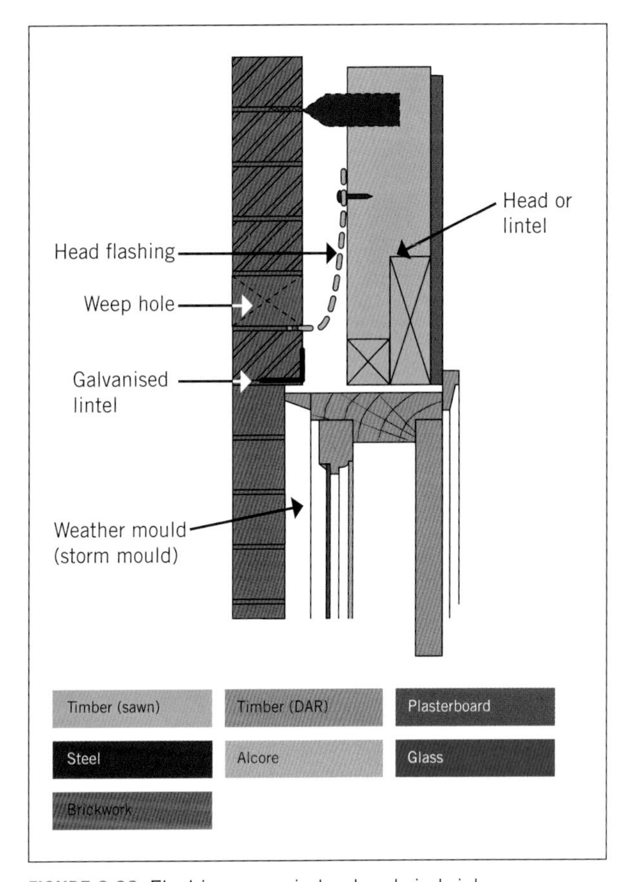

FIGURE 3.30 Flashing over window heads in brick veneer construction

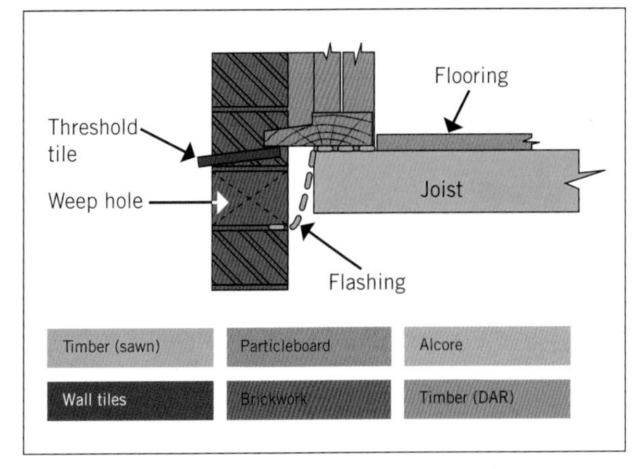

FIGURE 3.31 Tiled sill to threshold showing flashing

Flashings in aluminium windows

In aluminium constructions, the head flashing is usually set between the aluminium fin and the **reveal** lining at the factory (see Figure 3.32). It should be fastened to the window studs and jack studs with clouts or wide crown staples. If the window does not come with flashings, the alternative is to fasten flashing to the studs and embed the flashing into the weep hole course, as indicated as an alternative in Figure 3.33.

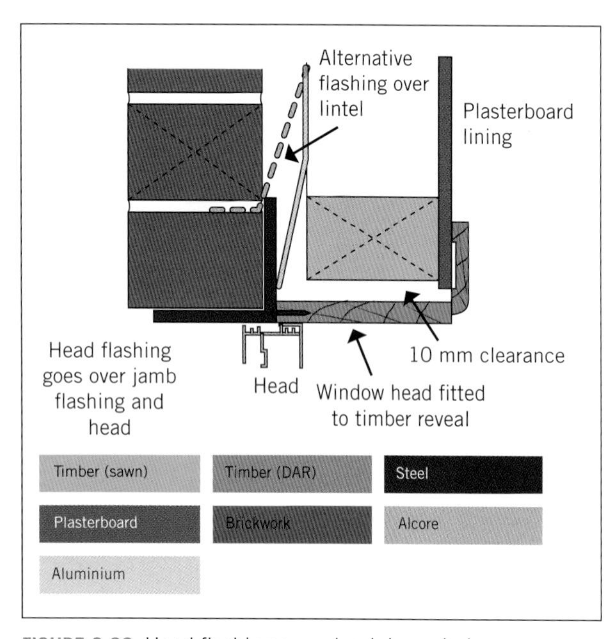

FIGURE 3.32 Head flashing over aluminium windows

Flashing below aluminium windows can be done similarly to timber windows where no flashing material has been attached at the time of manufacture. When the flashing is attached to a window, the flashing material should extend at least three brick courses below the bottom of the window and preferably be laid into the weep hole course (see Figure 3.34).

Openings that take aluminium door frames in brick veneer and timber-framed constructions using bearer and joist methods need to have a sole plate or fixing plate cut into the joists to provide an even bearing surface for the door frame to sit on; otherwise the aluminium frame can buckle under the weight of traffic, causing problems with door alignment. A timber threshold can be used, or a common alternative is to use a brick sill.

Note: A 10 mm gap must be left between the threshold or sill in ground floor construction and increased to 15 mm in first floor and above constructions.

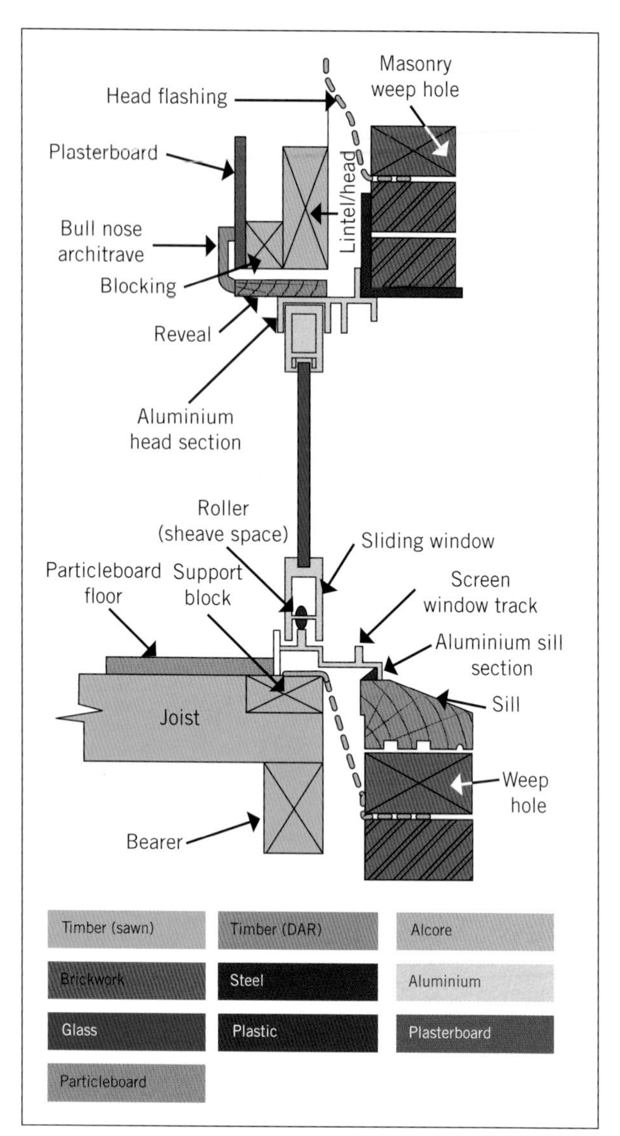

FIGURE 3.33 Head flashing in brick veneer construction for aluminium windows

In new constructions, thresholds must be designed to facilitate access for people with disability. This may include a small ramp.

During the construction stage, aluminium door sills and thresholds need to be protected from damage by foot traffic, wheelbarrows and general building grime (sand, cement, mud, etc.). Build a bridge or ramp to protect the full width of the sill (see Figure 3.35).

Flashings in cavity brick construction

In cavity brick construction, the flashing to the threshold is laid in place before the door frame is stood in place to ensure that it forms a tray around the threshold and sufficient width is allowed so that

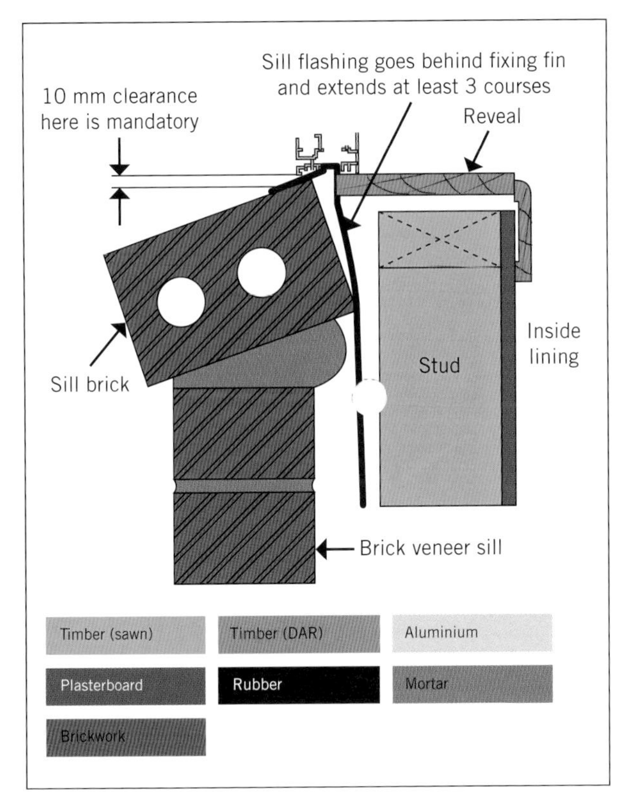

FIGURE 3.34 Sill flashing in brick veneer construction with aluminium windows

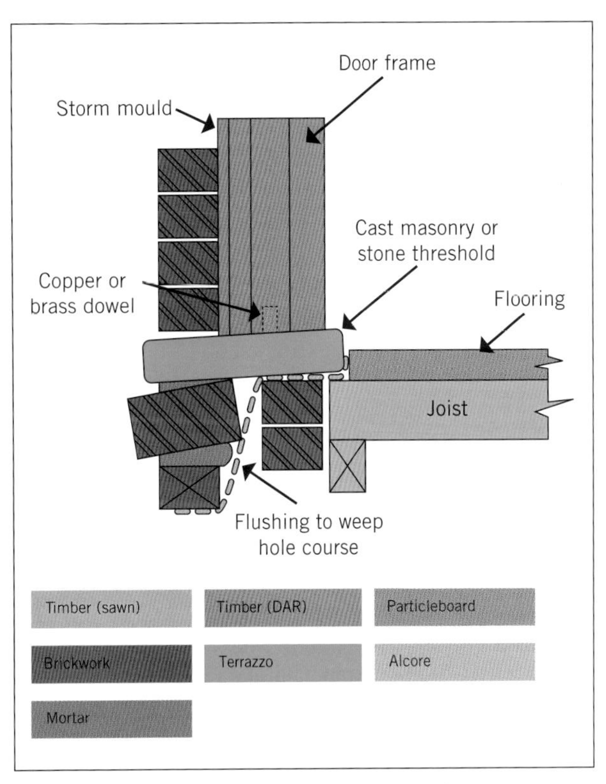

FIGURE 3.36 Flashing in cavity brick construction

the loose edge can be embedded in the exterior masonry face wall at weep hole level, as shown in Figure 3.36.

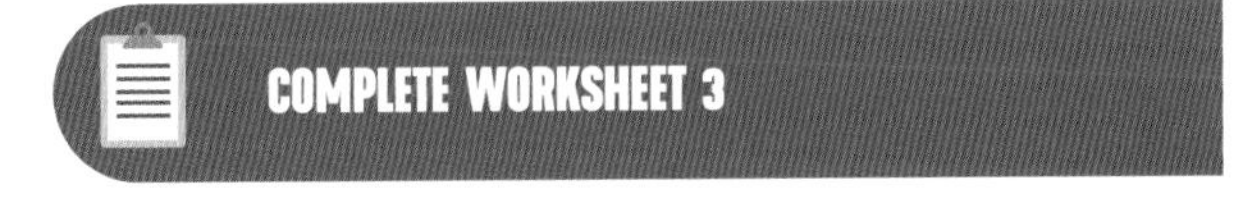

Door and door frame installation

There are two main forms of door frame for timber doors:

- door frames for external use
- door frames for internal use.

Their construction is described in Chapter 9.

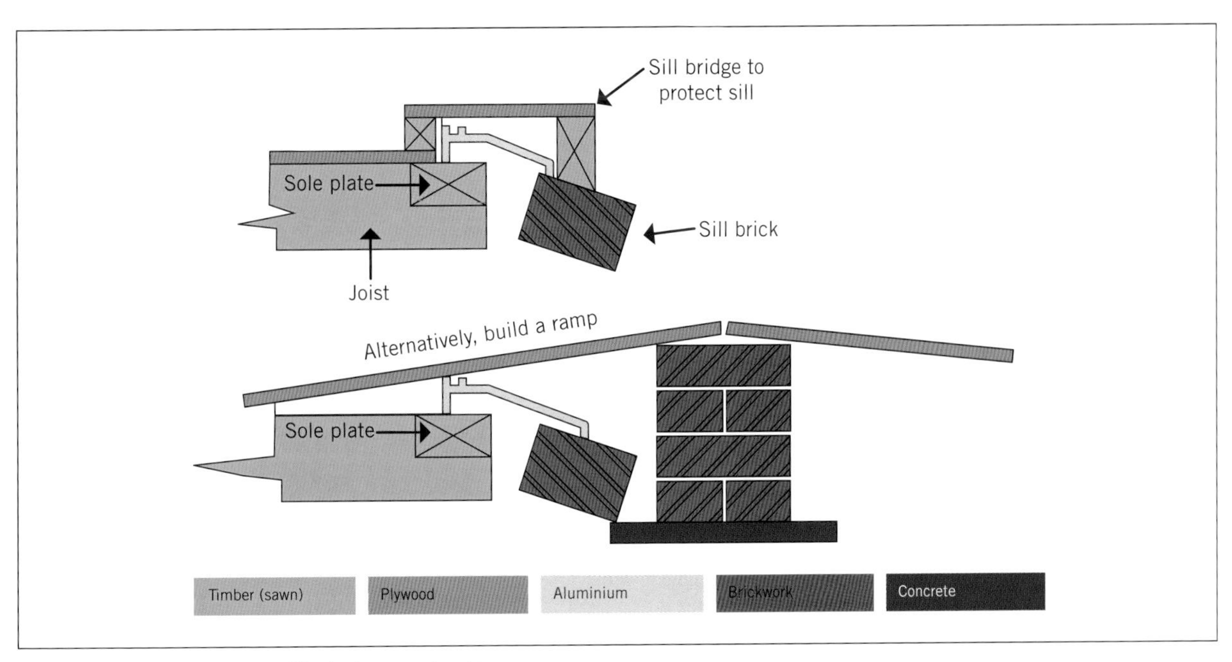

FIGURE 3.35 Protecting the sills during construction

Installing door frames will be different for each type of wall construction (e.g. timber framed, brick veneer, cavity brick and steel framed) and floor construction (e.g. suspended floor or concrete slab).

In this section, we outline the methods and processes used for each form of construction.

For guidance in relation to all types of door installation, including the type, gauge and quantity of fasteners the 'Australian Window Association – An Industry Guide to the Correct Fixing of Windows and Doors' should be used. The Australian standard 2688 *Timber and composite doors* should also be referenced when hanging doors.

Installing external door frames in timber-framed wall constructions

Installing external door frames is a straightforward process of standing and plumbing the frame in the opening and fixing the frame using nails. You must be sure the door frame material is wide enough to cover the full range of lining and cladding materials being used (see **Figure 3.37**). Otherwise an additional reveal lining material may be required.

The entrance door should always open inwards, and if fitted with a screen or security door, you should ensure that the rebate sizes are suitable (43–45 mm for the external door and 25 mm for the security door).

Note: A plant or stop can be used to form a rebate for a security door that is being retrofitted.

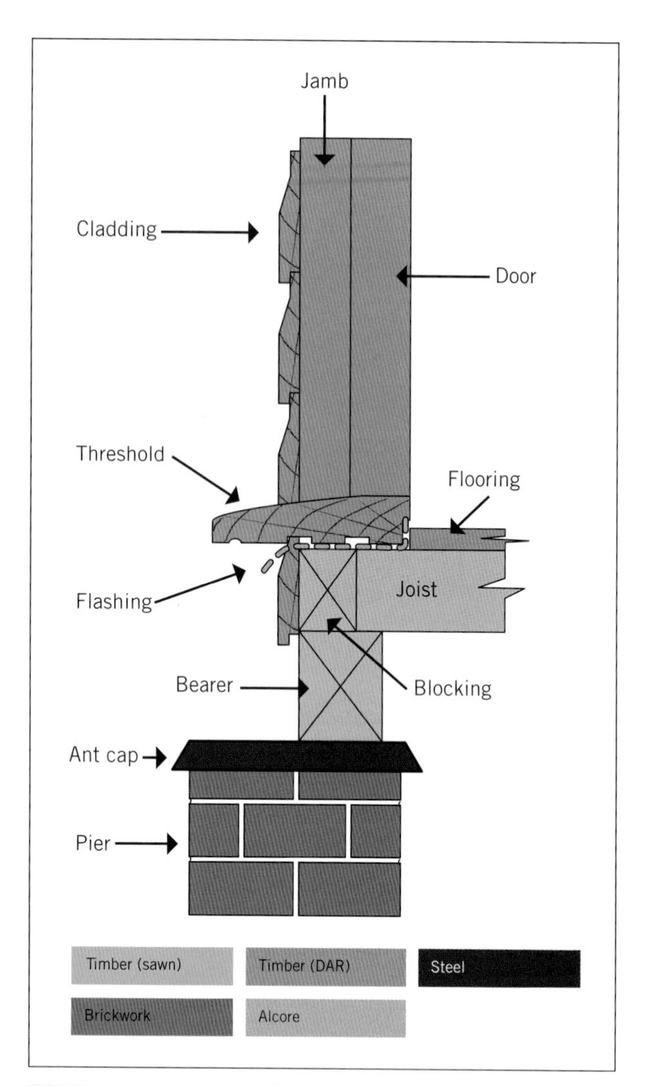

FIGURE 3.37 Door frame aligns with cladding and lining materials

Thresholds

Thresholds can be made from a range of materials such as solid timber, stone (including sandstone, marble and granite) and pre-cast masonry products such as terrazzo. Solid timber used in sills must be of durable quality and structural appearance grade or select grade and in bushfire-prone areas of a specific density. Some common species used for thresholds are:

- Sydney blue gum
- red river gum
- karri
- tallow wood
- blackbutt
- kapur
- grey gum.

Other species may be available in your state. The threshold must extend beyond the face of the cladding material (see **Figure 3.38**).

A timber threshold (see **Figure 3.39**) is fixed to the bottom of the jamb stiles and is used as a spreader to keep the jamb stiles parallel during fabrication, transport and installation.

Cast masonry or stone thresholds are generally used in masonry construction and are bedded in cement, packed level side to side, but raked towards the front to allow water to run away from the building and not back into it (see **Figure 3.36** earlier).

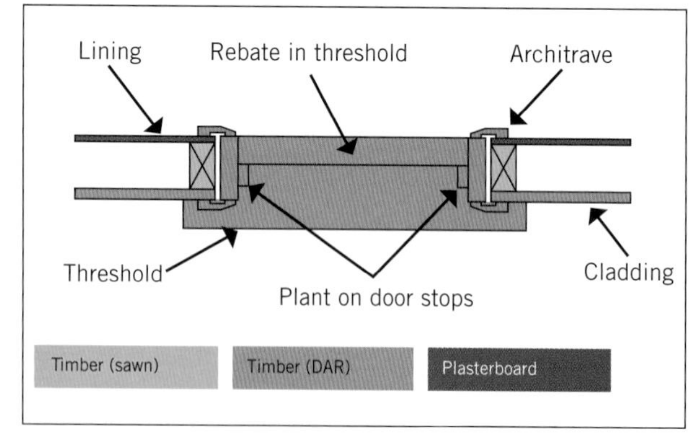

FIGURE 3.38 Lining up the jamb lining with lining and cladding

Note: If the door frame is not fitted with a threshold, a temporary spreader should be fitted about 75 mm up from floor level to keep the jamb stiles parallel.

If the door frame has been made to suit the opening in the wall frame, which has been constructed with the appropriate allowances for the door frame (including

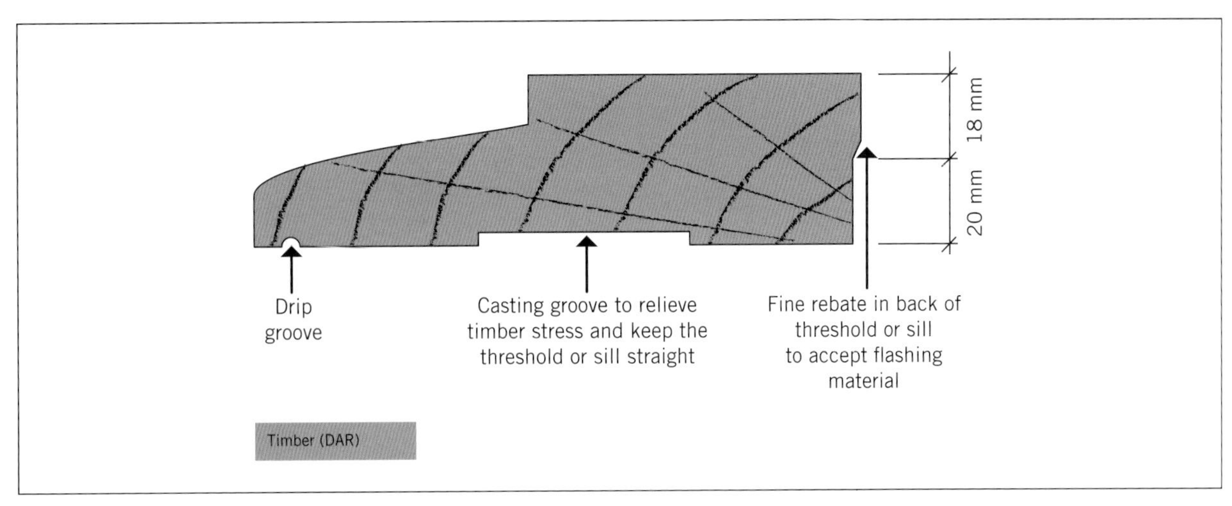

FIGURE 3.39 Typical timber threshold profile

jambs, head and threshold), locating external door frames in brick veneer and timber-framed buildings is a simple process.

Ensure the flashings for the threshold are in place. Leave the temporary door frame braces in place to help keep the door frame square during installation. Then follow these steps. See 'How to' on p. 134 for detailed instructions.

Step 1

Check the opening for size and plumb (in both directions) and stand the door frame in position.

Step 2

Prepare some packing for the frame. Wedges are best for timber-framed and brick veneer constructions as these can be inserted from both edges of each jamb stile and adjusted easily (see Figure 3.40). They can be made from off-cuts of framing material and should be about 50 mm wide with a long pitch (about 1:12). Plastic wedges, or packers, are often used in contemporary building practice, eliminating the need to source scrap material. They do not rot or shrink and can be easily cut and nailed off. If you do not use wedges, use solid timber, plywood or a similar material cut to the right thickness. A little construction adhesive will hold the packing in place when you are nailing. The packing should be placed at the hinge positions to provide the most support and midway down each stile.

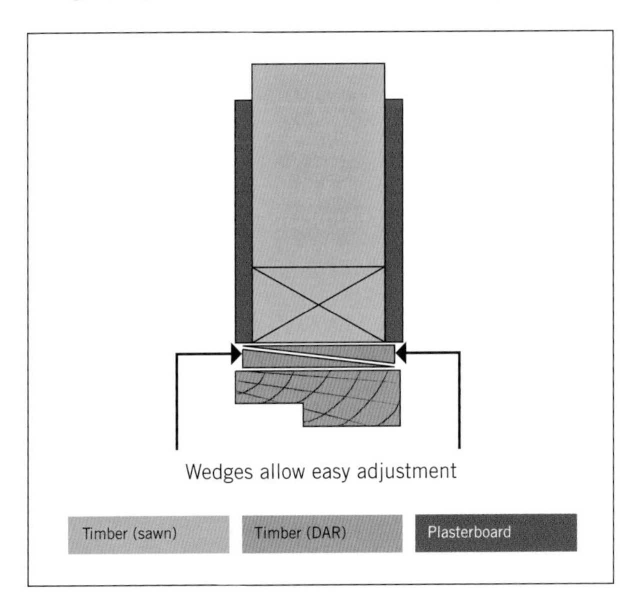

FIGURE 3.40 Using wedges

Step 3

Adjust the frame so that the stiles are plumb in both directions and align with the frame, allowing for lining and cladding materials on each side. In an external door frame, the threshold is usually attached to the stiles at this stage. Use a long straight edge (preferably 2 m long) to check the entire length of the jamb for plumb and straight. Crooked jambs will result in uneven gaps around the door. Adjust the frame so that the head is level. You may need to pack under the threshold to achieve level. Packing below the threshold should be between the flashing and the threshold.

Note: An internal door will not have a threshold so you must pack the jamb stile to get the head level and then scribe the reducing amount from one side to the other using a pair of compasses. Remove the temporary nails and trim the bottom end of the jamb stile so the head will sit level, and reinstall the frame plumb. Depending upon the floor covering and the amount to be trimmed off, you may not need to cut the long stile as the gap may be covered up by the floor covering.

Step 4

Nail off the door jamb using appropriately sized nails (75 mm is a good size for this job) (see Figure 3.41). Punch the heads below the surface at least 2 mm. If the entrance door is to have a polished finish, position the nails so they are unobtrusive. Trim any excess packing flush with the frame or lining.

Note: A similar method is used for nailing all door frames in all forms of construction. In steel frame construction, you need to insert blocking into the

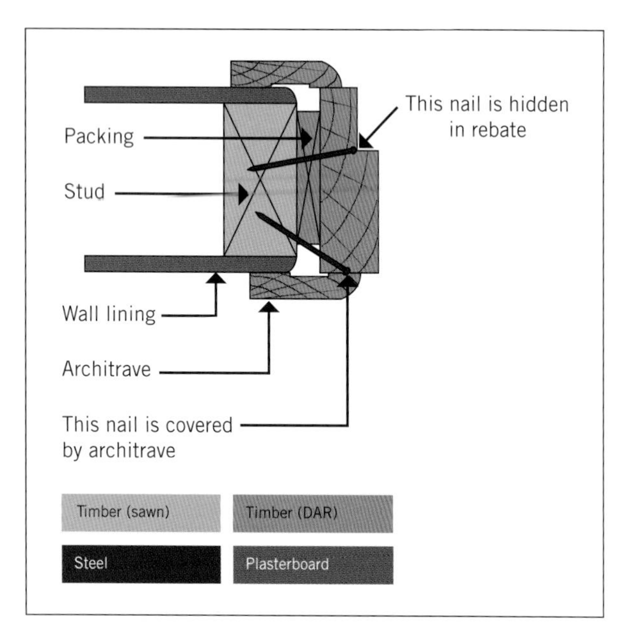

FIGURE 3.41 Nailing off the jamb

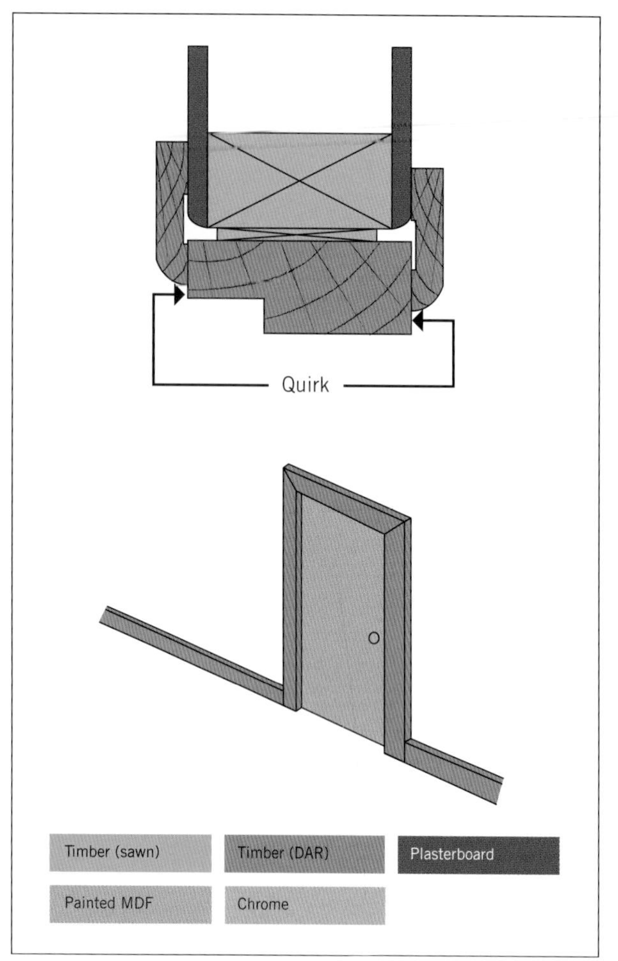

FIGURE 3.42 Fitting architraves

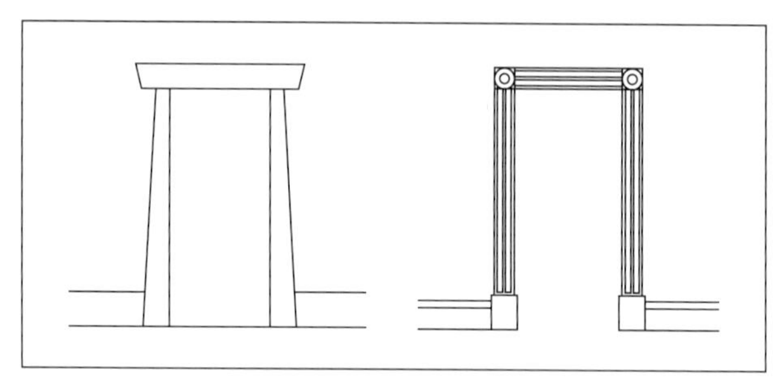

FIGURE 3.43 Alternative architrave styles

frames to take the nails or use newly developed nails for fixing to the steel stud.

Step 5

Architraves are generally mitred around the door frame. They hide the gap between the wall lining or cladding and the door frame. If the door frame has been put in out of plumb and level, the mitres will need to be hand-fitted. This can be very time-consuming and will invariably be noticeable.

Architraves are fitted leaving a **quirk** (5 mm set-back from the edge of the jamb stile) and should draw the eye to the opening (see Figure 3.42). This is why the splay, bull nose or heavily moulded edge is placed towards the opening, whether it is a window or a door. There are alternative styles for architraves (see Figure 3.43). Sometimes in modern construction the door jamb is square set to the wall lining with no architraves (see Figure 3.47 later).

Timber split door jambs

Timber split jambs are an alternative to regular door frames. They come partly assembled with architraves attached. Split jambs allow for simple installation fixing from each side, particularly if the wall frame has a wind in it (see Figure 3.44). The two halves are tongue and grooved together, making adjustment easy and providing complete gap cover. You still need to adjust the hinge-side plumb and level so that the door opens and closes properly.

The steps in installation are as follows.

Step 1

Pack the wall frames ready to accept the frame. Wedges cannot be used as they cannot be adjusted, so solid packing needs to be used. Insert the section with the pre-hung door still attached to the door frame into the opening, ensuring that the door is handed the correct way and using the door as a guide to adjust the head for level and the jambs for plumb and straight.

Tack the first side in position, using appropriate nails and making sure that you do not nail into the groove.

The door is usually fitted with loose pin hinges so it can be easily removed before installation of the split frame system.

Step 2

Insert the opposite half of the split frame and ensure that it fits tight against the wall lining (see Figure 3.44). Nail the side in position through the jamb and packing and finally through the architraves on each side of the frame (see Figure 3.45).

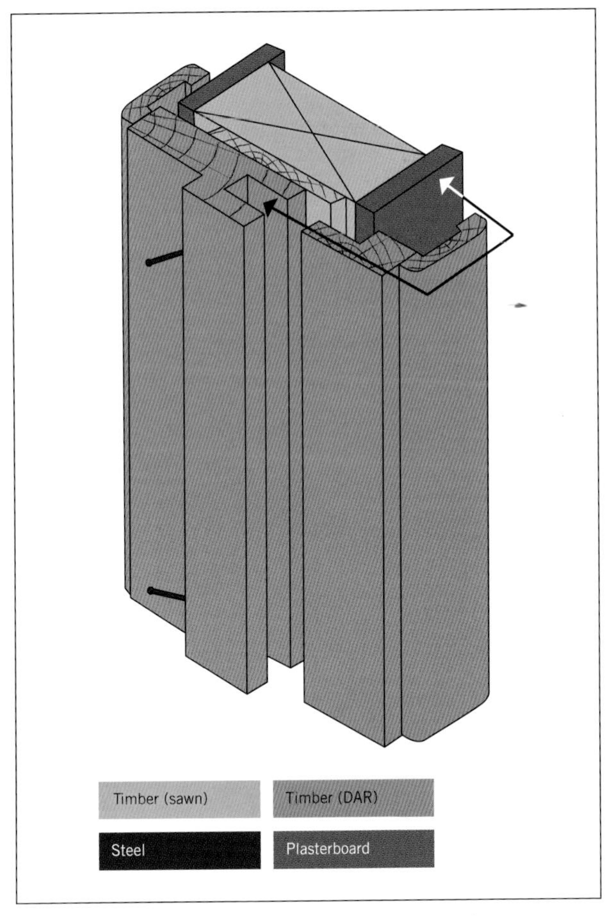

FIGURE 3.44 Insert the other half of the split jamb

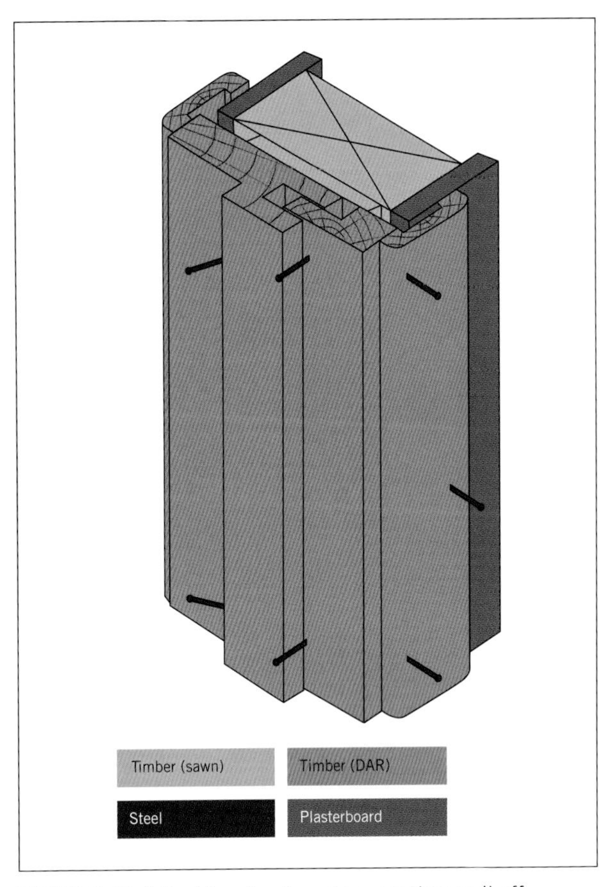

FIGURE 3.45 Adjust for plumb and square then nail off all components

Metal split door jambs

Metal split door jambs are a recent innovation and are used in modern architecture to maintain clean lines and provide a flush finished wall. They remove the need for architraves by incorporating a perforated plasterer's angle on each side. They are fitted to each side of the opening and fastened off using plasterboard screws.

Metal split jambs can be used in timber-frame and steel-frame construction. They allow the plasterboard to be set to the edge of the door frame, and are adjustable in width so they can handle a variation in wall thickness. A reveal-lining material is also available that can be used with cavity door frames and windows. These frames are not for external use or for use in conjunction with solid core or entry doors. Ezy Jamb is a brand name used by Studco Building Industries (see Figures 3.46 and 3.47).

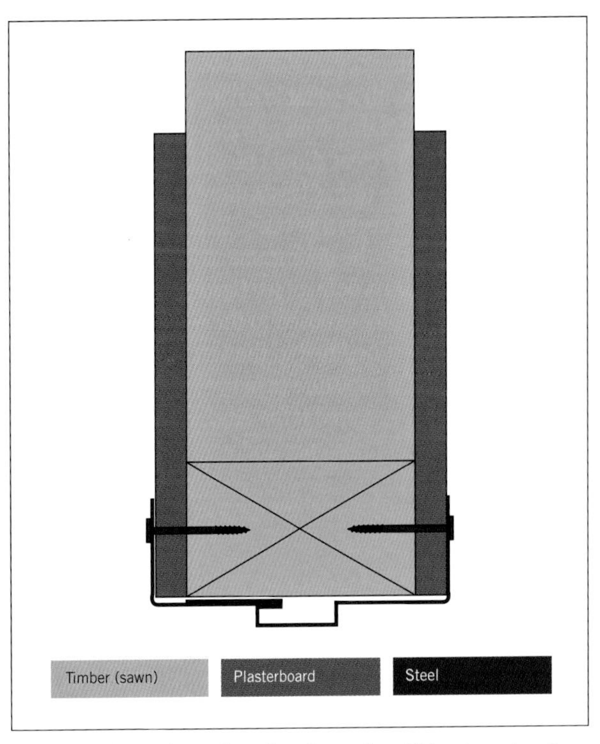

FIGURE 3.46 Horizontal section through split jamb and wall

Installing external door frames in cavity brick constructions

Masonry ties are used to fix the external door frame to cavity brick constructions. The masonry ties are either galvanised or stainless steel straps. The quantity and spacing requirements are as follows:

- A minimum of four ties must be fixed to the back of each jamb stile at 600 mm (maximum) spacings, using a minimum of two galvanised clouts or screws. Fasteners must penetrate at least 20 mm into the jamb stiles.

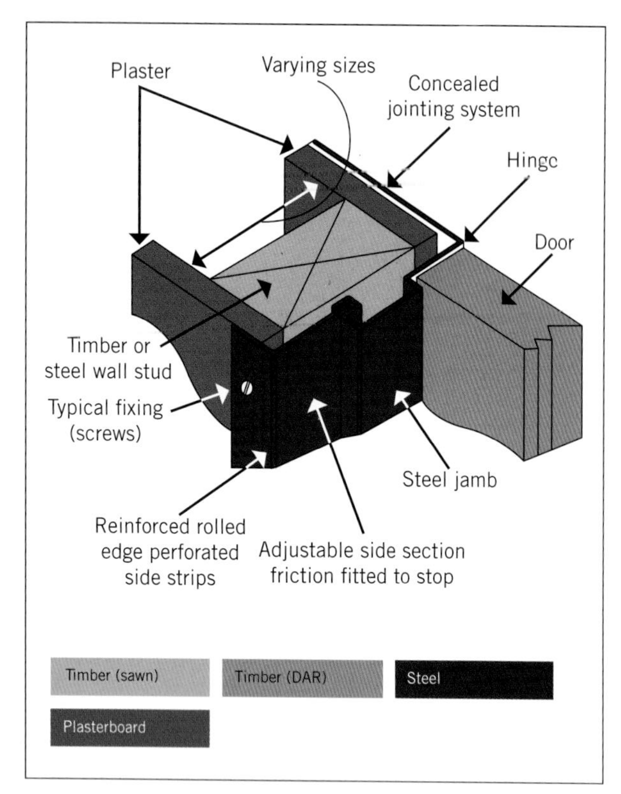

FIGURE 3.47 Ezy Jamb system from Studco

- Ties or straps must be at least 200 × 25 × 1.5 mm, with a corrugated or deformed end for proper keying into the masonry bed.

There are several methods used to install the door frames in cavity brick construction and to cover the cavity. When the door frame covers the full width of the cavity, it should be fitted with a suitable cavity cleat, which is fastened to the door jamb with galvanised or stainless steel fasteners. The cavity cleat should be 40 × 40 mm in dimension and made from Class 2 timber or suitably treated timber that meets this class. The cleat provides additional stiffness to the door frame, as well as greater security and weatherproofing (see Figure 3.48).

If the full edges are exposed to the weather, a full height flashing must be fitted to the vertical edges to prevent wind-driven moisture penetration through any gap. No flashing is required if the edges are covered by a minimum of 5 mm of masonry and a storm mould is used.

Installing external door frames in brick veneer constructions

When prefabricated wall frames are used, door openings are made significantly wider to allow adjustment of the positioning of the door frames so that brickwork can be gauged to suit the actual opening. This requires secondary studs to be fixed in place once the door frames are accurately positioned and the bottom plate is cut away between the studs.

In brick veneer construction, galvanised or stainless steel fasteners (nails and/or screws) should be used to fix the frame to the wall frame. They should be spaced not more than 600 mm apart. There are requirements in the NCC and in AS 1684 and AS 1720.3:2016 as to the design of timber frames and diameter and length of nails to be used. The Australian Glass and Window Association (AGWA) provides a handy guide to the correct fixing of windows and doors and selecting the right size nail (see AGWA website link at the end of this chapter).

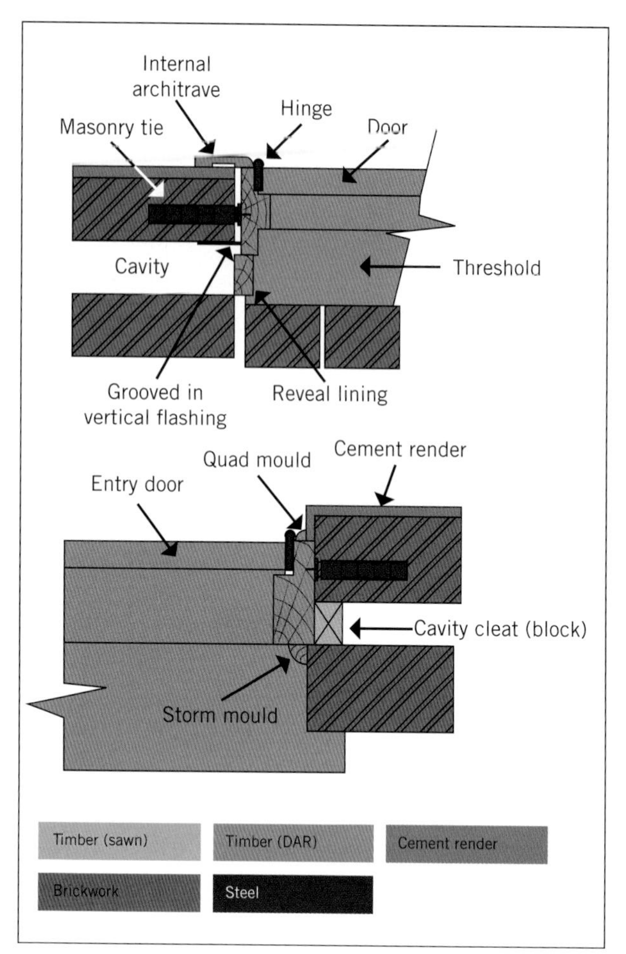

FIGURE 3.48 Alternative methods used to install door frames into cavity brick and masonry

The door frame will have a vertical flashing if it is set over the external brickwork. If butting up to the external brickwork (minimum 5 mm), as in Figure 3.49, a storm mould will be used.

Installing external doors in single brick constructions

In single brick construction, door frames are often used for rear entrances or entrance ways through attached garages. Where a building sits on a concrete slab, the door frame is usually assembled without a threshold as the slab provides the weathering surface; in some buildings the slab is rebated to accommodate a brick and tile sill. Metal door frame heads are supported using nails fixed to braces (see Figure 3.50). For some attached garage constructions, brick piers are built up on the garage side of each side of the door frame to provide additional support and allow a vertical flashing to be installed (see Figure 3.51).

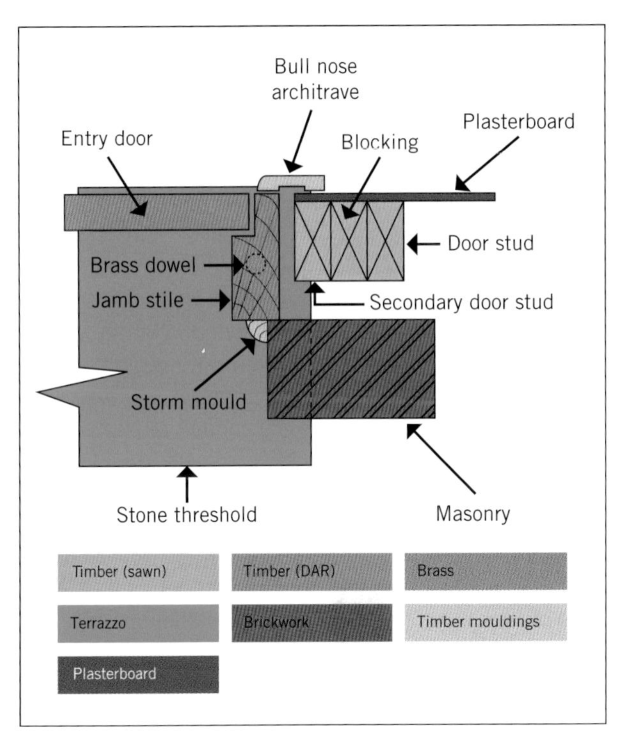

FIGURE 3.49 Entry door in brick veneer construction

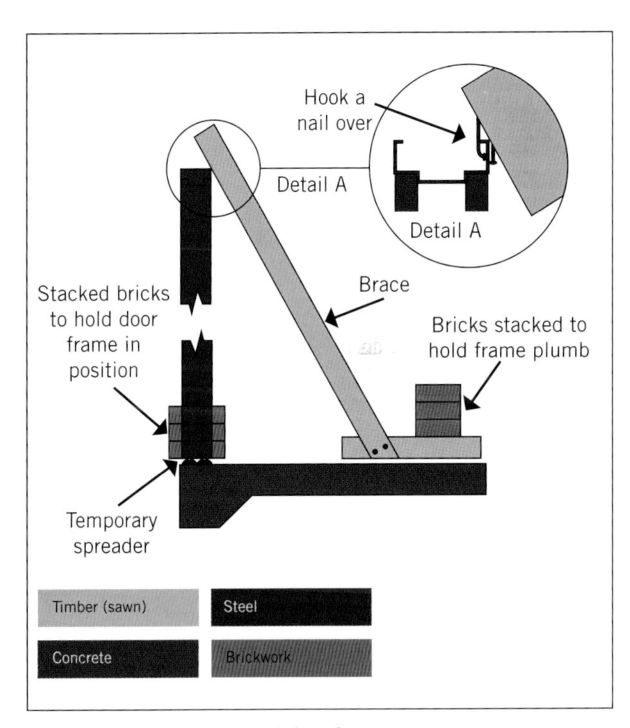

FIGURE 3.50 Bracing metal door frames

Where no threshold is used, a temporary spreader is attached to keep the jamb stiles parallel. The door frame is set out according to the brick bond or gauge. The frame is stood in position and braced back to the slab or joist from the head to keep the frame plumb. When bracing door frames in raft slab construction, do not nail braces to the slab; nail the braces to a suitable length block (450 mm), then pack bricks or sandbags on the brace to avoid movement. If the block is nailed to the slab, removal of the block will cause a hole in the concrete surface that will have to be repaired.

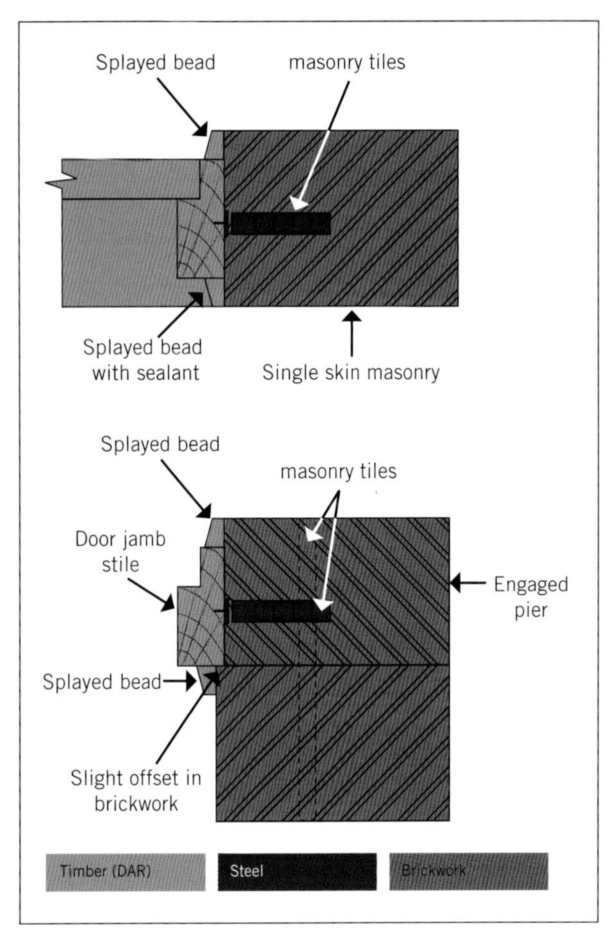

FIGURE 3.51 Method of door frame placement and fixing in single skin construction

Temporary diagonal braces remain in place to keep the frame square until the brickwork has been erected. The bricklayers will usually install the frame ties using the same regulations as for cavity brick construction (minimum of four ties $\leq$ 600 mm apart).

Installing metal door frames in single brick walls

Metal door frames are often used with single brick wall construction, particularly in commercial construction and home units where security and maintenance are a concern. Metal frames are available for single brick walls, cavity brick construction and cement block construction. Wire-type galvanised door frame ties are used to build steel door frames into the single skin of brickwork (see Figure 3.52). They are clipped into the back of the door frame and are able to slide up and down to position the bonding to the brickwork. The hollow back of the jamb stile is filled progressively with mortar to provide a solid jamb on completion. Where fire regulations apply, the hollow back can be filled with appropriate fire retardant filler.

External door frames installed into cavity brick, brick veneer and single brick constructions require some additional preparation, as the opening is usually placed to

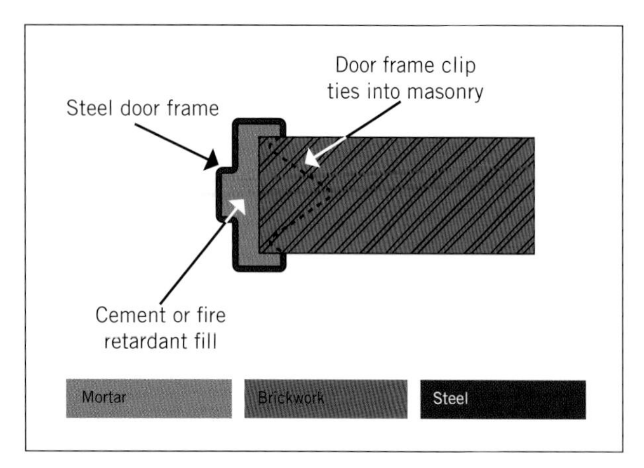

FIGURE 3.52 Method of building in metal door frames in single skin construction

suit brick bond. This involves the adjustment of brick perpend joints, in the horizontal plane, to allow the brick reveals to finish to the back of the storm moulds of the door frame while maintaining equal thickness perpend joints and full or half bricks for the wall.

Note: It may not always be possible to maintain full or half bricks where there are many door or window openings in one wall. Always check for level plumb and wind (twist) before final fixing of door frames (see Figure 3.53).

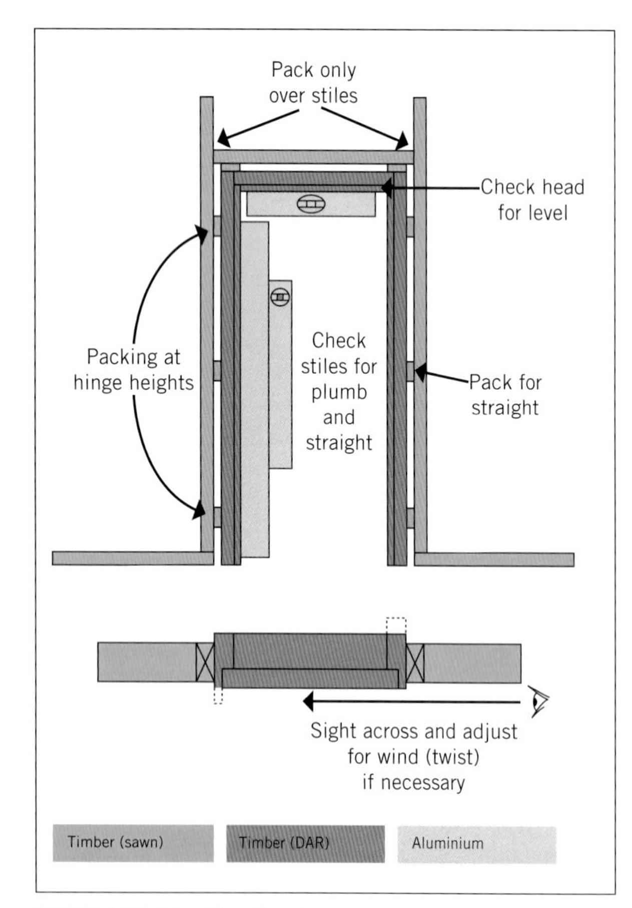

FIGURE 3.53 Checking door frames for level plumb and wind

Installing aluminium door frames

Aluminium door frames are delivered to the building site assembled and often with timber reveal linings and flashings attached. Most aluminium frames are manufactured with a fin section that is attached to the reveal lining. The frames are screwed together using stainless steel screws and silicone sealed joints. Flashing is fitted between the fin and the reveal on all sides of the frame. Most aluminium frames are built to hold aluminium sliding doors, but swing doors, bifold and other commercial forms of aluminium-framed construction are becoming more common.

These frame constructions can contain multiple window types, including awning, casement, double-hung and louvres, as well as door units, in the one frame.

Installation is the same as for other door frame installations, requiring appropriate packing and tying-in to wall frame and masonry constructions.

Note: In all external door frame installations, a suitable flexible sealant must be used to seal gaps between walls and frames. This is done by applying the sealant before fixing the storm moulds in position.

Installing sliding doors

Over the wall sliding doors

Internal timber sliding doors are usually mounted on tracks over the surface of the wall above the opening (see Figure 3.54). A guide is fitted to the floor to prevent the door from flying out. The track should be mounted level so that the door does not automatically close or open. A pelmet is used to cover the sliding mechanism. Mounting the doors this way has the advantages of easy mounting and increasing usable floor space, but has the disadvantage of taking up wall space.

Cavity sliding doors

These doors are concealed in the wall cavity via specially designed units that are integrated into the wall framing before linings are applied (see Figure 3.55). The doors can be made from a range of materials including timber or aluminium framed and frameless glass panels. They are available to suit a range of wall thicknesses.

Cavity sliders increase the usable wall and floor space. In contemporary homes, they can be made to run floor to ceiling with the track system finishing flush with the ceiling (see Figure 3.56). Door units can be automated for hands-free operation for clients with disability and provide security access when needed.

Note: Take care in setting out wall frames where cavity sliders are to be accommodated.

When fitting and finishing timber or hollow core doors, all edges must be well sealed with paint or primer, including the top and bottom edges, to prevent doors warping or surface deterioration due to moisture

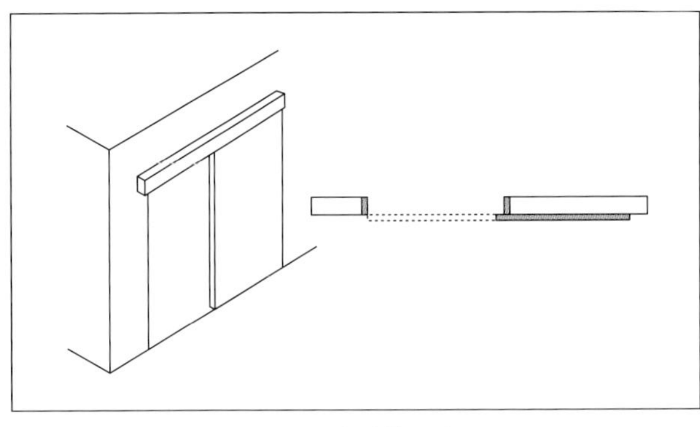

FIGURE 3.54 Surface mounted sliding door

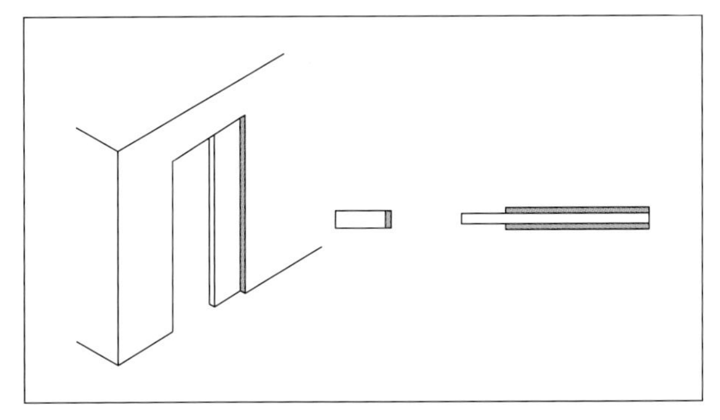

FIGURE 3.55 Cavity sliding door unit

Source: Alamy Stock Photo/caia image

FIGURE 3.56 Floor-to-ceiling cavity sliders

uptake, particularly in wet areas. Failing to seal all surfaces can void the warranty on doors should a claim need to be made on the manufacturer.

Installing a cavity sliding door

Cavity sliding doors are generally sold as a kit that needs to be installed in the wall before internal cladding is installed.

Depending on the kit requirements, you will need to assemble the frame on site.

When installing the cavity sliding unit, care must be taken to ensure the frame is packed to ensure the frame sides are straight and the door can close as designed.

The unit should be set back 10 mm from the face of the timber wall to allow for the plaster installation to finish flush.

Once the door frame is installed in the wall, the base plates can be attached to the top of the door. The base plates are what connects the rollers to the door.

When locating the base plates, it is important to review the manufacturer instructions, however, as a typical rule they are installed 80 mm from the sides of the top of the door.

To ensure the height of the door is correct and even, the locking nuts can be adjusted up and down once the door is in place.

Once the door is sliding evenly, door guides are installed to keep the door from swinging on the rollers in the cavity space.

Hanging doors

Hanging a door is the process of fitting the door to the opening and fitting the operating mechanism that allows the door to open and close. Hinges are used for this purpose. There is a vast range of hinges available to hang doors in door frames. Selecting the correct hinge and hinge size is critical to doing the job correctly.

Table 3.3 outlines various external and internal hinges and their uses.

Selecting the right hinge

External doors should be fitted with three hinges, while internal doors usually have two hinges. Table 3.4 will assist you in selecting the right hinge and hinge size for the job.

A general rule of thumb is the heavier the door the greater the number of hinges that should be used and the bigger the hinges should be.

TABLE 3.3 Hinges and their uses

Type of hinge	Material and finish	Use
Butt hinge Loose pin – internal use Fixed pin – external use	Steel (galvanised, nickel and cadmium plated), stainless steel, brass, aluminium (powder coated and anodised)	Internal/external doors, hollow core and solid core doors – standard opening

>>

>>

Type of hinge	Material and finish	Use
Broad butt hinge Loose pin – internal use Fixed pin – external use	Steel, stainless steel, brass, aluminium (powder coated and anodised)	Internal/external doors, hollow core and solid core doors – 180° opening allows doors to open back against the wall
Parliament hinge	Steel, stainless steel, brass	Internal/external doors, hollow core and solid core doors – 180° opening allows doors to open back against the wall
T hinge	Steel, wrought iron, galvanised steel, nickel and cadmium plated	Gates and external doors on sheds – use galvanised for external purposes
Strap hinge	Steel, wrought iron, galvanised steel, nickel and cadmium plated	Gates and external doors on sheds – use galvanised for external purposes
Concealed hinge	Brass, steel and stainless steel	Specialist hinge used when no hinge or hinge knuckle is to be seen
Sprung hinge	Steel and stainless steel	Self-closing action for privacy and security
Gravity hinge	Steel and stainless steel	Self-closing action for toilet partitions
Pivot hinge Heavy duty floor spring Aluminium track, Top guide, Top pivot set, Bottom pivot set, Rubber bumper Lightweight pivot system with tracks	Steel, stainless steel and plated	Various models depending on door types

TABLE 3.4 Door sizes and hinge selection

Thickness (mm)	Width (mm)	Height (mm)	Recommended hinge and quantity per door
30–45 mm	< 900 mm	Standard height 2040 mm	Internal doors: 2 × 100 (H) × 75 mm (W) External doors: 3 × 100 (H) × 75 mm (W)
30–45 mm	< 900 mm	From 2040 mm to 2340 mm From 2340 mm to 3000 mm	Internal doors: 3 × 100 (H) × 75 mm (W) External doors: 4 × 100 (H) × 75 mm (W)
45–55 mm	< 1050 mm	Standard door 2040 mm From 2040 mm to 2340 mm From 2340 mm to 3000 mm	3 × 100 (H) × 75 mm (W) 3 × 100 (H) × 75 mm (W) 4 × 100 (H) × 75 mm (W)

For applications where doors are 'oversized', heavy or regularly used, or for fire doors, stainless steel hinges with ball bearing knuckles should be used (see Figure 3.57).

Regular hinges don't allow doors to open flat against a wall, but broad butt and parliament hinges do (see Figure 3.58).

Source: Shutterstock.com/Francesco Milanese

FIGURE 3.57 Hinge with ball-bearing knuckle

Fast-fix hinges that require no mortising are available for internal doors and are the standard type used in commercial partition work and for aluminium-framed glass doors (see Figure 3.59).

Hinges are often bought in packs containing a pair of hinges and correctly sized screws. When ordering hinges separately, it is critical that you order the correct screws to suit the hinge type being used.

For regular internal doors, 20 mm, 8 gauge countersunk screws are normal. However, if the screws

FIGURE 3.59 Fast-fix hinge

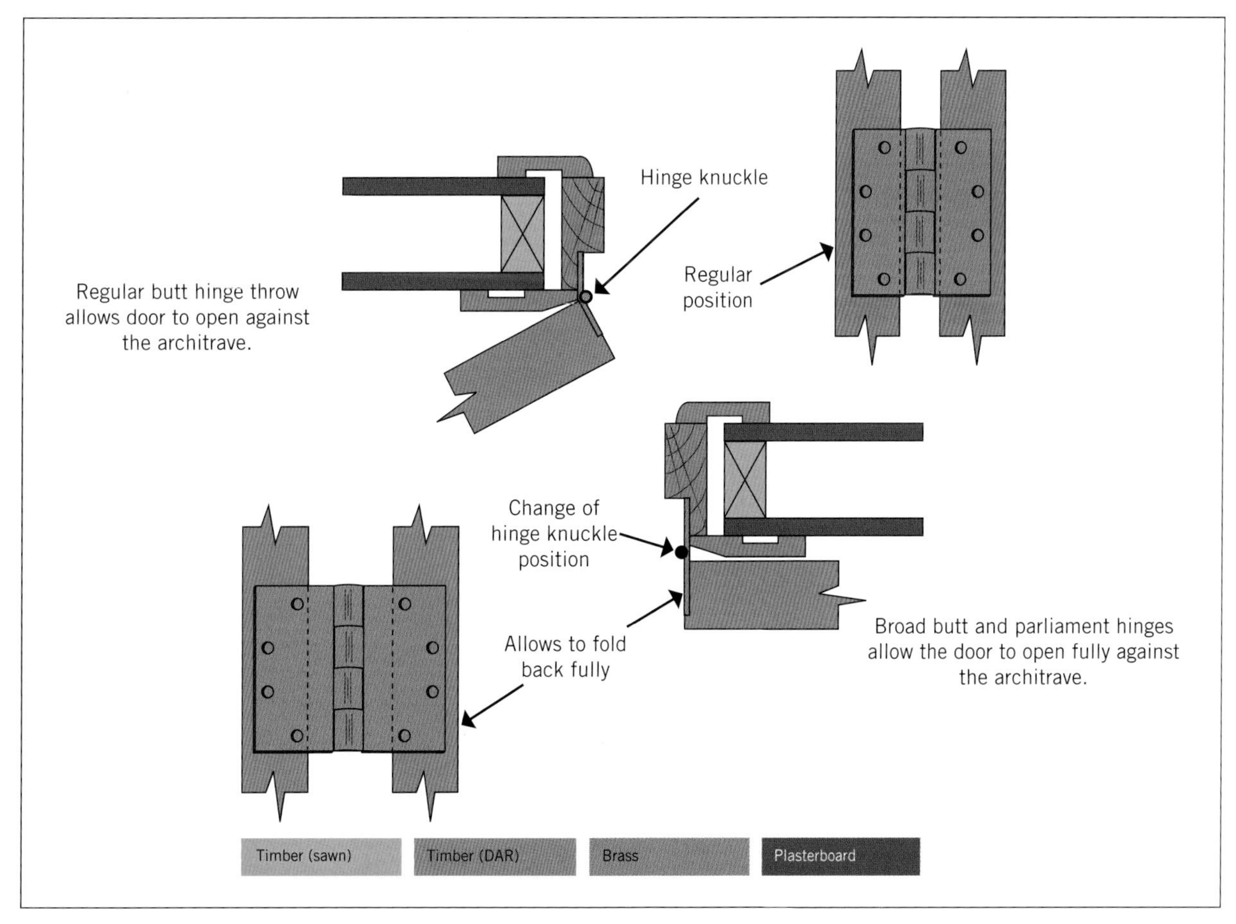

FIGURE 3.58 Hinge throw on broad butt and parliament hinges

do not come with the hinges, a rule of thumb is to use 8 gauge screws with 75 mm hinges and 10 gauge screws with 100 mm hinges. Make sure the screws are of a compatible material (e.g. stainless steel screws with stainless steel hinges). Ensure also that the heads of the screws finish flush with the hinge leaf. If the screws sit above the surface of the leaf, the hinge will bind on the screws and affect the fit of the door.

Special hinges include electric hinges, which are used in conjunction with electronic door locking and security systems.

Handles, knobs, locks, latches and accessories

Simple latches and bolts

The most basic door security systems include the hasp and staple and barrel bolt and pad bolt (see Figure 3.60). They are mainly used for external doors and, in particular, garages, sheds and gates. They are generally face fixed and are available in a range of finishes including galvanised and plated.

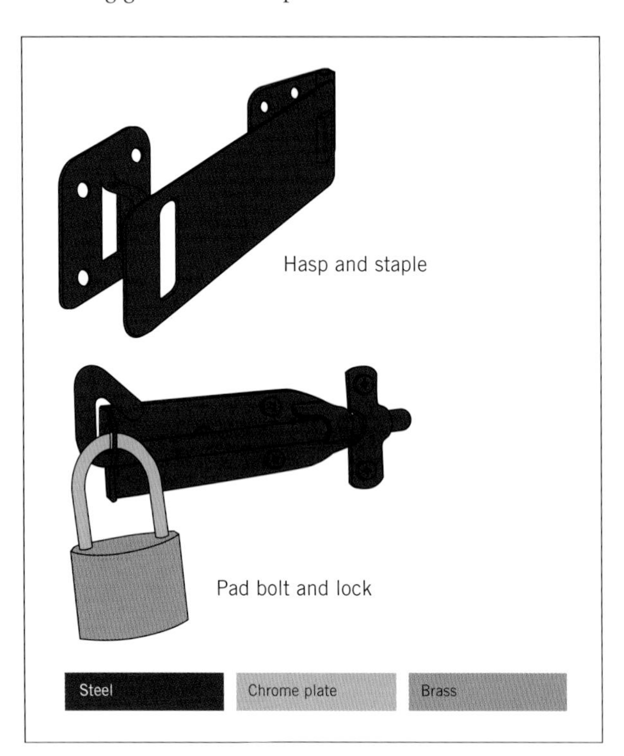

FIGURE 3.60 Simple latches with keyed padlocks

There are lockable patio bolts for French and sliding doors.

Flush bolts are mortised into the inside face or closing edge of the door at the top or bottom, or both.

FIGURE 3.61 Flush bolt

Mortise latch

Latches keep a door closed. Used both internally and externally, the mortise latch is cut into the appropriate closing edge of hollow core doors or the stile on a framed and panelled door. The majority of waste is cleared by drilling out the waste and cleaning up the mortise with a chisel if necessary. A hole is drilled from both faces of the door to take the spindle, which is then attached to the handle or lever. A strike plate is let in flush to the rebate on the door jamb.

Mortise latches are not often used today as they are costly to install; they have generally been replaced by the cylinder or tubular latch (see Figure 3.62). The cylinder or tubular latch requires a single 22 mm hole to be drilled in the edge to take the body of the latch and a single hole drilled from each face for the spindle to accept either a lever or a knob. Some modern tube latches are 'knock in', only requiring a hole drilled in the edge and one in the face for the spindle. There is no face plate to let in, but a strike plate for the jamb is required.

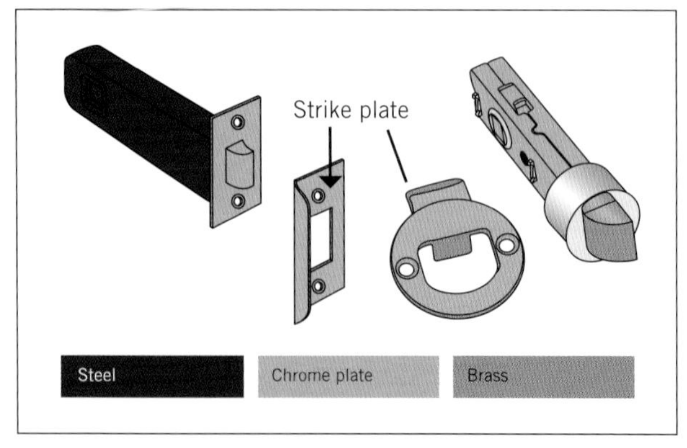

FIGURE 3.62 Tubular latch

Mortise lock

Mortise locks, as well as having a bevelled latch to hold the door closed, contain a key-operated bolt for security (see Figure 3.63). Being fully recessed into the edge of the door, these locks are time-consuming and expensive to install and have largely been replaced by face-fixed deadlocks.

Source: Greg Cheetham

FIGURE 3.63 Mortise lock and handle with plate

Rim lock and night latch

Rim locks and night latches are face fixed to the inside of the door. A knob allows them to act as a latch, holding the door closed. A keyed barrel is inserted from the external side that locks the latch for security. There is a broad range of styles and finishes.

Cylindrical lockset

The cylindrical lockset is the most common of the door passage locksets and it is available in a large range of styles and designs from many manufacturers (see Figure 3.64). A hole saw is used from both sides of the door to make a hole (usually 54 mm), the back-set distance from the edge of the door (see Figure 3.65), and a 22 mm hole for the latch fitting is made in the edge of

Source: Ed Hawkins

FIGURE 3.64 Cylindrical dead latch lockset

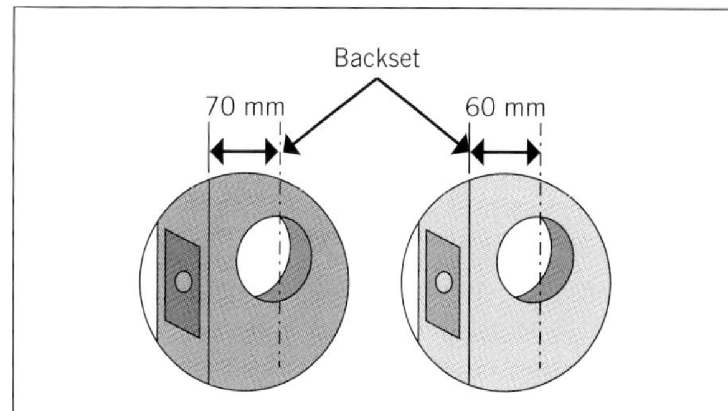

FIGURE 3.65 Back-set on locks

the door. A push-button locking mechanism fitted to the inside of the door prevents the external handle from being turned unless by a key. Latch extensions are available to fit the lock in the centre of the door width.

Note: A 60 mm back-set is the regular distance, but 70 mm back-set is available. Check the width of the door stile for the appropriate size back-set.

Deadlock and deadbolt

These locks are key-only operated and can be used in conjunction with latches for greater convenience. They are available in a range of finishes, including stainless steel and brass plated and other electroplated finishes.

Handles and knobs

Mortise locks are usually fitted with lever handles, while cylindrical locks are usually fitted with knobs (see Figure 3.66). The lever handle itself can be fitted with locks, but usually there is a backing plate with a keyhole fitted as part of the set. Knob handles can be plain or fitted with keyed locks or buttons. Lever handles must be used in fire-rated situations.

FIGURE 3.66 Door handle and spindle

Keying systems

Locks can be keyed separately or keyed alike. Entry points on buildings are usually keyed alike. In multi-residential and commercial applications, a combination of systems may be used (e.g. keyed-alike systems in combination with master-keyed systems). In domestic residential construction, electronic locking systems are becoming more common.

Door accessories

Other door accessories include:

- door closers that not only close the door but also hold the door open – most closers are surface mounted and can be adjusted to close at different rates
- weather strips, fitted to the edges of doors, particularly the bottom edge – there are many different types and styles suitable for different purposes
- door knockers
- spy holes
- door chains and security bolts, which allow the door to open only a fraction of the normal distance.

Hanging a door in an existing opening

The first step in hanging a door in an existing opening is to make a door wedge block. This is an essential part of a carpenter's tool kit – your teacher can provide you with a plan for one. The one shown in Figure 3.67 is the most common style.

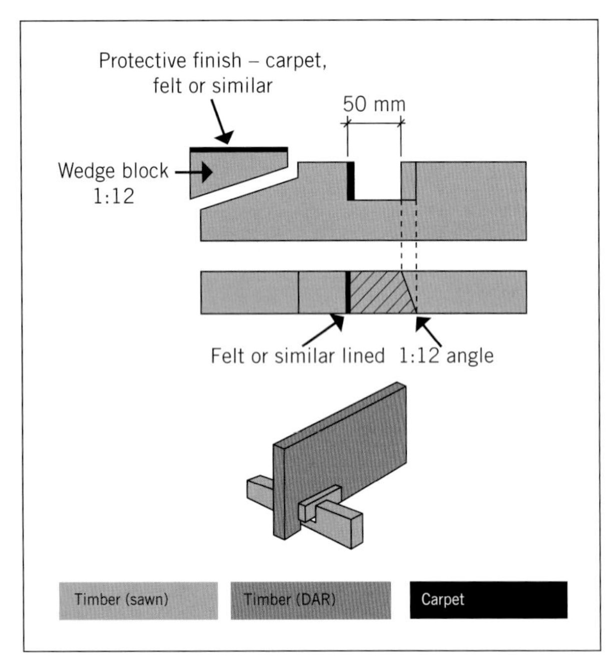

FIGURE 3.67 Door wedge block

Steps for hanging a door

Step 1

Once you have your tools and work area set up, select the appropriate door for the opening, making sure you have identified the face side (for external doors) and the lock stile.

Check for the manufacturer's sticker or stamp on hollow core doors. If you can't immediately identify the lock stile, tap the door gently about 90 mm in from the closing edge on the face side, and listen and feel for the solid sound.

Step 2

Check the door for twist and bow. If the door is too bowed or twisted it cannot be used. There are several good guides on what is acceptable in terms of acceptable construction standards for twist in doors. These are listed at the end of this chapter. If there is a slight bow, the door should be placed so that the top and bottom of the door touch the rebate of the top and bottom of the jamb. The latch or lock can be used to close the door properly. This is usually not possible with solid timber or solid core doors.

Step 3

Pack the door into the frame along the bottom and opening edges, ensuring that the hinge edge and the top of the door are butted firmly to the head and jamb stile. Use a pencil and a piece of scrap material the same thickness as the largest gap down the stile to scribe a line along the closing edge (see Figure 3.68).

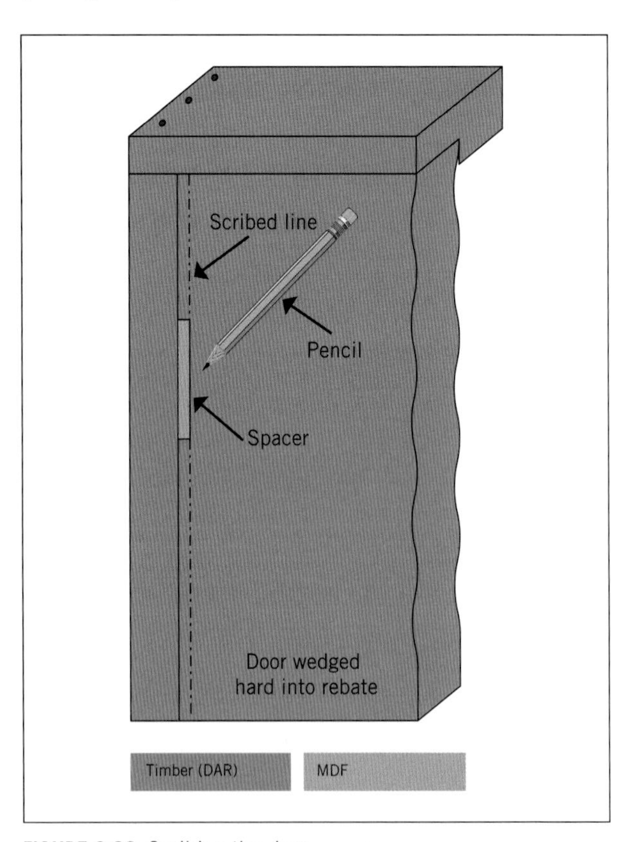

FIGURE 3.68 Scribing the door

Step 4

Remove the door from the frame and place it in the door wedge block, and then plane the edge and end of the door to the scribe lines. For a demonstration on making a door holding jig, see the Video resources section at the end of the chapter.

Step 5

Put the door back in the frame and pack evenly around the sides and top to 2–3 mm, depending on the finish, and check the gap at the foot of the door to ensure you have enough clearance to accommodate the floor finish (20–22 mm for carpet). Use a block of the appropriate thickness to scribe across the floor.

Step 6

With the door in position, measure down from the top of the door 150 mm (or 175 mm) to locate the top of the top hinge and then measure up 225 mm (or 250 mm) from the bottom of the door to locate the bottom of the lowest hinge. Make sure you mark both the door and the jamb stile together. If it is an entry door, or a heavy door requiring three hinges, the centre hinge is located in the middle of the top and bottom hinge. (If replacing a door, use the original hinge set-out.) The larger the hinge, the greater the distance the hinges are set down from the top and up from the bottom. The reason the spacing is greater at the bottom than at the top is because this creates an optical illusion of the door being slightly heavier at the bottom, and also because the visual line from eye to bottom is longer than eye to top, thus creating balance (see Figure 3.69).

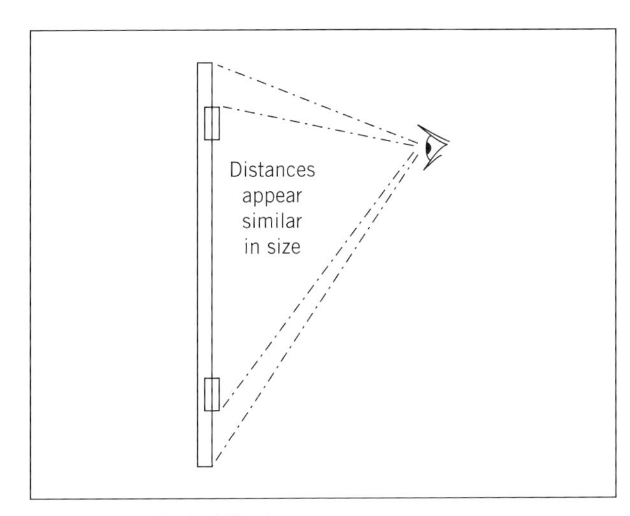

FIGURE 3.69 Optical illusion

Step 7

Remove the door from the frame and place in the door block wedge. With a jack plane, 'back off' the closing edge of the door so it does not bind on the door jamb during opening or closing – a 2 mm bevel is all that is required (see Figure 3.70).

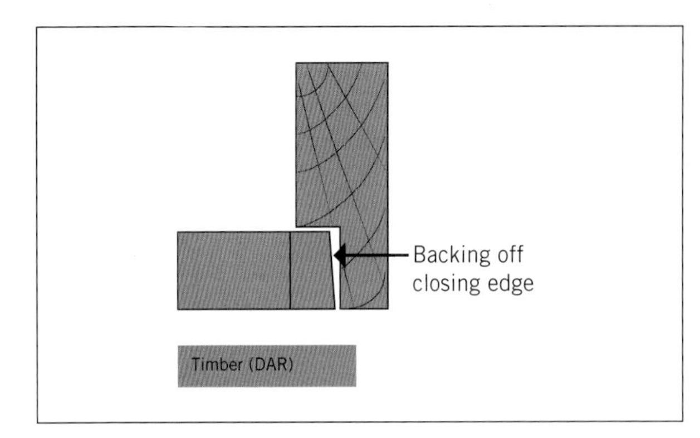

FIGURE 3.70 Backing off the door

Step 8

Using a combination square, mark out the hinges on the door frame. Align the knuckle of the hinge with the edge of the door frame (see Figure 3.71). Use a sharp pencil to trace the height and width of the hinge leaf position on the door frame and the door.

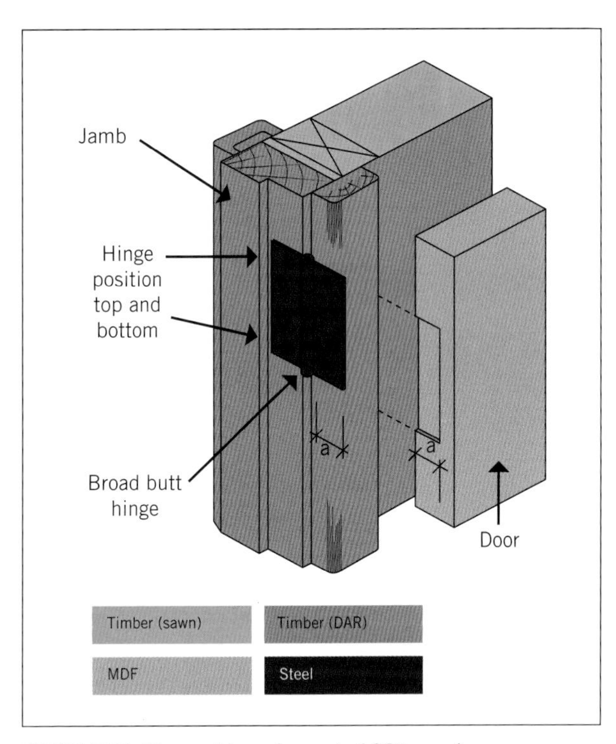

FIGURE 3.71 Narrow hinge for up to 120° opening

In Figure 3.72 a broad butt hinge has been used to show the offset knuckle. The knuckle of the broad butt must align with the centre of half the thickness of the architrave in order for the door to open to a full 180°.

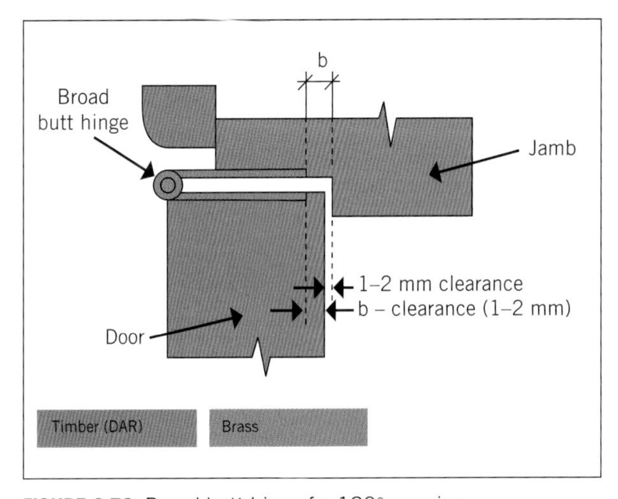

FIGURE 3.72 Broad butt hinge for 180° opening

A butt gauge is an alternative tool used for marking out hinges (see Figure 3.73). Ask your teacher to demonstrate the correct use of this tool. For a demonstration on using a butt gauge, see the Video resources section at the end of the chapter.

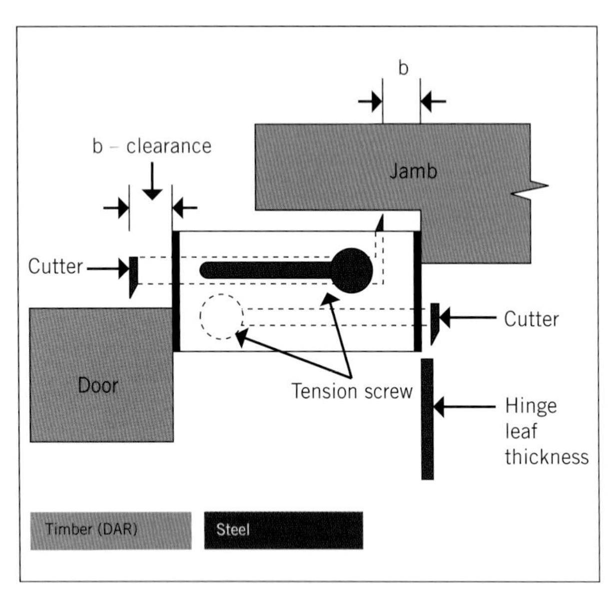

FIGURE 3.73 The butt gauge

Use a utility knife to cut the cross-grain ends of the hinge recess. Search 'making a hinge jig' on YouTube to see many ways to make these jigs.

Step 9

Remove the waste with a sharp chisel, router, or trimmer and jig to cut to the lines. When chiselling out waste, start by marking the outline with a light tap of the chisel, holding the flat side just inside the line so that when the chisel is driven in it does not push outside the line. This procedure will ensure a tight fit, with no ugly gaps around the butt hinge after fitting.

Step 10

After removing the waste, set the hinge in the recess and drill pilot holes for the screws. Drilling pilot holes will prevent the jamb stile and door from splitting. Use the correct length and gauge of screw to fasten the hinge to the door before lifting the door into position, engaging the leaves in the jamb recesses (pack the door to the height to make alignment easier). If slot-head screws are used, align the slots vertically for a professional finish (known as 'heading' the screws).

Fitting the lock or latch

The position of the lock or latch will usually be specified on the drawings or in the specifications. A regular height for door locks and latches is 900 mm (up to 1200 mm is common) from finished floor level. Locks, latches and handles in frame panelled doors are usually fitted in line with the mid-rail.

When fitting locks, it is a good idea to wedge the door into the open position so you have access to both faces and the edge of the door. You can make timber wedges or use a couple of wide chisels.

If it is a hollow core door, check again to make sure you have hinged the door on the right stile and the lock block is on the closing side. It is much easier to repair the edges than to repair a hole in the face that is in the wrong spot.

Remember to follow the manufacturer's installation instructions in all cases.

Following are the steps for fitting the lock or latch.

Step 1

Measure up from the floor the height of the centre line of the handle or knob. Depending on the type of lock being fitted, mark the back-set required. This is usually the distance from the face of the latch plate to the centre of the spindle. Don't forget to allow for the backing off measurement on the closing edge when marking the back-set position.

Step 2

Use a combination square or marking gauge to mark a centre line on the edge of the door that aligns with the back-set line (see Figure 3.74). For a tubular latch or mortise lock, select the appropriate-sized drill bit or hole saw to remove the waste material to accept the cylinder lock or spindles. Remember the old adage: measure twice, cut once.

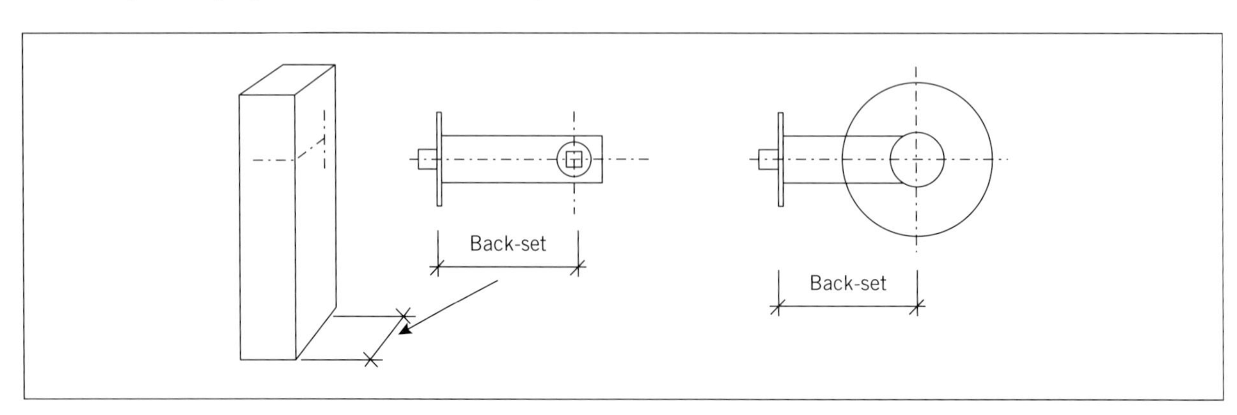

FIGURE 3.74 Back-set on a tubular latch and cylinder lock

Step 3

Slide the tubular latch into the edge hole and with a sharp pencil mark around the outer face of the face plate. Use a utility knife with a combination square to cut across the grain for a neater finish before removing the waste with a sharp mortise or butt chisel. Check that the spindle hole in the tubular latch or the joining lugs on the cylinder lock are accessible via the face holes you have drilled. Be sure to drill perpendicular to the edge or you can risk skewing the side of the door.

Step 4

Follow the manufacturer's assembly instructions for the cylinder lock and install the lock in position, ensuring the key action is on the correct side. For the tubular latch, use the appropriate screws to install the lock and handle set. Be sure the latch tongues are facing the correct way for the door to close.

Step 5

Close the door with the latch tongue against the jamb stile, and then mark the centre position of the tongue and the top and bottom of the rest of the latch tongue against the outside edge. Square the marks across the face of the rebate. From the inside of the rebate, mark the distance from the closing face of the door to the flat face of the tongue, adding 1–2 mm for clearance (see Figures 3.75 and 3.76).

Step 6

Lay the strike plate over the two marks and trace around the outside and inside of the plate, ensuring that the plate aligns with the tongue measurement (see Figure 3.76). Use a 22 mm spade bit to drill a hole or use a chisel to remove the waste to at least 2 mm deeper than the depth of the tongue. Finally, remove the strike-plate waste with a chisel to the depth of the strike plate and fasten the strike plate with the supplied screws. Try closing the door and make any necessary adjustments.

There are many different types of locks and latches. Check out some of the videos available on YouTube by searching 'install a lock on a door'. Today, new homes are being smart-wired to include electronic door-locking systems that are integrated with other security systems, which can be operated remotely. You may need special instructions on how to install these systems, or to bring in a licensed data cable installer.

Replacing an existing door

When replacing an existing door with a new one, you need to check the opening dimensions and the rebate depth carefully before you order the new door. Check the diagonals for square (see Figure 3.77) and stiles and head for plumb and level. If the opening is significantly out of square, you may need to order an oversized door in order to get a good fit.

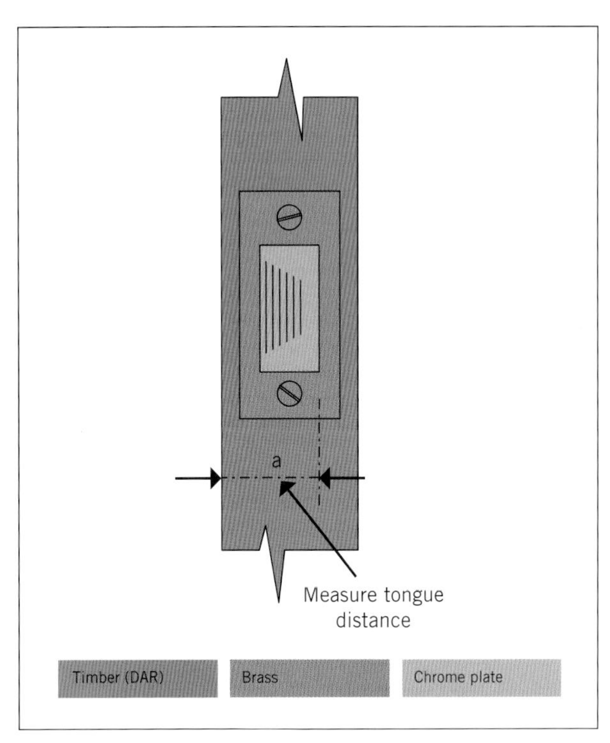

FIGURE 3.75 Measuring and marking the tongue

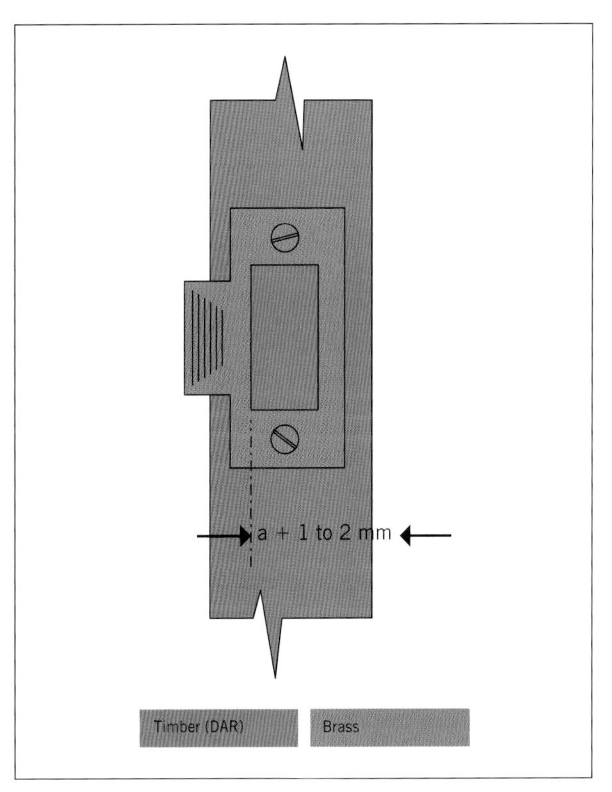

FIGURE 3.76 Marking the strike plate

Once you have obtained the new door, remove the existing door by taking the door and hinges from the jamb stiles. Fit the new door to the opening, as outlined earlier, and transfer the hinge positions to the new door.

Take the door from the opening and set it in the door wedge block, with the hinge edge up. Reuse

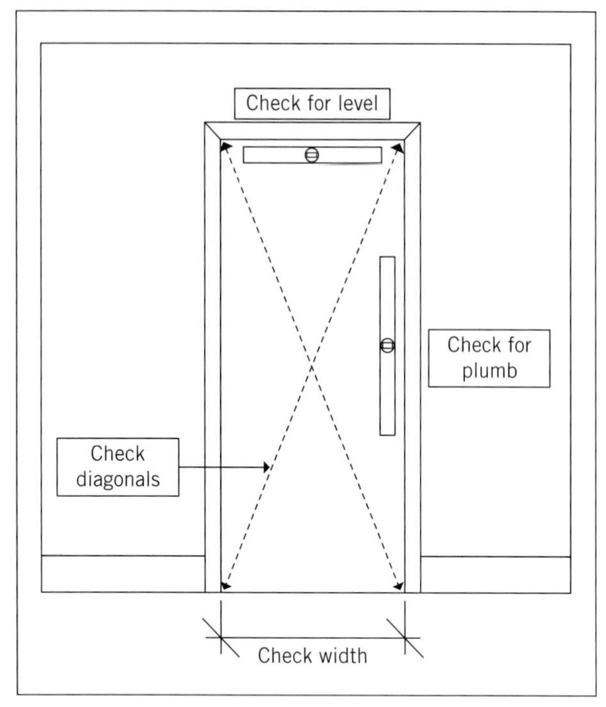

FIGURE 3.77 Check diagonals

one of the existing hinges wherever possible to mark out the hinge recesses on the edge of the door (see Figure 3.78). You can also set up a butt gauge to match the existing recesses and mark the hinge width and depths on the door.

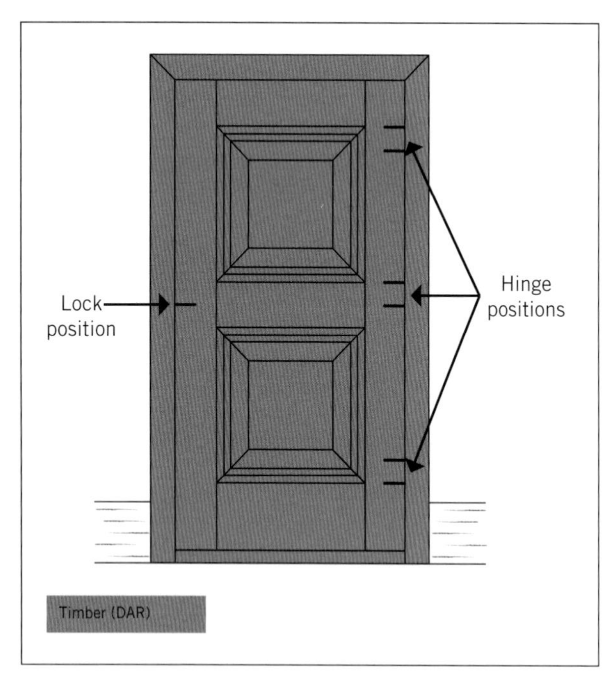

FIGURE 3.78 Marking hinge and strike plate positions

Window installation

This section will cover the various types of windows that you will likely come across in residential housing

LEARNING TASK 3.4 BUSHFIRE-PRONE AREAS

Discuss in class the various types and ways external timber doors and windows can be utilised in bushfire-prone areas, focusing on the meaning of different BALs. Consider:

- the types of timbers used
- construction methods
- installation methods, such as gaps and sealants
- impacts of design features
- the types of glass used.

Useful resources include:

- AS 3959 – Construction of buildings in bushfire-prone areas
- *Guide for Compliant Construction of Timber Windows and Doors in Bushfire-Prone Areas*, from the Window and Door Industry Council: https://wadic.org.au/knowledge-hub/resources and click on 'Bushfire Compliance Guide 2012'
- *A Guide to Windows & Doors in Bushfire Prone Areas*, from the Australia Glass and Window Association: https://www.agwa.com.au/ (follow the links on the 'Consumers' page of the website)
- Students are to consult AS 3959, and outline the simplified steps used to determine a building's BAL.

1 For a building with a BAL rating of 12.5, what sections of the Standard outline construction requirements?
2 For a building with a BAL rating of 19, what sections of the Standard outline construction requirements?
3 Answer True or False.
   a Windows in a building with a BAL rating of 12.5 must have Grade A safety glass minimum of 3 mm thickness.
   b External doors in a building with a BAL rating of 19 can have hollow core doors as long as they have a non-combustible kick plate extending 400 mm above the threshold.
   c Glazed external doors in a building with a BAL rating of 19 must use toughened glass 5 mm thick.

construction. *AS2047 Windows and external glazed doors in buildings* should be referenced and complied with when installing windows to ensure compliance with the National Construction Code.

Installing aluminium windows in timber-framed constructions

The cladding material should be in place to the height of the top of the window, allowing you to install the head flashing after window installation.

Having prepared the openings ready to receive the window frame with 10 mm clearance at the sides and top, it is simply a matter of slipping the window into the opening. Make sure that the factory-fitted flashing is not kinked or folded back under the window and that it sits in front of the frame.

Then pack the sill level if necessary, check the head for level and move the window to align with the external cladding or flush with the internal lining, or both if the reveal linings allow. If the reveal linings are too wide, they may have to be trimmed down. Finally, plumb the stiles, inserting solid packing (or using wedges), and nail at maximum 600 mm centres. Don't drive the nails all the way in until you have checked the frame for square and tested the window for operation. Once satisfied that the window will operate correctly, finish the nailing off by punching the nail heads below the surface by 2 mm to allow for suitable filling and finishing with architraves.

It is good practice to check the type, gauge and quantity of fasteners required for each window by accessing the window manufacturer's installation guides and recommended fasteners and fixings.

Installing aluminium windows in brick veneer constructions

Having prepared the openings to suit the window schedule, adjust any secondary jamb studs to suit the brick bond (if secondary jamb studs have not been used, allow 10 mm clearances each side and at the head).

Make sure the factory-fitted flashing is not kinked or folded back under the window and that it sits in front of the frame and down into the cavity.

Place the window into the opening, aligning it flush with the internal lining (see Figure 3.79). Pack the sill straight and check the head for level (see Figure 3.80). If it is a wide window and packed from the sill trimmer, insert a timber batten hard up under the outside of the window frame to prevent the frame sagging. Now plumb both stiles, insert solid packing and temporarily nail off (not more than 600 mm centres).

Make sure that the window operates properly before finishing the nailing off by punching the heads 2 mm below the surface.

Wait until the brickwork and sill bricks have been laid before applying a bead of sealant in the gaps to the external joint between the brickwork and the window and below the sill if no flexible seal is present. If necessary, install an aluminium angle storm mould to prevent moisture penetration.

Note: No head flashing is required if the window head aligns with the eaves lining.

Under no circumstances should nails be driven into the window sill as nails will act as a conduit for moisture and cause rot to the sill, as well as fixing the sill and restricting the ability of the window to move within the building envelope. This can place additional stress on the window and cause sashes to bind on frames.

Packing should not be placed over the head of the window as the building load could then be transferred to the frame, distorting the frame and causing the sashes to bind, and in the worst circumstance causing the glass to break.

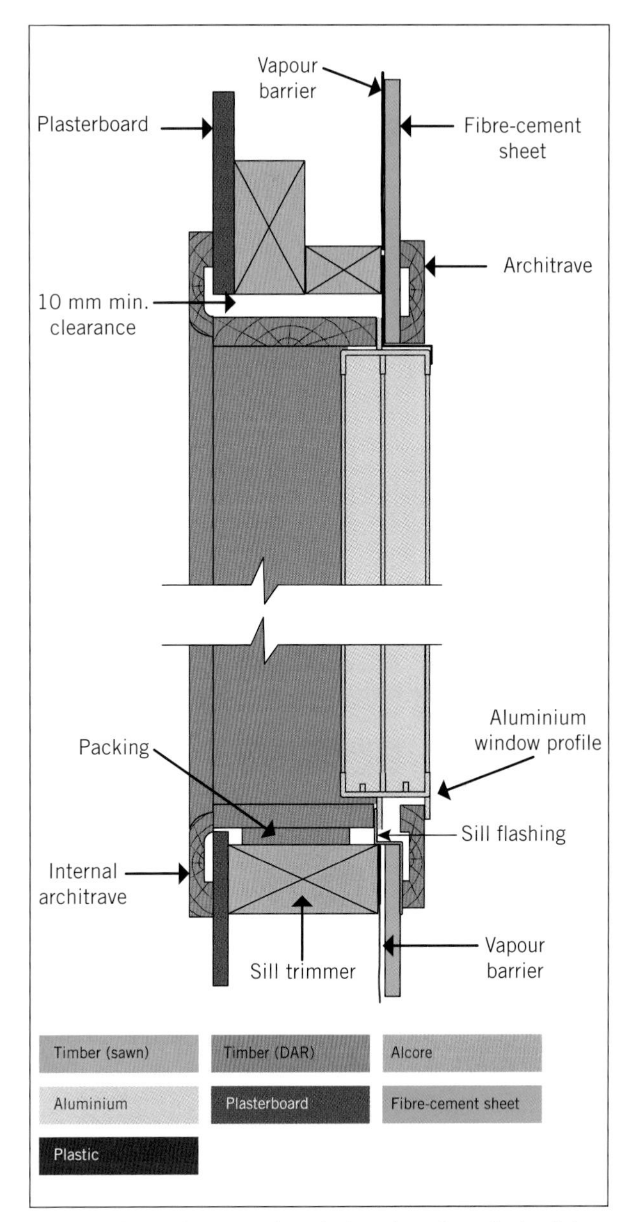

FIGURE 3.79 Horizontal and vertical sections through aluminium window with timber reveals

Installing aluminium windows in cavity brick constructions

Aluminium windows are installed in cavity brick constructions usually after the brickwork has been built (but before any rendering if it is to be applied to the reveal areas). Building-in lugs are clipped to the aluminium window fins and embedded in the brickwork as the walls go up (see Figure 3.81).

The outer wall opening is built to the overall height and width of the windows (less the fins), while the inner opening is built to the same size plus an allowance for building-in lugs (fixing clips). Head flashings are installed as the walls are built, while the

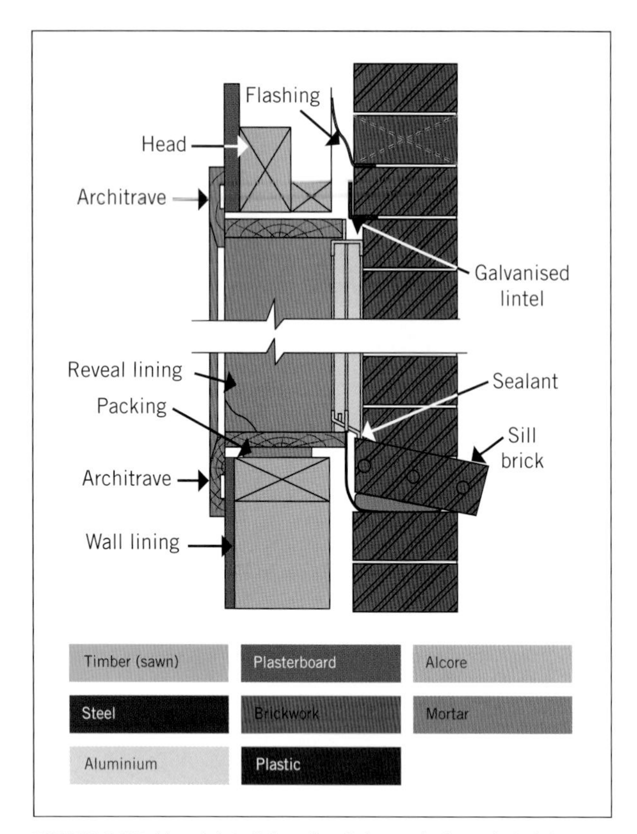

FIGURE 3.80 Head detail for aluminium windows in brick veneer construction

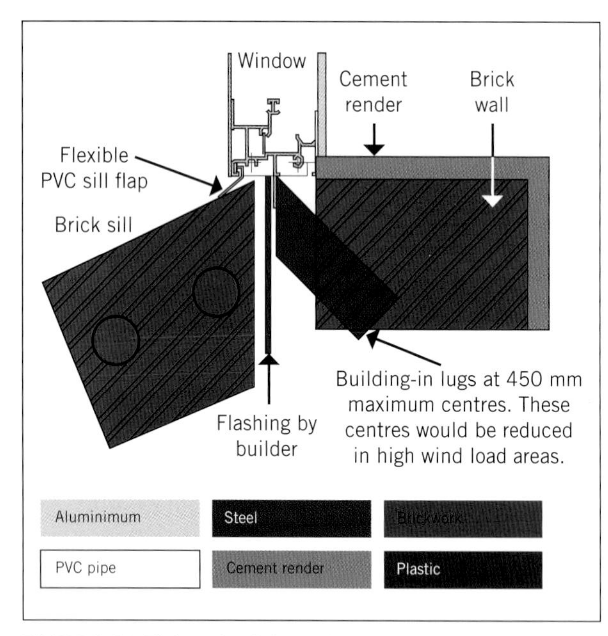

FIGURE 3.81 Fixing aluminium window without reveals in cavity brick construction with fixing clip or building-in lug

sill flashing is completed once the sills are installed (see Figure 3.82).

If the internal brickwork is to be face brickwork, the window is installed using building-in lugs embedded in the internal brickwork.

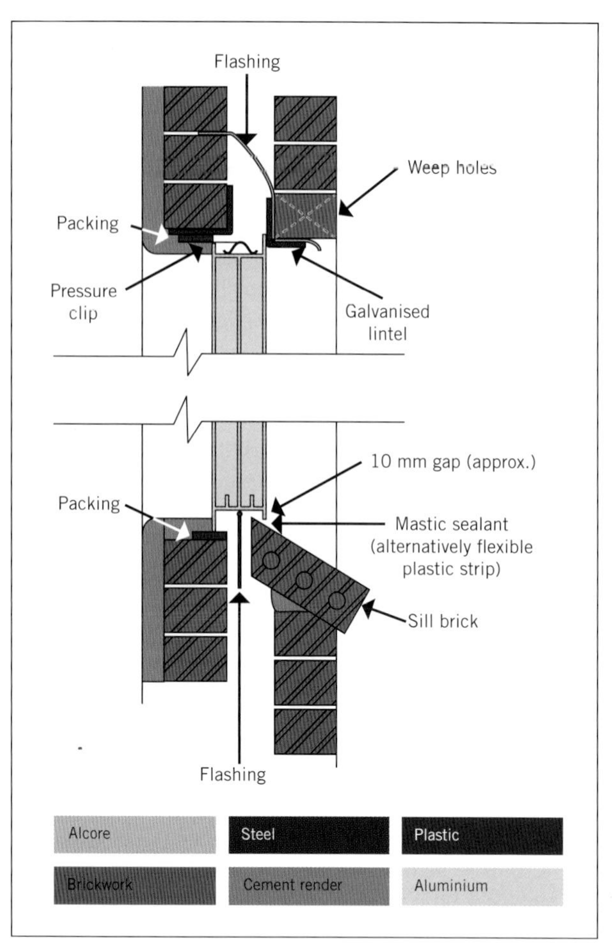

FIGURE 3.82 Vertical section through window and cavity wall

If the window is to cover the cavity, the brickwork butts to the frame on the inside and a storm mould is fixed to the external face of the window frame.

It is good practice to prevent direct contact between sill and outside brick work with an isolating material to stop the aluminium reacting with brick mortar. Direct contact can cause corrosion.

Installing windows with timber reveals in cavity brick constructions

To avoid staining or cement damage, windows with timber reveals can be installed after the walls have been constructed and cleaned. The frame is placed in the opening with the fin hard against the inside face of the outer brickwork. This allows the timber reveals to cover the cavity and finish flush with the internal face of the brickwork or render finish. The architraves are then fitted to the reveal linings with appropriate quirks (see Figure 3.83). The mortar joints can be raked clean and a slightly twisted timber plug is then inserted into the mortar gap. The twist is to help the timber plug hold in the gap. Use two 75 × 3.15 mm nails and punch them into each plug at least 2 mm below the surface and putty the heads.

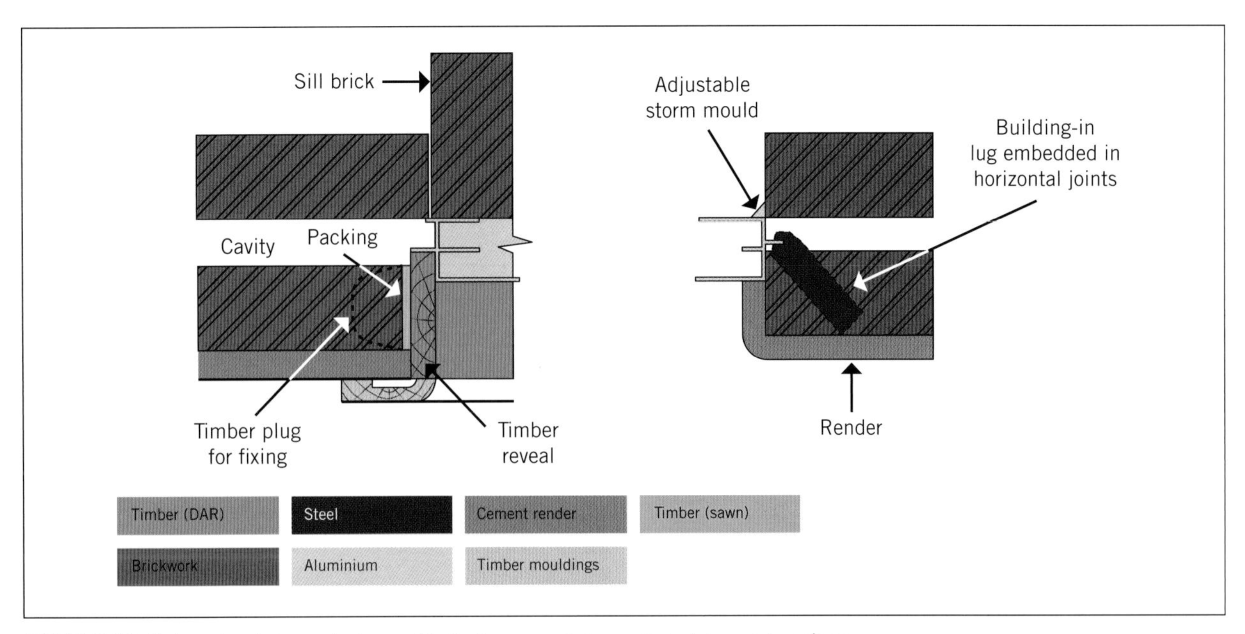

FIGURE 3.83 Fixing aluminium windows with timber reveals in cavity brick constructions

Installing timber windows in timber-framed constructions

Wall frames are prepared with the cladding installed to the top of the opening and the flashings are cut into the wall frames.

Cut and install the packing to both sides of the opening, making sure that the packing is plumb top to bottom and the horizontal distance is equal top and bottom (see Figure 3.84). Remember, the packing cannot be more than 600 mm apart. Check that the sill trimmer is level, and if not place appropriate packing below the stiles (and mullions, if necessary). Use a straight edge and level to ensure that it is level right across.

The sill may need to be trimmed to fit over the cladding before installation.

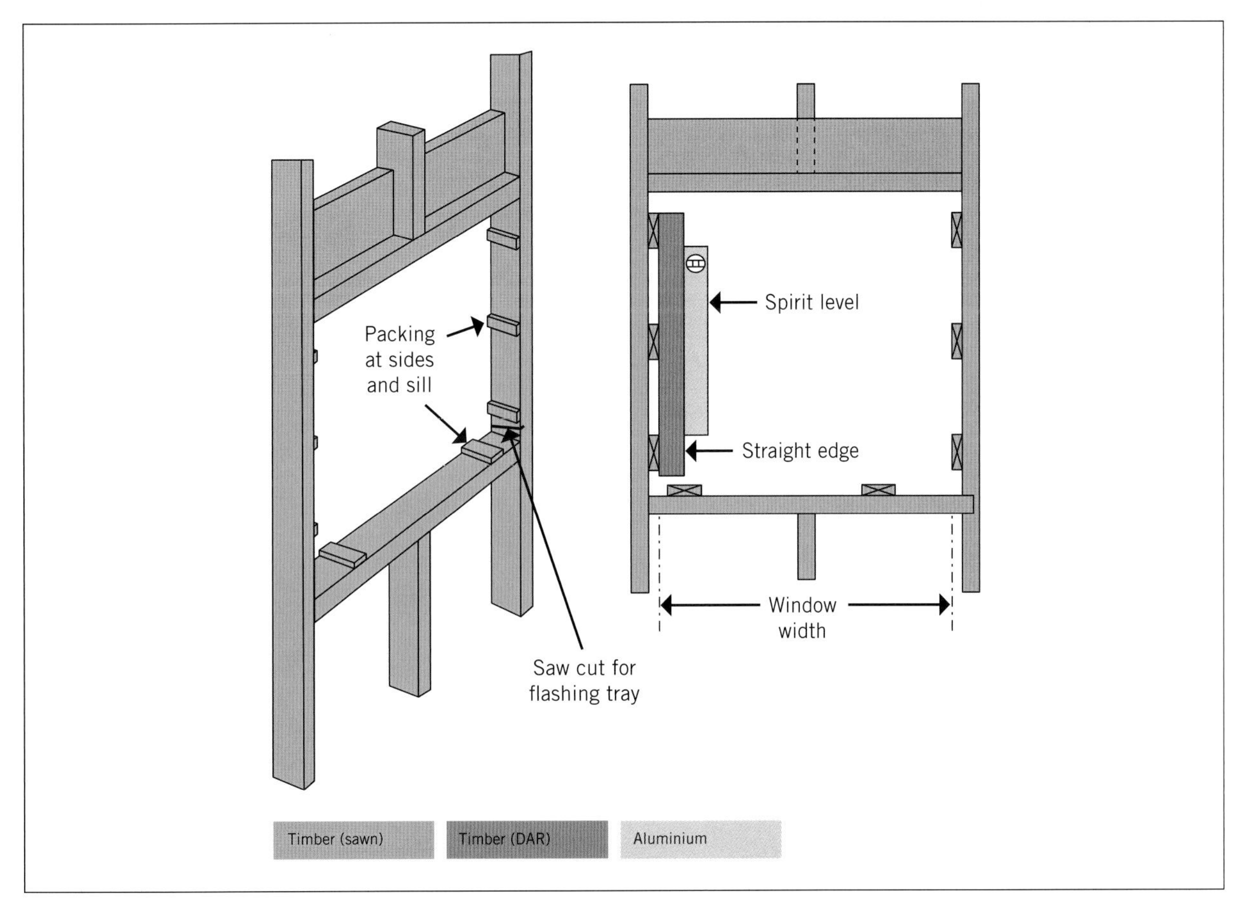

FIGURE 3.84 Packing the opening ready for the window

Sit the frame in the opening and push out until the frame is flush with the outside of the cladding surface. Check the window for wind before temporary nailing of the window through the stiles, through the packing and into the frame. Make a final check that the sashes operate correctly. Make any adjustments now before finishing the nailing by punching the heads below the surface.

Fix the external architrave by nailing into the frame, through the cladding and into the studs. (If using a sill, fit the apron piece below and trim off any excess flashing.) Fix the head flashing packing above the window. Install the head flashing and fold down over the packing block and trim off any excess – 10 mm overhang is normal (see Figure 3.85).

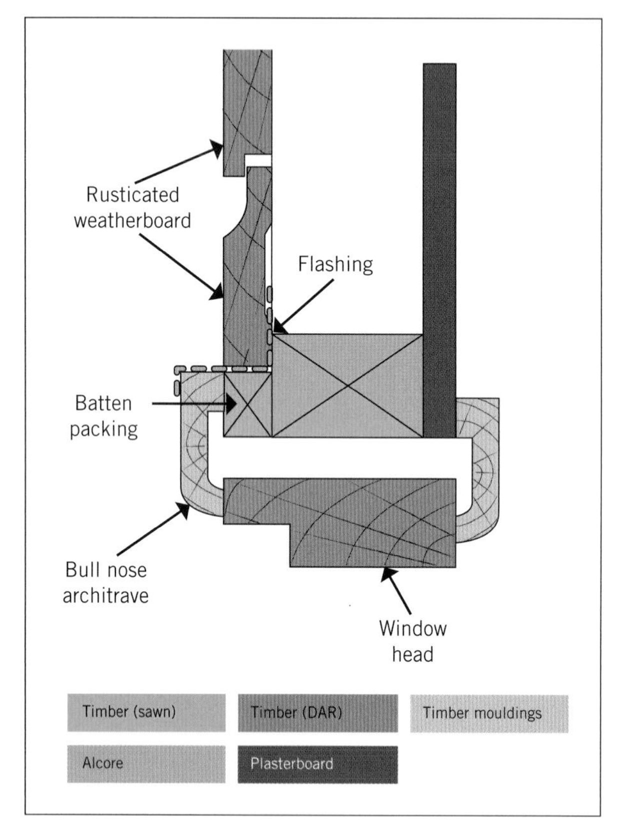

FIGURE 3.85 Position of head flashing

If necessary, you can cut and fix the external architrave to the frame and install the window from the outside.

After the internal lining is installed, trim around the frame, as well as the architraves, nosing and apron piece (see Figure 3.86).

Installing timber windows in brick veneer constructions

As with aluminium windows, timber windows are installed in cavity brick construction after the brickwork has been built (but before rendering, if any, is applied to the reveal areas).

FIGURE 3.86 Nosing and apron piece

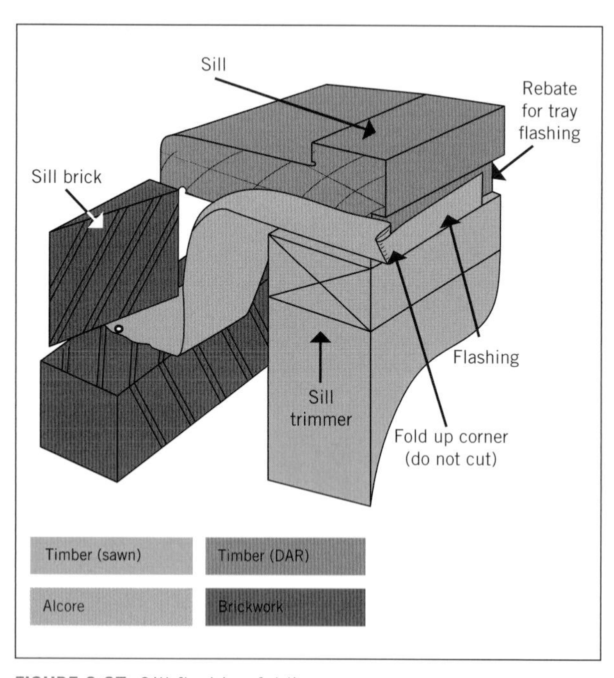

FIGURE 3.87 Sill flashing folding

A flashing tray is folded up around the sill and the window is inserted into the opening in the frame, similar to a door threshold (see Figure 3.87).

If vertical flashing is fitted to the window, the window should be installed before the brickwork is erected and the window positioned to align with the brick gauge.

The packing is then installed, or secondary floating window studs are fastened off top and bottom.

The window is pushed forward to align with the inner face of the brickwork and temporarily nailed off to the studs through the packing. Check the operation of the window before final nailing and punching of the heads.

Fit reveal linings, if necessary, and install nosing, architraves and apron pieces.

Once the brickwork is complete (see Figure 3.88), a sealant can be applied to any gaps and the storm moulds (quad, splayed or other moulds) can be fitted externally around the window to cover any gaps.

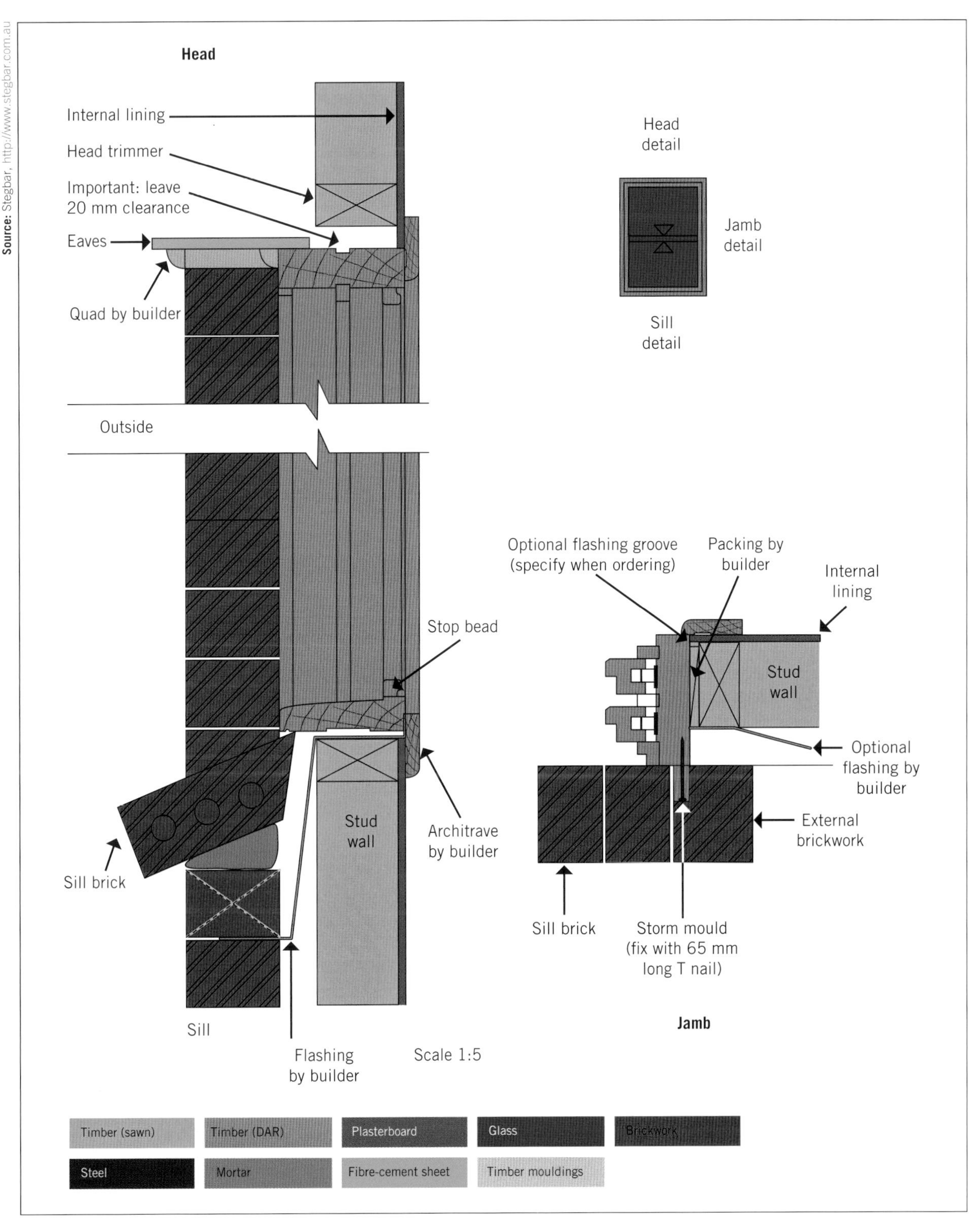

FIGURE 3.88 Windows in a brick veneer construction

Installing timber windows in cavity brick constructions

Installing timber windows in a cavity brick wall is carried out when the brickwork on the inner skin reaches inner sill height.

Before installing, make up the sill tray, as shown in Figure 3.87, and fit it to the window.

Sit the window in the opening and adjust horizontally for the brick bond (if the brickwork is not complete).

If the window is to have timber reveal linings, the window is moved forward so that the window frame aligns with the inner face of the external brick skin. If the frame does not have reveal linings, the inner face of the frame is positioned to align with the inner wall finish surface.

If the inner reveal is to be rendered, the window can be positioned as shown in Figure 3.89. The window sill should have a sufficient bearing surface to accept packing.

Note: Timber windows must be protected from any damage that may be caused by cement render. Timber reacts badly to cement render, causing staining that will be difficult to hide if the windows are to have a clear finish applied later.

If the window does not cover the cavity completely, an outer reveal lining can be applied to cover the void.

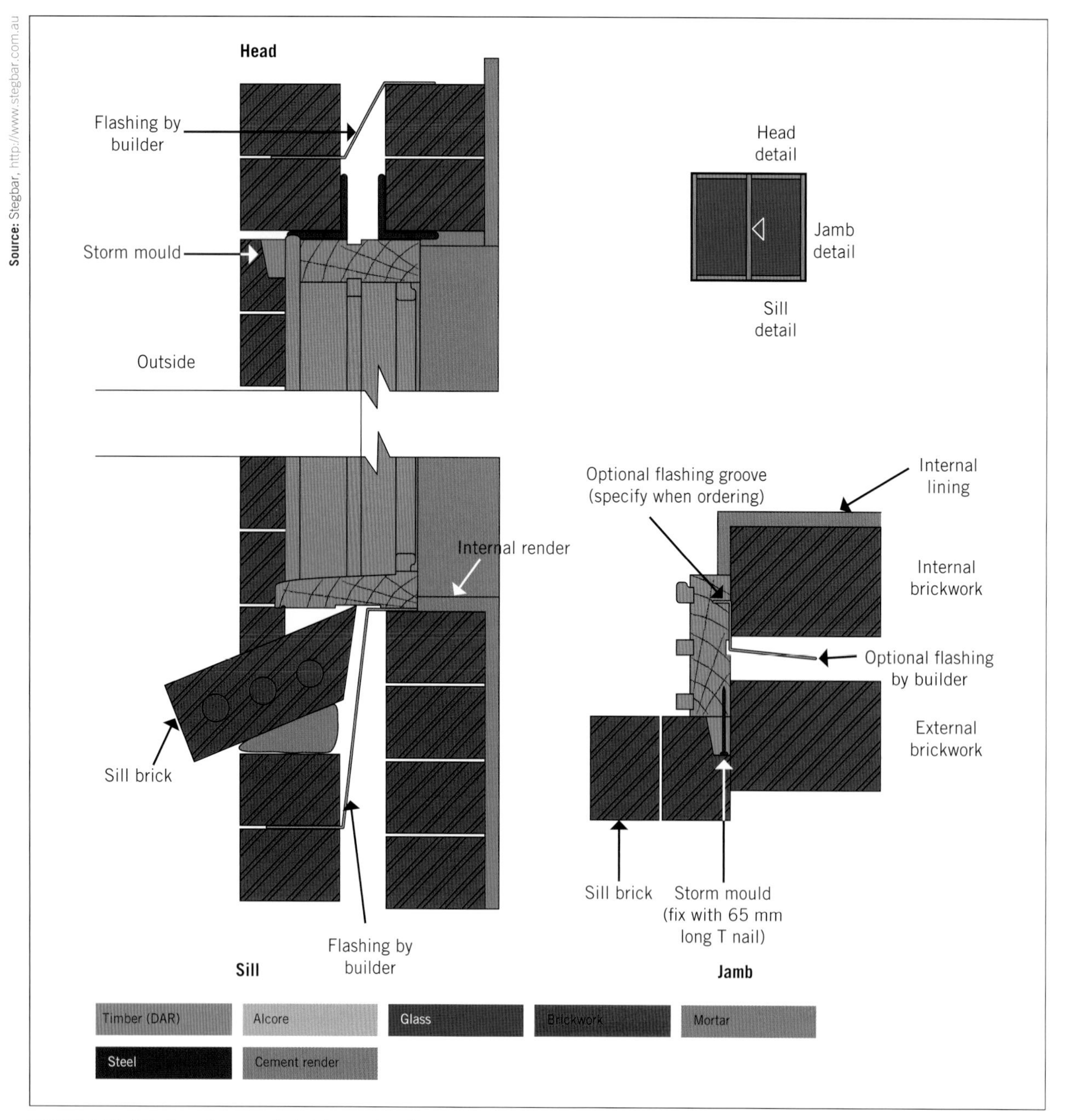

FIGURE 3.89 Windows in a cavity brick construction

If the windows are to be stood before the brickwork has gone up and a cavity cleat used to tie the window into the brickwork, the cavity cleat should be fastened to the frame and the frame packed level and plumb and braced in position.

Removing door and window frames

To replace a door or window, you need to first identify the surrounding wall construction method. The modern methods used to install windows and doors are outlined above, so you should have some idea of the process required to undo what has been done.

Method for old buildings

In the nineteenth and early twentieth centuries it was not uncommon for door and window frame heads to be extended beyond the frames and embedded into the brickwork or the cavity. If the sill and head sections of the door you wish to remove have been cut back into the cavity (see Figure 3.90), you need to cut through the head and sill to remove them or, alternatively, break out the masonry work and repair it when the new window has been installed, which is a more expensive method.

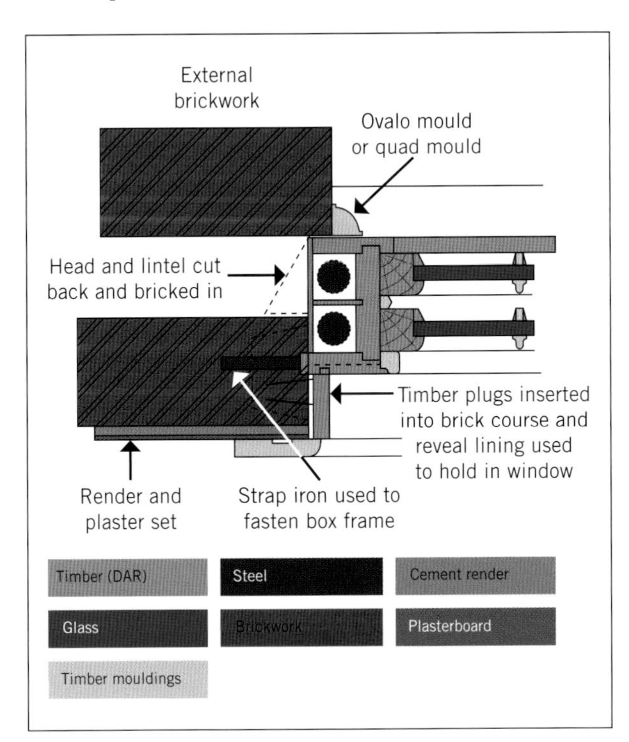

FIGURE 3.90 Three alternative methods used to fix box frame windows

Method for modern constructions

The first step is to cut around the joint between the architrave and the frame with a utility knife to break any paint layers that may bond the architraves to the window or door frame. Then remove the surrounding architraves, storm mould, window nosings and apron, and other joinery items that prevent you from accessing the gaps and cavities.

Take care to use a wide-bladed chisel or lever (**pinch bar**) to lever off the architrave on the frame side of the architrave first (see Figure 3.91). Then lever off the architrave on the wall side, using a backing block behind the chisel or lever (a metal cabinet scraper works well behind chisels); otherwise, you can easily damage the surrounding wall finishes by crushing plasterboard or damaging render and plasterwork. Work from the bottom up.

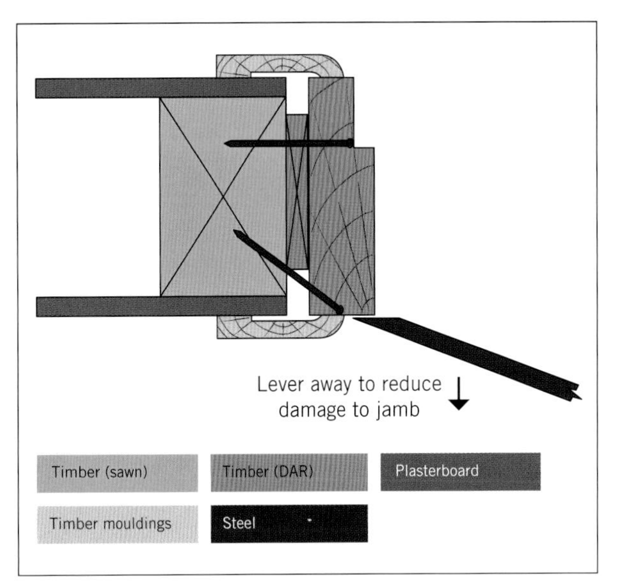

FIGURE 3.91 Removing architrave

Once the architraves have been removed, check the installation method and the position of the ties and fastenings.

Internal door jambs in timber-framed walls are usually nailed off, so you should be able to simply lever the door frame from the bottom of the wall towards the opening.

If the walls are masonry and the frames are tied in with brick ties or straps, you will need to use a reciprocating saw (see Figure 3.92) with a metal blade or a hacksaw blade to cut through the ties to release the frame from the opening. A reciprocating saw with a flexible metal blade will make the job a lot easier.

Finally, lift the door or window frame from the opening. Inspect and make good any flashings that may need repair or replacement before installing the new door or window frame.

At the end of the day, check the door or window is operating correctly by opening and closing

Source: Milwaukee Power Tools Australia

FIGURE 3.92 Reciprocating saw

several times. Whenever possible, speak with the customer and show them the completed installation so any concerns can be addressed before leaving the job.

Cleaning up

Once the job is completed you will need to clean up the work area. Your workplace will have specific requirements in place to ensure this task is completed properly. It is important you understand these and complete the task accordingly.

Failure to properly clean the area once you have completed work may create a hazard that other workers will need to deal with; for example tripping hazards left on the ground or dust that can cause irritation in the eyes and throat.

Leftover materials should be reused or recycled wherever possible. Heavy items should be moved using appropriate manual handling techniques.

Tools

Tools and equipment that were used during the installation must be inspected for damage and checked to make sure they work as designed and are fit for purpose. Any damage that is observed must be reported to the supervisor immediately, and the tool/equipment taken out of use and tagged, so other people do not use it before it is repaired or replaced.

COMPLETE WORKSHEET 5

SUMMARY

This chapter addressed the skills and knowledge required for installing windows and doors, including:

- understanding door and window styles and specifications; for example:
  - reading plans and specifications to determine doors as internal or external, their types (framed, hollow core or other) and their swing or hung direction and sizes
  - defining and determining the types of windows, glazing details, sizing and regulatory requirements.
- planning and preparing door and window schedules, including:
  - taking off quantities from plans and specifications
  - handling and storing doors and windows safely onsite
  - understanding the requirements in bushfire-prone areas (Bushfire Attack Levels, or BALs).
- understanding flashing details for doors and windows as they apply to:
  - different construction methods (e.g. full masonry, brick veneer and timber framed)
  - door and window framing materials (e.g. steel or timber door frames, and aluminium or timber windows)
- installing internal and external doors and door frames, including:
  - thresholds and split door jambs in timber and metal
  - cavity sliding and aluminium sliding.
- hanging doors, including:
  - selecting and installing the right hinge type
  - selecting and installing the right door furniture, including locks and latches to doors
  - fitting and hanging the door in the door jamb.
- installing windows, including:
  - different types of windows (e.g. timber framed and aluminium) in different types of walls
  - constructions such as:
    - timber framed and clad
    - brick veneer
    - cavity brick.
- Removing door and window joinery from different built forms.

REFERENCES AND FURTHER READING

Manufacturers' or suppliers' brochures for windows and doors web-based resources.

National Construction Code 2022

Copies of the NCC can be obtained free of charge from the Australian Building Codes Board website: **http://www.abcb.gov.au**.

- National Construction Code – Volume One has the requirements for multi-residential, commercial, industrial and public buildings and structures.
- National Construction Code 2022 – Volume Two is a uniform set of technical provisions for the design and construction of buildings and other structures allowing for variations in climate and geological or geographic conditions. Volume 2 contains the requirements for Class 1 (residential) and Class 10 (non-habitable) buildings and structures.
- National Construction Code 2022 – ABCB – Housing Provisions contains Deemed-to-Satisfy Provisions that are considered to be acceptable forms of construction that meet the requirements for complying with Parts H1 to H8 of NCC Volume Two.

Register or log in regularly to ensure you always have the latest versions at **http://www.abcb.gov.au**.

Texts

Acceptable Standards of Construction Committee (NSW). *Acceptable Standards of Construction – Class 1 & Class 10 Buildings*, Acceptable Standards of Construction Committee (NSW).

South Western Sydney Institute of TAFE NSW (2020). *Basic Building and Construction Skills* (6th edn), Cengage Learning Australia, Melbourne.

Victorian Building Authority (2015). *Guide to Standards and Tolerances*, Building Commission, Melbourne.

Web-based resources

Government and industry organisations

Architectural Door Hardware Association: **http://www.adha.net.au**

Association of Building Sustainability Assessors: **https://www.absa.net.au**

Australian Glass and Window Association: **https://www.agwa.com.au**

BASIX Certificate Centre: **http://www.basixcertificatecentre.com.au**

Building Products Innovation Council: **http://www.bpic.asn.au**

National Home Energy Ratings Scheme (NatHERS): **https://www.nathers.gov.au/**

Skylight Industry Association: **http://www.siai.com.au**

Window and Door Industry Council: **http://wadic.org.au**. Publishes the *Guide for Compliant Construction of Timber Windows and Doors in Bushfire-Prone Areas.*

Window Energy Rating System (WERS): **https://awa.associationonline.com.au**

Window Film Association of Australia and New Zealand: **http://www.wfaanz.org.au**

Corporate

Canterbury Windows and Doors: **http://www.canterburywindows.com.au**

Hume Doors: **http://www.humedoors.com.au**

Jeld-Wen Australia (owners of Corinthian, Stegbar, Airlite and other companies in the door and window industry): **http://www.jeld-wen.com.au**

Trend Windows and Doors: **http://www.trendwindows.com.au**

Vantage Aluminium Joinery: **https://www.awsaustralia.com.au/vantage**

Video resources

Australian Glass and Window Association videos playlist: **https://www.youtube.com/playlist?list=PLw25vx2FAb7K8-MOuZoNfGcKh5EP8Zh2Z**

Windows

How to cyclone proof windows and doors: **https://www.youtube.com/watch?v=q5CE17JVVHg&list=PLSBARC465CGaekhuXoL3KDvuPU7rKGca_&index=1**

How to install a timber awning window: **https://www.youtube.com/watch?v=F5YPcM7GXdw&list=PLSBARC465CGaekhuXoL3KDvuPU7rKGca_&index=9**

How to install an aluminium window: **https://www.youtube.com/watch?v=cUkwbxPrEq8&list=PLSBARC465CGaekhuXoL3KDvuPU7rKGca_&index=5**

How to remove and refit spiral balances: **https://www.youtube.com/watch?v=Se-YGNegpOo**

Doors

Assemble and install a door jamb: **https://www.youtube.com/watch?v=ICYwPa5_bFY**

Flush door parts: **https://www.youtube.com/watch?v=ejizLMk7FMw&list=PLHFT_d-5EhN5aQHw1ZzPeu52DSH3oQjJi**

How to hang a door: **https://www.youtube.com/watch?v=nNgA-8Bn-CQ&list=PLSBARC465CGbz0Ym3zvxlfdhl9luJRAda&index=7**

How to install a cavity sliding door: **https://www.youtube.com/watch?v=tsKC79aTd1I&list=PLSBARC465CGbz0Ym3zvxlfdhl9luJRAda&index=15**

How to install a door jamb: **https://www.youtube.com/watch?v=hdrnH95BwYk**

How to install sliding doors: **https://www.youtube.com/watch?v=kafj4PT9gJI&list=PLSBARC465CGbz0Ym3zvxlfdhl9luJRAda&index=9**

Pre-hung door jamb: **https://www.youtube.com/watch?v=XcauX3EvzXo**

Other useful videos

How to fit a tubular mortice latch and keep: **https://www.youtube.com/watch?v=ExzXomOEBzI**

How to make a lock jig: **https://www.youtube.com/watch?v=BqASVwYNO-I**

How to use a hinge butt gauge (Easy hinge mortises with the no.94 dual beam scribe): **https://www.youtube.com/watch?v=SO9UXgEOcmM**

Making a door-holding jig (How to make a basic door stand): **https://www.youtube.com/watch?v=QQHJ1jVd05Q**

Ryobi door hinge kit: **https://www.youtube.com/watch?v=Ip92O2i9IpA**

Relevant Australian Standards
AS 1288 Glass in buildings – Selection and installation
AS 2047 Windows in buildings – Selection and installation
AS 4420.0 Windows – Methods of test – General introduction and list of methods. **Note:** Other Standards in the series cover air infiltration tests, operating force tests, ultimate strength tests and water penetration resistance tests.
Australian Standards can be purchased from the Standards Australia website: **https://www.standards.org.au/**, or see your teacher or librarian for assistance in accessing Australian Standards online.

GET IT RIGHT

INSTALL AND REPLACE WINDOWS AND DOORS

Questions

Below are four pictures showing some examples of window and door details.

1 What could be the problem with the gaps around these older windows?

____

____

2 How should these older windows have been completed and maintained correctly?

____

____

3 What should be done to rectify the problem?

____

____

4 What could happen if nothing is done?

____

____

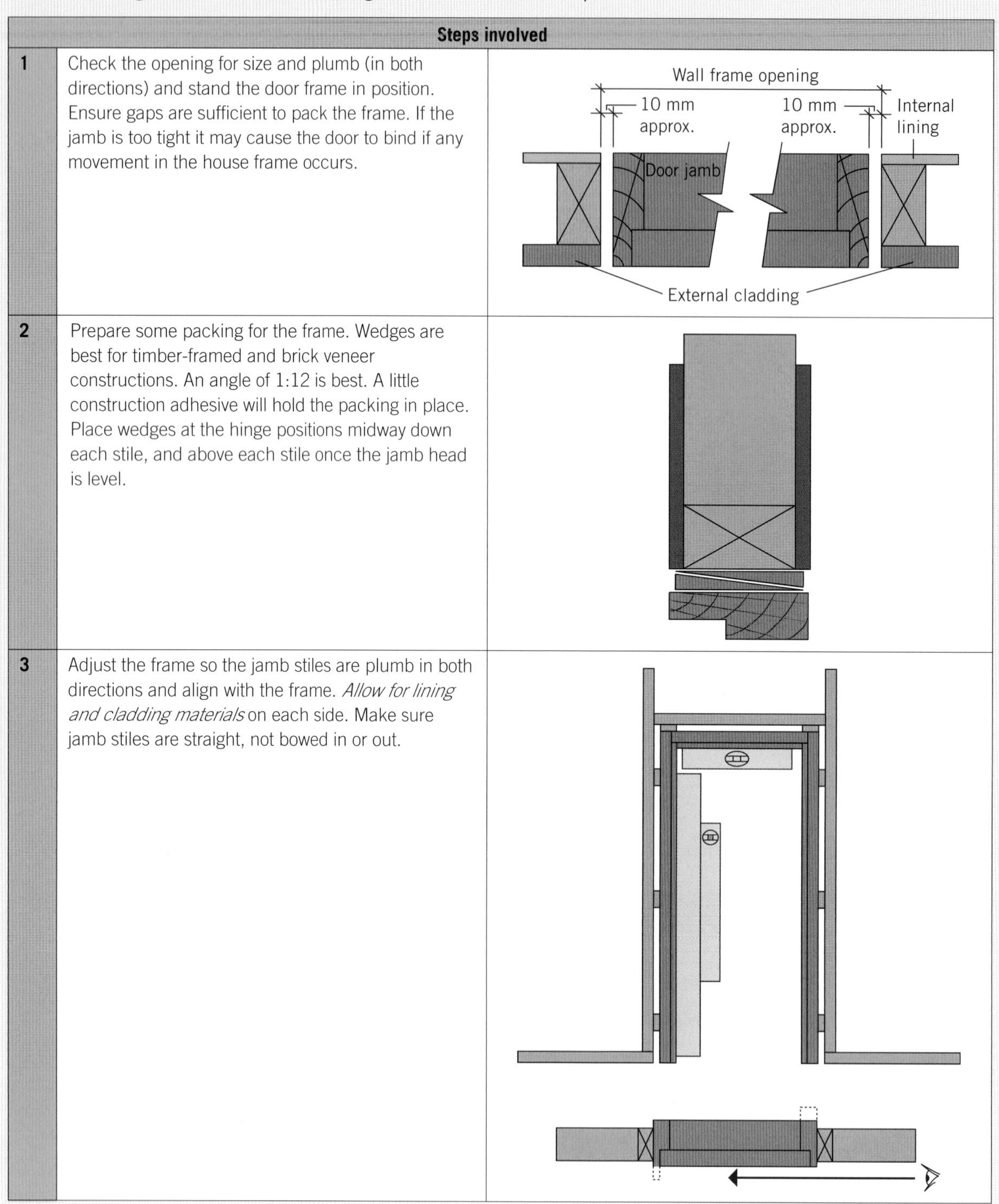

HOW TO

INSTALL EXTERNAL DOOR FRAMES IN TIMBER-FRAMED WALL CONSTRUCTIONS

Before installing frames, ensure the flashings for the threshold are in place.

Steps involved	
1	Check the opening for size and plumb (in both directions) and stand the door frame in position. Ensure gaps are sufficient to pack the frame. If the jamb is too tight it may cause the door to bind if any movement in the house frame occurs.
2	Prepare some packing for the frame. Wedges are best for timber-framed and brick veneer constructions. An angle of 1:12 is best. A little construction adhesive will hold the packing in place. Place wedges at the hinge positions midway down each stile, and above each stile once the jamb head is level.
3	Adjust the frame so the jamb stiles are plumb in both directions and align with the frame. *Allow for lining and cladding materials* on each side. Make sure jamb stiles are straight, not bowed in or out.

>>

>>

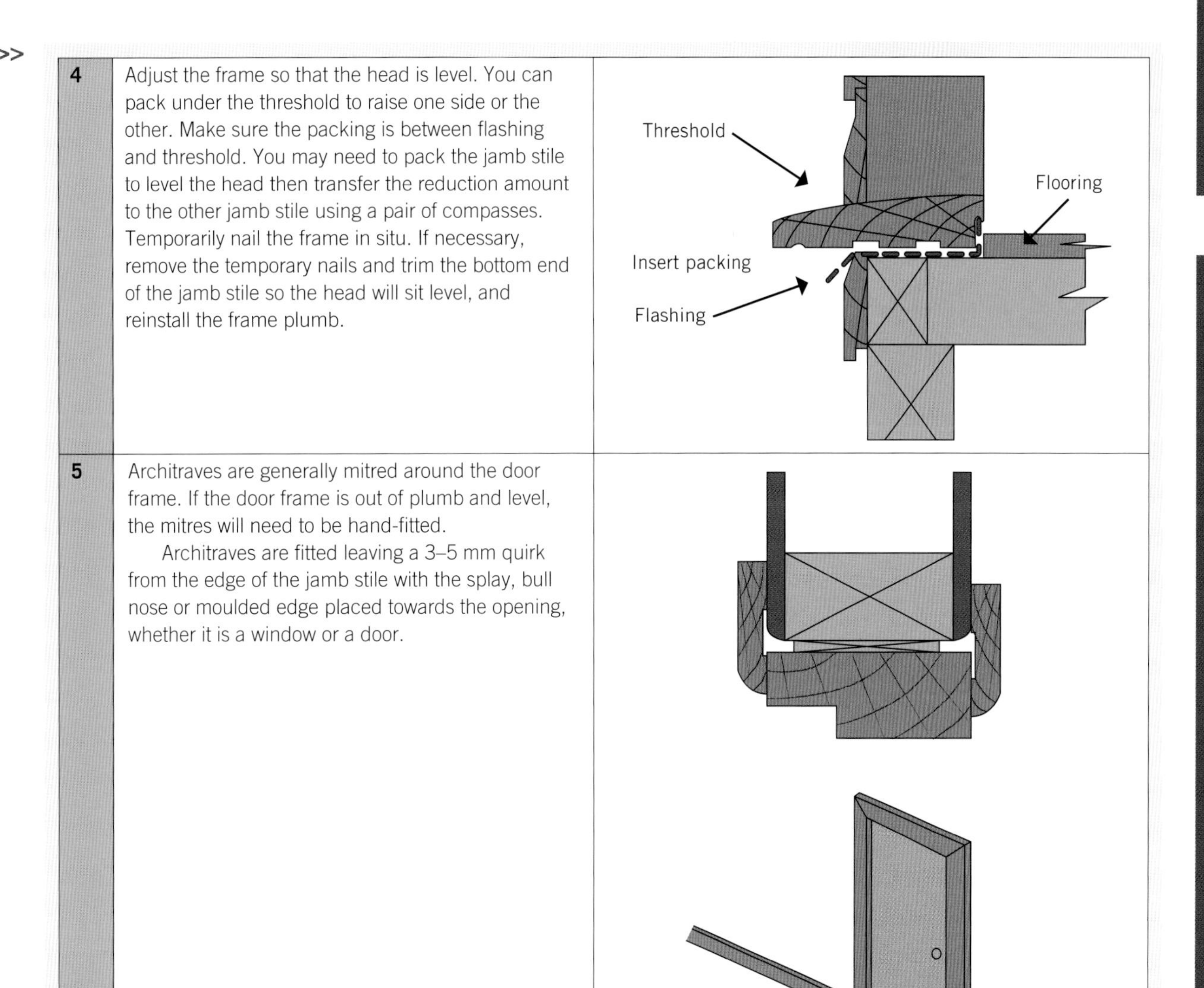

4	Adjust the frame so that the head is level. You can pack under the threshold to raise one side or the other. Make sure the packing is between flashing and threshold. You may need to pack the jamb stile to level the head then transfer the reduction amount to the other jamb stile using a pair of compasses. Temporarily nail the frame in situ. If necessary, remove the temporary nails and trim the bottom end of the jamb stile so the head will sit level, and reinstall the frame plumb.	
5	Architraves are generally mitred around the door frame. If the door frame is out of plumb and level, the mitres will need to be hand-fitted. Architraves are fitted leaving a 3–5 mm quirk from the edge of the jamb stile with the splay, bull nose or moulded edge placed towards the opening, whether it is a window or a door.	

WORKSHEET 1

To be completed by teachers

Student competent ☐

Student not yet competent ☐

Student name: ______________________

Enrolment year: ______________________

Class code: ______________________

Competency name/Number: ______________________

Answer the following questions.

1 From the drawn door symbols shown below, identify each one and write the correct name in the space below.

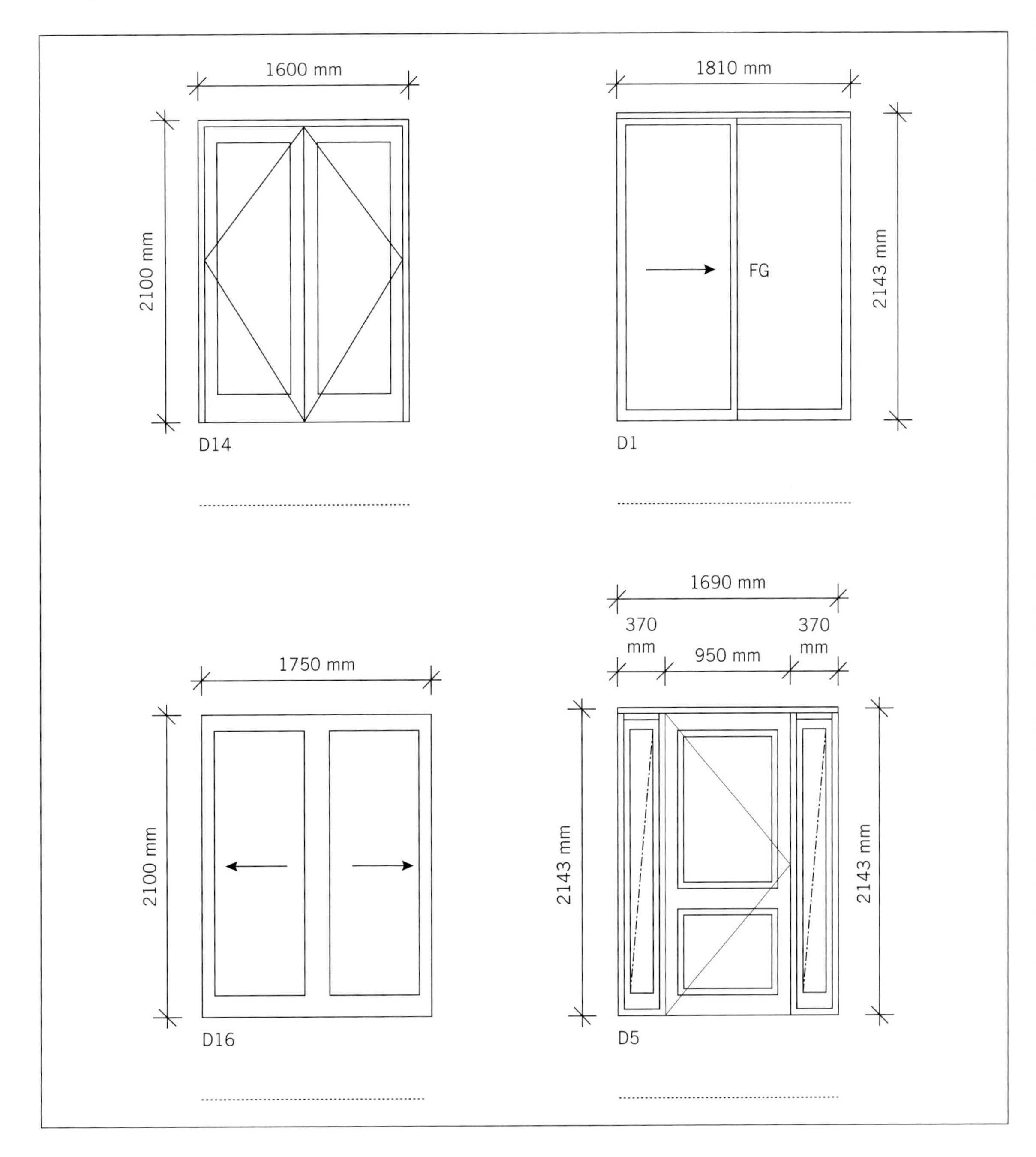

2 In the spaces provided beside the sketch below, indicate which is the hinged side and which is the lock side.

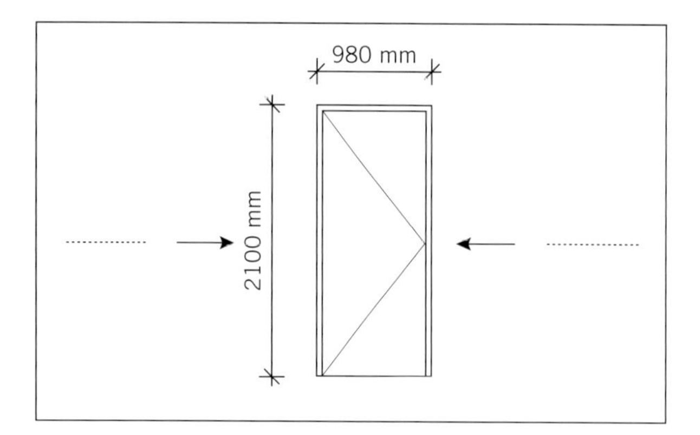

3 Explain the three basic functions of a threshold.

a ______________________________

b ______________________________

c ______________________________

4 Why is a double-rebated door jamb used in an entrance door frame?

____

____

5 How can a double rebate be formed in an entrance door frame without machining rebates?

____

____

6 List two species of timber suitable for thresholds.

a ______________________________

b ______________________________

7 List three materials that door frames can be made from.

a ______________________________

b ______________________________

c ______________________________

8 How do you calculate the opening size in a wall frame?

____

____

9 What is the purpose of fitting temporary braces to door frames and where are they fitted?

____

____

10 What are the standard height and width dimensions for internal doors?

____

11 Flashings to entrance door frames are placed in which positions?

____

____

12 Identify the form of construction and label the parts in the drawing below.

Form of construction: ______________________________

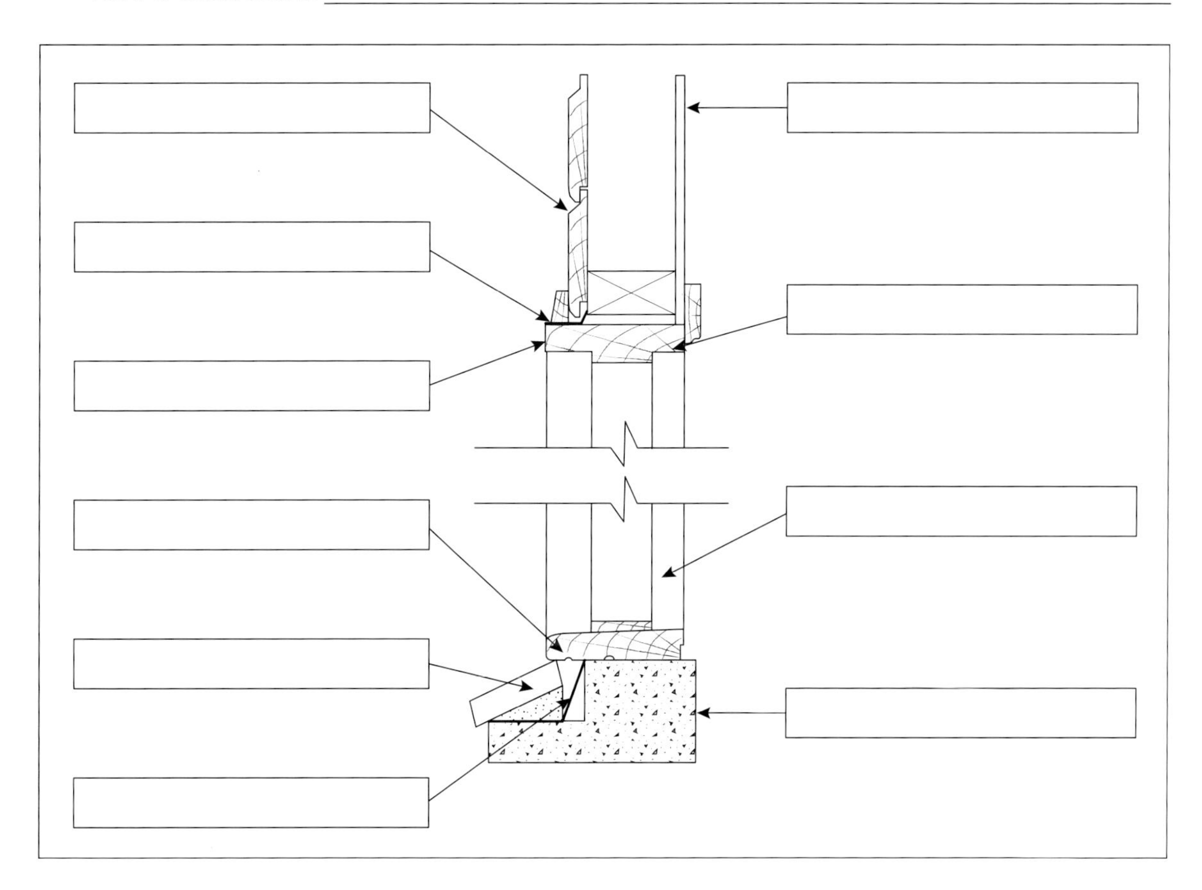

13 What is the Australian Standard for constructing buildings in bushfire-prone areas?

____

14 Explain the six Bushfire Attack Levels (BAL).

a ______________________________

____

b ______________________________

____

c ______________________________

____

d ______________________________

____

e ______________________________

____

f ______________________________

____

15 In Section 6.5.2 of AS 3959 Construction of buildings in bushfire-prone areas covering doors and windows:

a To what BAL does this section refer?

____

b What are the general requirements for windows?

____

c What are the general requirements for external doors?

____

16 The National Construction Code 2022 – Volume Two, Part H6 deals with energy efficiency in buildings. What is the primary objective of this Part?

____

____

17 Tick Yes or No.

Is the National Construction Code – Volume Two, Part H6 relevant in your state or territory?

Yes ☐ No ☐

If your answer is Yes go to Question 18.

If your answer is No, explain what alternative solution or requirement applies in your state.

____

____

18 Summarise the two means of achieving energy efficiency under National Construction Code – Volume Two, Part H6.

WORKSHEET 2

To be completed by teachers	
Student competent	☐
Student not yet competent	☐

Student name: ______________________

Enrolment year: ______________________

Class code: ______________________

Competency name/Number: ______________________

Answer the following questions.

1 What is the main influence in positioning door frames in brick veneer and cavity brick construction?

____

2 Fill in the missing words: In brick veneer construction, __________ or __________ fasteners (nails and/or screws) should be used to fix the frame to the wall frame.

3 Explain how door frames in brick veneer construction can be sealed to prevent the penetration of water.

____

____

____

4 What three checks to the door frame need to be made to ensure that the door will fit and operate properly?

a ______________________

b ______________________

c ______________________

5 What are the two methods used to fasten door frames in cavity brick walls?

a ______________________

b ______________________

6 What is the range of thicknesses used for an internal door and an entry (external) door?

Internal doors: ______________________

External doors: ______________________

7 Why is flashing folded up around thresholds?

____

____

8 The width of the rebate in a glazed panelled door is affected by the thickness of glass. Link the glass thickness to the correct rebate width with a line or arrow. Refer to AS 1288 Glass in buildings, Section 8, Table 8.1.

Glass thickness	Rebate width
3–4 mm thick	10–13 mm wide
15 mm thick	8–9 mm wide
10 and 12 mm thick	18–20 mm wide
5, 6 and 8 mm thick	13–15 mm wide

9 When nailing off a door frame that is to be clear finished (polished, stained or lacquered), where should the nails be placed?

____

____

10 Describe two methods that can be used to protect aluminium door frames from damage during construction.

a ______________________________

b ______________________________

11 What is the minimum clearance to be provided between the underside of the lintel, ring beams or lintel/ring-beam trimmer and the top of the window or door frame? (See AS 1684 – 2010 Section 6.2.3.)

____

____

12 Using AS 1684.3 for cyclonic areas and Table 6.3 – Size of windowsill trimmers, determine the section size for a windowsill trimmer when the window opening is 2200 mm wide, framing timbers are MGP12 and the wind classification is C2.

____

____

13 If a door stud is carrying a concentrated load (i.e. in a load-bearing wall), it is correct to have a single stud using the same standard framing material.

True ☐ False ☐

14 Answer True or False to the following question.

In a timber-framed floor, blocking is required between the joist below the door and window openings.

True ☐ False ☐

15 In terms of wind classifications for building, what do the following terms indicate?

a N1

____

b C1

____

c TC

____

16 Briefly explain the following terrain classes with an example of a specific location near you.

a Terrain Category 2 (TC2)

____

____

b Terrain Category 3 (TC3)

____

____

c Terrain Category 1.5 (TC1.5)

____

____

WORKSHEET 3

To be completed by teachers	
Student competent	☐
Student not yet competent	☐

Student name: ______________________________

Enrolment year: ______________________________

Class code: ______________________________

Competency name/Number: ______________________________

Answer the following questions.

1 Explain the term 'out of wind' as used to describe a situation in door frame installation.

____

2 The following statements are either True or False. Write either true or false in the boxes provided.

a Split door jambs are an alternative to standard rebated door jambs. ☐

b Flashing is required under a timber threshold only in an external door frame. ☐

c Exterior doors generally have two hinges. ☐

d Loose pin hinges must be used on external doors and emergency exits. ☐

e Parliament hinges allow doors to open flat back to the wall. ☐

f Deadlocks are key-locked from both sides. ☐

g A mortise latch is harder to fit to a door than a tubular latch ☐

h Storm moulds are only used on timber windows in cavity and brick veneer construction. ☐

i Timber windows should be protected from cement splash and diluted acid brick cleaning materials. ☐

3 Describe two advantages of split door jambs.

a ______________________________

b ______________________________

4 Describe one advantage of metal split door jambs.

____

5 Describe two methods you can use to establish the lock stile on hollow core internal doors.

a ______________________________

b ______________________________

6 Explain the term 'backing off a door' and why it is done.

____

____

____

7 Door hinges should be set down from the top edge of the door 150 mm or 175 mm. What distance from the bottom edge of the door should the bottom hinges be placed relative to the top hinges?

Down from top	Up from bottom
150 mm	________
175 mm	________

You can indicate on the sketch below:

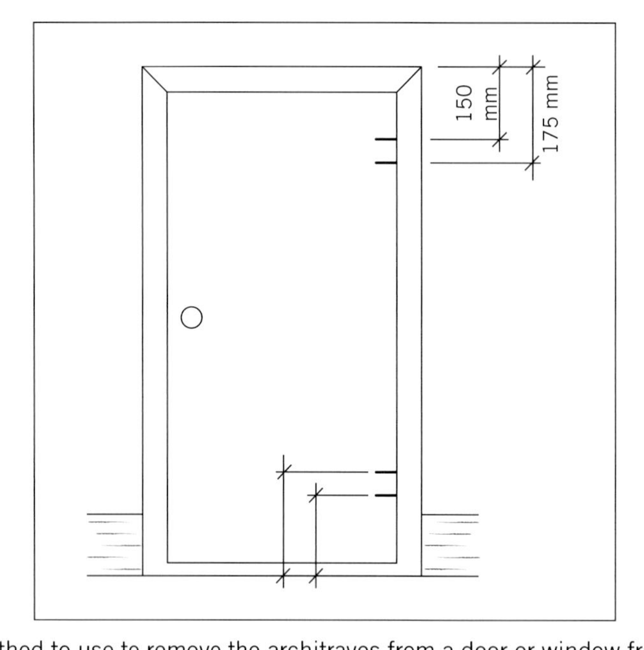

8 Outline the best method to use to remove the architraves from a door or window frame.

____

____

____

9 Describe how the back-set on a door lock is determined.

____

____

____

10 Describe the difference between a lock and a latch.

____

____

____

WORKSHEET 4

To be completed by teachers	
Student competent	☐
Student not yet competent	☐

Student name: ______________________

Enrolment year: ______________________

Class code: ______________________

Competency name/Number: ______________________

Answer the following questions.

1 From the drawn window symbols shown below, identify each type of window and write the correct name in the space below. Obtain a copy of the *Stegbar Timber Windows and Doors Installation Guides* from the company's website (**https://www.stegbar.com.au/information-centre/resources**).

Source: Stegbar, http://www.stegbar.com.au

____

2 The National Construction Code specifies how much natural light should be provided to a habitable room as a percentage (%) of the average daylight factor.

a What is the average daylight factor as a percentage?

____

b Where in the National Construction Code – Volume Two will you find the calculation?

____

3 The selection and installation of windows in buildings is covered by which Australian Standard?

____

4 Aluminium windows are constructed using what type of screws?

____

5 What treatment should be applied to the joints in aluminium frames?

____

6 Name four types of glass that can be used in windows.

a ____

b ____

c ____

d ____

7 What information must be included on the labels on windows to comply with the Australian Standard?

____

____

____

8 Describe the purpose of reveal linings in windows.

____

____

____

WORKSHEET 5

To be completed by teachers	
Student competent	☐
Student not yet competent	☐

Student name: ______________________

Enrolment year: ______________________

Class code: ______________________

Competency name/Number: ______________________

Answer the following questions.

1a According to AS 1684, are you able to notch or house window studs within the window frame opening? Answer yes or no.

____

1b Where in the code did you find the answer?

____

2a Are you able to use common studs at the side of a door opening?

____

2b Explain why or why not.

____

____

2c Where did you find the answer?

____

3a Where can you find sizing information for windowsill trimmers in openings greater than 1500 mm?

____

3b What size sill trimmer is required for an opening 2400 mm to 2700 mm wide when using MGP10 grade timber where the wind classification for the location is N3?

____

4 List the five materials that may be used as glass sealant materials according to AS 1288.

____

____

____

____

____

5 What type of glass must be used in a framed window sash where the sill is 1100 mm above the floor level?

___

6 What type of glass is Grade B safety glass?

___

7 Under what conditions can you use a standard annealed (non-safety) glass mirror?

___

___

8 What sort of glass must be used for louvres in a potential human impact zone?

___

___

___

9 To what does Section 6 of AS 1288 apply? Give an example.

___

___

10 Access to an open window must be restricted in Class 2 to 9 buildings when window opening is less than 1.7 m from the floor. Provide two examples of how access may be restricted.

___

___

INSTALL WET AREA FIXTURES

4

This chapter aligns with the unit of competency, 'Frame and fit wet area fixtures' and also aligns with aspects of the wall framing units of competency. The chapter outlines the knowledge and skills required to build and install the framework required in wet areas so fittings such as baths, showers, vanities and railings along wall linings, and waterproofing systems, can be installed successfully.

In particular the chapter addresses the key competencies of:

- planning and reading job specifications and locating information such as Australian Standards (AS), council regulations and National Construction Code requirements
- reading, developing, interpreting and applying work health and safety/occupational health and safety (WHS/OHS) regulations, site safety plans, state and territory legislation and regulations, and organisational policies and procedures, including quality requirements
- using tools, plant and equipment safely and effectively
- communicating with and working effectively with others and in teams.

Practical projects as a minimum should involve the complete framing and finishing of at least one bathroom complete with linings including a bath framing, and preparation for cabinetry and railing fixing. The project should include flashings, waterproofing and penetrations.

Learning objectives

By the end of this chapter you will be able to:

- identify, plan and prepare for wet area constructions, including reading plans, organising tools, calculating quantities and costs, and understanding the need for barricades, quality control and waste management requirements
- construct suitable timber floor and wall framing and concrete slabs, including the need for control joints in tiled or non-tiled wet areas
- install baths, showers and spas in wet areas, including framing up and venting around baths and spas; and install showers, shower trays, grates and hobs in wet areas
- install appropriate flashings in preparation for waterproofing around baths, spas and showers
- select and install appropriate wall lining materials, including fibre-cement sheeting and wet area plasterboard and alternatives; this includes applying sealants around penetrations such as floor wastes and tap heads and installing joinery items
- apply basic waterproofing techniques, including selecting compatible materials, choosing the right tools, and understanding the steps in applying sealants and membranes.

What is a wet area?

A wet area is defined as any area within a building where water is used, including bathrooms, kitchens, en suites, water closets and laundries.

Water can travel vertically by capillary action to penetrate even the smallest spaces.

It has the capacity to:

- destroy the structural integrity of many materials
- cause decay, mould growth, rust in metals
- create swelling and breakdown of manufactured building boards
- cause some adhesives to fail
- impact personal health and hygiene.

Note: AS 3740 Waterproofing of domestic wet areas does not technically include kitchens as wet areas, but when dishwashers, sinks or taps leak they can cause considerable damage to joinery.

Source: Greg Cheetham

FIGURE 4.1 Drainage works

LEARNING TASK 4.1

Obtain a copy of the current AS 3740 (your college library should be able to assist). In small groups:

1 identify where kitchens are excluded from the Standard
2 discuss what can cause water leaks or water damage in bathrooms, laundries, water closets and kitchens.
3 discuss ways in which damage caused by water leaks can be eliminated or reduced.

During discussion, use any examples from your experience that could assist others in their understanding.

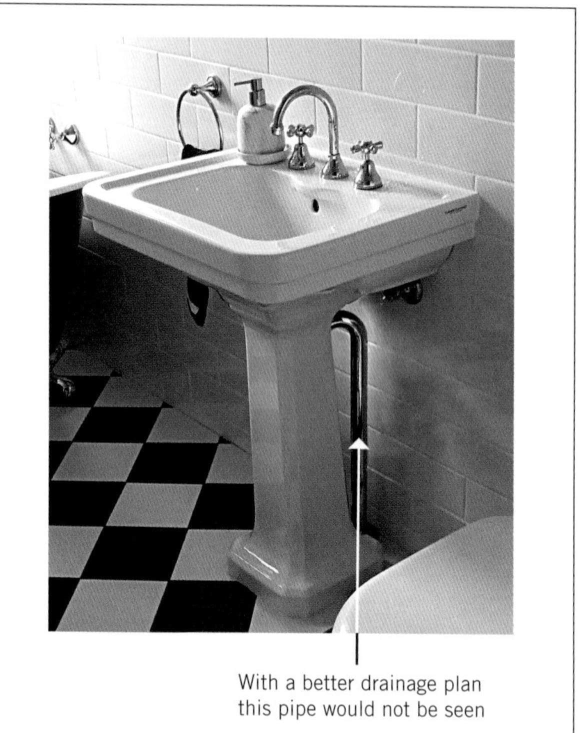

Source: Greg Cheetham

FIGURE 4.2 Hand basin drain pipe

Planning and preparation

In this section we will outline the importance of interpreting plans and specifications and planning for wet area framing and lining.

Well-drawn plans, properly interpreted, can reduce construction time and eliminate costly errors and reworking. In wet areas it is essential to have an accurate drainage plan to be carried out carefully by plumbers. For example, if the drain pipe for the hand basin in Figure 4.1 had been located just 200 mm further off the wall, the drain pipe would not have been visible (see Figure 4.2).

Reading plans and specifications

It is important to read all plans and specifications carefully (see Figure 4.3), identifying:

- materials required for framing, lining and waterproofing the wet area
- fixtures and fitments; for example, baths, **vanity** units and hand basins (ensure they will be available when required)
- finishes specified for walls, ceilings, floors and fitments
- location of the fixtures and fittings (these can be marked out in situ as the work progresses)
- environmental and quality control issues, such as when specialist chemicals are used – read the safety data sheet (SDS) before use and ensure storage and disposal of wastes/residue is carried out correctly
- WHS/OHS issues – risk assessments and safe work method statements (SWMS) need to be completed, recorded and stored before work commences, and the SDS for all materials being used in the construction need to be available onsite. Different materials are used in construction and need to be safely handled using the correct PPE and manual handling techniques. Organise the work area to reduce foot traffic to and from work vehicles and

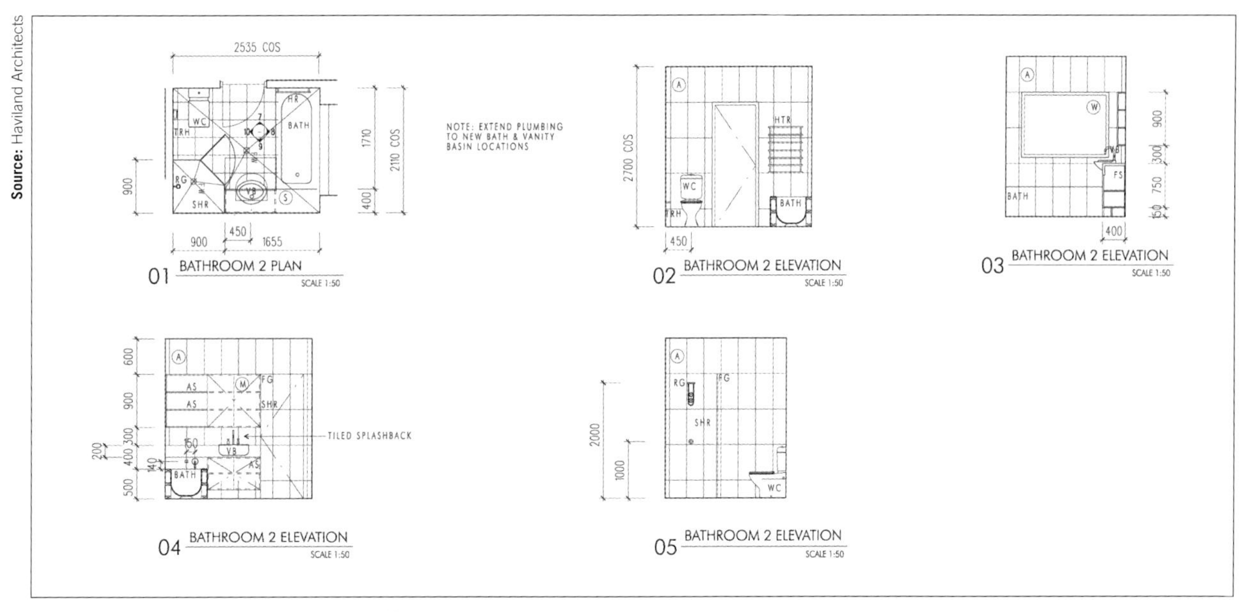

FIGURE 4.3 Example of plan and specifications

storage areas to reduce trip hazards and maximise efficiency.
- different PPE and safety equipment that will be needed at various times depending on the stage of construction. Always be alert and aware of the surroundings. Review the correct PPE and safety equipment needed for different work activities. Masks will be needed for waterproofing or when cutting timber but may not be used at other times. The SWMS and JSA will provide WHS safety information to comply with laws, regulations and local requirements. Ensure these are followed and understood before work commences. PPE must be worn at all times, especially the use of gloves when handling timber.

Use the hierarchy of hazard control to reduce identified risks and safety hazards in construction work.

Calculating quantities and costs

During the planning and preparation process, you will need to calculate and cost the quantity of materials required from the plans and specifications. These are the steps:

1 Make a list of all tools and materials required.
2 Calculate material quantities accurately.
3 Research current prices for materials.
4 Make and check calculations.

Note: Special tools like wet saws, grinders and wall chasers may need to be hired to complete work in wet areas. Hire costs need to be taken into consideration early in the construction process.

The majority of materials quantities will have been calculated for the entire construction during the original quoting and ordering process. However, you may need to calculate the amount of timber required from site stocks when constructing the wet area framework. You may also need to calculate the additional materials for bath support framework, additional noggins and frame trimming, for example.

The basic calculations you will need to work out include the following.

Calculation of **solid timber quantities** in lineal metres:

Total lineal metres ÷ stock length

Calculation of **sheet materials** in m^2:

Perimeter of room × wall height (minus door and window openings) ÷ m^2 of the appropriate sheet size.

You will also need to be able to calculate the surface area to be waterproofed and convert this information into the quantity of waterproofing materials required. This will require you to be able to calculate **coverage rates for liquids** such as membrane materials, fibreglass matting (if required) and paintable surface sealants; that is:

total m^2 ÷ coverage rate of the liquid
e.g. 5.0 m^2 ÷ 10.0 m^2 = 0.5 litres

LEARNING TASK 4.2 READING PLANS

From the plan in Figure 4.3 identify the following features:

1 Drawing 01 Bathroom 2 Plan – what scale is used?
2 What is the height of the vanity top?
3 What do the letters HTR stand for on elevation 02 bathroom 2 elevation?
4 What would the shower screen need to be made from if it is to comply with AS 1288 Glass in buildings – Selection and installation and AS/NZS 2208 Safety glazing materials in buildings?

COMPLETE WORKSHEET 1

Barricades and signage

While the use of **barricades** is not generally necessary during wet area construction, signage should be used when wet area **flashings**, **membranes** and lining material finishes are being applied (or are drying) or are being sanded. During renovations of wet areas, barricades for sealing off the work area may be necessary to prevent the escape of dust and fumes to other areas. When renovating you will need to protect those parts of the building that are not being renovated, so it may be necessary to put down drop sheets. A range of self-adhesive films have been developed to protect finished floors, especially carpets (see Figure 4.4).

Standards and environmental and quality controls

Under the National Construction Code, wet areas must be designed and constructed to prevent damage from water brought into or taken from the construction via plumbing and drainage works. AS 3740 Waterproofing of domestic wet areas provides guidance on how this is to be achieved via a combination of materials and procedures. This Standard needs to be read in conjunction with AS 1684 Residential timber framed construction. You also need to follow the manufacturer's specifications and installation guidance for any materials or products being installed in order to maintain warranties. Other Standards and state regulations will need to be considered during wet area construction (see the waterproofing resource links at the end of the chapter).

Prepare a quality control checklist so you can check the quality of materials and practices during the construction and finishing stages. An example of a simple checklist is shown in Table 4.1.

Source: Protective Trade Group

FIGURE 4.4 Stick 'n' protect floor film

Tools

The tools required for the process will depend upon the construction materials being used. Timber-framed construction will require the standard carpentry tools, while steel-framed construction will require a range of additional specialist tools like metal shears or tin snips to form cut-outs in the studs.

Cutting of **fibre-cement lining** materials will require sheet guillotines (more commonly called fibro cutters) or cutting may be carried out using the score and snap method (see Chapter 8). Time-saving specialist power tools and adapters have been developed to work with

TABLE 4.1 Quality control checklist

Note: Safe work method statements to be completed, checked and filed before work commences.		
1	Floor frame in place and to AS 1684	☐
2	Compressed fibre-cement sheeting in place and correctly fastened	☐
3	Wall frames in place, plumb and square, and all noggins and trimmers in place	☐
4	Plumbing and drainage in place and leak tested	☐
5	Electrical roughed in correctly	☐
6	Flashings, yokes, collars installed correctly	☐
7	Lining materials installed according to specs and manufacturer's guides	☐
8	Waterproofing applied according to specs and manufacturer's guides	☐
9	Penetrations sealed	☐
10	Floor and wall finishes completed to specs and tested for correct falls	☐
11	Fitments installed in correct positions and sealed to standards and specs	☐
12	Plumbing work commissioned and completed	☐
13	Painted surfaces completed to specs	☐
14	Electrical fitments commissioned and installed	☐
15	Quality control checklist and defects list completed and distributed	☐
Site supervisor signature: ____________		
Date: ________		

some of the difficult materials found in wet area construction; for example, mechanical fibre-cement shears and metal shears are available in corded and battery operated models.

Complete a visual inspection of all tools and equipment to ensure safe operation. Regular checks and maintenance of tools and equipment should be done each time they are used. Tools and equipment should operate as per the manufacturer's guidelines. If faults are found, follow workplace procedures to record problems, tag the tools or equipment, and relocate them away from other functioning tools.

Wet area construction

In this section timber and steel framing for wet areas, including control, is discussed.

Timber-framed floors

Timber-framed floors have the advantage that plumbing can be retrofitted and aligned accurately after the shower base has been fitted. To identify the best orientation for the base, an under-floor examination before cutting is a good idea.

Simply set the base on the floor, before the bedding mix has been placed, and mark the centre of the floor waste, drill and cut-out for the waste pipe.

Prepare the bedding mix and set the tray in position, making sure sufficient bedding mix is used to provide full support for the bottom of the tray. Use a spirit level to level the tray in both directions.

Timber-framed wet area floors consisting of bearers and joists will usually have a compressed fibre-cement floor. This material is difficult to work with, and usually requires a diamond saw to cut the sheets and masonry bits for drilling and forming penetrations. Angle grinders with masonry wheels are used for bevelling and jointing. The sheets are screw fixed using stainless steel or (at minimum) cadmium-plated screws. Do not glue fix the sheets as any movement in the joists when glued can cause waterproofing materials to delaminate.

Many bathrooms today are designed with a reduced floor level so that tiling will finish flush with other connecting floor finishes, and to avoid trip and fall hazards. This is done by cutting down floor joists in the bath area to allow for cement toppings and the tile thickness. You need to ensure that additional supports (extra piers) are provided before laying wet area flooring. An alternative is to reduce the height of the joists but increase their thickness (see Figure 4.5). In any

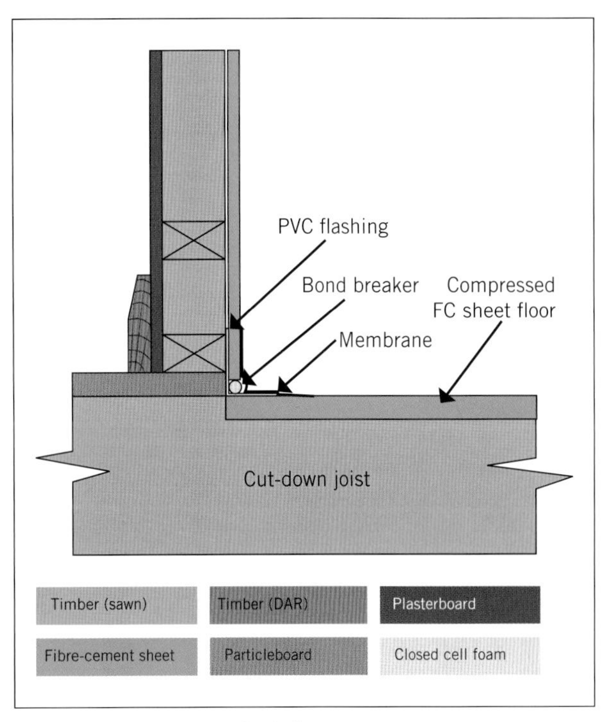

FIGURE 4.5 Cutting in a flush floor

circumstances, the subfloor framing must comply with AS 1684 Residential timber-framed construction and the National Construction Code.

Concrete slabs

Concrete slabs should be formed up to incorporate a step-down in the slab to allow the finished floor levels to be seamless throughout the building.

LEARNING TASK 4.3

CONCRETE IN RESIDENTIAL CONSTRUCTION

Obtain a copy of Cement Concrete & Aggregates Australia's *Guide to Residential Floors* (1998, updated 2003) and answer the following questions:

1. Identify the Australian Standard that covers the design principles for domestic house floors and footings.
2. Describe three different design types of concrete raft slabs used in residential construction.
3. Discuss the reasons for the different design types.

Timber wall frames

Wall frames are typically made from timber or steel. Well-seasoned timber should be used to avoid excessive shrinkage. Section sizes should conform to specifications and AS 1684. Poorly seasoned timbers shrink significantly and create stresses in the structure that can cause membranes and joints to crack, thus allowing the penetration of moisture into cavities and potentially causing rot to timber frames and rust in steel frames.

Standard framing practices should be maintained; that is, 600 mm maximum stud spacing for non-load-bearing walls and 450 mm maximum for load-bearing walls.

Do not stagger noggins behind sheet material laid horizontally (see Figure 4.6). These joints need full support as any movement caused by vibration or impact can cause the joint to allow water penetration into the cavity.

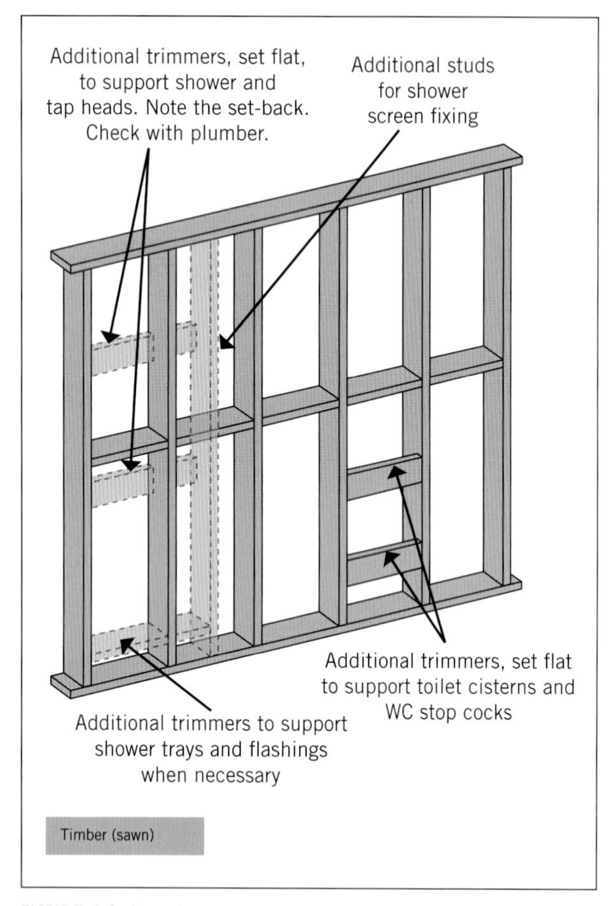

FIGURE 4.6 Noggin and trimmer positions

Where baths, **spas** and shower trays are being installed, studwork may need to be checked out (to a maximum of 20 mm) to allow the lip of the unit to fit in far enough to permit the wall lining to cover the lip and provide appropriate flashing (see Figures 4.7, 4.8, 4.12 and 4.13).

A three-stud cluster should be used at all internal and external wall frame corners. Additional studs and noggins should be incorporated into the frame where fittings and fixtures are being installed; for example:

- trimmers on edge to support plumbed-in fixtures such as toilet cisterns, vanity basins, showerheads, taps and laundry tubs
- extra studs to support shower screens and shower modules
- a row of noggins 25 mm above the shower tray or bath to provide sheet fixing.

Ensure that support noggins are sufficient and properly located to accept the range of fitments that will be incorporated into the project. This requires consultation with the various contractors and specialist trades that will come onsite when the wet area is ready.

Steel stud framing

Steel framing must comply with AS 3623 Domestic metal framing.

Construction techniques are similar to those for timber-framed construction, with additional noggins and cut-outs required to accept fitments (see Figure 4.8).

A more modern approach in steel-framed wet areas is to not cut stud work or any additional noggins for fixing of fittings, other than those required to support **stop cocks** for tap heads and toilet cisterns. The alternative is to line the walls with 12 mm waterproof plywood (Type A bond) before lining with 6 mm fibre-cement sheeting or wet area plasterboard to allow for random fixing points for fittings and fixtures (see Figure 4.9). In fixing out this way there are no restrictions on fitting or retrofitting railings, screens, mirrors or cabinetry to walls.

In extreme climates you may have to install insulation material between the steel frame and the lining to prevent problems associated with thermal movement (see Figure 4.10).

Control joints

Control joints must be installed in large spans of wall to allow for structural movement. In non-tiled areas, control joints should be no more than 7.2 m apart.

Where a tiled area is to be constructed, the control joints should be placed at not more than 4.2 m spacings. The joints are constructed using double studs plus an allowance for expansion and contraction, and should incorporate the lining material and tiles (see Figure 4.11).

COMPLETE WORKSHEET 3

Baths and spas

There is a large range of modern baths and spas. Some are free-standing, which eliminates a lot of the work in checking out frames (see Figure 4.12).

Modern baths and spas are generally moulded from acrylic or fibreglass; some are pressed metal, while others are porcelain finished and with a lip around the top to provide a finish and a flashing. They can be ordered in different configurations to suit different bathroom layouts. Most are made to fit into a corner position. They have closed sides and/or ends that are recessed into the wall framing, as discussed earlier, to

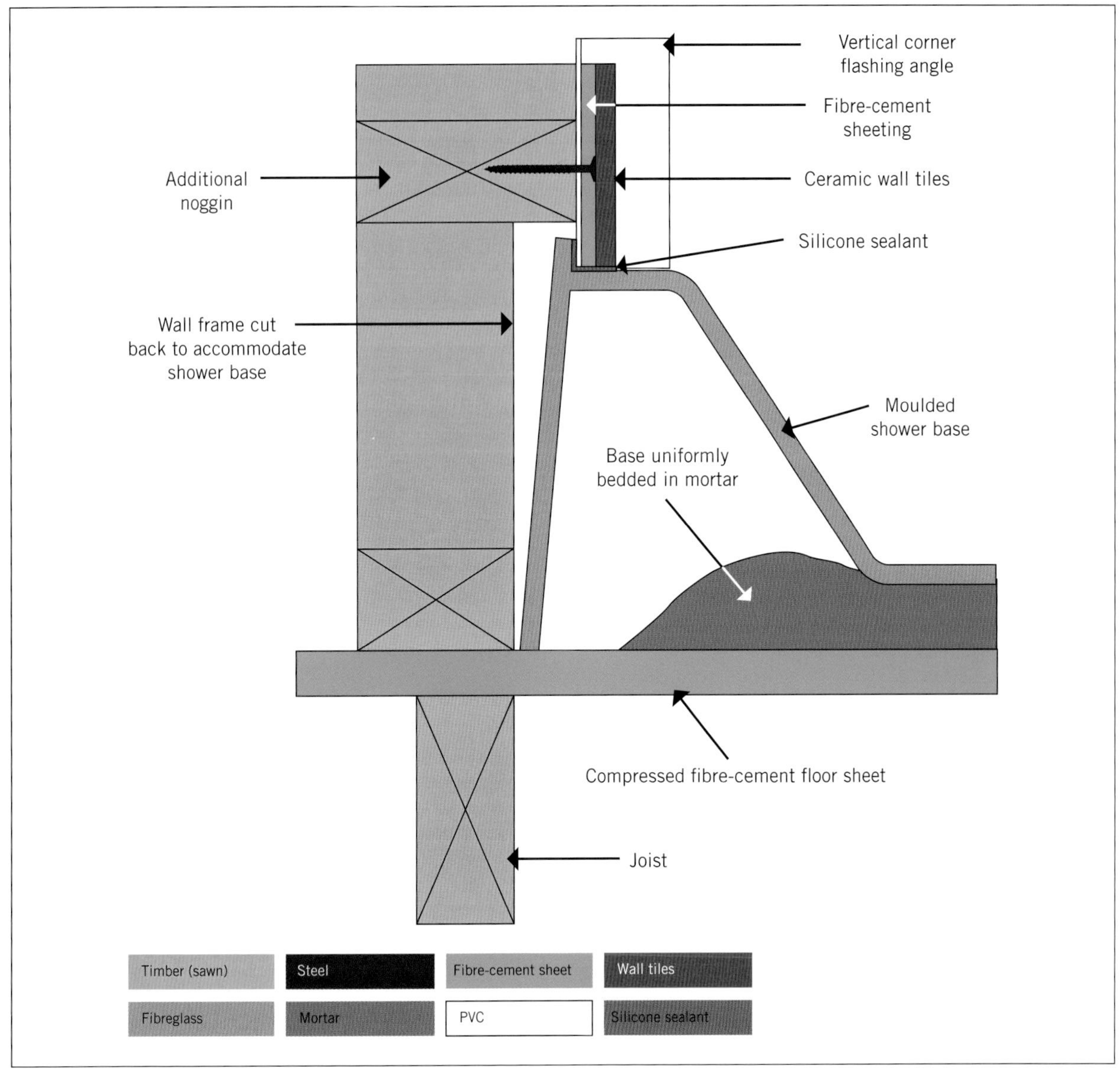

FIGURE 4.7 Cutting in a shower tray to a timber-framed wall

allow the wall lining materials to fit over the lip and seal the wall side and cavity. This makes them a built-in fixture as opposed to a fitting that can be removed.

Baths and spas can be made from a range of other materials such as:

- pressed metal with a baked enamel surface (these are relatively thin metal and require perimeter support)
- stainless steel
- cast iron with a baked enamel finish.

Fibreglass and acrylic baths and spas require additional timber perimeter support, while the base can be supported using a dry sand cement mix, and occasionally masonry materials; alternatively, expanding foam can be sprayed into the void while the bath is in place (see Figures 4.13 and 4.14). Do not apply the foam into enclosed voids.

Protecting fixtures and finished surfaces

There are some simple precautions that should be taken to protect fixtures. You should:

- protect the surfaces of baths and spas from damage, so prior to installation be sure to keep them stored safely in the manufacturer's packaging
- take care after installation to prevent chipping, scratching or cracking by not removing any manufacturer's applied coatings; and provide additional cover when necessary by completely covering the unit with plastic or other packaging materials

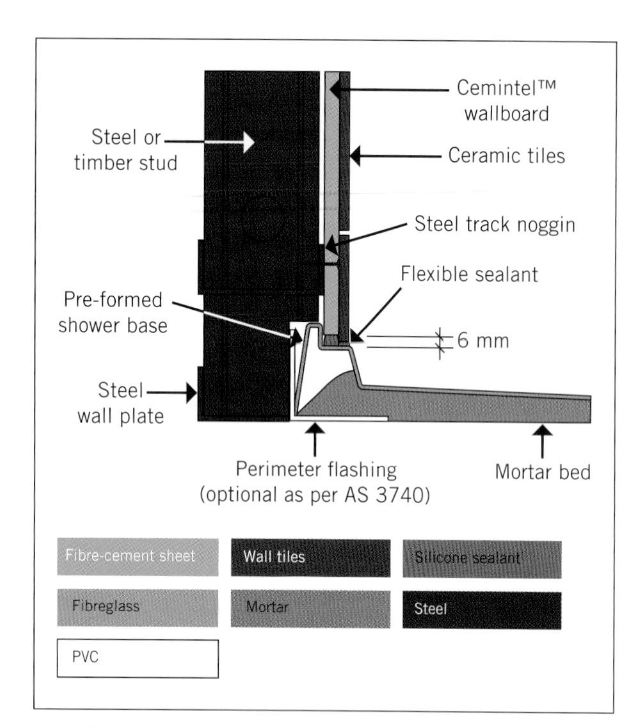

FIGURE 4.8 Pre-formed shower base on concrete floor

FIGURE 4.9 Wet area walls fully lined with 12 mm A bond plywood before fibre-cement sheeting

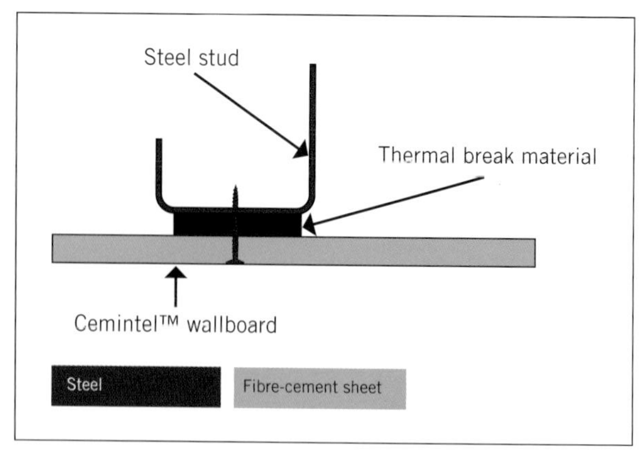

FIGURE 4.10 Thermal bridging on steel framing

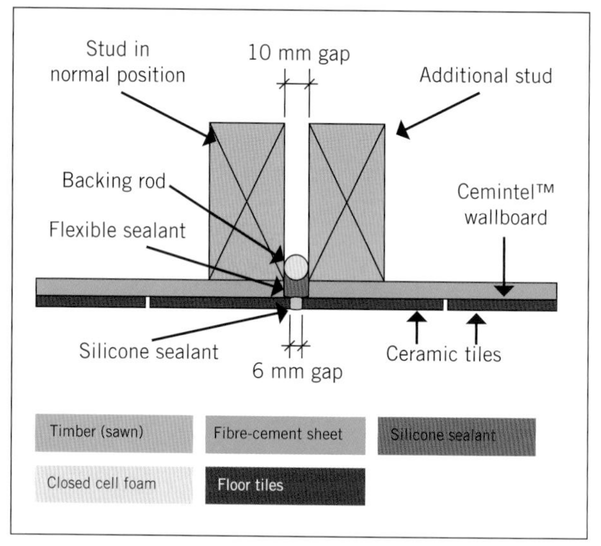

FIGURE 4.11 Control joint for tiled areas

FIGURE 4.12 Highgrove free-standing acrylic bath

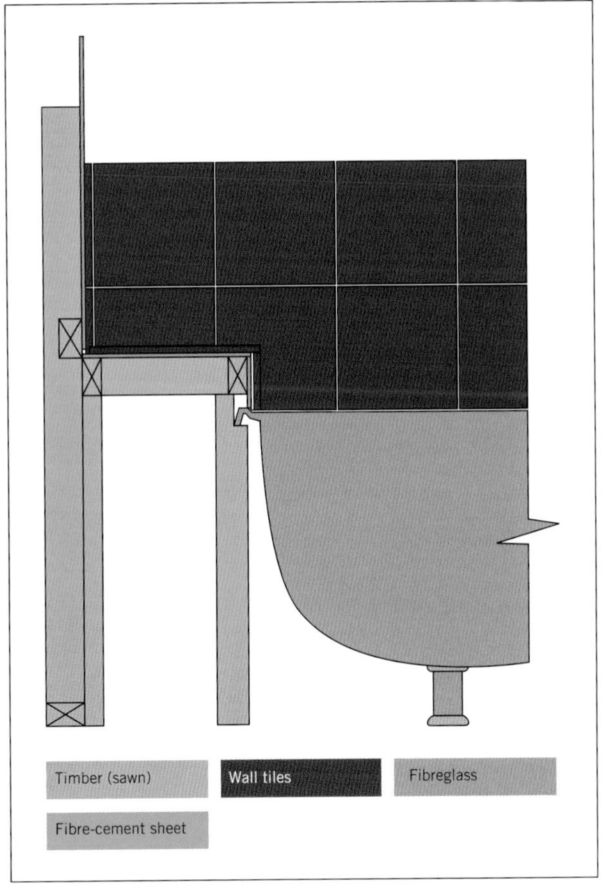

FIGURE 4.13 Flat hob end

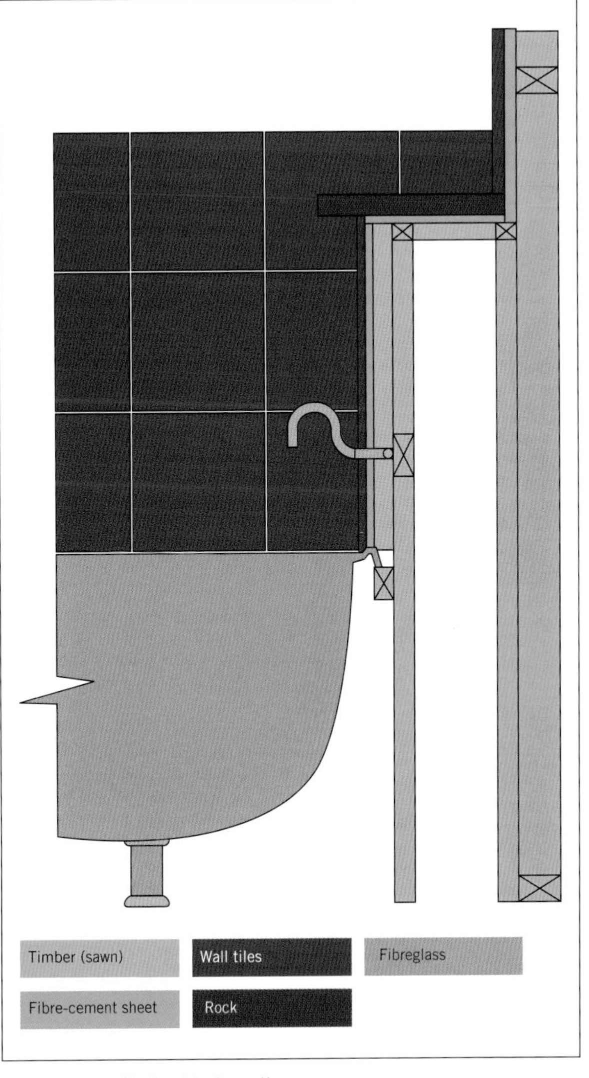

FIGURE 4.14 Raised hob walls

- never stand on or in the bath during the finishing processes.
- Use plywood or 8 mm Coreflute to protect the finished surfaces (see Figure 4.15).

Framing around baths and spas

Timber framing around baths and taps follows typical AS 1684 Timber-framed construction practices.

Most regular baths require approximately 450 mm in height for framing, but check the depth requirements with the manufacturer or measure the depth. Allow an additional 50–100 mm for **sparging** (bedding in dry mortar mix) underneath.

Continuous edge support needs to be provided for baths on all sides. There are a couple of methods used (see Figure 4.16).

The old method is to cut a notch in each stud no deeper than 20 mm and insert a continuous (ribbon) trimmer around the perimeter to support the bath or spa on the wall sides. Make sure the bath or spa fitting is level. A noggin or continuous (ribbon) trimmer will be required above the bath height to provide a means of fixing the wall lining.

The modern method is to build a complete frame surround for the bath to sit in. This has some advantages:

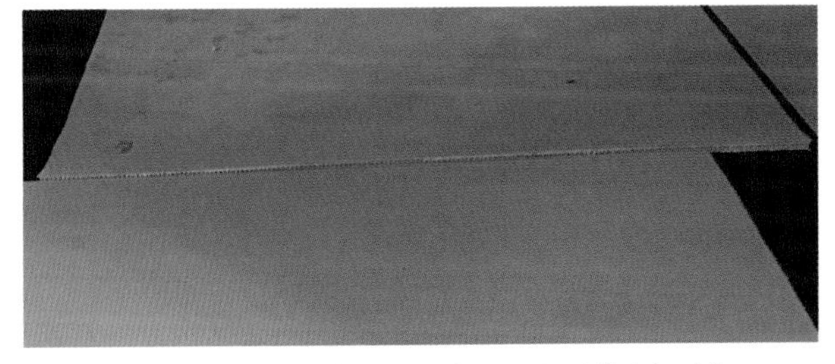

FIGURE 4.15 8 mm Coreflute used to protect finished floors

- No studs are cut, no additional noggins need to be inserted and no support trimmer has to be inserted, although a fixing trimmer is inserted into the bath frame.
- The frame is lined and a membrane is applied according to AS 3740 Waterproofing of wet areas within residential buildings.
- Walls can be fully lined to the floor, removing the potential for moisture to enter the cavity.

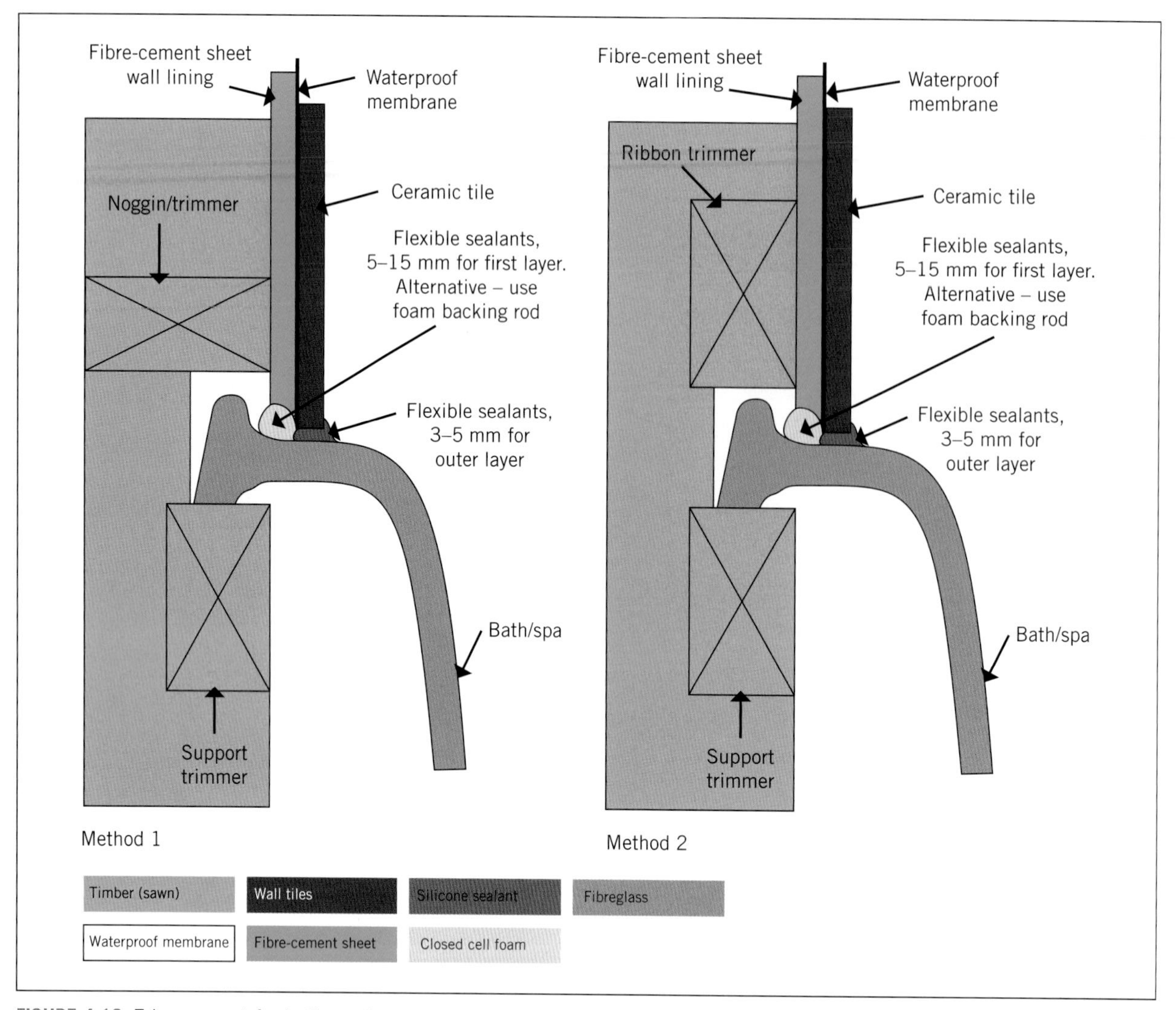

FIGURE 4.16 Edge support for baths and spas cut into wall

An alternative is to enclose the bath in masonry product. A membrane is applied to the Standard AS 3740 Waterproofing of wet areas within residential buildings.

Alternatively, the bath can be supported by trimmers, a continuous batten inserted into the frame on the enclosed side (see Figure 4.17) and by masonry products such as bricks, cement blocks or autoclaved aerated cement (See Glossary of terms) on the exposed faces.

To determine the height and lengths of the frames exactly, set the bath in position on the wall trimmers and level the bath in both directions. Plumb down the position of the edge of the bath to locate the frames on the floor. Decide if the bath is to sit on top of the hob frame, finish flush or hang over the edge slightly (see Figure 4.20). From your plumbed marks set out the outline of the frames on the floor. Be sure to allow for the thickness of the lining materials, the tiles and the adhesive. Set out your frame sizes using 450 mm stud spacing.

You may need to make adjustments to the top plate or allow for a support batten if the rim of the bath is not wide enough to use full-sized framing materials at the top (see Figures 4.18 and 4.19).

Venting

Venting around baths and spas needs to be considered in any frame design. Ventilation should comprise two vents built into the lining materials to provide sufficient ventilation to eliminate a build-up of condensation in the cavity space around the bath.

When spas are installed, pumps will also be installed and these will require maintenance from time to time. The location of access panels needs to be considered if it is not possible to locate the pumps externally.

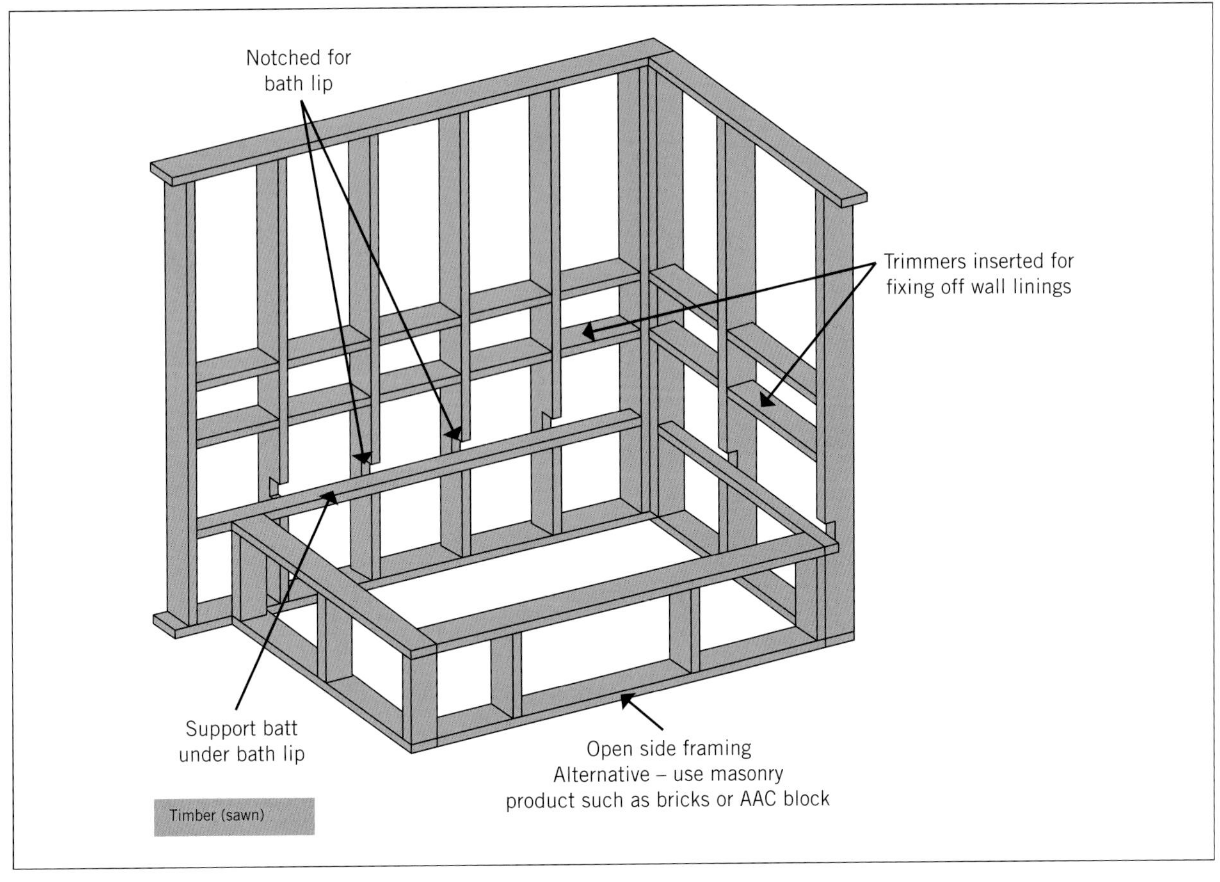

FIGURE 4.17 Bath fully framed

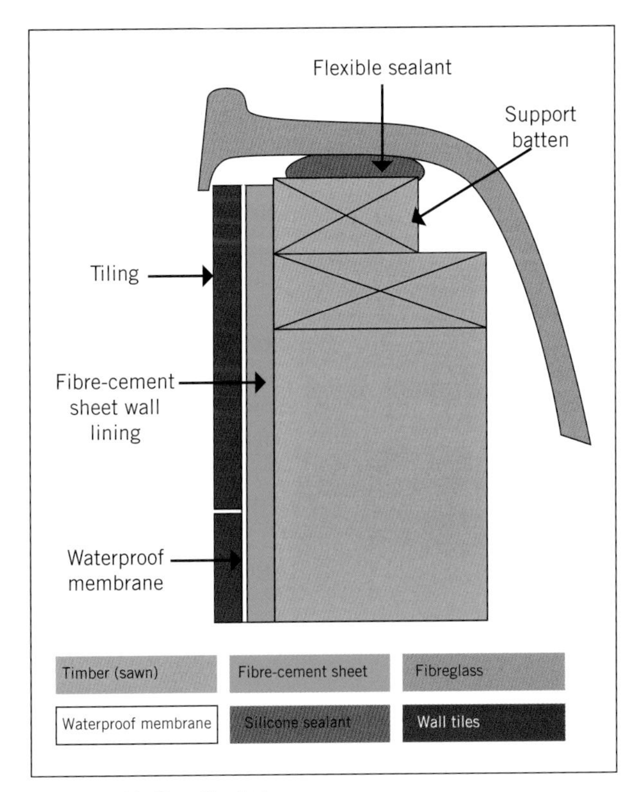

FIGURE 4.18 Over the hob

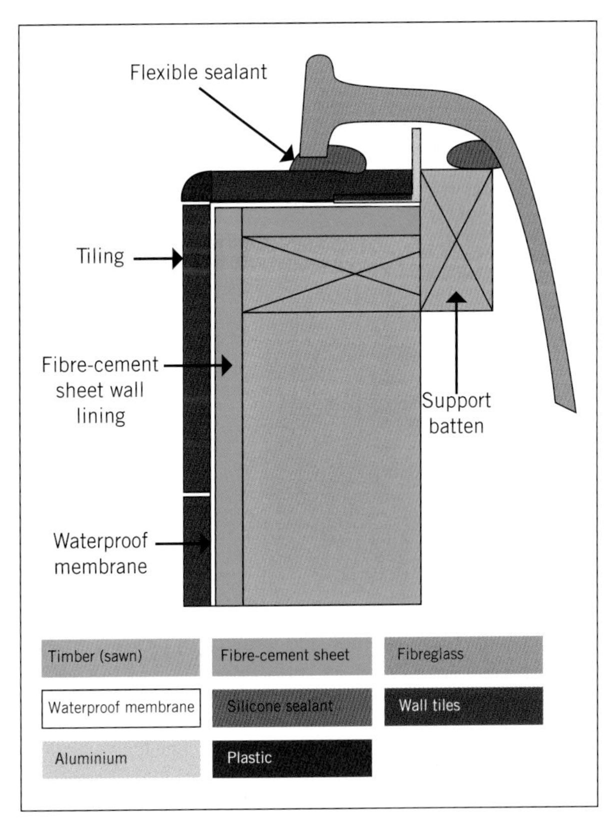

FIGURE 4.19 On top of the hob

Hob walls for baths

Hob walls for baths are walls, usually tiled, at the open end (or both ends) of a bath. Their purpose is to close off the area between the bath and the walls or to create a wall to support plumbing or to separate the bath from the shower recess. Shower screens can be supported on hobs or hob walls.

The height of the hob should be set out to full tile height using a gauge rod (see Figure 4.20). The hob can be timber framed and lined or of masonry construction (see Figure 4.21).

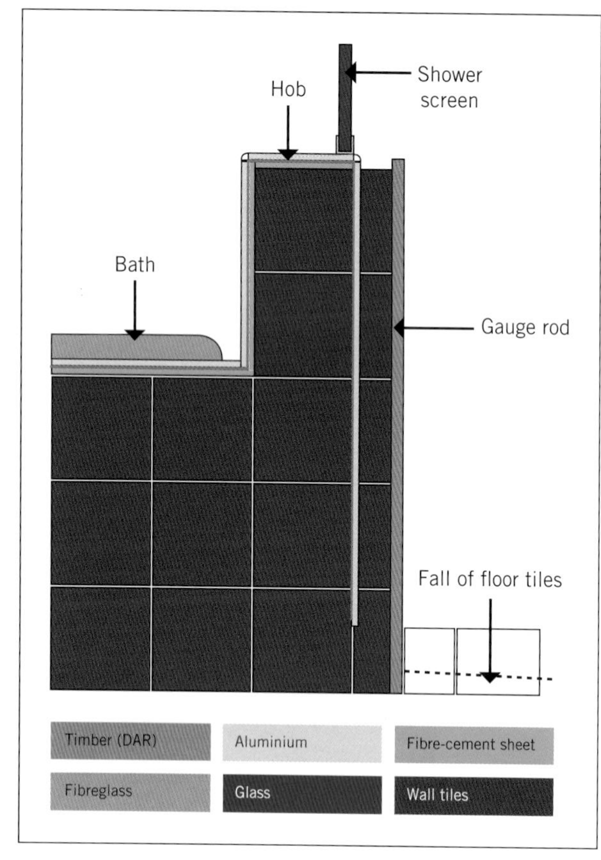

FIGURE 4.20 Hob wall

Note: It is not recommended practice to use timber for shower hobs.

Showers

Shower areas

Shower areas can be enclosed or open (unenclosed). In the regulations, an unenclosed shower area extends 1500 mm horizontally from the shower connection on the wall and up the wall to a height of 1800 mm from the finished floor (see Figure 4.22).

An enclosed shower area is bounded by walls and screens up to a height of 1800 mm from the finished floor. Screens with hinged or sliding doors contain the spread of water within the enclosure.

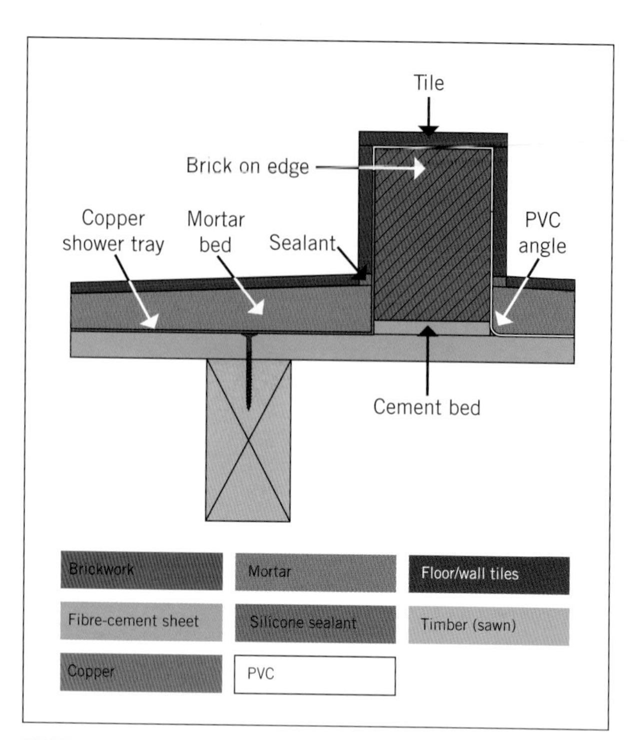

FIGURE 4.21 Shower hob showing

The provision of shower hobs in renovations of older homes is governed by AS 3740 and can only be used in bathrooms not subject to the Standard for Livable Housing Design (i.e. new residential construction in at least one bathroom). See the National Construction Code – Volume Two, H8P1 Livable housing design.

Shower base

A shower base is a pre-formed, pre-finished platform or module that is installed as the finished floor of a shower (see Figure 4.23), or a complete shower module or compartment that is then connected to appropriate drainage. Shower bases must sit on a bed of mortar for support. They can be made from acrylic or enamelled pressed metal.

The internal corner of the frames should be notched out accurately, as shown earlier in Figure 4.7, when the frames are being constructed and before they are erected.

When preparing the frames, ensure that an appropriate allowance (15–20 mm) is made for the thickness of the shower base and the shower-base bedding mix.

The thickness of the lip of the tray is marked from the face of the framing, ensuring that the lining material will fit over the lip of the tray when it is installed. Make sure that the bottom plate is trimmed back to width only in the area where the tray is to fit. Studs are notched in and noggins prepared and fitted to allow for wall lining fixing.

Positioning of trays on concrete slabs requires accurate setting of the drainage connection penetrations before the slab is poured. A special offset fitting (see Figure 4.24) can be used in conjunction with the 100 mm sewer pipe to allow correct alignment.

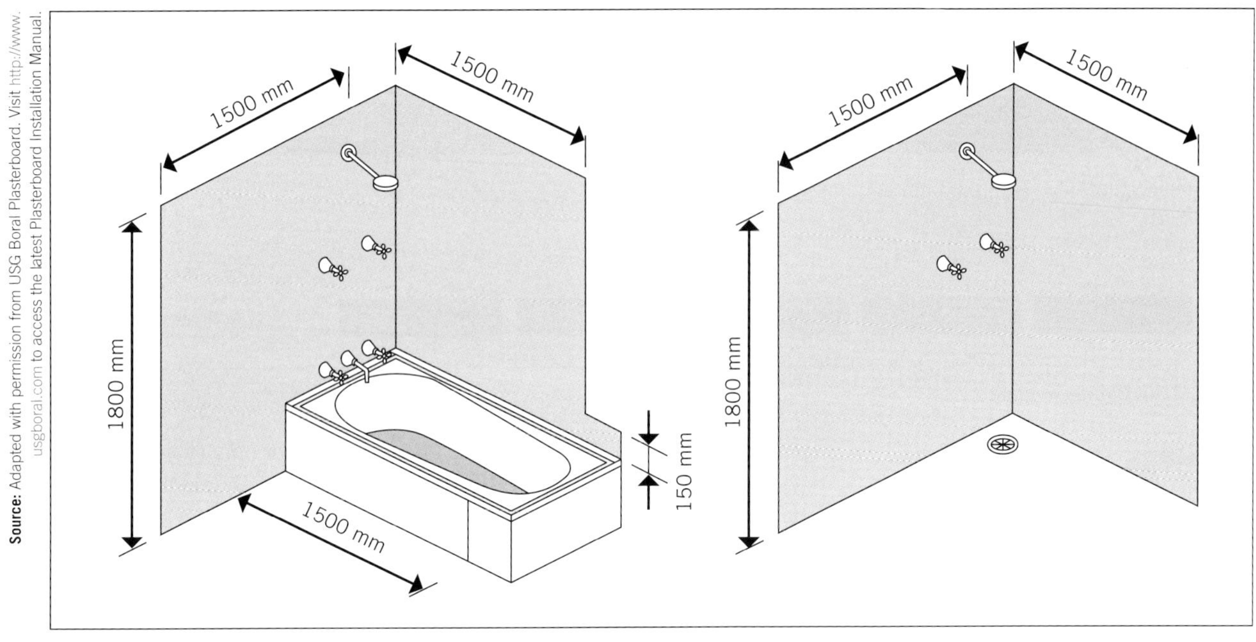

Source: Adapted with permission from USG Boral Plasterboard. Visit http://www.usgboral.com to access the latest Plasterboard Installation Manual.

FIGURE 4.22 Unenclosed shower over bath and unenclosed shower

Source: Ed Hawkins

FIGURE 4.23 Typical shower tray

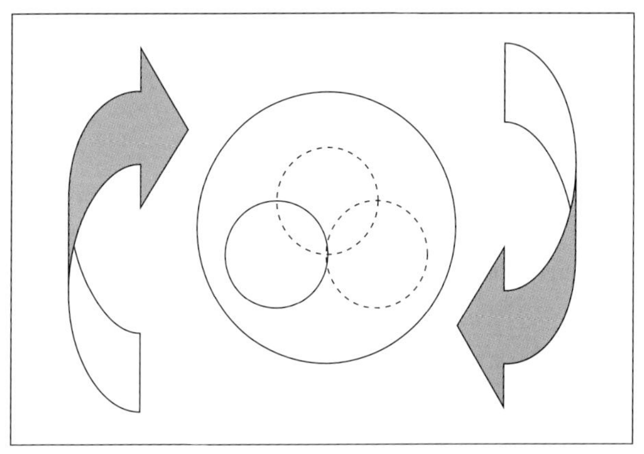

FIGURE 4.24 Adjustable cap

When installing the tray, ensure that it has sufficient fall to the waste.

Complete shower modules are an alternative to shower bases. The modules can combine wall linings, multiple shower fittings and doors in one complete unit. These units need to be fitted to the manufacturer's specifications and require flashing in different ways. They can eliminate the need for membrane waterproofing in shower areas as they are completely sealed units. However, some waterproofing is still required.

Shower grates

Shower grates are again becoming a common feature in bathrooms. The concept has been around for centuries in the form of open drains, but in more recent times they dropped out of favour for hygiene reasons. Their redevelopment and inclusion in bathrooms comes from the need for access for people with disability, and they are now being

Source: Greg Cheetham

FIGURE 4.25 Bathroom grate styles

incorporated into hospitals and aged care facilities (see Figure 4.25). Shower grates allow wheelchair access, reduce the risk of trips over hobs and edges, and fit with modern design concepts of clean lines and minimalism.

Care needs to be taken when installing shower grates as they require a fall to be built into the tray for run-off. The tray is set in a mortar bed or packed up and sealed (see Figure 4.26).

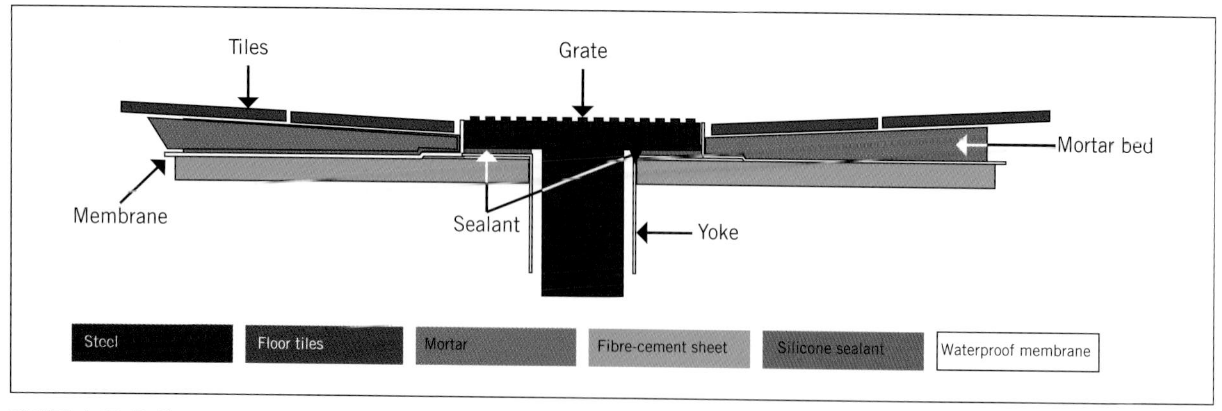

FIGURE 4.26 Bathroom grate

Flashings

In wet areas, flashings are usually required at wall and floor intersections. As a minimum, the flashing must extend to 25 mm above the finished floor level. There are various types of flashing available. The most common is a PVC (polyvinyl chloride) angle flashing, which is available in a range of sizes; 50, 75 or 100 mm × 50 mm are the most commonly used. The appropriate size depends on the floor finish and the height to the bottom of the wall lining. You will need to be familiar with AS 3740 Waterproofing of wet areas within residential buildings.

Figure 4.27 shows the typical requirements for installing a pre-formed shower base.

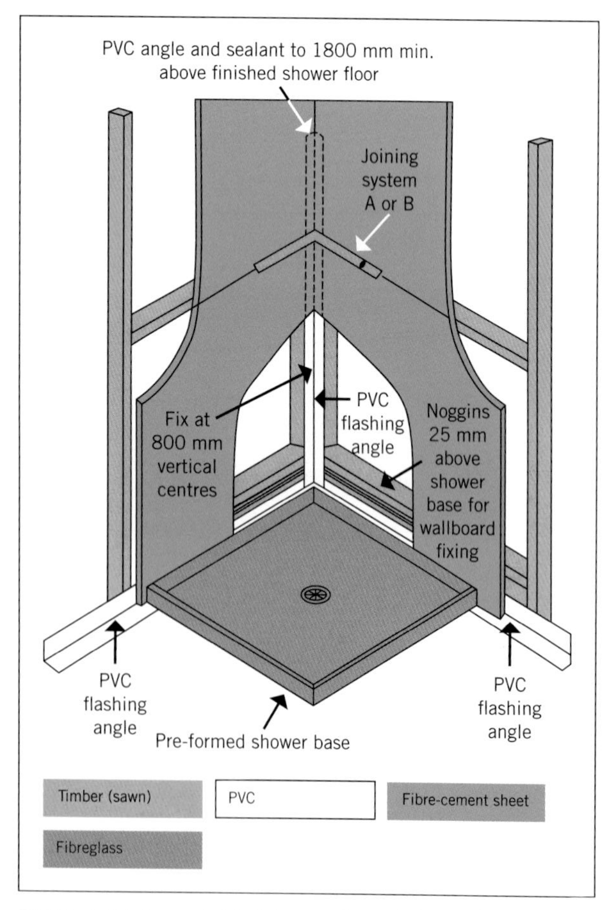

FIGURE 4.27 Typical detail for pre-formed shower base

The flashing is glued to the floor only to maintain its waterproof integrity. Joints in length should be lapped and corners are cut and folded in a similar way to window flashings to maintain a continuous barrier at internal and external corners (see Figure 4.28).

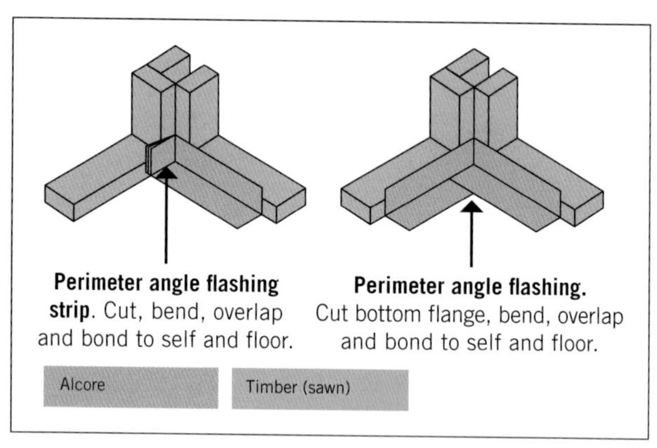

FIGURE 4.28 Perimeter flashing

A synthetic, rubber-like material may be adhered to the floor using a water-based two-part epoxy adhesive. An alternative is a fibreglass matting glued around the perimeter, up vertical corners and over floor joints using the manufacturer's recommended adhesive or sealant.

Vertical corners can also be fastened using plasterer's metal angle or PVC flashing. Using plasterer's metal angle secures the corner and reduces the risk of the corner joint (tiled grout lines) cracking with movement. Make sure it has a corrosion-resistant coating. Ensure the angle extends down over the floor PVC angle and also over and into the shower tray where applicable.

Note that there are two types of shower trays:

- modern pre-formed shower bases, which include waste fittings
- copper shower trays with folded and soldered corners. In the past, these were installed and tiled over. They can still be used but are not recommended as they are difficult to seal.

The modern method is to tank the entire recess with a membrane material. Either a painted or fibre-reinforced membrane is generally used; alternatively, a clear vinyl membrane can be constructed in situ, with all corners and joints being plastic welded to form an impervious membrane.

Linings

In most circumstances, wet areas are lined with water-resistant plasterboards or fibre-cement sheeting and are waterproofed and tiled. Other materials such as timber, solid plastic laminate sheeting (e.g. Lamipanel from Laminex Industries) and galvanised iron are suitable alternatives, but will also have special installation requirements to stop water penetration.

The main advantages of plasterboard are its cost and ease of installation. The disadvantage is that some brands require additional surface treatment to seal them to improve their water resistance. Fibre-cement sheeting is more difficult to cut but has a tougher surface that is more water resistant than plasterboard.

As mentioned earlier in this chapter (see 'Steel stud framing') it is common practice to line framed-up wet areas with 12 mm A bond plywood prior to the installation of other lining materials to make fixing of fittings easier and reduce the need for measuring, cutting and installing noggins.

The sheets will be fixed in line with normal practice. Use a wallboard adhesive applied to studs, plates and noggins and then screw (for metal frames) or nail (timber frames) to finish. Consult with your board manufacturer for recommended fixing methods.

Plasterboard

The major manufacturers of wet area plasterboards in Australia and their products are USG Boral (Wet Area Board) and CSR (Gyprock Aquachek).

All of the manufacturers have a range of sheet sizes from 2400 mm to 4800 mm in length by 1200 mm or 1350 mm wide. Typical thicknesses are 10 mm and 13 mm. All makers use a distinctive blue or green colouring in the sheet linings to distinguish the moisture-resistant forms from other plasterboards.

Most boards are made with recessed edges for jointing in wet areas (see Figure 4.29). All boards have a water-resistant face side and back lining and a water-resistant gypsum core.

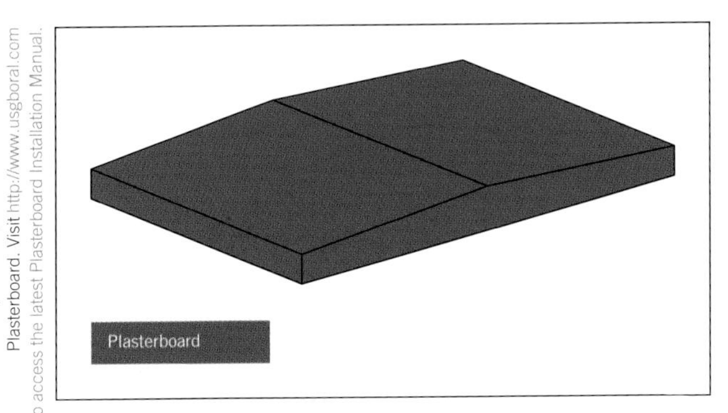

Source: Adapted with permission from USG Boral Plasterboard. Visit http://www.usgboral.com to access the latest Plasterboard Installation Manual.

FIGURE 4.29 Recessed edge

Fibre-cement sheets

Fibre-reinforced cement-based products should conform to AS 2908.2 Cellulose cement products – Part 2: Flat sheets, making them impervious to water penetration and rot.

The main manufacturers of wet area fibre-cement sheeting in Australia are James Hardie (Villaboard®) and CSR (CeminSeal™). Typical sheet sizes are as follows:

Width (mm)	Length (mm)
900	2400, 3600, 4200
1200	1800, 2100, 2400, 2700, 3000, 3600, 4200
1350	2400, 3000, 3600

Typical sheet thickness is 6 mm, with a weight of approximately 9 kg/m^2 at 6 mm.

Fixing plasterboard

Both plasterboard and fibre-cement sheeting are fixed using similar methods, spacings and fastenings. Basic fastening positions are indicated in Figure 4.30.

You should only use plasterboard in wet areas when the surface is to be tiled to a minimum height of 1800 mm. It should be installed to comply with AS/NZS 2589 Gypsum linings – Application and finishing.

When fixing plasterboard in tiled areas, use screw fasteners only (without adhesive) to allow for movement. Electric or battery-operated screwdrivers can be torque adjusted to prevent screws breaking the paper face lining. The risk of damaging the paper face lining is too great to allow for nail fixing – damaged face linings expose the inner gypsum core to moisture.

Note: Nails can only be used to fix plasterboard in areas where it is **not** going to be tiled.

Nails and screws should be placed at 150 mm centres for edges and 200 mm centres for sheet bodies. However, when tiles more than 6.5 mm thick are to be applied, nails and screws should be placed at 100 mm centres.

All joints and junctions must be waterproofed before any other surface finish is applied.

When fixing sheet lining, vertical joints should be staggered.

To waterproof the surface and joints of plasterboard, all joints must be set with a proprietary brand of wet area jointing compound and jointing tapes (see Figure 4.31), following the manufacturer's instructions carefully. Normal jointing compounds can be used in wet areas but not in shower recesses. Shower recesses have special requirements.

Before installing boards:

- check the noggins, tap heads and roughed-in plumbing, and that electrical components have been completed
- make sure all PVC and metal angles are in place before starting to fix the sheets
- lay the sheets horizontally, wherever possible, starting with the bottom sheet. Ensure that each sheet is fixed 10 mm up from the floor level to allow

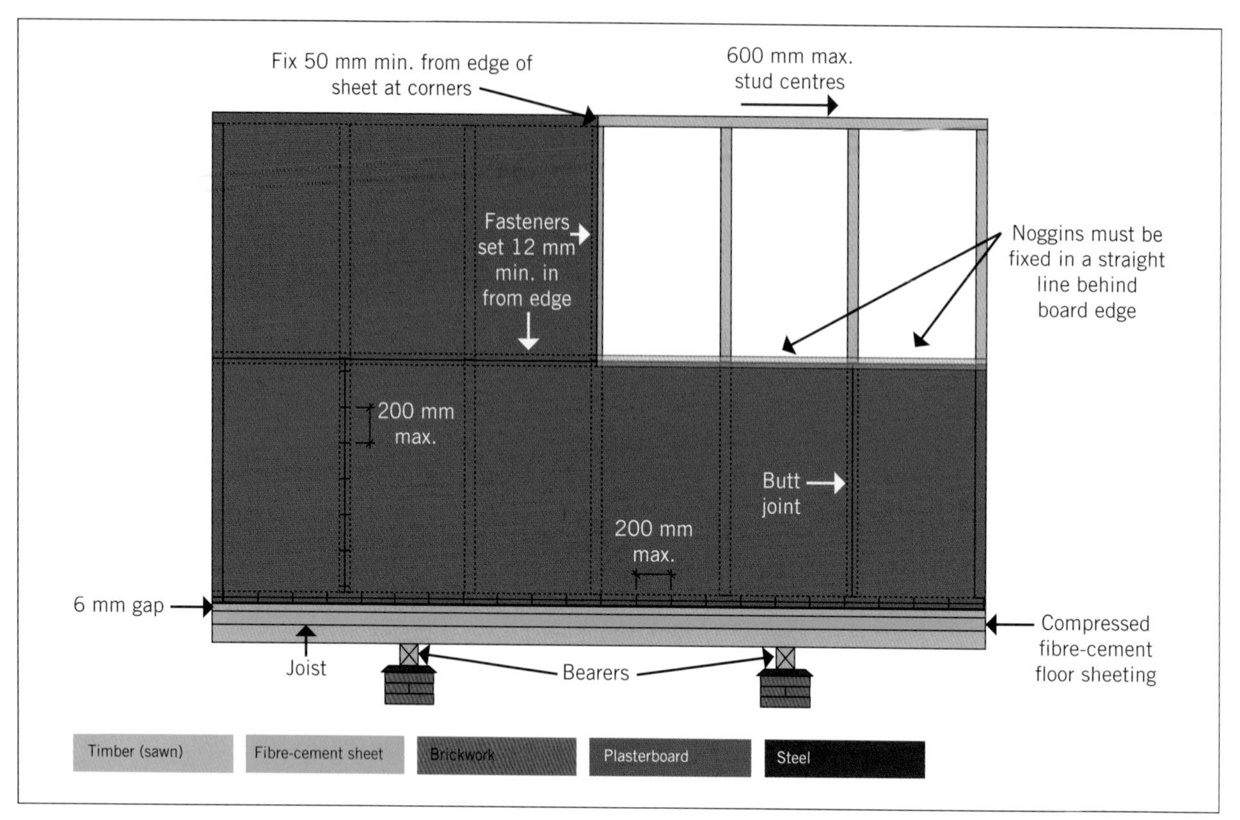

FIGURE 4.30 Fastening positions for plasterboard and fibre-cement sheeting

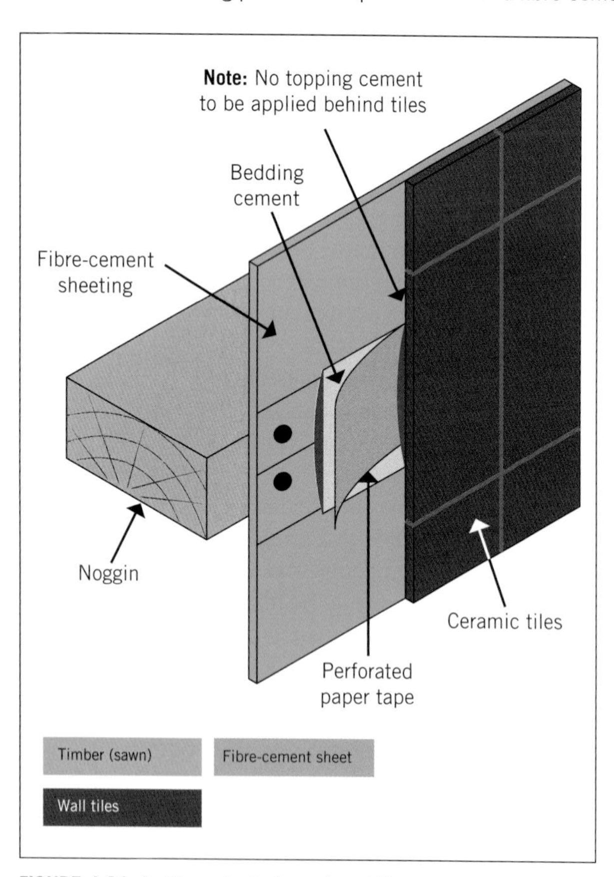

FIGURE 4.31 Setting plasterboard and fibre-cement sheeting joints

for later application of waterproofing. Use temporary blocks to support the sheets

- mark out and cut penetrations with a utility knife or plasterboard saw. Do not punch holes with a hammer (see Figure 4.32) – these are too hard to seal if the edges of the penetration are ragged.

FIGURE 4.32 Making penetrations in plasterboard

- a renovator saw (see Figure 4.33) or multi tool can be used to cut neat holes in plasterboard and fibre-cement sheeting.

Note: In South Australia, installers should also refer to the *Development Act 1993* – Minister's Specification SA F 1.7 Waterproofing of wet areas in buildings for additional waterproofing requirements in wet areas.

FIGURE 4.33 Renovator saw

Fixing fibre-cement sheeting

Cutting fibre-cement sheeting is more difficult than cutting plasterboard, and requires special tools and techniques. A hand guillotine (fibro cutter) is the most common tool used for making straight cuts (see Figure 4.34), but an electric guillotine and a power drill adaptor can also be used.

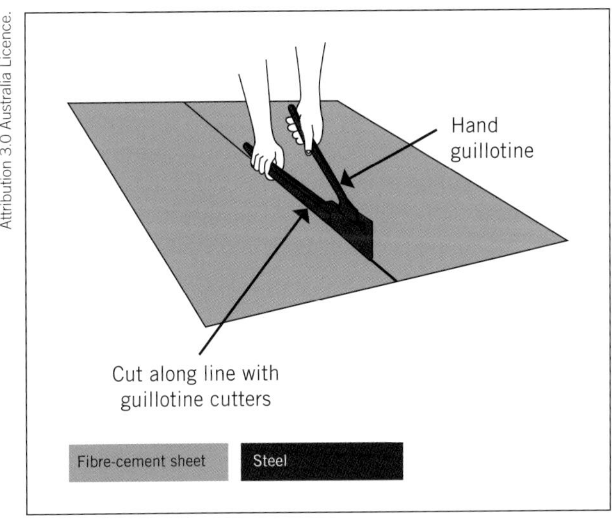

FIGURE 4.34 Hand guillotine or fibro cutter

Mark the sheets carefully, including penetrations, and then check before cutting.

Lay the sheet flat on 100 × 50 mm timbers, saw horses or similar supports to allow the guillotine or powered shears to slide across the sheet without obstruction. Don't forget to cut on the waste side of the line.

You can also use a score and snap action with a tungsten carbide-tipped scoring tool combined with a straight edge. **Score** on the face side of the sheet only and bend the sheet upwards to ensure a clean cut.

Penetrations can be made by drilling small holes around the marked-out penetration and punching out the waste with a hammer from the face side (see Figure 4.35). The edges can be cleaned up with a rasp.

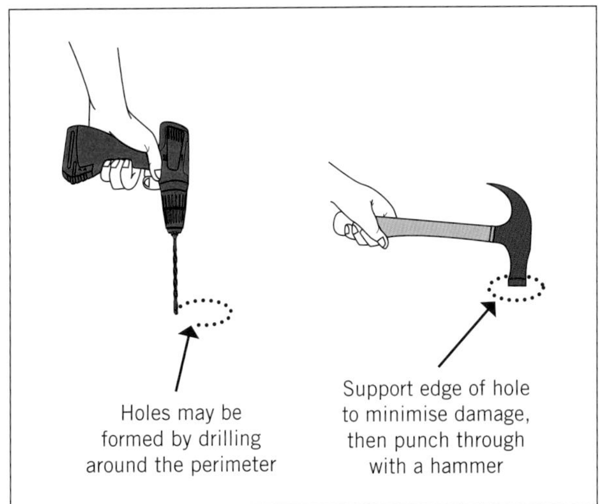

FIGURE 4.35 Making small penetrations in fibre-cement sheeting

A number of power tool manufacturers have oscillating blade tools, both corded and battery powered, that are ideal for making cut-outs in fibre-cement sheeting. Where edges need to be bevelled for jointing and setting there are special power tools available to grind the edge to the correct width and depth.

Large penetrations (such as those needed for vents) can be cut out using a score and snap method. Score around the penetration; make a hole in the centre of the penetration; saw cut to the edges of the penetration; then snap face side up to release the waste (see Figure 4.36).

When cutting plasterboard or fibre-cement sheeting, it is recommended practice to use an appropriately rated respirator mask. A P1 grade filter is the minimum recommended, while a P2 filter offers even greater protection against breathing harmful dust.

Sealants

Sealants must be used around wet areas:

- at wall and floor junctions
- around shower and bath fitments
- at all penetrations to walls (generally below 1.8 m), including taps (if no PVC aprons are used) and floors.

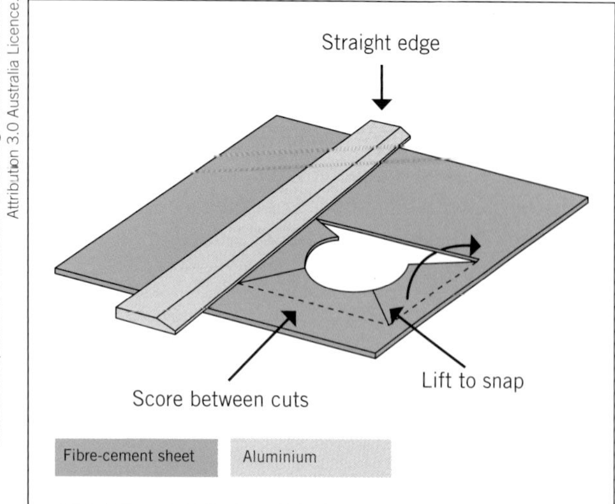

FIGURE 4.36 Score and snap large penetrations

Suitable types of sealants include silicone sealants, mastics and water-based acrylics. They must be used strictly according to the manufacturer's instructions; otherwise the warranty may be void.

There are mould-resistant grouts and silicones, as well as epoxy grouts, to help keep grout lines clean, particularly when white grouts are used on floors.

Penetrations

Penetrations are any building elements passing through a wall, floor or ceiling; for example, plumbing heads for taps and spouts, and floor wastes for drainage.

Floor wastes

It is important that all plumbing wastes are sealed, particularly in a shower recess. A PVC waste yoke must be bonded to the flooring and the waste pipe, using an epoxy-based resin, before fitting grates and other fixtures (see Figure 4.37). A leak control waste system should be used to enable moisture to drain if it gets below the floor tiles.

Tap head wall penetrations

Figure 4.38 shows details for a tap head wall penetration.

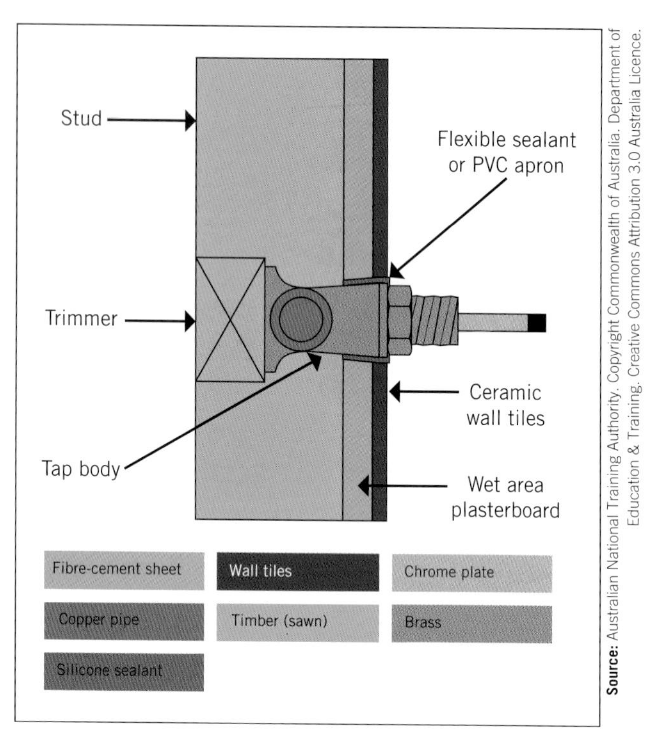

FIGURE 4.38 Tap head wall penetration

Vanities and tubs

Check the plans carefully to determine the quantity, size, position, type and other details needed for the installation of vanities and cabinet accessories such as mirrors, lighting and venting bulkheads. This needs to be done at an early stage in the building work so that the necessary noggins and supports can be built into the frames.

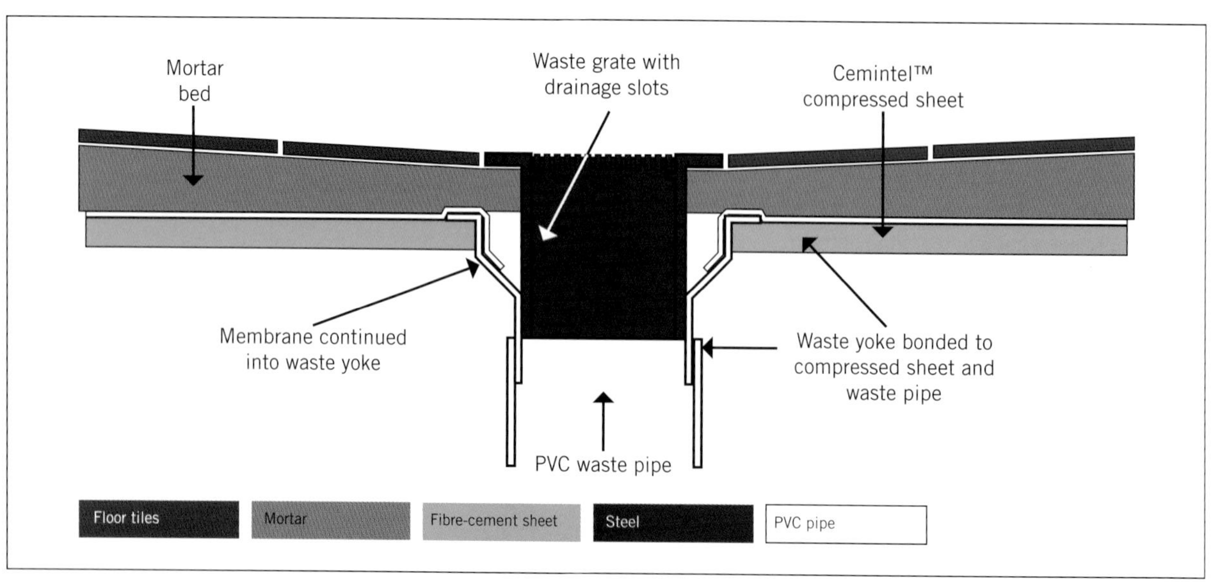

FIGURE 4.37 Waste detail

Careful setting out of tiling and high-quality tiling can also eliminate a lot of unnecessary work. Marking out cabinet positions accurately on walls before tiling makes for closer fitting of tiles and avoids scribing to tiles. Setting out so you achieve even lines on each side of windows and other openings will improve the finish. Small holes around plumbing fitments will reduce caulking and potential water leaks. (See Figures 4.39, 4.40 and 4.41.)

Source: Greg Cheetham

FIGURE 4.39 Cabinetry

Source: Greg Cheetham

FIGURE 4.40 Setting out tiling around windows

Source: Greg Cheetham

FIGURE 4.41 Penetrations

Installing vanities and benches

Vanities, laundry tubs and benches are usually installed after the floor finish has been applied.

Some units come with separate bases that are levelled by packing or scribing and then fastened to floors and walls. Cabinets are then set on top and fastened back to the walls with screws in timber- and steel-framed construction or masonry **anchors** in brick construction. Some manufacturers are using height-adjustable stainless steel legs as an alternative to scribe fixed bases.

Cabinets should be made from high moisture-resistant (HMR) particleboard or medium density fibreboard (MDF), or other materials that will not be affected by contact with moisture. There are now waterproof PVC polymer boards available in standard sheet sizes of 2400 × 1200 × 12, 16 and 18 mm thick. They are good for carcass work but are not recommended for doors. Bases should be at least 75 mm high.

A modern trend is to eliminate bases and suspend cabinets directly from the walls. This requires additional noggins to be placed in the walls for fixing or a support batten fixed below the cabinet on the face of the wall lining material (see Figure 4.42). Alternatively, the walls can be lined with 12 mm A bond plywood behind fibre-cement sheeting or plasterboard so that the fixing of cabinets and fittings is easier.

Check that the cabinet sits plumb and level in both directions. Ensure all exposed edges are sealed to prevent moisture penetration.

Splashbacks are often part of the design of the cabinet. Walls behind sinks and basins should be waterproofed to a height of 150 mm above the finished bench top. If the cabinet does not have an integrated splashback, a row of tiles should be fixed to the wall above the bench top. A mirror extending down to the bench top is an alternative.

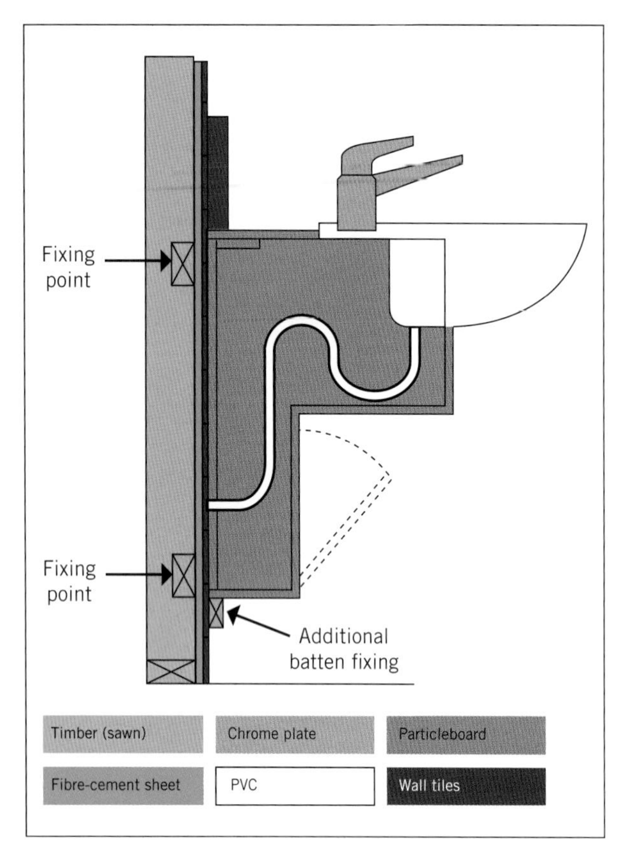

FIGURE 4.42 Fixing wall-suspended vanities

LEARNING TASK 4.4

TESTING FIXING METHODS AND WET AREA LINING SYSTEMS

Carry out the following practical exercise to test fixing methods and wet area lining systems.

On a section of wall frame, fasten a section of wet area plasterboard. Leave the adhesives to set, and then:

1. Below the first piece of wet area plasterboard, fix a small section of 12 mm A bond plywood and cover the section with a piece of 6 mm fibre-cement sheeting.
2. Fix a bathroom fitting, such as a towel ring, using:
   a. hollow wall anchors to the wet area plasterboard
   b. a screw fixing to a ply-backed wall lining.
3. Use your hands only to pull each fitting from the wall.
   Identify which method provides the stronger fixing method.

COMPLETE WORKSHEET 4

Handy hints for wet area construction

Following are some handy hints you will find useful in wet area construction:

- Keep waste holes capped when constructing wet areas to prevent any rubbish entering the drainage system.
- When using silicone sealants to seal intersections or joints between fitments and tiled finishes, apply the silicone in a thin bead. Before cleaning off excess silicone, spray it over with a non-ammonia-based glass cleaner, then wipe off the excess. The spray cleaner prevents the silicone from adhering to adjoining surfaces. Alternatively, use masking tape to prevent the spread of excess silicone on to surfaces. This makes clean-up much easier.
- Clean up dust using a vacuum cleaner to prevent the spread of dust to other areas of the building.
- Clean up tools and equipment after use and dispose of any waste water according to environmental guidelines.

GREEN TIP

Many of the materials used in wet areas contain chemicals that can contaminate waterways and sensitive environments.

LEARNING TASK 4.5 BATHROOM DESIGN

Teacher-led activity: Students are to do an internet search using the search string (term) 'modern bathroom designs'.

In groups, select two modern bathroom designs and discuss the following:

1. What are the common construction features in each bathroom?
2. What are the most commonly used materials for wall linings?
3. What are the critical points you will need to consider at the pre-construction planning stage?

In this exercise students will need to break into groups, conduct research and prepare a group presentation on their findings. The presentation must include photographs with labels and arrows indicating key features. Students can use PowerPoint or similar software for the presentation.

Waterproofing

In residential construction all waterproofing practices must comply with AS 3740 Waterproofing of domestic wet areas, and in some states must be carried out by licensed waterproofing contractors. The Standard gives minimum requirements for materials, designs and installations and waterproofing wet areas within residential constructions.

Poor waterproofing practices in residential construction can cause significant defects in buildings and in multi-residential constructions. Waterproofing defects are created because of:

- poor workmanship – lack of material and technology knowledge and skills
- quality control – including supervision, inspection and testing
- waterproofing maintenance during renovations – the fixing of fittings after waterproofing and applied finishes
- poor attitudes – towards quality, performance and contractual obligations and compliance.

Make sure all flashings are installed correctly at junctions of intersecting walls and floors. There are different requirements for different circumstances. These are outlined in AS 3740 Waterproofing of domestic wet areas – 3.9 Junctions. Also see the section on flashings.

Material selection

The key points to selecting the right waterproofing materials include:

- understanding the material's compatibility – especially around movement, expansion and contraction of different materials
- ambient temperature variations in the location
- exposure to other materials – alkalis in cement mortar and perhaps cleaning chemicals.

Make sure membrane materials comply with AS/NZS 4858 Wet area membranes. If in doubt, ask the supplier for proof. Some imported products may not comply.

Note: The use of stainless steel, copper and vinyl trays with welded or soldered joints may be an alternative in some situations, but these must be installed correctly.

Tools required are:

- broom or banister brush, pan and dust masks
- scraper to remove any lumps of render or building waste from floor and wall surfaces
- paint brush and paint roller kit
- caulking gun
- gloves – surgical gloves are best
- face mask/respirator – some products may be harmful if breathed (check the SDS for requirements)
- film depth guide – to ensure the correct thickness of membrane is applied.

Step 1

The first step is to make sure all surfaces are clean and free from dust or loose materials. Clean off any cement render splatter.

You need to consider whether the membrane will be applied above or below the cement screed so you know the fall to the drainage points. Either method is acceptable as long as the materials are compatible. Always check the manufacturer's specifications and guides first and don't get tempted into mixing materials and systems.

Generally, where big expansion, contraction, building movement or construction joints are present a 'below screed' method is appropriate and may mean using welded PVC or bitumen (torched on) membrane.

When using a bonded 'above screed' membrane provide a proper bond breaker system.

Note: Above screed membranes are *not* appropriate when a solvent-based polyurethane membrane is applied because of incompatibility with tile adhesives and some other finishes.

Screeds should be a minimum of 15 mm for bonded screed or 40 mm for an unbonded screed at their thinnest point.

Before applying any waterproofing ensure all floor finishes allow water to drain without pooling to floor wastes. This can be done with a:

- spirit level to check for low points and fall
- bucket of water (you need to be careful)
- marble or ball bearing
- combination of above.

The fall should be at least a ratio of 1:100 outside of showers, and 1:80 in the shower recess.

Step 2

Use a brush or roller to apply the primer to the surfaces of the area to be waterproofed (see Figure 4.43).

Source: www.bunnings.com.au

FIGURE 4.43 Applying the primer

Step 3

All joints and penetrations (floors and walls) must be fully sealed with a compatible sealant before applying waterproofing materials. Check with membrane manufacturers for the right sealant type.

Sealant fillets and bond breakers are critical to the integrity of waterproofing and must be selected and applied correctly (see Figure 4.44).

Where differential movement is greatest (e.g. between building frames, linings and shower trays or baths) the use of a foam **backing rod** or flexible tapes are critical to the integrity of the seal.

Think of it like an elastic band. They stretch best along the length but not the width; and the thicker they are, the less elastic they are in width (see Figure 4.45).

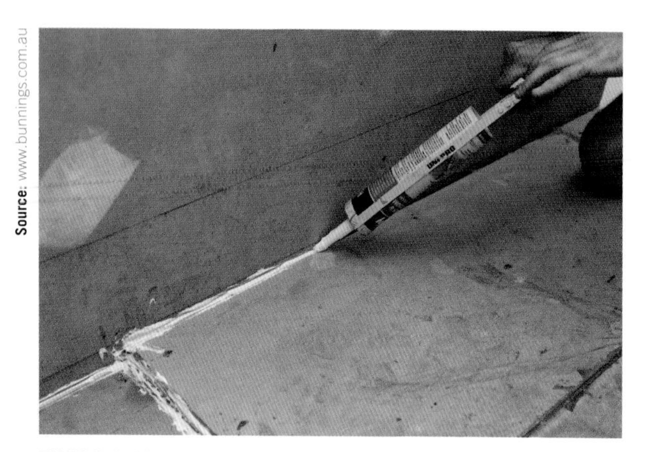

FIGURE 4.44 Applying the sealant

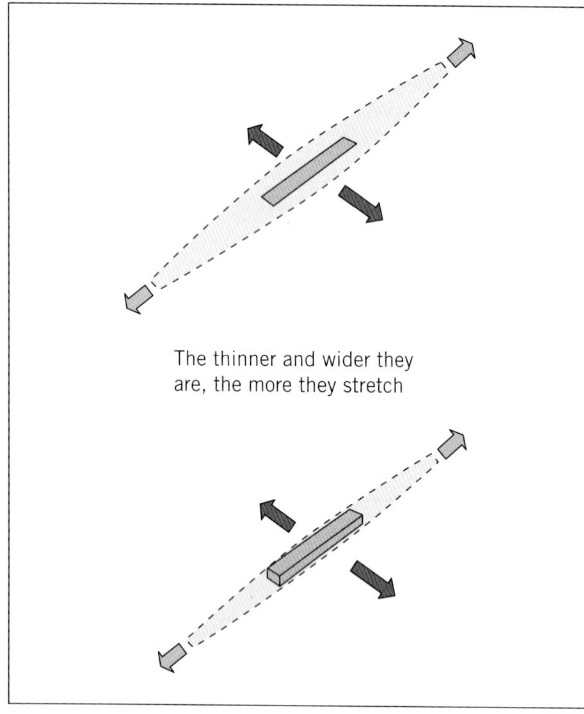

FIGURE 4.45 Elastic band effect

Polyurethane and silicone-based sealants *are not* proper bond breakers.

Step 4

Apply the first coat of membrane material to joints (150 mm minimum). Use masking tape to help minimise waste, see Figure 4.46.

Step 5

Install reinforcing fibre or sealant tape to all joints by applying membrane material with a brush or roller. Make sure there are no air bubbles in or under the flexible tape or sealant tape. Allow to cure fully. See Figure 4.47.

Step 6

Finally, apply a full coat of membrane material to all surfaces.

FIGURE 4.46 Applying the first membrane coat

FIGURE 4.47 Installing the flexible joint reinforcement tape

Note: If the shower has a hob, the membrane must extend over the hob and extend a minimum of 50 mm past the outside face of the hob. The top of the hob brick should be splayed inwards to assist any water penetrating the tiles to flow back into the shower recess. When tiled, no plastic decorative edge can be used on the internal joint of the hob. The top tile should overlap the internal face tile (see Figure 4.48).

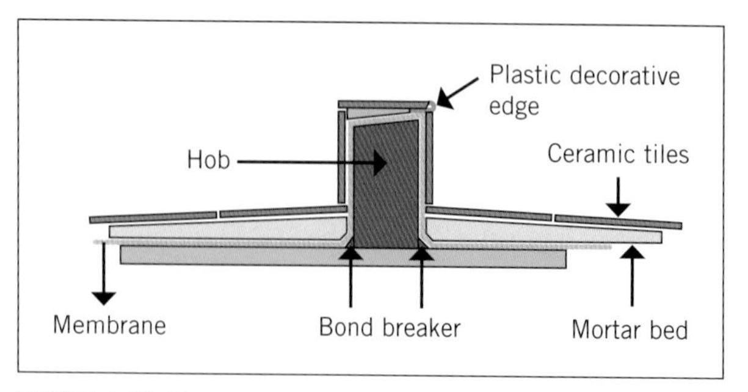

FIGURE 4.48 Shower hob

All internal and external wall junctions and horizontal joints must be waterproofed to 1800 mm above the floor with a minimum width of 40 mm either side of the junction.

LEARNING TASK 4.6

Teacher is to divide the class into three-person groups.

Students are to investigate different types of waterproofing materials and methods and discuss the best method and materials to use in these situations.

- Masonry and rendered walls to compressed fibre-cement sheet flooring.
- Timber-framed walls with wet area fibre-cement sheeting to a recessed concrete slab before topping.
- Timber framed walls with wet area fibre-cement sheeting to a recessed concrete slab after topping has been applied.
- Whether to waterproof above or below cement screed on floors.
- Renovating a timber-framed cottage with timber floorboards.
- Renovating a period home with masonry-rendered walls, and an old tessellated tiled floor laid on a mortar screed on top of rubble-filled substrate.

You need to consider issues around:

- compatibility of materials
- complexity to build and the impact on building costs
- environmental and sustainability considerations.

You can use drawings and sketches to demonstrate what is required.

Articles to read and consider: Building Connection, 'Waterproofing – joint sealants to connect or to release?' by Andrew Golle, 2 September 2015, https://buildingconnection.com.au/2015/09/02/waterproofing-joint-sealants-to-connect-or-to-release/. And Building Connection https://buildingconnection.com.au – look in the Articles drop-down menu and click on Columns for Andrew Golle's waterproofing videos.

Clean up

When finished, the work the area needs to be cleaned, and any materials need to be sorted for landfill, recycling and reuse according to the work site requirements and workplace procedures.

Clean up the work area, removing any debris, tools and excess materials. Follow the manufacturer's guidelines when disposing, storing and handling hazardous chemicals used for waterproofing and sealing.

Complete a visual inspection of all tools and equipment to ensure safe operation. Regular checks and maintenance of tools and equipment should be done each time they are used. If faults are found, follow workplace procedures to record problems, tag the tools or equipment, and relocate them away from other functioning tools.

Restore the surroundings to their original condition, ensuring that any damage caused during installation is repaired.

Obtain the necessary sign-off and approvals from the client, project manager, or any relevant authorities to confirm the completion of the installation. Ensure that all necessary documentation has been obtained and filed appropriately.

COMPLETE WORKSHEET 5

SUMMARY

In this chapter we have looked at installing wet area fittings and fixtures in bathrooms, including framing and lining materials for walls and floors and the requirements of the Australian Standards and National Construction Code for waterproofing.

The main points in planning and preparation are:

- accurate reading of plans and specifications to ensure that the right materials for framing, lining and waterproofing are selected
- quality control: use a checklist to ensure all aspects of the job have been checked at different stages of the construction
- environmental and sustainability considerations such as water use and waste management have been taken into account
- calculating material quantities and costs is critical to managing budgets and ensuring a profit is made
- choosing the right tool for the tasks – keep up to date with the latest tool technology.

In wet area construction:

- timber-framed floors require special treatment to frame and waterproof them. Compressed fibre-cement sheet is laid, glued and screwed down over the joists
- concrete raft slabs incorporate a step-down to allow for tiling and level finishing
- additional noggins and trimmers may be inserted for the fixing of plumbed items. Alternatively, sheet the entire wall with wet area plywood before fibre-cement sheeting is applied
- frames can be checked out to a maximum of 20 mm. Any cut-outs have to be strengthened
- steel stud frames must comply with AS 3623 Domestic metal framing
- control joints are required in long walls to allow for structural movement
- most baths, showers and spas are supported on a bed of sand cement mix or a combination of masonry materials. Some are set on legs.

Shower areas

- Shower areas can be open, or enclosed with framing. An unenclosed shower area extends to 1500 mm horizontally from the shower head.
- The wall frames are cut back to accept preformed shower trays and allow wall linings and finishes to overlap it. In modern construction, grates are installed directly into the floor to create hobless shower areas, eliminating trip hazards.

Flashings

- Flashings must extend a minimum of 25 mm up the wall/floor intersections and to a height of 1800 mm in vertical corners. Flashing can be PVC, fibreglass or rubber based, with epoxy resin or similar waterproof adhesives.

Linings

- Wall lining materials include wet area plasterboard, marine ply, fibre-cement sheeting and solid timber lining boards.
- Plasterboard and fibre-cement sheeting are nailed or screw-fixed to studwork, and joints are taped and set using special wet area jointing compounds.
- Sealants (silicone, mastic, etc.) must be used in wet areas at wall and floor junctions and shower and bath fitments and penetrations.

Installing vanities and benches

- Vanities, laundry tubs and benches are usually installed after the floor finish has been applied.
- They are fixed to walls with screws in timber and steel-framed construction or with masonry anchors in brick construction. Adjustable legs are an alternative.
- Cabinets should be made from high moisture-resistant (HMR) particleboard or medium-density fibreboard (MDF), or materials not affected by moisture.

REFERENCES AND FURTHER READING

Texts

CemintelTM (2016). *CemintelTM Wet Area Systems*, **http://www.cemintel.com.au** > search term 'wet area systems'.

James Hardie (2016). *Application Guide – Interiors – Wet Area Construction*, **http://www.jameshardie.com.au** > look under 'Resources' drop-down menu/Technical Guides/ Internal linings.

Rondo (2015). *Professional Manual*, **http://www.rondo.com.au**. Look under 'wall product manuals'.

Web-based resources

Many of these companies have web-based resources such as technical application and installation guides, videos and even e-learning platforms you can download or access for free.

Cement Concrete & Aggregates Australia: **https://www.ccaa.com.au** – *Guide to Residential Floors*. Under the Publications drop-down menu, click on Technical publications/Guides

Construct NSW: **https://www.nsw.gov.au/building-commissioner/construct-nsw** – website provides advice and guidance on industry and regulatory reforms to the residential apartment building construction industry

CSR Cemintel™: **http://www.cemintel.com.au** – how-to videos and technical data and BIM drawings

CSR Gyprock™: **https://www.gyprock.com.au**

James Hardie Ltd: **http://www.jameshardie.com.au** – how-to guides and other technical information.

Siniat: **https://siniat.com.au/technical-manuals** – Look under Resources and Downloads drop-down menus.

TAFE NSW: **https://store.training.tafensw.edu.au/product-category/construct-nsw/** – online learning module Waterproofing Design Principles, targeting Class 2 buildings. Offered by Construct NSW, it's fee-based but recognised as meeting Continuing Professional Development (CPD) requirements

USG Boral: **http://www.usgboral.com/content/usgcom/en_australia.html** – how-to guides, technical data and a BIM tool to assist in designing or selecting wall or ceiling system for projects. Look in the Resources & Tools drop-down menu.

Waterproofing:

- Building Connection: **https://buildingconnection.com.au** – look in the Columns drop-down menu for Andrew Golle's waterproofing articles.
- D&R Henderson: **https://drhenderson.com.au/distributed-products/pvc-polymer** – PVC polymer board
- Shalex industries: **https://www.shalex.com.au/** – application guides and safety data sheets (SDS).

National Construction Code

Copies of the National Construction Code can be obtained free of charge from the Australian Building Codes Board website: **http://www.abcb.gov.au**. In particular:

National Construction Code 2022, Building Code of Australia – Volume One, Class 2 to Class 9 Buildings, Section F – Part F.2 – Wet areas and overflow protection.

National Construction Code 2022, Building Code of Australia – Volume Two, Class 1 to Class 10 Buildings, Part H4 – Health and amenity.

National Construction Code 2022 – Housing Provisions Standard – Section 10 Health and amenity – Part 10.2.

In South Australia only, the *Development Act 1993* – Minister's Specification SA F 1.7 Waterproofing of wet areas in buildings – Additional requirements for materials and techniques for waterproofing wet areas (i.e. toilets, laundries, bathrooms and kitchens).

In South Australia only, the Minister's Specification SA F1.7 Waterproofing of wet areas in buildings: additional requirements should be referred to in addition to the above.

Video resources

Bathroom waterproofing: **https://www.youtube.com/watch?v=vSDCZkYa7ZE**

Categories of water in water-damaged buildings: **https://www.youtube.com/watch?v=m7sJXY5z4AM**

Crommelin how-to guides: **https://www.crommelin.com.au/crommelin-how-to-guide/**

Dunlop Express Wet Area Waterproofing: **https://www.youtube.com/watch?v=Eq_ACJivK34**

GripSet high performance waterproofing membranes: **https://gripset.com/sealed-for-good/videos/**

Hardie glaze lining – wet area installation: **https://www.youtube.com/watch?v=nsgoCxmTGsw**

How to identify water leaks in walls: **https://www.youtube.com/watch?v=YtNxM-6-7fY**

Prepare for sheeting a bathroom: **https://www.youtube.com/watch?v=XnR477yKOCg**

Relevant Australian Standards
AS 1684.2 Residential timber-framed construction – Non-cyclonic areas – with amendments
AS 1684.3 Residential timber-framed construction – Cyclonic areas – with amendments
AS 2908.2 Cellulose cement products – Part 2: Flat sheets
AS 3740 Waterproofing of domestic wet areas
AS 3958.1 Ceramic tiles – Guide to the installation of ceramic tiles
AS/NZS 2269.0 Plywood – Structural – Specifications
AS/NZS 2589 Gypsum linings – Application and finishing
AS/NZS 4386 Cabinetry in the built-in environment – Commercial and domestic
AS/NZS 4858 Wet area membranes
AS/NZS 1859 Series on reconstituted wood-based panels – i.e. particleboard, dry and wet process fibreboard and decorative overlaid wood panels
Australian Standards can be purchased from the Standards Australia website: **https://www.standards.org.au/**, or see your teacher or librarian for assistance in accessing Australian Standards online.

GET IT RIGHT

INSTALL WET AREA FIXTURES

Questions

Below are two pictures showing a wet area installation.

1 What is missing in this wet area installation?

____

____

2 How should this job have been completed correctly?

____

____

3 What should be done to rectify the problem?

____

____

4 What could happen if nothing is done?

____

____

Source: Greg Cheetham

Source: Greg Cheetham

HOW TO

WATERPROOF A WET AREA

In this method we are applying the membrane below the cement screed. Always follow the manufacturer's instructions with any waterproofing product.

Steps involved		
1	Make sure all surfaces are clean and free from dust or loose materials. Clean off any cement render splatter with a scraper. Use a vacuum cleaner for best results.	 Source: Greg Cheetham
2	Use a brush or roller to apply the primer to the surfaces of the area to be waterproofed.	 Source: www.bunnings.com.au
3	All joints and penetrations (floors and walls) must be fully sealed with a compatible sealant before applying waterproofing materials. Sealant fillets and bond breakers are critical to the integrity of waterproofing and must be selected and applied correctly. Where differential movement is greatest (e.g. between building frames, linings and shower trays or baths) the use of a foam backing rod or flexible tapes is critical to the integrity of the seal.	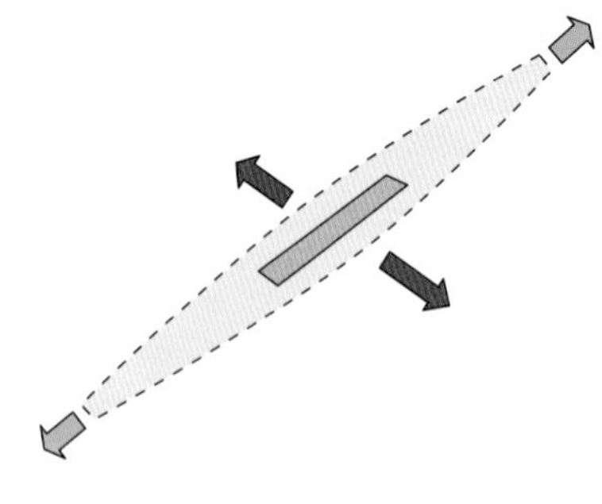 The thinner and wider they are, the more they stretch 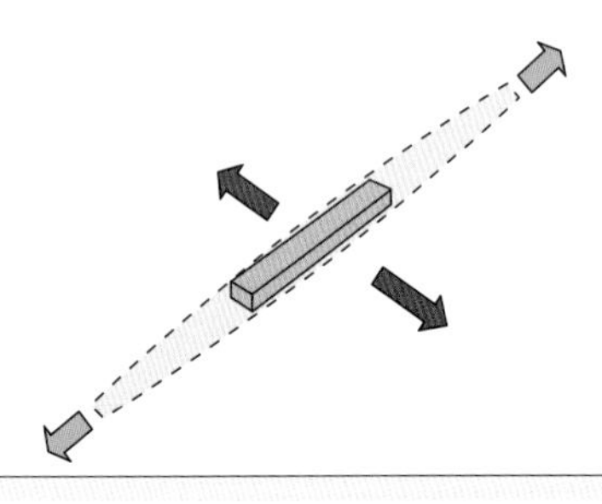

>>

>>

		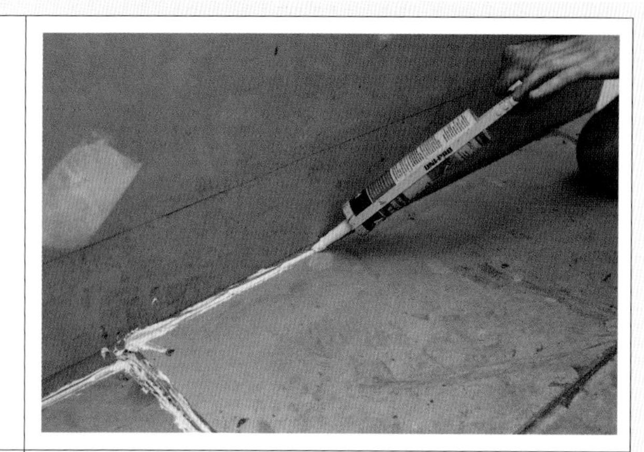
4	Apply the first coat of membrane material to joints (150 mm minimum). Use masking tape to help minimise waste.	
5	Install reinforcing fibre or sealant tape to all joints by applying membrane material with a brush or roller. Make sure there are no air bubbles in or under the flexible tape or sealant tape. Allow to cure fully.	
6	Finally, apply a full coat of membrane material to all surfaces. If the shower has a hob, the membrane must extend over the hob and extend a minimum of 50 mm past the outside face of the hob. The top of the hob brick should be splayed inwards to assist any water penetrating the tiles to flow back into the shower recess. When tiled, no plastic decorative edge can be used on the internal joint of the hob. The top tile should overlap the internal face tile. All internal and external wall junctions and horizontal joints must be waterproofed to 1800 mm above the floor with a minimum width of 40 mm either side of the junction.	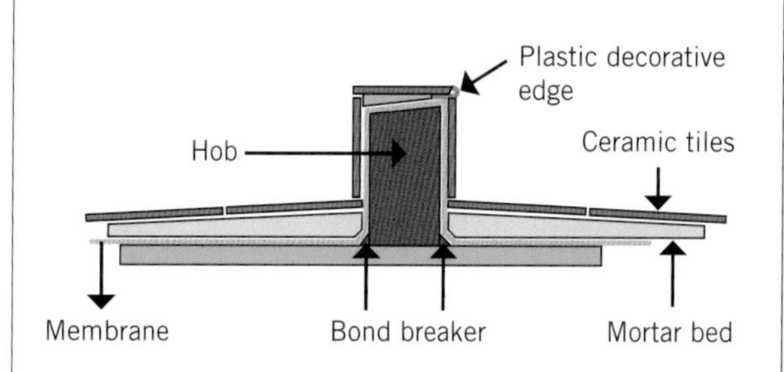

WORKSHEET 1

To be completed by teachers

Student competent ☐

Student not yet competent ☐

Student name: ______________________

Enrolment year: ______________________

Class code: ______________________

Competency name/Number: ______________________

Answer the following questions.

1 From the detailed drawing below, calculate the amount of 90 × 45 mm stud material required to frame the walls and write a timber order for the amount. The wall height is 2.4 m.

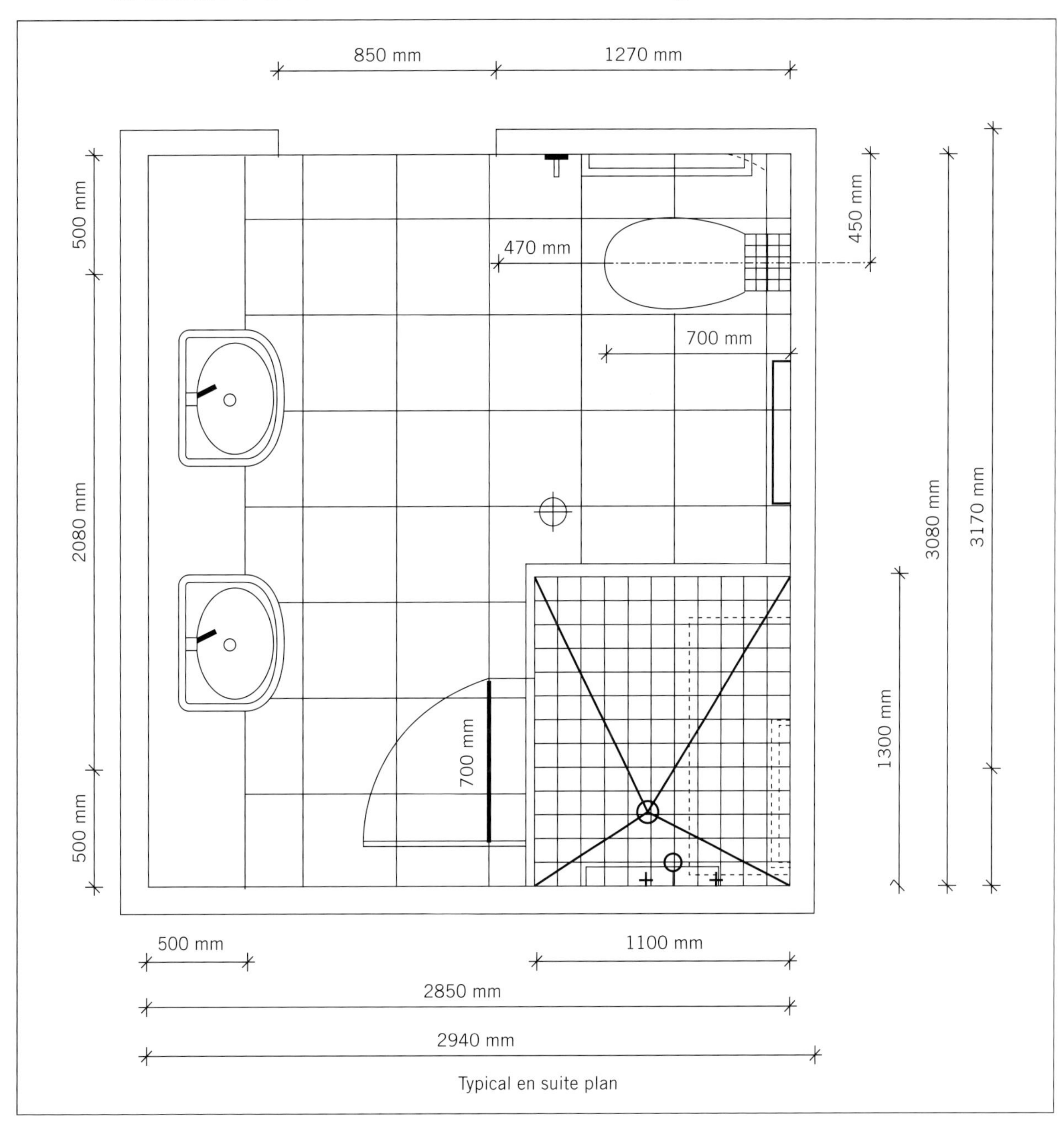

Typical en suite plan

Timber frame calculations

2 From the plan in Question 1, calculate the number of sheets of 6 mm wet area fibre-cement sheeting required to line the walls. Assume that the door is standard height and allow 2060 mm for opening head and 900 mm for opening width. A combination of sheet sizes is practical.

WORKSHEET 2

To be completed by teachers	
Student competent	☐
Student not yet competent	☐

Student name: ______________________

Enrolment year: ______________________

Class code: ______________________

Competency name/Number: ______________________

Answer the following questions.

1 Which Australian Standard was developed as a result of complaints regarding leaking shower recesses. When was it updated?

____

____

2 What should be used in conjunction with the Australian Standard mentioned above?

____

____

3 What is the maximum stud spacing in wet areas?

____

____

4 Where should noggins be included in wet area framing?

____

____

5 An alternative to inserting noggins to support fixtures and fittings is to line the walls with 12 mm A bond plywood behind other wall linings and finishes.

What are the advantages and disadvantages of this method?

Advantages	Disadvantages
a ______________	d ______________
b ______________	e ______________
c ______________	f ______________

WORKSHEET 3

To be completed by teachers	
Student competent	☐
Student not yet competent	☐

Student name: ____________________

Enrolment year: ____________________

Class code: ____________________

Competency name/Number: ____________________

Answer the following questions.

1 At what intervals should control joints be included in framing for wet areas?

____

____

2 Why are control joints used?

____

____

3 What types of flashing materials can be installed at wall and floor intersections in wet areas?

____

____

4 What flashing material is commonly used in vertical internal corners and what alternatives are there?

____

____

5 Where are sealants used in a wet area?

____

____

6 What materials can be used for wall linings in wet areas?

____

____

WORKSHEETS 4

7 What companies are the major manufacturers of wet area lining materials and what are the common names of these lining materials?

WORKSHEET 4

To be completed by teachers	
Student competent	☐
Student not yet competent	☐

Student name: ______________________

Enrolment year: ______________________

Class code: ______________________

Competency name/Number: ______________________

Answer the following questions.

1 Name four methods of cutting fibre-cement sheeting.

a ______________________

b ______________________

c ______________________

d ______________________

2 Describe the method used to cut small rectangular holes in fibre-cement sheeting.

____

____

3 How should sheets be fixed in wet areas – horizontally or vertically?

____

____

4 When is stud adhesive not used for wall linings in wet areas?

____

____

5 What size gap is left along the floor when installing fibre-cement sheets?

____

____

6 What is the recommended spacing for nailing fibre-cement sheets?

____

____

7 Where do you fix sheets at the floor/wall intersection in wet areas?

____

____

8 What is fixed at the internal angle when lining wet areas with plasterboard?

____

9 How are holes cut in wet area plasterboard for tap penetrations?

____

____

10 What gap should be left at the floor when fixing wet area plasterboard?

____

11 What fastener spacings are used when tiles more than 6.5 mm thick are fixed to wet area plasterboard?

____

____

12 What type of reinforcing tape is embedded in fibre-cement sheet or plasterboard joints?

____

13 How should jointing in other than tiled areas be carried out?

____

____

14 What material is finally applied over the board and joints before tiling?

____

WORKSHEET 5

To be completed by teachers	
Student competent	☐
Student not yet competent	☐

Student name: ______________________________

Enrolment year: ______________________________

Class code: ______________________________

Competency name/Number: ______________________________

Answer the following questions.

1 Write the correct response (True or False) in the box provided for the following statements relating to the use of primers in waterproofing:

a Only the junctions of walls and floor in a shower need to be primed. ☐

b Only old substrates require priming. ☐

c Primer should be dry before waterproofing. ☐

d Primers speed up the curing of concrete. ☐

2 Where do you find correct information on disposing of excess liquid membrane?

____

____

3 List four safety precautions you must use when applying solvent-based primer in waterproofing activities.

____

____

____

____

4 Correctly fill in the missing measurements in the spaces provided:

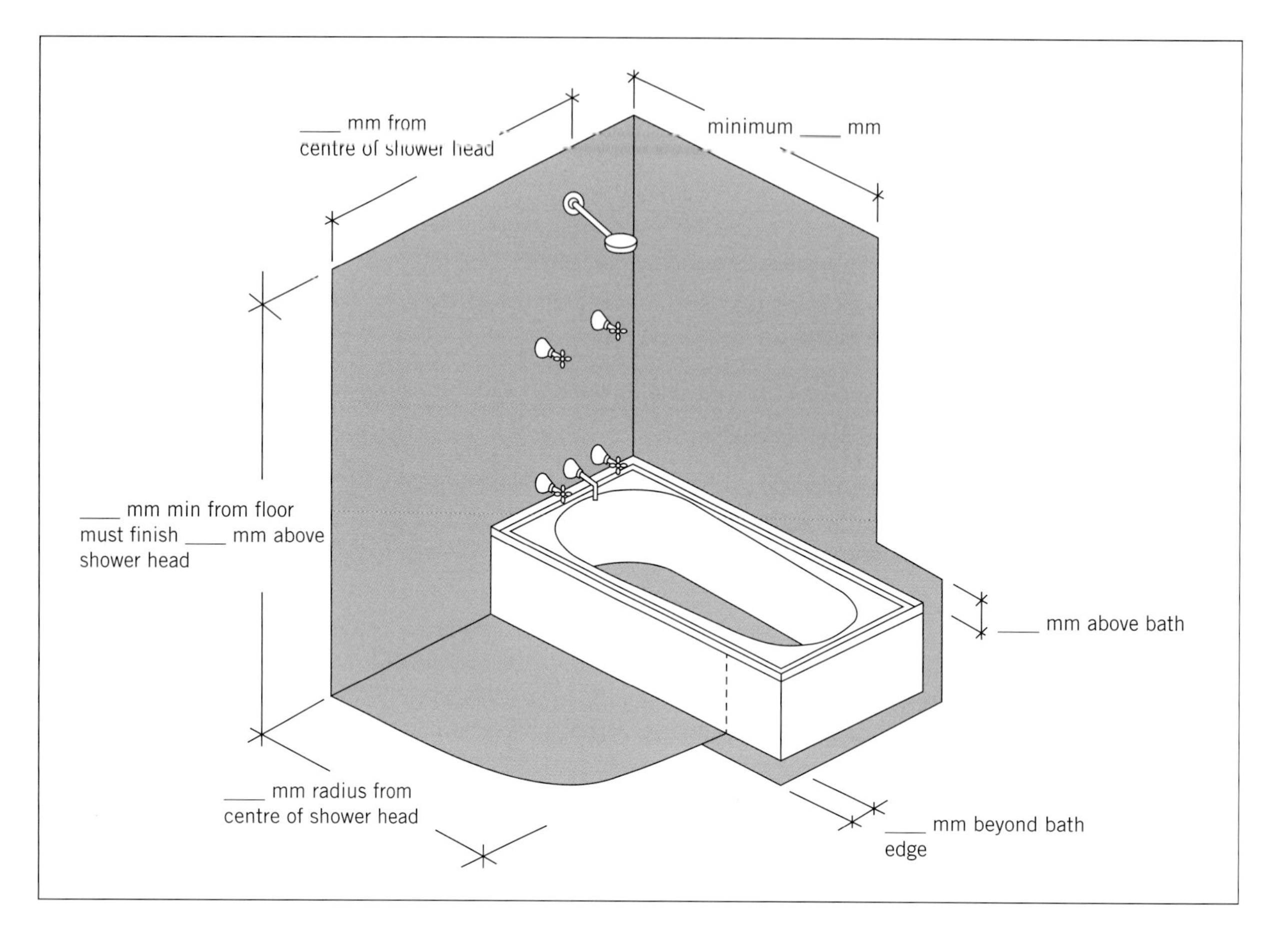

5 Which of the illustrations below best demonstrates the correct method to seal a movement or expansion joint: A, B or C?

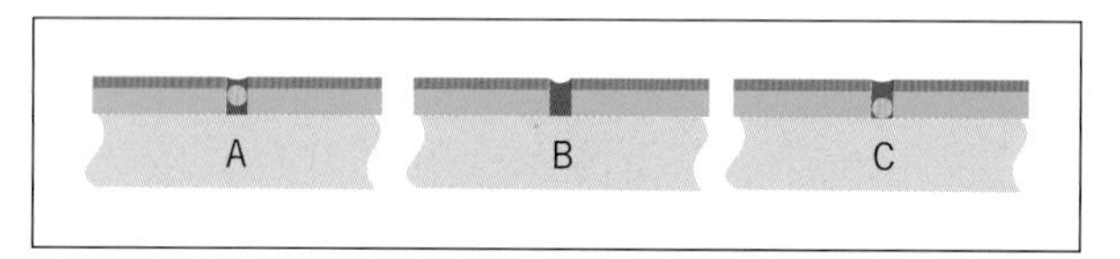

6 When will a waterproof membrane be ready to tile over?

___

___

7 For general bathroom floor area, the minimum fall to the waste should be:

a 1:50

b 1:80

c 1:100

d 1:150

8 Fill in the missing words from AS 3740 3.9.1.1:
Perimeter flashing to wall/floor surfaces shall be ____________ to the horizontal surface and have a vertical leg of a/an ____________ above the finished floor level, except across doorways, and the horizontal leg shall be a minimum width of ____________.

9 Select the correct answer for the question below by placing a cross in the right box.
When a preformed shower base is used, wet area plasterboard must be kept a minimum of how many millimetres above the shower base?

a 5 mm ☐

b 10 mm ☐

c 15 mm ☐

d All of the above are OK ☐

10 Match each architectural symbol in the first column with the correct corresponding term in the second column. All relate to bathroom plans.

Architectural symbol	Architectural term
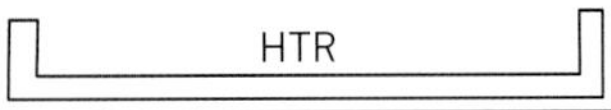 HTR	Toilet
 HWC	Basin or sink
	Heated towel rail
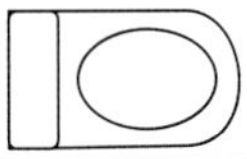	Bath
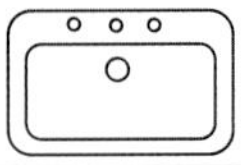	Light fitting
	Hot water heater
	Urinal
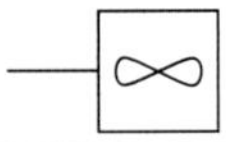	Shower tray
	Double power outlet
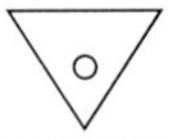	Exhaust fan

INSTALL LINING, PANELLING AND MOULDING

This chapter aligns with the unit of competency, 'Install lining, panelling and moulding'. The unit outlines the knowledge and skills required to install decorative lining, panelling and mouldings to masonry and timber- or metal-framed walls. Some of the key learnings include the range of materials available to line a building, and some simple techniques to achieve the best possible finishes in fit-outs and finishing.

In particular the chapter addresses the key competencies of:

- planning and preparing for installing lining, panelling and moulding, including:
  - selecting the right materials, tools, fixing methods and safe work practices
  - calculating quantities and costs
- installing lining and panelling, including methods of jointing and finishing panelling. This includes measuring, marking and cutting and fixing the lining and panelling materials
- cutting and fixing skirtings, architraves, mouldings and cornices at internal and external corners. This includes the setting-out geometry for raking moulds
- cleaning up once the job is done, including cleaning tools, disposing of waste materials and recycling whenever possible.

Learning objectives

By the end of this chapter you will be able to:

- identify suitable panelling, lining and moulding materials, and panelling components including timber lining boards, dado panelling, mouldings and beads used in heritage-style wall lining applications and current construction methods. You will also be able to identify and select plasterboard and other building boards for different wall lining applications.
- prepare walls for installation of linings, including straightening frames, crippling studs and calculating material quantities. You will also be able to understand the need to locate and safely deal with hidden services such as electricity, gas, water and data cabling. The need for incorporating energy efficiency through the use of effective insulation is also covered.
- set out lining materials using different application methods; that is, horizontal, vertical and diagonal where:
  - walls are constructed of timber frames or masonry where extra noggins or battening is required to provide suitable fixing points
  - lining walls with metal studs are used
  - roof trusses are used in construction, requiring the use of metal furring channel to provide a framework for battening before long materials are fixed.

- install lining materials by selecting the appropriate fixing method, tools and fasteners, including installing matched veneered products to achieve the desired aesthetic.
- use geometry to set out true shapes and mitre joints for raked moulds, and set out hopper bevels for crown mouldings.
- use the most appropriate finishing method for internal and external corners depending on materials used, including scribed joints, rebates and mitre joints for skirting boards and other joinery applications such as dado panelling, spandrels and soffits and raked mouldings.

Lining, panelling and moulding materials

The most common and economical wall lining material is plasterboard; however, a number of other materials are available and commonly used, including:

- solid timber lining boards
- timber veneered panelling:
  - plywood
  - medium density fibreboard (MDF)
  - particleboard
- hardboard
- fibre-cement sheeting
  - especially in wet areas like bathrooms, laundries and kitchens
- specialist plasterboards
- laminated panelling PVC polymer boards.

LEARNING TASK 5.1 WALL LININGS

Students are to collect and report on a range of wall lining information, including but not limited to:

- plasterboard – regular and wet area
- decorative plywood or timber veneered product
- fibre-cement sheeting.

  Information should include:
- installation information
- cost per sheet or per square metre.

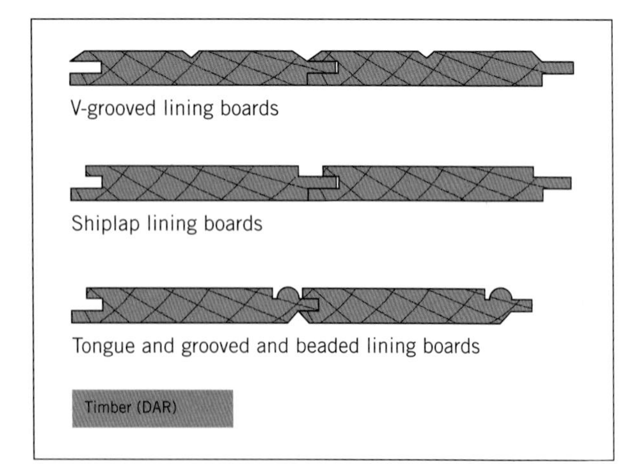

FIGURE 5.1 Timber lining board profiles

Solid timber

Solid timber is used as a decorative feature on walls. Traditionally, it was installed vertically as tongue and grooved lining boards, but it is also made up into frames and panels and fixed as dado panelling.

Lining boards

Lining boards are available in a range of timbers and profiles, with the most common being V-grooved, shiplap or beaded (see Figure 5.1). In this age where time is critical, prefinished boards are available. These boards require no sanding, and have UV stabilised applied finishes and are often end matched. A soft faced mallet should be used, not the usual claw hammer when installing the materials as faces can be easily damaged. If you must use a claw hammer, use a block to protect the surface.

Lining boards can also be mounted diagonally (at 45°) or horizontally, or arranged to form different geometrical shapes and designs (see Figure 5.2).

Lining boards range from 86 to 137 mm wide, and between 10 and 15 mm thick depending on the timber supplier. **Note:** These measurements refer to effective cover. Check with suppliers for accurate information.

(a)

(b)

(c)

(d)

Source: (a), (b), (d) Ed Hawkins; (c) Gordon Slater

FIGURE 5.2 Lining boards: (a) diagonal, (b) geometric pattern, (c) raked ceiling, (d) horizontal

Wall panelling

Wall panelling is more often found in commercial buildings as a means of hiding services and building components like columns, although there are now some more cost-effective plasterboard systems that imitate the real thing.

Panelling can be used to line walls from floor to ceiling, but often only extends from the floor to a nominal height between 700 and 1500 mm high. This form is referred to as **dado** panelling (see Figures 5.3 and 5.4).

The panelling is made up of frames much like doors, using rails and stiles, frieze rails and muntins (or mullions).

Pilasters are intermediate constructions used to break up the run of panelling where columns or doors or other fitments interrupt the flow of the panelling.

The built-in panel can be made from plywood, solid timber or other material such as veneered particleboard or MDF and metal sheeting.

Dado frames are packed level and straight, usually to a datum line, and can be fixed directly to timber framing with appropriate sized nails or with construction adhesive. The capping rail and skirting boards secure the frames in place and allow for some movement.

When masonry walls are to be panel lined, it is not uncommon for a groundwork of battens to be fitted to the walls to ensure a straight plumb base for the panelling. The battens should be fixed horizontally using masonry plugs and screws at 600 mm centres. This allows for the secret nail fixing of the panelling. Vertical battening can be done whenever panels are joined or a pilaster is required. This will be detailed on the drawings.

Source: Ed Hawkins

FIGURE 5.3 Dado panelling

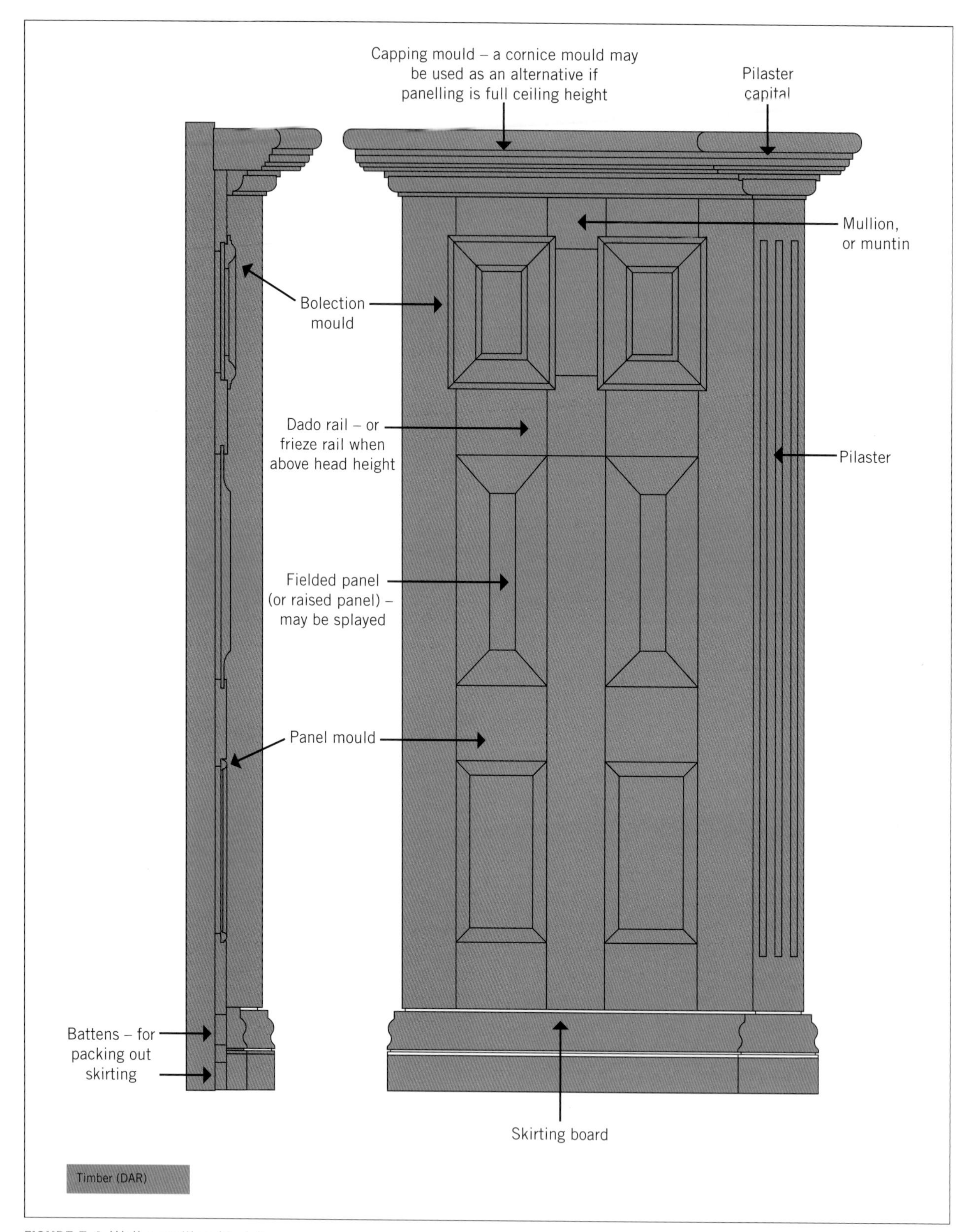

FIGURE 5.4 Wall panelling (dado)

The dado panels can be directly scribe fitted to the floor gaps, or highs and lows in the floor, and are supported on blocks above the skirting. The panelling is often finished off using ornate **cornices**, frieze rails or picture rails.

Timber mouldings

Skirting and architraves

Skirting boards (see Figure 5.9) are used internally to cover gaps between floors and walls and **architraves** are used internally to cover the gaps between window and door frames and walls. Some common moulding profiles are described next.

Bull nose and pencil round profiles

Bull nose (see Figure 5.5) is a typical moulding profile that was used extensively between the 1950s and the 1980s in Australia for skirting boards and architraves. The bull nose radius is typically the thickness of the materials; that is, 19 mm. Standard sizes are 42 × 19 mm, 67 × 19 mm and 91 × 19 mm. Special sizes can be made to order, and MDF boards can be cut and run.

Pencil round mouldings (see Figure 5.6) gained popularity as skirting and architraves in the late 1980s and continue to be used in modern residential construction. In recent years, section sizes have shrunk in thickness to 12 mm to keep costs low. The radii of pencil rounds are smaller than for the bull nose: 10–12 mm radius for 18 mm thick stock, and 6–8 mm for the modern 12 mm thick stock.

Splayed moulding (see Figure 5.7) is an alternative to bull nose and pencil round and is also available in combination with pencil round.

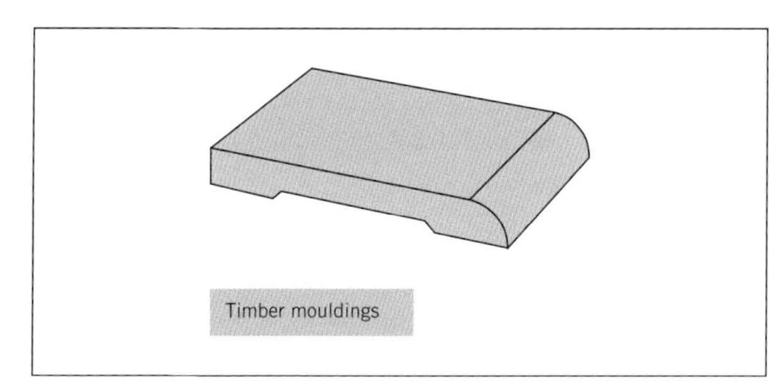

FIGURE 5.5 Bull nose

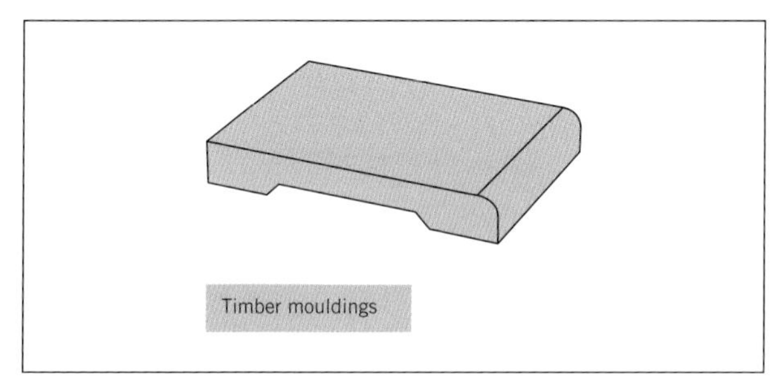

FIGURE 5.6 Pencil round

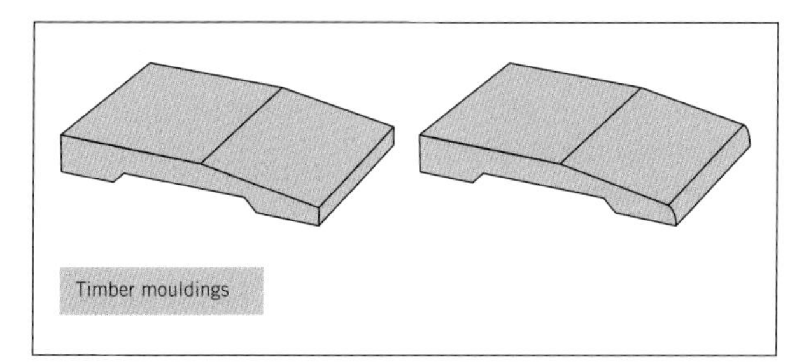

FIGURE 5.7 Splayed mould; splayed with pencil round

Casting grooves

A wide groove is often run on the back side of many skirting boards, architraves, window sills and lining boards. Wide boards can have a tendency to cup and twist because of stress. A wide groove, known as a **casting groove**, will reduce the stress in a board and can prevent twisting and cupping. The groove also allows for easier fitting to uneven surfaces where a tight fit is required, such as between door and window frames and linings (see Figures 5.5, 5.6, 5.7 and 5.9).

Note: window sills may have other grooves, such as a 'drip' groove and a groove to take lining boards.

Beading – flat, quad, cover strips and nosing

Beading is generally used to cover gaps between building elements but can also be used as decorative elements (see Figure 5.8).

- Beadings are generally small in section: less than 19 mm square.
- Beadings are usually decorative in nature and can be flat or square faced.
- Quad moulds (and scotia mould) are used to cover gaps in corners and elements that adjoin at 90°.
- Mouldings are decorative and functional (e.g. as a picture rail or skirting board).
- Regular mouldings are usually greater than 19 mm square.
- Cover strips are used to cover joints in wall cladding and lining materials such as fibre-cement sheeting.
- Nosings are used at the bottom of window frames to cover gaps between frames and wall lining materials, particularly in older style buildings.

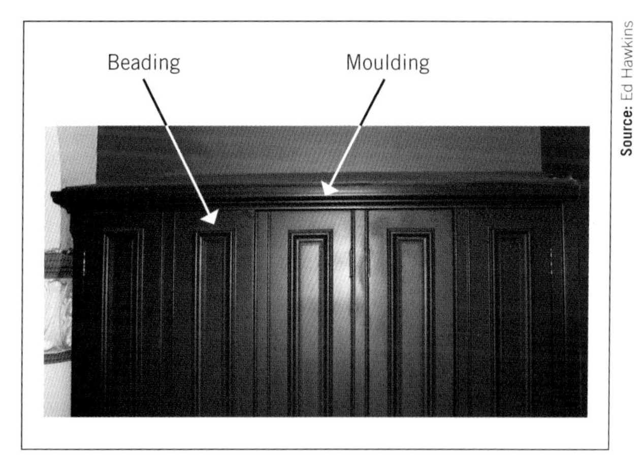

Source: Ed Hawkins

FIGURE 5.8 Beading and moulding

Ornate period profiles

These are produced to match heritage profiles (see Figure 5.9). Careful consideration must be given to the age and period of the building in order to select the correct moulding. The primary timbers used in the manufacture of these mouldings are clear finger-jointed hoop or radiata pine, or typically western red cedar. Alternatives are California redwood or Australian cedar, which was commonly used in heritage buildings. It was

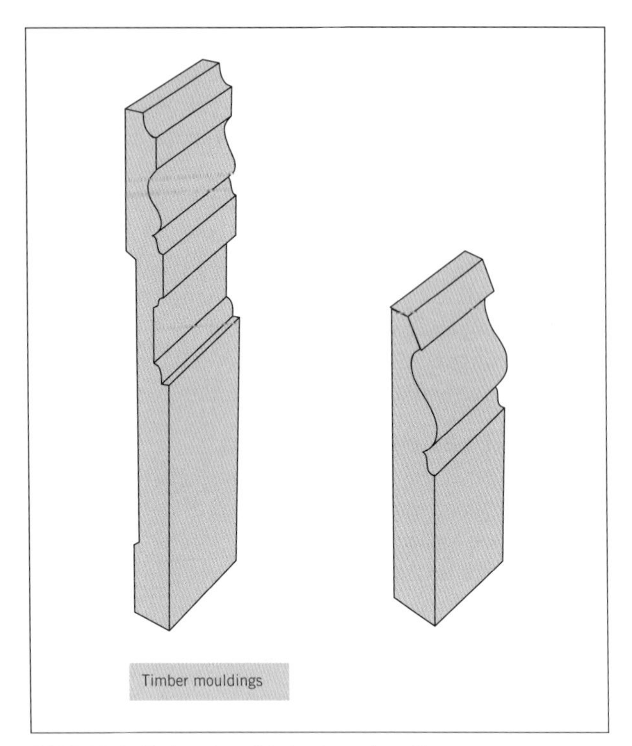

FIGURE 5.9 Federation skirting board profiles

not uncommon for high-quality Australian red cedar to be filled and stained to look like Queensland silky oak or English oak.

Some timber suppliers reproduce original mouldings and match existing mouldings when the specifications call for matching to be done. However, it is very expensive to match existing mouldings. Using recycled architraves and skirting is an economical alternative.

Many of the original mouldings are now reproduced using MDF, which can be bought pre-primed ready to paint.

For more information on the range of skirting and architrave profiles available, see the websites listed at the end of this chapter.

Picture rails

The picture rail is a special moulding that is placed at a level height, usually in period homes and on walls over 2.4 m high (see Figures 5.10 and 5.11). In days gone by, ornate plaster moulds were used to decorate timber battens that were then fixed to the masonry wall using wedge plugs and nails (see Figure 5.12).

Quad, scotia and other mouldings

Quad and scotia are moulds used to cover gaps and provide a neat finish to boards and panels in corners (see Figure 5.13). They are available in a range of sizes from beads, 9.5 × 9.5 mm, to larger sections, 42 × 42 mm, with offset variations such as 32 × 19 mm to cover and finish materials of different thicknesses.

They can be used to cover gaps between walls and floors when shrinkage occurs in bearers and joists. The back corner is usually chamfered to allow the moulding to fit tightly into corners.

Typical joints used to join quad and scotia moulds include **scarf joints** for joining lengths and scribe joints for corner intersections.

Crown mouldings

Crown moulding is the description given to a range of mouldings designed to finish off the top edge of walls (also known as cornice mould – see Figures 5.14 and 5.15), columns and cabinets (see Figure 5.15). As cornice

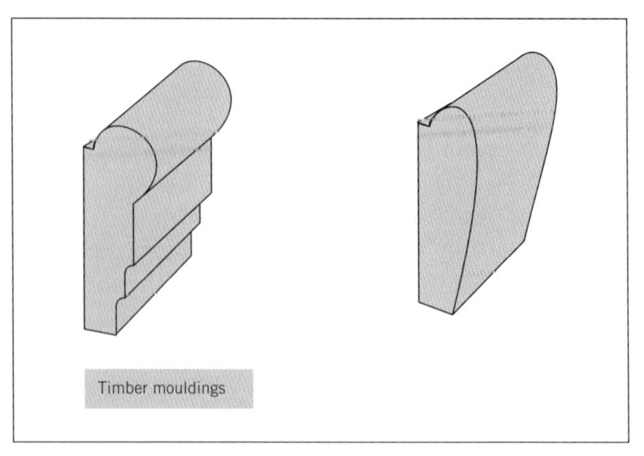

FIGURE 5.10 Picture rail profiles

Source: Ed Hawkins

FIGURE 5.11 Picture rail

Source: Greg Cheetham

FIGURE 5.12 Plaster picture rail

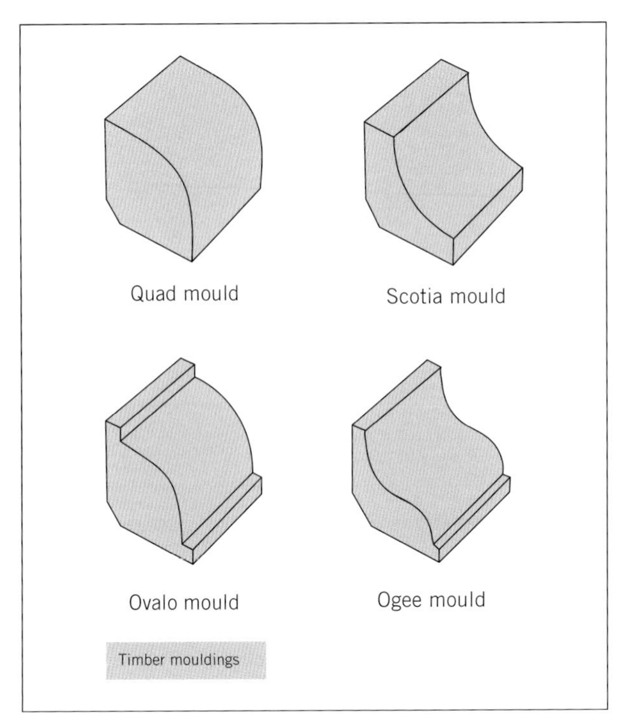

FIGURE 5.13 Typical mouldings

Source: Greg Cheetham

FIGURE 5.14 Plaster cornice

mould, **crown mouldings** are generally decorated plaster or wooden trim used where walls meet ceilings.

Other special moulds

There is a range of other special mould shapes used in the industry for different purposes; for example:

- *Splay bead* is used as bead to fix glass or panels into rebated frames (see Figures 5.16 and 5.19).
- *Staff (or stop) bead* is used to form a channel for the bottom sash or casement sash to run in (see Figure 5.17 and Figure 9.44 in Chapter 9).
- *Fly mould* is used to secure fly-screen wire to timber fly screens for doors and windows (see Figure 5.19).
- *Cover strip* is used to cover the gaps and joints in sheet panels used to clad and line walls (see Figures 5.18 and 5.19).

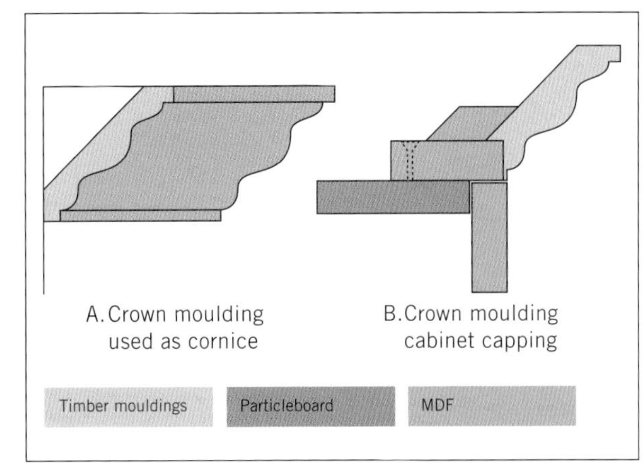

FIGURE 5.15 Crown moulding used as cornice and cabinet capping

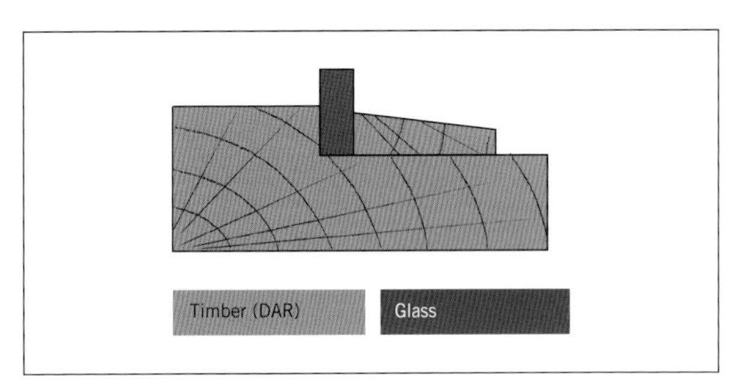

FIGURE 5.16 Splay bead

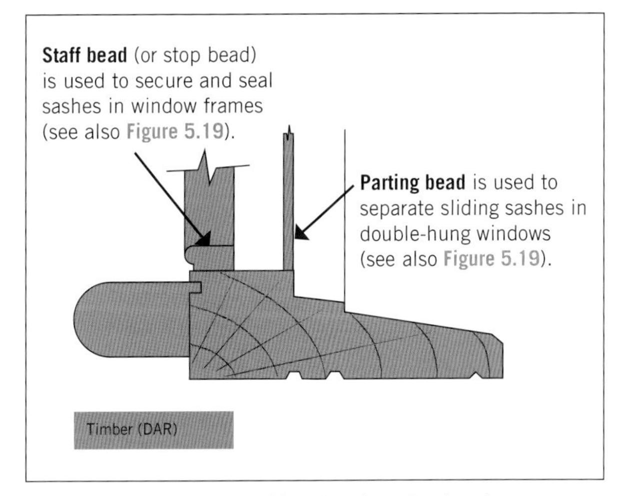

FIGURE 5.17 Staff (or stop) bead and parting bead

Source: Ed Hawkins

FIGURE 5.18 Cover strip

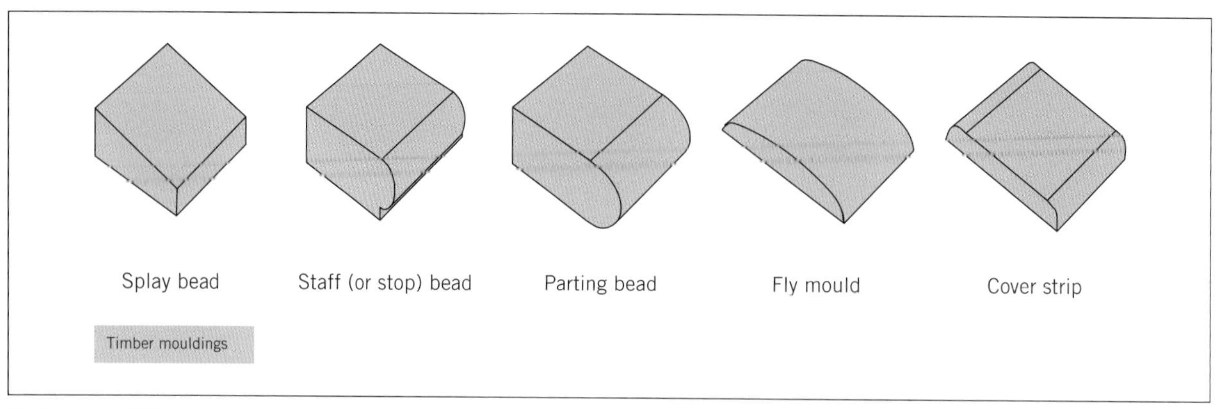

FIGURE 5.19 Other special moulds

- *Nosing* is used to cover gaps between window sills and internal walls and provides a ledge on which things can be rested (see Figures 5.20, 5.21 and 5.23).
- An ***apron piece*** is a piece of moulded timber placed below the nosing board to cover the gap between the wall finish and the nosing board (see Figure 5.22). It can be a scotia, ovalo or quad mould, or alternatively, a piece of architrave. Return mitres are used at the ends of the nosing and apron pieces so that a constant moulded profile is maintained (see Figure 5.23).

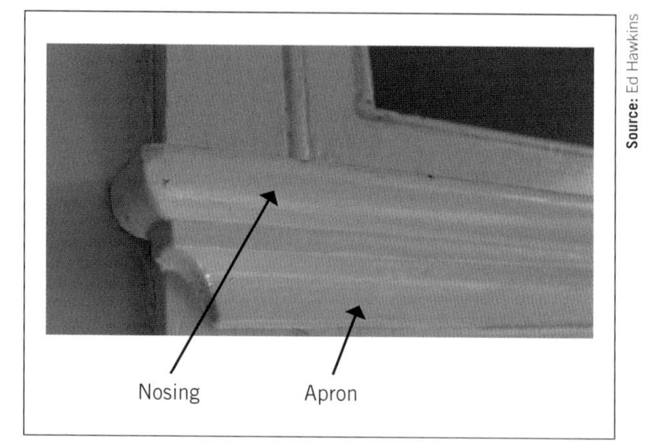

Source: Ed Hawkins

FIGURE 5.21 Window nosing and apron

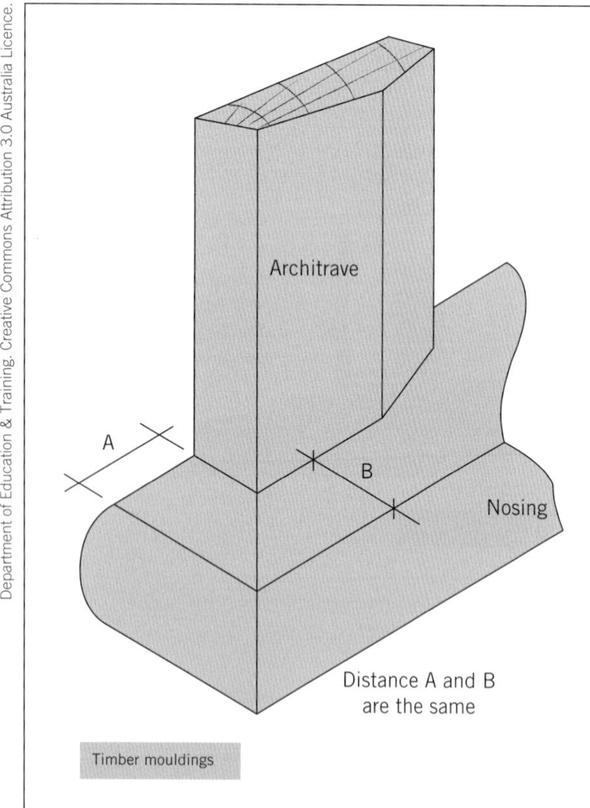

FIGURE 5.20 Determining nosing overhang

Source: Ed Hawkins

FIGURE 5.22 Bull-nose nosing and scotia apron piece

In modern construction, nosings and apron pieces are not common, having been replaced by simply mitring the architraves all around the window frame (see Figure 5.24).

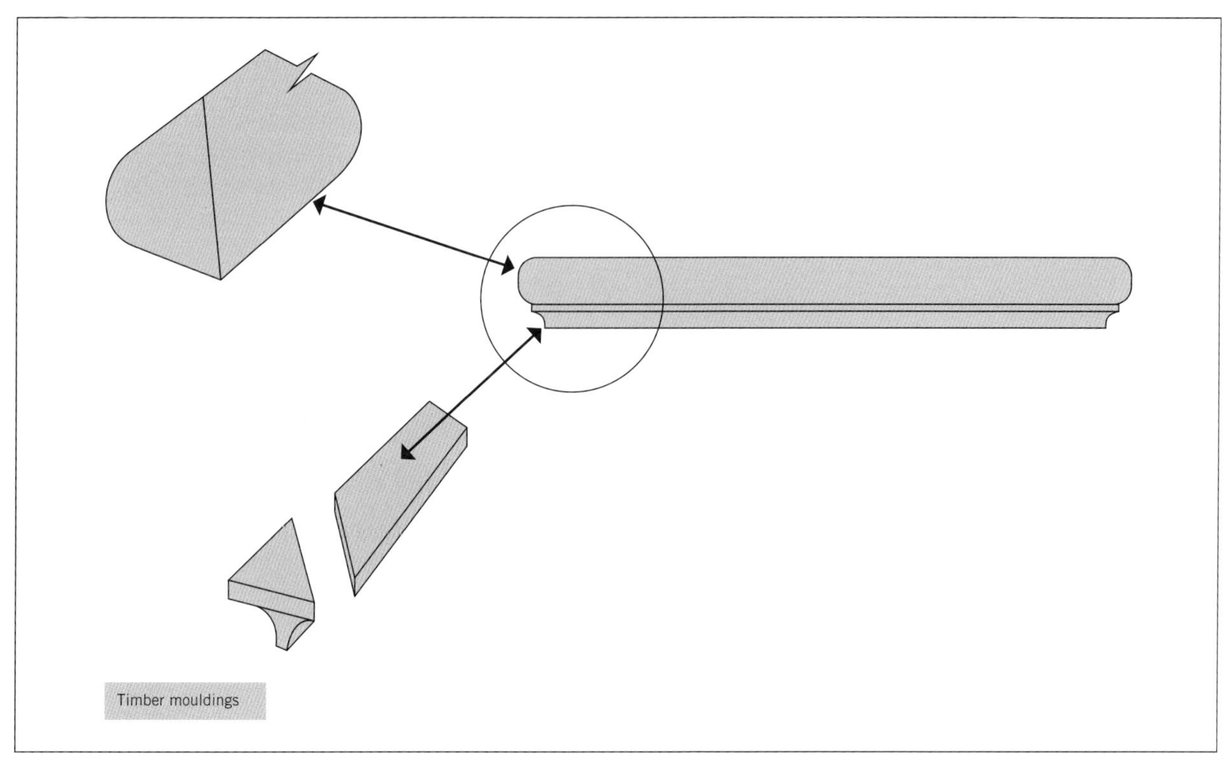

FIGURE 5.23 Bull-nose nosing and scotia apron piece – diagram

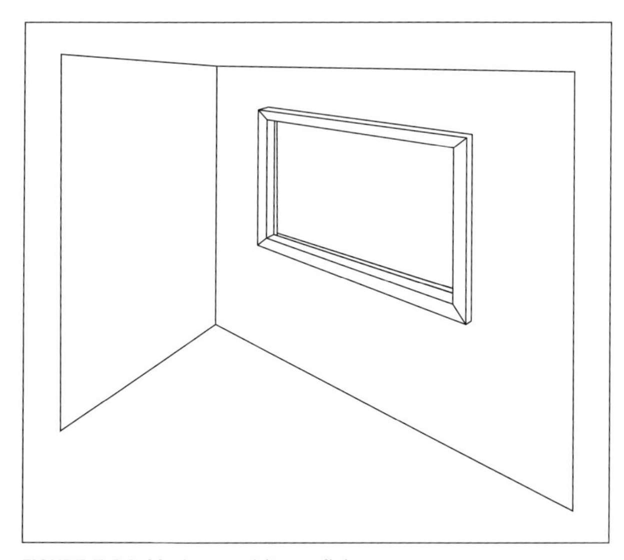

FIGURE 5.24 Modern architrave fixing

LEARNING TASK 5.2 TIMBER MOULDINGS

Using the internet, investigate a number of timber moulding specialists in your region. In doing this, prepare a report that:

1 provides the companies' contact details
2 outlines their product range
3 provides a *labelled* sample piece/section of:
   a architrave mould – indicating the mould shape
   b ovalo mould
   c cornice mould
   d bolection mould.

Each piece must be 90 mm long and presented on a board to your teacher with your name and class number on the back.

To help with your research, try using this search string: timber moulding specialist + <insert your town>.

Sheet material

Plasterboard

Plasterboard is made from gypsum with a strong paper-based material covering suitable for accepting a range of finishes including paint and wallpaper. It is the most common form of wall lining material used. Plasterboard is available in lengths from 2400 mm to 6000 mm and regular widths of 1200 mm and 1350 mm to suit wall heights and stud spacings. The typical thickness for standard wall partitions is 10 mm and 13–16 mm for fire- and acoustic-rated wall finishes. The board is made with recessed edges to allow for flushing, using typical plasterboard setting techniques.

Sheets are fixed to timber wall frames using lattice head nails and stud adhesive, with perimeter nailing at 150 mm centres, 10–15 mm from the edge and no closer than 50 mm from the end. Nails should be spaced 300 mm apart around door and window openings.

Plasterglass is plasterboard that has been reinforced with fibreglass filaments for extra strength. It is usually heavily patterned and much heavier than plasterboard. It is commonly found in period houses up until the major uptake of plasterboard in the 1960s in Australia (see Figure 5.25).

Note: Plasterboard is usually installed and set by qualified tradespersons.

FIGURE 5.25 Fibre-reinforced plaster ceiling

Fire-rated plasterboard

Plasterboard and plaster products are considered non-combustible materials. Special plasterboard and fixing systems have been developed to meet fire and acoustic (noise transmission) rating levels.

The issues around fire-rated plasterboard walls are too many and complex to be covered in detail in this chapter. However, in general, fire and acoustic ratings using plasterboard are achieved by careful placement and layering of plasterboard sheets. See the Web-based resources section at the end of this chapter for links to fire-rated plasterboard installation guides.

Fire-rated sheet layers must have staggered joints and, like plywood, are often required to be laid in alternate directions with joints taped and set in each layer.

Plasterboard used to achieve acoustic ratings is available from major manufacturers. In order to reduce noise transmission significantly, wall frames are generally specially built with cavity dampers (insulation), split frames and dual frames (in multi-residential accommodation), and in some cases require a specialised frame to be fixed to existing walls to eliminate transferred vibration.

The relevant Australian Standards (AS) applicable to fire and acoustic properties for plasterboard and other wall lining materials are:

- AS/NZS 2589 Gypsum linings – Application and finishing
- AS 1530.1 Methods for fire tests on building materials, components and structures – Combustibility test for materials
- AS/NZS ISO 717.1 Acoustics – Rating of sound insulation in buildings and of building elements – Airborne sound insulation.

LEARNING TASK 5.3

FIRE RATINGS FOR WALL LINING SYSTEMS

Fire ratings for plasterboard and other wall lining systems are used in medium density and fire-prone areas.

1. Explain the term 'fire resistance levels (FRL)' in an Australian context.
2. How would you recognise the requirements in a set of specifications or plans and explain each element of the example accurately? Try using the search string: 'Fire resistance levels + Australia' in your web browser.

Plywood

Plywood is often used as an internal lining (see Figure 5.26). It comes in stained and varnished, painted and low-pressure laminate surface finishes. Interior plywood is manufactured to AS 2270 Plywood and block board for interior use.

Plywood is available in a wide range of sheet sizes and veneer finishes. Usually an A grade face finish is required on timber veneers where a clear or stained finish is required. The back surface can be finished with a downgraded C or D grade finish.

Plywood lining has three main surface finishes:

- timber veneered (Good 1 Side – G1S; plain)
- V-grooved (smooth veneered)
- low-pressure laminated (melamine formaldehyde) (usually timber grained).

The most common plywood sheet sizes are 2400 × 1200 mm and 2700 × 1200 mm. Sheets 900 mm wide are also available.

Sheet thicknesses vary between manufacturers; however, standard panel thicknesses range from 3 mm up to 25 mm, with 3 mm to 9 mm being the usual thicknesses used for external cladding.

There are a number of appearance grade veneer qualities specified for plywood in AS 2269 Plywood – Structural – Specifications, including:

- A: high-quality appearance grade veneer suitable for clear finishing of decorative surfaces such as furniture and internal wall panelling

Source: Ed Hawkins

FIGURE 5.26 Plywood wall lining

- S: an appearance grade permitting natural timber defects (e.g. knots) as a decorative feature
- B: used under high-quality paint finishes.

Type C glue bond is suitable for most interior lining applications, but if you are using plywood in damp or wet areas such as kitchens, bathrooms and laundries, a Type B bond is recommended.

More information on grading and specifications on plywood can be found at the Engineered Wood Products Association of Australasia website: https://ewp.asn.au/.

Plywood can be fixed to walls using concealed nailing techniques, or it can be hung on split battens, cut and laid in a pattern or simply slip matched in the same way as particleboard and MDF. The advantages of plywood over other similar materials are its weight and structural strength factors.

Hardboard

Not often seen as a decorative finish for walls and ceilings, hardboard is a useful lining material that does not generally split or crack. Made from highly compressed hardwood fibres, it has a uniform density throughout and a smooth face for finishing, and is suitable for wall and ceiling linings. There are special types, such as pegboard, available for other purposes such as noise suppression and displays. It also comes in pre-primed and finished form.

Hardboard can be fixed directly to studs or to ceiling joists, but trussed roofs require the installation of furring channel systems before ceilings can be installed.

Reduced density hardboard is a similar product but lacks the surface toughness of hardboard; however, it is of sufficient surface strength to be used as a wall and ceiling lining material. It is also used as an underlay for floor coverings, such as vinyl and cork, and as bracing for wall frames. Both hardboard and reduced density hardboard are available in sheets with long bevelled edges ready for taping and setting with regular plasterboard finishing materials.

Tempered hardboard can also be used to line walls, particularly in semi-exposed areas such as carports and in wet areas like bathrooms and kitchens. This material is available in an oven-cured polyurethane paint finish or a laminated finish using DAP (diallyl phthalate) resin or vinyl overlay, as well as in an embossed spray finish simulating a tiled finish (see Figure 5.27).

Source: Ed Hawkins

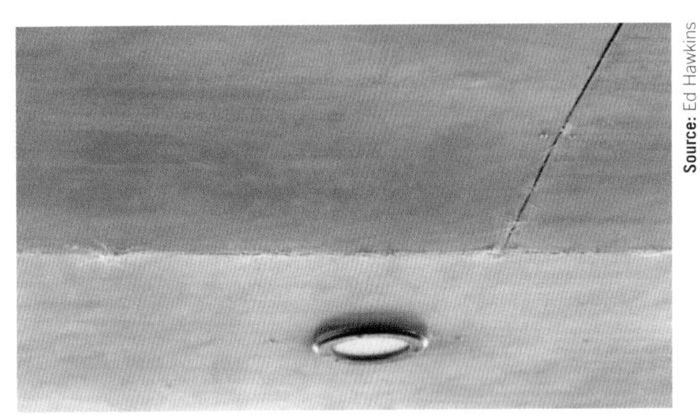

FIGURE 5.27 Hardboard pre-finished lining material

Hardboard is fixed directly to studs or ceiling joists, or walls and ceilings are battened off to take the sheets (see Figure 5.28). Recessed edges can be taped and set similarly to plasterboard.

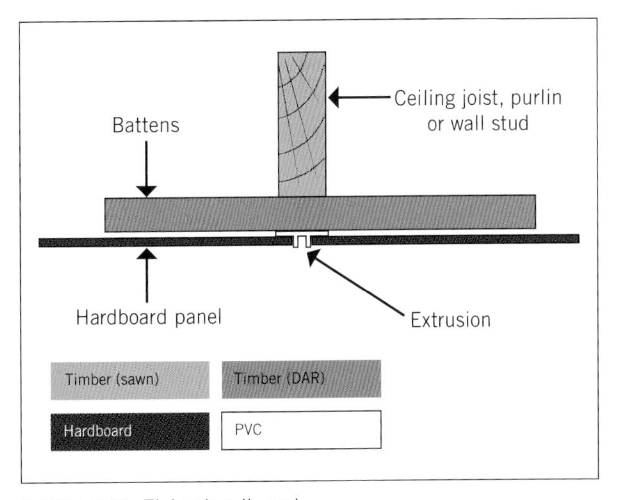

FIGURE 5.28 Fixing hardboard

A modern alternative is to use a powder-coated or anodised aluminium extrusion as a joining strip and then to nail the sheets off to the extrusion, leaving a clean 10 mm reveal between sheet edges.

Wet area plastic laminate

Wet area plastic laminate is a high-pressure laminate sheet product that has a tough melamine formaldehyde surface finish that is suitable for use in wet areas. Sheet thickness ranges from 2.7 mm to 16 mm (depending on the manufacturer), providing a solid flat surface laminate that resists bowing and has excellent moisture resistance. Stud spacings should not exceed 450 mm centres. This sheet provides a grout-free bathroom wall surface finish that is easy to maintain and can easily withstand high temperatures (> 100° C).

Fibre-cement sheeting

Fibre-cement sheets should conform to Australian Standard 2908.2 Cellulose cement products – Part 2: Flat sheets, making them highly resistant to water penetration and rot. They are used extensively as wall linings in wet areas such as kitchens, bathrooms and laundries. They can be tiled over or set and finished similarly to plasterboard and have a paint finish applied. There are also special 'wet area' fibreboards available from most manufacturers. Common brand names are Villaboard and Duraliner.

Fibre-cement sheets can be installed horizontally or vertically. If the joints are to be taped and set for a flush finish, order sheets the height of the wall (nearest over-height sheet) so that the vertical joints are factory recessed; otherwise you will spend additional time grinding the joints for setting.

Typical sheet sizes are as follows:

Width (mm)	Length (mm)
900	2400, 3600, 4200
1200	1800, 2100, 2400, 2700, 3000, 3600, 4200
1350	2400, 3000, 3600

Typical sheet thickness is 6 mm.

Refer to Chapter 4 for more information on cutting, fixing and finishing fibre-cement sheeting in wet areas.

Handling, stacking and storage

When storing sheet materials onsite, including adhesives and fillers, store them out of the weather. Sheets should be stacked flat and off the ground on full-width bearers (or gluts) at minimum 600 mm centres (see Figure 5.29). Edges should be protected from damage. Alternatively, the sheets can be stacked on edge, off the ground on bearers, and leaning against a straight wall or frame (see Figure 5.30).

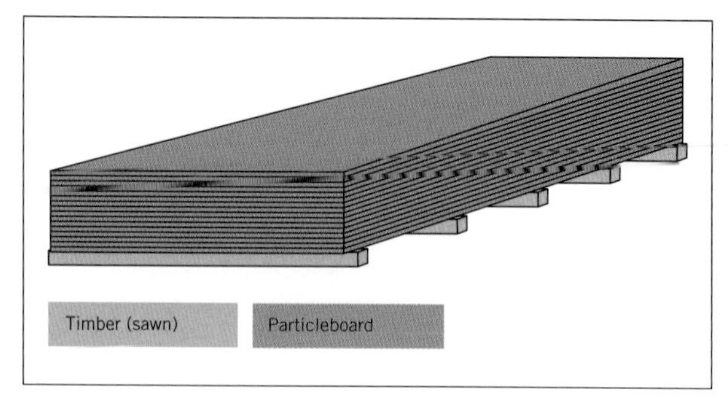

FIGURE 5.29 Flat stacking

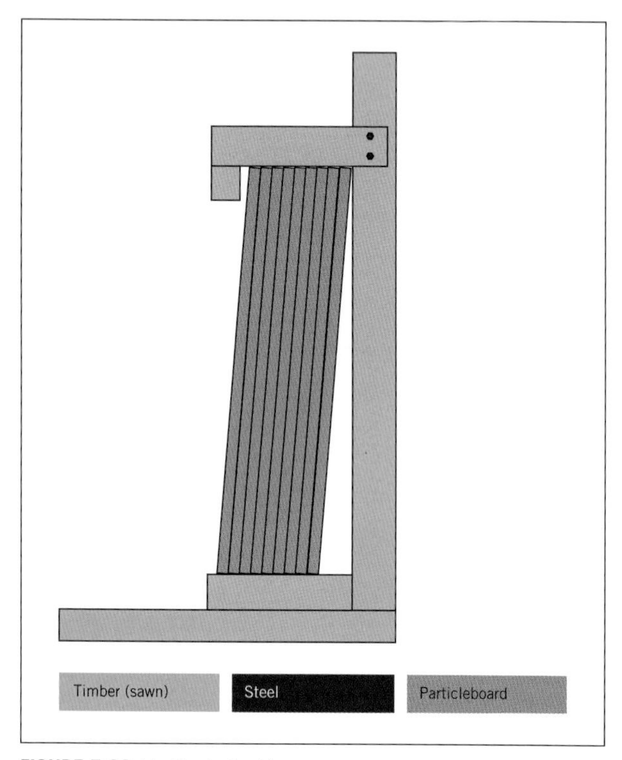

FIGURE 5.30 Vertical stacking

Tie or cleat the stack back against the wall to prevent the possibility of the sheets falling over. A batten, cleat or a clamp and block can be used to hold the stack in place.

Note: All wall and ceiling lining materials should be fire tested, as outlined in the National Construction Code. The appropriate material will depend on the class of building, the location within the building (e.g. common wall in multi-residential building) and whether or not the building is sprinkler system protected.

If the building is **not** fire protected, the material should have a smoke growth rating of < 100. More details on this can be obtained from the National Construction Code and the web links at the end of this chapter.

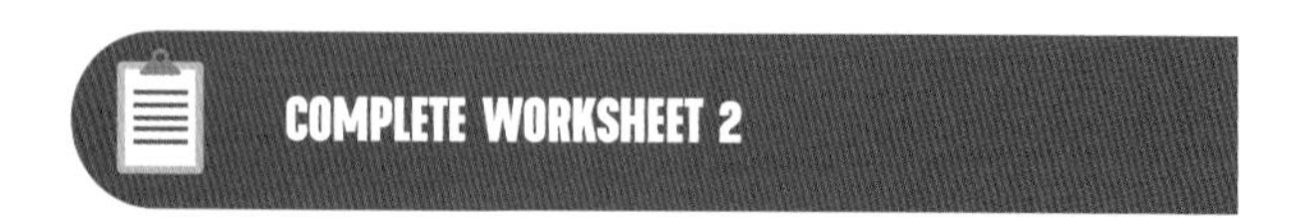

Preparing walls for installation of linings

Framed walls are prepared for internal linings in the same way that external walls are prepared for cladding; that is, the studs are checked for:

- straight – checked for highs and lows with a straight edge or laser level
- plumb – vertical alignment, using a plumb bob, spirit level or laser level
- straight – alignment with other studs, using a straight edge, laser level or string line and block method.

Sprung studs are either planed straight if they are high or packed if they are low or not plumb. If studs are too far sprung to make planing practical, crippling the stud by cutting, wedging and cleating is an acceptable alternative. This process is frowned upon today as it can weaken the stud if not done correctly, but it is sometimes necessary (see 'Straightening wall frames' in Chapter 8 for more details and check AS 1684 Residential timber-framed construction for allowable measures in practice).

When installing lining materials, particularly to ceilings, you may need to organise scaffolding for the site, depending on the ceiling height. The type of scaffolding will be determined by the job requirements. Scaffolding requirements are regulated by AS/NZS 1576.1 Scaffolding – General requirements. You will need to determine the type of scaffolding to be used and the quantity required.

In most circumstances, a mobile scaffold with outriggers will be the correct type. Where a large hall space, for example, is to be lined, consideration should be given to using a scissor lift or other form of elevated work platform. In some states you may need a licence to erect the scaffold and a licence to operate the elevated work platform. Check with your state's WorkSafe organisation to identify these requirements, and make sure only trained operators and erectors perform the work. Ladders and trestles are covered by the AS/NZS 1892 suite of standards and AS/NZS 1576.5 Scaffolding – Prefabricated splitheads and trestles. Elevating work (and similar) platforms are covered under AS 1418 Cranes, hoists and winches suite of standards. The safe use of mobile elevating work platforms is covered by AS 2550.10.

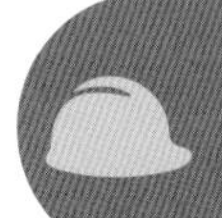

You will need a 'high risk work licence' to carry out scaffolding work. Check with your state or territory's WorkSafe agency to determine the right licence classification for the work involved.

You need to determine a safe method for lifting the sheets or lining materials and supporting the materials while they are being fixed. Gin wheels and block and tackle are useful lifting aids for slinging quantities of timber lengths, while specialised sheet lifters and scissor lifts are commonly used to raise sheet material to suitable working heights.

PPE requirements will vary depending on the job task. Ensure gloves, work boots, hearing protection and eye protection are used when handling materials and working with tools and equipment.

Dealing with hidden services

All hidden services such as electrical, water, gas, ducted air-conditioning and telecommunications need to be identified and roughed in before wall linings are installed.

When renovating, all such services must be disconnected before work can commence safely.

Appropriately licensed tradespeople must be employed to do the electrical and plumbing work – make sure you have allowed for them in your quote and work schedules.

Environmental considerations

When planning to install wall linings, consideration should be given to the inclusion of a suitable insulation material inside the wall cavities or between masonry walls and lining when battens are used. Rockwool batts are the most common material for timber- and steel-framed walls, while polystyrene foam sheeting is used in the batten cavity in masonry walls.

The *National Construction Code 2022* – Volume Two, Part H6 Energy efficiency, provides further guidance on the thermal efficiency of walls and wall systems.

It is important to assess the task and determine what hazards may be presented while completing the task and how they can be controlled. Always refer to the Safe Work Method Statement (SWMS) and the Job Safety Analysis (JSA) for each job.

A site induction may be required that will address specific safety requirements of a site. Hazards should be identified, and where possible controls and safety signage installed. Safety concerns should be monitored and reviewed during construction, especially if the work is carried over several days or when other trades are also working on site.

COMPLETE WORKSHEET 3

Calculating material quantities

Never assume that a room is perfectly symmetrical, rectangular or square. The following factors can also influence material calculations:

- surface area to be covered
- type of lining material
- direction of lining
- door and window opening sizes
- effective cover of timber lining boards
- sheet sizes available – all dimensions.

As the basic calculations for establishing material quantities and associated costs are the same as those used for calculating quantities of external cladding (see Chapter 8), use the same formula for your calculations:

Square metre calculations = Length × Width (or height) = m^2

Perimeter calculations = (Length + Width) × 2 = Perimeter distance for basic squares and rectangles

You may need other formulas such as:

Surface area of triangles = $\frac{1}{2}$ base × perpendicular height

Perimeter of a circle = π × diameter (for arches and arcs)

To find the hypotenuse of a triangle (for example, the length of a raked ceiling), use **Pythagoras' theorem** (see Figure 5.31):

$$C = \sqrt{A^2 + B^2}$$

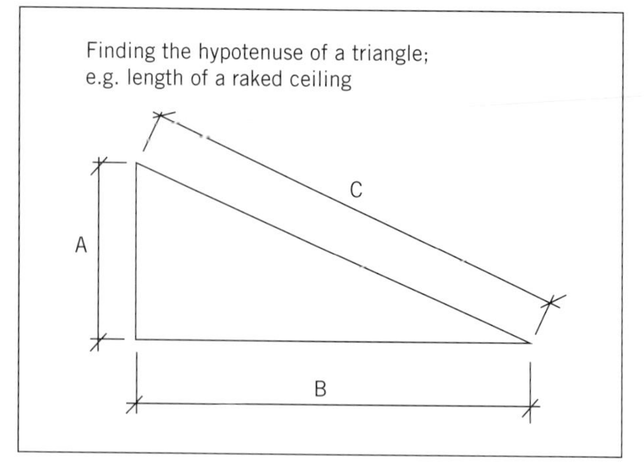

FIGURE 5.31 Right-angled triangle

Setting out lining materials

There are several factors that influence the setting out of sheet and timber lining materials. These factors are:

- direction of the lining materials (i.e. horizontal, vertical or diagonal – see Figures 5.32 and 5.33)
- lines of fastening, such as stud or batten positions
- starting point for fixing
- visual or aesthetic balancing of materials.

In general, sheet lining is installed at 90° to the stud or ceiling joist direction. When this is not possible, battens or additional rows of noggins must be installed at specified spacings.

If the building has a truss roof system, and a ceiling lining is to be installed, a furring channel should be installed or, alternatively, the trusses need to be battened off (see Figure 5.34).

For more information on plasterboard fixing details, you can obtain a technical manual and view instructional videos on the websites of CSR Gyprock, USG Boral and Rondo, which are listed at the end of the chapter.

Lining boards placed vertically on the wall create an illusion of higher ceilings.

Fixing to steel studs requires the wall to be battened using timber battens so that secret nailing techniques

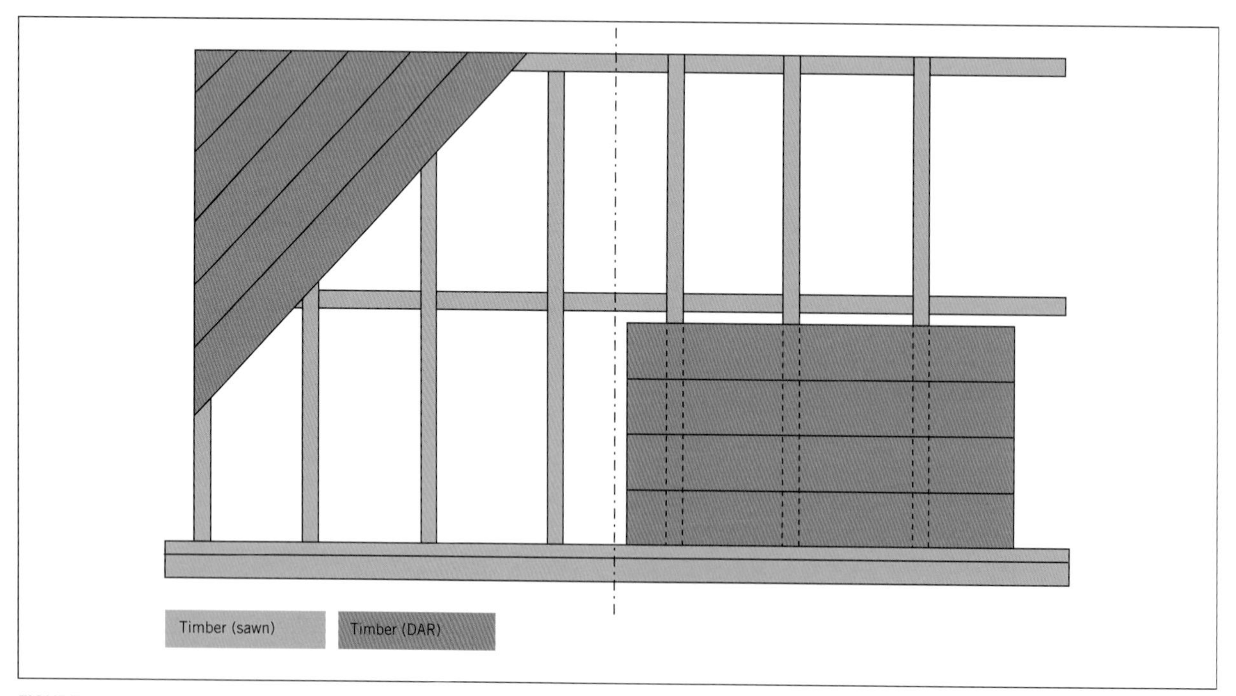

FIGURE 5.32 Horizontal and diagonal fixing – no additional battening required

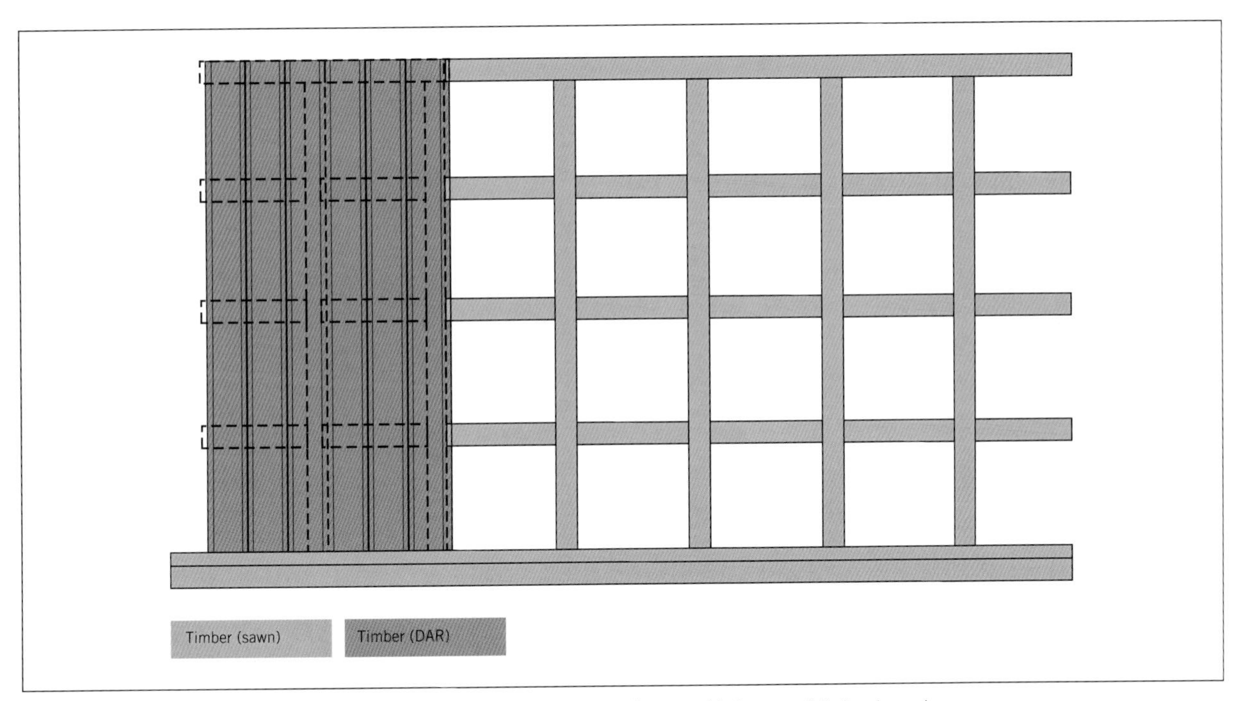

FIGURE 5.33 Vertical fixing – additional noggins required, depending on thickness of lining boards

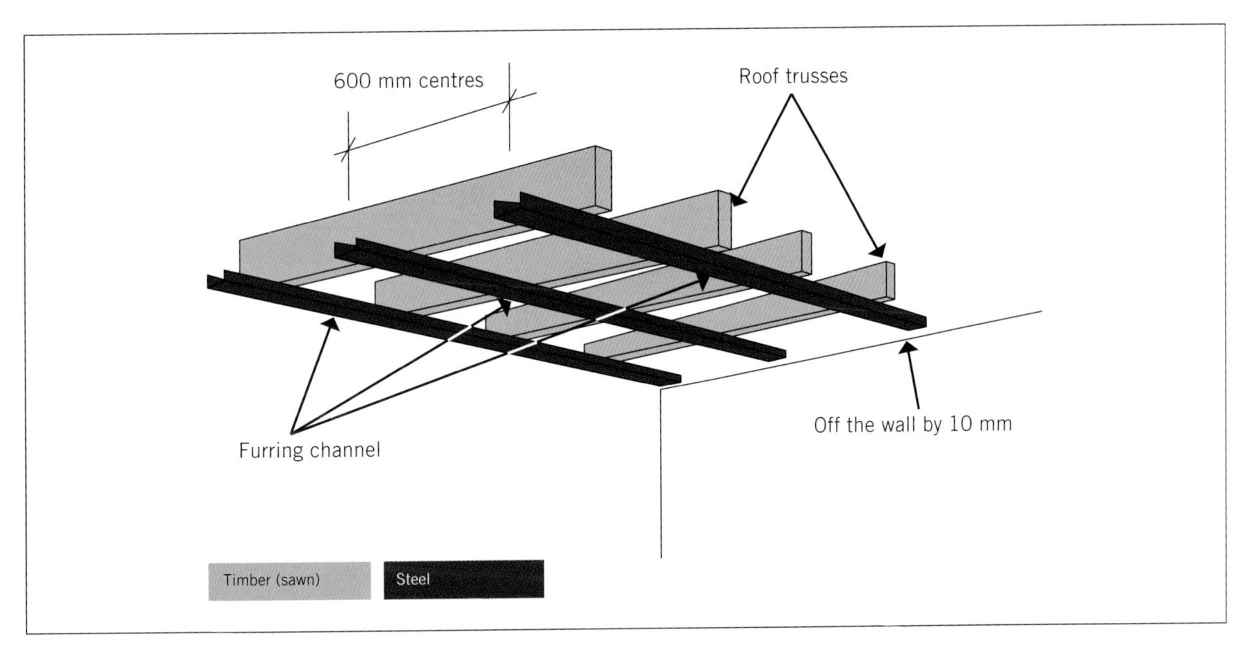

FIGURE 5.34 Furring channel for truss roof systems

can be used (see Figure 5.35). Horizontal battening is common practice, but battens can also be fixed vertically flush with the edge of the studs. Additional timber noggins can be inserted into the steel frame.

The visual appearance of wall linings can determine the way wall panelling is applied. For example, when veneered sheet panelling has a plain grain pattern, simple slip matching techniques can be applied, as in Figure 5.36 (A); however, when using a highly figured grain pattern, you may wish to centre the pattern on the wall, as in Figure 5.36 (B).

Diagonal fixing requires the use of centre lines to achieve a balanced visual aesthetic (see Figure 5.37).

Fixing battens to a wall surface

Plain battens are used to support panels and lining boards when:

- no studs are available
- masonry walls are to be covered by materials that cannot be adhered directly; that is, dado panelling
- walls are too rough for direct fixing.

Split battens are a useful means of attaching wall panelling to wall frames where there is no visible means of fixing. This system is useful if panelling needs to be removed to gain access to hidden services, to be cleaned or to be replaced regularly.

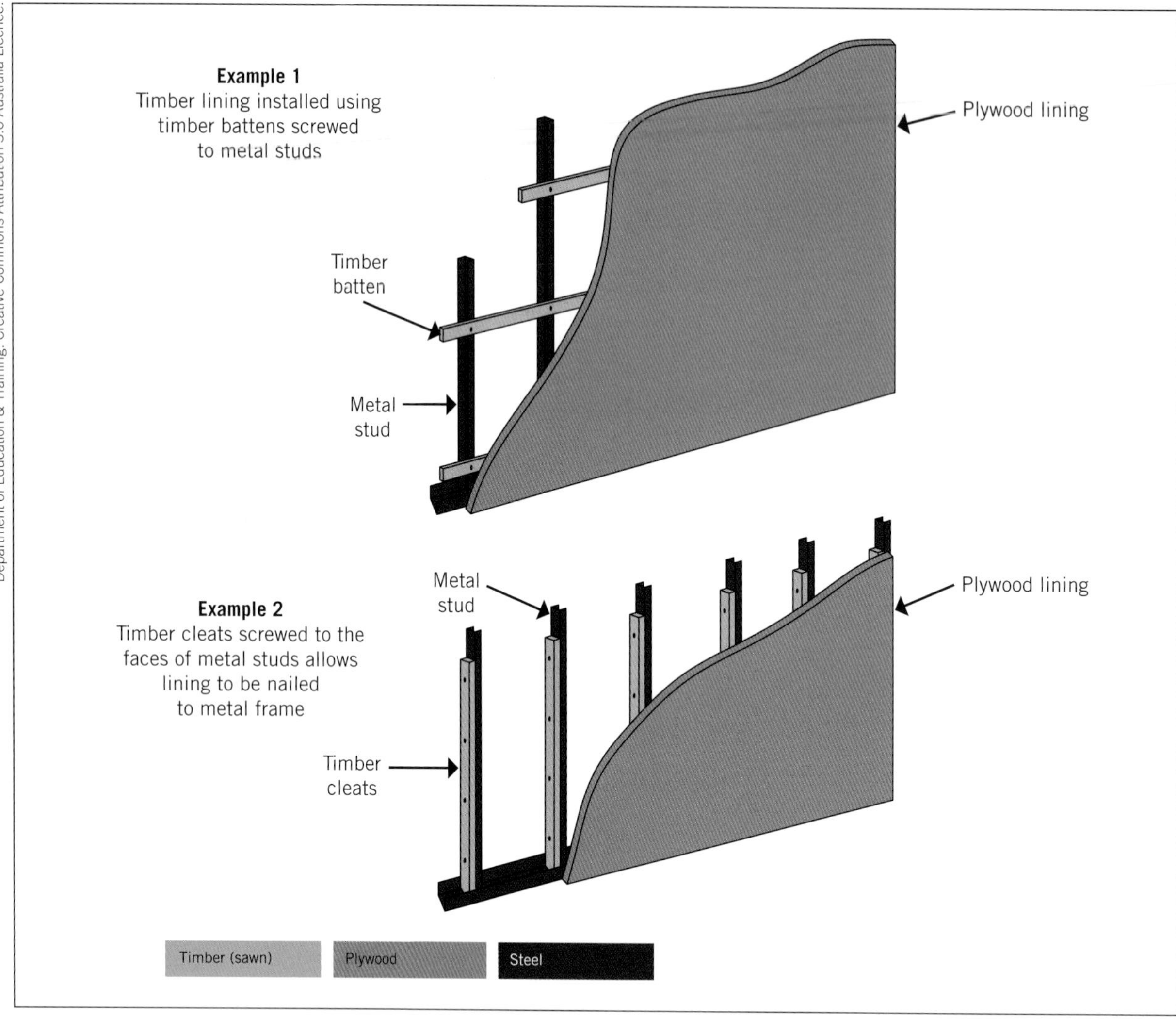

FIGURE 5.35 Fixing lining boards to steel studs

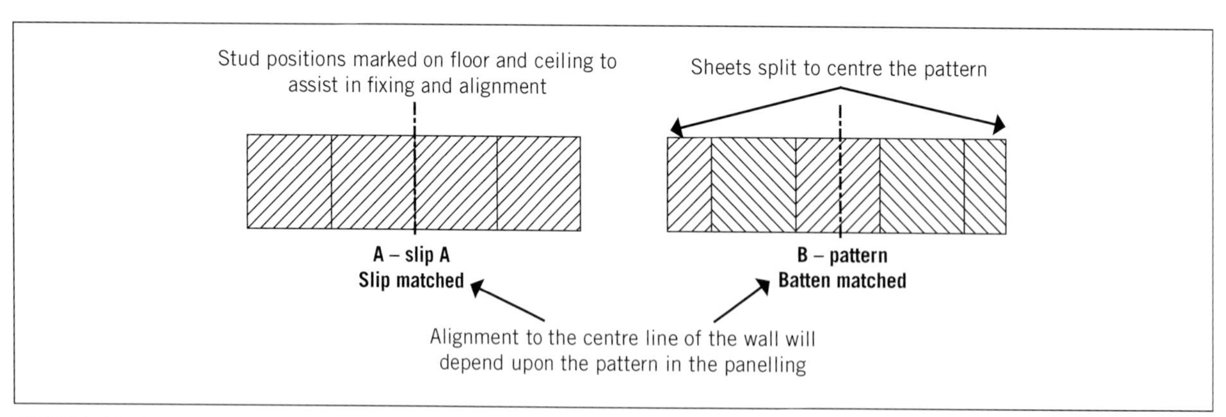

FIGURE 5.36 Grain and pattern matching techniques

Make sure that the wall battens are fixed level and at the correct height in relation to the panel battens to ensure the correct alignment.

As panel battens need to be attached to the back of each panel at exactly the same height, a jig is required to ensure even and accurate spacing (see 'Direct fixing of sheet materials to masonry' later in the chapter for more set-out details).

Fixing lining boards to masonry walls requires the installation of *timber or steel battens*. The battening process ensures that irregularities in the surface are corrected and that linings are protected from moisture penetration. It also allows nail or screw fixing and provides a cavity for services behind the lining.

Battens can be fixed vertically or horizontally (see **Figure 5.38**), but this will depend on the

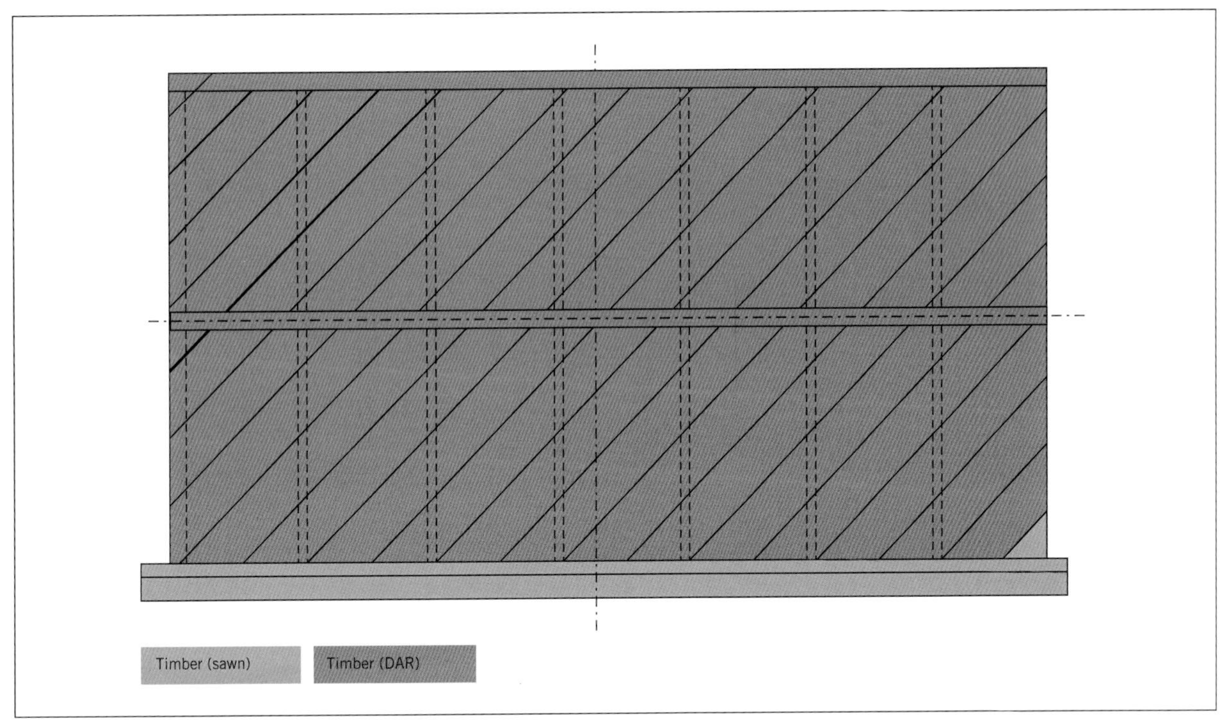

FIGURE 5.37 Balanced visual aesthetic

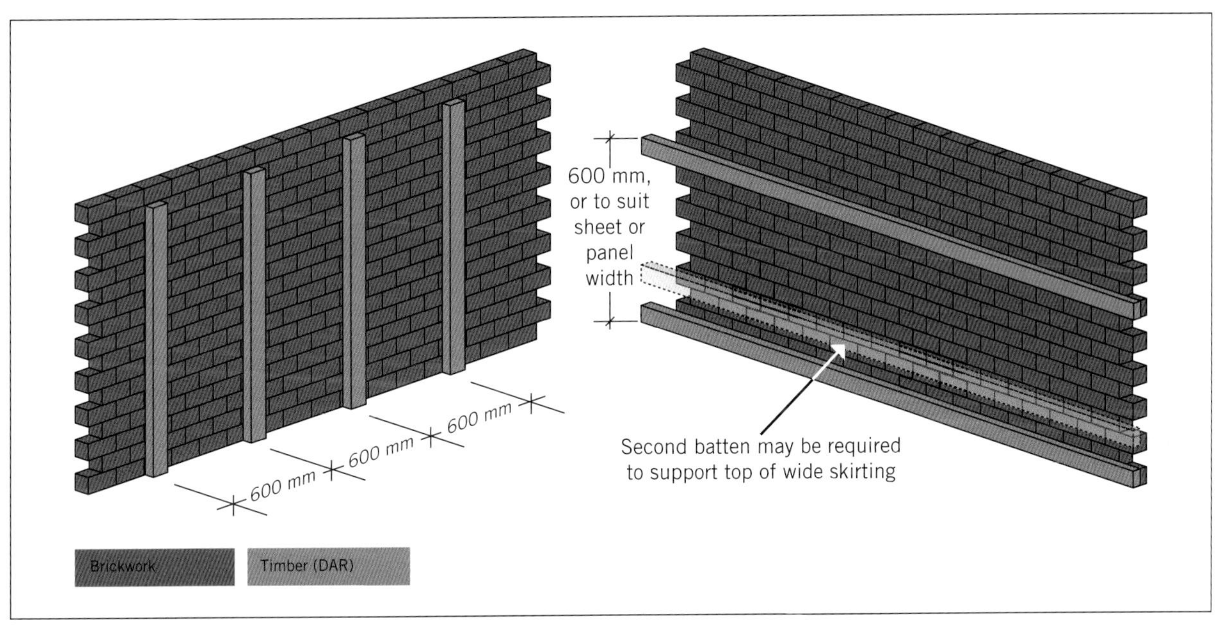

FIGURE 5.38 Vertical and horizontal batten fixing

panelling material being used and the requirements of the job.

Battens are fixed to a masonry wall using a combination of adhesive and masonry anchors, or screws and star plugs (see Figure 5.39). Masonry anchors that usually sit proud of the surface must be counter-bored or sunk into the battens in order for the lining materials to sit flush against the battens and not show the lumps and bumps of the anchors (see Figure 5.40).

For more information on selecting the appropriate masonry anchor system, see the masonry anchor websites listed at the end of the chapter.

Packing may be required behind battens to align them, keep the fixing surface flat and plumb, and to hide irregularities in the surface (see Figure 5.41).

Installing lining materials

A range of tools are required to fix lining materials depending on the materials being used. Some of those tools include hand tools, pneumatic tools (see Figure 5.42) and both corded and battery-operated tools. For fixing timber architraves, mouldings and

FIGURE 5.39 Masonry plug

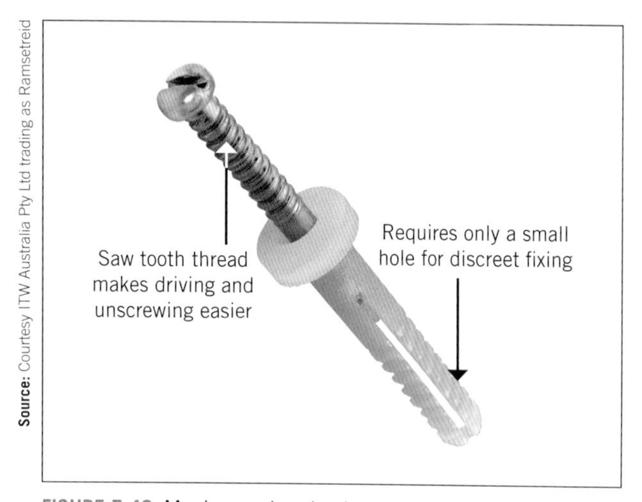

Source: Courtesy ITW Australia Pty Ltd trading as Ramsetreid

FIGURE 5.40 Mushroom-head nylon masonry anchor

Source: Australian National Training Authority. Copyright Commonwealth of Australia. Department of Education & Training. Creative Commons Attribution 3.0 Australia Licence.

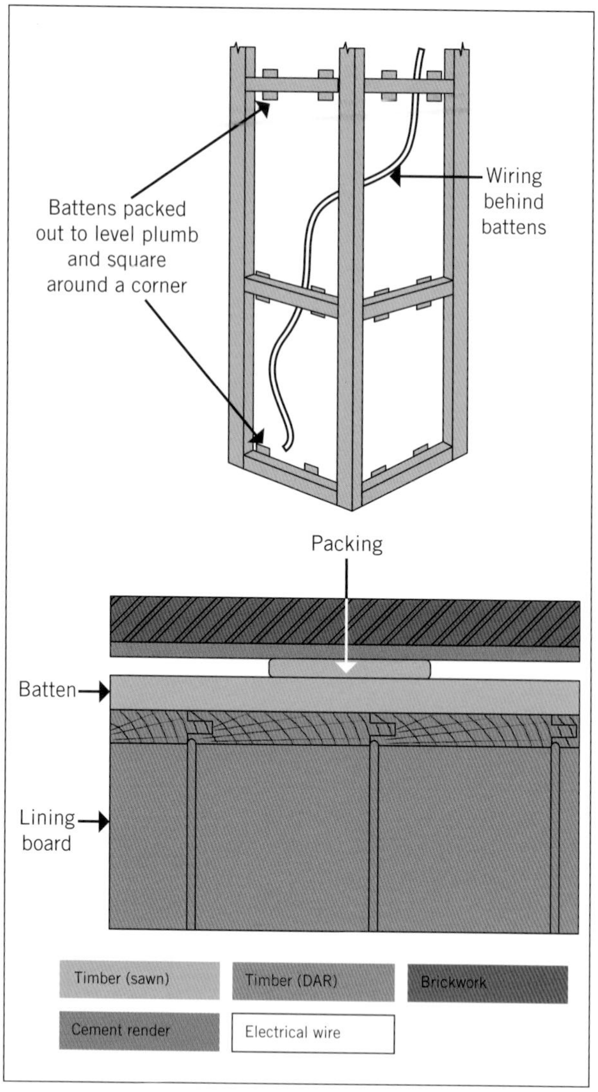

FIGURE 5.41 Packing-out battens

lining boards a brad gun is the most common. Brad guns are available in both pneumatic and battery-operated models.

Other tools that will be required include:

- hand tools:
  - hand saws including tenon and coping
  - hammers – soft faced and claw
  - squares (combination and tri) and bevels
  - chisels
  - caulking guns
  - cramps – G clamps and quick release
  - measuring tapes, rules, marking equipment, spirit levels, straight edges and string lines
  - planes – smoothing planes and rebate planes

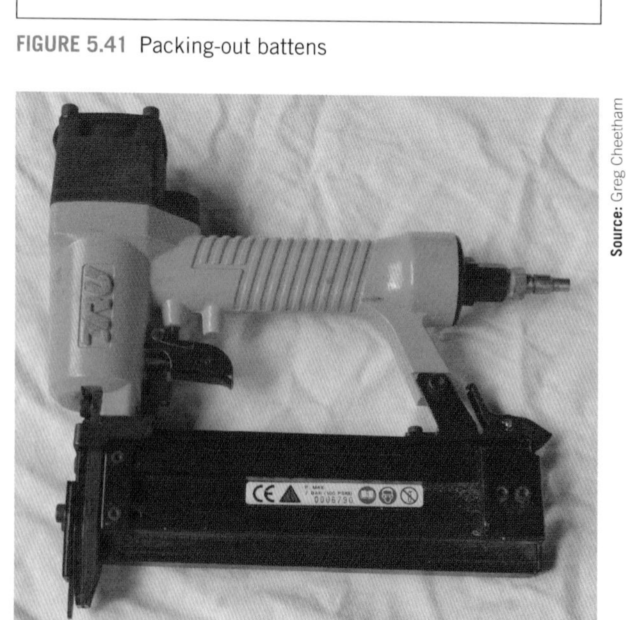

Source: Greg Cheetham

FIGURE 5.42 Brad gun

- power tools:
  - air compressors and hoses
  - drills – battery and corded and power leads
  - power planers
  - power saws – circular saws and jigsaws
  - routers and trimmers
  - laser levels.

Look for obvious signs of damage or wear and ensure they are clean, as this could interfere with their functioning. Check for cracks in handles, buttons, sharpness of blades, hinges or other components to ensure they are functioning correctly. Check power cords have been tagged and tested.

Remove any tools and equipment with obvious faults away from functioning tools. Label and report them as faulty so that other tradespersons won't use them.

Always check work instructions for the type of tools, equipment and PPE that will be required throughout the job. It may vary depending on whether or not the activity requires working at height, or using specialist equipment.

Fixing lining boards

Generally, if the lining boards are to be used in wall frames, the lining fixing centres should be no greater than 600 mm. The board thickness and the timber species may have a bearing on this measurement.

When the boards are being laid over exposed rafters, the fixing centres can vary between 600 mm and 1200 mm, depending on the species and timber thickness.

Lining boards are usually fixed directly to studs using concealed nailing techniques in combination with construction adhesive (see Figure 5.43). End joints can be undercut square across and slightly over length and sprung into position for a neat joint. Some lining boards can be purchased with end-matched joints for off-stud jointing. All other end joints should be pre-drilled and then nailed to reduce the risk of splitting.

Proprietary metal clips are also available for totally concealed fixing of some lining boards (see Figure 5.44).

Vertically aligned lining boards need to have the first or last board, or both boards (depending on the set-out), ripped and scribe fitted to the adjoining wall. Cut and hold the board plumb against the adjoining wall, and using a 50 mm block of timber, scribe up or down the adjoining boards marking the profile on the board (see Figure 5.45). Alternatively, you can use a compass to scribe a line along a wall or floor (see Figure 5.46).

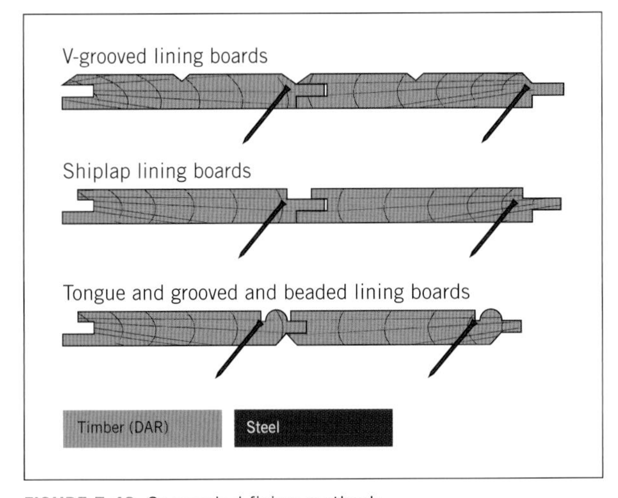

FIGURE 5.43 Concealed fixing methods

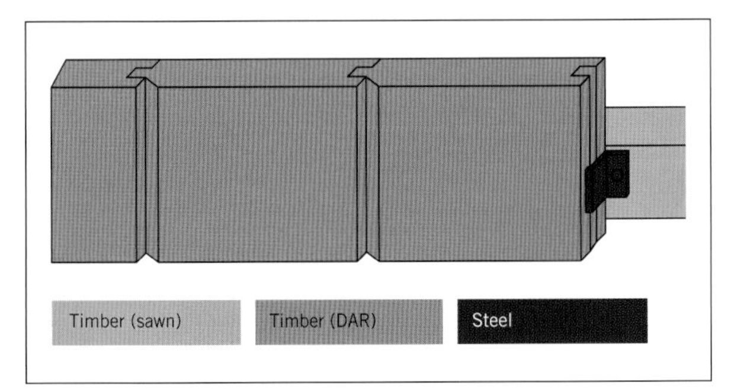

FIGURE 5.44 Proprietary metal clips

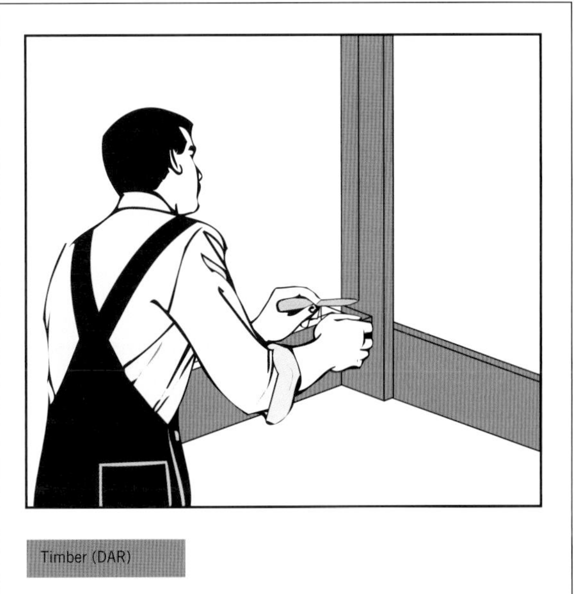

FIGURE 5.45 Scribing the first board

FIGURE 5.46 Scribing with a compass

Permanently fix the first board to the battens with 25 mm panel pins or 30 mm nails for stud walls (depending on the lining board thickness), using concealed fixing methods and nailing at an angle through the **tongue**. Take care not to split the tongue. Check before nailing that you have a neat fit and adjust the fit using an off-cut held against the edge. Check regularly that the boards are being installed plumb.

When you come to a window or door, temporarily nail the last board to the second last board and use an off-cut of cladding or compass to scribe it. Continue cutting short boards for above or below windows and doors as necessary.

Finally, finish off any decorative trims that may be required; for example, skirting, architraves, window aprons, capping moulds and beads, quads or scotia.

Once all the pieces are in place, punch any protruding nail heads below the surface if not done by your nailing tool. Putty the heads of the nails and any major defects, using a colour-matched putty. Lightly sand the project with 150 grit abrasive to remove pencil marks and other blemishes.

Fixing plywood, MDF and veneered particleboard sheeting

Timber-based sheet materials can be nail fixed to timber and steel stud walls using techniques similar to those used for lining boards. As outlined previously, there are two basic methods for aesthetically setting out timber-veneered panels; however, in larger jobs the process can be more complicated and panels from a supplier may need to be matched in different ways:

- **Sequence-matched panel sets** (see Figure 5.47). These work best when walls have no doors or windows and a uniform panel width and height. In expensive jobs, the veneers may be selected from a **flitch**. When more than one flitch is required, careful matching and sequencing is needed to achieve the correct alignment. Doors are not matched and corners may not match.
- **Factory matched sets** (see Figure 5.48). These numbered sequenced sets of sheets are usually 2400 × 1200 mm in size. They can be from one flitch or several near-matched flitches. As they are mass-produced, no two sets will be the same and doors cannot be matched. They are best installed in full-width sheets, with adjustments made at corner sheets.
- **Blueprint matched panels** (see Figure 5.49). These are custom-made to match from a flitch that provides consecutive veneers and a perfect match in grain and colour across the wall. All panels, including doors, match. To ensure that the panel sizes are correct, scaled site drawings may need to be sent to the board manufacturer.

The panels can be installed directly to the frames with the edges closely butted together and nail fixed. Alternatively, an internal spline can be used to attach the panels (see Figure 5.50), or the panels can be suspended on split battens (see Figure 5.51).

It is not uncommon for sheets to be tongue and grooved (see Figure 5.52) or biscuit jointed together to get a flush face.

Sometimes board veneers are laid on special core materials like black MDF. The sheets are then given a slight rebate to reveal the black edge and provide a shadow line when installed (see Figure 5.53).

When boards are installed on split battens, allowance has to be made at the top for lifting the panels on and off (see Figures 5.54 and 5.55). The gap can be covered with a lighting pelmet or a cornice or crown mould.

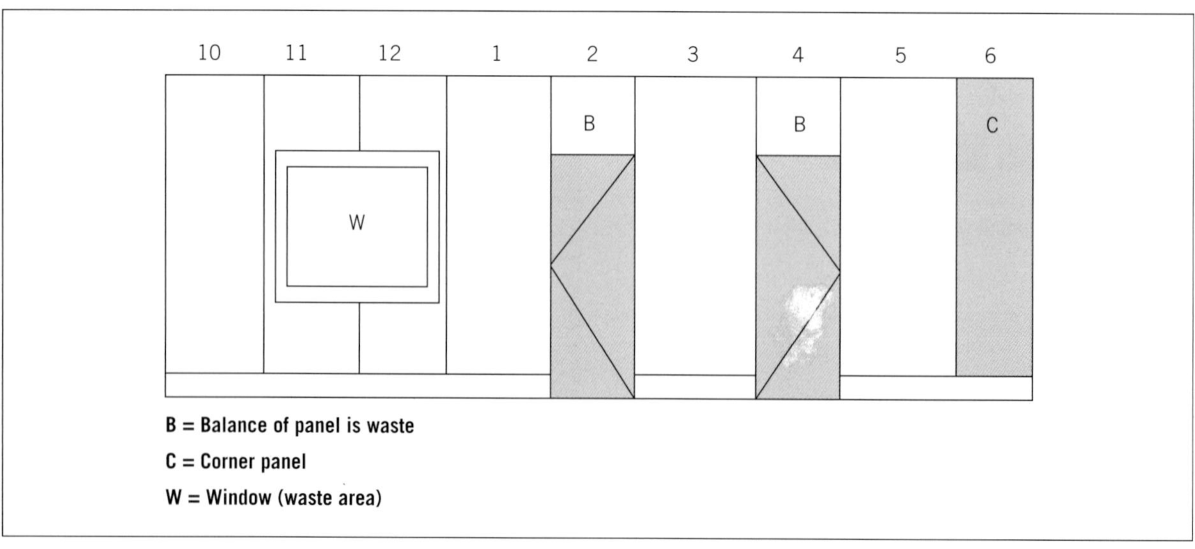

FIGURE 5.47 Sequence matching

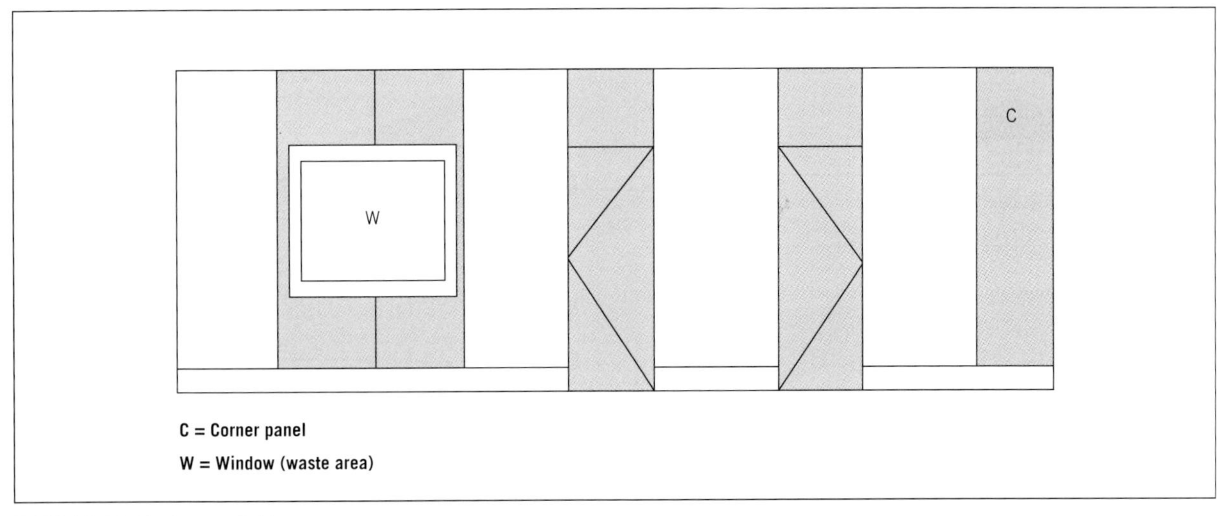

FIGURE 5.48 Factory matching

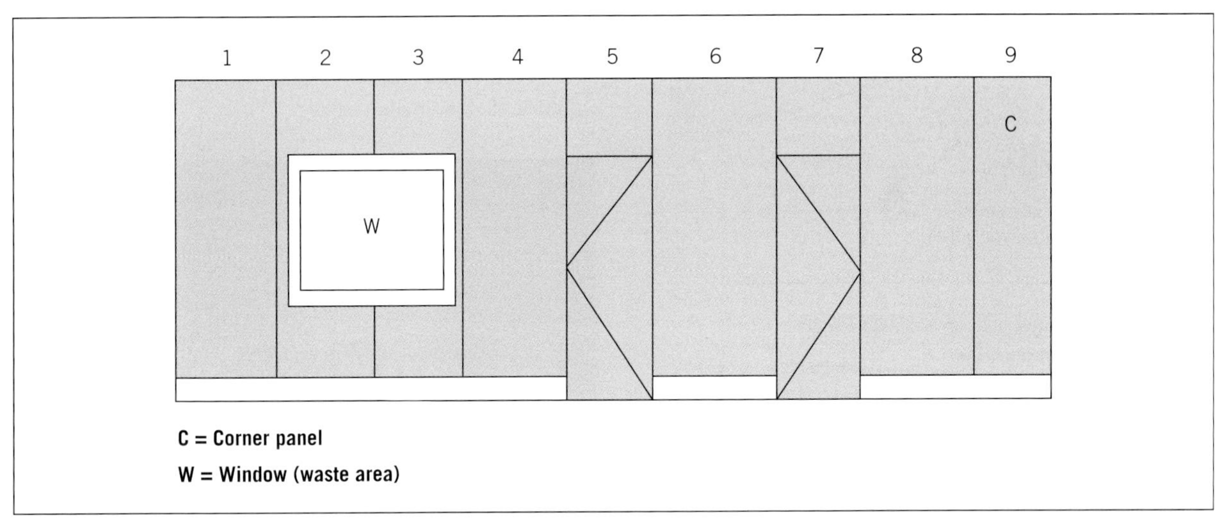

FIGURE 5.49 Blueprint matching

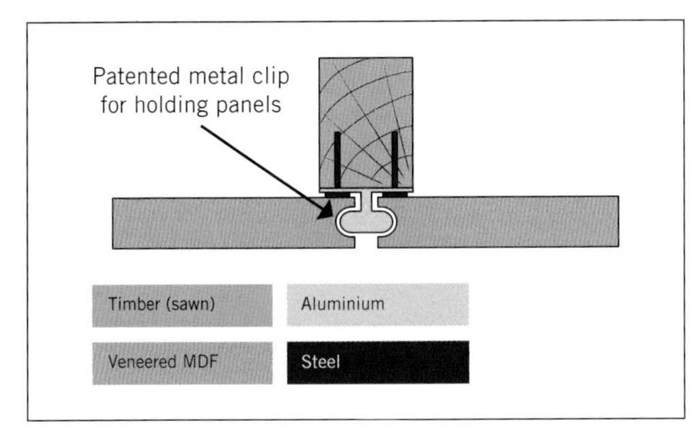

FIGURE 5.50 Patented metal spline fixed to stud

When setting up for split battens, you will need to work from a datum line set up using a water level, spirit level and straight edge, or laser level (see Figure 5.56).

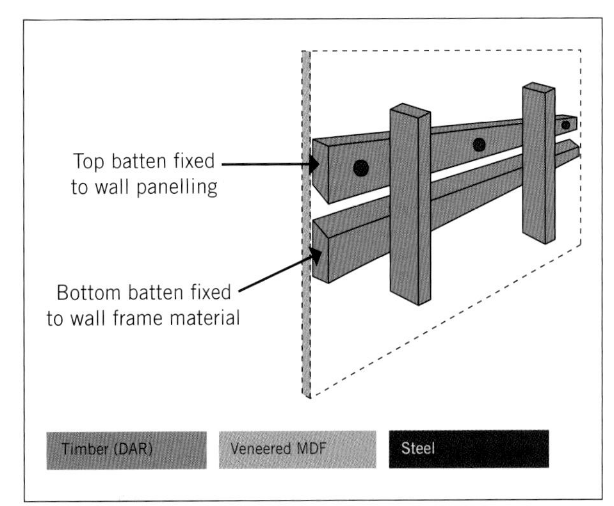

FIGURE 5.51 Split batten

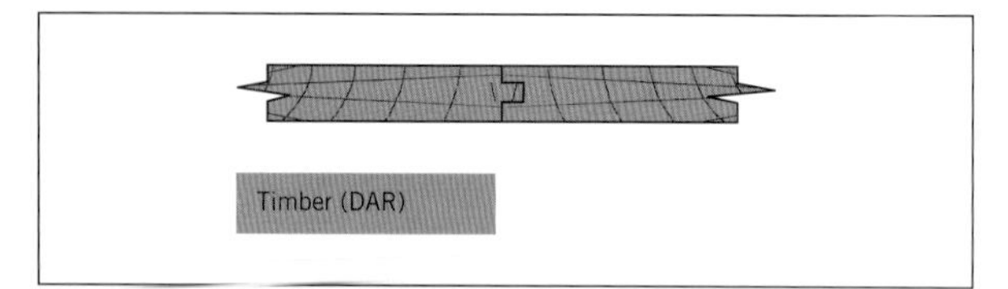

FIGURE 5.52 Tongue and groove joint

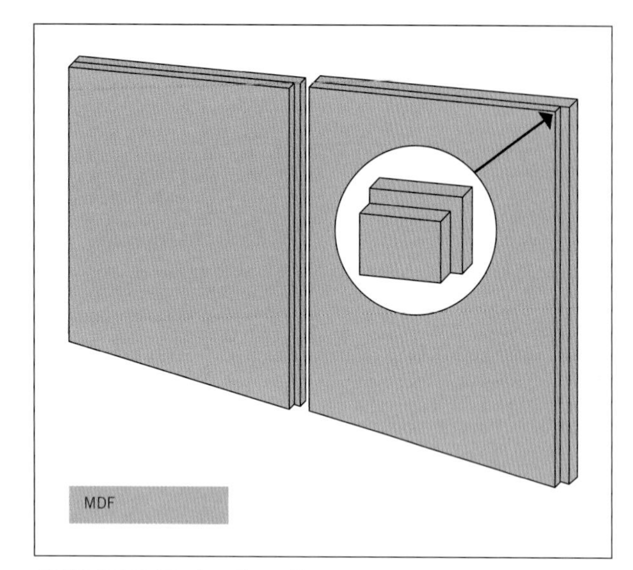

FIGURE 5.53 Shadow line effect

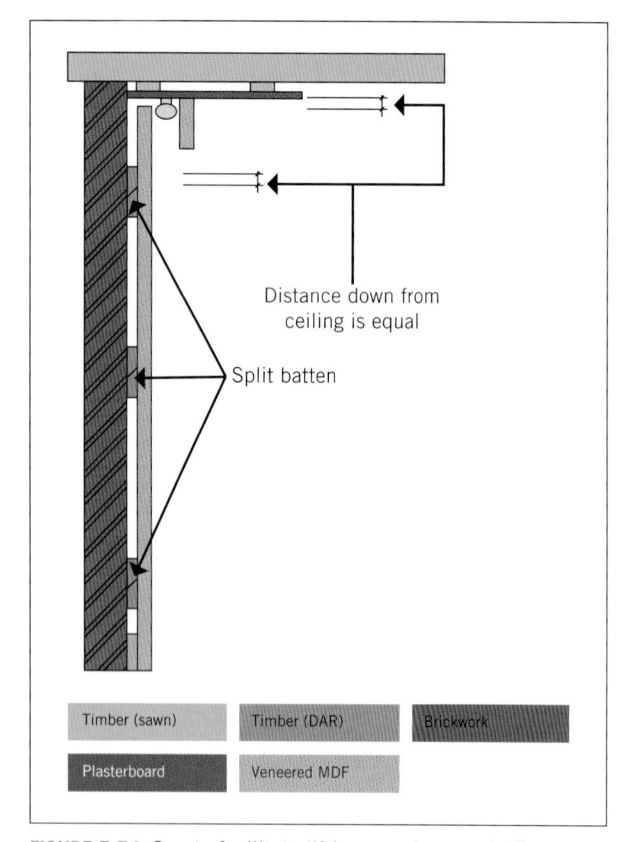

FIGURE 5.54 Gap to facilitate lifting panels on and off

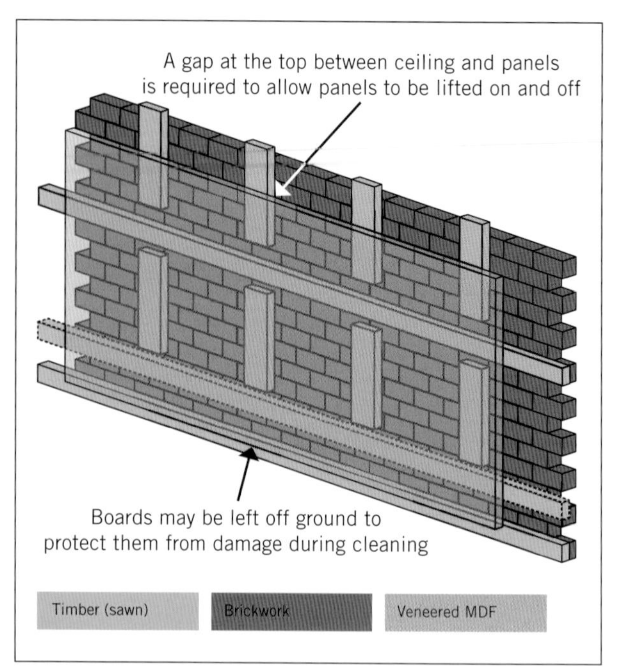

FIGURE 5.55 Split batten fixing

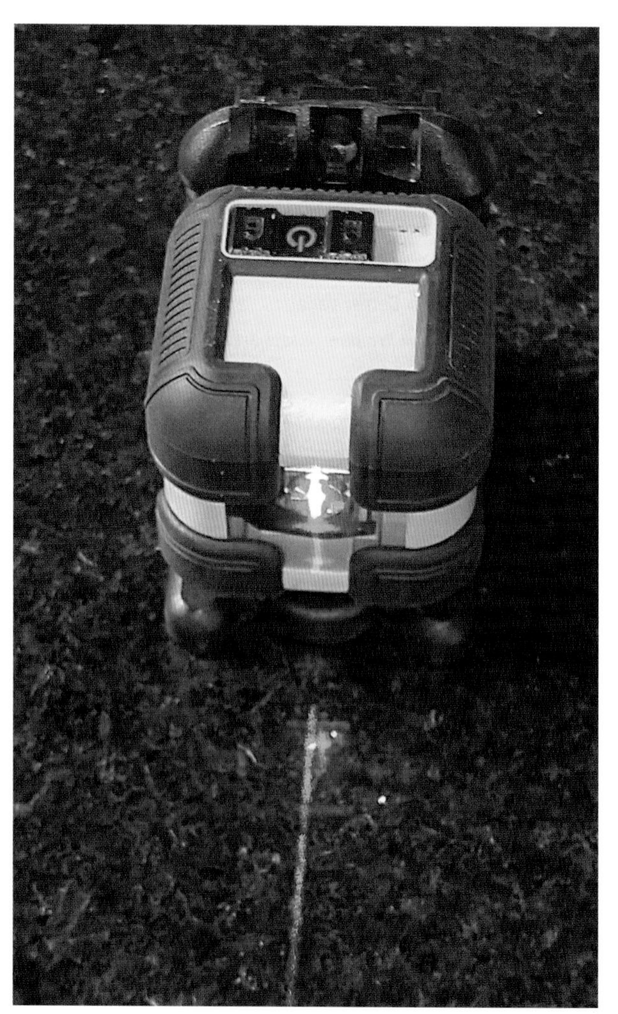

FIGURE 5.56 Laser level

Direct fixing of sheet materials to masonry

Sheet materials can be directly adhered to masonry walls as long as the walls are properly prepared and free from dust, oil and other materials such as formwork release agents.

Sheets can be installed horizontally or vertically, with staggered butt joints where necessary. They are temporarily held (for 90 to 120 minutes) in position using masonry nails at the sheet edges and temporary blocks in the middle of the sheets if necessary (see Figure 5.57). Sheets should be braced with stud material and battens if necessary. A 10 mm gap should be left between the floor and the bottom edges of the boards. When the masonry wall surface is irregular (15–25 mm), 75 × 50 mm packing pieces (see Figure 5.58) are used to pack the fixing line out to an in-line plane. A string line or straight edge can be used to ensure that the pieces are level. The packing pieces are set in a daub of the adhesive at 600 mm centres. Additional daubs of the adhesive are applied at 450 mm centres.

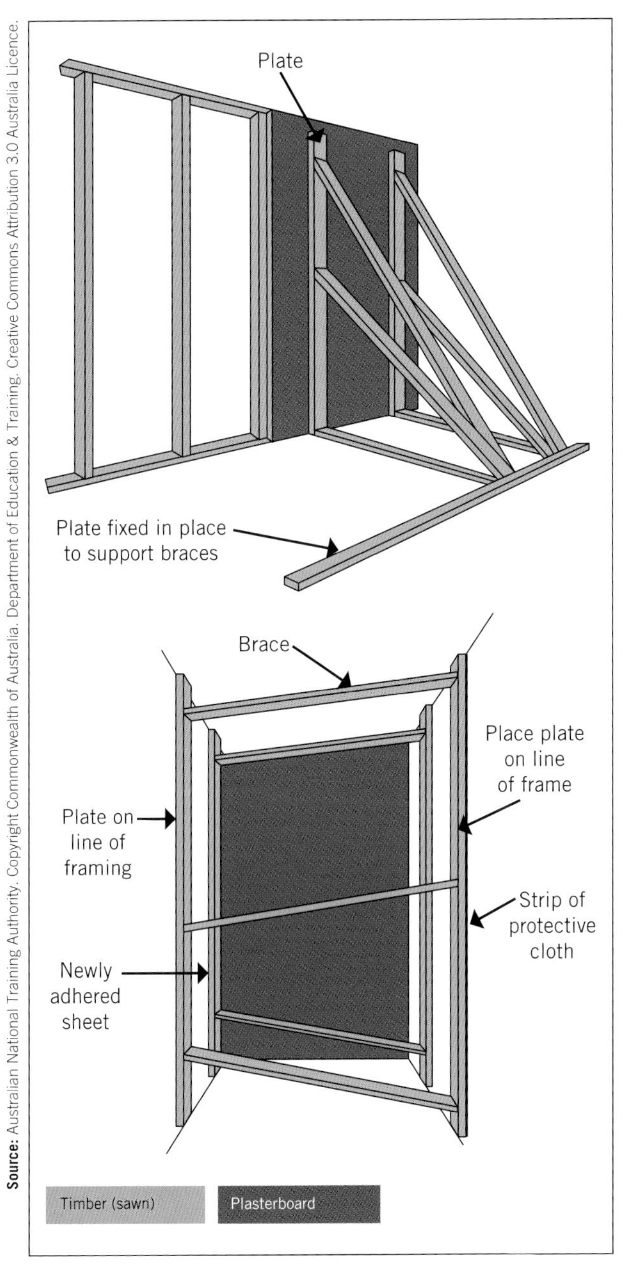

FIGURE 5.57 Direct adhesive fixing – temporary bracing

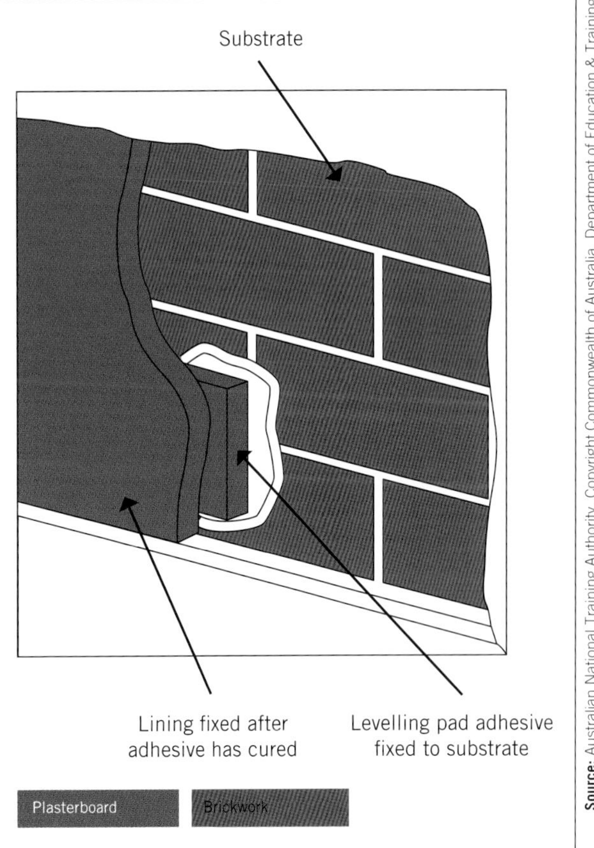

FIGURE 5.58 Levelling pads, 75 × 50 mm

Direct-fixing sheet materials to wall panels eliminates battening and wet trades like plastering and rendering – it is faster too. A disadvantage is that control joints need to be incorporated because of movement between dissimilar materials, and masonry fixings must be used to fasten all fixtures.

Finishing

Finishing internal and external corners and ceiling lines

Internal corners of lining boards can be scribe finished (see Figure 5.45 earlier in the chapter), or quads or scotia (see Figures 5.59 and 5.60), or square or triangular beads can be added. The size of the material used can affect the aesthetic finish of corners and consideration needs to be given to how the moulding is finished at the top and bottom. Quads can be scribe fitted over bull nose and similar skirting but not over ornate period skirting. Scotia works well with period skirting but not so well with the bull nose.

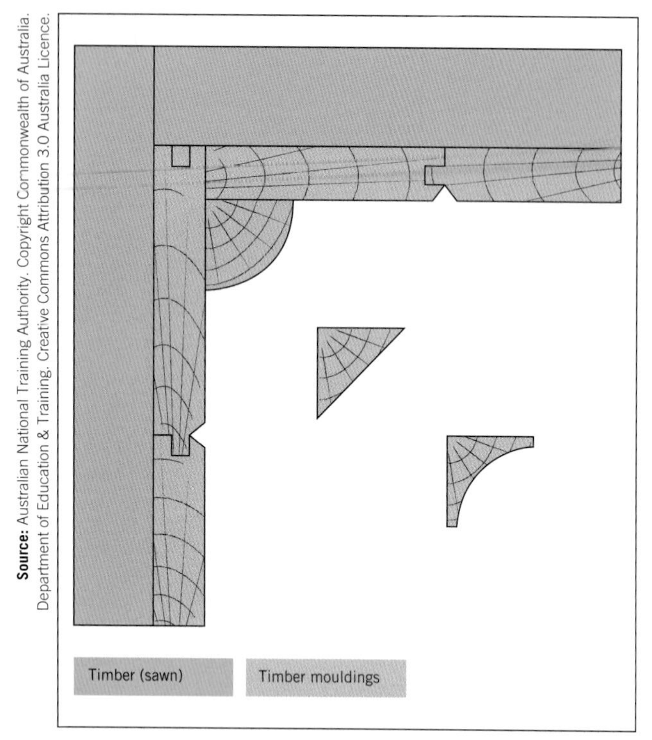

FIGURE 5.59 Scribe finishing internal corners

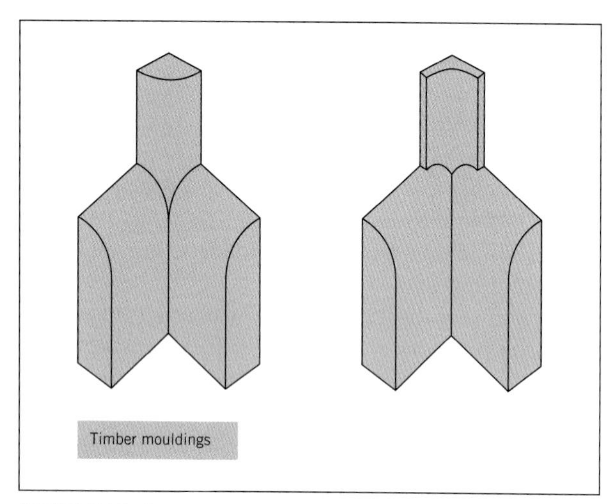

FIGURE 5.60 Quad and scotia

External corners of panels and lining materials need to be protected from day-to-day knocks and damage (see Figure 5.61). Setting with metal angles can protect plasterboard and fibre-cement sheeting. Panelling requires moulding or extrusion to be used. There really is no end to the materials that can be used. Simple squared and dressed battens can be applied to corners, or rebated beads or mouldings can be used to protect the corners of panelling. It is not uncommon to use aluminium angles to finish the corners in a modern panelled wall (see Figure 5.62). Another alternative is to machine specialised joints to complete the finish.

Corners in traditional wall panelling

In traditional wall panelling, a number of different methods are used to join the panels at external corners (see Figure 5.63) and internal corners.

Figure 5.63 shows the use of a fluted mould in combination with a scotia mould to build up a corner that protects and beautifies the corner and at the same time allows you to cope with what might be rough groundwork.

At internal corners, bare-faced tongue and groove joints traditionally have been used to join panelling. The benefit is that panels can be adjusted in and out of the groove, which allows adjustment and movement in the building structure and in the panelling material. Modern alternatives include biscuit joints and using plant-on mouldings and beads to hide the internal corner joints.

Mitre joints for internal and external corners, over the length of panels, are too difficult and **always** result in shrinkage and movement gaps.

Installing skirting boards

Depending on the floor finish and the quality of the finished skirting boards, these boards may have to be scribe fitted to the floor to achieve a decent quality finish. If the floor is to be carpeted, you probably won't need to scribe the boards unless the gaps are greater than 20 mm.

Often in older homes, bearers and joists may have shrunk to such an extent that significant gaps have appeared, creating draft points and vermin access. Unless you intend to remove, replace or refit the skirting, it is often easier and faster to use a bead or quad to cover any gaps (see Figure 5.64). Use a bead of construction adhesive when nailing the moulding to the skirting, and be sure to nail the moulding to the skirting and not to the floor. If you nail to the floor, the moulding will move up and down with the floor, causing paint lines to crack.

If you decide to remove skirting boards, take care not to damage the existing wall finish as this can often be difficult and time-consuming to repair. Use a backing block behind a chisel (see Figure 5.65), a wrecking bar and a hammer to keep the damage to a minimum.

When planning to fit lining materials over existing lined walls, it is best to leave the existing material intact and batten the wall out to the thickness of the skirting. This allows you to leave the skirting in place. Make sure all the battens are the same thickness. You may have to

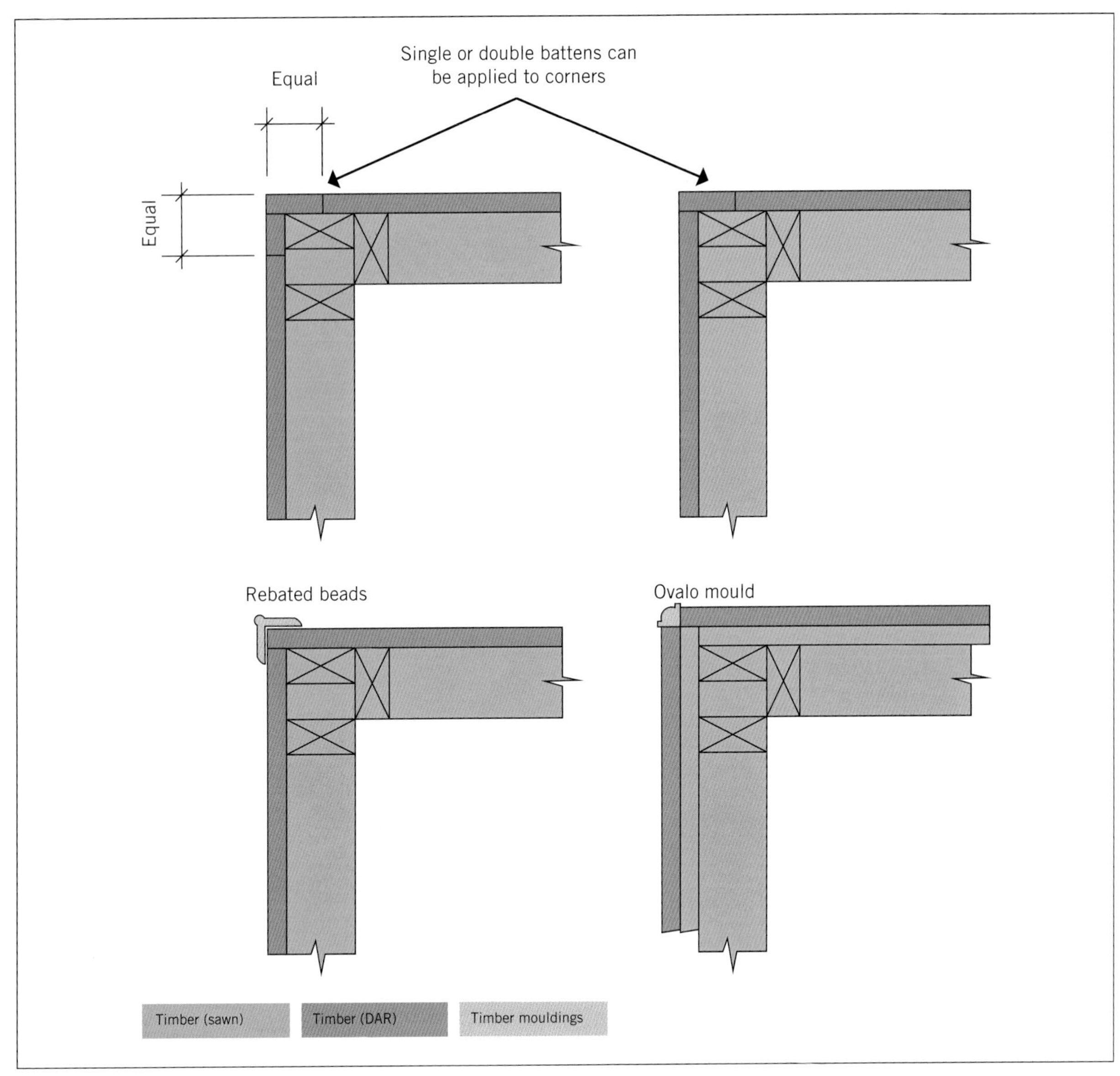

FIGURE 5.61 Finishing external corners

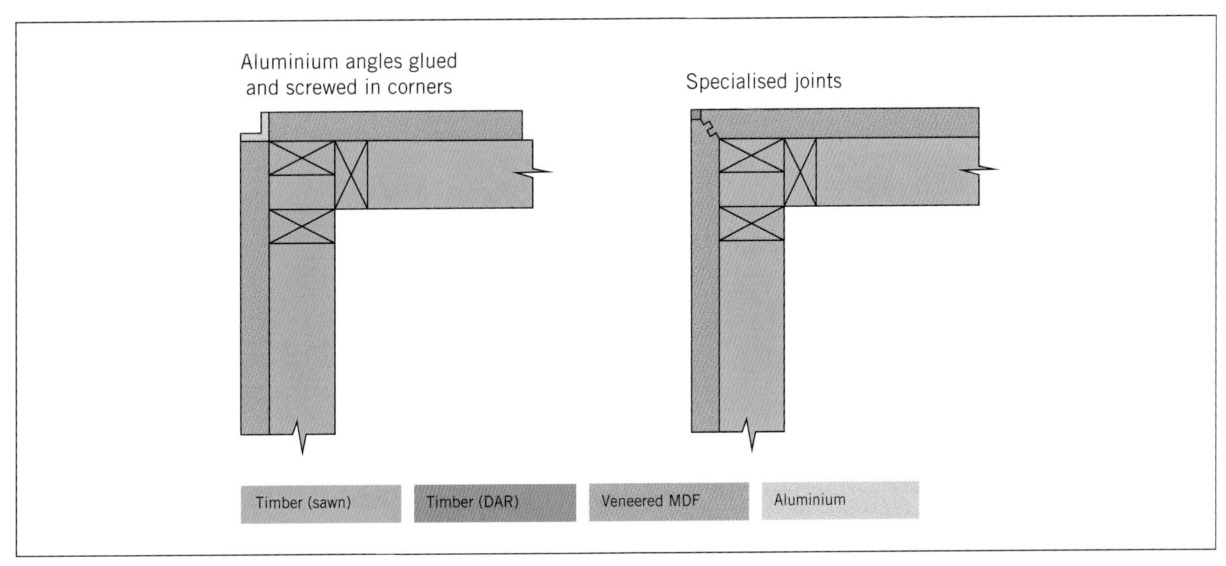

FIGURE 5.62 Aluminium angles and specialised joints used with modern panelling

Source: Ed Hawkins

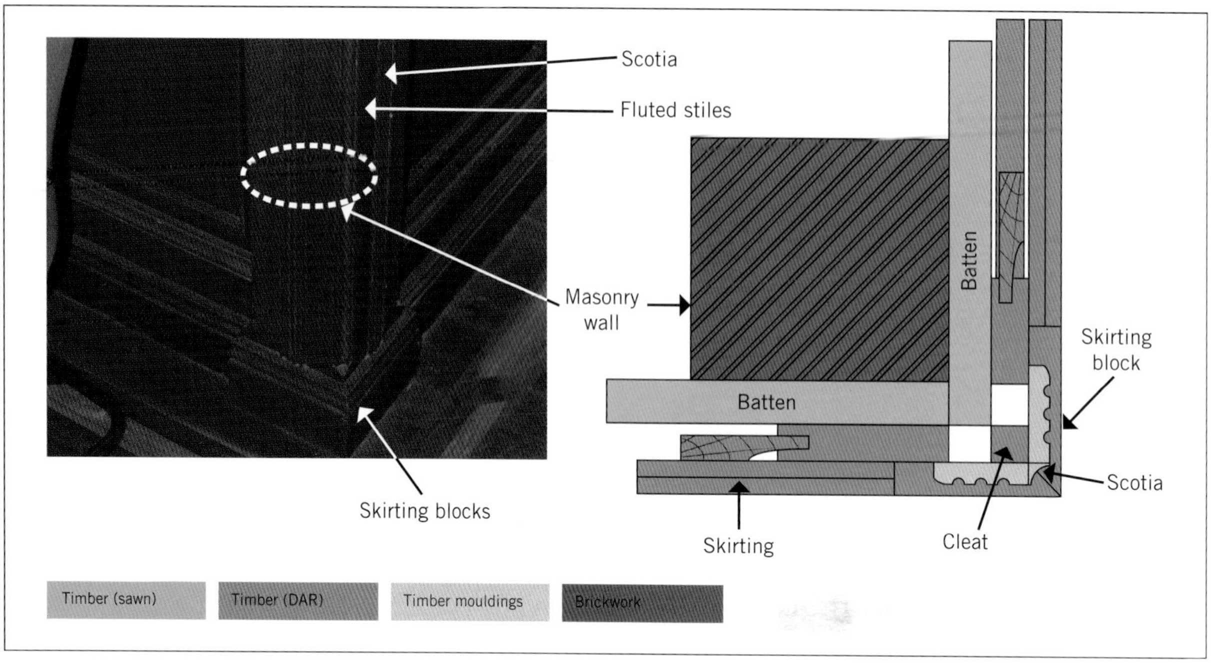

Scotia
Fluted stiles
Masonry wall
Batten
Batten
Skirting block
Scotia
Skirting blocks
Skirting
Cleat

Timber (sawn) | Timber (DAR) | Timber mouldings | Brickwork

FIGURE 5.63 Section views of corner details

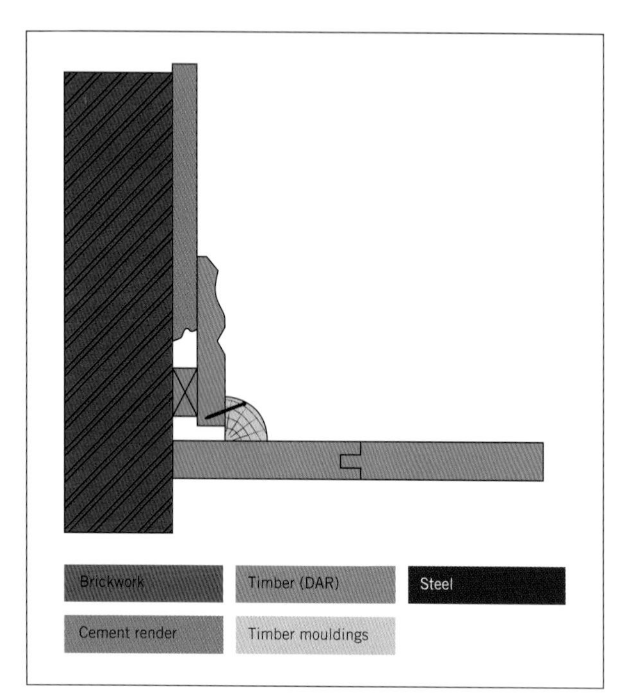

FIGURE 5.64 Quad mould as gap filler

FIGURE 5.65 Protecting the wall finish

remove and pack out the skirting board with short lengths of the lining material if it is thinner than the battens.

Fixing skirting to masonry walls

Where masonry walls have been render finished, you will often need to pack out the bottom edge of the skirting to keep it plumb and prevent it from collapsing in (see Figure 5.66).

Traditionally, every third or fourth perpendicular joint in masonry walls was plugged with twisted timber plugs so that skirting boards could be nailed off; 75 mm

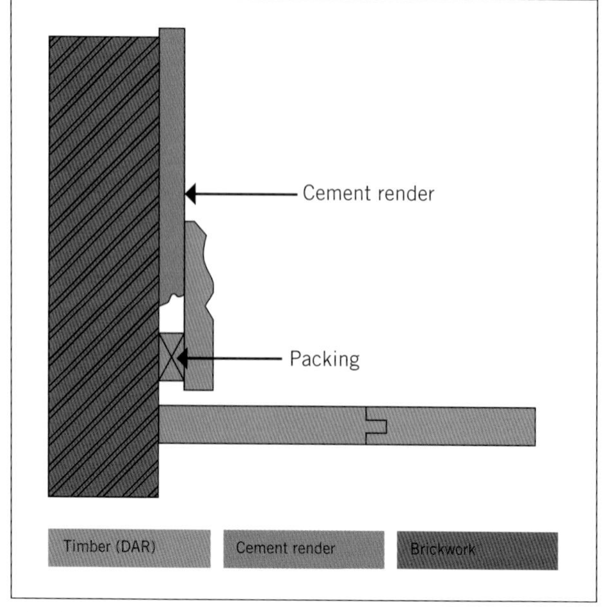

FIGURE 5.66 Packing out skirting

nails were used to fix the board to the walls. This is still a good method, but may require the scraping out of the joints so that timber plugs can be inserted. Alternatively, you can mark out the brickwork using a chalk line every 600 mm and drill 12 mm holes 50 mm deep (65 mm in rendered walls) with a masonry bit and insert timber dowels into the holes (see Figure 5.67). Use two rows of nails to fix skirting that is 65 mm or wider. One nail at each point is usually sufficient for 50 mm skirting.

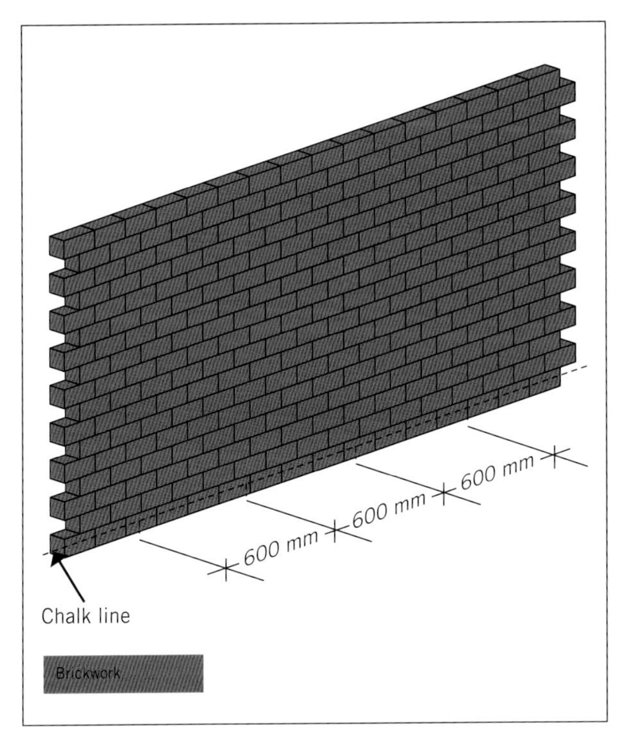

FIGURE 5.67 Setting out for plugs

Another option is to use construction adhesive. Follow the adhesive manufacturer's instructions and brace the skirting against the wall until the adhesive has set.

Note: Adhesive fixing to plasterboard will make the plasterboard difficult to remove and replace or repair should it be necessary.

Joints

Mitre joints are used for external corners (see Figure 5.68). These joints should be undercut to allow the mitred edge to fit neatly, even if the wall is slightly out of square. This will result in a perfect mitre joint. The back of the mitre joint is removed with a plane, chisel or wood rasp (see Figure 5.69).

Scribe joints are used to fit skirting boards at internal corners (see Figure 5.69). Before installing the skirting, work out a sequence of installation that will allow the scribe joints to be viewed from side-on instead of head-on when people enter the room. The skirting board opposite the doorway should be

Source: Ed Hawkins

FIGURE 5.68 Joints for internal and external corners

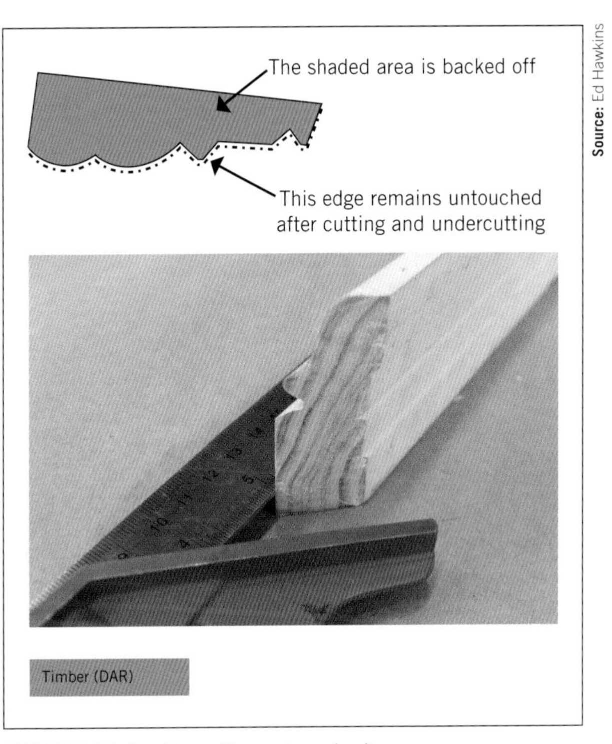

Source: Ed Hawkins

FIGURE 5.69 Backing off an external mitre

square cut between the walls with the side lengths scribed over.

Steps for creating a scribe joint

Step 1

Mark the overall length of the skirting board, including the thickness of the boards to be scribed over at each corner (see Figure 5.70).

Step 2

Use a drop saw, or similar, to make a 45° cut on the end of the board to be scribed (see Figure 5.71). You can turn

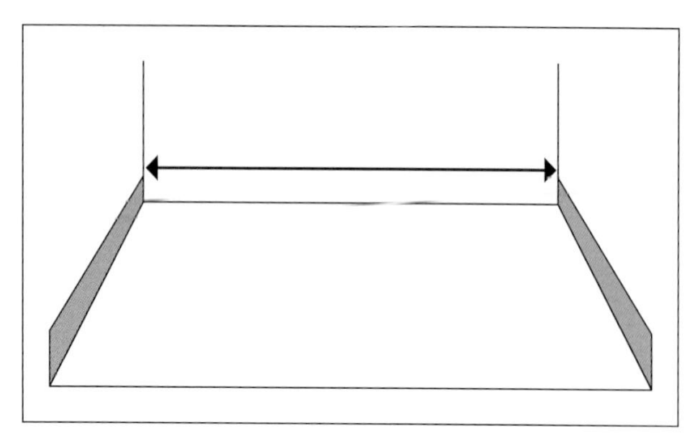

FIGURE 5.70 Marking the length of the skirting board

FIGURE 5.71 Starting a scribed joint

the mould bottom edge up and make a square cut down the mitre line (see Figure 5.71).

Step 3

When fragile areas will be created during the scribing process, use your forefinger to support the timber and prevent breakout (see Figure 5.72). Use a coping saw to remove the remaining waste, making sure you undercut the timber, leaving a fine edge to fit neatly against the adjoining board.

Step 4

Finally, clean up using a chisel, wood rasp file and abrasive paper. Figure 5.73 shows the finished scribe joint.

Architraves

Architraves are mouldings used around windows and doors to cover gaps and provide an aesthetic frame (see Figures 5.74, 5.75, 5.76 and 5.77).

Architraves around windows are installed after the nosing and apron have been fitted to the window sill if nosings and apron pieces are part of the design.

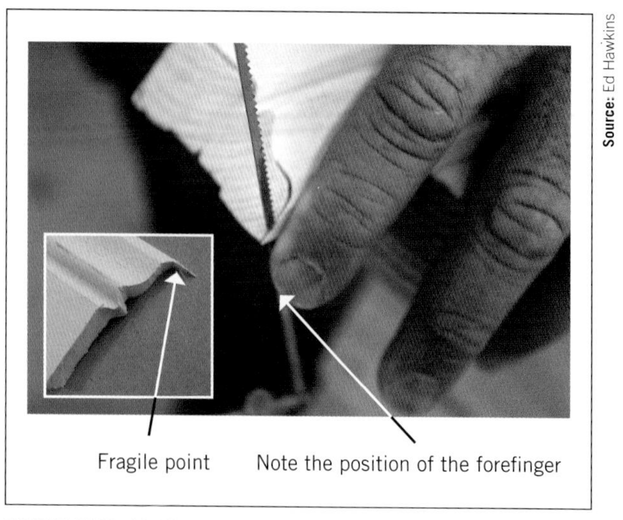

FIGURE 5.72 Making the scribed cut

FIGURE 5.73 Finished scribe joint

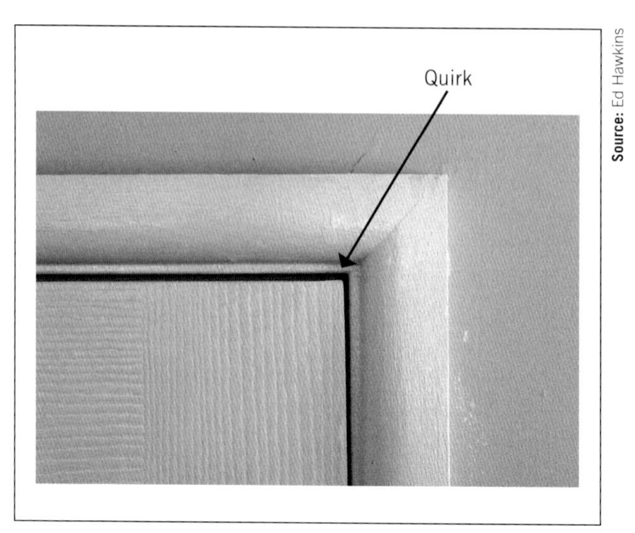

FIGURE 5.74 Quirk

FIGURE 5.75 Californian bungalow style architrave

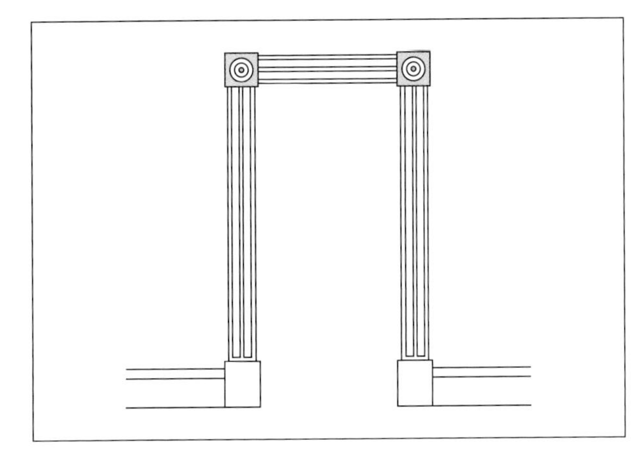

FIGURE 5.76 Victorian style architrave

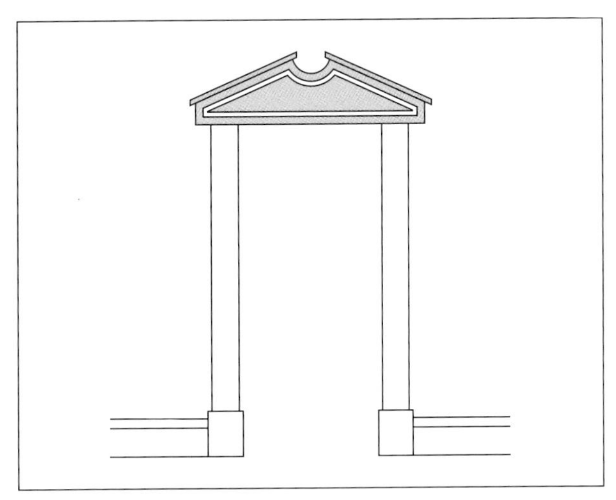

FIGURE 5.77 Georgian style architrave

Both door and window architraves are fitted with a quirk (see Figure 5.74). Do not forget to allow for the quirk when making calculations for architraves and skirtings. The quirk distance is usually between 3 and 5 mm.

Start by installing a vertical stile. Then mark the quirk distance at the top and bottom of the frame.

Measure the distance to the quirk mark across the top window head, then add the width of the skirting to find the length of the side pieces. Measure, mark and cut the mitres for the top two side architraves. Undercut the mitres slightly.

Pin the architrave in place without completely driving the nails home as you may have to adjust the fit later. This may be difficult if using a power nailer or brad gun, so get it right first time.

Measure and mark out the mitres for the head section. The distance over the outside of the architraves is used.

With architraves carefully mitred and fitted, remove the head section, apply a little adhesive to both mitres, and glue and nail the head section back into place. Finish off nailing the two sides, using one nail into the window frame and one nail through the wall lining material (on framed constructions or when fixing battens/blocking in steel frames) and into the studs. Finally, arris on all edges and fill any nail holes and other blemishes before sanding, ready for finishing.

If the faces of the architraves do not fit flush, pack the back of the architraves until the faces are flush and then drill through the top edge of the head architrave into the side architrave (see Figure 5.78). Nail off the joint.

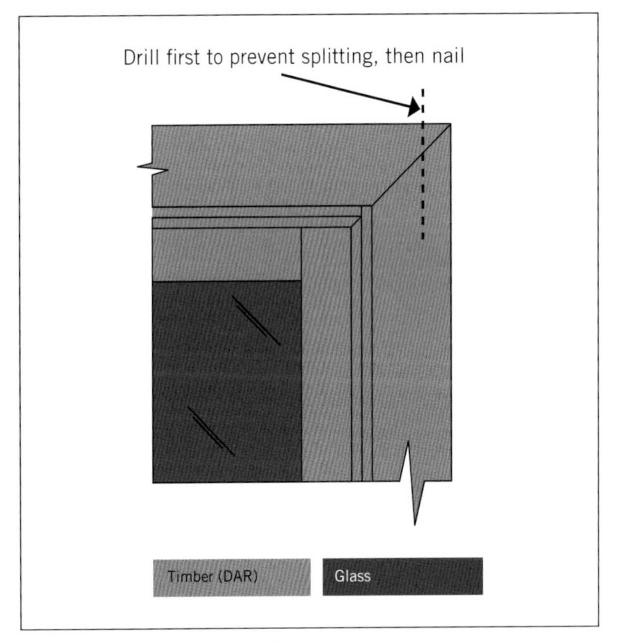

FIGURE 5.78 Nailing of a mitre joint

Mitre joints are not the only means of finishing off around windows and doors, but joints in combination with shaped head pieces can dress up an entrance.

Pelmets provide cover for curtain tracks and fittings (see Figure 5.79).

Source: Gordon Slater

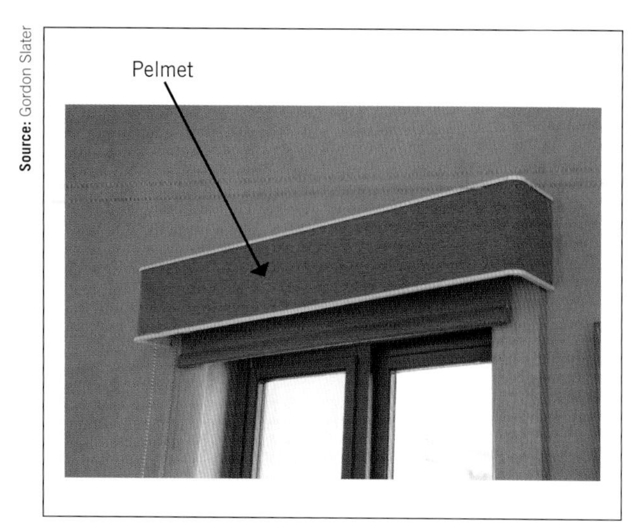

FIGURE 5.79 Window pelmet

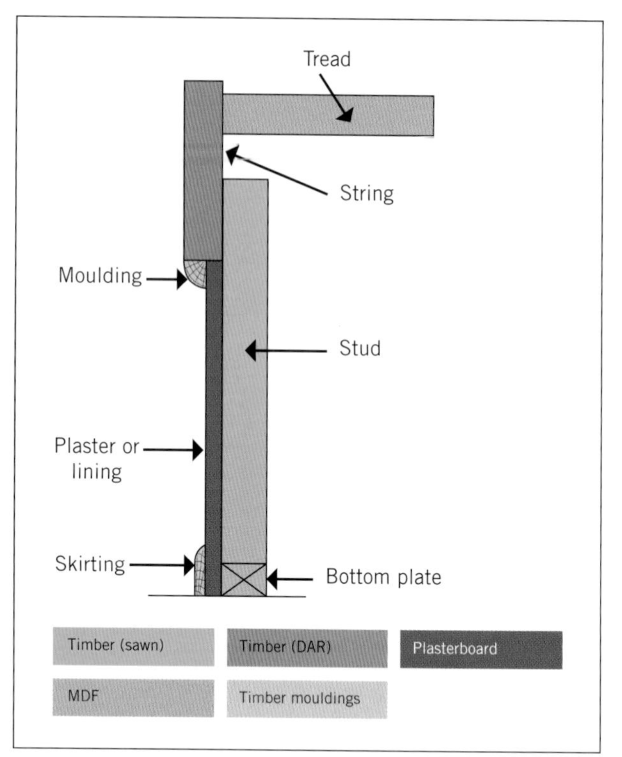

FIGURE 5.81 Spandrel panel

Spandrels and soffits

Often the triangular area beneath a staircase is called a **spandrel** (see Figures 5.80 and 5.82). A soffit in this instance is the lined underside of a flight of stairs (see Figures 5.82 and 5.83). The spandrel area is used for storage and so is often framed, panelled and fitted with doors for access. Or it may be done simply with plasterboard or plywood over a simple frame (see Figure 5.81).

Source: Greg Cheetham

FIGURE 5.80 Spandrel and door

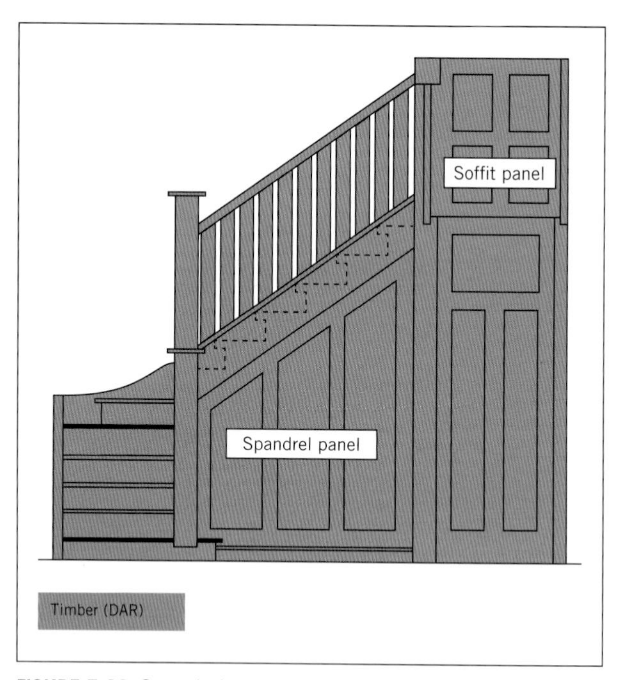

FIGURE 5.82 Spandrel and soffit

Sometimes the area immediately below the treads and risers, known as the soffit, is covered in to hide the means of fixing the strings, treads and risers (see Figure 5.83).

The most economical approach is to cover the entire area below the stairs with plasterboard and use beading or moulds to cover gaps and protect soft edges. Other methods include making up a frame and panel construction or using lining boards to finish the soffit (see Figure 5.83).

The alternative is to rebate the bottom edge of the string (see Figure 5.84).

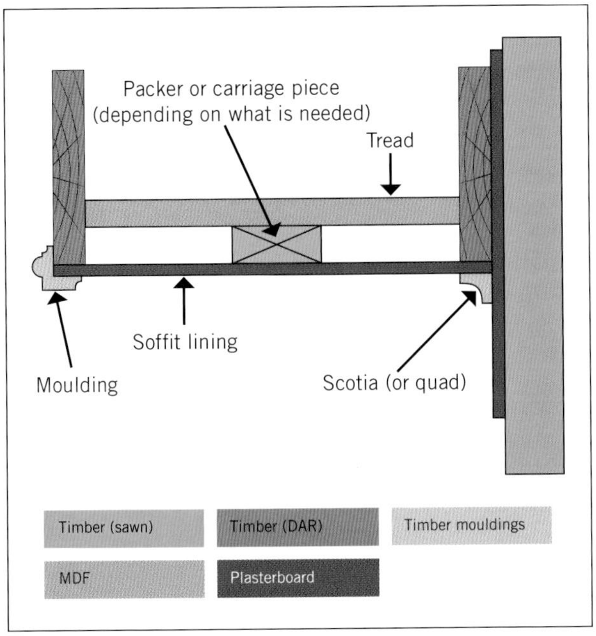

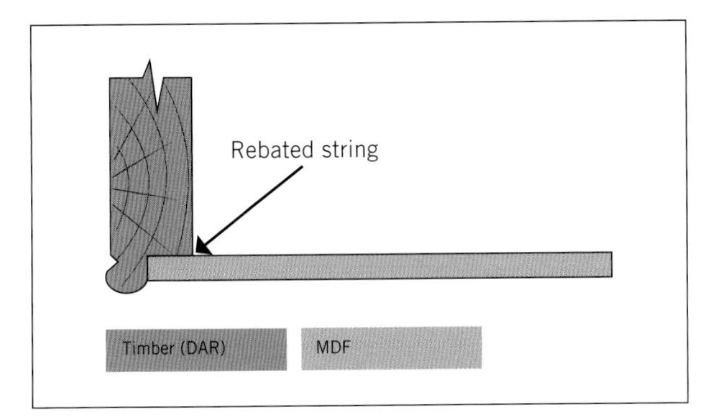

FIGURE 5.83 Soffit lining

FIGURE 5.84 Rebated string

Raked mouldings

Raked mouldings are used to cover gaps between walls and raked ceilings and other inclined constructions such as stairs and ramps. In order for the mouldings to be installed correctly, three different moulds must be used: a mould at the top (acute angle), a standard mould along the sides and an obtuse angle mould at the bottom.

Steps for developing raked mouldings

Step 1

Set out the raked angle elevation then a section of original mould with square corners as it will sit in the square corner of the side wall and raked ceiling (see Figure 5.85).

Step 2

Extend the moulding profile lines to the acute angle and the obtuse angle as indicated on the set-out board (see Figure 5.86).

Step 3

Redraw the squared mould at acute and obtuse angle ends (see Figure 5.87).

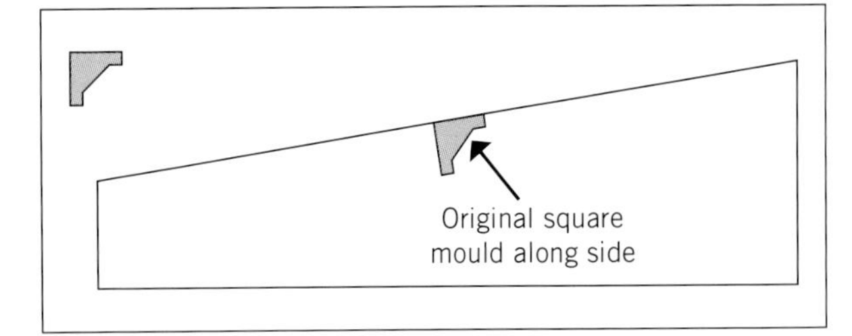

FIGURE 5.85 Create an elevation

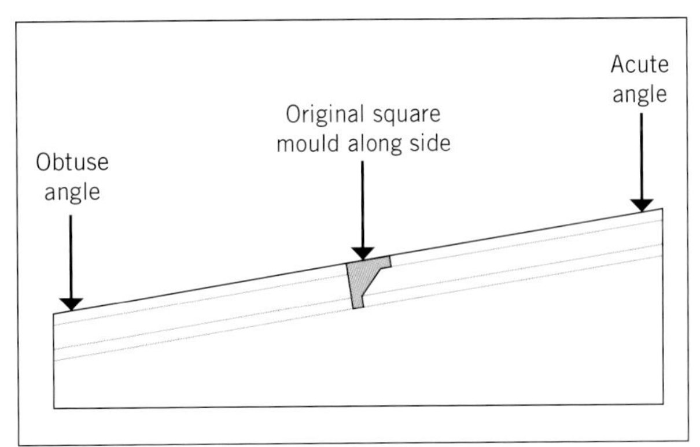

FIGURE 5.86 Extend the moulding profile

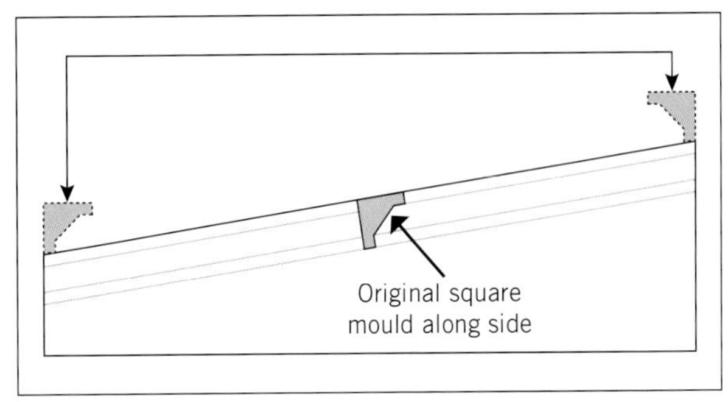

FIGURE 5.87 Redraw the moulding at each angle

Step 4

Drop down the true widths of the mould, as shown in Figure 5.88.

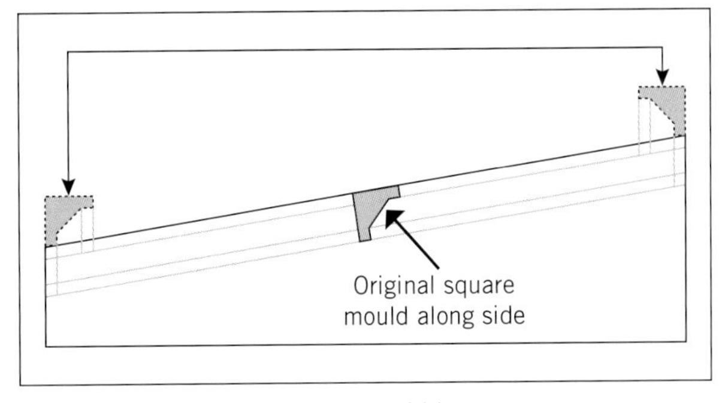

FIGURE 5.88 Drop down the plan widths

Step 5

Draw in the true top and bottom mould shapes (see Figure 5.89).

These two new mould shapes will have to be machined by a specialist joinery company with the

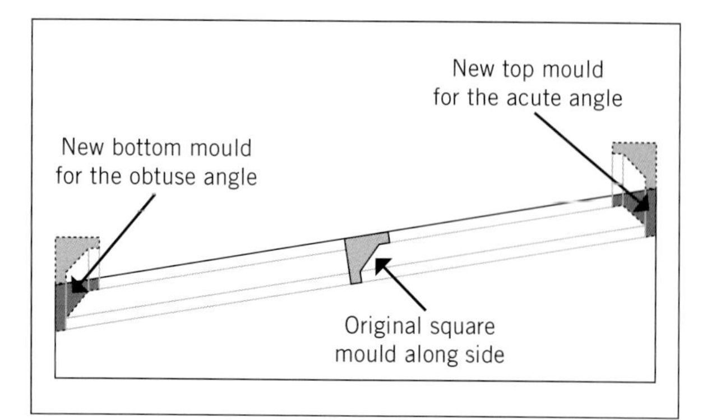

FIGURE 5.89 True top and bottom mould shapes

appropriate machines. You will need to pay for the grinding of special cutters and machine set-up costs.

The next step is to develop the true mitre angle in raking moulds.

From a plan view, the angle itself appears to be a regular 45°; however, because the two pieces are sloping you need to develop the true mitre angle.

Developing the true mitre angle in raked moulds involves being able to see the angle from a 90° angle. To do this, set out a plan view of the project angles using a sliding bevel set to the top and bottom angles and transfer these marks to a suitable set-out board.

The steps below are for setting out the obtuse or bottom angle and the theory is the same for the top angle. You just need to change the mould shape.

Step 1

Set out the rake angle and develop a plan view of the mitre angle (see Figure 5.90).

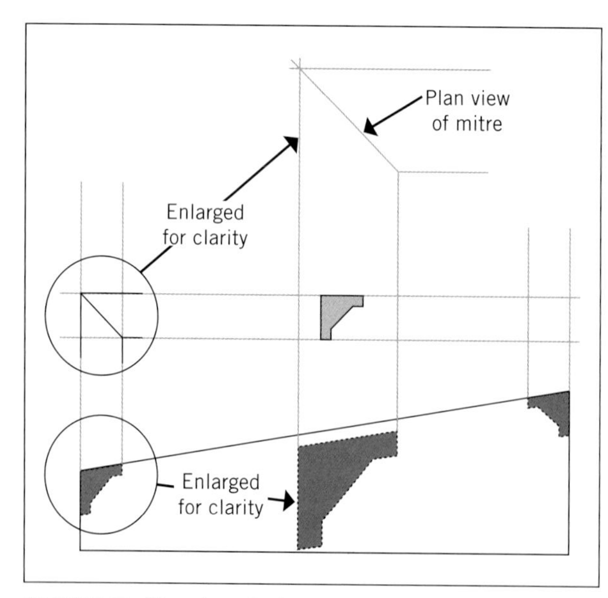

FIGURE 5.90 Plan view of mitre angle

Step 2

Use a compass, from the bottom corner of the mould (as indicated in Figure 5.91), to drop a line representing the true length (width of the mould to the horizontal plane).

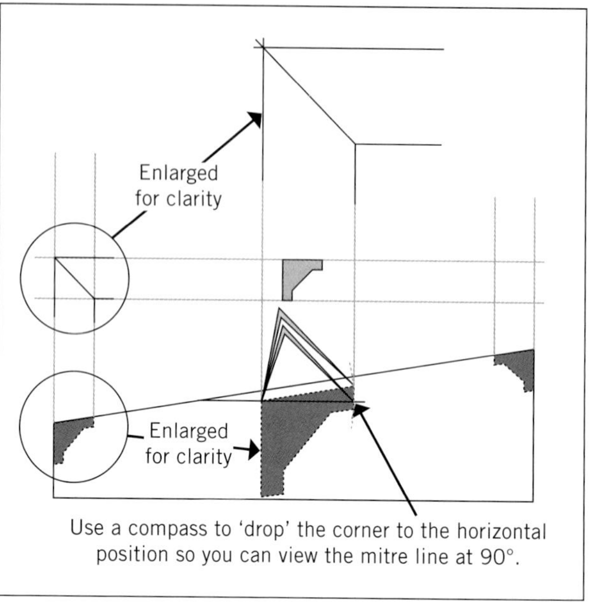

FIGURE 5.91 Drawing in a horizontal plane

Step 3

Project the new point up to the plan view (see Figure 5.92).

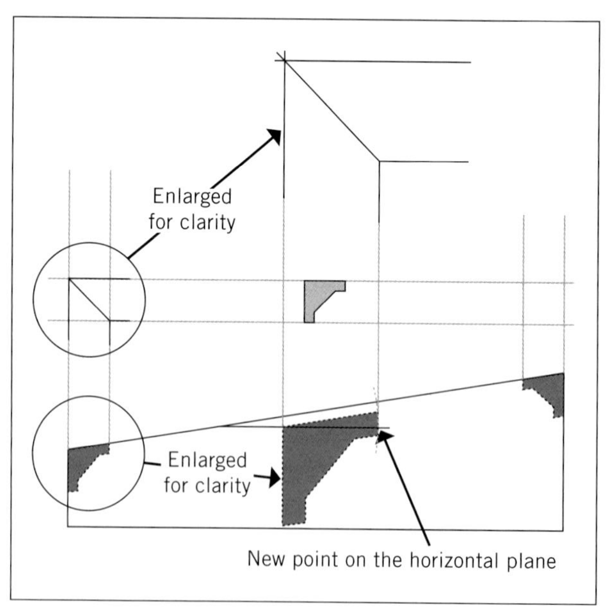

FIGURE 5.92 New point on the horizontal plane

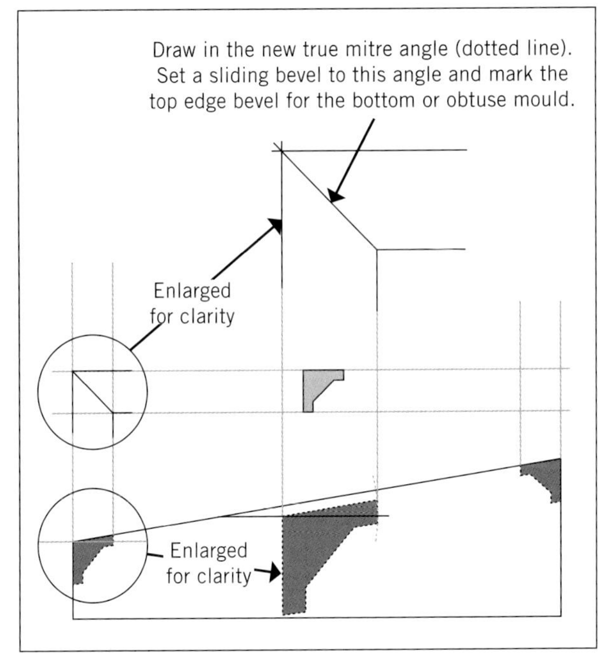

FIGURE 5.93 True mitre angle

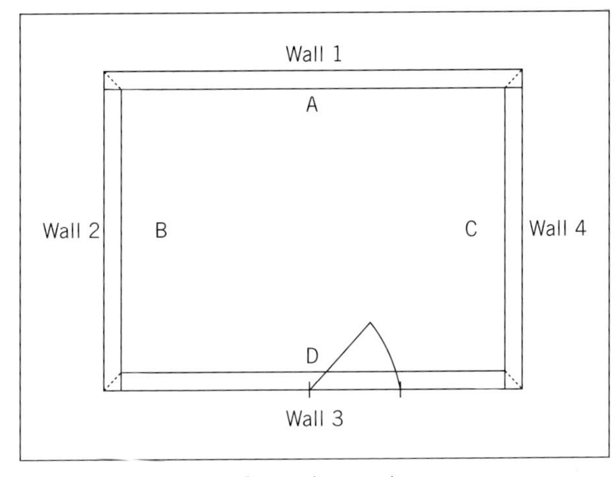

FIGURE 5.94 Sequence for cutting cornices

Step 4

Draw in the new true mitre angle. A sliding bevel is then used to transfer the new angle to the timber, or to set up the drop saw for cutting (see Figure 5.93).

Fitting cornices and crown mouldings

Cornices are fitted in a sequence that allows the scribe joints to face away from the line of sight when you enter the room (see Figure 5.94).

The sequence for cornice installation is as follows:

1 Square cut and fit a length of cornice A neatly between walls 2 and 4.
2 Scribe fit cornices B and C to cornice A and then cut square to fit wall 2.
3 Measure and scribe cut cornice D to fit cornices B and C.

Standard cove plaster cornice is scribed in a similar way to timber except that, because of its curved shape, you need to build a mitre box to enable you to get accurate cuts (see Learning task 5.4). This is even more important when irregular moulded cornices are being installed (see Figure 5.95).

The easiest way to think about the cutting of the mitres for scribing is to construct the mitre box and consider that the perpendicular leg (A) is the wall, while the horizontal leg (B) represents the ceiling line (see Figure 5.96).

It is critical that the moulding be held in exactly the correct position while the mitre is being cut. First, determine the position of the crown moulding or cornice in relation to the walls and ceilings. How far down will the cornice sit on the wall, and how much of the ceiling will the cornice or crown moulding cover?

Cut some suitable scrap material and pin it in position on the wall and ceiling legs to hold the material in place while you make the mitre cut. These same

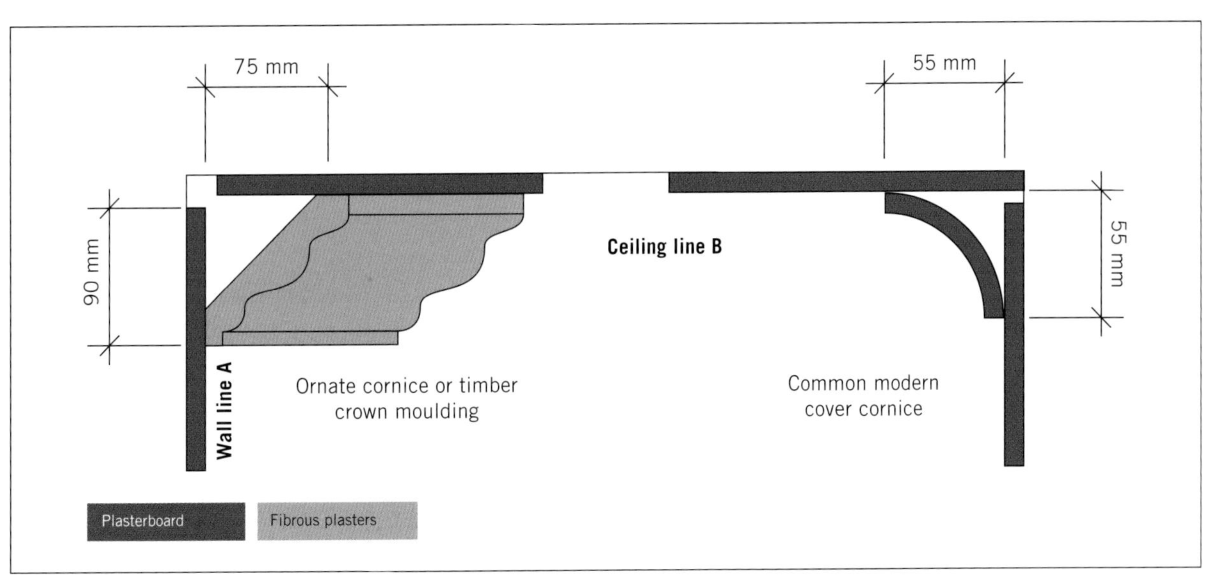

FIGURE 5.95 Wall and ceiling line

LEARNING TASK 5.4 MAKING A CORNICE MITRE BOX

Make a standard 90 × 90 mm plaster cornice mitre box (see the image below). Use 90 × 45 mm for the base and 135 × 19 mm for the sides.

Screw the pieces together using 3/8 g × 45 mm screws.

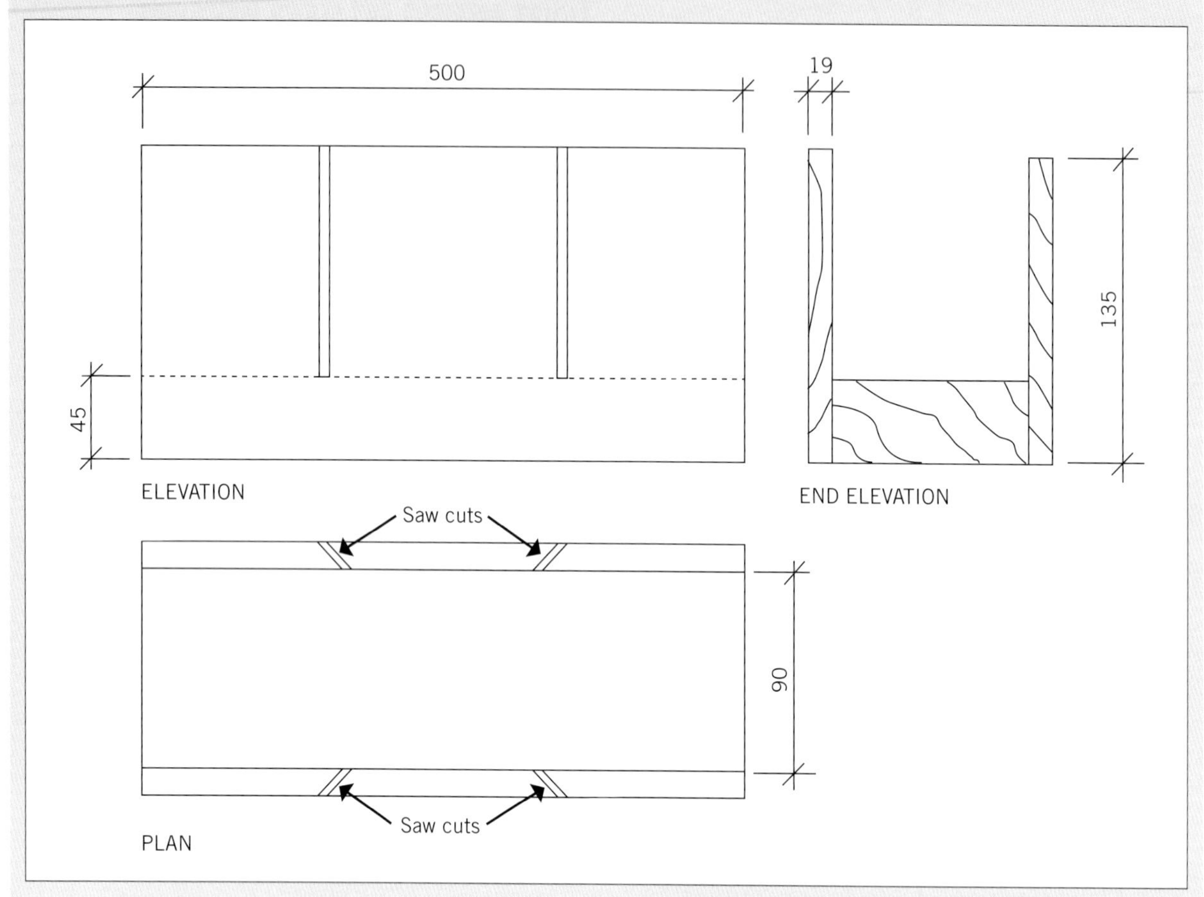

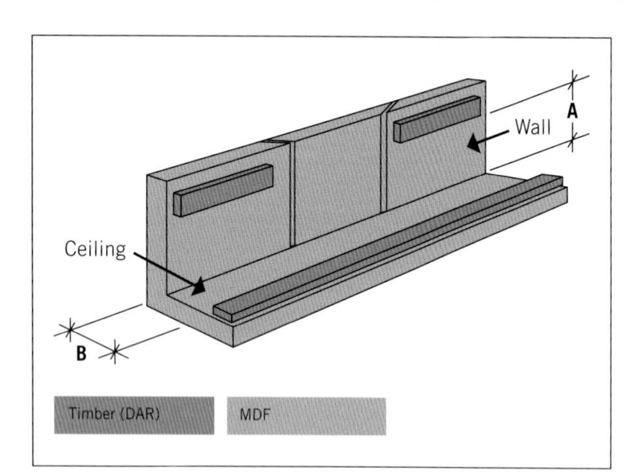

FIGURE 5.96 Cornice and crown moulding mitre box

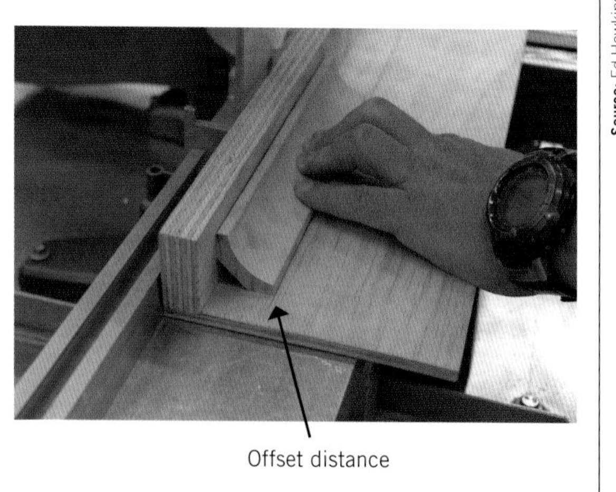

Source: Ed Hawkins

FIGURE 5.97 Set up the fences

fences and stops can be set up on a drop saw to make more accurate cuts. Follow these steps:

1 Set up the fences to take the crown moulding – remember that the vertical fence is the wall and the bed of the drop saw is the ceiling (see Figure 5.97). Make sure that the moulding is going to sit flush against the fence (see Figure 5.98) and work out the offset distance.
2 Use a spacer to set the distance (see Figure 5.99). This makes sure it is even along the fence. Use screws to

Source: Ed Hawkins

FIGURE 5.98 Crown moulding flush against the fence

Source: Ed Hawkins

FIGURE 5.99 Set the width of the fence

fasten the fence but take care not to place them near the saw cut area.

3 Remove the waste material, making sure the fences remain fastened in their original position (see Figure 5.100). The saw cuts act as registration points for your pencil marks on the crown mould lengths.
4 Measure and mark the lengths of the crown moulding. It is not a bad idea to lightly mark or draw the angle required on the face of the crown mould. Insert it into the drop saw, aligning the correct registration mark with the saw cut in the fence (see Figure 5.101). Before cutting, make sure you have the ceiling edge and the wall edges in the correct alignment.

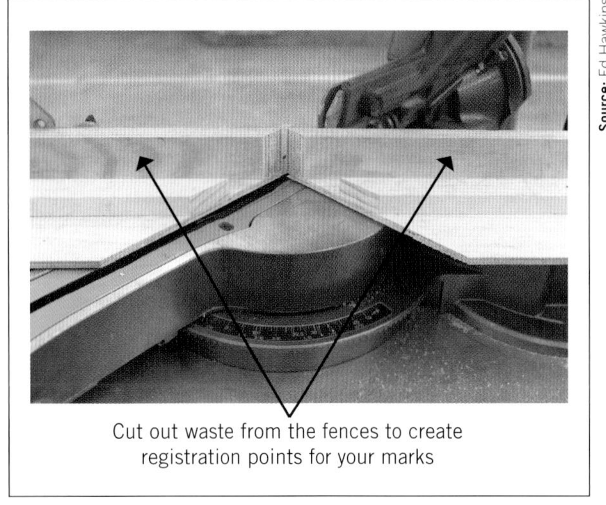

Source: Ed Hawkins

FIGURE 5.100 Cut out the waste

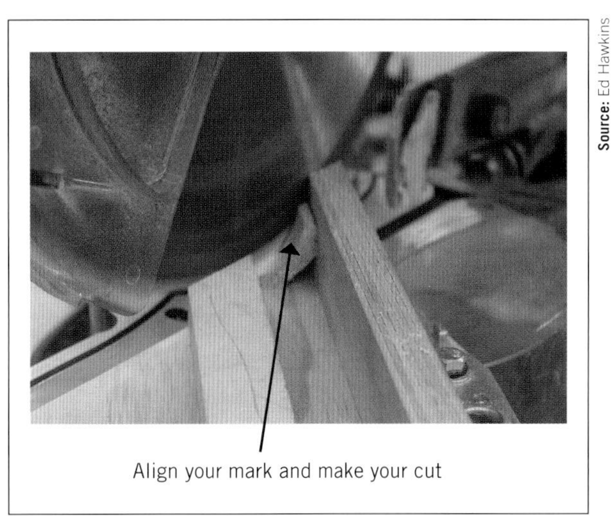

Source: Ed Hawkins

FIGURE 5.101 Align the marks and make the cut

Geometrically setting out compound mitres

The alternative to setting up a drop or compound mitre saw is to cut the material flat and develop the correct angles and bevels. This may be the necessary method if the material is too large to fit below the blade in the sloped position. The angles formed then become compound mitres, being angled in two directions.

Setting out for compound mitres requires the use of a compass and other drawing equipment. The basic steps are similar to those for developing raking moulds in that you need to be able to view the joint from 90°.

1 Set out the angle and shape of the crown moulding in situ. Make sure you identify a suitable pivot point, which is a line parallel to the back of the moulding, and measure the true thickness of the material from the pivot line to the back of the moulding (see Figure 5.102). This becomes important later on. It is a good idea to number the corners of the original crown mould.

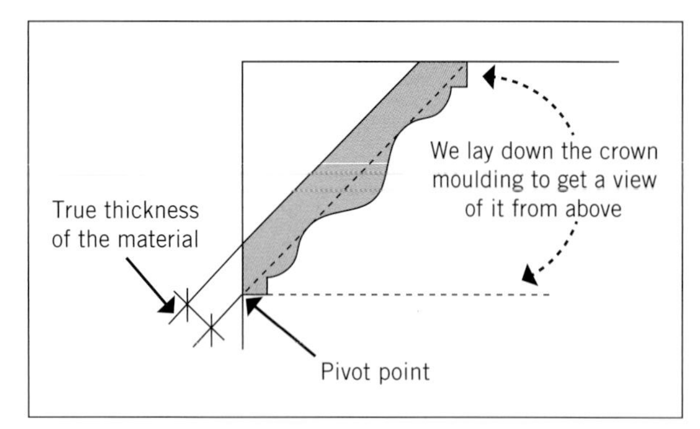

FIGURE 5.102 Lay down the crown moulding

2 A: Drop down an elevation of the original crown moulding with the compass from the pivot point. Measure up the true thickness of the material from the horizontal plane and draw in another horizontal line representing the thickness of the material. Use the compass to transfer the mitre points down (see Figure 5.103).

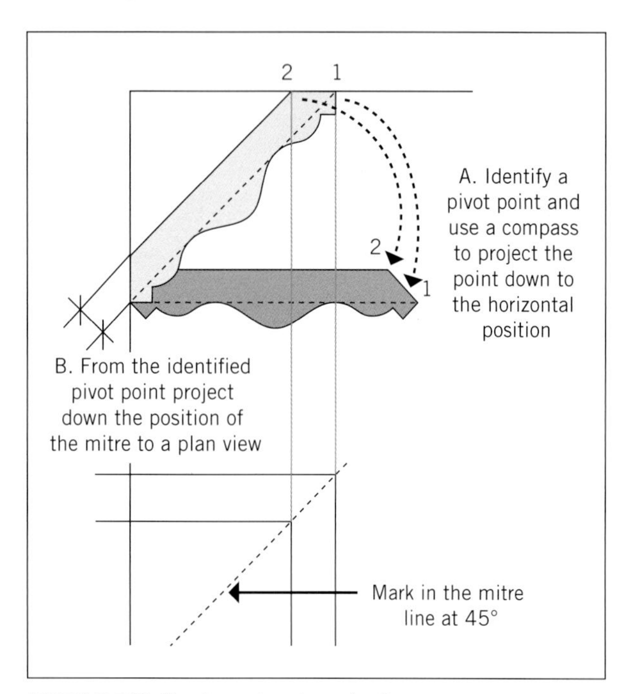

FIGURE 5.103 Create a plan view of mitre

B: Create a plan view of the original mitre by dropping down the lines from the elevation as shown (see Figure 5.104).

3 Drop down the new mitre lines and then project, across from the original mitre lines, lines that intersect the new mitre/face bevel lines (see Figure 5.104).

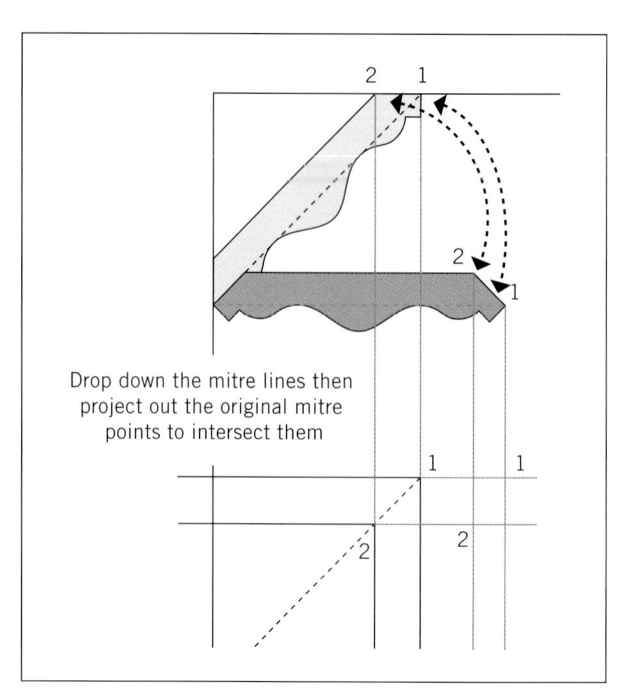

FIGURE 5.104 Creating the face bevel

4 Now join up the original pivot point (it hasn't moved) to the new outside mitre point and from the outside mitre point to the inside mitre point (see Figure 5.105). This forms the true face bevel (see Figure 5.106).

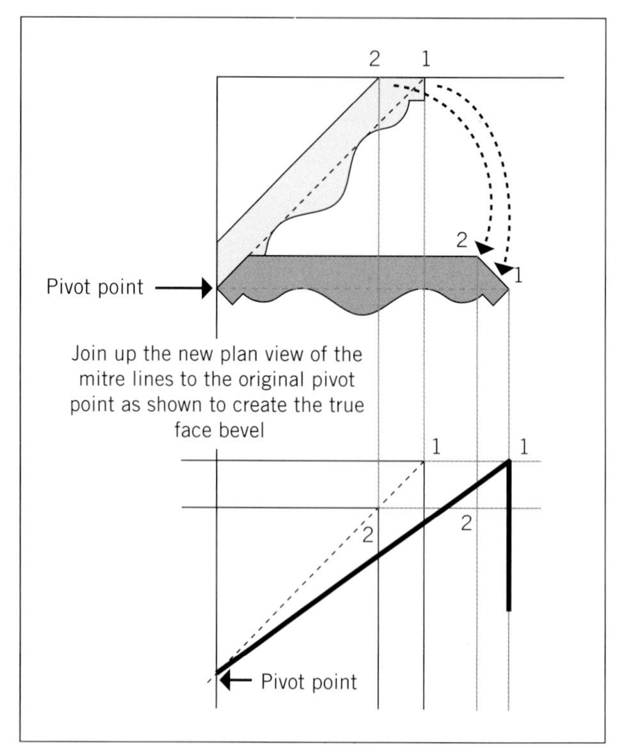

FIGURE 5.105 Create the face bevel

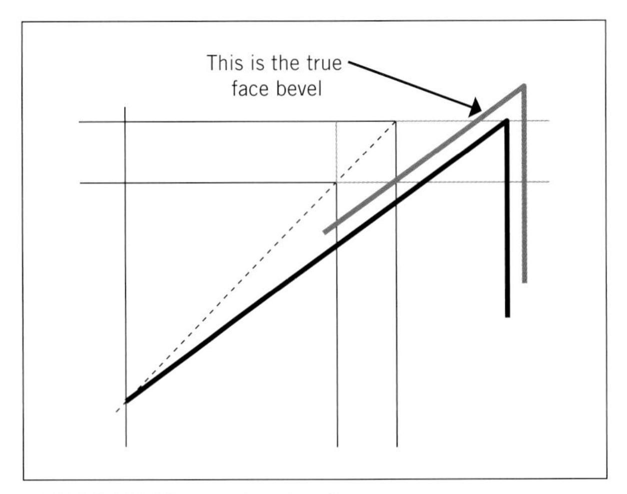

FIGURE 5.106 The true face bevel

5 To set out the true mitre angle, draw in a parallel line to the true face bevel from the inner mitre line (see Figure 5.107). Then overlay the true thickness of the material (as shown) at 90° (see Figure 5.108).

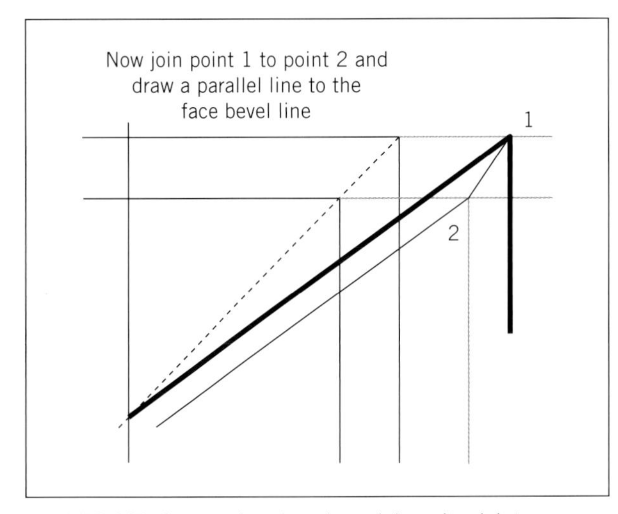

FIGURE 5.107 Set out the plan view of the mitre joint

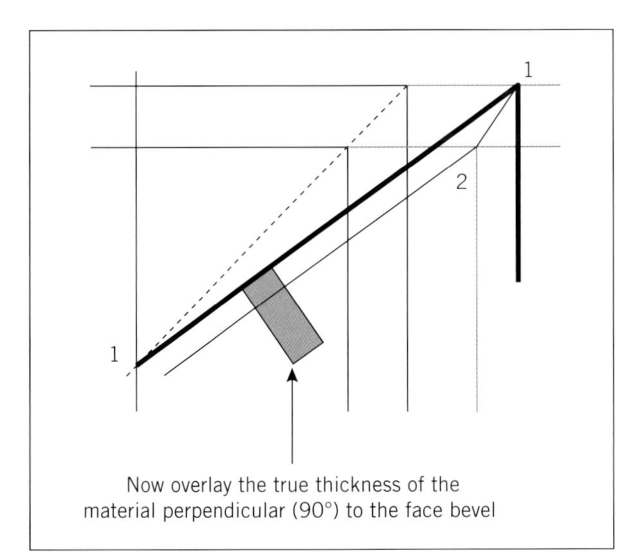

FIGURE 5.108 Overlay the true thickness of material

6 Mark across the diagonal as shown in Figure 5.109. You have created the true mitre angle. This is the angle at which you tilt the saw blade on the compound mitre saw.

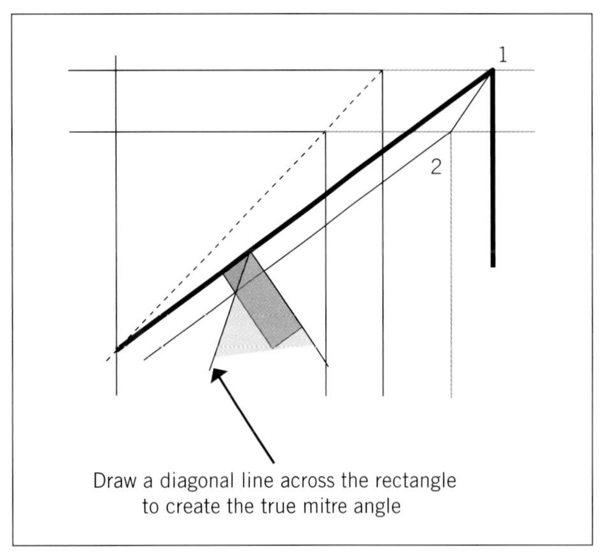

FIGURE 5.109 Creating the true mitre angle

Clean up

When finished, the work the area needs to be cleaned, and any materials sorted for landfill, recycling and reuse according to the worksite requirements and workplace procedures.

Clean up the work area, removing any debris, tools and excess materials. Complete a visual inspection of all tools and equipment to ensure safe operation. Regular checks and maintenance of tools and equipment should be done each time they are used. If faults are found, follow workplace procedures to record problems, tag the tools or equipment, and relocate them away from other functioning tools.

Restore the surroundings to their original condition, ensuring that any damage caused during installation is repaired.

Obtain the necessary sign-off and approvals from the client or project supervisor. Ensure that all necessary documentation has been obtained and filed appropriately.

SUMMARY

This chapter looked at the various materials and techniques used in the processes of lining walls and installing skirtings, architraves, cornices and crown moulds. Important points covered include:

- identifying basic materials, including solid timber lining boards; plasterboards and fibre-cement sheeting; and timber-based panelling such as hardboard, plywood, veneered MDF and particleboard
- preparing framed (timber and steel) and masonry walls to take linings. This includes:
  - straightening techniques (crippling bowed studs)
  - adding extra noggins if required
  - identifying and dealing with hidden services such as electricity, water, gas and data cabling
  - considering sustainability measures such as installing thermal and noise insulation
- using safe and effective materials-handling techniques including scaffolds, mechanical materials, lifting devices and elevating work platforms
- setting out lining boards and panelling materials accurately to design requirements and specifications, which may include:
  - installing materials horizontally, vertically and diagonally
  - matching veneers (book matched and slip matched)
- understanding the common terms used in wall panelling and mouldings (e.g. dado, frieze and pilaster skirting, and raked)
- naming a range of common moulds used for skirtings and architraves (e.g. bull nose, splayed and colonial) and mouldings and beading (e.g. quad, scotia and ovolo), which are all used to hide gaps
- knowing how to mark out and cut mitre joints and scribed joints
- applying the basic calculations required to determine material quantities for surface area and perimeter (e.g. Pythagoras' theorem)
- applying geometry techniques to develop hopper bevels and true mitre angles for raked ceilings, cornices and crown mouldings.

REFERENCES AND FURTHER READING

Web-based resources

Masonry anchors

Bremick: **http://www.bremick.com.au**
Hilti: **http://www.hilti.com.au**
Macsim®: **http://www.macsim.com.au**
Ramset: **http://www.ramset.com.au**

Plasterboard

BGC Plasterboard: **http://www.bgc.com.au/plasterboard**
CSR Gyprock: **https://www.gyprock.com.au/resources/redbook**
Knauf Australia: **https://www.knauf.co.uk/systems-and-products**
Siniat: **https://siniat.com.au/products**
James Hardie: **http://www.jameshardie.com.au/products/**

Fire-rated plasterboard installation guides

Gyprock Fyrchek: **https://www.gyprock.com.au/products/plasterboard-fyrchek**
Knauf – Firestop: **https://www.knaufapac.com/en_au/products/interior-linings/technical-boards/firestop-fire-resistant-plasterboard.html**

Skirting and architrave profiles

Australian Moulding Company: **http://www.australianmoulding.com**
Heritage Profiles and Mouldings: **http://www.heritageprofiles.com.au**

Timber

Chippendale Restorations: **https://www.chippendalerestorations.com.au/products-category/joinery/**
Hudson & Sons Illustrated Joinery Catalogue: **https://cavrep.com.au/J/joineryHudson1910.pdf**
The Woodworkers – restoration catalogue: **https://www.woodworkers.com.au/catalogues/QLD/complete/3%20-%20restoration%20catalogue.pdf**
Tilling: **http://www.tilling.com.au**
Wood Solutions: **https://www.woodsolutions.com.au/**

Wet area plastic laminates

Laminex®: **http://www.laminex.com.au**

Environmental sustainability

There are many places to go for information on environmental sustainability.
Green Building Council of Australia – **https://www.gbca.org.au**
Supply Chain Sustainability School – **http://www.supplychainschool.org.au/**
Wood Solutions: **https://www.woodsolutions.com.au/**
Your Home: **http://www.yourhome.gov.au/**

Audiovisual resources

Buildsum YouTube channel – videos on mould enlargement and reduction, raking moulds and installing skirting: **https://www.youtube.com/channel/UCHIkUPCm0eUnhZ5R1gA-6BA**

- Rondo wall and ceiling systems (instructional videos): **http://www.rondo.com.au**
- Cutting crown mouldings – two of the best videos demonstrating how to cut crown mouldings are:
  - Set up a mitre saw to cut crown moulding: **https://www.finewoodworking.com/2005/10/25/set-up-a-miter-saw-to-cut-crown-molding**
  - How to cut crown moulding: **https://www.thisoldhouse.com/how-to/how-to-cut-crown-molding**
- Ramset product education videos on masonry fastening: **http://www.ramset.com.au/Document/Section/213/Product-Education-Videos**

National Construction Code

Copies of the National Construction Code can be obtained free of charge from the Australian Building Codes Board: **http://www.abcb.gov.au**.

National Construction Code 2019, Building Code of Australia – Volume Two, Class 1 and Class 10 Buildings, Australian Building Codes Board.

Section 3.7 Fire separation in multi-residential constructions

Section 3.12 Energy efficiency and thermal ratings

Relevant Australian Standards
AS 1530.1 Methods for fire tests on building materials, components and structures
AS 1684.2 Residential timber-framed construction – Non-cyclonic areas
AS 1684.3 Residential timber-framed construction – Cyclonic areas
AS 2270 Plywood and block board for interior use
AS 2550.10 Cranes, hoists and winches – Safe use – Mobile elevating work platforms
AS 2796.1 Timber – Hardwood – Sawn and milled products
AS 3959 Construction of buildings in bushfire-prone areas
AS 4785.1 Timber – Softwood – Sawn and milled products
AS/NZS 1418.10 Cranes, hoists and winches – Mobile elevating work platforms
AS/NZS 1576.1 Scaffolding – General requirements
AS/NZS 1576.3 Scaffolding – Prefabricated and tube-and-coupler scaffolding
AS/NZS 1859.2 Reconstituted wood-based panels – Specifications – Dry process fibreboard
AS/NZS 1859.4 Reconstituted wood-based panels – Specifications – Wet-processed fibreboard (hardboard, medium density fibreboard)
AS/NZS 1892.5 Portable ladders – Selection, safe use and care
AS/NZS 2098 – Methods of test for veneer and plywood – testing includes moisture content, formaldehyde emissions, bond quality and water absorption/swelling
AS/NZS 2589 Gypsum linings – Application and finishing
AS/NZS 2908.2 Cellulose-cement products – Flat sheet
AS/NZS 2924.1 High pressure decorative laminates
AS/NZS ISO 717.1 Acoustics – Rating of sound insulation in buildings and of building elements
Australian Standards can be purchased from the Standards Australia website: **https://www.standards.org.au/**, or see your teacher or librarian for assistance in accessing Australian Standards online.

GET IT RIGHT

INSTALL LININGS, PANELLING AND MOULDING

Questions

Below are two pictures of skirting fixed to the wall, and floor tiles butting up to the skirting. The initial normal viewing position is from the left-hand side in both pictures.

1 What is less than desirable with the method used?

____

____

2 How should it be done correctly?

____

____

3 What should be done to rectify these skirtings?

____

____

HOW TO

SCRIBE JOINT AN ARCHITRAVE

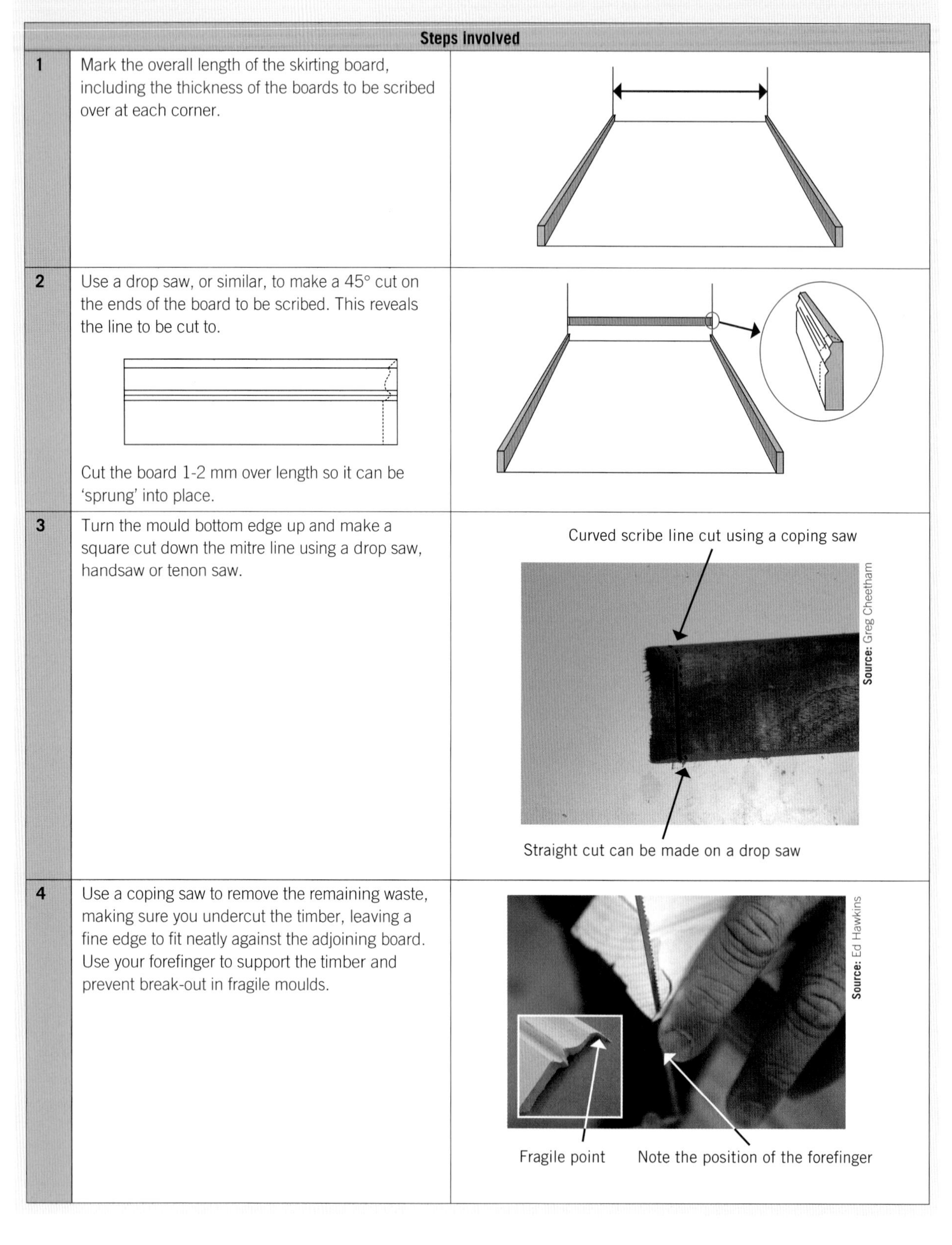

Steps involved	
1	Mark the overall length of the skirting board, including the thickness of the boards to be scribed over at each corner.
2	Use a drop saw, or similar, to make a 45° cut on the ends of the board to be scribed. This reveals the line to be cut to. Cut the board 1-2 mm over length so it can be 'sprung' into place.
3	Turn the mould bottom edge up and make a square cut down the mitre line using a drop saw, handsaw or tenon saw.
4	Use a coping saw to remove the remaining waste, making sure you undercut the timber, leaving a fine edge to fit neatly against the adjoining board. Use your forefinger to support the timber and prevent break-out in fragile moulds.

>>

>>

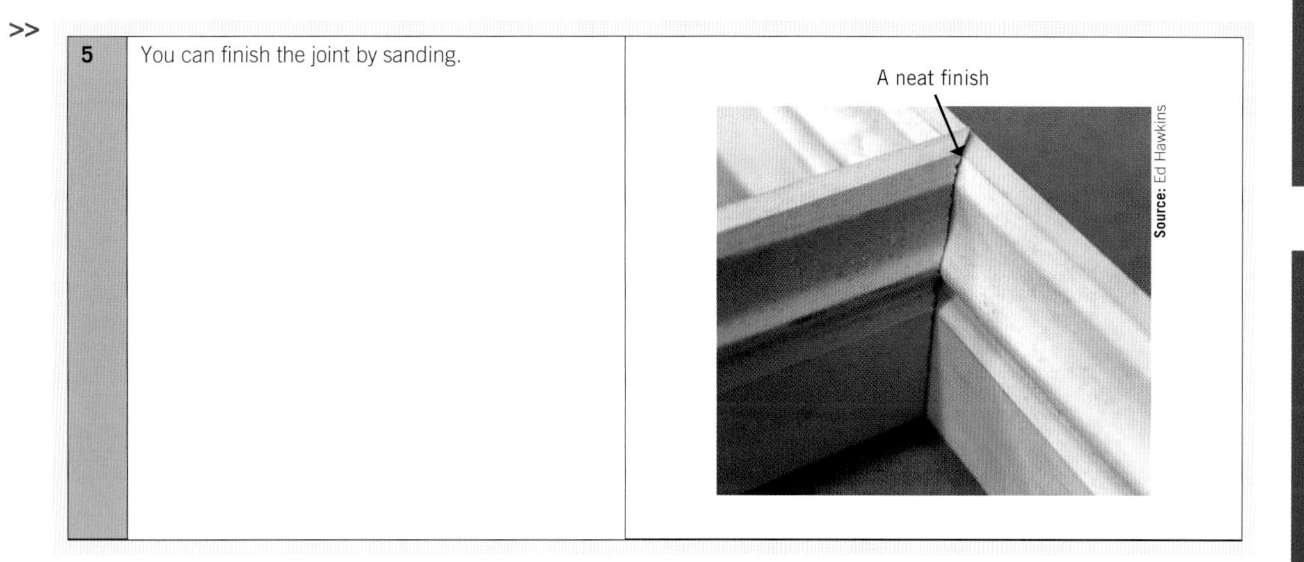

5	You can finish the joint by sanding.

WORKSHEET 1

To be completed by teachers

Student competent ☐

Student not yet competent ☐

Student name: ______________________

Enrolment year: ______________________

Class code: ______________________

Competency name/Number: ______________________

Note to teachers: Prepare skills bays where students can practise installing wall lining materials.

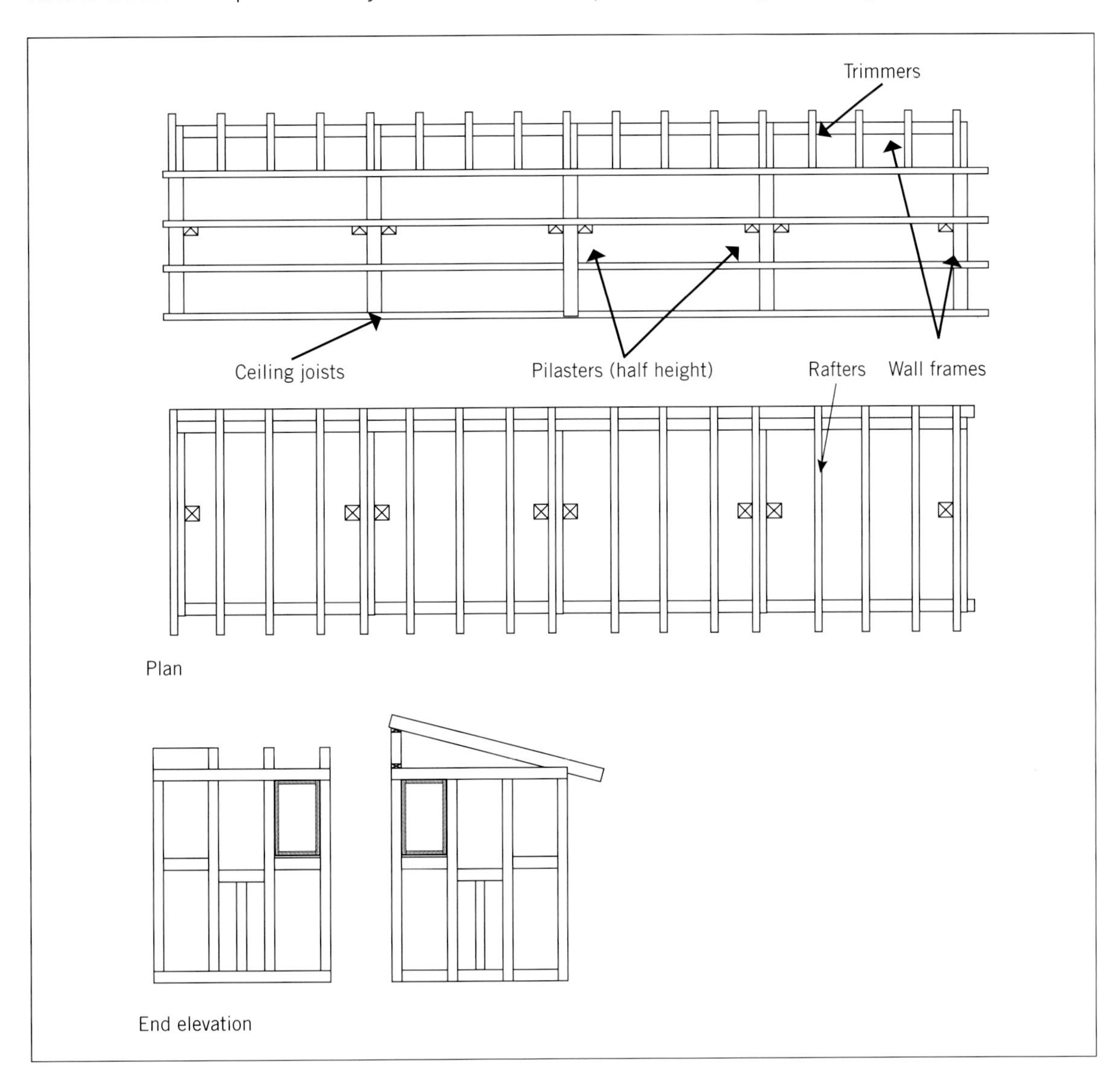

Not to scale: This is a sketch only and is indicative of the type of skills bay that may be used in a formal vocational education institution for the delivery and assessment of skills related to this topic.

Install wall lining

Before starting, ensure that you have your personal protective equipment (PPE), all the necessary tools and all safety equipment, including barricades and signage where necessary.

Note: All your work must be carried out according to the following task specifications. After each task, your assessor must check your progress and provide feedback.

Full details of materials and specifications will be provided by your teacher/assessor.

Task A

Complete diagonal lining of walls with lining boards, including scribing and mitred scotia, quad and skirting, with a minimum of one internal and one external corner for each moulding.

Task B

Complete the lining of walls and raked ceiling with sheet panelling, including the set-out, fabrication/machining and installation of raked moulds (not plasterboard cornice) at top and bottom, with a minimum of one mitre/scribed joint at the top and bottom of the raked ceiling.

Task C

Complete the installation of window nosing, apron and architraves to one window frame. The nosing and apron must have return moulding fitted at each end.

WORKSHEET 2

To be completed by teachers	
Student competent	☐
Student not yet competent	☐

Student name: ______________________

Enrolment year: ______________________

Class code: ______________________

Competency name/Number: ______________________

What to consider when installing wall linings

Answer the following questions.

1 List some important considerations when installing fibre-cement sheets to wet areas.

____

____

____

____

____

2 List some important considerations when installing tongue and groove lining boards.

____

____

____

____

____

3 List some important considerations when installing veneered board or plywood to walls.

____

____

____

____

____

4 From the list below, select the best tool to assist when installing tongue and groove lining boards:

a soft-faced hammer

b rubber mallet

c claw hammer

d wooden mallet.

5 What are the three main benefits of using prefinished lining boards?

6 What is the definition of a 'pilaster'?

7 Name three modern forms of the architrave mould.

a ______________________________

b ______________________________

c ______________________________

8 What is the difference between a panel mould and a bolection mould?

a ______________________________

b ______________________________

c ______________________________

WORKSHEET 3

Student name: ______________________

Enrolment year: ______________________

Class code: ______________________

Competency name/Number: ______________________

To be completed by teachers	
Student competent	☐
Student not yet competent	☐

WORKSHEETS 5

Wall insulation materials

Student task

Investigate and report on a range of three (minimum) ceiling and wall insulation materials suitable for a timber-framed wall, noting their R rating, and outlining the advantages, disadvantages, features and benefits of each.

The report should include at least a half page (200 words) on each material.

Below is a list of potential materials:

- fibreglass batts
- natural wool fibres
- recycled paper insulation (blown)
- foil products (corrugated and air cell).

WORKSHEET 4

To be completed by teachers	
Student competent	☐
Student not yet competent	☐

Student name: ______________________

Enrolment year: ______________________

Class code: ______________________

Competency name/Number: ______________________

Answer the following questions.

1 When fitting cornices and crown mouldings, what is the primary sequence of installation?

____

____

____

____

____

2 Explain the term 'raking moulds' and provide an example of where you might find one.

____

____

3 Explain with an example the term 'spandrel'.

____

____

4 What are split battens used for?

____

____

5 What are the two most critical factors when installing split battens?

a ______________________

b ______________________

6 What methods can be used to fix lining boards to walls for invisible fastening?

____

____

____

____

7 Explain how you would prepare masonry walls for direct fixing of wall panelling.

8 Fill in the missing words.

An apron piece goes ________ the ________ board to ________ between the wall and the nosing board.

9 Indicate on the sketch below the correct method to finish off the ends of a nosing board and apron piece. A simple plan view of a nosing is shown.

10 List three characteristic differences between a regular moulding and a bead.

11 Draw the profile of an ovalo mould and an ogee mould.

12 This question has two parts. Please answer both.

a Which Australian Standard covers plasterboards as an internal wall lining?

b According to this Standard, what is the minimum bearing surface of a timber stud or noggins where plasterboard sheets are butt jointed?

13 According to the Standard for plasterboard lining and finishing, three different levels for finishing linings are applicable. Give a brief summary of each level of finish and where/how it is applied.

Answers:

a ______

b ______

c ______

14 Develop the true face bevel and true mitre angle for the crown moulding drawn below. **Note:** use the drawing to show your solution.

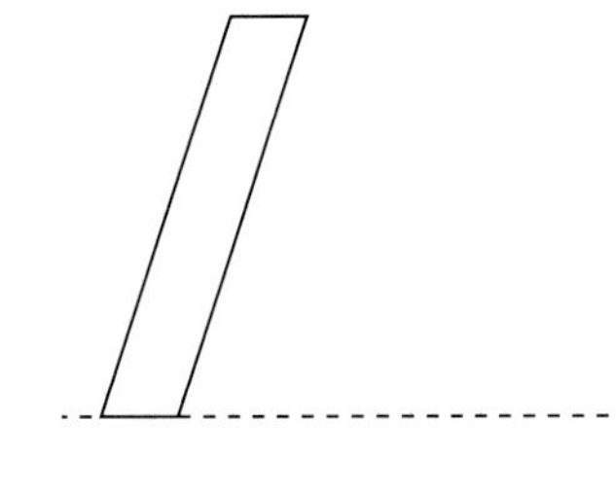

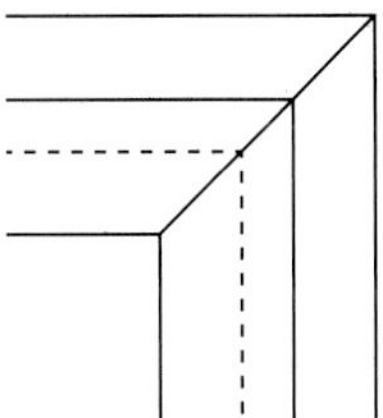

WORKSHEET 5

To be completed by teachers

Student competent ☐

Student not yet competent ☐

Student name: ____________________

Enrolment year: ____________________

Class code: ____________________

Competency name/Number: ____________________

Answer the following questions.

1 Where do you get information regarding the type and layout of lining material?

____

____

2 Explain why one lining material would be preferred over another.

____

____

3 What is the most common material used for ceiling lining and why?

____

____

4 The relevant Australian Standard for scaffolding is AS/NZS 1576.1 Scaffolding – General requirements. Mobile scaffold is often used in the installation of ceilings in buildings. What sections deal specifically with the operation of mobile scaffold?

____

____

____

5 Which other Australian Standard covers work platforms? Provide an outline of what that Standard covers.

____

____

____

6 List three frame or surface defects that require fixing before installing linings.

a ____________________

____

b ____________________

____

c

7 Three critical building items should be checked for completion before linings are installed. List them below.

a

b

c

8 What type of joint is used at internal corners for quad and scotia?

9 List four factors that should be considered when calculating quantities for wall lining.

a

b

c

d

10 Where would you find casting grooves, and what benefit do they provide?

WORKSHEET 6

To be completed by teachers	
Student competent	☐
Student not yet competent	☐

Student name: ______________________________

Enrolment year: ______________________________

Class code: ______________________________

Competency name/Number: ______________________________

1 A room requires timber lining boards installed to wall A and 10 mm plasterboard to the remaining walls and ceiling.

   a Part A: Use the following specifications to calculate the total metres of timber lining boards (+ 5 per cent waste) and the number of plasterboard sheets required, as well as the appropriate sheet sizes required to line the walls.

      i Direction of timber lining boards: vertical effective cover 133 mm.

      ii Calculate the total lineal metres required to line the wall.

      iii To help calculate sheet sizes, visit the CSR Gyprock website: http://www.gyprock.com.au.

      iv Wall A – solid timber lining boards. Remaining walls are plasterboard sheets.

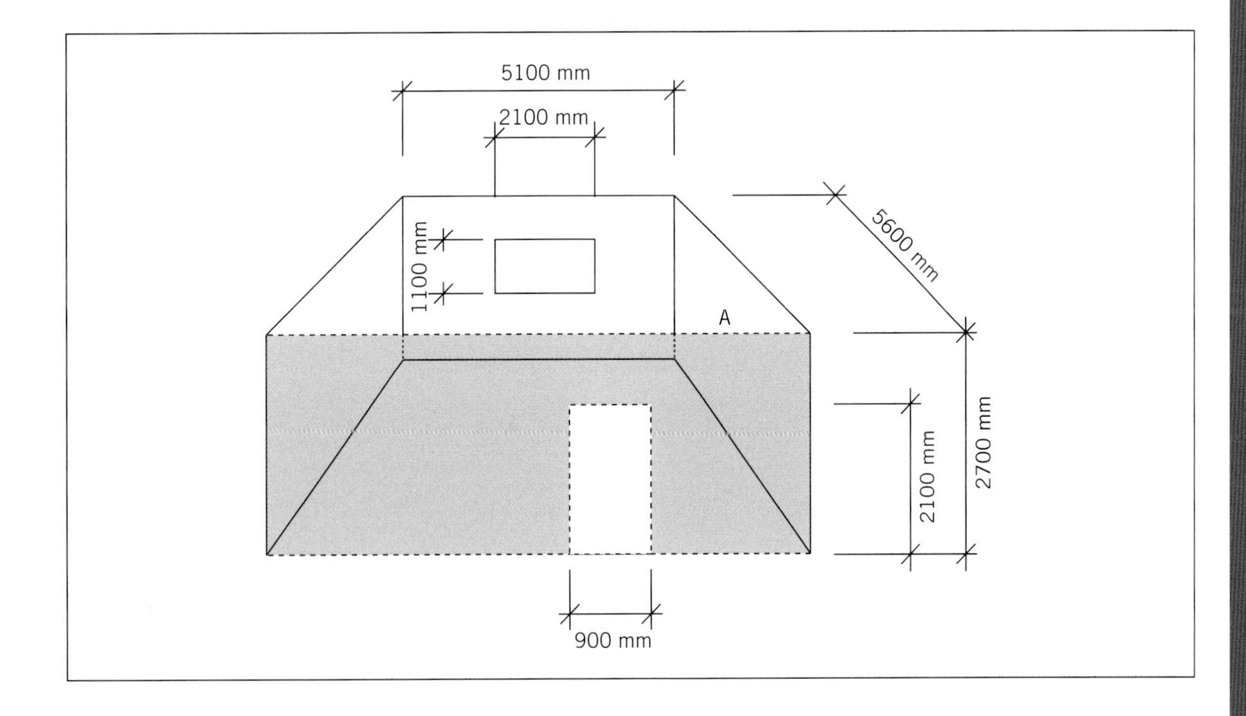

b Part B: Calculate the total cost of material to line the walls and ceiling, and finish with cornice architraves and skirtings to doors and windows. **Note:** Windows have no nosing.

Item	Lengths	Rate	Cost
Lining boards 133 mm pine	Stock sizes	\$3.60/m	
Plasterboard	Stock sizes	\$6.77/m^2	
Cornice – 90 mm		\$3.50/m	
Skirting – 66 mm colonial		\$6.90/m	
Architrave – 92 mm splayed		\$9.80/m	

6

CONSTRUCT, ASSEMBLE AND INSTALL INTERNAL AND EXTERNAL TIMBER STAIRS

This chapter aligns with the unit of competency, 'Construct, assemble and install timber external stairs' and a range of internal stair types. It will focus on timber stair construction, different types of handrails and balustrades, and other National Construction Code design requirements for a range of stair types.

In particular the chapter addresses the key competencies of:

- designing and setting out straight flights with landings and winders in accordance with the National Construction Code
- installing straight flights both internally and externally, including handrails and with changes in direction
- designing and setting out stairs and assembling and installing stairs.

This edition also provides further resources on the use of modern computer numerical control (CNC) wood-machining technology to produce stair components.

Essential to stair construction is consideration of the environment in which they will be placed, as well as knowledge of the Australian Standards as they relate to timber grading, preservative treatments and appropriate construction for fire-prone areas.

Learning objectives

By the end of this chapter you will be able to:

- know and understand the terminology associated with stair construction
- identify different types of stairs by name
- apply work health and safety/occupational health and safety (WHS/OHS) and quality assurance requirements in making and installing stairs
- set out straight flight stairs accurately according to the National Construction Code, use the correct formulas for calculating the pitch so as to meet a range of design requirements, and calculate material quantities accurately
- identify the appropriate tools, equipment and machinery used to make stairs
- make and join basic handrail sections and fit balusters and balustrades to straight flight stairs
- install straight flight stairs and landings to support stair flights for both internal and external use.

Stair terminology

A set of stairs is called a flight. A number of flights can be joined together by **landings** to increase the height or to change direction.

A flight is the part of a stair having a continuous set of **risers**, uninterrupted by a landing. This applies also to flights that change direction throughout the rise by using winders such as spirals and quarter and half turns – where the intersection of the tread and the rise or nosing creates a continuous slope (see **Figures 6.1** and **6.2**).

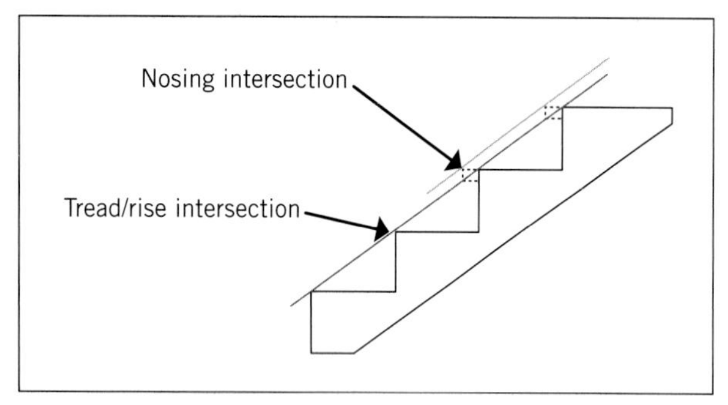

FIGURE 6.1 Simple stair set-out showing tread and rise and nosing

Special note: According to the National Construction Code, a long flight of stairs should be interrupted by a landing or change of direction to limit the distance a person might fall.

Balusters are vertical members between the string or landing and the handrail. The handrail and the balusters are referred to as the *balustrade*. The **handrail** is fixed parallel above the **string** to provide support when climbing or descending the stairs. The **newels** are vertical posts, generally at the ends of a flight of stairs, to which the string and handrail are sometimes connected (see **Figure 6.3**).

Winders are tapered treads within a straight flight, used to change direction of the stairs around corners while continuing the slope (see **Figure 6.4**). They allow the stairs to gain height where the horizontal distance (or total going) is limited.

Types of stairs

A *straight flight with landing* (see **Figure 6.5**) is used in areas with a large length of floor space at ground floor level (going).

Note: For flights over 18 risers, a landing must be incorporated.

Quarter turn stairs use a minimum amount of floor space at both levels and a minimum width for both flights. A quarter space landing is used to make the turn (see **Figure 6.6**), or alternatively, winders can be used to make the turn and shorten the flights.

Half turn or dog-legged stairs need a minimum length of floor space (total going) but need sufficient width for two stair flights (see **Figure 6.7**). A half space landing is used when installing a dog-legged stair.

A **spiral stair** is a flight of stairs, circular in plan, winding around a central post, with steps that radiate from a common centre. *Tapered treads* are used on spiral and geometrically (elliptically) curved stairs, tapering towards the inside of the curve (see **Figure 6.8**).

Cut strings have the top edge cut to the lines of the treads and risers (see **Figure 6.9**). A variation on the cut string is the cut and bracketed string.

Closed strings have the treads and risers housed into the strings and the top edge is parallel to the bottom edge (see **Figures 6.3** and **6.4**).

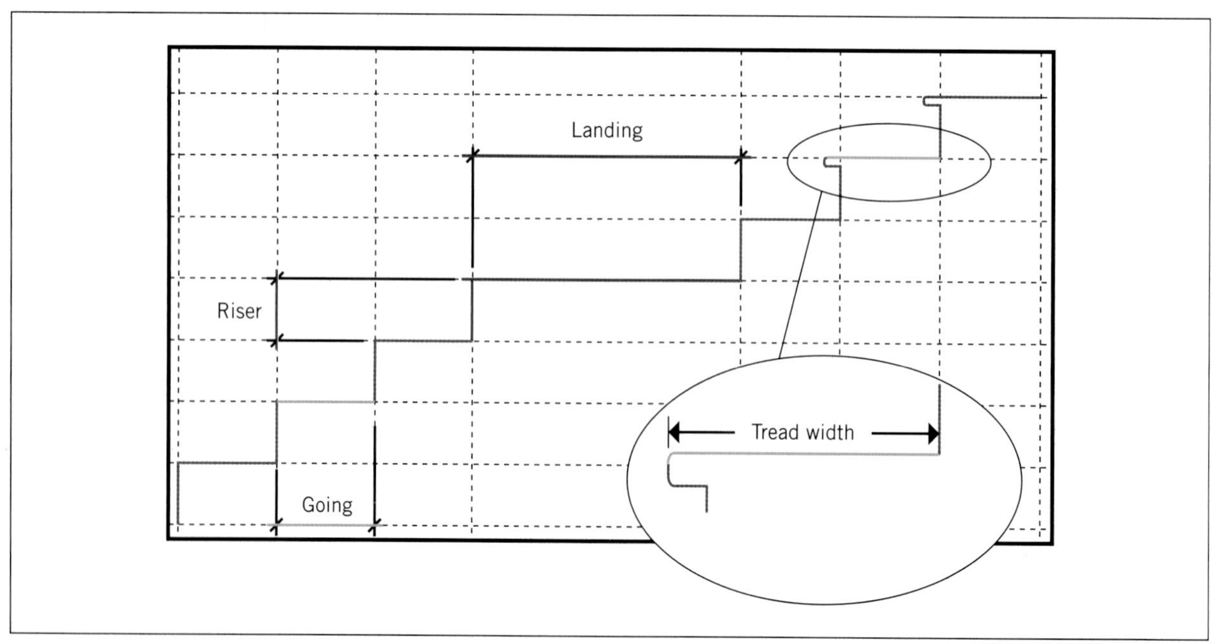

FIGURE 6.2 Riser, tread (going) and landing

Source: Ed Hawkins

FIGURE 6.3 Balusters, handrails and newels

Source: Ed Hawkins

FIGURE 6.4 Winders

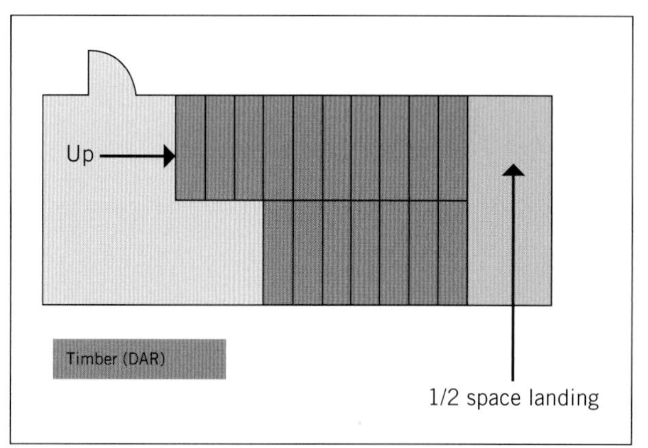

FIGURE 6.7 Half turn or dog-legged stairs

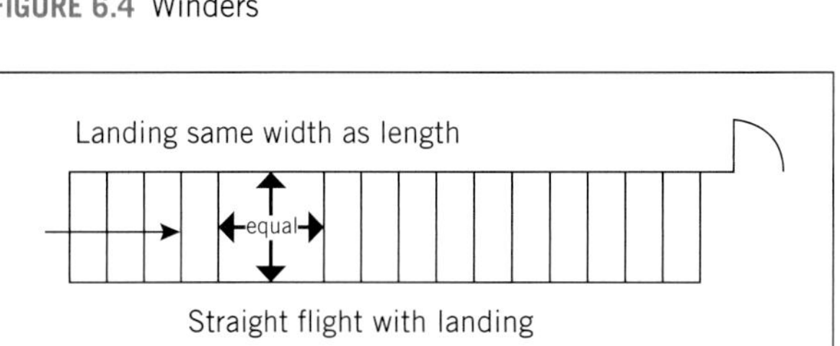

FIGURE 6.5 Straight flight with landing

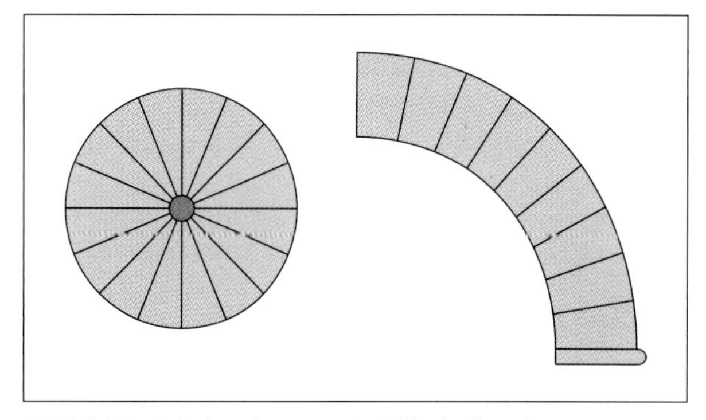

FIGURE 6.8 Spiral and geometric (elliptical) stairs

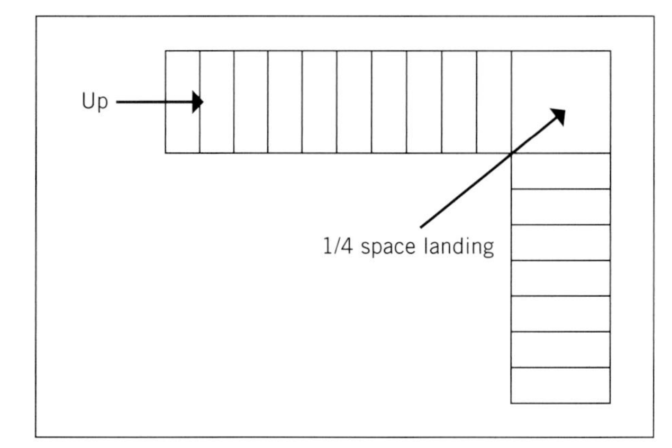

FIGURE 6.6 Quarter turn stairs

Wall strings are located on the wall side of the flight and are usually closed, while the **well string** is located on the open side of the flight and may be closed or cut strings (see Figure 6.10).

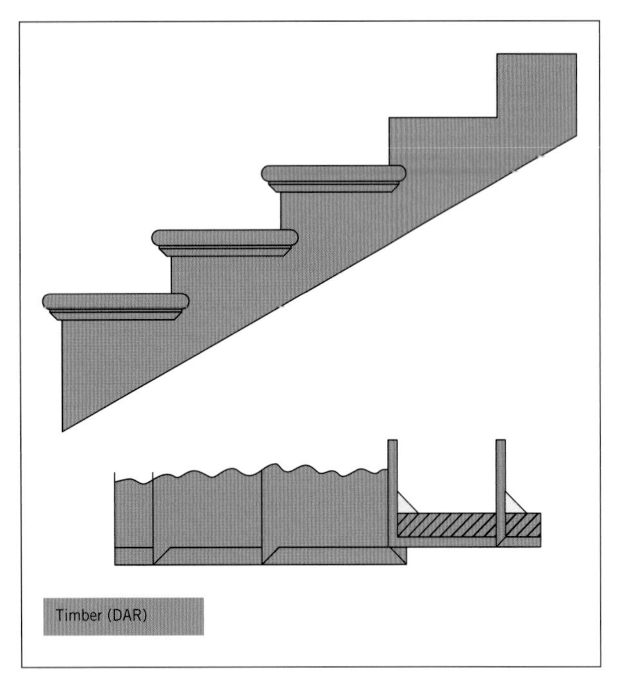

FIGURE 6.9 Cut strings

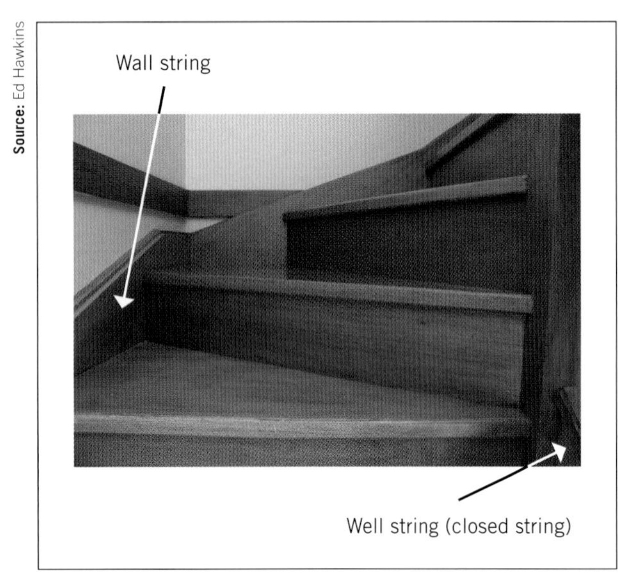

Source: Ed Hawkins

FIGURE 6.10 Well string (closed) and wall string

Quality assurance, work health and safety

Quality assurance covers the set-out and manufacture of stairs in workshops and on building sites. The selection of quality materials and effective materials handling, stacking and storage, as well as the care and maintenance of machines and hand and power tools, are critical.

When handling rough sawn timber, it is wise to wear leather (not calf skin) work gloves to protect your hands from splinters.

Complete a site risk assessment before commencing work onsite. You may need to complete a site induction. Refer to the SWMS and JSA to ensure compliance with all WHS measures on site. Ensure hazards are identified and apply risk-control measures, including signage and barricades where required. Before starting work, make sure the area is clean and free from obstacles and that materials are stacked safely and ready for use. Check the Environmental Plan for requirements in relation to dust, vibration and noise. Locate the waste and recycling points in order to maintain a clean workspace, reduce trip hazards and comply with environmental laws.

Assembled stairs can be particularly heavy so use team-lifting strategies to reduce potential injuries, or use a mechanical lifting aid like a pallet jack or forklift to move materials around the site. Other types of lifting devices include:

- block and tackles
- engine lifters
- gantry cranes.

Maintenance of tools, equipment and machinery should include:

- checking blades for sharpness and belts for wear and tear
- having power tools tested and tagged safe for use when necessary
- correctly adjusting blades, fences and drive belts
- checking guards, fences and stops are in place and operating properly.

Always refer to the manufacturer's instructions for care and ongoing maintenance, especially for tools and equipment that require calibration. Personal protective equipment (PPE) must be worn. PPE includes foot protection, eye and hearing protection, dust masks, suitable clothing, hard hats and gloves. It depends on where the work is being carried out as to what PPE is necessary. For example, in a joinery shop a hard hat is not normally worn, but a respirator-type dust mask may be necessary when applying surface coatings and sanding all components.

The large, section-size, solid timbers required for stair strings are no longer sustainably available. Sustainable substitute materials for long strings include laminated timber beams or laminated veneer lumber (LVL). These can be decorated with a range of sustainably sourced veneered products and capped to hide lesser quality timbers.

Wear a dust mask or respirator when working with MDF or any other materials that produce dust or sawdust.

Setting out stairs

Before commencing work, always check the plans and drawings for the location and measurements of the stairs.

When interpreting the plans and drawings, stairs will be represented by horizontal or vertical lines. Check the plans for both Topdown and Sectional views.

Check the measurements from the plans and drawings at a site inspection. Ensure the most current version of the plans and drawings are used. Changes and variations are often made to plans and drawings, and sometimes updates are made during construction. In some instances, stairs are made offsite, so it is important that they will fit into the space when completed.

After checking measurements, calculate the quantities and materials required to complete the work. Plan and sequence the work activities to ensure work is completed in an efficient manner.

Stair set-out must comply with the National Construction Code – Volume One, Section D Access and egress, and Volume Two, Part H5 Safe movement and access, as well as the Australian Building Codes Board's Standard for Livable Housing Design. The slope, called a pitch, for a set of stairs cannot be too high (or it will be a ladder). And the pitch cannot be too shallow (because fatigue from climbing too many steps will be a problem).

The principal factors governing stair design are:

- horizontal floor space available
- location and height of any obstructions, such as a door and/or windows; for example, a doorway cannot open directly on to stairs – it must open on to a landing
- available headroom, which is measured plumb from the nosing to the underside of the ceiling or soffit above (see **Figure 6.11**). The head room clearance height is 2 m, but check the Australian Building Codes Board's Standards for Livable Housing Design for clarification.

In most cases, the design of a set of stairs is generally a compromise between the pitch and the floor space or landing space available.

The critical dimensions to be considered when designing stairs are:

- the vertical **rise** of each step (see **Table 6.1**)
- the **going** of each step (see **Table 6.1**)
- the maximum number of risers in a flight (18)
- the handrail heights measured plumb from the tread and riser intersection point on the staircase. The minimum height is 865 mm (see **Figure 6.12**). At balconies higher than 3 m, the balustrade must be $\geq$ 1.0 m.
- The width of stairways – See National Construction Code – Volume One, Section D and your state or territory's schedule of requirements. When deciding the width of stairs you need to consider who or what is going to travel the stairs. For example, how wide does it need to be to enable carrying a double bed, fridge or bookcase up or down the flight? Some old inner-city terrace buildings have very steep and narrow staircases, making it very difficult to carry furniture
- the dimensions of landings (minimum 750 mm) – should be equal in length and width (see **Figure 6.12**)
- the minimum space between balusters in a balustrade – should not be greater than 125 mm, so that a baby cannot get their head caught or their shoulders through (see **Figure 6.12**)
- the gap between the undersides of one tread and the top of the tread below – should not be more than 125 mm in an open rise flight (see **Figure 6.12**).

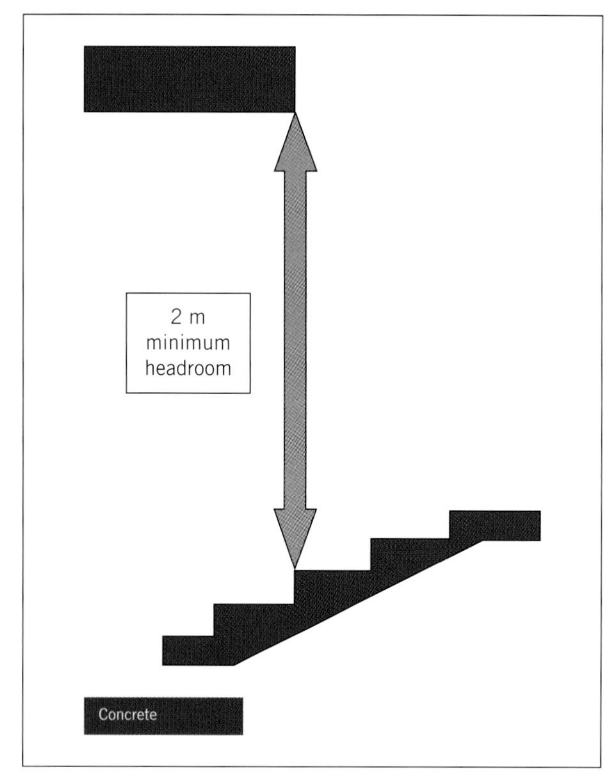

FIGURE 6.11 Minimum headroom measured from nosing and underside of ceiling or soffit above

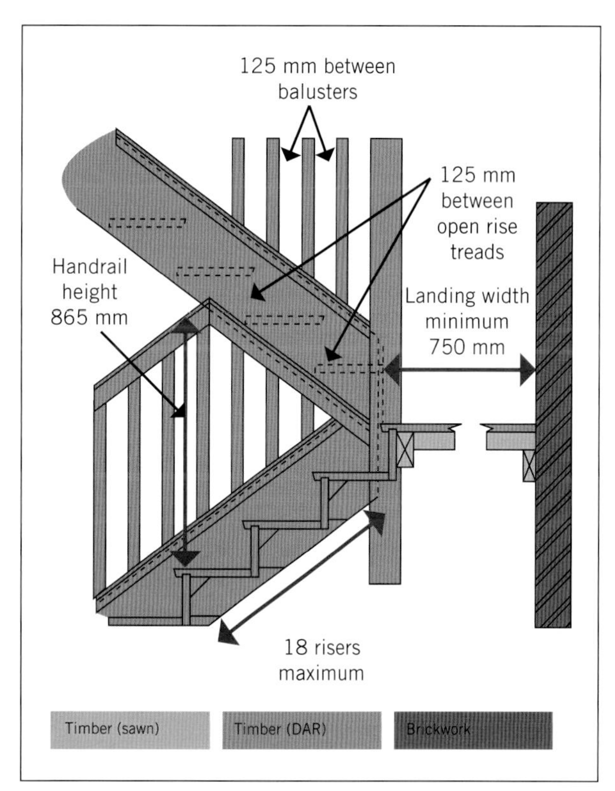

FIGURE 6.12 Basic design dimensions of stairs

TABLE 6.1 National Construction Code – formula for pitch

Stair riser and going dimensions (mm)						
	Riser (R)		Going (G)		Formula for pitch = 2R + G	
	Max.	Min.	Max.	Min.	Max.	Min.
Stairs other than spiral	190	115	355	240	700	550
Spiral	220	140	370	210	680	590

Table 6.1 outlines the basic formula for checking that a flight of stairs complies with the National Construction Code.

Applying the formula for pitch

The formula for testing the compliance of the pitch of a straight flight of stairs is: **2R + G = 550 to 700 mm.**

For example:

Rise = 150 mm

Going = 310 mm

Pitch = 300 + 310 = 610 mm

Therefore, a set of stairs using the rise and tread sizes (see Figure 6.13) given in the previous example would comply because the final calculation falls between 550 and 700 mm.

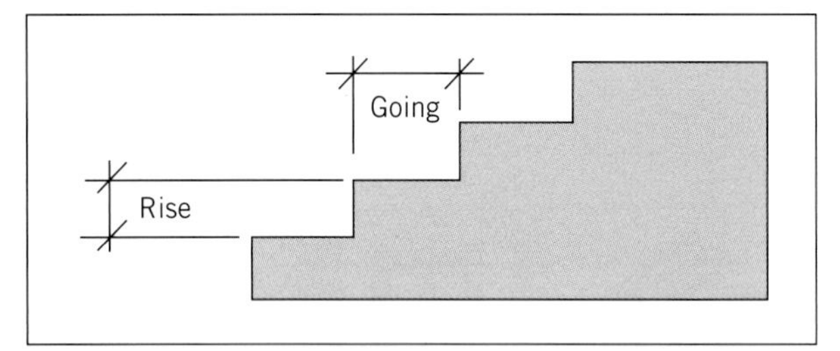

FIGURE 6.13 Going and rise

The following example would not comply:

Rise = 185 mm

Going = 350 mm

Pitch = 370 + 355 = 725 mm (2R + G is greater than 700 mm)

For spiral stairs, the same formula is used, but there are different maximums and minimums. Spiral stairs are usually built to save space and tend to have a steeper pitch than regular flights. For safety reasons, the formula restrictions have been reduced to prevent manufacturers from making spiral stairs too steep. If a steeper flight is required than that available under the National Construction Code, an attic ladder can be used.

LEARNING TASK 6.1 SETTING OUT STAIRS

Do the following rise and going dimensions meet the National Construction Code? Explain why or why not.

- Stair 1: going = 305 mm; rise = 157 mm
- Stair 2: rise = 120 mm; going = 250 mm
- Stair 3: going = 235 mm; rise = 190 mm

Installing attic ladders will require cutting into or repositioning ceilings timbers and possibly roof timbers and roof trusses as well. Seek professional help to determine the installation requirements for attic ladders.

Stairs with winders

Stair flights have winders in them to reduce the amount of floor space the staircase uses. Winder treads are tapered, and are built using the same rise as for the rest of the staircase. You cannot use more than three winders in a 90° turn or six winders in a 180° turn.

For flights less than 1 m wide, the going should be the same as for the adjoining parallel treads when measured at the centre of the unobstructed flight width (see Figure 6.14).

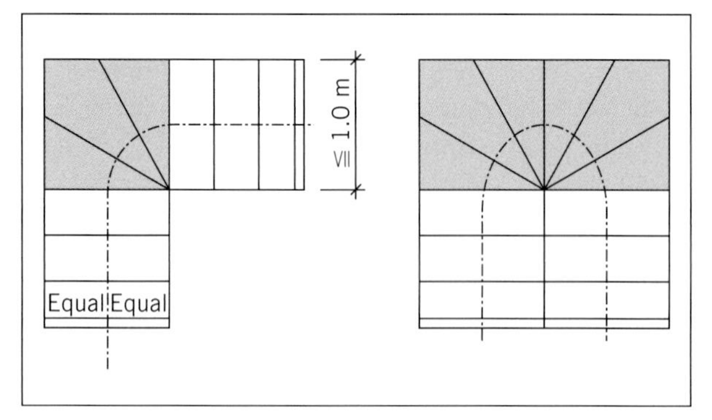

FIGURE 6.14 Quarter and a half turn stair flight with winders set out less than 1 m wide

Note: For flights > 1 m wide, the going should be measured at 400 mm out from the inside of the handrail side (see Figure 6.15).

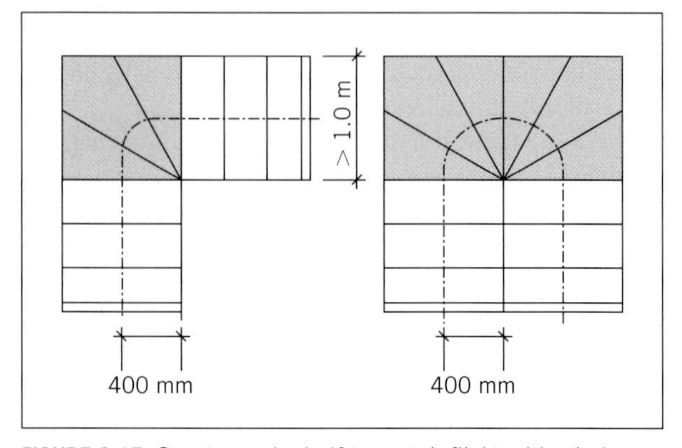

FIGURE 6.15 Quarter and a half turn stair flight with winders set out more than 1 m wide

For curved stairs, the pitch measurement rules are the same. For full details refer to the National Construction Code.

LEARNING TASK 6.2 SOLVING A STAIR PROBLEM

Design a legal set of stairs that meets the National Construction Code and fits into the floor plan below.

Calculate a legal rise and going for the set of stairs.

Note: you will need a landing: use a 900 mm × 900 mm landing.

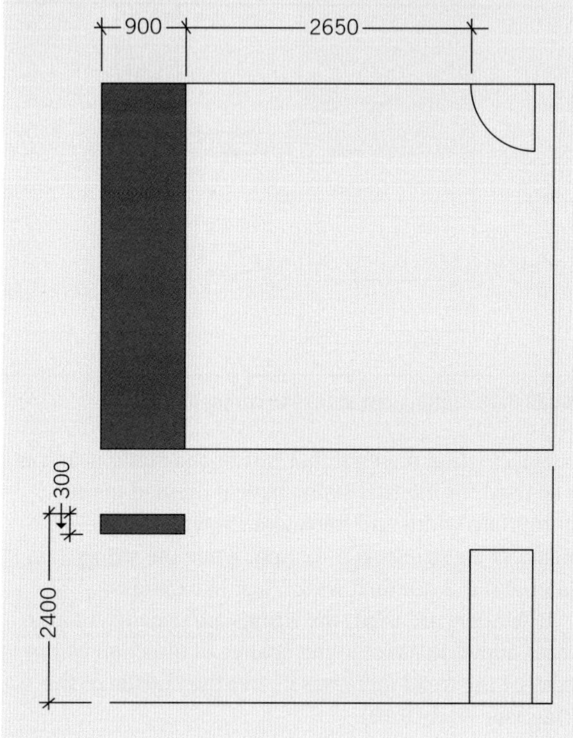

Site measuring for timber stairs

When measuring up onsite, have a couple of lengths of timber or board that can be used to make rise and going rods. Record the information on the boards as you take the measurements. The boards can also be used as straight edges to check the floors and walls for level and plumb.

The critical information required is as follows:

- The height from finished floor to finished floor sets out the total rise of the flight in order to establish whether or not a landing is required. Watch out for out-of-level floors as the dimensions may cause the rise height calculations to be incorrect (see Figure 6.16). Use a spirit level and straight edge to get the overall rise height as close as possible to where the stairs will finish.
- Check the walls for parallel, particularly if the flight is to fit neatly between them, and allow for cover strips and moulds to finish off the stairs neatly.
- Check for obstructions and restrictions that may affect the stair design, such as windows, doors and head clearance in both the overall going and rise.
- Check the wall and floor constructions to make sure you can fasten the staircase or install sufficient bearers, joists and landing timbers to support the flight.

Calculating material quantities

String length

We use Pythagoras' theorem to calculate the string length (see Figure 6.17).

In this example, we will use an average rise height of 150 mm and a typical going of 310 mm, with both rise

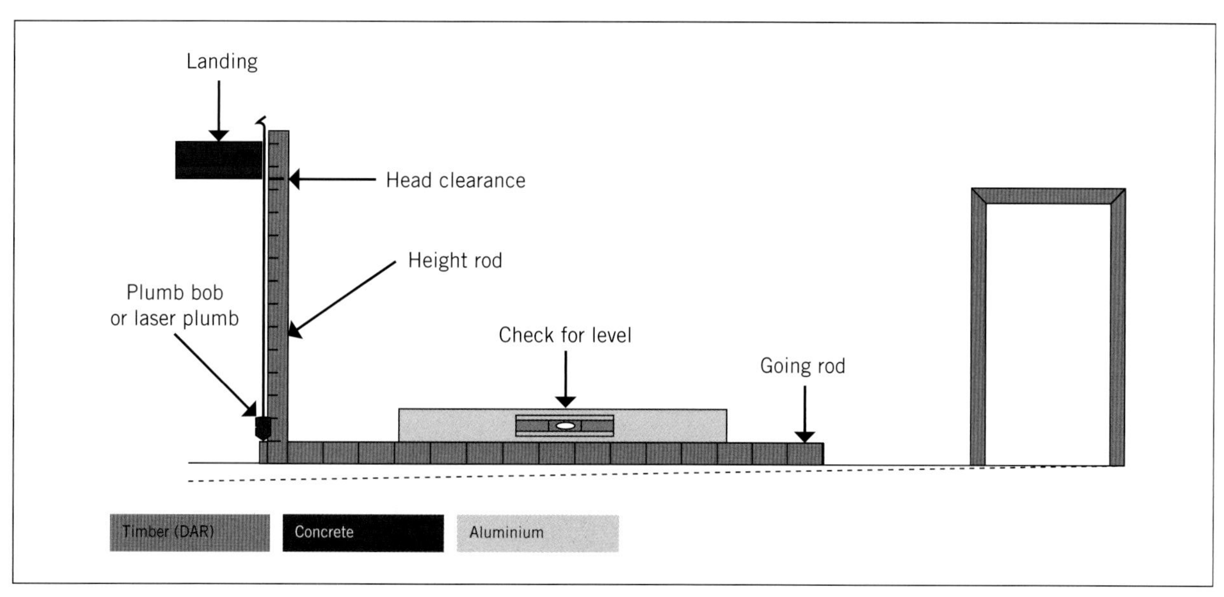

FIGURE 6.16 Measuring the rise height with out-of-level floors

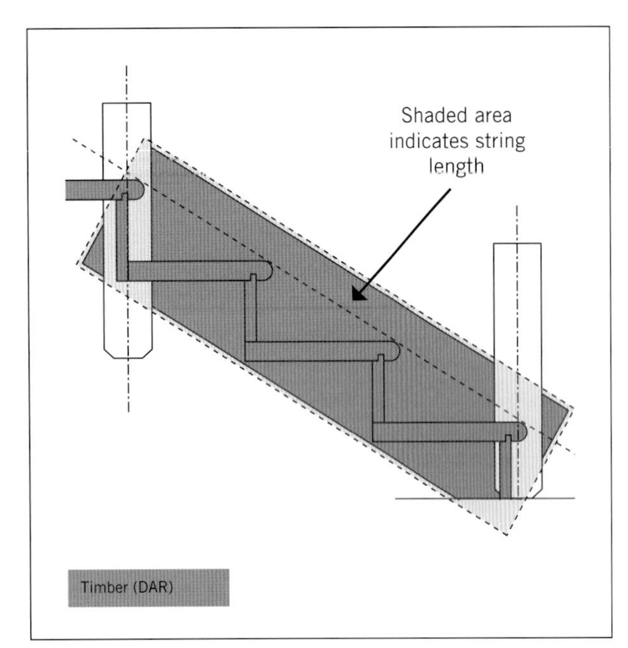

FIGURE 6.17 String length

and going complying with the National Construction Code requirements:

2R + G = 550–700 mm

(2 × 150) + 310 = 610 mm

Now, using Pythagoras' theorem, $c = \sqrt{a^2 + b^2}$, we calculate the approximate length of the string; that is:

$$c = \sqrt{0.6^2 + 0.93^2} = 1.106 \text{ m}$$

Therefore, the length of timber needed for the string is 1.106 m. Taking it to the nearest stock length, it will be 1.2 m (depending on the timber species and the lengths the timber merchant supplies).

It is critical that allowances be made for joints, the fixing of other components, how skirting boards are to be aligned at landings and how the string is to be fixed off to the floor joist or landing materials.

Length of treads

Treads are calculated by:

Number of treads × length of each tread = total metres

A best stock length calculation is done to arrive at the purchase quantities.

Note: If winders or split landings are used, you may need to allow for the edge jointing of boards to make wider treads.

An alternative method is to use 'Pete's rule of thumb'. This method uses the **nominal width** of the tread material (i.e. to the nearest 25 mm, inclusive of nosing) multiplied by the number of risers plus 1. From the previous example, that is:

330 mm finished tread size = > 350 mm nominal
4 risers + 1 = > 5
350 × 5 = 1750 mm

So you would order a 1.8 m stringer.

Note: 'Pete's rule of thumb' comes from Peter Costin, a renowned Geelong specialist joiner.

COMPLETE WORKSHEET 2

Setting out the strings

The most common forms of stairs are *set out to centre lines* on strings and newel posts. The well string and newel are located on the centre line (see **Figure 6.18**).

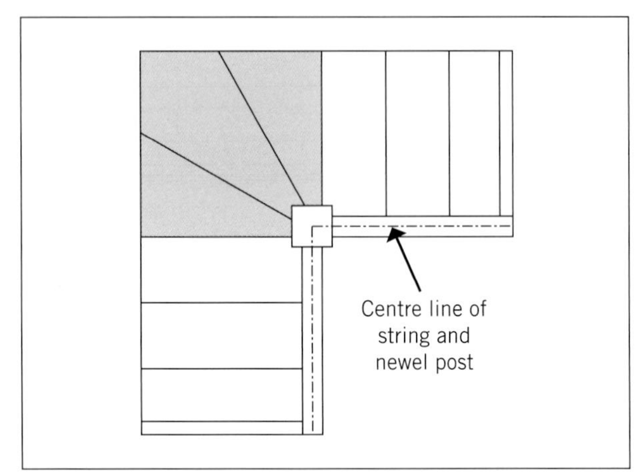

FIGURE 6.18 Setting out strings to centre lines

Select string material that is true and straight. If it is to be used for internal stairs, have it dressed to the appropriate width and thickness – a typical material thickness for strings is > 32 mm, while the string material has a nominal width, just like treads.

If winders are used, the strings are usually cut and pieces added to assist in the change of direction as the depth of the tread increases nearer the outside of the string (see **Figure 6.19**).

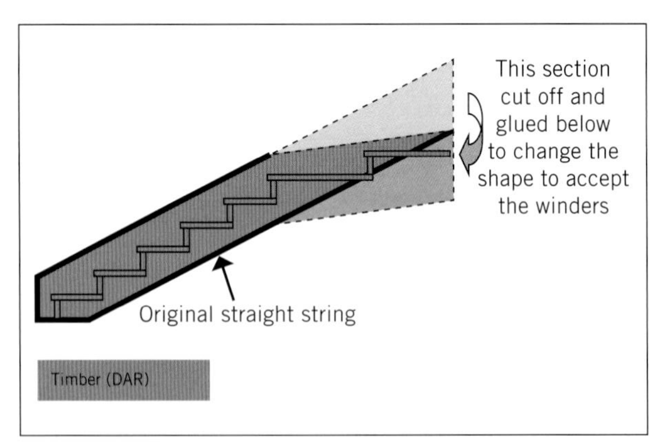

FIGURE 6.19 String for winders

The top edge is dressed and a **margin line** is drawn on the strings (see **Figure 6.20**). This keeps the **nosing** and/or tread and riser intersection aligned. It is a good idea to set the strings up as a pair on a set of saw horses or trestles so that they can be marked out together. A rule of thumb for the margin line is the nominal thickness of the string material (e.g. 38 mm nom. = 32 mm margin; 50 mm nom. = 42 mm margin).

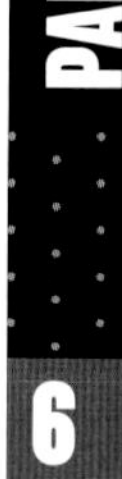

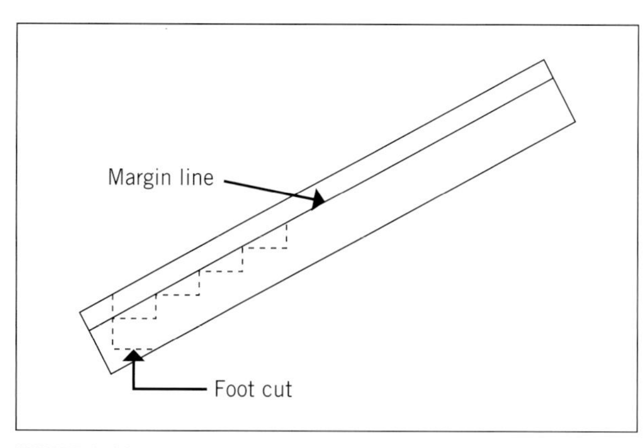

FIGURE 6.20 Margin line

Setting out the treads and risers

Make the rise and going calculations (2R + G) and set up a steel square or jig and templates.

A number of different tools can be used to mark out the treads and risers when you are setting them out.

A **pitch board** can be made from plywood. The pitch board shows the margin, the rise height and the tread width (see Figure 6.21).

An adjustable router template can also be used to mark out the treads and risers and this will be used again if routing out for treads and risers (see Figure 6.22).

Note: When making this type of jig, you need to allow for the difference between the cutter diameter and the diameter of the template guide (see Figure 6.23).

When wedges are used, they should have a pitch of no less than 12:1 (see Figure 6.24). This has to be allowed for in the construction of the jig. The wedge should have a thickness of 5 mm at the narrow end and be long enough to support the full width of the tread without interfering with the insertion of the wedges for the risers.

A builder's square (carpenter's square, roofing square, steel square) and fence are also used to set out

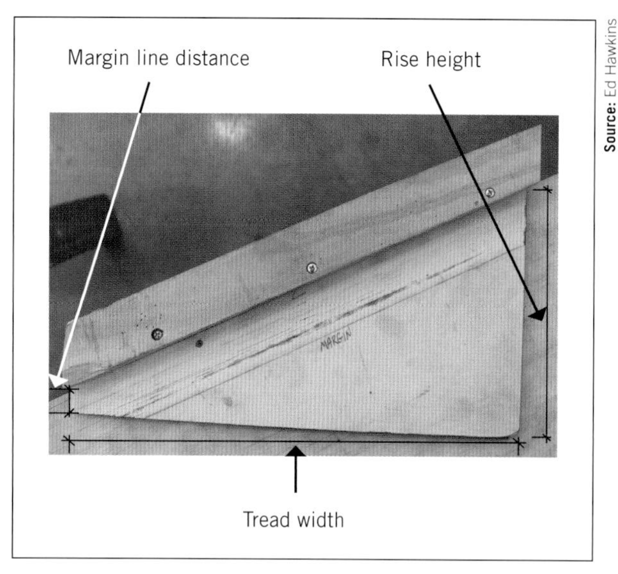

Source: Ed Hawkins

FIGURE 6.21 Pitch board

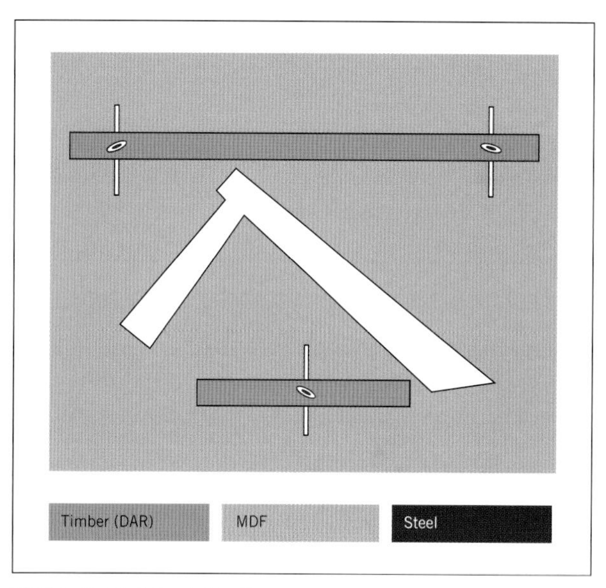

FIGURE 6.22 Adjustable jig

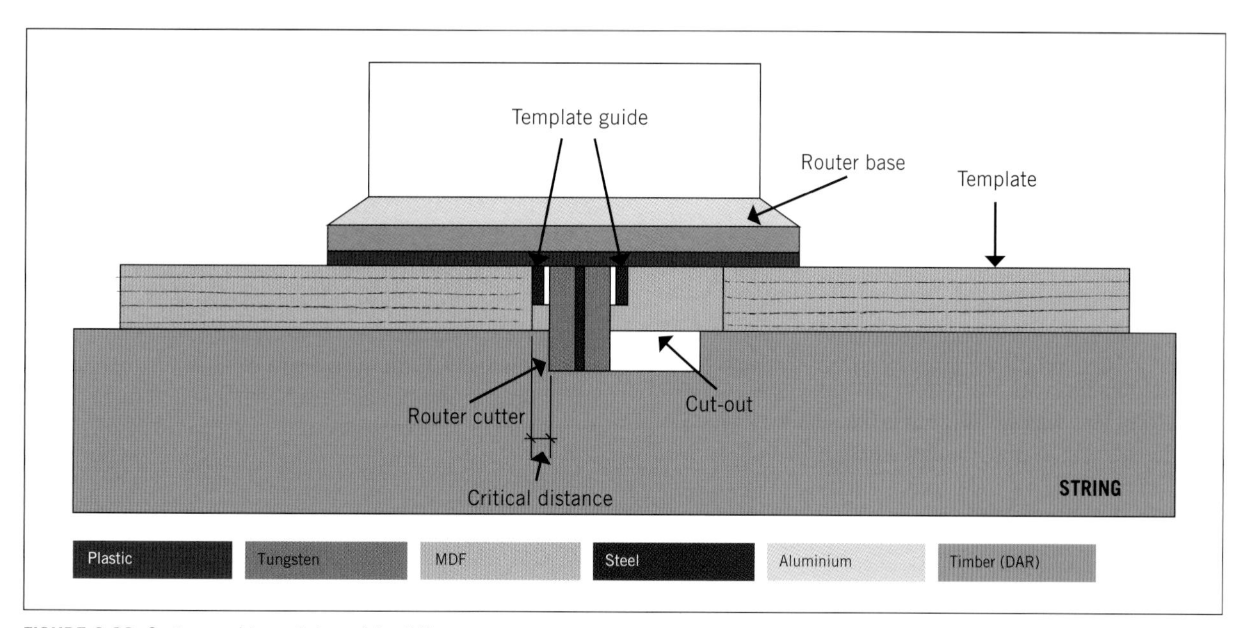

FIGURE 6.23 Cutter and template guide difference

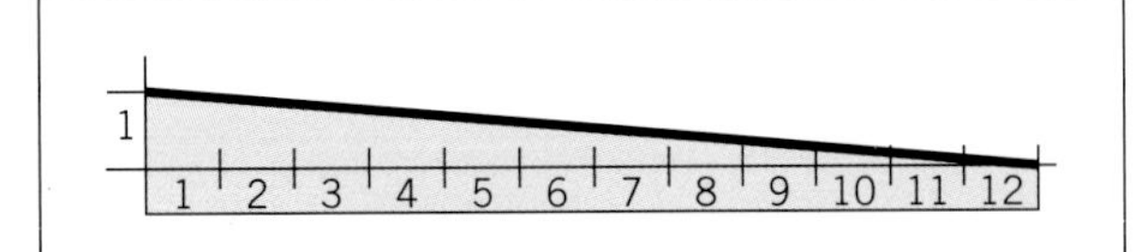

FIGURE 6.24 Pitch for wedges

the strings for treads and risers (see Figure 6.25). The steel square can also be used in conjunction with stair buttons to mark out the treads and risers (see Figure 6.26).

Source: Ed Hawkins

FIGURE 6.25 Builder's square and fence

Source: Ed Hawkins

FIGURE 6.26 Builder's square with stair buttons

Using closed strings means that the tread and riser are set out from the margin line. Treads and risers intersect at the margin line (see Figure 6.27).

Foot and plumb cuts

The **foot** and plumb cuts are located at the bottom end of the string, and these are marked out at the same time as the treads and risers.

The plumb cut on the wall string is traditionally finished to align with the face of the newel post on the well string (see Figure 6.28), but is also made to blend into the skirting in high-quality work.

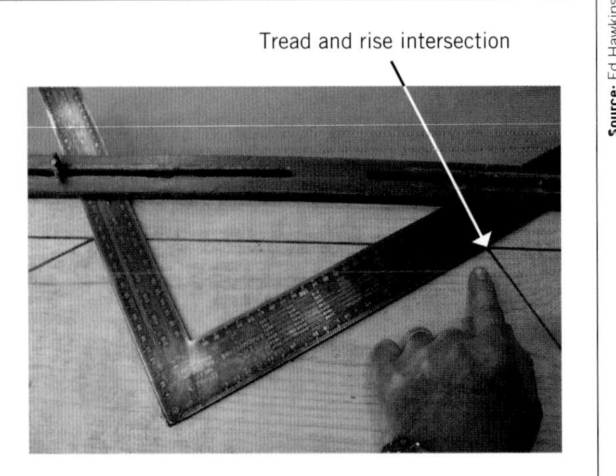

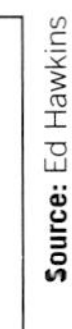

FIGURE 6.27 Treads and risers meet at the margin line

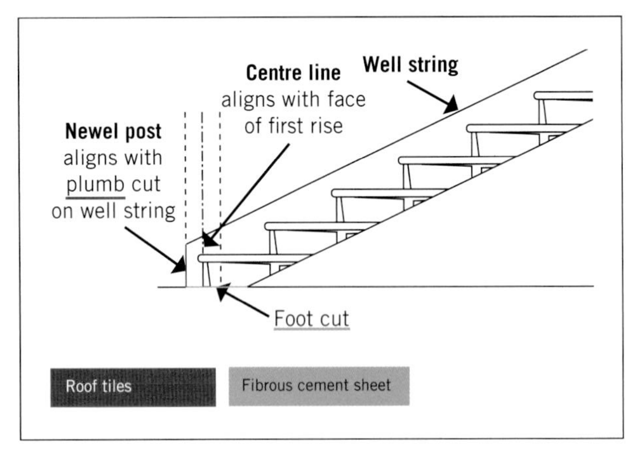

FIGURE 6.28 Foot and plumb cut

For external stairs, it is not uncommon for the newel post to be screwed or bolted to the strings, with the handrails bolted to the newel posts and the balusters fixed to the outside of the handrails and strings (see Figure 6.29).

At the top of each string, allowances for joining to landings, newel posts or wall winder strings must be made.

If the string is to join to a landing, it is common for a hook to be cut at the top to allow for the string to be supported at the top by the landing (see Figure 6.30).

Sufficient material has to be left on the end of the wall string so that the architrave can butt neatly to it.

The last tread is actually a 'half tread' and is supported by a riser board that is not fastened with a wedge. This allows for some adjustment back and forth when fitting the stairs.

Alternatively, if the string is to be attached to a newel post at the top and bottom, sufficient material must be retained for cutting of the tenons (see Figure 6.31).

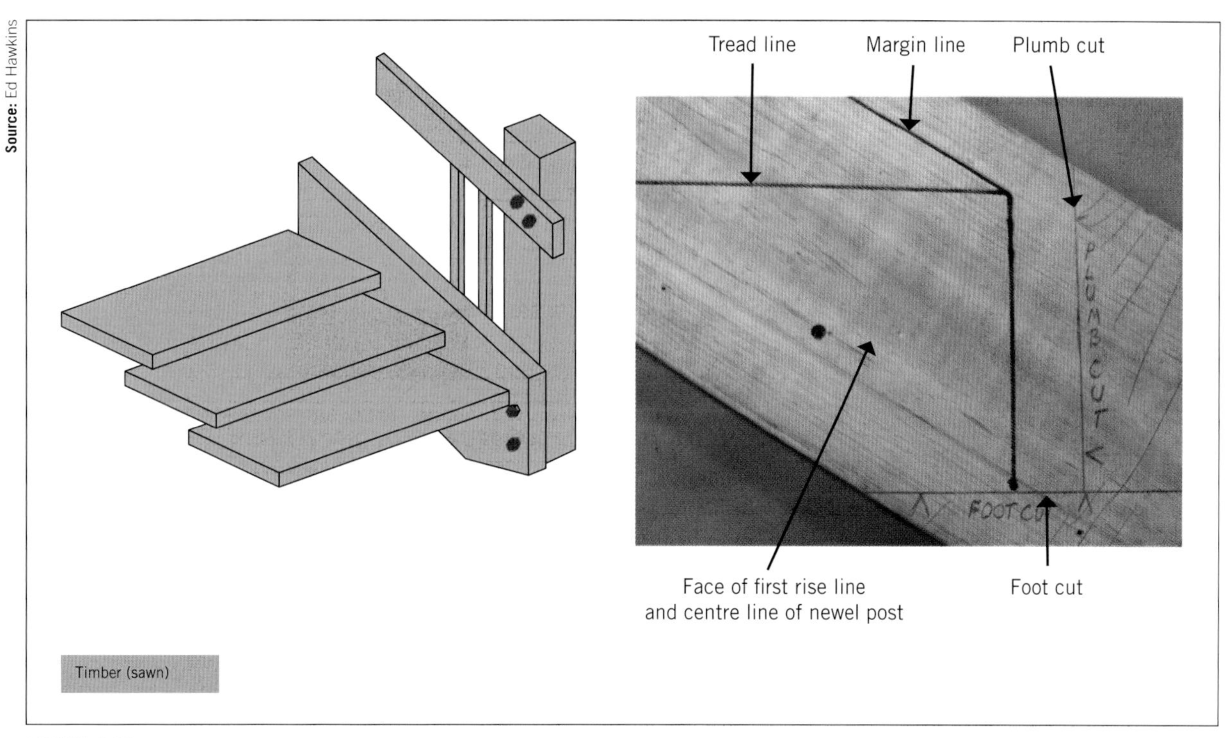

FIGURE 6.29 Alternative means of attaching newel posts and handrails

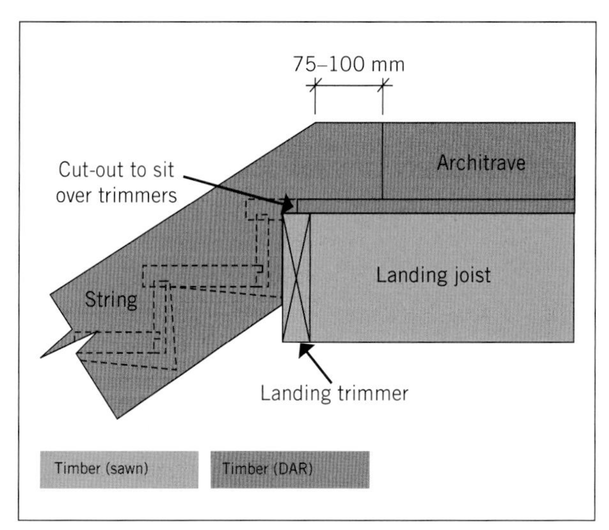

FIGURE 6.30 Wall string hooked over landing trimmer

When setting out cut strings, allowance has to be made for the thickness of the rise material if it is to be planted over the face of the string (see Figure 6.32). This is also important in flights over 1.2 m wide, which require a central support (called a **carriage piece**) under the treads and behind the rise.

Joints

Double-haunched mortise and tenon joints are used to join newels and strings together at the top and bottom (see Figure 6.33). Dowel joints are an alternative.

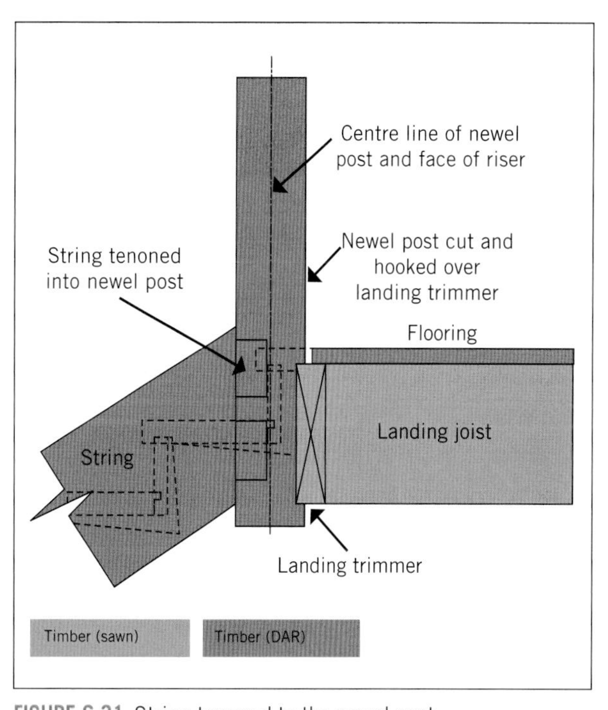

FIGURE 6.31 String tenoned to the newel post

When stairs take a change in direction at a landing, the newel post must be machined to accept the rise for the continuing flight, as indicated in Figure 6.34. (Note that in this figure, for clarity, the notching for the landing timbers is not shown.)

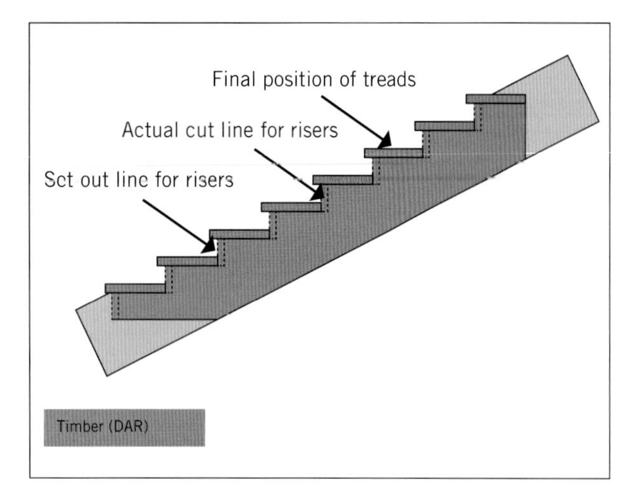

FIGURE 6.32 Cut string set-out

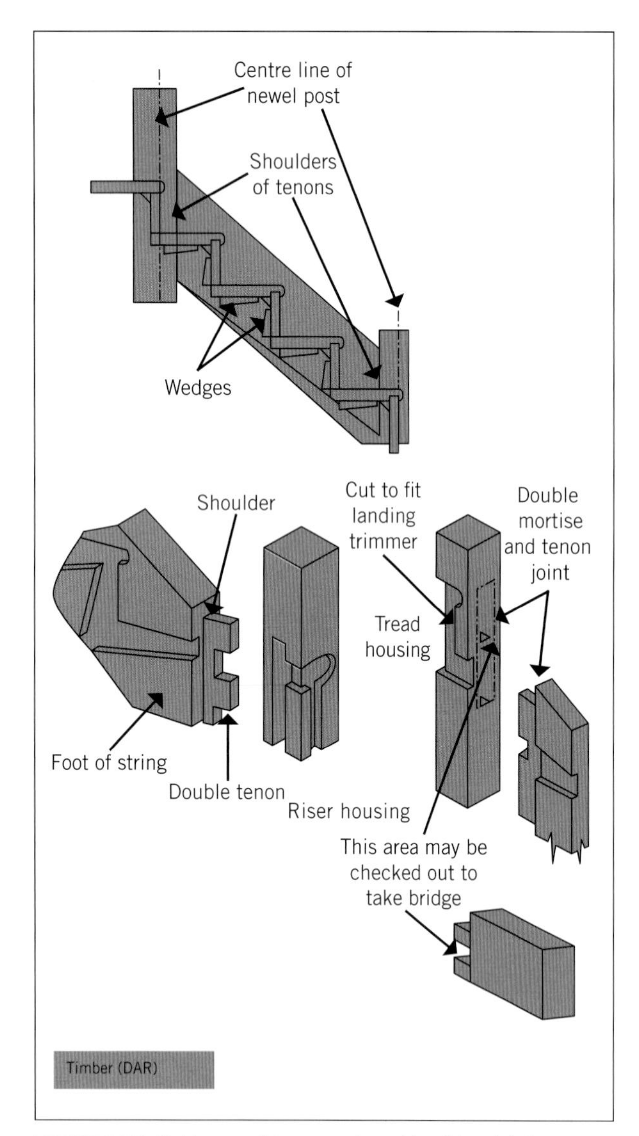

FIGURE 6.33 Bottom and top newel post to string joints

Tread and riser joints

Treads and risers are often fitted into closed strings in housings 10–12 mm deep. These housings are made using a router and template, as described earlier.

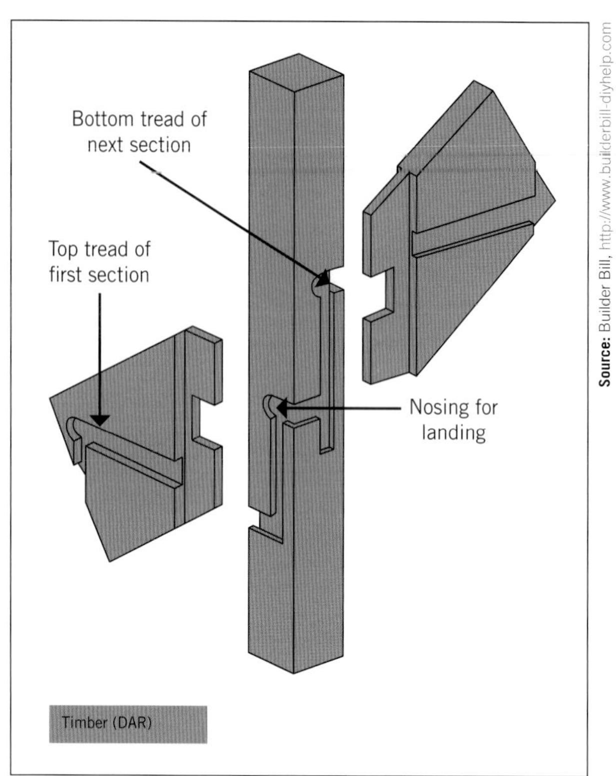

FIGURE 6.34 Newel post joints at landing intersections

When making the jig, it is essential to allow for the wedges. The wedges keep the faces of the treads and risers pushed hard to the shoulder of the housing and provide additional strength to the joint by increasing the gluing surface. Glue blocks are also inserted behind the rise and tread joint (see Figure 6.35).

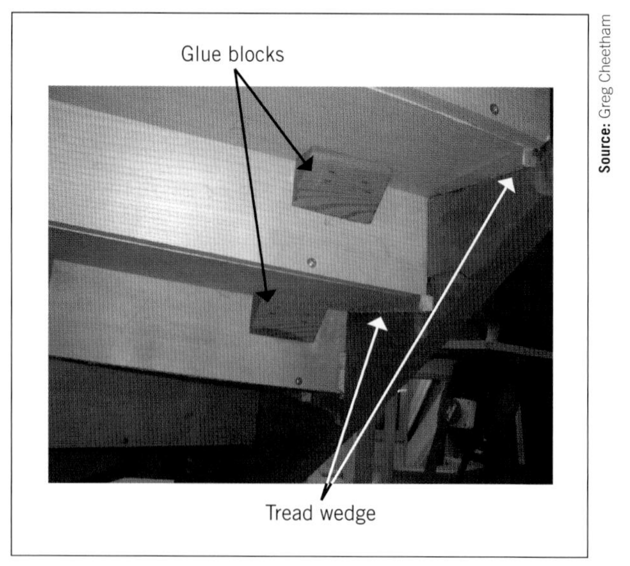

FIGURE 6.35 Wedge and glue block

Because of the movement inherent in stairs from human traffic, treads and risers need to be joined. The joints reduce gaps and increase the strength of each member by continuous support. The joints should be machined and assembled to prevent gaps appearing when movement is present.

Typical joints (see Figure 6.36) include:

- housed
- bare-faced tongue and groove
- screwed
- housed with scotia mould.

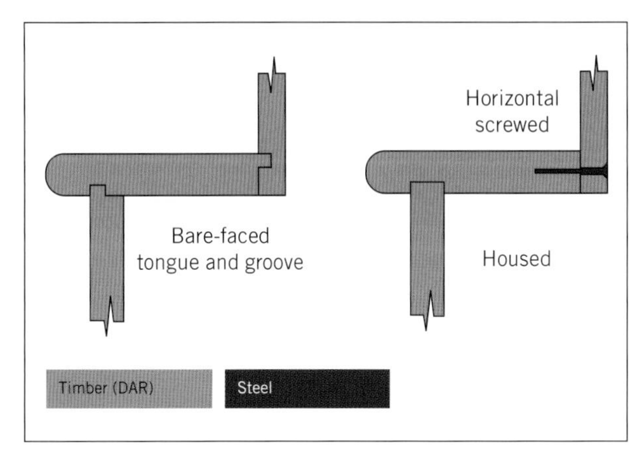

FIGURE 6.36 Typical joints for treads and risers

When horizontal screwed, the tread is backed off (about 1–1.5 mm maximum), which makes the tread to riser joint fit tightly when screwed. The riser is shortened by 2–3 mm to ensure that the riser does not foul the tread wedge (see Figure 6.36).

When manufacturing is done off-site, the treads and risers can be joined together and easily packaged to speed up installation (see Figure 6.37).

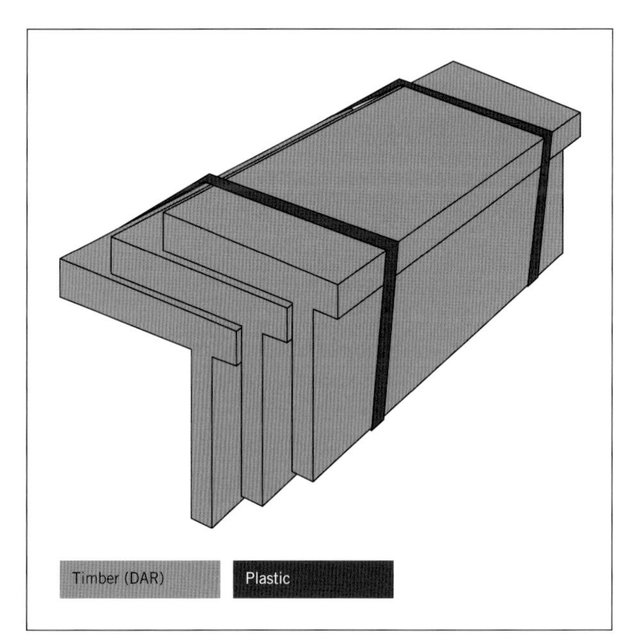

FIGURE 6.37 Treads and risers supplied assembled

Other joint methods

Bolts with nuts and washers, **coach bolts** and threaded rod are used in conjunction with stair brackets, cleats or nailing to keep the treads, strings and risers connected (see Figure 6.38). In external stairs, treads shrink as a result of being out in the weather and housings expand; as a result, the narrower treads and larger housings make for very loose joints. Threaded rod is therefore used to keep the joints tight, particularly in external stair construction. The number of rods depends on the number of treads. As a minimum, one rod is set below the second tread and one below the second top tread, with other rods being used every third tread in between. Internal stairs may not need them if the treads are housed and seasoned timber has been used. If the treads are not housed, threaded rod is about the only successful method of keeping the joints between strings and tread tight over time.

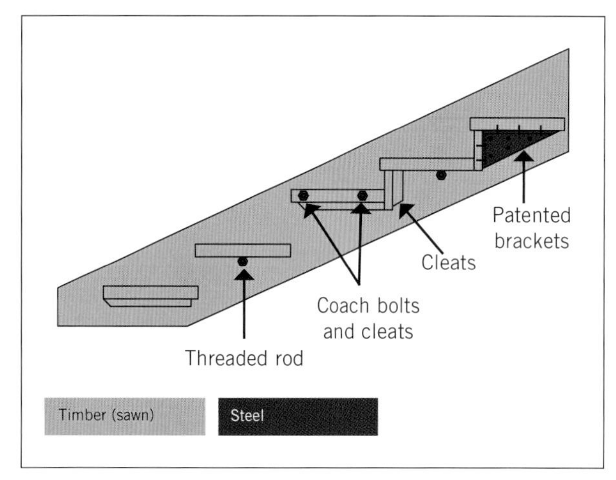

FIGURE 6.38 Other joining methods used for stairs

Coach bolts should not be used as the sole means of attaching treads to strings. The treads (and risers if used) must be supported on cleats. The closed rise form of construction is not recommended for external stairs because of the problems with timber shrinkage and movement.

Where coach bolts and threaded rods are used, it is good practice to recess the bolt heads or, alternatively, use dome nuts on the end of the threaded rod (see Figure 6.39).

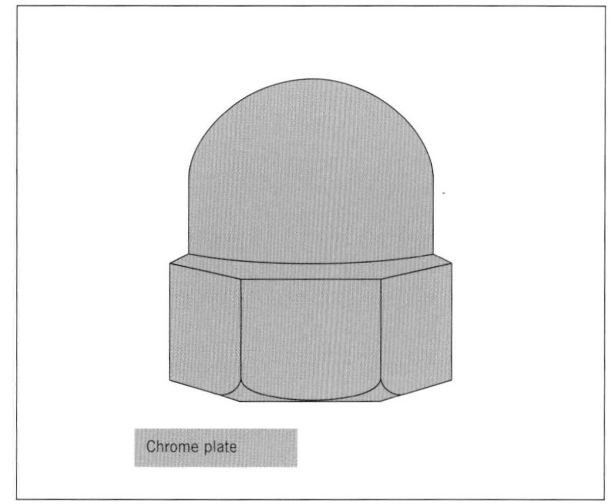

FIGURE 6.39 Dome nut

LEARNING TASK 6.3 SETTING OUT STRINGS

With the teacher's assistance, students are to set out (at half scale) a set of stairs with four rises, with a closed string on one side and a cut string on the opposite side, and with open rises.

- Materials: 240 × 45 mm treated pine – planer gauged
- Overall rise: 720 mm
- Overall going: 800 mm
- Width: 750 mm
- Margin: 20 mm
  A piece of ply or MDF can be used for the set-out.

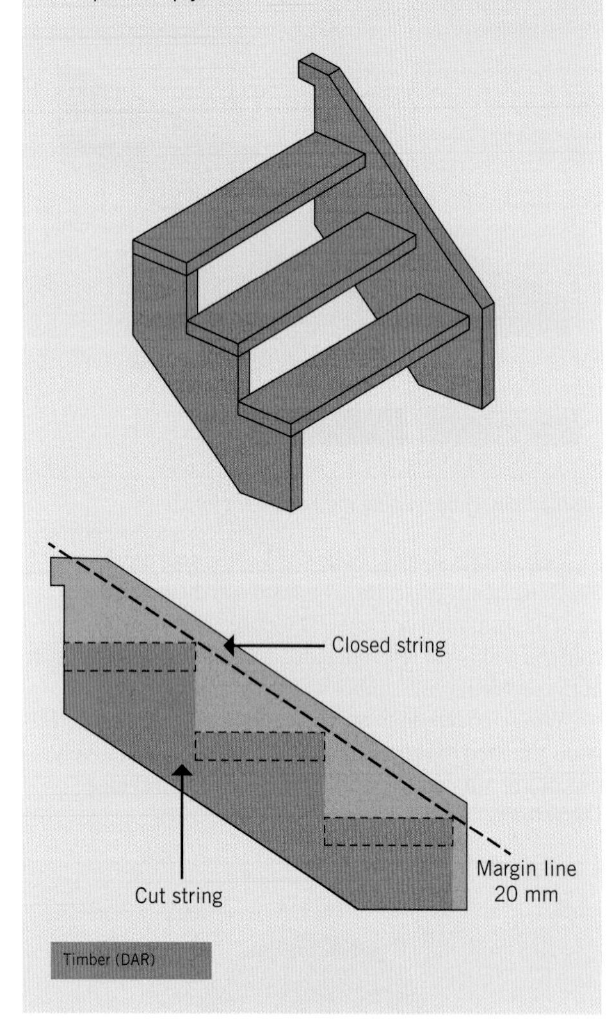

Machines used in stair production

Some of the different types of static wood machines used in stair production include:

- sliding compound mitre saw (see Figure 6.40) or radial arm saw – for rough docking of timber to length
- rip saw – used to rip timber to width and thickness before jointing
- tenoner – used to produce tenons
- mortiser (chisel or chain type) – used for mortising

FIGURE 6.40 Sliding compound mitre saw

- jointer (or buzzer) – used to dress timber true, straight and square; also to taper material for balusters
- thicknesser (panel planer) – used to dress timber to width and thickness
- spindle moulder – used to produce handrail sections and other moulded timber materials.

Note: AS 4024.1 Safety of machinery includes standards detailing controls, guarding, safety stops and interlocks, as well as ergonomic and safe working environment around machinery.

Make sure that all fences and guards are in place and operational, switches operate properly and cutters are adjusted and sharp. Keep the work area clean and clear of obstacles.

Computer numerical control machinery

CNC machinery is fast overtaking traditional machinery in the production of basic stair components such as tapered treads, strings and handrails.

As outlined in Chapter 9 (see 'CNC machine centres'), there are three basic forms of CNC machinery that are suitable for joinery. These are the basic CNC machinery for flat panel work, three-axis machines allowing for greater variety of custom joinery, and the latest model five-axis machines providing the greatest scope of works imaginable (see Figure 6.41).

CNC machinery for nested production has the capacity to produce many of the basic staircase components, such as strings, treads, risers and winders. The three-axis machines are capable of producing the flat panel components as well as some of the more difficult components such as newel posts and handrail sections. The five-axis machines can produce and reproduce exactly the same complex compound curved components found in geometrical staircases such as wreathed strings

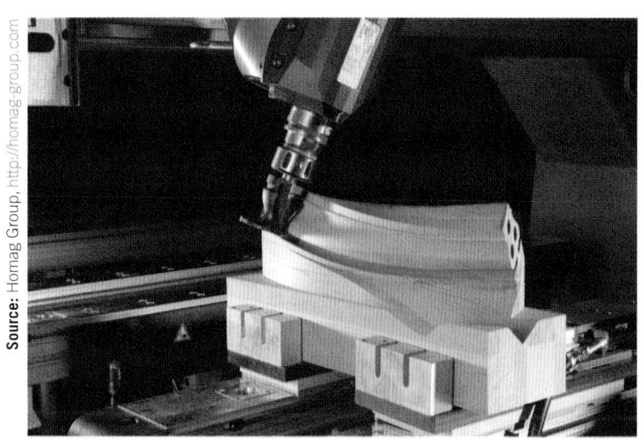
Source: Homag Group, http://homag-group.com

FIGURE 6.41 CNC machinery

and volutes. The purchase of CNC machines should be based on sound investment criteria including:

- cost of machinery plus software and tooling
- maintenance costs – programmed services and replacement parts
- usage – whether it will be used daily, weekly or occasionally
- financing costs – lease or buy decisions
- opportunity cost; for example, what profit would have been earned if the money were spent on a dedicated stair string CNC machine compared to the profit funds earned from leaving the money in the bank
- increased production capacity
- staff capability – training and staff retention costs
- integration in the entire production system – for example, common design and programming software.

Advantages of CNC machines are:

- quick-change tooling and machining
- work health and safety advantages
- quality reproduction of standardised product.

Disadvantages of CNC machining are:

- cost of machinery, particularly five-axis machinery
- programming times for complex components
- can create costly mistakes and significant downtime if not properly set up or maintained
- often underutilised or not used to take advantage of full capability.

Visit the websites at the end of this chapter to view a demonstration of the use of CNC machines in stair manufacture.

Handrails and balustrades

Handrails are an integral part of the construction of stairs. They are supported on balusters or on additional supports fixed to the walls adjoining the stairs and supported on brackets (see Figure 6.59 later in the chapter).

All stairs rising more than 1 m must have a handrail installed on at least one side.

Handrail heights are governed by Australian Standards and the National Construction Code, Volume Two. The recommended height of 865 mm is measured plumb from the front of the intersection of the tread and riser. For balconies higher than 3 m, the balustrade must be ≥ 1 m; for balconies less than 3 m high, the handrail height should be no less than 865 mm (see Figure 6.42).

For floors more than 4 m above the surface beneath, any horizontal elements within the balustrade or other barrier between 150 mm and 760 mm above the floor must not facilitate climbing.

Handrails can be fitted to newel posts using several different methods, as shown in Figure 6.43.

Handrail and balustrade materials

Handrails are manufactured from many different materials and are fitted using many different methods. Some are attached to stairs via balustrades, but when stairs are bounded by walls the handrails are fitted directly to the wall.

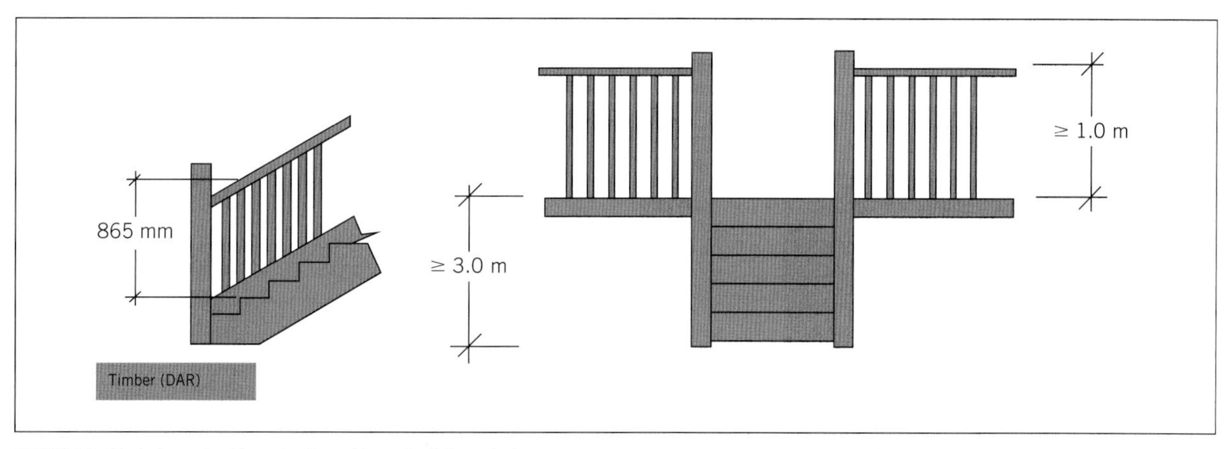

FIGURE 6.42 Integrated handrail and handrail bracket

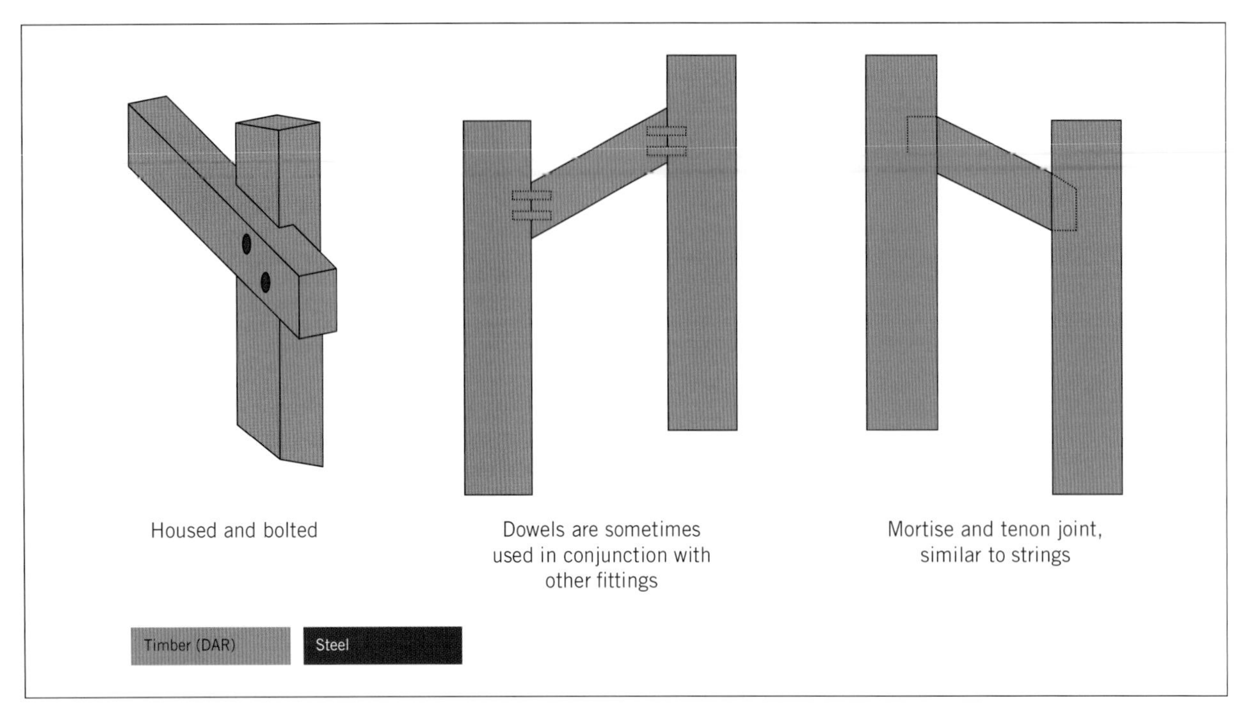

FIGURE 6.43 Methods of fitting handrails to newel posts

Timber is made in many different profiles; for example, moulded to traditional shapes with hand and finger grooves or grooved below to cater for fillets or simple rectangular shapes. Figure 6.44 shows an elaborate timber and metal balustrade. More information on shapes is available from the websites listed at the end of this chapter.

Source: Ed Hawkins

FIGURE 6.44 Elaborate timber and metal balustrade

Timber is also used in its rectangular form, either on the flat or on edge (see Figure 6.45).

Metal handrails are quite common. The main materials used are wrought iron, steel, stainless steel and aluminium. Polished brass is also used, mainly in high-quality internal joinery work. Bronze castings are sometimes used in commercial work and high-quality homes for balusters, handrail ends and short lengths of handrail at entrances.

Stainless steel wire has become popular in modern stair construction. The wire must be 2.5 mm thick (minimum) and must be spaced at a maximum of 100 mm apart horizontally and a maximum of 110 mm apart vertically. The tension on the wires is also critical – the tension will depend upon the spacing of the wires and the distance between posts.

The National Construction Code – Volume Two, Part H5 Safe movement and access provides important advice on incorporating stainless steel wire in balustrades and should be referred to whenever stainless steel wire is specified.

DIY kits can be purchased, but because special tools such as swages and wire cutters are required to work with this material, it is best to hire an expert to achieve the right result.

Glass balustrades are popular in modern architectural constructions (see Figure 6.46). This material requires special fitting, which is outside the scope of this chapter. However, you can find more information on materials and selection by referring to AS 1288 Glass in buildings – Selection and installation and AS 2208 Safety glazing materials in buildings.

Source: Ed Hawkins

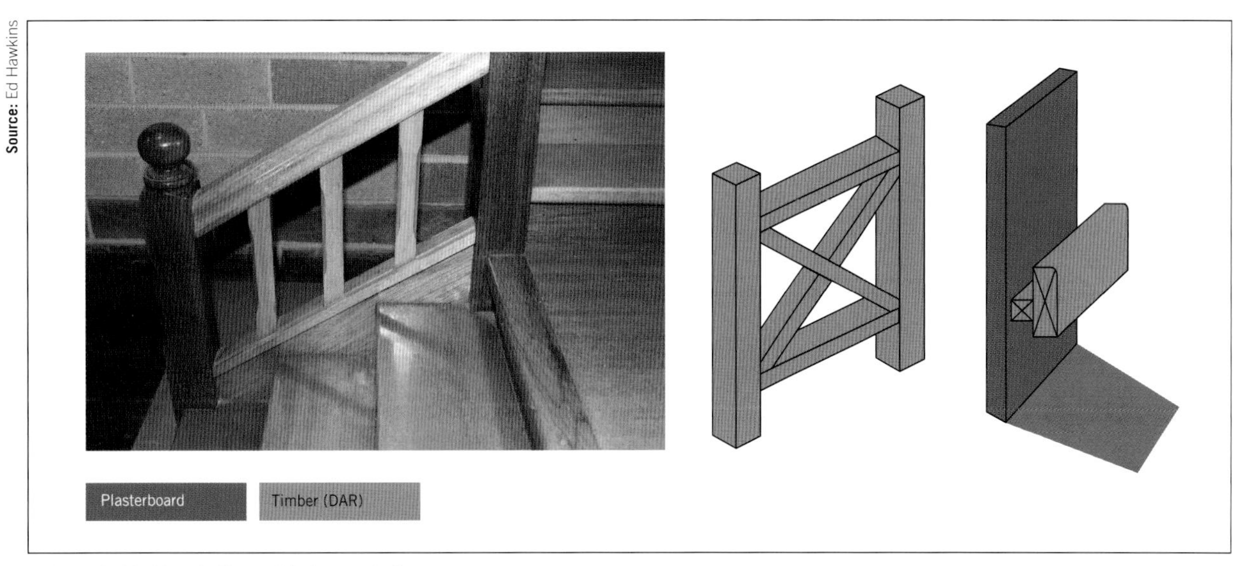

FIGURE 6.45 Handrails and balustrade forms

FIGURE 6.46 Glass panel balustrade

Installing timber balustrades

In this section we will concentrate on the making and installing of a straight handrail and balustrade. Making curved and continuous handrails is a specialist joinery task requiring advanced drawing and geometry skills.

Pythagoras' theorem is used to calculate the length of handrail material needed. It will be roughly the same length as for the strings. However, you will need to calculate the material quantities for the balusters.

To calculate the number of balusters required, first consider how they will be installed. If the stair strings are cut and bracketed and the balusters are to be directly fixed to the tread face, it is common practice to use two or three balusters per tread. So you multiply the number of treads $\times$ 2 (or 3). Some manufacturers produce balusters to different lengths so that they appear to step down evenly with each tread at the handrail end. This looks more professional (see Figure 6.47).

If the balusters are to be fitted above the cut string in the grooved channel, the same length will be required for them.

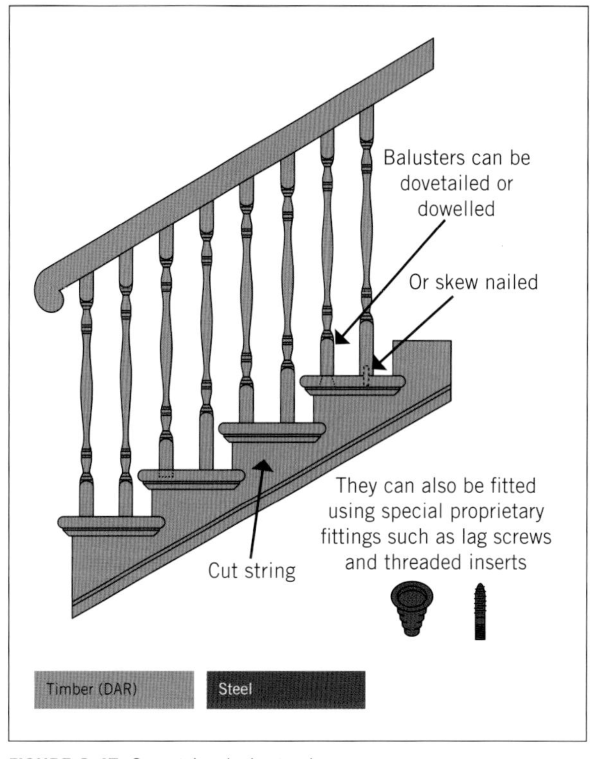

FIGURE 6.47 Cut string balustrade

It is very important to regularly check for plumb when installing balusters with fillets because incorrectly sized fillets can lead to a cumulative error, and the balusters will appear more out of plumb the more of them you install.

Fillets are used to evenly space out balusters and keep them separated (see Figures 6.48 and 6.49). They are easier and faster to install than skew nailing or other forms of fastener (see Figure 6.50).

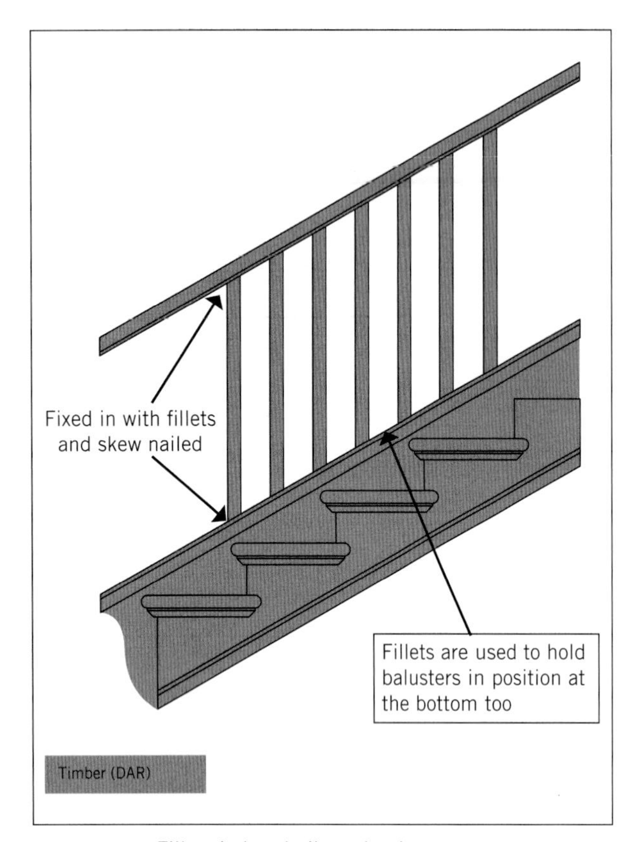

FIGURE 6.48 Fillets in handrails and strings

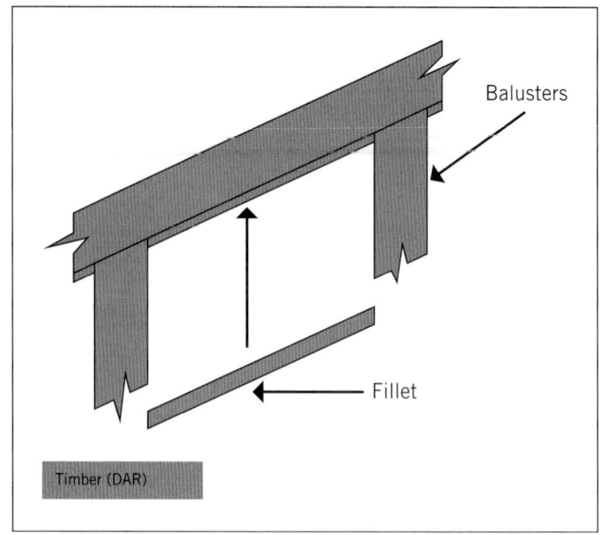

FIGURE 6.50 Fillets evenly spaced and separating balusters

Source: Ed Hawkins

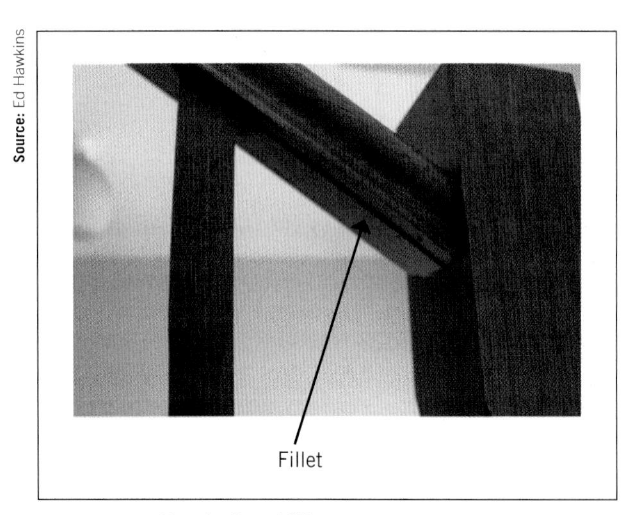

FIGURE 6.49 Handrail and fillet

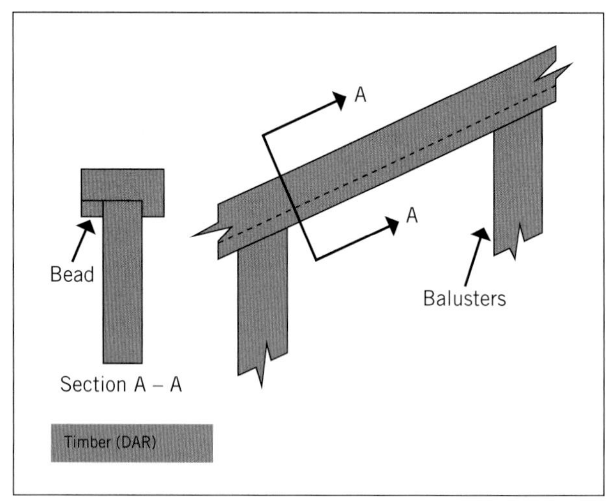

FIGURE 6.51 Balusters fitted with beads and rebates

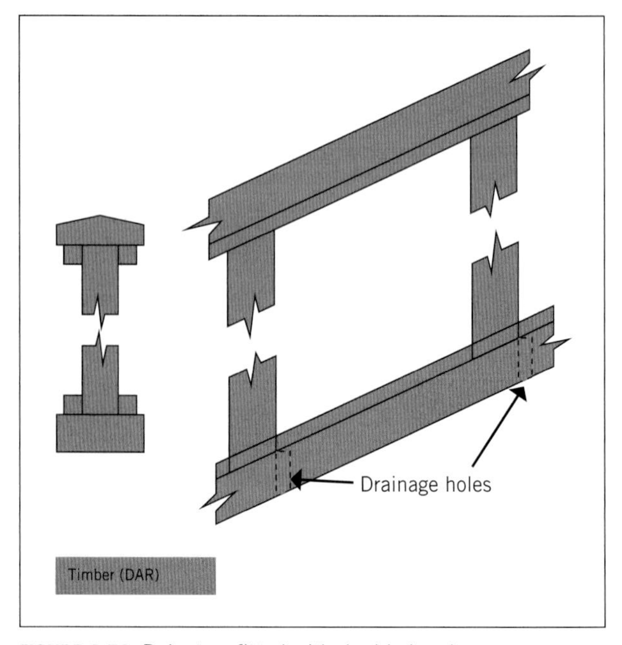

FIGURE 6.52 Balusters fitted with double beads

Balusters can also be fitted using bead and rebates (see Figure 6.51). Or they can be fitted using double beads (see Figure 6.52).

Note: A fillet should be used at the bottom between balusters to prevent water ponding in the recess of external stairs where the balusters are connected to the string via capping. Alternatively, a drainage hole should be drilled close to the bottom newel to relieve water ponding (see Figure 6.52).

The top of the handrail should be bevelled on a horizontal balustrade to allow water to run off.

Closed strings can have a grooved capping fitted and balusters inserted using fillets (see Figure 6.53).

FIGURE 6.53 Balusters fitted using fillets

Handrail joints

Handrails sometimes are joined end to end to make up long lengths, to continue around a corner or to change direction at landings. Each of these joinings has a particular name (see Figure 6.54).

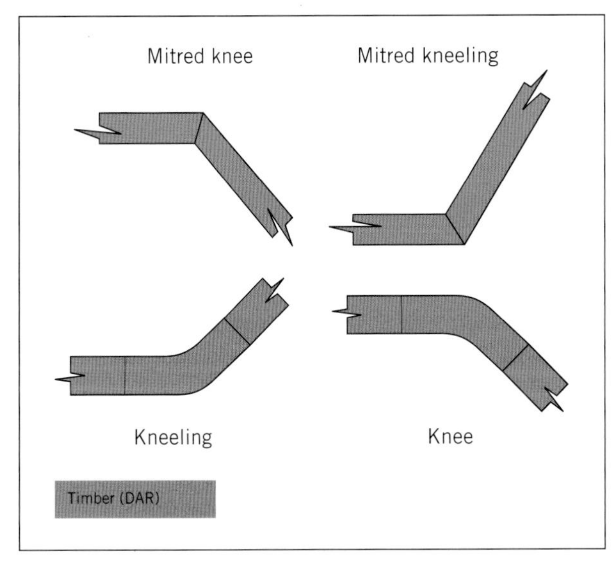

FIGURE 6.54 Basic handrail joints at landings

Gooseneck handrail joints are used to go around a corner when winders and newel posts are being used (see Figure 6.55). The **wreath** is the curved section used in continuous handrails to change direction (see Figure 6.56).

Handrail bolts (see Figure 6.57) are used with dowels to join lengths of handrail together to form a continuous railing. Jigs are used to locate the centres for dowels and handrail bolts.

Installing handrails directly to walls

As indicated earlier, some handrails are fitted directly to walls. The handrails need to conform to the National Construction Code – Volume Two and be no less than

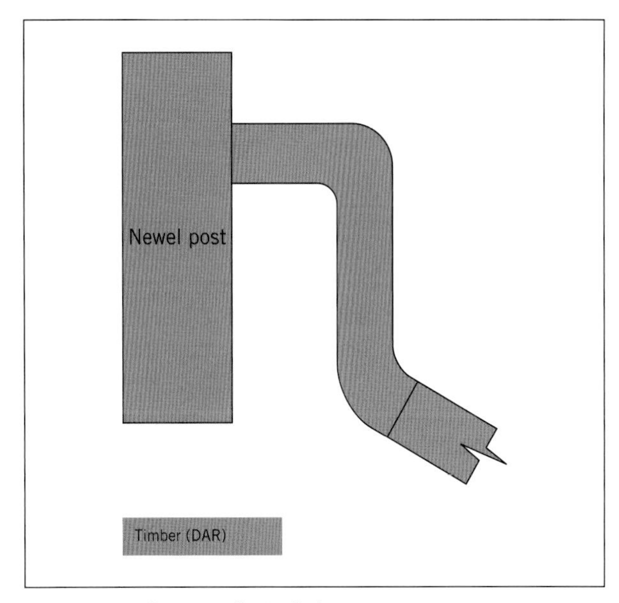

FIGURE 6.55 Gooseneck at winders

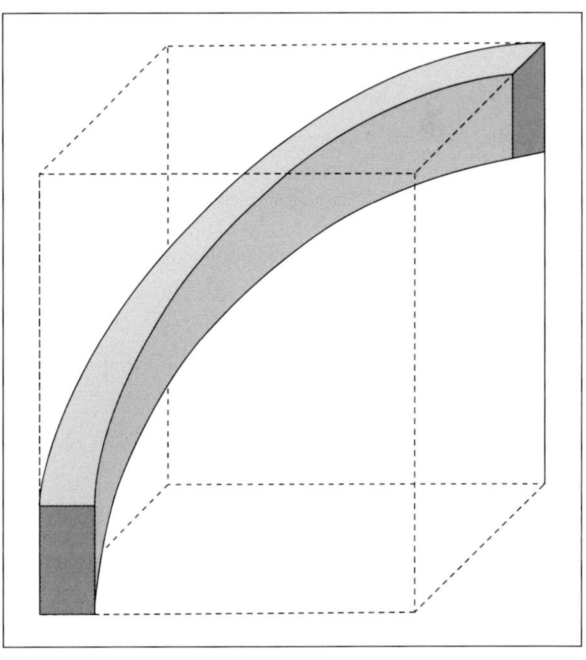

FIGURE 6.56 Wreath

865 mm to the top of the handrail, measured vertically from the tread/riser intersection (see Figure 6.58).

Handrail brackets are the most common means of fixing handrails to walls (see Figure 6.59).

The most important thing is to ensure they are well fastened using the appropriate fasteners; that is:

- screws and plugs for masonry, or masonry fasteners (DynaBolts™)
- screws into studs or specially fitted noggins behind plasterboard.

Metal-framed construction should be backed with timber noggins or infill timbers, as metal is often not strong enough to support a person's weight should they fall.

Source: Zipbolt Pty Ltd

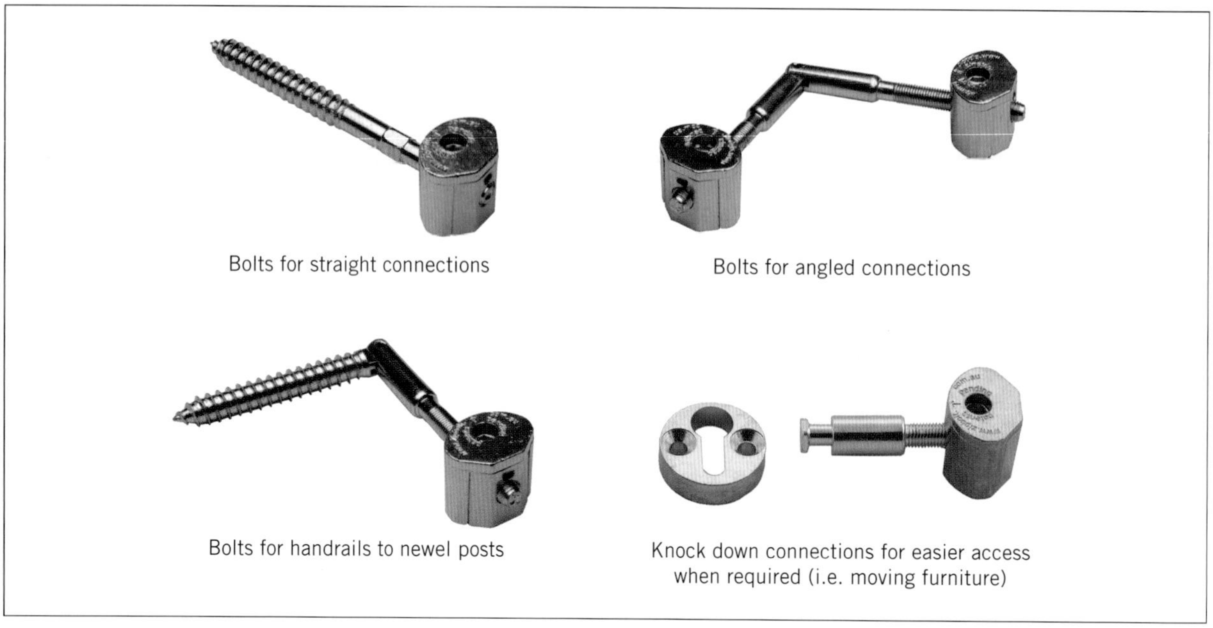

Bolts for straight connections
Bolts for angled connections
Bolts for handrails to newel posts
Knock down connections for easier access when required (i.e. moving furniture)

FIGURE 6.57 Handrail bolts

865 mm
865 mm

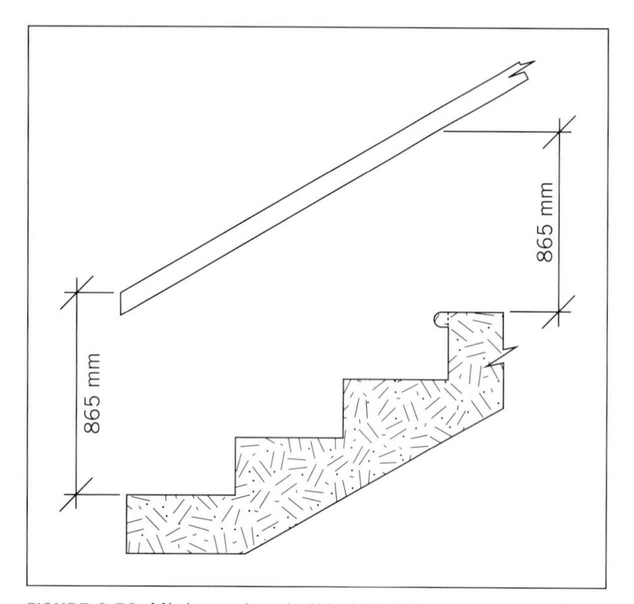

FIGURE 6.58 Minimum handrail height 865 mm measured to top of handrail from tread/riser intersection

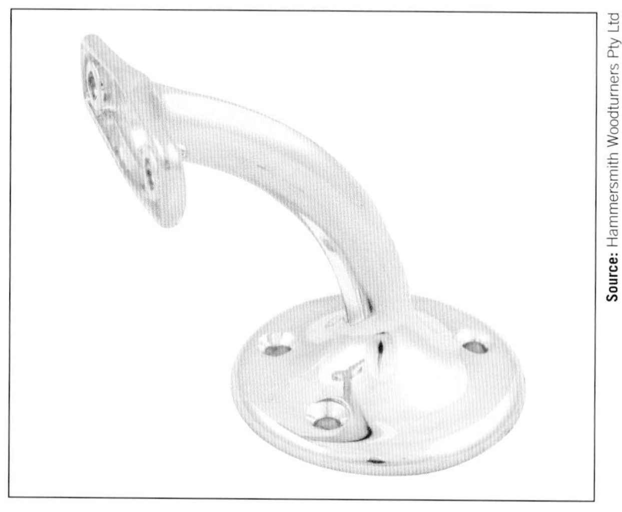

Source: Hammersmith Woodturners Pty Ltd

FIGURE 6.59 Handrail bracket

Calculating baluster spacing

To calculate the baluster spacing (see Figure 6.60), measure the horizontal distance between the newel posts and divide this amount by 125 mm (maximum baluster spacing).

For example:

Horizontal distance	=	1500 mm
Baluster spacing	=	1500 ÷ 125
	=	12 balusters

Now calculate the width of 12 balusters if each baluster is 40 mm wide:

12 × 40 mm = 480 mm

480 mm
1020 mm
1500 mm

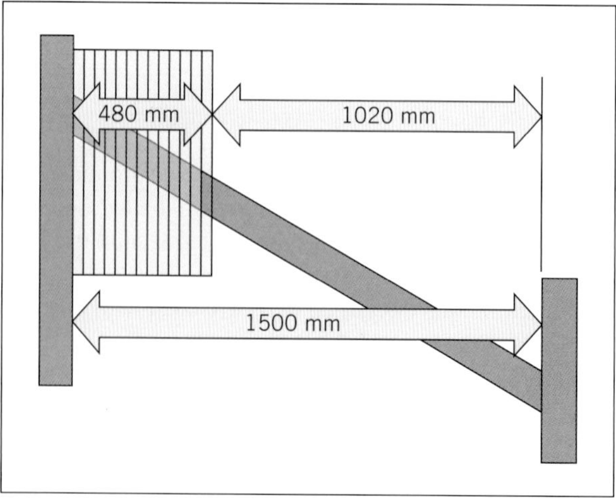

FIGURE 6.60 Calculating baluster spacing

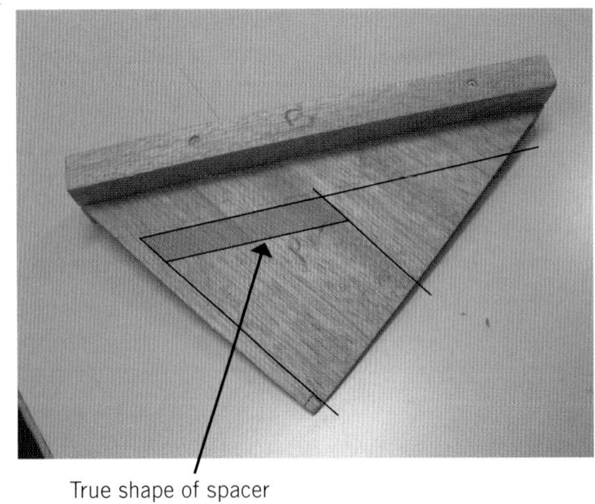

FIGURE 6.61 Setting out fillet size

Having 12 balusters means there will be 13 spaces (like the number of studs in a wall calculation), so reduce the horizontal distance by the number of baluster widths; that is:

1500 – 480 mm = 1020 mm

Divide the amount left by the number of spaces to find the horizontal distance between balusters, that is:

1020 ÷ 13 = 78.46 (say, 78.5) mm (horizontal distance)

Set this distance out on a pitch board and draw in the true shape of the spacer or fillet (see Figure 6.61). Having marked off the true length of the spacer or fillet, these can then be cut with a drop saw from the appropriate sized stock.

Wherever possible, use the same spacings on balconies and landings as used on the flight. This means that you will need to set out the posts carefully to ensure the gaps are even.

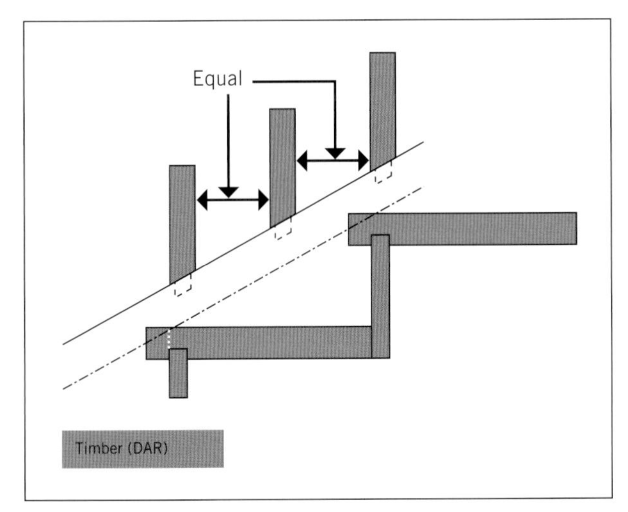

FIGURE 6.62 Alternative method of calculating baluster spacing

Alternative method of calculating baluster spacing

An alternative set-out is to align the face of the balusters to the face of the risers, and then position the intermediate balusters centrally between them (see Figure 6.62). This approach works well when all balusters are of the same width and complies with the National Construction Code.

Landing timbers

Landing timbers used in wall frames are sometimes supported on a diagonal bearer (see Figure 6.63) using

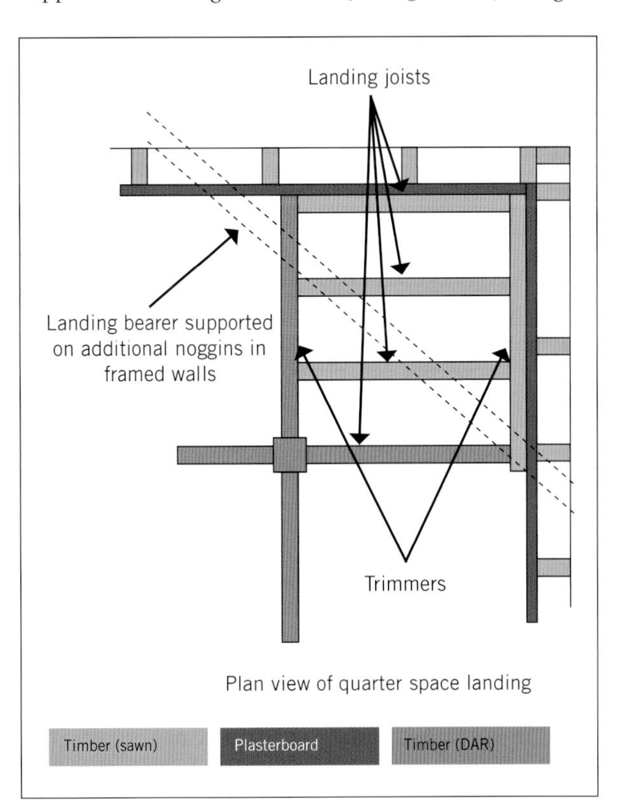

FIGURE 6.63 Quarter space landing timbers

typical framing timber measuring 90 × 45 mm. It is possible to do away with the diagonal brace by increasing the width of the joist timber to 200 mm, as with deep floor joist construction when stairs have newel posts going to the floor.

The frame is built on the ground, then lifted into position.

The frame may be supported temporarily at the wall by using nails set to the correct height, as determined by the rise calculations, and the open corner is clamped to a **tom**, or temporary prop, if a full height newel post has not been used (see Figure 6.64). When a full height newel post is used, the frame is supported by the newel post. The frame is then fastened off with coach bolts or screws to the stud work.

The landing bearer can be set into masonry walls (see Figure 6.65) or supported by timbers fixed to the wall using high strength masonry fasteners such as DynaBolts™ or Loxins™.

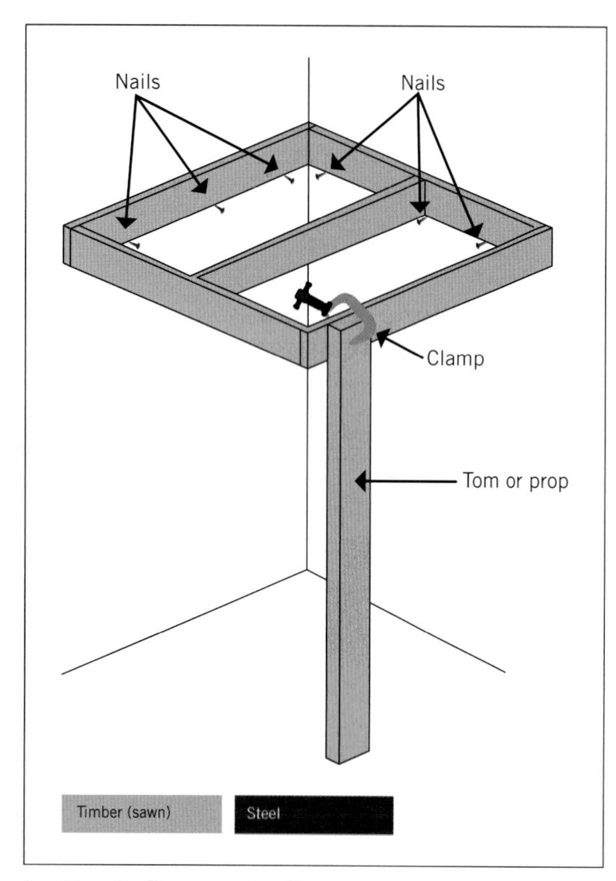

FIGURE 6.64 Propping a landing

LEARNING TASK 6.4

DESIGNING STAIRS FOR EASY ACCESS

Teachers are advised to obtain a copy of *The good, the bad and the ugly – design and construction for access* published by the Australian Human Rights Commission (available at https://www.humanrights.gov.au/publications – use the keyword search term 'access').

In a teacher-led discussion, students should consider the following issues:

- *Issue 1:* Tactile ground surface indicators (TGSIs) – public staircases and the installation of TGSIs. Where, when and why are TGSIs required?
- *Issue 2:* Handrails on stairways – newel posts on landings and change of direction points make access for people with disability difficult. How can the design be improved to remove these difficulties?
- *Issue 3:* Nosings on stairways – sight-impaired people may see a short set of stairs without nosing highlighters as ramps rather than stairs. How can the stairs be made more visible?
- *Issue 4:* Open risers and overhanging treads on stairways – these types of stairs also cause difficulty for sight-impaired people. How can the design be improved to reduce accidents?

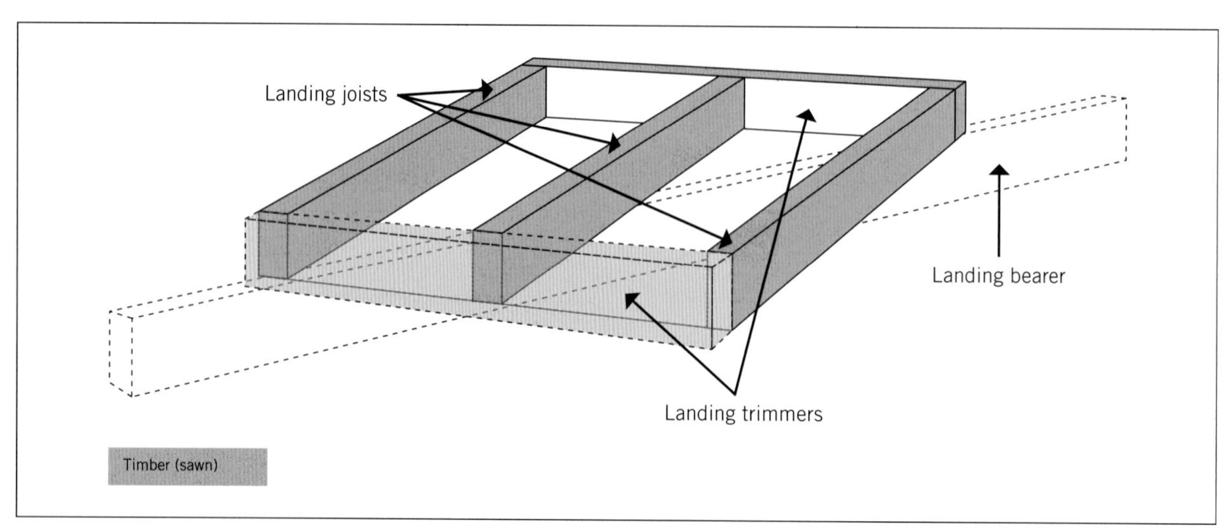

FIGURE 6.65 Landing framing with diagonal landing bearer

COMPLETE WORKSHEET 3

Installing stairs

Installing stairs in low-rise residential buildings often requires a specialist tradesperson. As there are many different types of stairs – such as open rise, closed rise, cut string, closed string and spiral, to name a few – there will naturally be many ways to install stairs. As long as you have taken accurate measurements as outlined in the section 'Site measuring for timber stairs' earlier in the chapter, you are a long way towards creating a set of stairs that will fit the position required.

The tools required to install stairs include:

- levelling tools such as spirit level, plumb bob and/or laser level
- saws – circular saw, compound mitre saw and multi tool
- measuring and calculating tools – four fold rules, laser, measuring tapes and straight edges
- electric drill with a range of bits, including spade bits
- chalk line, string lines and squares
- saw stools, ladders and scaffolds
- hammers and nailers
- screwdrivers, screw guns, spanners, sockets and wrenches
- chisels and planes – power planers
- torches or similar.

Installing external stairs

External timber stairs should be made from Class 1 or 2 timbers such as merbau, blackbutt, jarrah or spotted gum, or an H3 preservative-treated timber.

External stairs are installed using galvanised iron post brackets, carriage bolts, nuts and washers to connect the posts to the strings and landings. In Figure 6.66, the strings are bolted to the posts and deck bearers. The galvanised post brackets are bedded into 300 × 300 × 450 mm concrete piers or fastened to a concrete slab using masonry anchors.

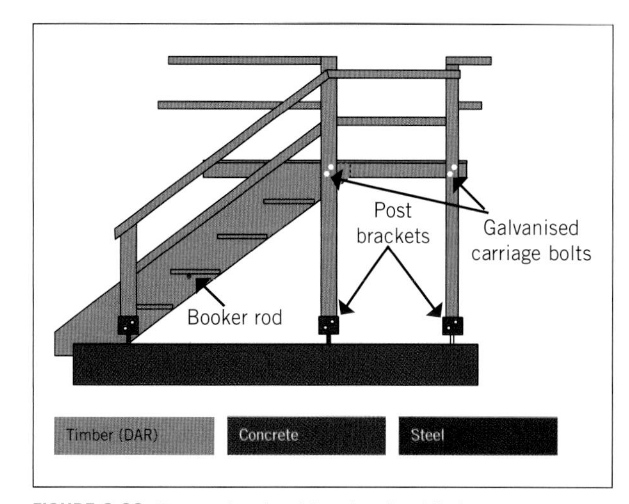

FIGURE 6.66 External stair with galvanised fittings

An alternative to post brackets is to build a concrete pad at the foot of the stairs and insert brass dowels to prevent the strings from moving. This method should be used only in well-weatherproofed areas as the timber will pick up moisture and may rot prematurely. If this method is used, it is good practice to treat the foot of the strings with a paint or sealer before installing them.

Cut strings can be supported using a bearer fixed to the ground with masonry anchors and the foot of the string cut to fit over the bearer (see Figure 6.67).

Depending on the size and construction methods of the staircase, the flight may be assembled on the ground and lifted into position by two people, or it may require the standing and temporary bracing of the strings and newel post (see Figure 6.68). If bracing is required, use

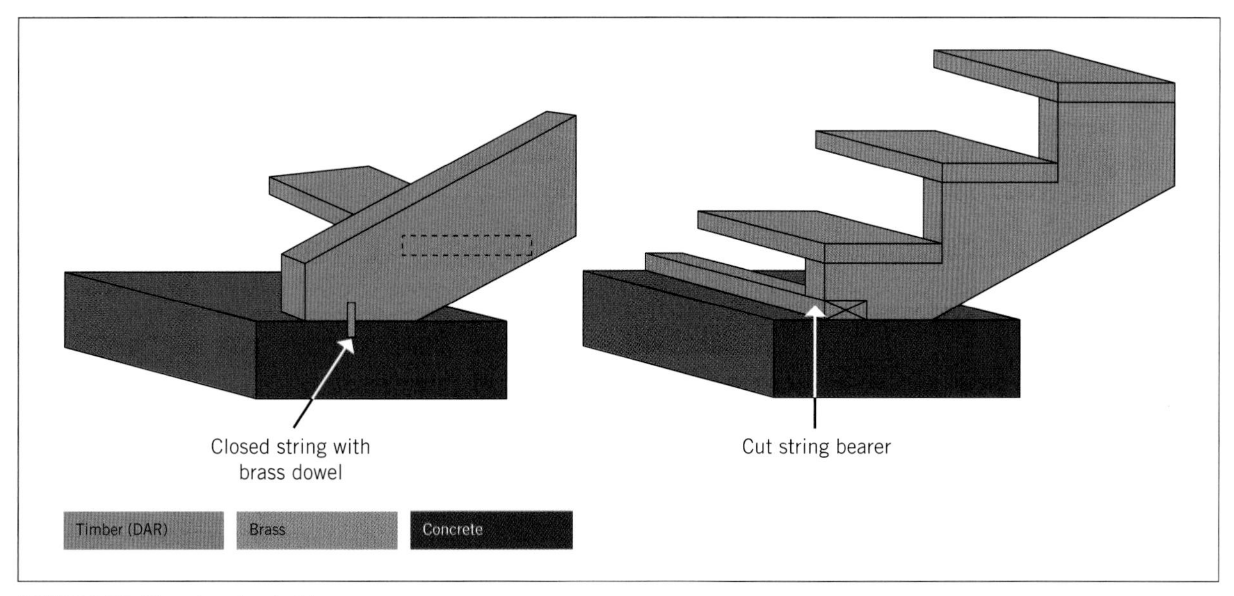

FIGURE 6.67 Closed and cut strings

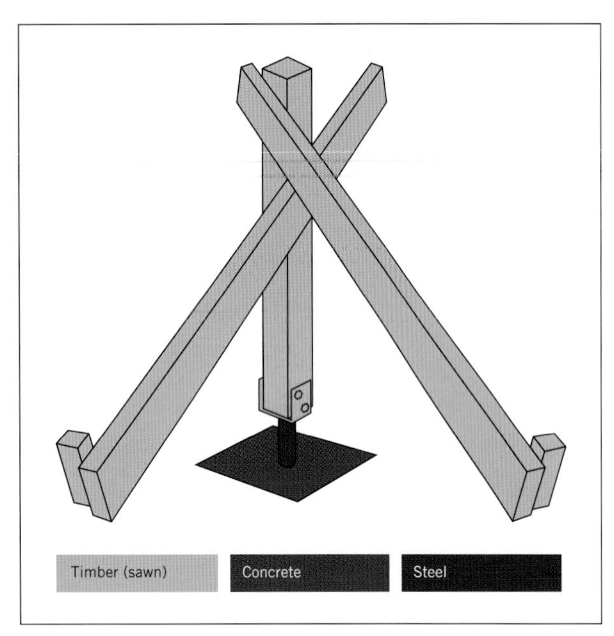

FIGURE 6.68 Temporary bracing of newel post

quick-release clamps and backing blocks so you do not damage the timbers, and fix off the foot of the brace to temporary pegs in the ground. Make sure the posts are plumb in both directions.

Check the strings are parallel and that the tread housing or cleats, or cut string tread cuts, are level.

If the strings are housed, the treads can be slid in from the front or back depending on the shape of the housing. Make sure you line up the front edges of the treads (see Figure 6.69).

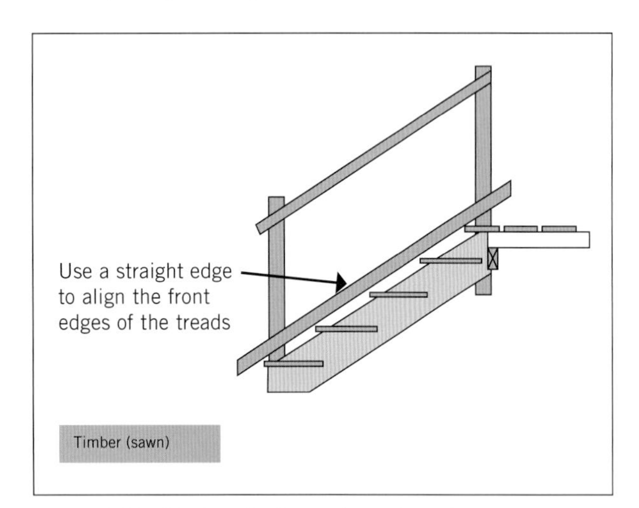

FIGURE 6.69 Aligning the front edges of treads

Nonslip tread

Nonslip finishes provide a safe landing and are often used for external stairs. The stairs need to be clean, dry and free from debris. Prefabricated nonslip materials are commonly used. For a more sturdier tread, metal brackets with nonslip tread can be installed on the edge of the tread. These will need to be screwed into the tread.

When using prefabricated nonslip tape, measure the required amount of tread and follow the manufacturer's guidelines for application.

Clean up

When finished, the work area needs to be cleaned, and all materials sorted for landfill, recycling and reuse according to the worksite requirements and workplace procedures.

Clean up the work area, removing any debris, tools, and excess materials. Stack and store excess materials according to workplace procedures.

Complete a visual inspection of all tools and equipment to ensure safe operation. Regular checks and maintenance of tools and equipment should be done each time they are used. If faults are found, follow workplace procedures to record problems, tag the tools or equipment and relocate them away from other functioning tools.

Restore the surroundings to their original condition, ensuring that any damage caused during installation is repaired.

Obtain the necessary sign-off and approvals from the client, project manager, or any relevant authorities to confirm the completion of the installation. Ensure that all necessary documentation has been obtained and filed appropriately.

Site stairs

When working on building sites, it is common practice to build access steps and ramps to assist the movement of people and materials around the site (see Figure 6.70). These should always be built according to the same regulations and standards as for normal stair flights.

Source: Ed Hawkins

FIGURE 6.70 Site stairs

SUMMARY

In this chapter we looked at the basic knowledge and skills required to construct, assemble and install straight flight internal and external timber stairs involving one or more flights, including:

- understanding basic stair construction terminology, including treads, risers, cut and closed strings, well and wall strings and balustrades, and types of stairs (including straight flights, dogleg, half and quarter turn, spiral and geometric)
- selecting and applying safe work practices and using quality materials
- setting out stairs and stair components compliant with building codes and Australian Standards using a range of methods and tools and equipment, including routers, roofing squares, templates and guides
- housing strings for treads and risers using different joint techniques such as double mortise and tenon, bare-faced tongue and grooves
- calculating material quantities for strings, treads, risers, balusters and waste allowances
- understanding modern manufacturing processes including CNC machinery and multi-axis routers
- making and installing balustrades, balusters, newel posts and handrails, and understanding specialist terminology such as fillets, wreath, knee, gooseneck and kneeling
- installing straight flights for both internal and external stairs, including basic landing construction methods, foot and plumb cuts, and newel post and handrail connections.

REFERENCES AND FURTHER READING

Texts

Blanc, Alan (1996), *Stairs, Steps and Ramps*, Butterworth Architecture, Oxford.

Engel, Andy (2007), *Building Stairs*, Taunton Press, Newtown, CT.

Hasluck, Paul N. (2009), *Practical Staircase Joinery*, Stobart Davies Ltd, Ammanford, Wales.

Spence, William Perkins (2000), *Constructing Staircases, Balustrades & Landings*, Sterling Publishing, New York.

Web-based resources

EeStairs – a visual display of stunning and novel staircases: **https://www.eestairs.com/en**

The Design Library – The Ultimate Guide To Stairs: Part 1 (Stair design, Part 2 (Stairs regulations), Part 3 (Stairs materials): **https://www.designlibrary.com.au/ultimate-guide-stairs-stairs-design-part-1-3/**

Wood Solutions – Guide 8: Stairs, balustrades and handrails class 1 Buildings – construction: **https://www.woodsolutions.com.au/publications**

Audiovisual resources

Buildsum – series of videos on setting out for restricted and unrestricted run and rise of stairs: **https://www.youtube.com/user/Buildsum/videos**

Benedict Millworks – CNC cutting out stair stringers: **https://www.youtube.com/watch?v=3-cPxU9cTfg**

First time building stairs – Everything you need to know, Part 1: **https://www.youtube.com/watch?v=UFIqiwtaAYA**

First time building stairs – Everything you need to know, Part 2: **https://www.youtube.com/watch?v=jx7-uJ_eVoE**

How to build a large outdoor staircase: **https://www.youtube.com/watch?v=pCFsroLfTaI**

MAKA 5 axis machining – PE70 wreath handrail: **https://www.youtube.com/watch?v=pON5kbiHjcs**

Robot Master – milling of circular staircase stringers: **https://www.youtube.com/watch?v=m_sQPcMgrc0**

Routing stringers: **https://www.youtube.com/watch?v=YObQXCdxRNs**

Staircase: **https://www.youtube.com/watch?v=BW29YddwogA**

Timber

Britton Timbers: **https://brittontimbers.com.au/**

Fedwood – handrails, balusters and newel posts: **http://www.fedwood.com.au/**

Wood Solutions – Stairs, Exterior: **https://www.woodsolutions.com.au/applications-products/stairs-exterior**

Stainless steel wire balustrades

Australian Wire Solutions – balustrade and wire solutions: **http://www.awrsolutions.com.au/**

Miami Stainless: **http://www.miamistainless.com.au/index.php?route=common/home**

Ronstan – tensile architecture: **https://www.ronstantensilearch.com/**

CNC machine manufacturers

Multicam CNC Routing Machinery: **https://multicam.com.au/**

SCM – distributes Morbidelli, SCM machinery: **https://www.scmgroup.com/en_AU/scmwood**

Weinig Australia – distributors of Holz-Her machinery: **http://www.weinig.com.au**

These companies have many videos on YouTube that you can watch to gain an appreciation of what can be done by these new-generation machines.

National Construction Code

Copies of the National Construction Code can be obtained free of charge from the Australian Building Codes Board website: **http://www.abcb.gov.au**. In particular:

Relevant Australian Standards
AS 1288 Glass in buildings – Selection and installation
AS 1604.1 Specification for preservative treatment – Sawn and round timber. Relates to the requirements for preservative-treated timbers.
AS 1657 Fixed platforms, walkways, stairways and ladders – Design, construction and installation. The Standard does not apply to situations where special provision is made in appropriate building or other regulations such as the National Construction Code.
AS 1684 Residential timber-framed construction. Sets out the appropriate building practices and the selection, placement and fixing of construction materials as defined in the National Construction Code.
AS 2796.1 Timber – Hardwood – Sawn and milled products
AS 3959 Construction of buildings in bushfire-prone areas. Sets out requirements for the construction of buildings in bushfire-prone areas with regard to improved performance during bushfires.
AS 4024.1 Safety of machinery. Includes standards detailing controls, guarding, safety stops and interlocks as well as ergonomic and safe working environment around machinery.
AS 4785.1 Timber – Softwood – Sawn and milled products
AS 5068 Timber – Finger joints in structural products. Relates to requirements for the manufacture of finger-jointed timber building components.
AS/NZS 1170.0 Structural design actions – General principles
AS/NZS 1328.1 Glued laminated structural timber. Provides specifications on the timber, adhesive and strength of the load in laminated timber constructions and joints
AS/NZS 2208 Safety glazing materials in buildings
Australian Standards can be purchased from the Standards Australia website: **https://www.standards.org.au/**, or see your teacher or librarian for assistance in accessing Australian Standards online.

GET IT RIGHT

CONSTRUCT, ASSEMBLE AND INSTALL INTERNAL AND EXTERNAL TIMBER STAIRS

Questions

Below are two pictures of stairs showing the method used to secure them to the ground.

1 What is wrong with the method?

____

____

2 How should it be done correctly?

____

____

3 What should be done to rectify these stairs?

____

____

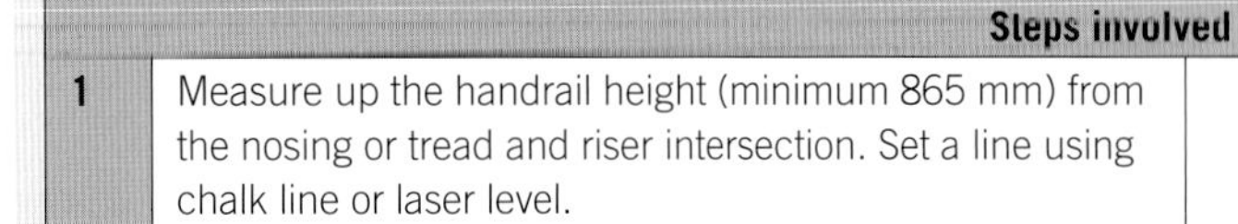

HOW TO

INSTALL HANDRAILS DIRECTLY TO WALLS

Steps involved		
1	Measure up the handrail height (minimum 865 mm) from the nosing or tread and riser intersection. Set a line using chalk line or laser level.	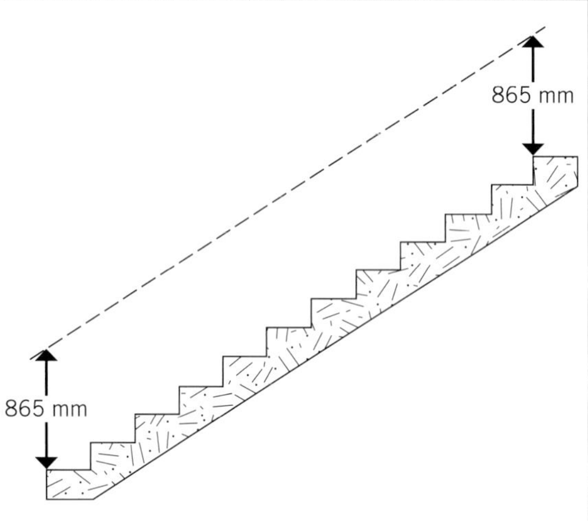
2	Use a section of handrail attached to a bracket to mark the fixing line of the brackets. Fix the top and bottom brackets in position, ensuring the angle matches the pitch of the stairs.	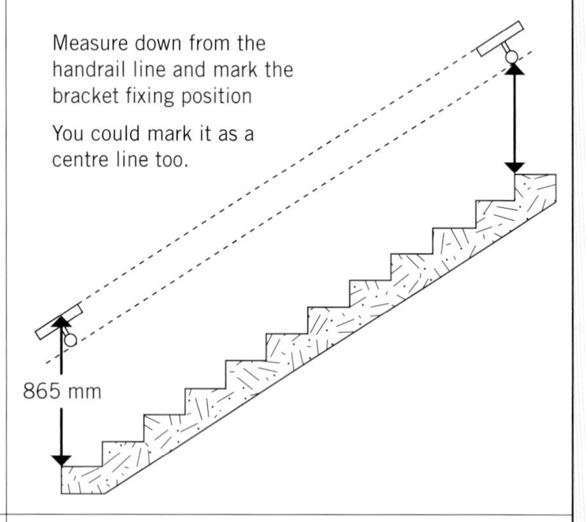
3	Measure and mark the length of handrail required and mark the position of the top and bottom brackets. Divide the distance between top and bottom marks evenly for masonry construction, or if it's a stud wall, locate the studs and plumb the stud centres to the bracket line to ensure you have fixing points. If you're building the wall, ensure you put in noggins as you build it. **Note:** Allow for joining handrail bolts. End handrails with plumb cuts. Use the original pitch board to mark and set the angles.	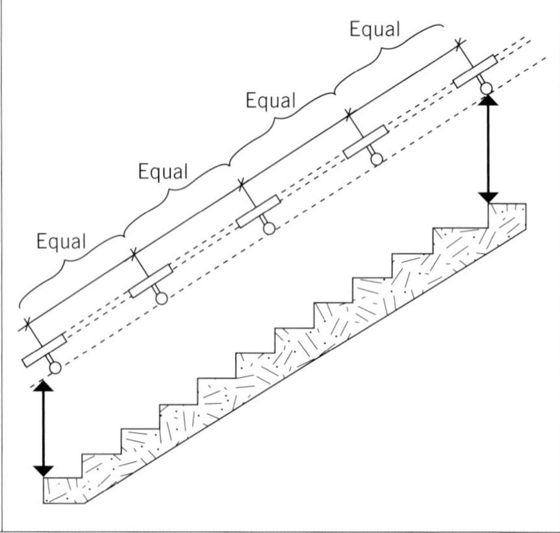

>>

>>

4	Temporarily fix the handrail brackets in position, allowing some tolerance for adjustment. Use a string line to align the bracket heads where the handrail will sit, and finally fasten the brackets.	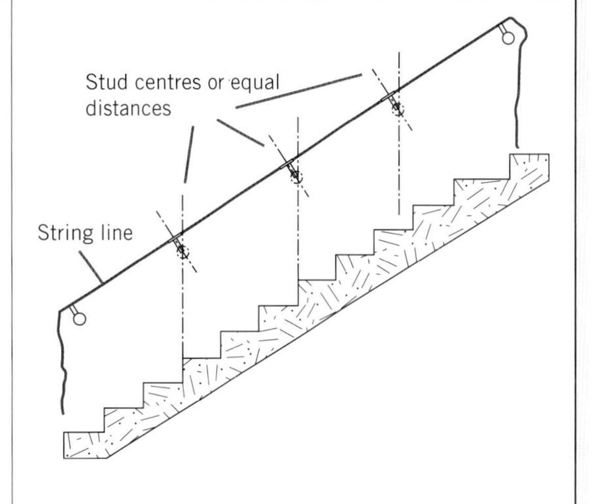
5	Align the handrail on the brackets allowing extra length for mitres and plumb cuts. Mark and cut the joints required before final fixing. 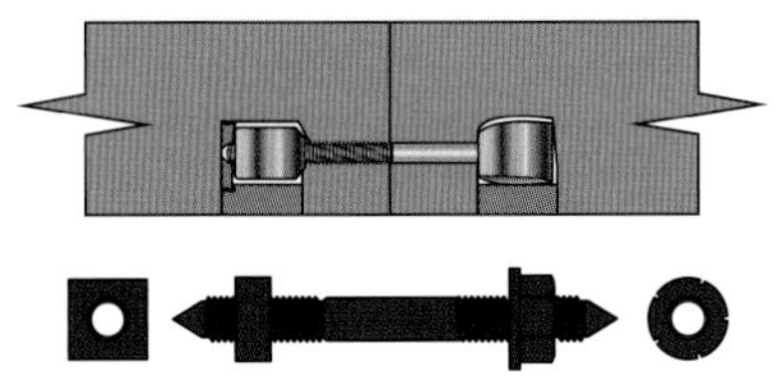 Allow for joining handrail bolts. End handrails with plumb cuts or alternatives. Clean up with sandpaper.	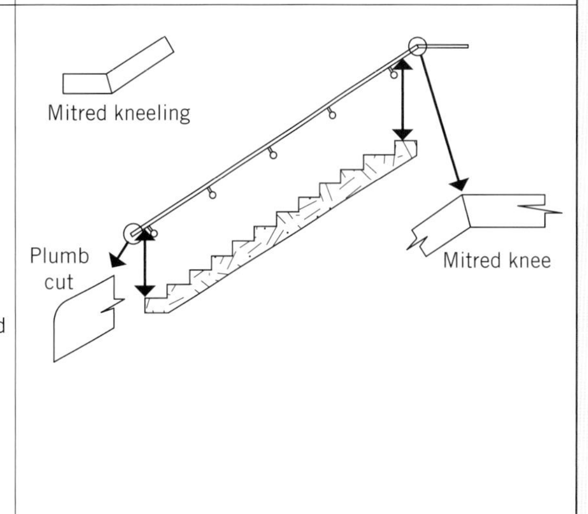

WORKSHEET 1

Student name: ____________________

Enrolment year: ____________________

Class code: ____________________

Competency name/Number: ____________________

To be completed by teachers	
Student competent	☐
Student not yet competent	☐

Task 1

From a copy of the National Construction Code, find the correct answers and include them in the spaces provided below.

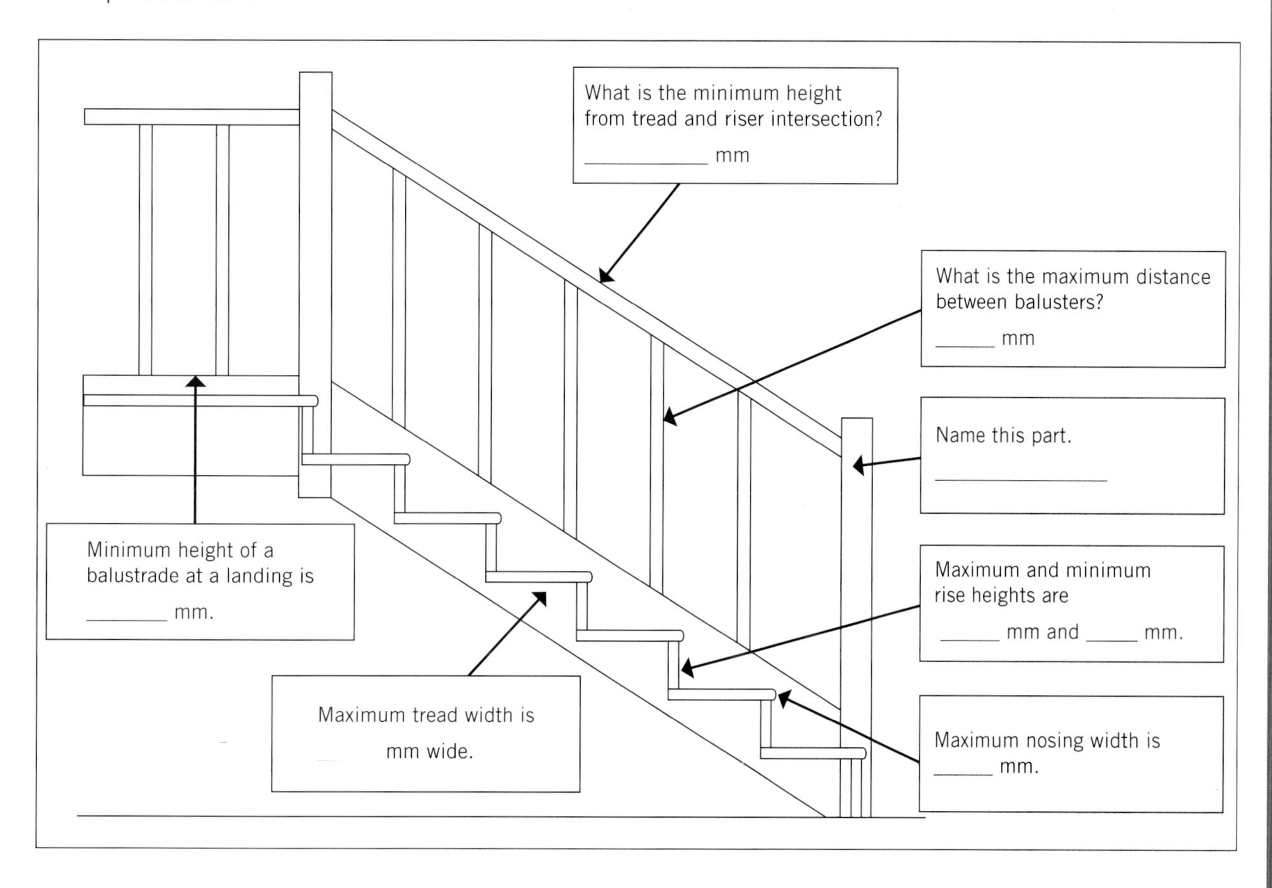

Task 2

Answer the questions below, and if required, use a diagram to show what to measure, and how and where to measure from.

a What is the minimum headroom height allowed between treads and landing above?

b What is the maximum riser height for spiral stairs?

c When measuring up for stairs, what points would you measure the going distance from and to?

d What is the maximum number of winders you can have in a 90° turn?

e How close can the last rise finish to a doorway at the top of a flight of more than three stairs? Indicate your answer on the drawing below (see the National Construction Code – Volume Two and the Housing Provisions Standards).

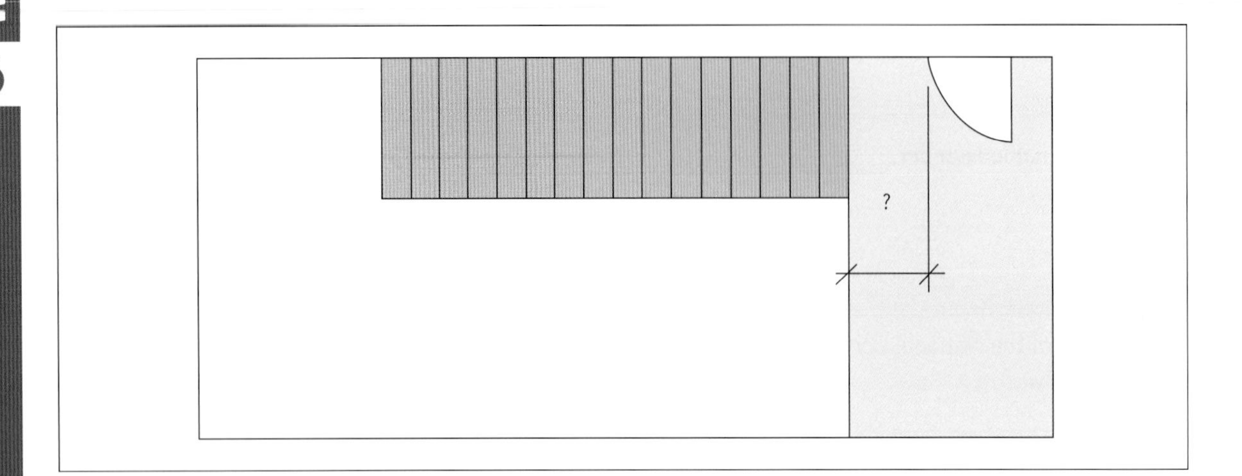

WORKSHEET 2

Student name: ____________________

Enrolment year: ____________________

Class code: ____________________

Competency name/Number: ____________________

To be completed by teachers	
Student competent	☐
Student not yet competent	☐

Answer the following questions.

1 Briefly define the following trade terms associated with stair work.

Carriage piece ____________________

Wall string ____________________

Tread ____________________

Riser ____________________

2 Sketch a 'closed stringer' and a 'cut and bracketed stringer'.

3 Answer True or False: 19 risers is the maximum number of risers allowed in a single flight of stairs.

☐ True ☐ False

4 Describe a landing with a straight flight of stairs.

____

____

5 How many winders can you use in the following situations?

a In a 90° turn

b In a 180° turn.

____

6 In stair construction describe the terms 'rise' and 'going'.

____

____

7 Why is a margin line required when setting out stairs?

____

____

8 Briefly describe what is meant by the following terms associated with stair work.

Nosing ______________________________

____

____

Spandrel ______________________________

____

____

Baluster ______________________________

____

____

Well ______________________________

____

____

Handrail balustrade ______

______

______

9 What is the allowable height for a handrail, according to the National Construction Code?

______

______

10 In which part of the National Construction Code would regulations for stairs be found?

______

______

11 List five items of importance the National Construction Code covers relating to stairs.

a ______

b ______

c ______

d ______

e ______

12 How is an open tread flight of stairs strengthened to prevent the treads from being dislodged from the stringers due to sideways movement?

______

______

13 Name the two types of handrails used in stair work.

a ______

b ______

14 What hardware item would be used to connect timber handrail sections?

______

______

15 List four different handrail joints.

a ______

b ______

c ______

d ______

16 When installing stairs, describe a method of fixing the bottom edge of the stringers to a reinforced concrete floor.

17 What is the top edge of the stringer fixed to during installation where no newel post is used?

18 When setting out stringers, what is the first step in the setting-out process?

19 What tool can be used in place of a steel roofing square when setting out stairs?

WORKSHEET 3

To be completed by teachers	
Student competent	☐
Student not yet competent	☐

Student name: ______________________________

Enrolment year: ______________________________

Class code: ______________________________

Competency name/Number: ______________________________

Your task is to set out and construct a staircase with six risers incorporating cut and closed stringers, a landing and winders with balustrade, as per the drawing below. A 1:2 scale may be used.

Overall rise:	930 mm
Overall going:	1800 mm
Width of flight:	900 mm
Specifications:	Treads to be housed in one string and the opposite string is to be a cut string

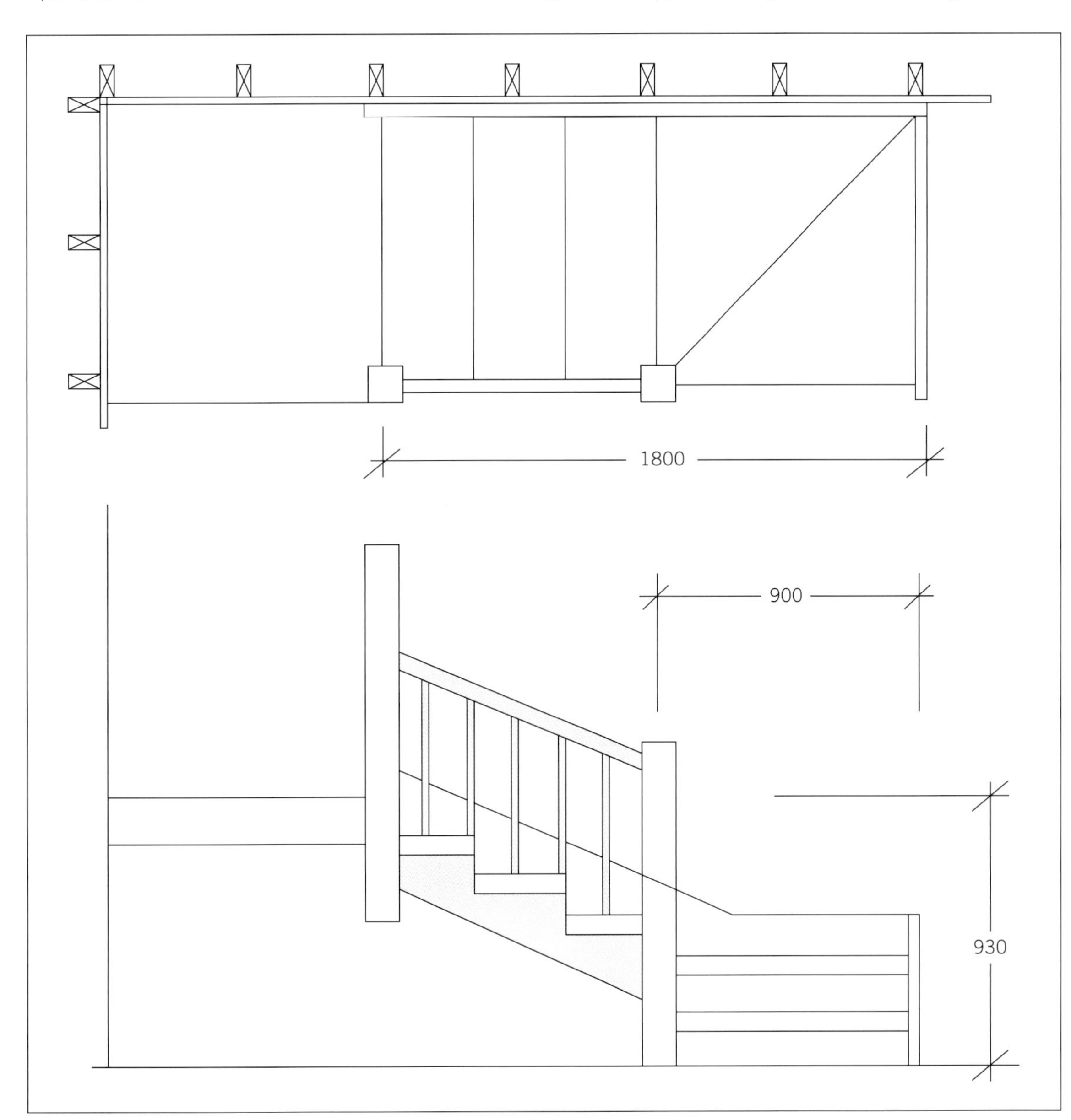

1 Calculate a suitable rise and going for each tread using the correct formula.

2 Calculate the length of material required for the strings and nominate a suitable stock length.

3 Calculate how many metres of tread material will be required; nominate a suitable material for use in external stairs; and nominate a suitable stock length to purchase.

a suitable material

____

b section size

____

c metres required

____

d suitable stock lengths

____

WORKSHEET 4

To be completed by teachers

Student competent ☐

Student not yet competent ☐

Student name: ______________________________

Enrolment year: ______________________________

Class code: ______________________________

Competency name/Number: ______________________________

This is an alternative practical exercise for joinery students.

Your task is to set out and build the set of stairs shown below. The unit may be made as a half-scale model by individual students or as a full-scale set of stairs by a team of three. The number of treads and risers can be varied. The sizes provided are nominal and can be varied to suit the needs of the training facility.

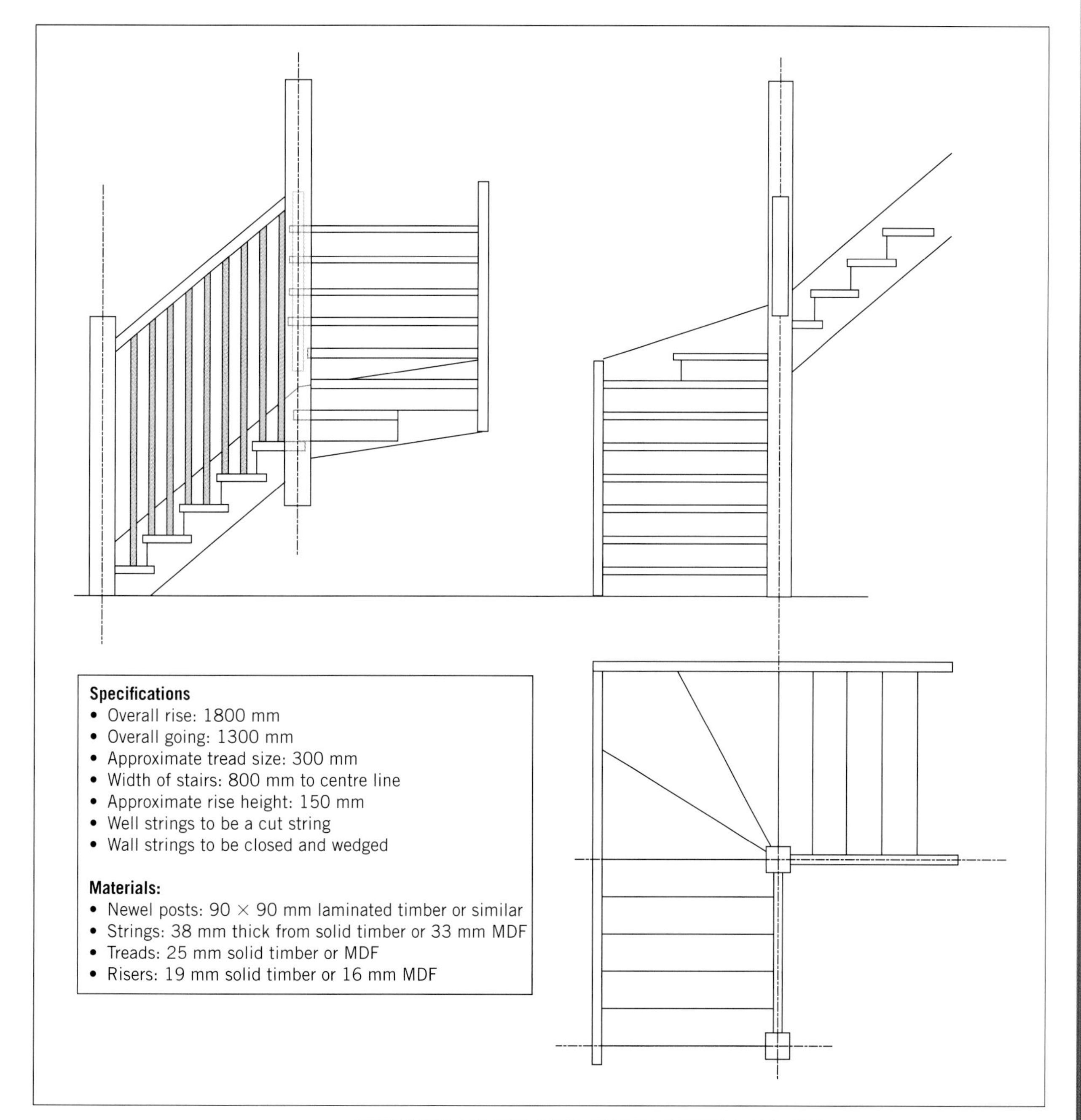

Specifications

- Overall rise: 1800 mm
- Overall going: 1300 mm
- Approximate tread size: 300 mm
- Width of stairs: 800 mm to centre line
- Approximate rise height: 150 mm
- Well strings to be a cut string
- Wall strings to be closed and wedged

Materials:

- Newel posts: 90 × 90 mm laminated timber or similar
- Strings: 38 mm thick from solid timber or 33 mm MDF
- Treads: 25 mm solid timber or MDF
- Risers: 19 mm solid timber or 16 mm MDF

PART 3

GENERAL ELECTIVES

Part 3 of this textbook deals with the complexities of setting out, cutting and pitching advanced roofs and installing wall cladding materials.

In particular it covers the geometry and mathematics required to set out the bevels and positioning of roof members in a range of different roof forms, and the installation of wall cladding materials to framed and masonry walls.

It is based on a number of units of competency and the content described in Chapters 7 and 8:

7 Construct advanced roofs
8 Install external cladding.

The key elements in Part 3 are:

- Planning and preparations associated with constructing advanced roofs and installing cladding, including plan and specification interpretation, job planning and work health and safety/occupational health and safety (WHS/OHS) requirements.
- Roof types, their geometry and setting out associated with complex roof shapes.
- Tools, equipment and techniques required to develop angles and bevels for roofs, and preparing exterior framing and substrates for cladding.
- Calculating material quantities, lengths and allowances for shortenings and other variations.
- Erecting and installing roof members and support framing, and preparing exterior wall frames for cladding.
- Ensuring the building is weatherproof by the appropriate fixing of flashings, vapour barriers and insulation according to plans, specifications, codes and standards.

Additional material on dormer window construction and installation has been included in Chapter 7 to further expand the knowledge and skills associated with advanced roof construction.

The learning outcomes for each chapter are a good indicator of what you will be required to perform, know, understand and apply on completion of each chapter. Teachers and students should discuss the full practical and knowledge evidence requirements of each unit of competency before undertaking any activities.

CONSTRUCT ADVANCED ROOFS

7

DR GLENN P COSTIN and GREG CHEETHAM

This chapter aligns with the unit of competency, 'Construct advanced roofs'. The unit outlines the knowledge and skills required to construct nine different roof forms that fall within the very broad advanced-roofing category. The roof forms include:

- gambrel (Dutch gable)
- jerkin head
- skewed gable
- oblique hip
- unequal pitch
- dormer windows
- octagonal roofs
- intersecting roofs of unequal pitch
- extended eaves on roofs.

In particular the chapter addresses the key competencies of:

- Planning and preparing for roof construction involving reading plans, specifications, tool and material selection, and WHS requirements
- Setting out, preparing and erecting roof members, involving geometric development of bevels, creating a pattern rafter, and correct erection sequences
- Installing roof supports such as purlins and struts.

For each of these, the mathematical, geometric and construction techniques are described in detail, which will give you the ability to explore other advanced roof forms. This chapter expands on – and at times challenges – the basic principles developed in *Site Establishment, Formwork and Framing* 4e (Laws 2020), which covered the setting out and construction of basic roofs, such as the gable, broken hip, valley and Scotch Valley.

Learning objectives

By the end of this chapter you will be able to apply:

- basic roofing techniques to more advanced roof structures by applying the 'Seven Pillars' principles, and the basic geometry required to calculate and/or set out the required angles, bevels and plumb cuts
- the principles of setting out a gambrel (or Dutch gable) roof using an extended ridge and waling piece to support jack rafters. This will include the various measurement and calculations for lengths of jack rafters and various reductions and adjustments required to align roof members
- the principles of setting out a jerkin head roof using extended ridge, reduced hips and soldier wall to support shortened rafters. This will include the various measurement and calculations required to set out the soldier wall and reduced hip lengths and required adjustments

- the principles of setting out a skewed gable roof, noting that each jack rafter in the skewed gable is different in length on each side of the skewed gable end, but that bevels and plumb cuts remain the same
- the principles of setting out unequal pitched roofs, noting that each jack rafter in unequal pitched roof is different in length, but that bevels and plumb cuts are different on each side of the ridge
- the knowledge of purlins and struts in different forms and configurations to support roof structures and transfer loads from roofs to load-bearing walls and ultimately building foundations
- the principles and knowledge of roofing to the construction of dormer windows
- the principles of setting out octagonal and semi-octagonal roof shapes using two different methods depending on design
- the principles required to construct intersecting roofs with different pitches.

National Construction Code 2022

The revised National Construction Code 2022 (NCC) has structural requirements relating to roofs. These requirements change depending on the region in which the building exists; for example, cyclonic or non-cyclonic, alpine, and bushfire- or earthquake-prone areas. The structural design considerations are outlined in NCC – Volume One, Part B Structures, and may be dependent on the building's use (e.g. habitable or shed).

NCC Volume Two provides the technical design and construction requirements for mainly Class 1 and 10 buildings. Section A stipulates mandatory NCC governing requirements and Schedules 4 to 11 have each state and territory's ancillary requirements.

The NCC Housing Provision Standards provide the 'Deemed to Satisfy Provisions' that are considered to be acceptable forms of construction that meet the requirements for complying with code. Alternative solutions can be applied as long as they meet the 'performance requirements' and are validated by an 'appropriate authority'.

The NCC references a number of Australian Standards (e.g. AS 1684 – Residential timber-framed construction) as a 'deemed to satisfy' compliance pathway for different aspects of construction.

Sustainability in advanced roof constructions

No matter which form of pitched roof you are constructing, it's important to consider the benefits that come using this construction method and how to best build sustainability principles into the design.

In Australia it's important to consider the following factors that will impact the design and construction of a pitched roof:

- The performance requirements of different timber elements determine the properties; that is, appearance, structural performance, service life, initial cost and ongoing costs.
- Environmental concerns impact the species selection; for example, a timber's natural durability may be an important factor in selection, influencing the pre- and post-construction treatments required. Durability can determine suitability of a timber species for a specific application. The durability rating of a species is based on the natural ability of the heartwood to resist decay and insect pests. This may be very important in termite-prone regions. Less durable species may be used, but the application of less environmentally friendly chemical treatments may be necessary.
- Using locally sourced materials from a managed forest is recommended, as opposed to imported species.
- The zone in which the building is being built is important. The Australian Standard AS 3959 Construction of buildings in bushfire-prone areas provides guidance on the specific design requirements for each zone.
- Where section sizes in some species are not available, it may be practical to use a laminated section. These can be made using lesser quality timbers and can produce extremely strong roof members.

Wood Solutions provides a lot of material on timber selection, use and guidance on solutions for timber materials based on the National Construction Code and Australian Standards requirements: https://www.woodsolutions.com.au/.

Inspect the worksite area before construction and locate services. Complete a risk assessment, apply signage and barricades and implement control measures. You must adhere to the codes and practice, or guidelines, provided by your state's WHS/OHS authority. Consider the following when planning roof construction:

- consult – with designers and engineers if applicable with the view to eliminating or significantly reducing the risks associated with working at heights
- manage risks – identify, assess, control and review with the aim of eliminating unsafe work wherever practical.

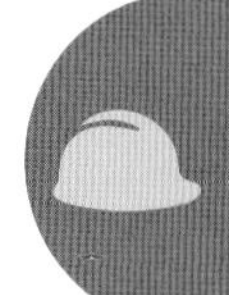

Building roofs is high-risk work. Obtain a copy of the *National Code of Practice for the Prevention of Falls in General Construction*. Section 7, 'Working at heights of 2 metres and above' is most relevant. Always follow the code when working on roofs.

The National Construction Code (NCC) and Australian Standards must be adhered to during planning and construction. This includes Volume Two, Section A, Governing requirements of the National Construction Code (NCC). The structure must also comply with Australian Standards 1684 for Residential timber-framed construction. Review workplace documents such as:

- Safe Work Method Statement (SWMS)
- Job Safety Analysis (JSA)
- Manufacturer's specifications (especially for components where load bearing is a concern)
- The Environmental Plan (which outlines the site information for dust, noise, vibration, wastage requirements and recycling).
- Plans, drawings and specifications.
- Engineering specifications are important to review, especially if construction is taking place in cyclonic climate zones.

Source: Synergy Scaffolding & Access

FIGURE 7.1 Narrow scaffoldings

When erecting a pitched roof, the use of a harness is required. There are several additional options available to reduce the risk of falling from a roof under construction:

- Mobile scaffolds (see Figure 7.1) are suitable for internal use and provide adequate working platforms. They need to fit through a door opening without being fully dismantled. A harness must be worn when working from a scaffold.
- Safety nets (see Figure 7.2) not only prevent falls from heights but also stop tools and debris such as larger off-cuts from falling through.

Source: © Copyright Safety Nets Australia Pty Ltd

FIGURE 7.2 Safety nets

- Perimeter scaffold provides both fall protection and a safe work platform around the perimeter of the building.
- Edge protection (see Figure 7.3) must be fitted to the side frames or structure. Make sure they don't interfere with roof members being erected.

Source: Australian Scaffolds

FIGURE 7.3 Edge protection and scaffold combined

Basic principles of roofing

The purpose of this revision is to reacquaint you with the basic principles of roofing, including the underpinning mathematics and geometry; and to highlight key assumptions within these basic principles that are challenged in advanced roofing, particularly by roofs with unequal pitches or tapering spans.

We suggest that you revisit Chapter 10 of *Site Establishment, Formwork and Framing* (Laws 2020).

Having developed an understanding of basic roofing, the following points mostly hold true for all roofs:

1 Roofs are made up of right angled triangles and are set out to centre lines (including hip and valley rafters, centring rafters and ridges).
2 Ridges run parallel and level to wall plates.
3 Regardless of the roof shape, rafters run at 90° to the wall plates.
4 Hips and valleys **bisect** internal and external corners regardless of the angles of those corners.

With the exception of point 3, all of the points mentioned will be challenged at some point in this chapter. And point 3 cannot be viewed as a given in all roofing, for there are occasions in some more complex architecture where it may be rational to forgo this 'rule' (though generally in small areas where only light loads apply).

Plans, drawings and workplace documents will provide you with the required measurements. Specifications for the roofing can be located on the plans or provided in the work documents. After considering all the work instructions, plan and sequence the work activities. Planning and organisation are key when working in teams. By following a logical order, workers can complete tasks efficiently and avoid unnecessary backtracking or rework. This saves time, reduces delays and increases overall productivity.

The mathematics: the 'Seven Pillars' revisited

From the text referred to earlier (Laws 2020), the following outlines the basis of the mathematics underpinning roofing. While there are some notable changes with regards to **hip** and **valley** length calculations, these basic principles still hold, as does the geometry they are derived from.

This sample calculation for a basic hipped roof is based upon a building with the following characteristics:

Pitch = 25°
Span = 8010 mm
Eave width = 600 mm
Rafter sectional size = 125 × 45 mm
Hip sectional size = 175 × 35 mm

It is important to develop a clear understanding of this system of calculations and, most importantly, the application of the common rafter length per metre run (CR factor) and, similarly, the hip length per metre run of common rafter (hip factor). It is by using these factors that lengths of some of the seemingly more difficult components are most easily found.

Pillar 1 Common rafter rise per metre run (CR rise/m run)

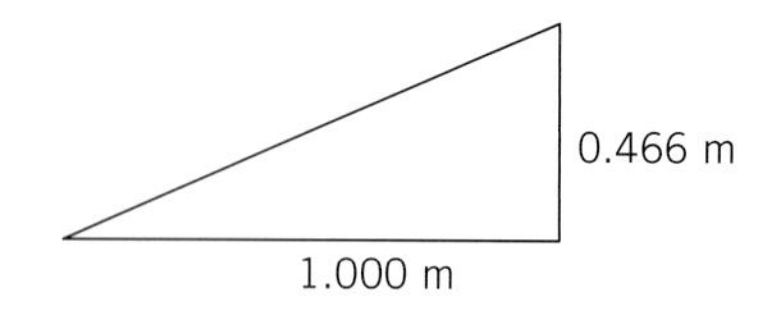

$$\text{CR rise/m run} = \tan\theta$$

$$\frac{Rise}{}$$

$$\tan 25° = 1.000$$

$$0.466 = \text{rise}$$

Pillar 2 Common rafter length per metre run (CR factor)

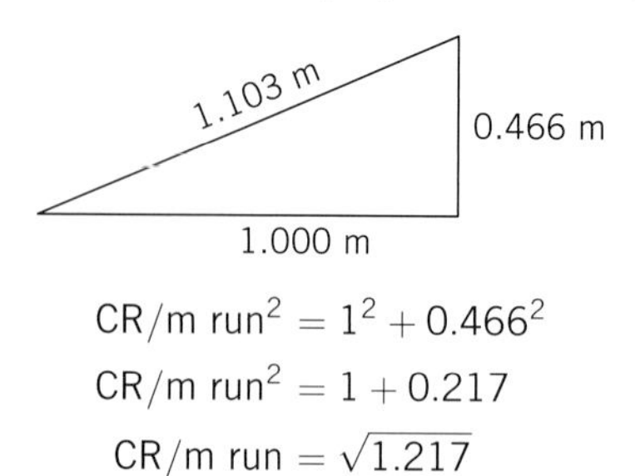

$$\text{CR/m run}^2 = 1^2 + 0.466^2$$

$$\text{CR/m run}^2 = 1 + 0.217$$

$$\text{CR/m run} = \sqrt{1.217}$$

$$\text{CR/m run} = \mathbf{1.103\ m}$$

Pillar 3 Common rafter set-out length (CR set-out length)

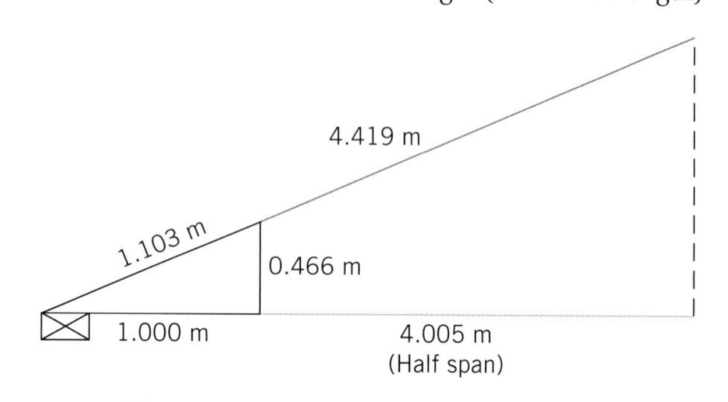

$$\text{CR set-out length} = \text{CR factor} \times \text{half span}$$

$$\text{CR set-out length} = 1.103 \times \mathbf{4.005}$$

$$\text{CR set-out length} = \mathbf{4.419\ m}$$

Pillar 4 Common rafter order length (CR order length)

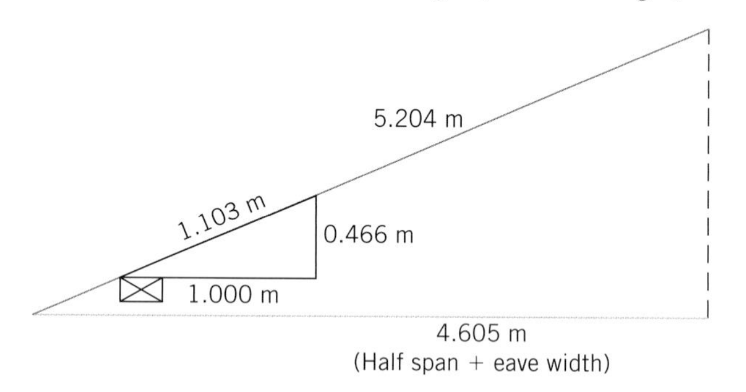

$$\text{CR order length} = [\text{CR factor} \times (\text{half span} + \text{eave width})] + \text{rafter depth}$$

$$\text{CR order length} = [1.103 \times (\mathbf{4.005} + \mathbf{0.600})] + \mathbf{0.125}$$

$$= [1.103 \times 4.605] + 0.125$$

$$= 5.079 + 0.125$$

$$= 5.204$$

$$\text{CR order length} = \mathbf{> 5.4\ m}$$

Remember to add the depth of the rafter material (in this case 125 mm) to allow for the bevel cut (see the 'Seven Pillars' PowerPoint for clarification if you are unclear on this issue).

Pillar 5 Hip length per metre run of common rafter (hip factor)

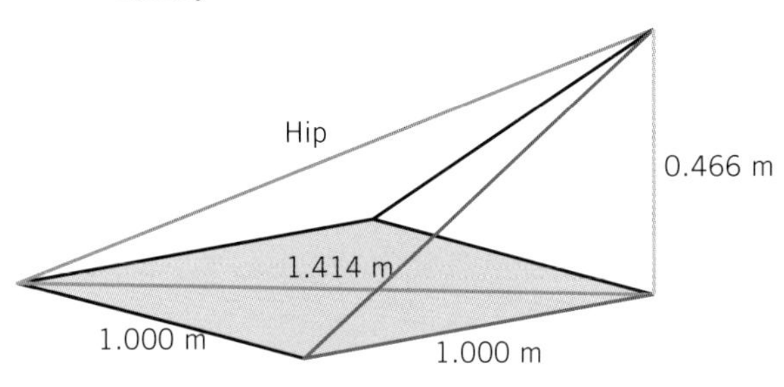

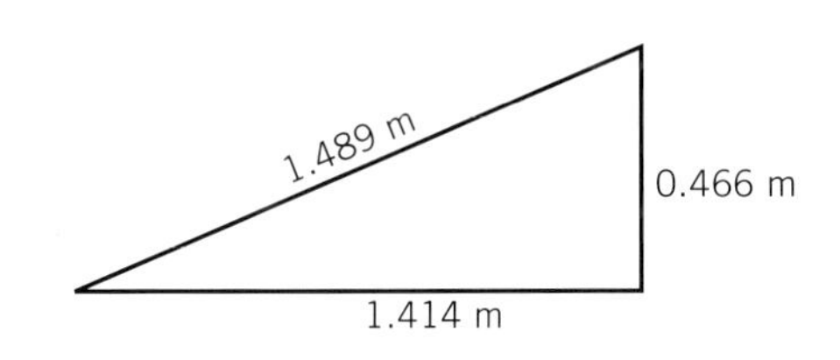

$$\text{Hip factor} = \sqrt{\text{hip run}^2 + \text{rise}^2}$$
$$\text{Hip factor}^2 = 1.414^2 + 0.466^2$$
$$\text{Hip factor}^2 = 2 + 0.217$$
$$\text{Hip factor} = \sqrt{2.217}$$
$$\text{Hip factor} = \mathbf{1.489\ m}$$

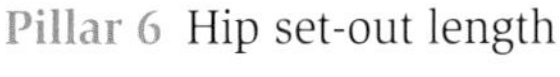

Pillar 6 Hip set-out length

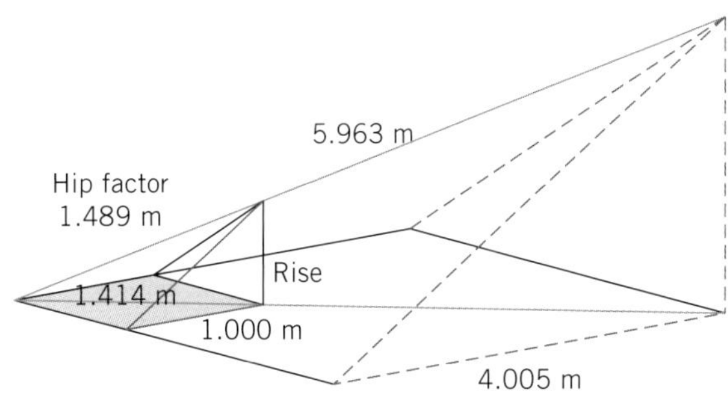

$$\text{Hip set-out length} = \text{hip factor} \times \text{half span}$$
$$\text{Hip length} = \mathbf{1.489 \times 4.005}$$
$$\text{Hip length} = \mathbf{5.963\ m}$$

Pillar 7 Hip order length

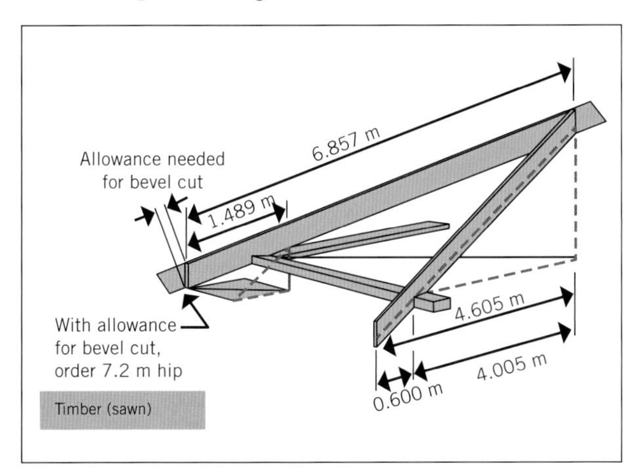

$$\text{Hip order length} = [\text{hip factor} \times (\text{half span} + \text{eave width})] + \text{hip depth}$$
$$\text{Hip order length} = [\mathbf{1.489} \times (4.005 + 0.600)] + 0.175$$
$$= [\mathbf{1.489} \times \mathbf{4.605}] + 0.175$$
$$= \mathbf{6.857} + 0.175$$
$$= 7.032$$
$$\text{Hip order length} = > \mathbf{7.2\ m}$$

Basic roofing geometry

From Chapter 10, 'Construct pitched roofs', in *Site Establishment, Formwork and Framing* (Laws 2020), you will be aware that, for basic roofing, eight bevels are required. The development shown in Figure 7.4 has been taken directly from that text. These eight bevels are required for our first two roofs (gambrel and jerkin head), and the basic principles hold for all the others. You need to focus on the development of the hip edge bevel. This is based on the level line (LL) principle (see Figures 7.5 and 7.6). This principle works on the basis that any line running at 90° to the component being considered is level. And as long as the top edge of that component is square to a line running plumb or vertically through it (generally the case with square or rectangular rafters, etc.), then this line is also running along that surface.

The skill in developing roof bevels is in determining where to find the relevant right angled triangle; that is,

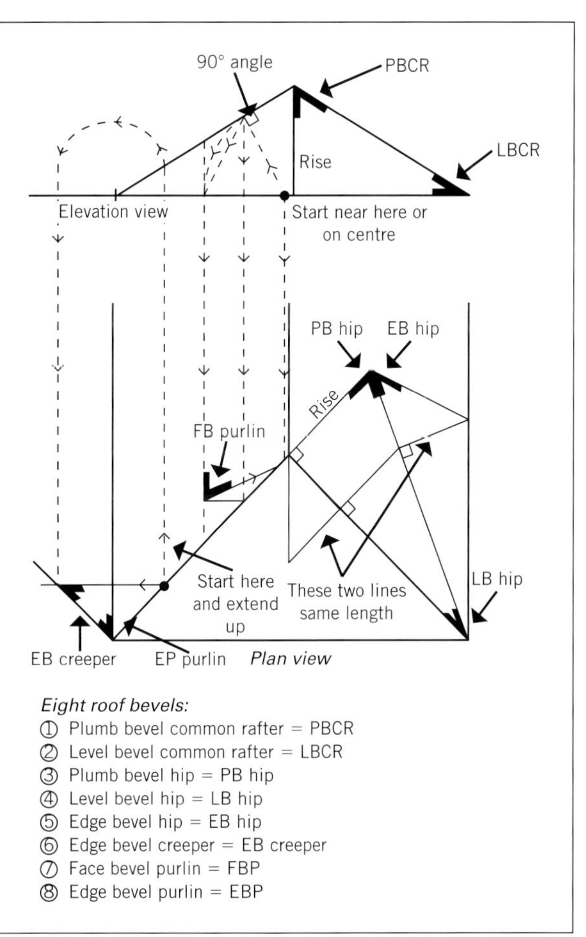

FIGURE 7.4 One approach to the development of roof bevels

LEARNING TASK 7.1 REVISION OF BASIC PRINCIPLES OF ROOFING

Use the diagram below to draw the basic plan view of a hip and valley roof.

Answer the following questions:

1 Calculate the CR rise/m run of the following common angles:
   a 37° =
   b 42° =
   c 33° =
2 Calculate the common rafter factor for the following CR/m run:
   a 26° = CR/m =
   b 36° = CR/m =
   c 40° = CR/m =
3 Calculate the common rafter set-out length for the given half spans:
   a 5500 (half span) × 35° =
   b 6200 (half span) × 39° =
   c 4800 (half span) × 43° =

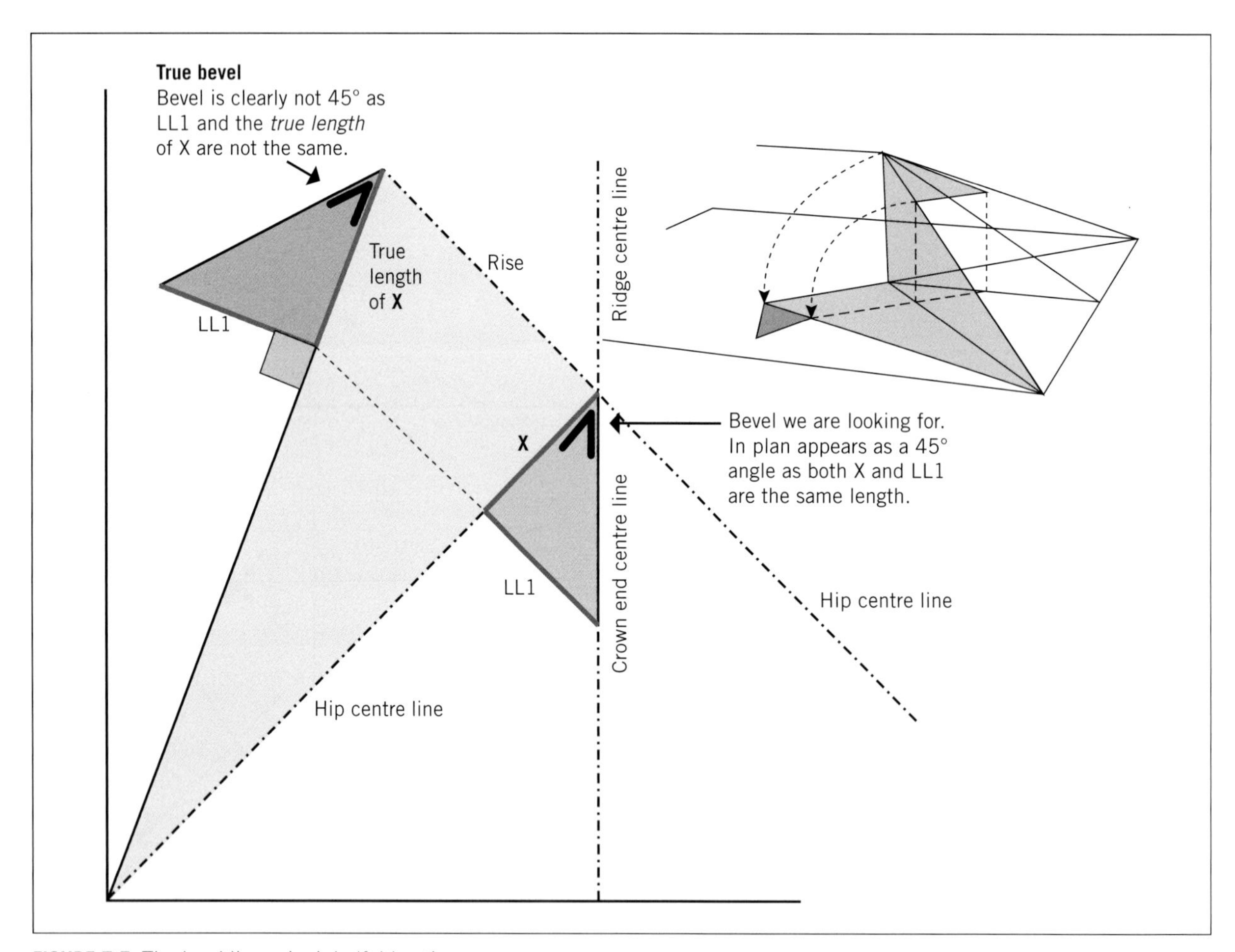

FIGURE 7.5 The level line principle 'fold out'

where to place the level line, and where to find the true length of the other side of that triangle. After that it is simply a matter of putting the two lengths together as a right angled triangle to form the 'true' bevel. It may be worthwhile to review a hip roof geometric development video. Buildsum has a useful video (https://www.youtube.com/watch?v=NfSkTT35UoI&t=21s).

In Figures 7.5 and 7.6 the component we are dealing with is a hip. Given that a level line in the plan view is a true length, it is the side running up the hip that is raking

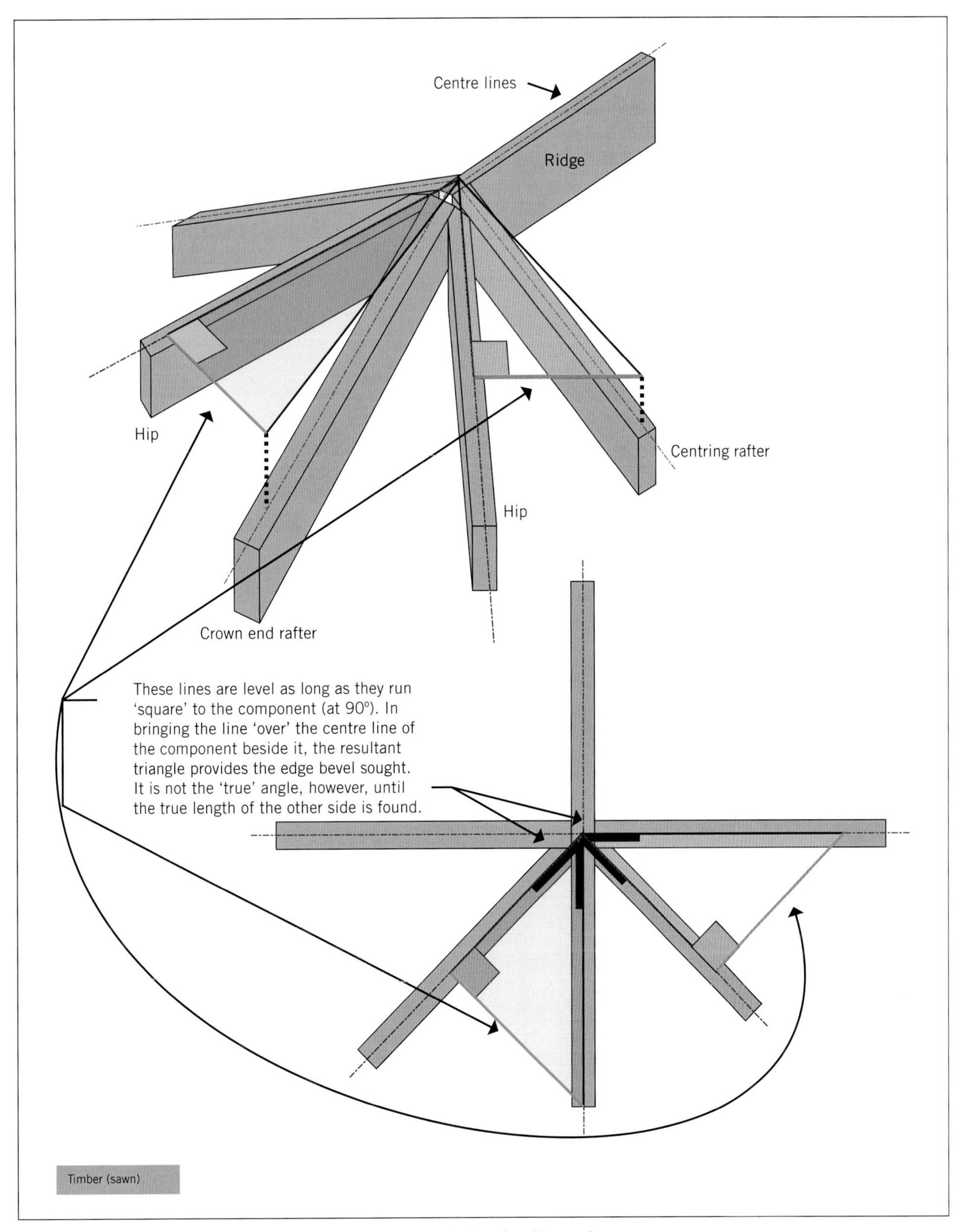

FIGURE 7.6 The level line principle – pictorial view (top) and plan view (bottom)

and so it is actually longer than it appears. The true length of this side is found in the side elevation of the hip.

Using a layout similar to that modelled in Figure 7.4, we can see one way of graphically putting this into practice (Figure 7.5).

Study this concept carefully until you can fully visualise what is being done. This concept will occur repeatedly in developing the bevels for each of the roofs that follow. It is also important to realise there are many ways the development of the above bevel may be laid out. However, each is working on the same principle of recognising when a line is being seen in its true length, and knowing how or where to find the true length of those lines that are raking towards or away from us.

Gambrel (Dutch gable) roofs

The gambrel or Dutch gable roof is basically a hipped roof with an extended **ridge** forming small gables at each end; that is, the hips start from the plates at the corners but do not reach the apex (see Figure 7.9). Sometimes the gable end is built as a vent or has vents let into it. This is a style that continues to be popular in contemporary home design.

In a gambrel roof the ridge is extended, as shown in Figures 7.7 and 7.8. This means extra rafters are installed and the hips shortened. The result is a vertical gable in the hip end, located wherever the builder, client or architect desires: usually at the most convenient common rafter position. A waling plate is fixed to the last set of common rafters to pick up the ends of the **jack rafters** that fill out the end of the roof (see Figure 7.7). Then the hips and any creepers are cut in to finish the framing.

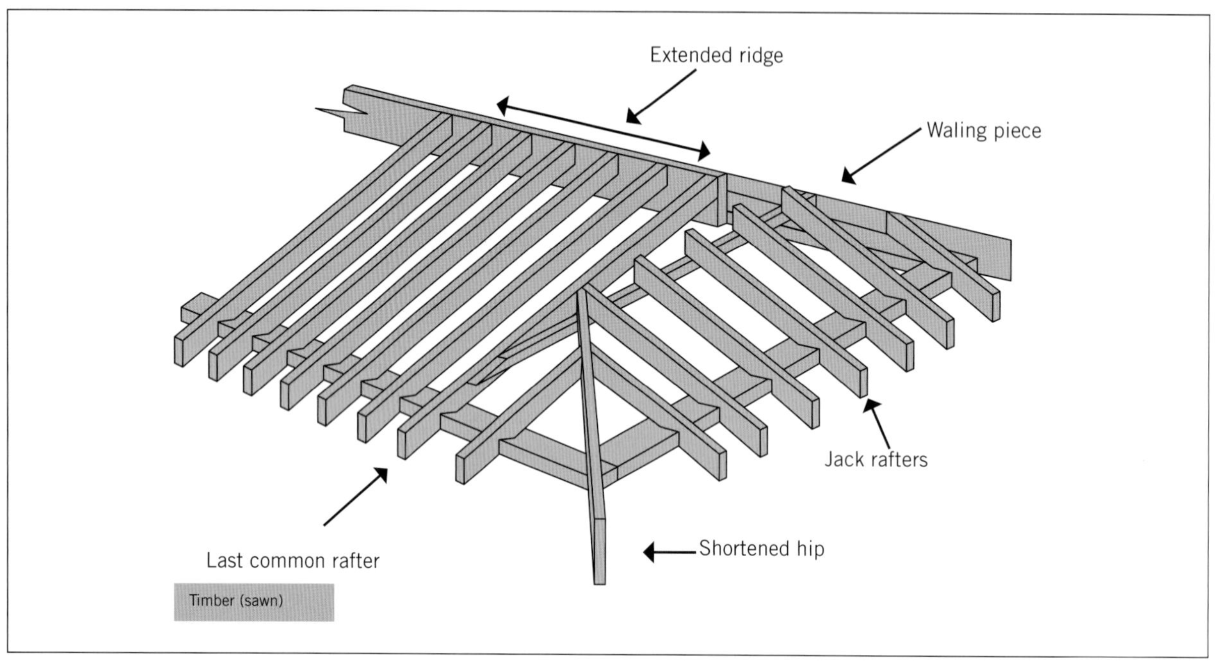

FIGURE 7.7 Gambrel roof frame

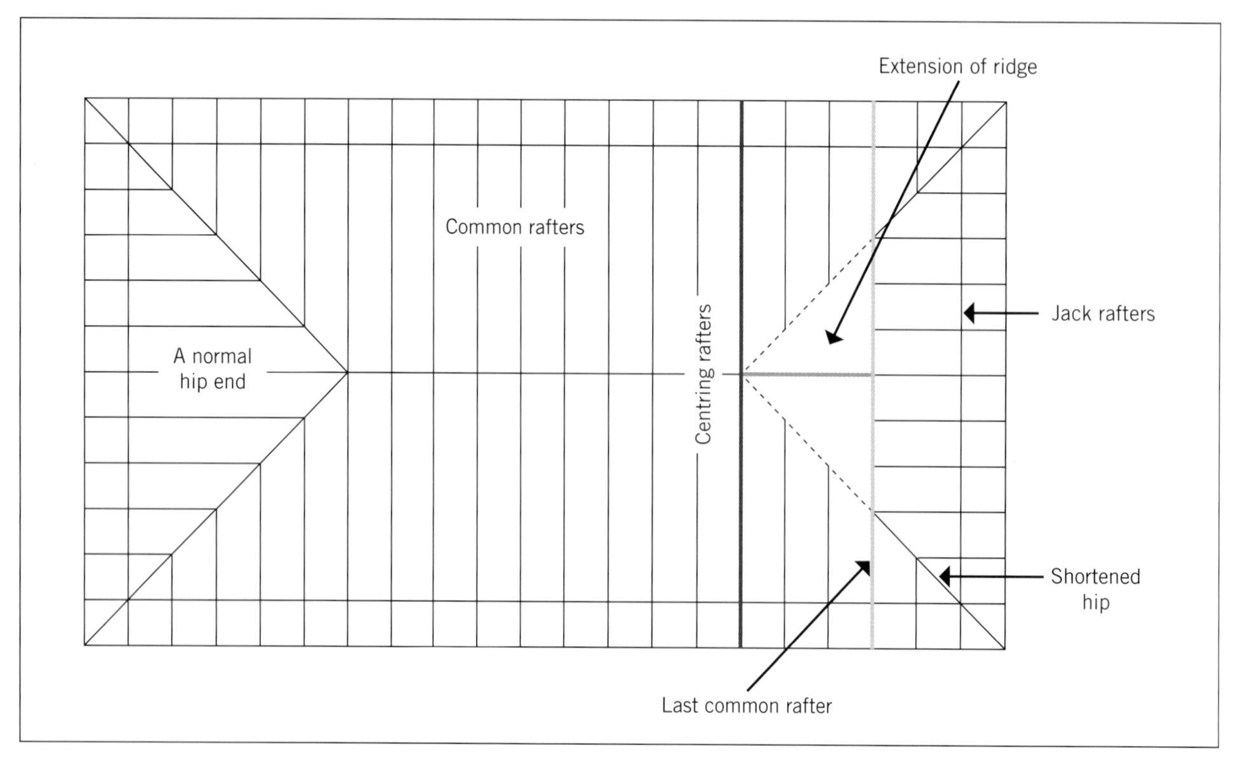

FIGURE 7.8 Plan of typical gambrel roof framing

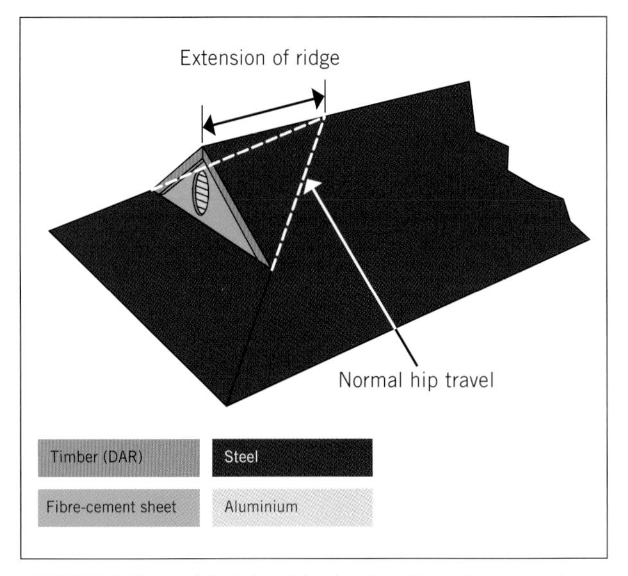

FIGURE 7.9 Typical Dutch gable showing the ridge extension

Setting out and constructing the gambrel roof

For simplicity, the roof being considered will have similar characteristics to that used in the revision exercises, but a narrower span; that is:

Pitch = 25°
Span = 4500 mm
Eave width = 600 mm
Rafter sectional size = 125 × 45 mm
Hip sectional size = 175 × 35 mm

Begin by marking out the top plates with the centre line of the centring rafters set back from the end of the build by the half span; the centre line of the crown end at the half span on the end wall plate; then all other rafter positions moving away from these centre line marks at the normal rafter spacings.

At this point you should do the development of all your basic bevels (the eight bevels shown in Figure 7.4) and the calculation of the first five of the 'Seven Pillars', as shown in Table 7.1.

TABLE 7.1 The first five roofing pillars

	First five roofing pillars	In this case (see pp. 298–9 for calculations).
1	Rise/m run of CR	0.466 m
2	CR length/m run (CR factor)	1.103 m
3	Set-out length of CR	2.482 m
4	Order length of CR	3.144 m
5	Hip length/m run of CR (hip factor)	1.489 m

From this information, you should create a pattern rafter as per normal (see Laws 2020, Chapter 10), but exclude the creeper rafter lengths for now.

Setting out the ridge

Ridge set-out is done by marking it off the wall plate as you would in normal roofing practice, the only difference being the allowance for the extension (see Figure 7.10). This is simply a matter of determining which rafter is going to be the last common one.

In this case, the plan calls for an extension of three rafters spaced at 450 mm centre to centre (see Figure 7.11). If in your plan no dimension is given, scale the length and then check with the client or architect.

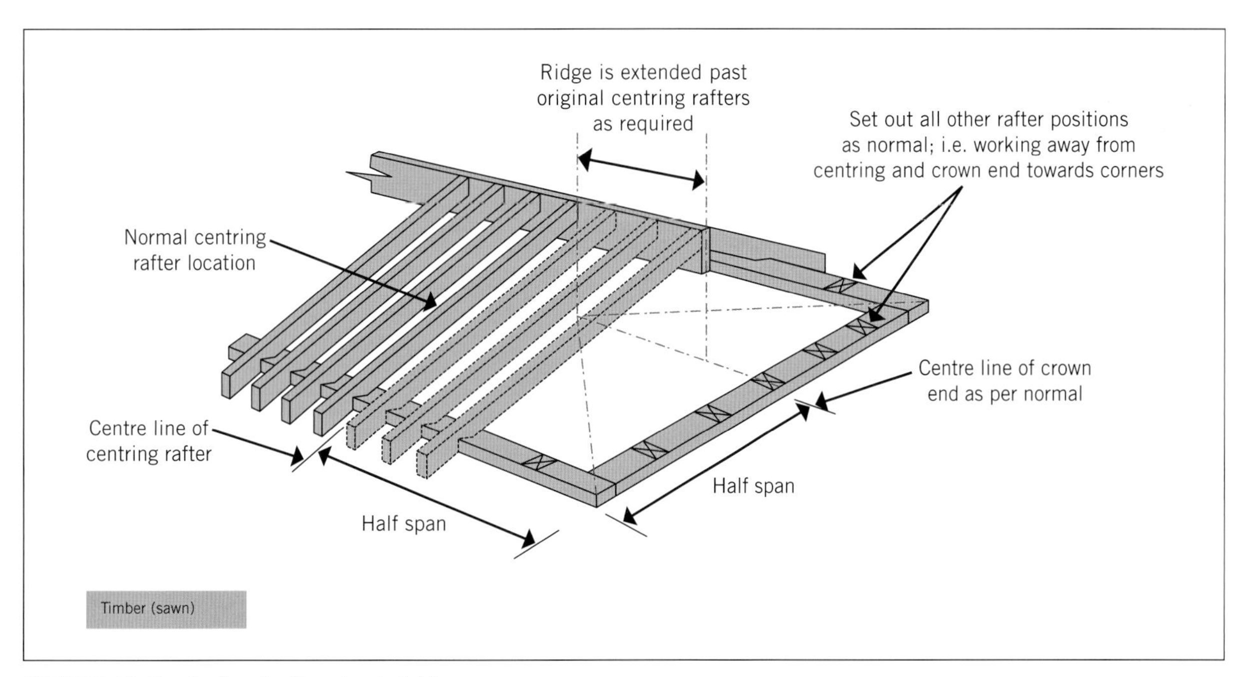

FIGURE 7.10 Gambrel roof with extended ridge

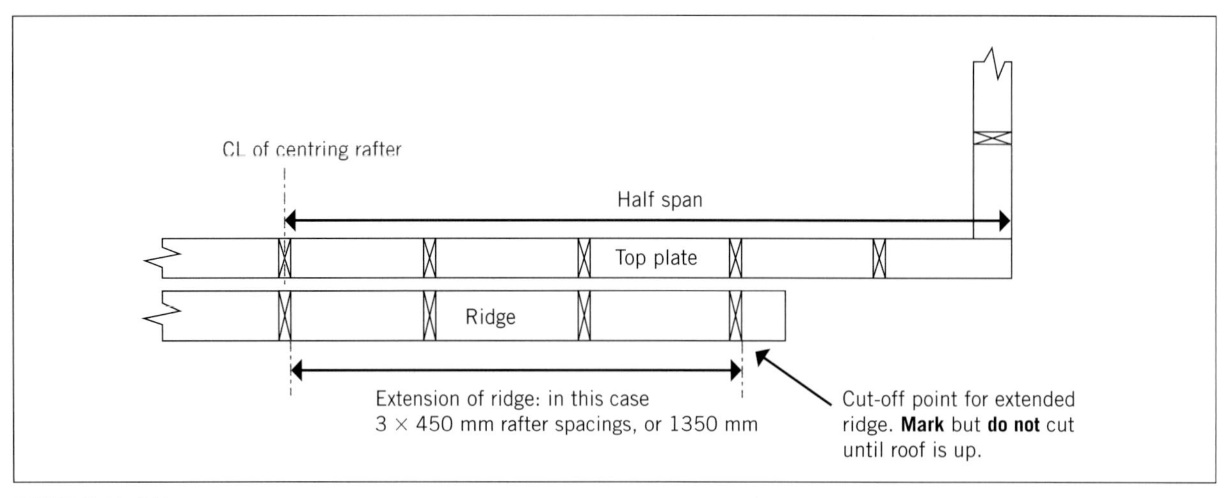

FIGURE 7.11 Ridge set-out

Standing the main roof section

Having marked out the ridge and pattern rafter, the first part of the roof may be cut out and assembled. Be sure to install a temporary brace at this point in case of unexpected high wind gusts (see Figure 7.12).

The jack rafters and location of waling piece

To find the location of the waling piece, we will need to set out and cut one of the jack rafters (see Figure 7.13). This will be used to mark where the top edge of the jacks will finish on the common rafters (see Figure 7.14).

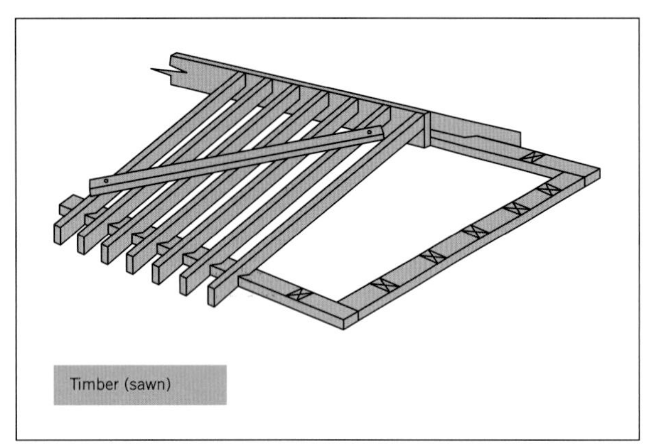

FIGURE 7.12 Main roof section erected and braced

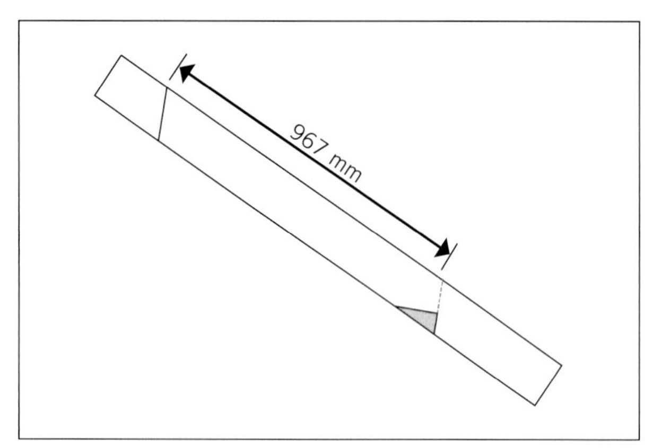

FIGURE 7.13 Jack rafter set-out

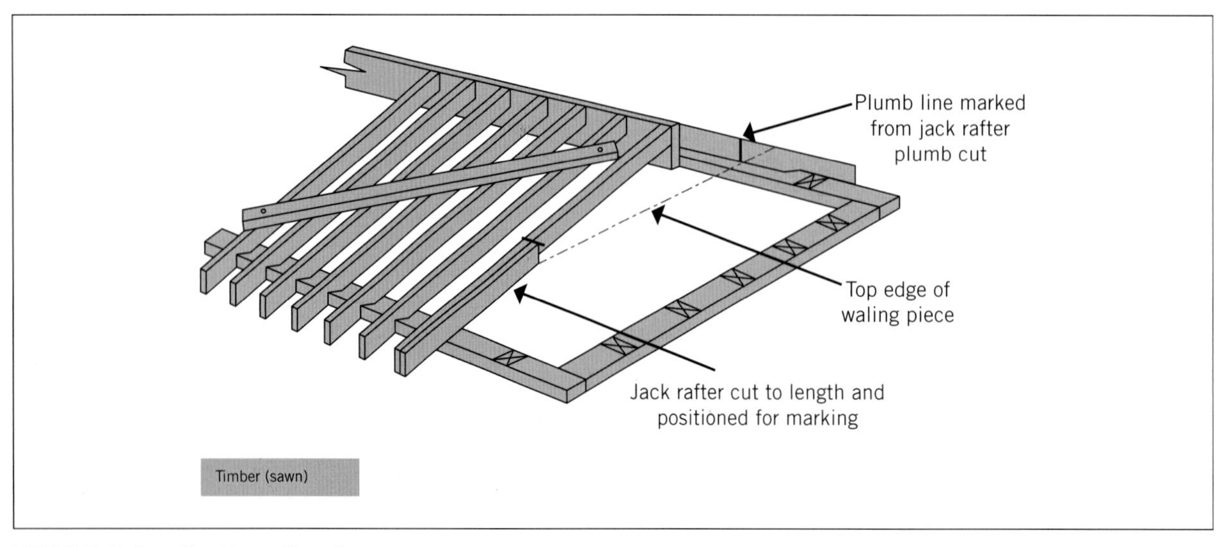

FIGURE 7.14 Locating the waling piece

The jack rafter is simply a shortened common rafter (hence the name 'jack'). To calculate the length of any component that has the same pitch as a common rafter, you need two easily found pieces of information: the run, or plan length, of the component; and the CR length/m run, or rafter factor.

The run of the jack is found by:

Half span – (ridge extension + half the thickness of a common rafter*)

In this case:

Jack run = 2250 – (1350 + 23*)
Jack run = 877 mm

The cut-off length of the jack rafter is then simply:

Cut-off length = component run × CR length/m run

In this case:

Cut-off length = 877 × 1.103
Cut-off length = 967 mm

Having cut at least one jack rafter as a pattern, place the rafter as shown and mark down the plumb cut (see Figure 7.15). Do this on both sides of the roof. Where this plumb line meets the bottom edge of the common rafters is where you will position the top edge of the waling piece.

***Note:** This reduction can be done later, as you would for a crown end rafter. In taking this alternative approach, you will be finding the 'set-out' length of the jack rafter. You must then be sure to take off the half thickness of CR (**horizontally**, and **not** down the length of the rafter).

Alternatives for locating waling piece

Locating the waling piece as described requires you to notch the underside of each jack rafter as shown in Option A. Options B and C offer alternatives to this approach (see Figure 7.16 earlier). Figure 7.17 shows the waling piece installed.

The jack rafters can now be installed. Note that the top outer edge of the two outer rafters should meet the edge of the common rafters as shown in Figure 7.18.

Hip rafters

To calculate the length of hips, we use the run of the common rafter that is required to obtain the same height. In basic hipped roofing this is simply the half span. In this case, as is clearly evident in Figure 7.19, the jack rafter obtains the height we require. So it is the run of the jack rafter that we need. We then simply multiply this length by the hip length/m run of CR (the hip factor).

Hip set-out length = run of jack rafter × hip length/m run of CR

In this case:

Hip set-out length = 0.877 × 1.489
Hip set-out length = 1.306 m

As we are using a jack run that already allows for the half thickness of a common rafter, no further reductions are required when setting out the hip. The hip set-out is shown in Figure 7.19.

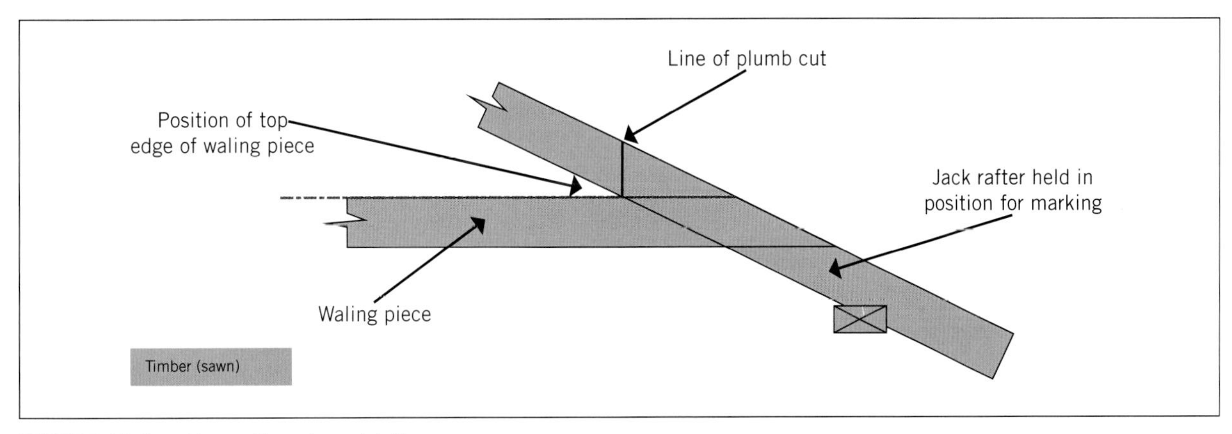

FIGURE 7.15 Locating waling piece detail

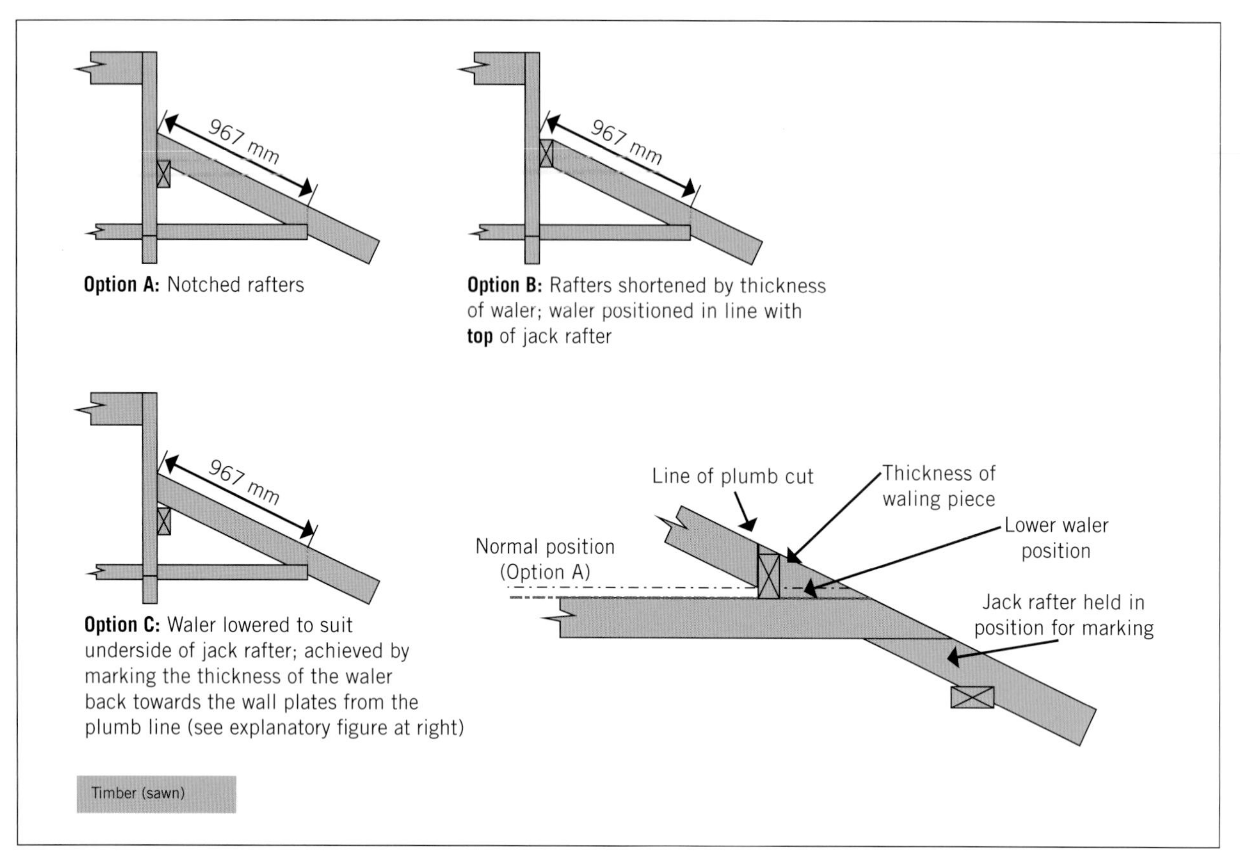

FIGURE 7.16 Alternative waling piece positions

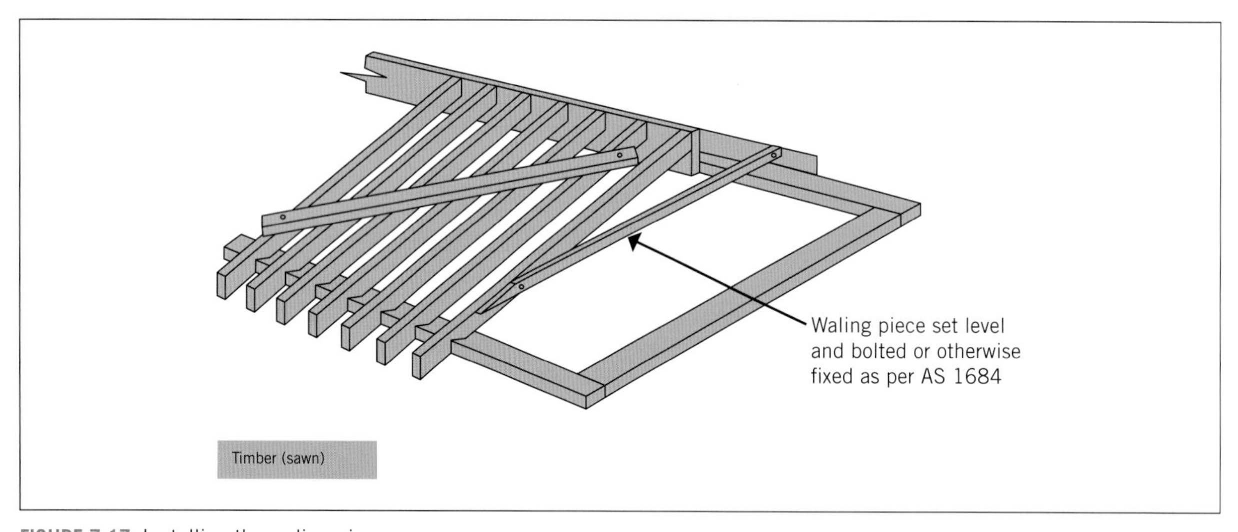

FIGURE 7.17 Installing the waling piece

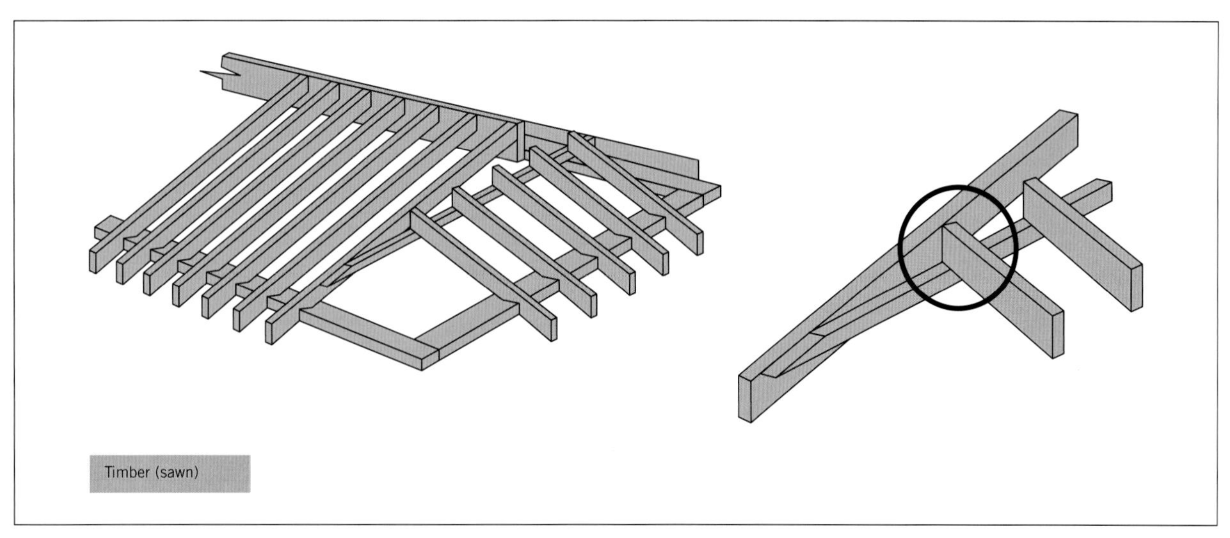

FIGURE 7.18 Jack rafters installed

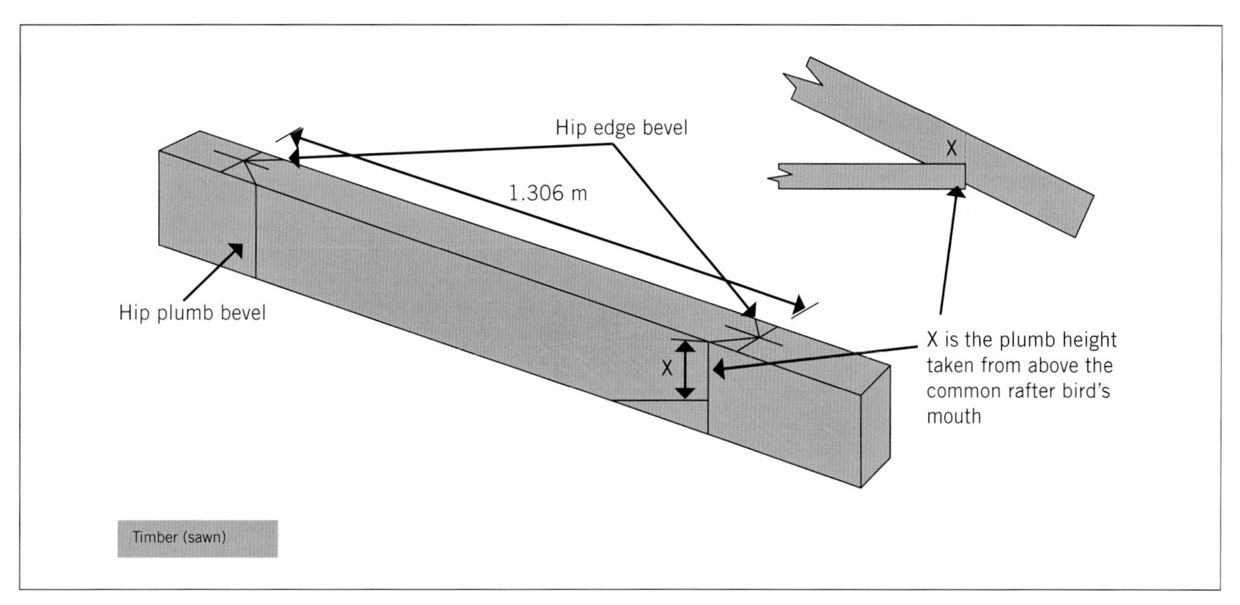

FIGURE 7.19 Hip set-out

Finishing the framing

Installation of the hips is much the same as normal, though you will need to notch the top end of the hip over the waling piece. This is most easily done by direct measurement and use of the normal hip plumb and level bevels. As with the jack rafters, be sure that the top edge of the hip aligns correctly with the edges of the jack and common rafters (see the detail in Figure 7.20).

Set-out, cutting and installation of the creepers is now done, as for a standard hipped roof (see Figure 7.21). As always, take care that all surfaces are true (in wind and straight).

LEARNING TASK 7.2

RESEARCHING GAMBREL (DUTCH GABLE) ROOFS

In class, discuss the advantages and disadvantages of gambrel roofs compared to gable end and hip and valley roofs. Key points to consider include:

- roof (attic) space
- climate considerations
- architectural factors.

To assist your research, conduct an internet search for images of Dutch gable roofs, and look at Buildsum https://www.youtube.com/watch?v=VSUsnhxjsxU\.

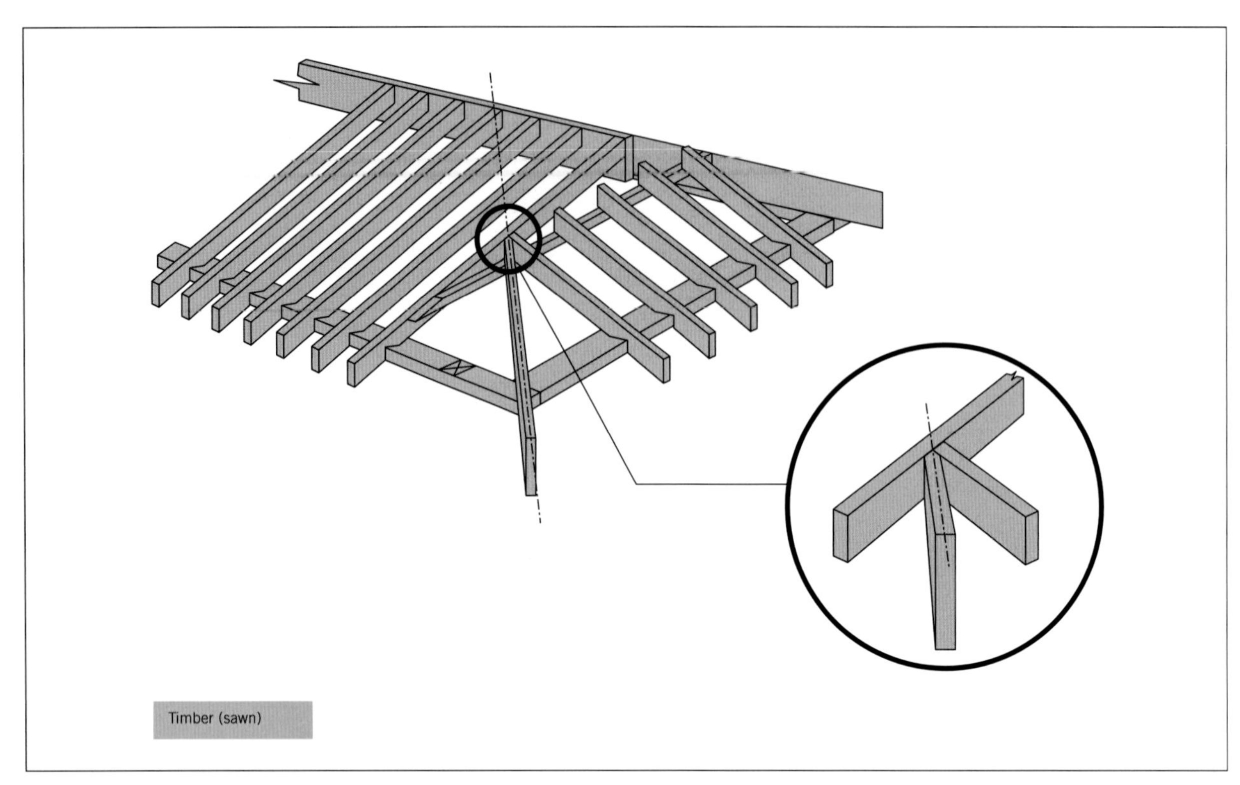

FIGURE 7.20 Hip position and detail

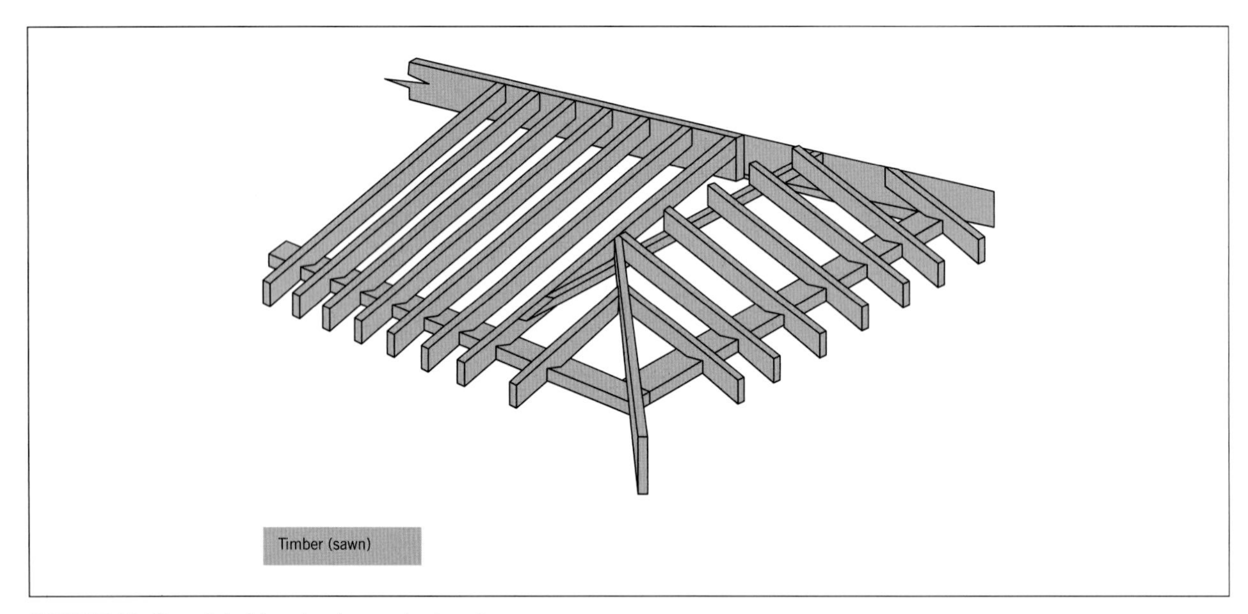

FIGURE 7.21 Completed framing for gambrel roof

Jerkin head roofs

A jerkin head roof, like the gambrel, is an adaptation of the hipped roof. This roof form is useful where the attic or roof space of a house is required as the living area, or where the architect desires to reduce the imposing nature of a large gable. While an interesting and pleasing roof from a construction point of view, it is not much used in contemporary architecture. Its inclusion in this chapter is based on its value in developing skills useful for more advanced forms.

Like the gambrel, in constructing a jerkin head roof the ridge is extended past the half span point where the centring rafters would normally be positioned. Unlike the gambrel, however, the hips and centring rafters move with the end of the ridge. The result may be

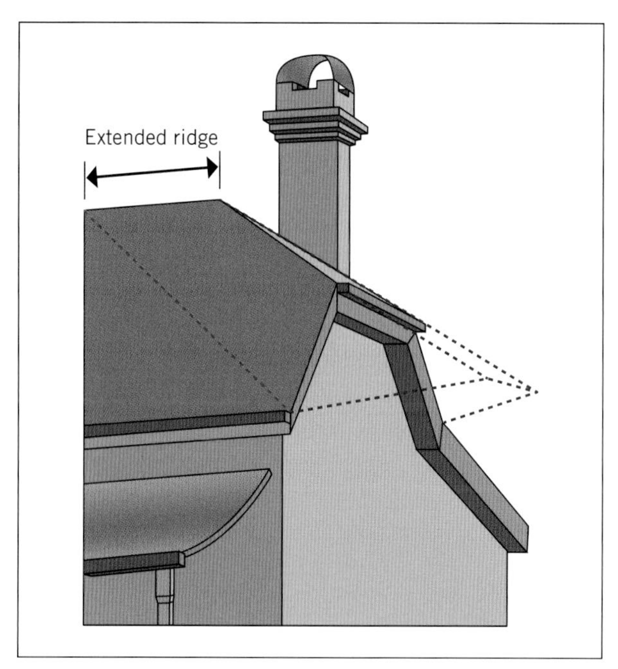

FIGURE 7.22 Ways of visualising the jerkin head roof

described as a cut-off hip end (see Figure 7.22, blue lines), or a truncated gable end wall. The hips do not intersect the corner of the building: instead, a soldier (or shortened) wall is constructed to support the crown end, creepers and hips (see Figure 7.23).

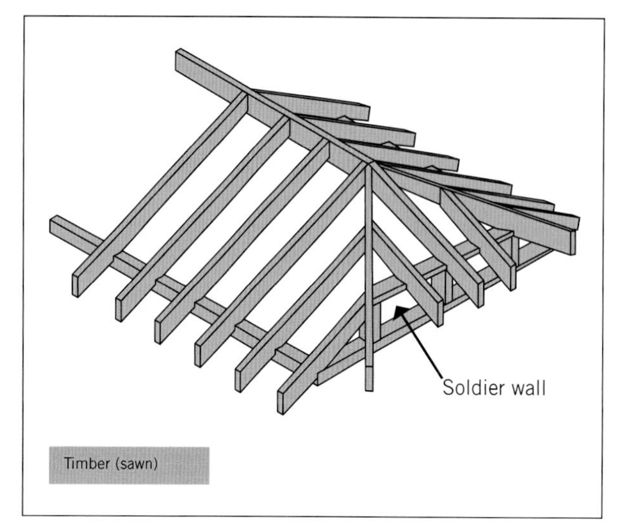

FIGURE 7.23 Pictorial view of jerkin head roof framing

As can be seen from the diagram in Figure 7.23, the set-out and development of all the various rafter components will be familiar to those with an understanding of basic hipped roofing.

The skill that you need to develop for the jerkin head roof is how to determine the height and width of the soldier wall. Figure 7.24 offers a comparison between the framing of a 'normal' hip end and the jerkin head. The length and position of the soldier wall is shown as a red line.

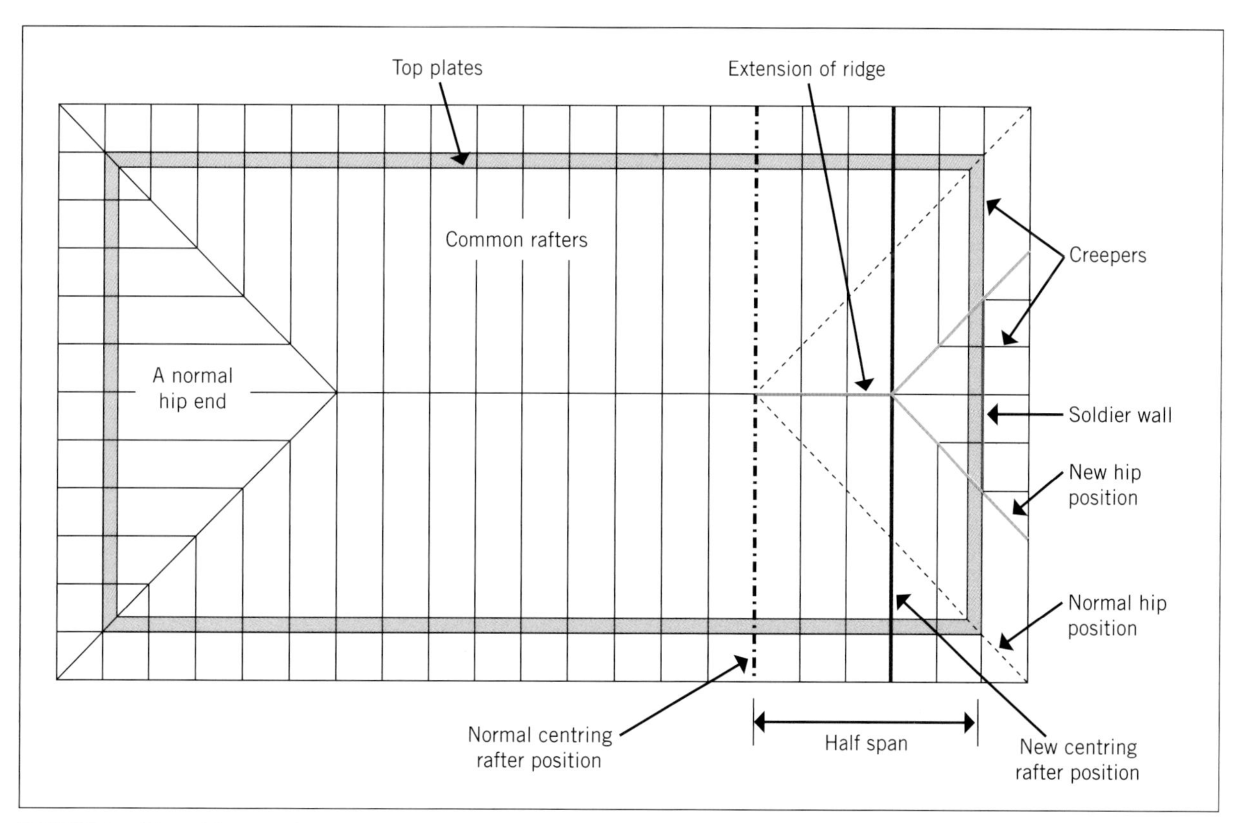

FIGURE 7.24 Plan of framing for a jerkin head roof compared to a 'normal' hip end

Calculating the height and width of the soldier wall

This is a very simple calculation once you understand the basic geometry from which it derives. For this discussion we will use the same characteristics as in the gambrel roof example:

Pitch = 25°
Span = 4500 mm
Eave width = 600 mm
Rafter sectional size = 125 × 45 mm
Hip sectional size = 175 × 35 mm

And once more the ridge will be extended by 3 × 450 mm rafter spacings, or **1350 mm**.

Now look closely at Figure 7.25. In this figure you can see that the new hip and the normal hip run parallel to each other. This means that the red line (ridge extension) and the green line (the horizontal run of the jerkin head rafter) are equal; that is, the run, or plan length, of the jerkin head rafter will always be equal to the amount by which you extend the ridge.

In this case:

Run of jerkin head rafter = 1350 mm

As Figure 7.26 shows clearly, the width of the soldier wall may now be determined by the following:

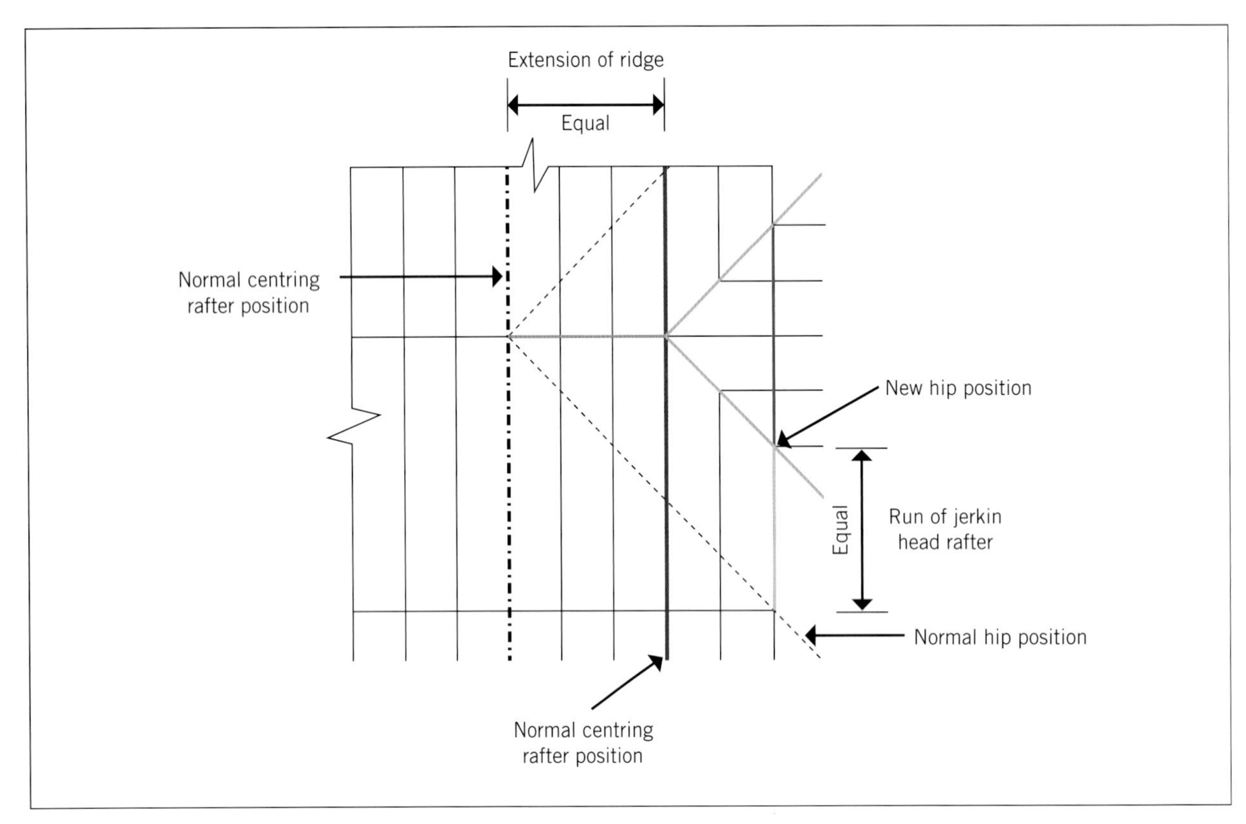

FIGURE 7.25 Determining the length of the soldier wall

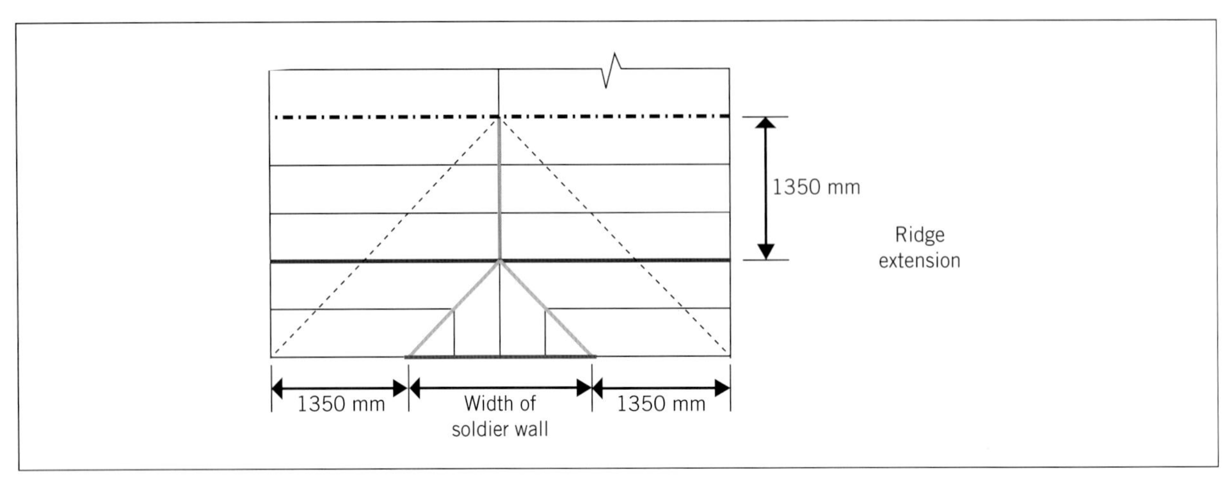

FIGURE 7.26 Width of the soldier wall

Soldier wall width = span – (2 × ridge extension)

Or in this case:

$$\text{Soldier wall width} = 4500 - (2 \times 1350) = 4500 - 2700 = 1800\,\text{mm}$$

As with the gambrel, at this point you should do the development of all your basic bevels (the eight bevels shown in Figure 7.4) and the calculation of the first five of the 'Seven Pillars'. Because we are using the same roof characteristics, these will be the same as those for the gambrel roof (see the earlier section 'The mathematics: The "Seven Pillars" revisited').

The height of the soldier wall is now determined using the first of the 'Seven Pillars'; that is, the rise/m run of CR. Figure 7.27 shows how this works.

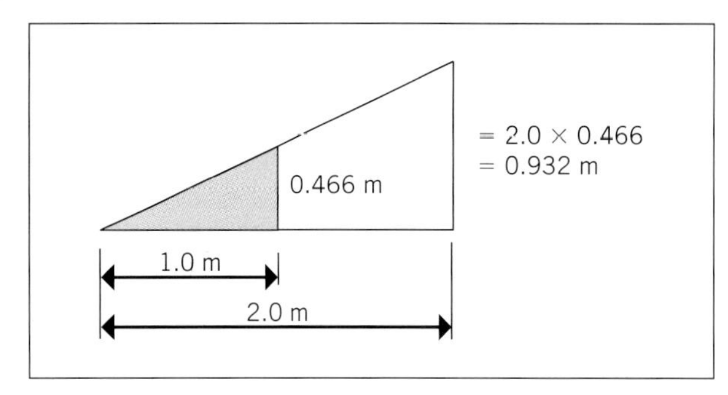

FIGURE 7.27 Rise/m run principle

That is, for every 1.0 m of run, a rafter at 25° will rise 466 mm. For 2.0 m, the rise will be twice as high (2 × 0.466). For 1.5 m, the rise will be 1.5 times as high (1.5 × 0.466), and so on.

The height of the soldier wall (see Figure 7.28) is therefore:

Height of the soldier wall = rise/m run of CR × ridge extension

In this case:

$$\text{Height of the soldier wall} = 0.466 \times 1.350 = 0.629\,\text{mm}$$

Framing out the roof

From this point on, the setting out and cutting of the crown end, hips and creepers is very much the same as for a normal hipped roof: working, as always, with plan lengths, and multiplying by the appropriate factor (hip or rafter). Figure 7.29 shows how to determine these lengths.

Crown end plan length = half span – ridge extension

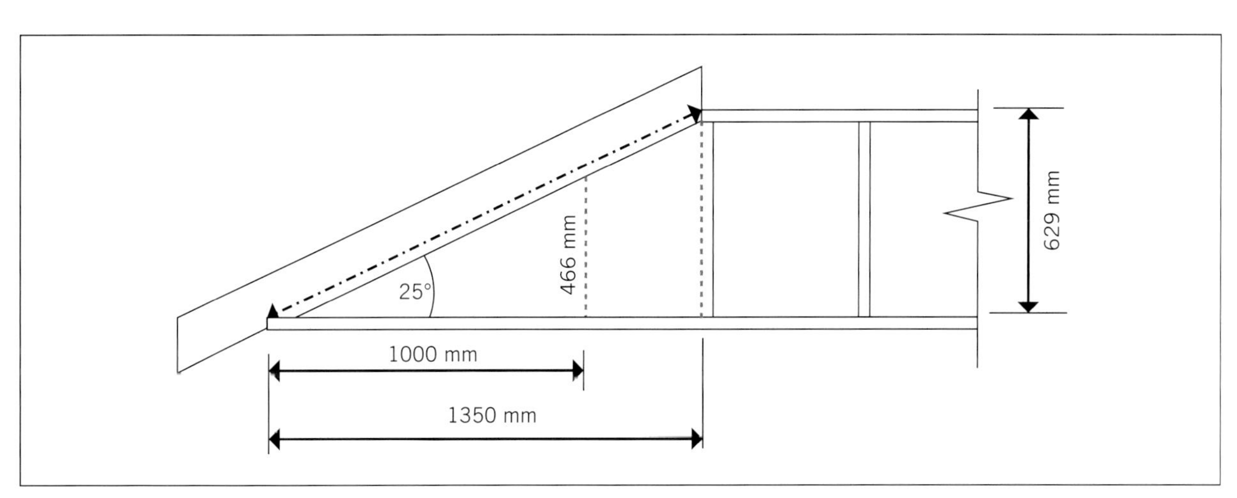

FIGURE 7.28 Height of the soldier wall

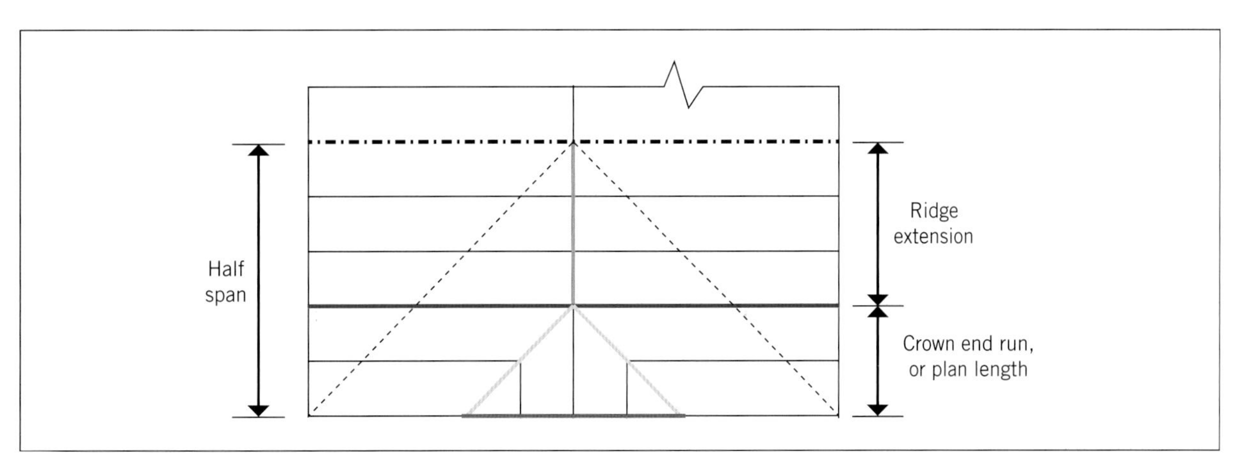

FIGURE 7.29 Finding the crown end plan length

In this case:

Crown end plan length = 2.250 − 1.350
= 0.900 m

Crown end set-out length = plan length × CR length/m run (rafter factor)

Crown end set-out length = 0.900 × 1.103
= 0.993 m

Remember, this is your set-out length. As with a normal hipped roof, you must make a horizontal reduction on your pattern of half the thickness of the common rafter material (see Figure 7.30). Positioning of the crown end rafter is shown in Figure 7.31.

The hips

As with a standard hipped roof, the hip factor is multiplied by the run of the appropriate rafter to gain

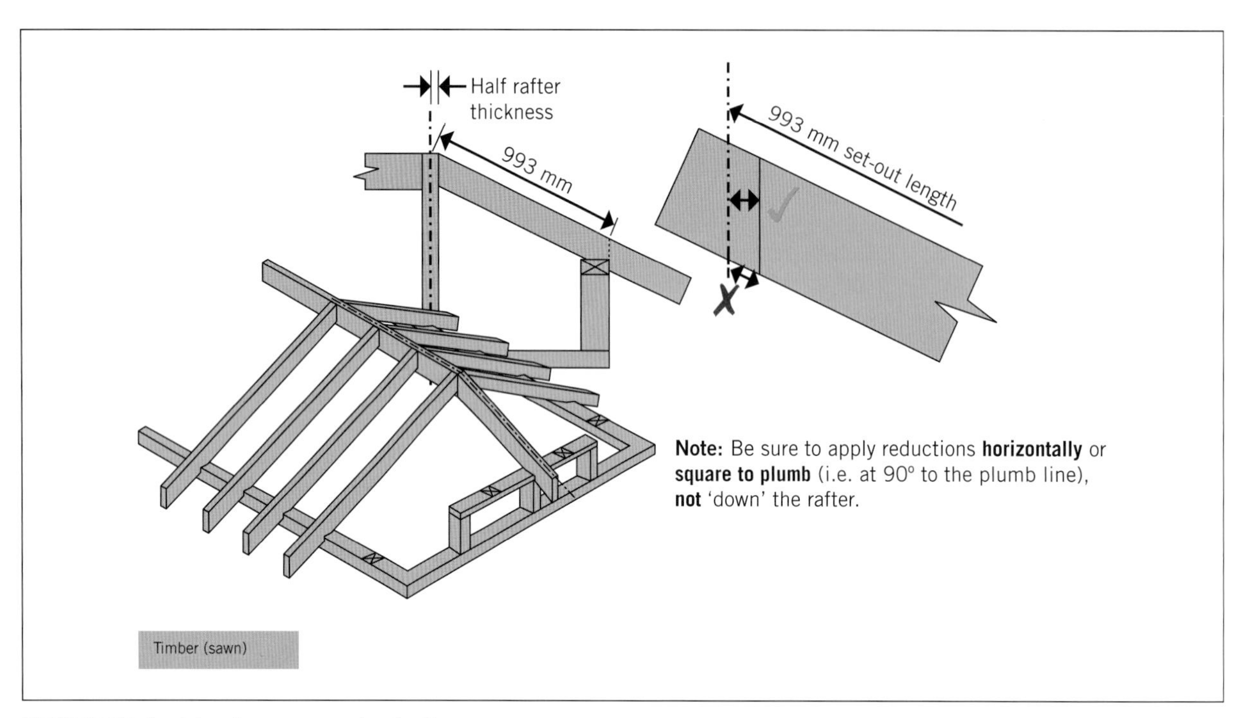

FIGURE 7.30 Applying the crown end reduction

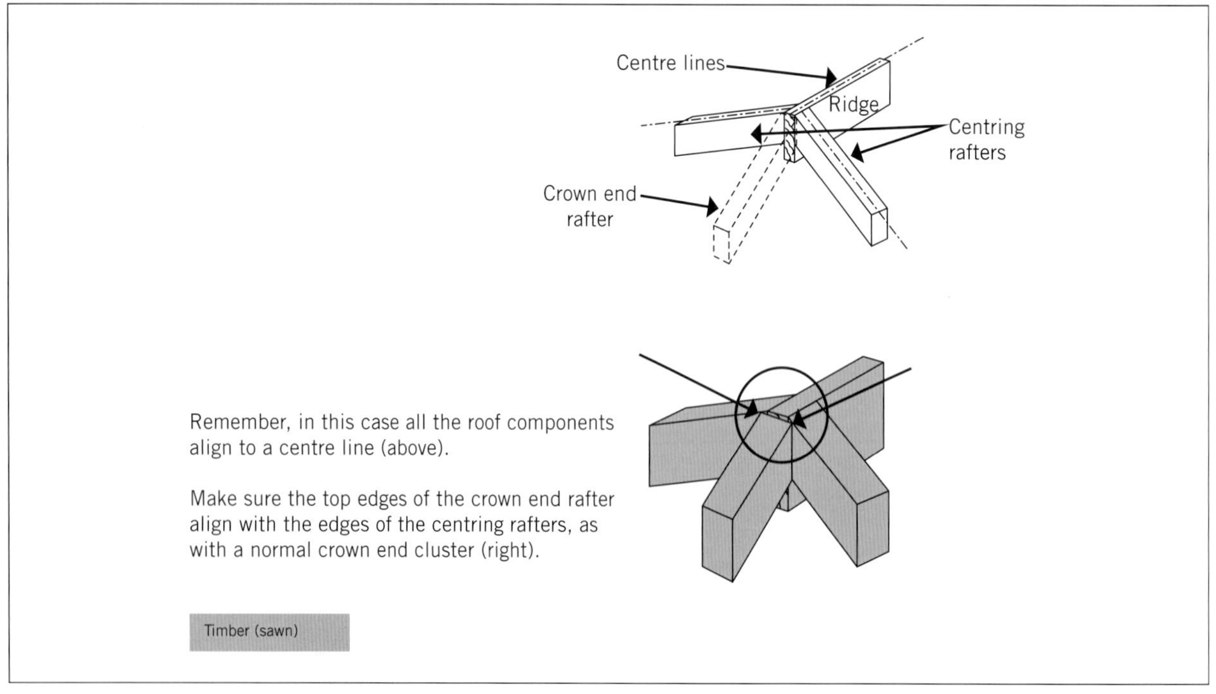

FIGURE 7.31 Positioning the crown end rafter

the hip set-out length, or our sixth pillar. In a normal hipped roof, we would multiply by the half span; that is, the run of the common rafter. In this case, the appropriate run, or plan length, is that of the crown end rafter found previously:

Hip set-out length = crown end plan length × hip factor

Hip set-out length = 0.900 × 1.489

Set-out is as per a normal hipped roof and, unlike the gambrel, you must make your standard reduction at the top of the hip of half the mitre thickness of the common rafter. See Figure 7.32 for further explanation.

The application of reductions and the full hip set-out are shown in Figure 7.33. As stated earlier, this is the same approach as would be used with a hip rafter for a standard hipped roof. Positioning of the hips is shown in Figure 7.34.

Cutting and installing creepers

Traditionally creepers are set out using a complicated reduction method to locate the long point of the first (longest) creeper. Only after this may the standard creeper reduction be used. The author has developed a method that makes finding the lengths of creepers for all roofs far simpler, and for jerkin head roofs, particularly so. Once more, it is about determining the plan length, or run, of a component, and then multiplying this distance by the appropriate factor (in this case, the rafter factor).

Figure 7.35 shows the measurements required for this approach. Normally these would be taken directly off the wall plates; however, it is possible to calculate them. Calculating these lengths is not difficult and it is important to know when dealing with larger roofs. The formulas for calculating these lengths are given after

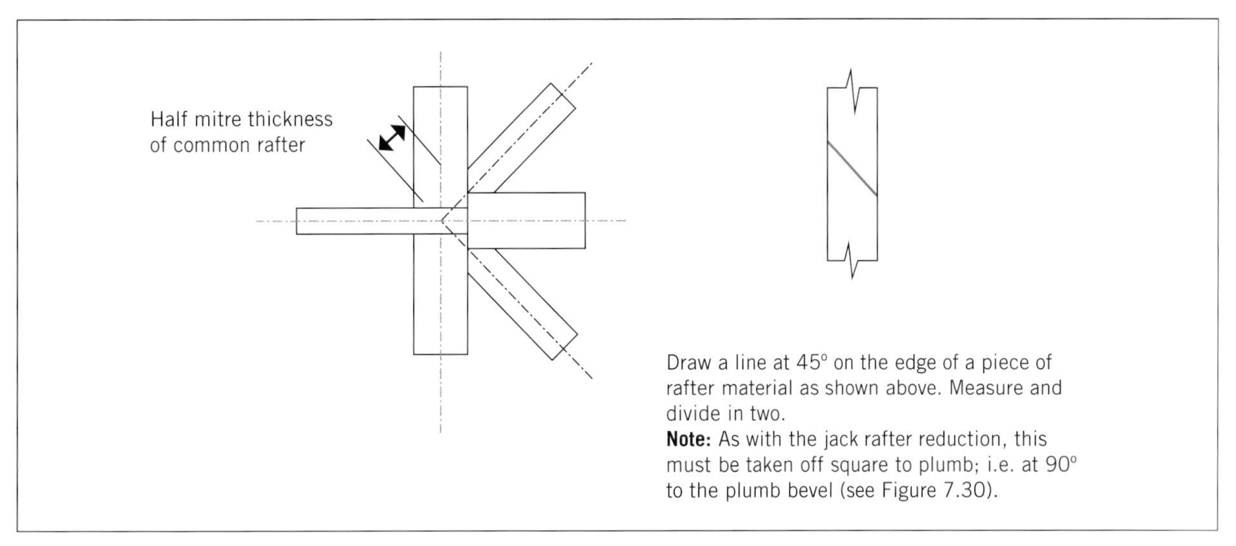

FIGURE 7.32 Hip reductions

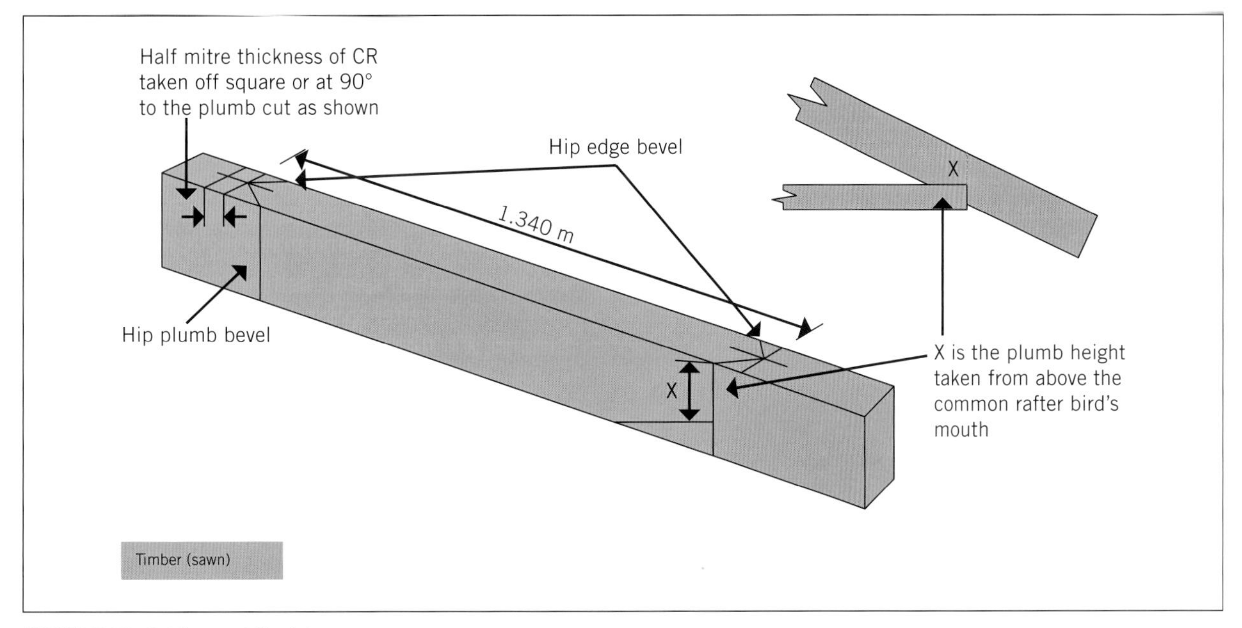

FIGURE 7.33 Setting out the hip

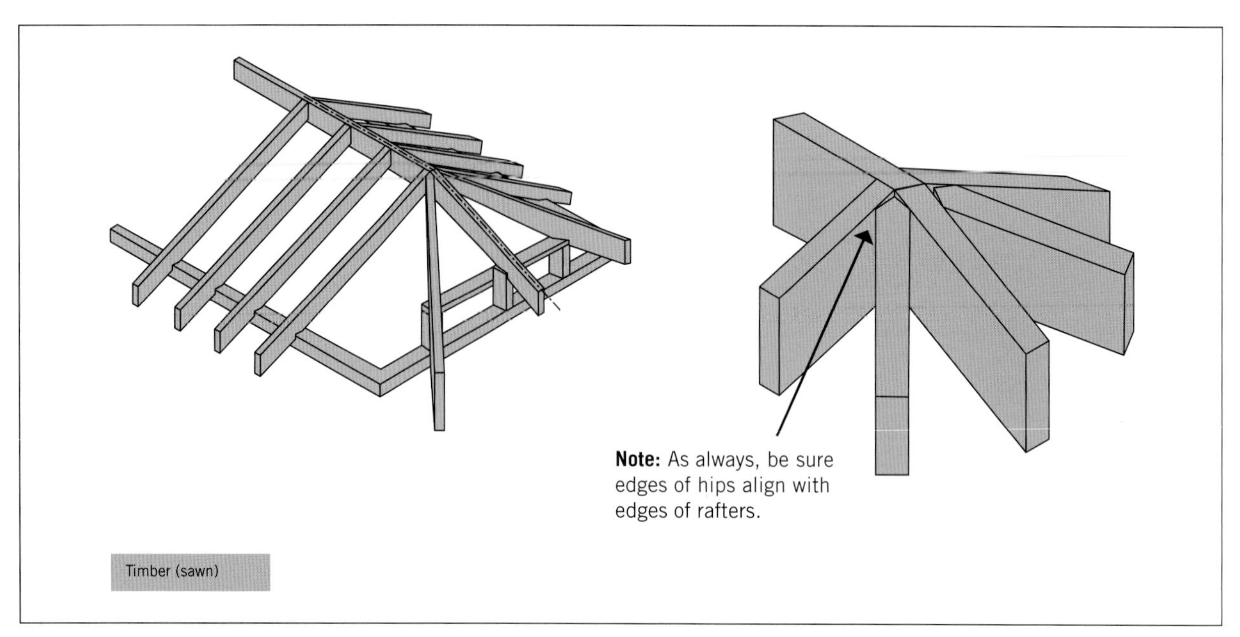

FIGURE 7.34 Positioning the hips

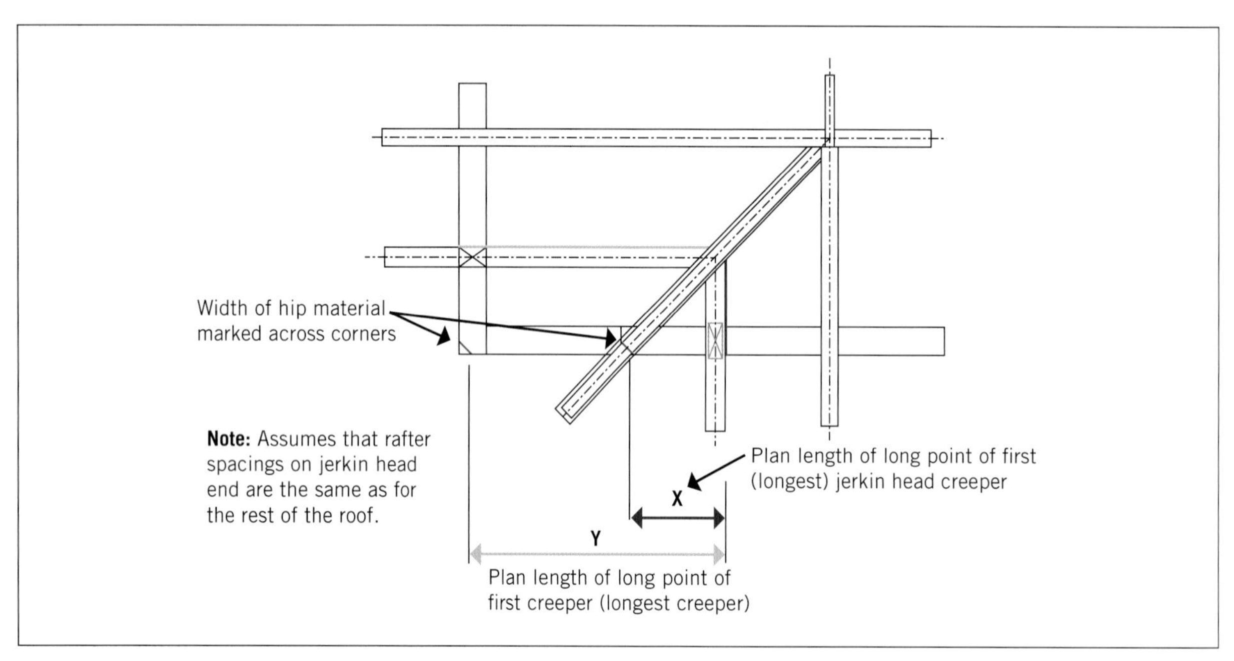

FIGURE 7.35 Determining the plan length, or run, of the long side of the first creeper

those for finding the set-out length for the first creeper. All other creepers may be set out from these long points (see Figure 7.36) using the standard creeper shortening:

Creeper shortening = rafter spacing × rafter factor

In this case:

Creeper shortening = 0.450 × 1.103
Creeper shortening = 0.496 m

Note: This distance is applied 'down' the rafter (see Figure 7.36).

Set-out length of first creeper = Y × rafter factor
Set-out length of first jerkin head creeper = X × rafter factor

These distances are usually direct measured; however, they may be found mathematically by:

Distance Y = (half span – half mitre thickness of hip) – (rafter spacing – half rafter thickness)
Distance X = (half soldier wall width – half mitre thickness of hip) – (rafter spacing – half rafter thickness)

It is not necessary to carry out these calculations here, as you are adequately equipped from the previous workings to undertake this yourself. However, it may be helpful to know the shorthand method of finding the half mitre thickness of a component when not direct measuring it:

Half mitre thickness = component thickness ÷ 1.414

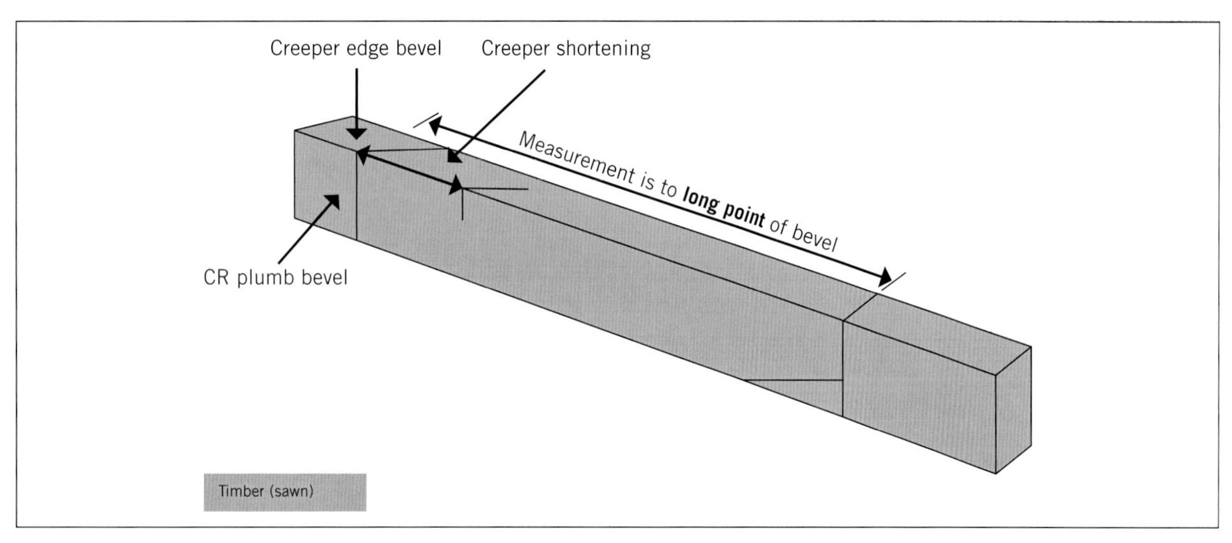

FIGURE 7.36 Setting out the first (or longest) creeper and applying the creeper shortening

This completes the main framing of the jerkin head roof (see Figure 7.37). All other components, such as purlins, strutting, barge trimmers and the like, should be familiar to you.

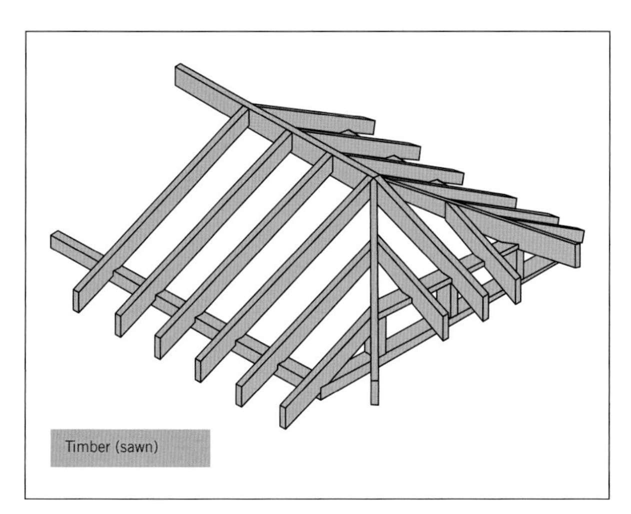

FIGURE 7.37 Completed roof framing for the jerkin head

LEARNING TASK 7.3

RESEARCHING JERKIN HEAD ROOFS

Investigate and report on the jerkin head roof form and, if possible, provide some examples of heritage-listed houses with jerkin head roofs in your state.

Key points include:

- when they became popular in Australia
- alternative names
- key design features
- examples of heritage-listed houses.

To assist with your research, conduct a search using the keyword phrase 'jerkin head' at: http://www.environment.gov.au/cgi-bin/ahdb/search.pl.

Also look at the Buildsum YouTube video at: https://www.youtube.com/watch?v=McIttRIFtB4.

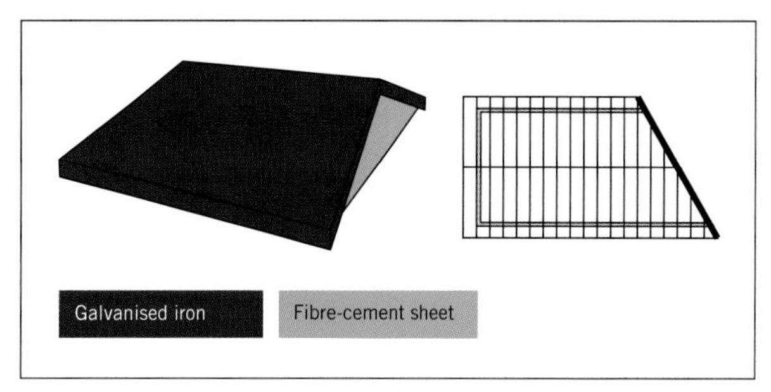

FIGURE 7.38 Skewed gable

Skewed gables

Buildings are not always built as square or rectangular structures. Sometimes the wall at the end of a building will run at an angle other than 90°, making what is known as a 'splayed' or '**oblique**' end. This has implications for roof design, and requires more thought in the cutting of either gable or hip rafters and creepers.

The **skewed gable** (see Figure 7.38) and the oblique hip (see later in this chapter) fall under the category of 'splayed ended roofs'. In this form, the skewed gable effectively has creeper rafters to be considered. For the moment, only roofs with equal pitches will be dealt with, leaving the issues arising from unequal pitches to a later section of this chapter.

As the diagram in Figure 7.39 demonstrates, the skewed gable roof is aptly named as the gable end is out of square to the building proper; that is to say, it is skewed. The amount of skew is not relevant to the setting out or construction as the mathematics and geometry remain the same. Likewise, the main roof may

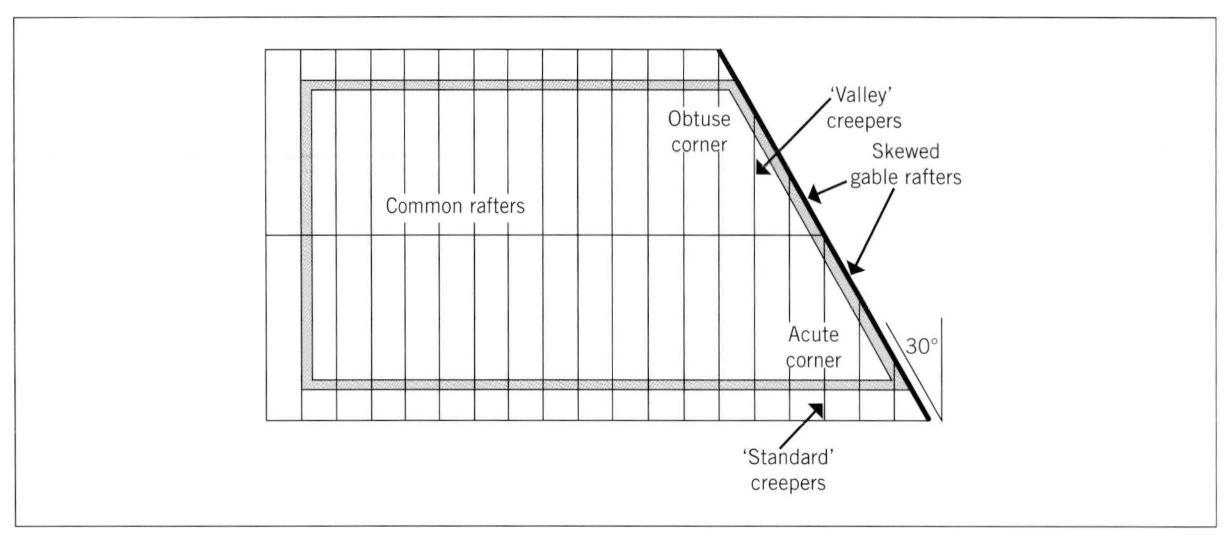

FIGURE 7.39 Framing plan for skewed gable

be constructed of unequal pitches without altering the geometry that will be shown here.

The main part of this project is simply a gable roof, which has been described previously in Laws (2020). This section will therefore focus upon the challenges offered by the skewed end.

As can be seen in Figure 7.40, for the ridge to remain level the rise at both gable ends must be equal. However, while the gable rafters at both ends have the same rise, the skewed gable rafters have a greater run, or plan length; that is, the blue line is longer than the green one. This means that different plumb and level bevels need to be found for these components.

In addition, the skewed gable rafters are not square to the ridge or wall plates. This requires us to find an edge bevel, much as we would for a hip or valley. Likewise, we have creeper edge bevels to find, as shown in Figure 7.39.

Doing the calculations

Once more, the roof being considered will have characteristics similar to that used in the revision exercises:

Pitch = 25°
Span = 4500 mm
Eave width = 600 mm
Rafter sectional size = 125 × 45 mm
Hip sectional size = 175 × 35 mm
Skew angle = 30°

In this exercise only the first four pillars need to be determined, along with the common rafter plumb and level bevels (see the earlier section 'The mathematics: The "Seven Pillars" revisited').

From the plan we must now calculate the following information:

- total rise of the roof
- run, or plan length, of the skewed gable rafters.

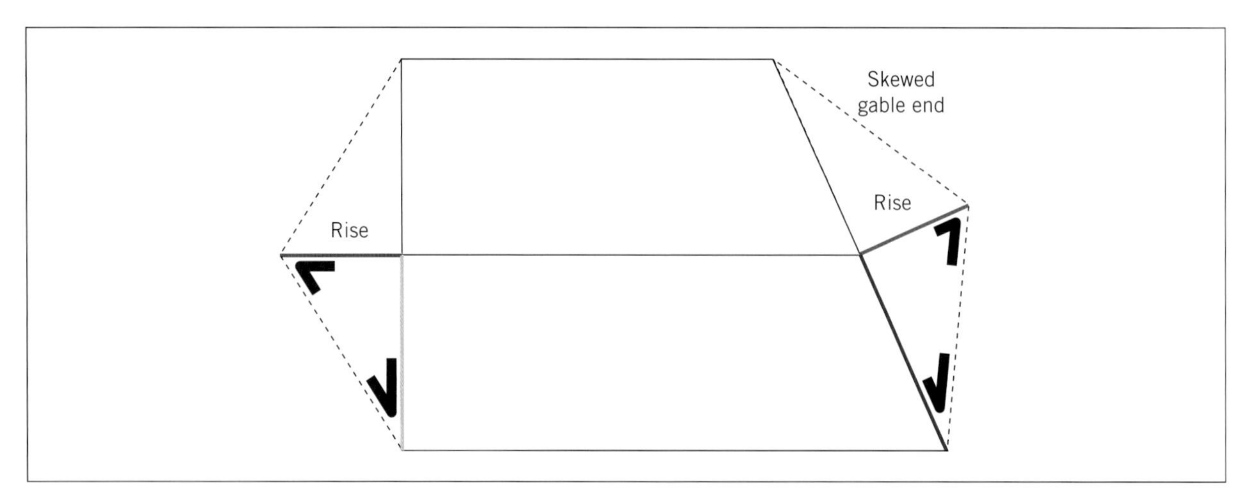

FIGURE 7.40 Skewed gable end elevations

From this information we will then go on to develop:

1 skewed gable length/m run of CR (effectively the skewed gable factor)
2 skewed gable set-out length
3 skewed gable order length
4 skewed gable plumb and level bevels
5 skewed gable edge and backing bevels
6 creeper edge bevel.

Total rise

Total rise = half span × CR rise/m run

In this case:

$$\text{Total rise} = 2.250 \times 0.466 = 1.049$$

Run of skewed gable rafters

The run is often directly measured off the roof. However, it is important, particularly for quantities and estimating purposes, to be able to do this mathematically. There are numerous mathematical paths that may be taken, of which two will be shown here. Method 1, where the angle of the skew is known, is the shorter route. Method 2 is used when only the skew offset length is known.

Method 1 (skew angle known)

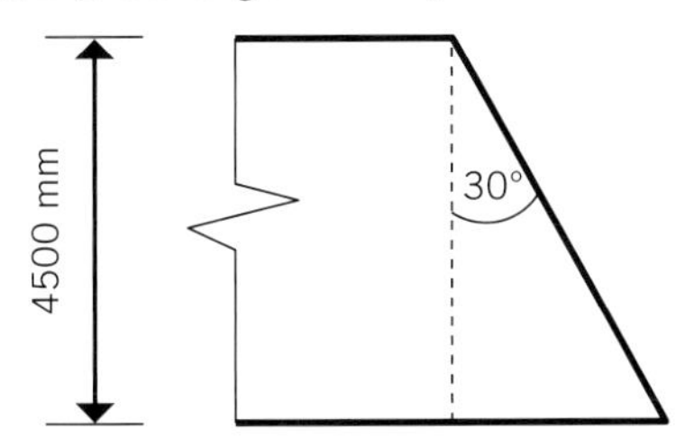

Length of skewed gable end = span ÷ cos θ°

In this case:

$$\text{Length of skewed gable end} = 4500 \div \cos 30^\circ = 5.196\text{ m}$$

Method 2 (skew offset length known)

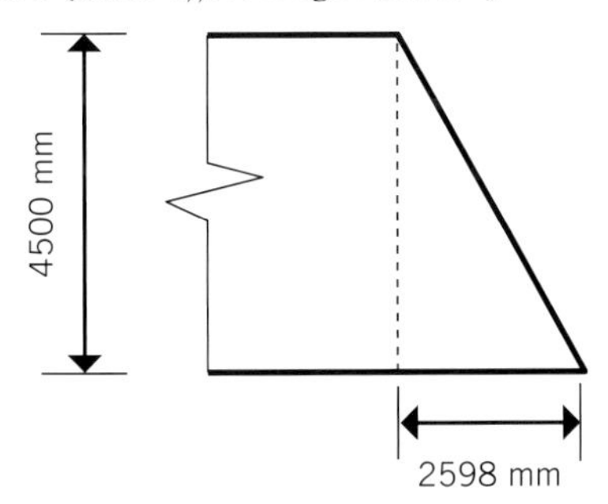

$$\text{Length of skewed gable end} = \sqrt{\text{span}^2 + \text{offset}^2}$$

In this case:

$$\text{Length of skewed gable end} = \sqrt{4.500^2 + 2.598^2} = 5.196\text{ m}$$

Divide by two to get the run of a single skew gable rafter; that is, 2598 mm. In this particular case, the answer acts as its own 'proof': see if you can figure out why.

Finding the other lengths

It is possible to forgo some of what follows and use a single series of calculations for the length of each component. The issue with this approach is that you need to repeat the full calculation on each occasion. For a one-off, this is acceptable, but for more extensive work the adaptation of the 'Seven Pillars' is preferred. It helps to think of the skewed gable rafters as hips or valleys; in fact, our first action is effectively to find the hip factor.

1 *Skewed gable length/m run of CR* (see the earlier section 'The mathematics: The "Seven Pillars" revisited' Pillar 5: hip factor)
We know that the skewed gable travels 5.196 m in plan, while the CR travels 4.5 m.
Therefore:

Skewed gable run/m run of CR = length of the skewed end ÷ span

In this case:

$$\text{Skewed gable run/m run of CR} = 5.196 \div 4.5 = 1.155\text{ m}$$

CR rise/m run = 0.466 m

Therefore the calculation for skewed gable length/m run of CR (see Figure 7.41) is:

$$\text{Skewed gable length/m run of CR} = \sqrt{1.155^2 + 0.466^2} = 1.245\text{ m}$$

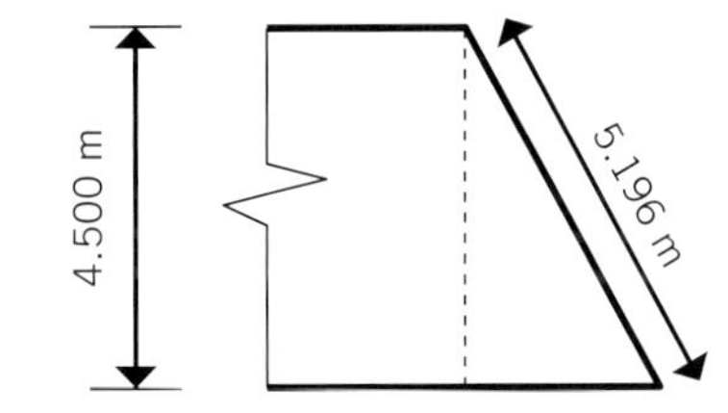

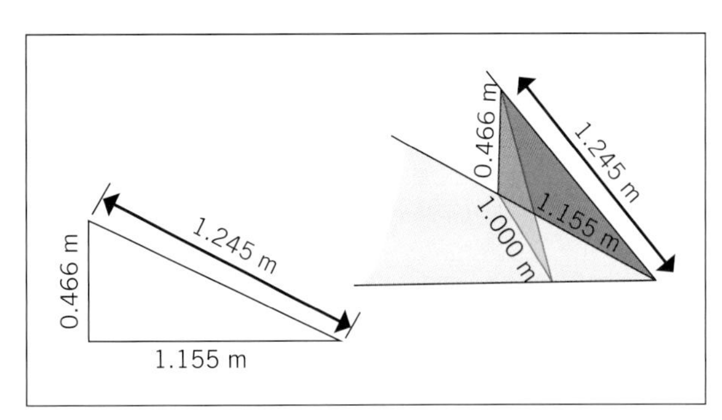

FIGURE 7.41 Finding skewed gable length/m run of common rafter

2 *Skewed gable set-out length*
As with a hip, the set-out length is:

Skewed gable set-out length = half span × skewed gable length/m run of CR
= 2.250 × 1.245
= 2.801 m

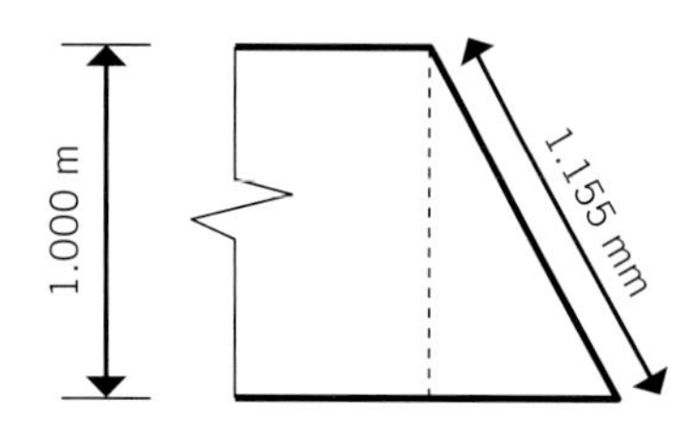

3 *Skewed gable order length*
Like a hip, the order length is:

Skewed gable order length
= [(half span + eave width) × skewed gable length/m run of CR] + material depth
= [(2.250 + 0.600) × 1.245] + 0.125
= 3.673 m (order 3.9 m)

4 *Skewed gable plumb and level bevels*
Figure 7.42 shows these bevels are found by a right angled triangle, matching the rise to the run of the skewed gable. By scaling these dimensions, this may be done on a steel square or a piece of timber, or a full-sized set-out on a concrete floor. If you need to know the angle in degrees, then use:

Level bevel = $\tan^{-1}$ (rise ÷ skewed gable run)
Plumb bevel = 180° – level bevel

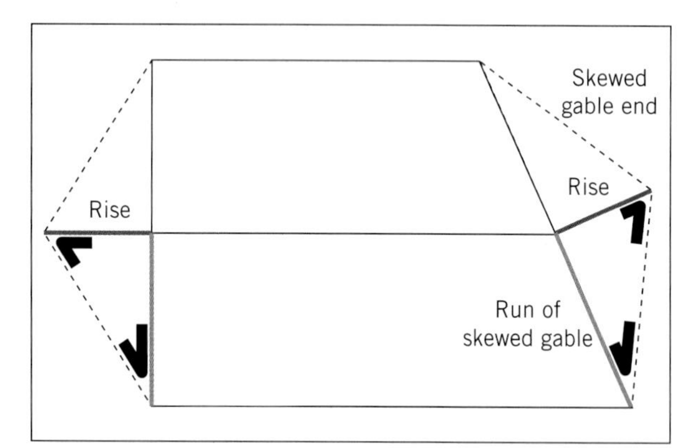

FIGURE 7.42 Skewed gable rafter plumb and level bevels

5 *Skewed gable edge and backing bevels*
As mentioned previously, the skewed gable can be thought of as a hip, so refresh your understanding of the hip edge bevel described in the earlier section 'The mathematics: The "Seven Pillars" revisited'. With this concept in mind, the edge bevel is found geometrically as shown in Figure 7.43: the two blue lines represent the skewed gable edge and, as with the hip edge bevel, a line drawn 'square' to these is

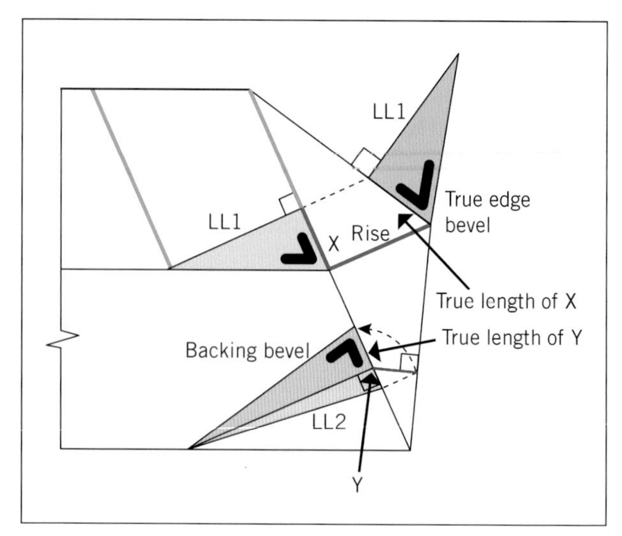

FIGURE 7.43 Edge and backing bevel geometric layout

level (LL1). The red line (X) is not a true length as it is actually running up the rafter. The true length of X is seen in the elevation of the skewed gable end as shown. This being the case, the bevel shown in grey is not true. The true length (LL1) and the true length of X are now matched together at 90°, thereby providing the edge bevel required.

Figure 7.44 shows how this concept is applied directly to a piece of rafter material. Setting your power saw to the angle of the skewed end, in this case 30°, and cutting to the plumb line will also achieve the same result.

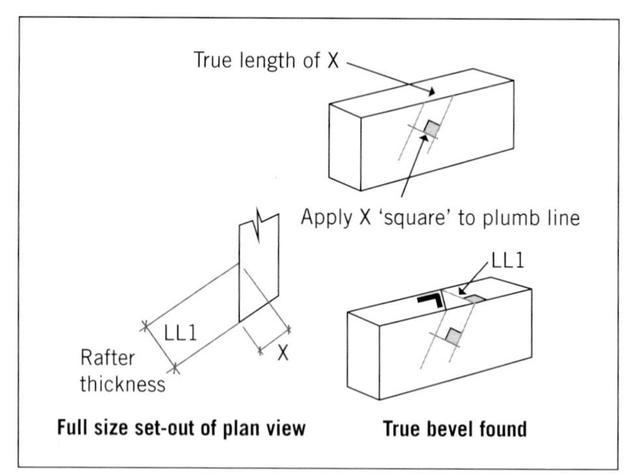

FIGURE 7.44 Obtaining the edge bevel – direct method

The skew gable backing bevel is used to back off the edge of the rafter so that it lies in line with the roof surface. This is found graphically by cutting a section through the rafter, then rolling it flat so that you can see the angle in its true form. In Figure 7.43, the pale, slimmer triangle is a plan view of this section. The darker *recoloured* one is the true shape of this triangle laid flat.

6 *Creeper edge bevel*
This is like a standard creeper meeting a hip. Graphically, its development is very similar, as Figure 7.45 demonstrates. Alternatively, it may be done on a steel square by matching the true length of CR to the skewed gable offset or, practically, by adapting the skew gable exercise using the information in Figure 7.46 (just two of the many means available).

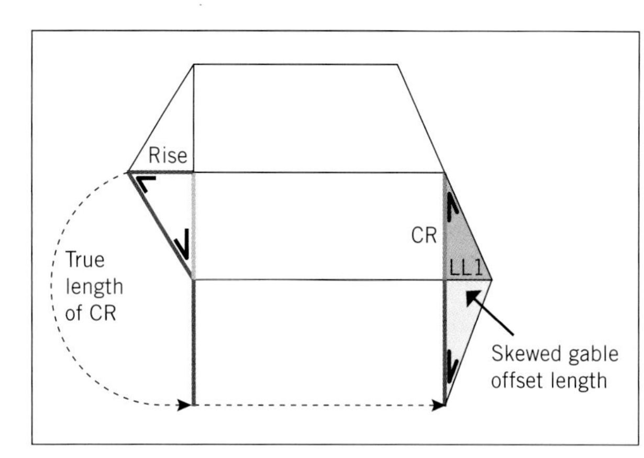

FIGURE 7.45 Edge bevel geometric layout

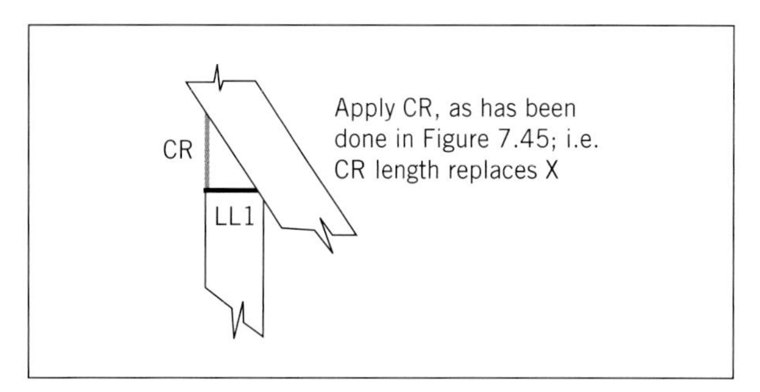

FIGURE 7.46 Full-sized plan view of creeper and skewed gable junction

Setting out and cutting components

First, set out the wall plates. The longest wall plate should have half the skewed gable offset distance clearly marked. Be sure to position the last common rafter at least two rafter thicknesses back from this point. This is transferred, along with all the usual rafter positions, to the ridge, which provides the centre line of the skewed end cut (see Figure 7.47).

Do not cut the ridge. Cutting the ridge leaves no margin for adjusting to plumb should some minor error have crept into your work.

Now cut and fit your common rafters and ridge as per normal practice (see Figure 7.48). Be sure to brace.

Next, set out the skewed gable rafters. Set out the **acute** rafter first (the one that fits to the narrow or pointed end of the gable). Set out on the inside face, as

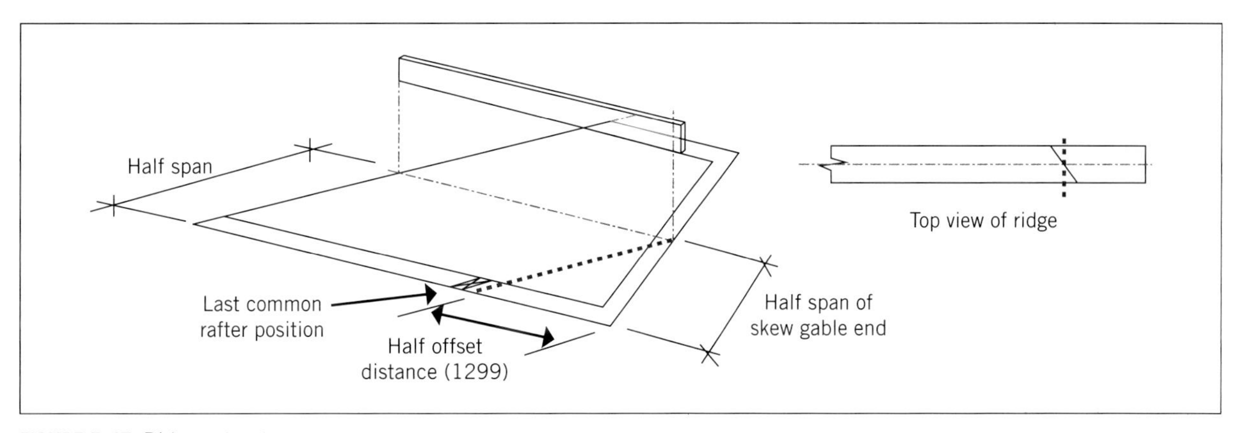

FIGURE 7.47 Ridge set-out

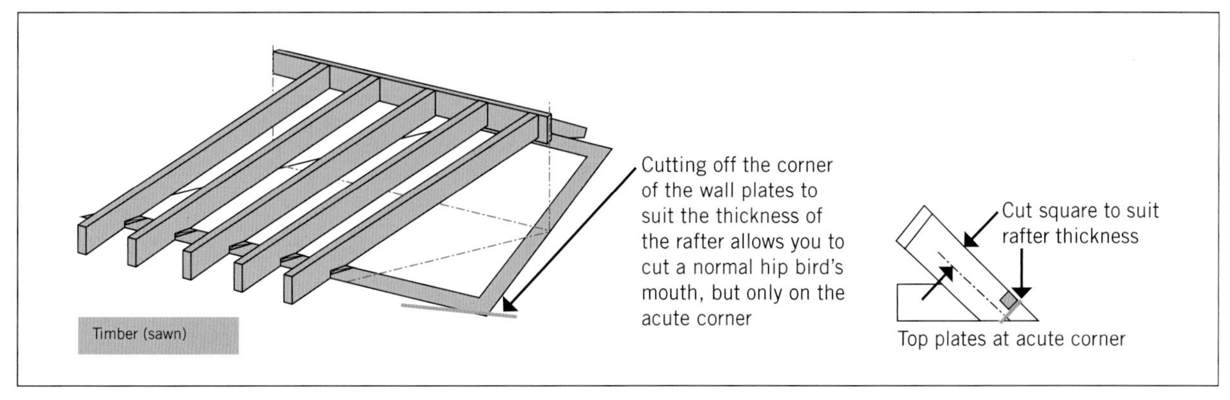

FIGURE 7.48 Establishing common rafters and ridge; trimming the acute corner to accept the hip bird's mouth

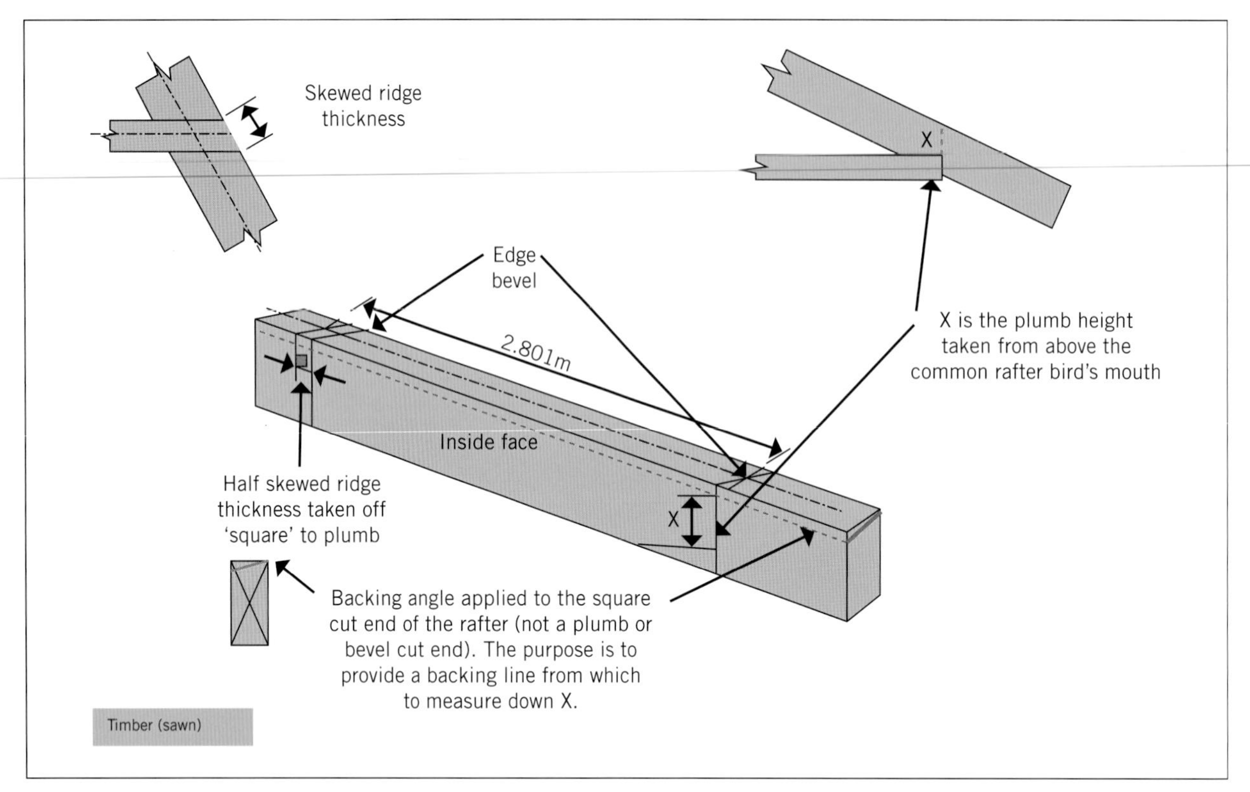

FIGURE 7.49 Acute skewed gable rafter set-out

shown in Figure 7.49. The bird's mouth to this rafter is done as you would for a hip. Cut your bird's mouth square across the rafter as normal. Then cut the acute corner of the top plates, as is done for a hipped roof (see Figure 7.48).

The **obtuse** gable rafter (the gable rafter to be fitted to the wide-angled side) is now set out in exactly the same way as for the acute rafter, but with one subtle, yet important, addition. This rafter must have a skew cut bird's mouth. To achieve this, mark the plumb height X on both sides of the rafter. On the outside face of the rafter (the side facing away from the building), X is measured down from the backing bevel line, as with the previous rafter. On the inside face, X is measured down from the top edge of the rafter (see Figure 7.50).

Install these rafters and proceed to setting out the creepers.

To set out the creepers, the length to the longest point of the first creeper, and the creeper shortening, must be found. In this case, however, the horizontal reduction in rafter length does not equal the rafter or creeper spacing. But we can still use the rafter spacing to help us find the horizontal reduction by means of a ratio factor. This factor, or multiplier, tells us how much a creeper runs horizontally for every metre it moves further away from the corner. Once we have that measurement, it is a simple matter of multiplying the run found by the rafter factor (CR length/m run).

First, the ratio between the span and the offset of the skewed end is found.

Creeper skew factor = span ÷ offset distance

In this case:

$$\text{Creeper skew factor} = 4.500 \div 2598 = 1.732\,\text{m}$$

To find the horizontal run to the long point of the longest creeper, measure directly off the plate from the face of the skewed gable rafter to the long side of the creeper position (see Figure 7.51). Multiply this by the creeper skew factor. This is the run of the long side of the first (or longest) creeper.

Run (to long point) of first creeper = Creeper skew factor X × Y

Assume measured Y in this case to be 0.950:

$$\text{Run (to long point) of first creeper} = 1.732 \times 0.950 = 1.645\,\text{m}$$

The long point length of the first creeper is found by:

$$\text{Long point length} = \text{run of first creeper} \times \text{rafter factor} = 1.645 \times 1.103 = 1.815\,\text{m}$$

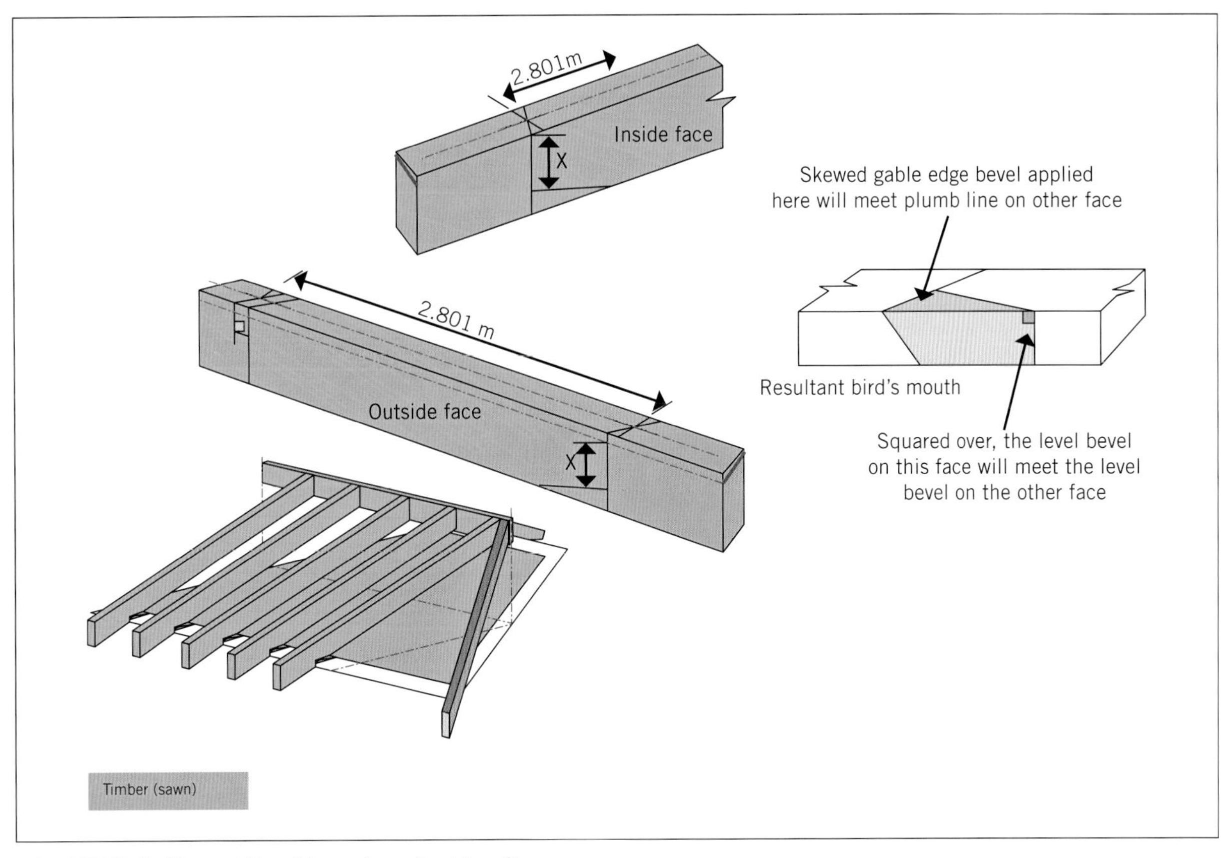

FIGURE 7.50 Setting out the obtuse skewed gable rafter

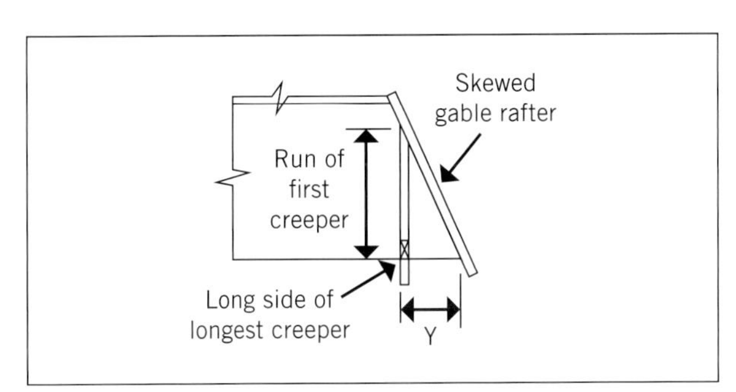

FIGURE 7.51 Measuring Y – the offset of the first creeper

The creeper shortening is now found in a similar manner:

Creeper shortening = rafter spacing × creeper skew factor × rafter factor

In this case:

$$\text{Creeper shortening} = 0.450 \times 1.732 \times 1.103 = 0.860\,\text{m}$$

Use this information to set out the creepers for the acute corner, as shown in Figure 7.52.

The same approach is used to find the lengths of the creepers for the obtuse corner. To find the long side of the longest creeper, direct measure the offset from the inside face of the skewed gable rafter (distance Y*), multiply by the creeper skew factor (in this case 1.732 m), and then by the rafter factor (in this case 1.103 m). Once again, all the other creepers are reduced by this length in multiples of the creeper shortening (in this case 0.860 m). No reduction is required for the ridge thickness as your measurement was taken along the face of the ridge to the face of the skewed gable rafter (see Figure 7.53).

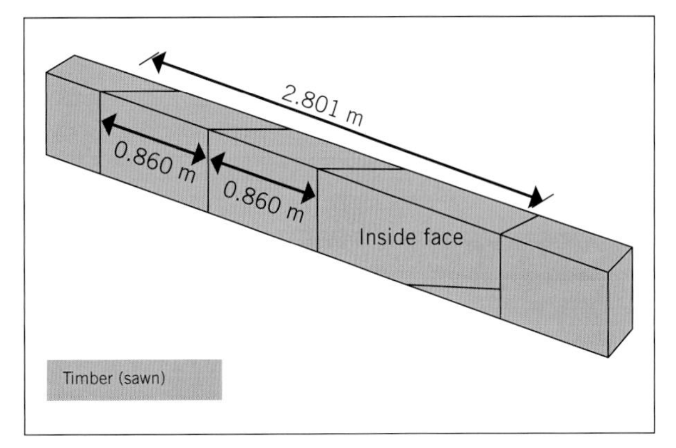

FIGURE 7.52 Setting out the acute creepers

*Note: Distance Y can be calculated if necessary. While space prohibits it being covered here, an adaptation of the calculations shown earlier in Figure 7.35 will assist you in calculating it.

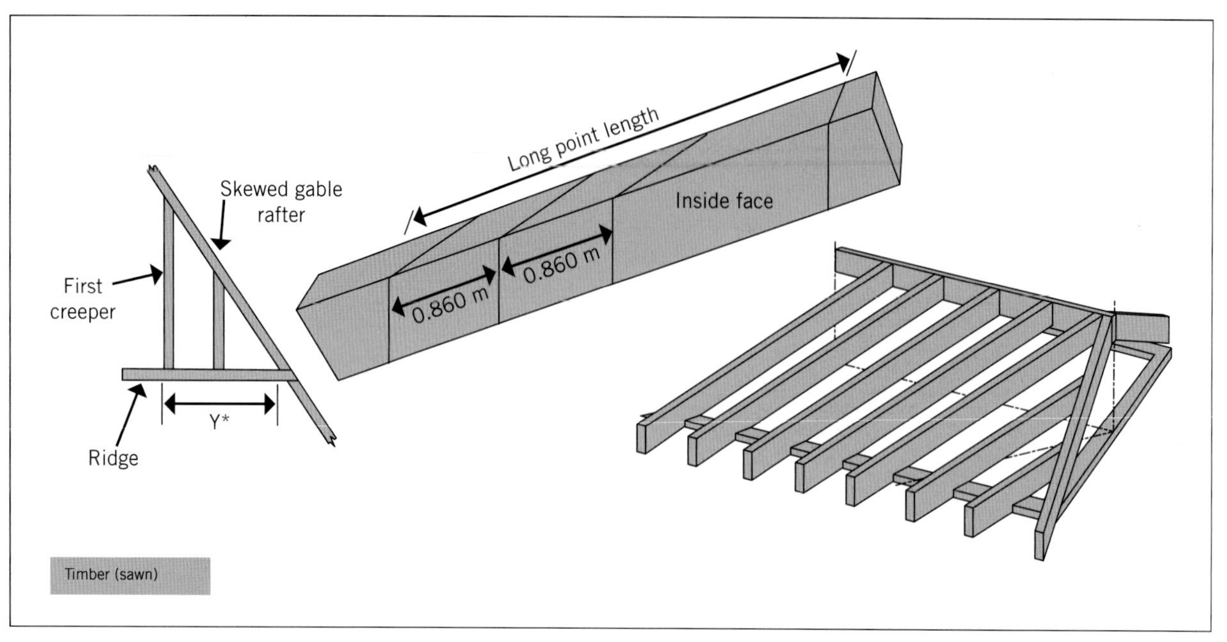

FIGURE 7.53 Setting out the obtuse creepers

LEARNING TASK 7.4

RESEARCHING SKEWED GABLE ROOFS

Watch these videos on skewed gables on the Buildsum YouTube channel and answer the following questions:

- Skew gable members and bevel positions: https://www.youtube.com/watch?v=8KY4yNicNus
- Skew gable bevel development: https://www.youtube.com/watch?v=pEJKQ50XApQ.

1 Apart from the plumb and level bevels required for common rafters, what is the extra bevel called in setting out and cutting creeper rafters?
2 Why is it necessary to have a backing angle on the top edge of the skewed gable rafter?
3 On the sketch below, set out the plumb cut and level bevel for the skewed rafters.

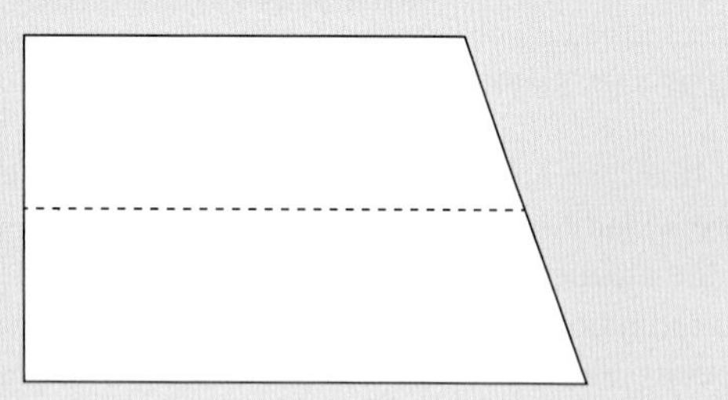

Oblique hips

The oblique hip (see Figure 7.54), the second splayed ended roof, is built over a floor plan similar to the skewed gable, so we will use the same plan, with the same roof characteristics, for our discussion (see Figure 7.55):

Pitch = 25°
Span = 4500
Eave width = 600
Rafter sectional size = 125 × 45
Hip sectional size = 175 × 35
Oblique angle = 30°

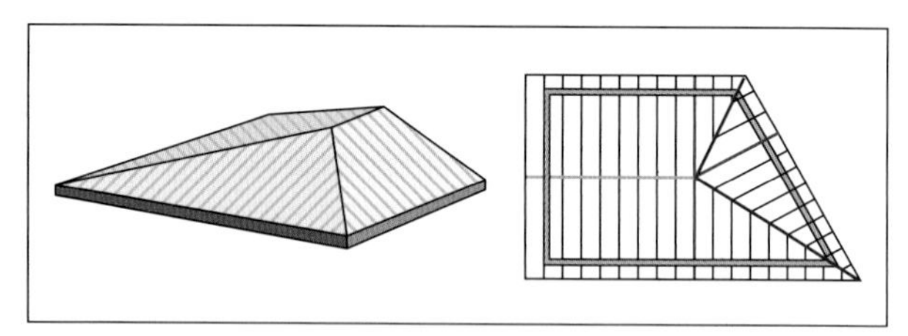

FIGURE 7.54 Oblique hip

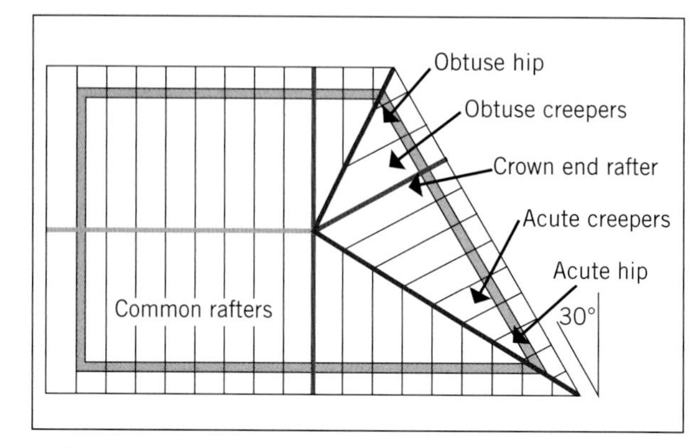

FIGURE 7.55 Oblique hip roof framing

Once again, only the first four pillars need to be determined, although, as previously, the other three pillars offer guidance for the rest of the calculations (see the earlier section 'The mathematics: The "Seven Pillars" revisited'). You should also develop the common rafter plumb and level bevels at this point.

From the discussion of the skewed gable roof, we know the following:

$$\text{Offset distance} = 2598\,\text{mm}$$

$$\text{Oblique end length} = 5196\,\text{mm}$$

We also know that the hips on an equal pitched roof (the rafters to each roof surface are pitched at the same angle) bisect (i.e. cut in half) the corners they arise from. In this case, our offset angle is 30° and therefore our acute internal corner is 60° (90° – 30°): bisected, this means our acute hip runs at 30° to the wall plates (see Figure 7.56).

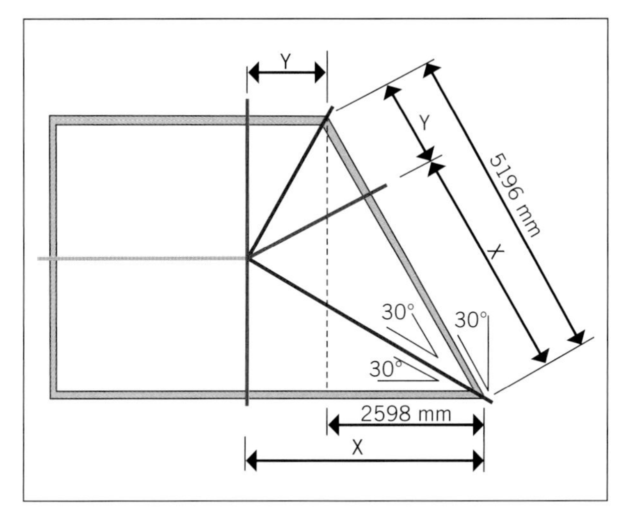

FIGURE 7.56 Practical set-out of oblique hip

Being equally pitched, the run of the crown end rafter (red line) is equal to that of the centring rafters (pink line).

With this information we are able to determine our three critical set-out positions: the two centring rafters (shown in pink); and the crown end (shown in red). Figure 7.56 plots this information on the plan.

Finding the centring rafter and crown end positions

Multiple mathematical and practical paths are available to solve this exercise.

Videos are available on the internet that demonstrate bisecting of angles for oblique hip roofs. Try using the search terms 'oblique hip geometry + video'.

A practical approach

A common practical approach is to bisect the two corners using timbers spanning the plates. The ends of these timbers are set at exactly equal distances from the corners, as shown in Figure 7.58. The distance from plate to plate is measured, then bisected (halved), and a clear mark made at this point. A string line is then fixed to the corner and passed under this mark to the wall plates opposite (see Figures 7.57 and 7.58). This is done both for the obtuse and acute corners. A further string line is set at the mid-span of the end wall (or any other convenient, or contrived, point if no end wall is as neatly located as shown here) and pulled through to the mid-span of the oblique end. These three lines should cross exactly at the one mid-point. Should any minor error be present, check the accuracy of the ridge string line, then adjust the two 'hip' lines to suit.

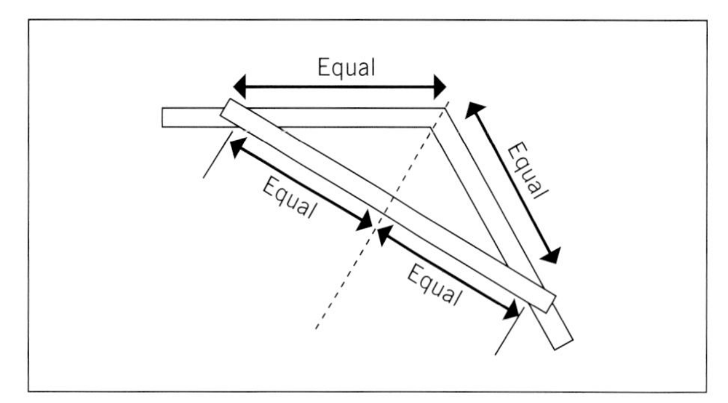

FIGURE 7.57 Bisecting the oblique hip – obtuse corner

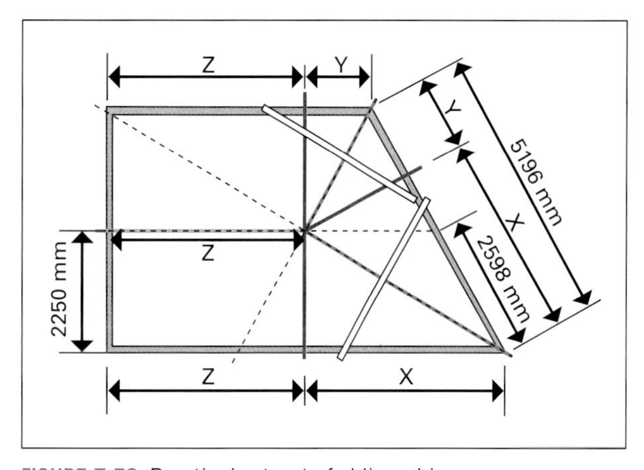

FIGURE 7.58 Practical set-out of oblique hip

The distance Z is now physically measured and transferred to both side wall plates, as shown in Figure 7.59. This locates the centre of the centring rafters.

The distances X and Y can now be measured, which locates the centre of the crown end rafter.

A mathematical approach

Figure 7.59 offers us several right angled triangles and their angles (given or derived). From this information we can look for simple paths by which to locate the centre lines of both the centring rafters and the crown end. The approach shown here is quick, simple and accurate.

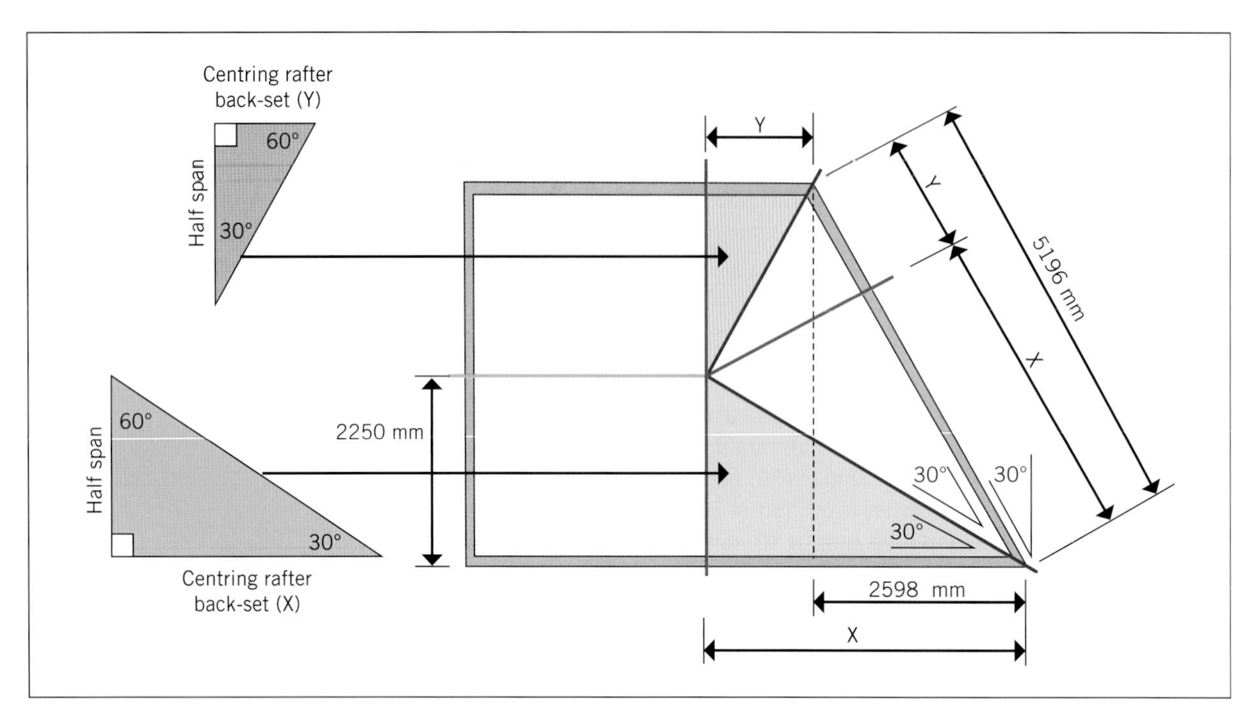

FIGURE 7.59 Mathematical approach to oblique hip set-out

The centring rafter back-set (X **or** Y) is found by solving **either one** of the triangles shown above (or both as a means of providing a check). We will solve for X.

$\tan\theta$ = opposite ÷ adjacent

=>

$\tan\theta$ = half span ÷ X

=>

Acute centring rafter back-set (X) = half-span ÷ $\tan\theta$ where θ = bisected angle of acute corner (in this case 30°)

In this case:

$$\text{Centring rafter back-set (X)} = 2.250 \div \tan 30^\circ$$
$$X = 3.897\,\text{m}$$

From Figure 7.60, X is also the centre line position of the crown end rafter. Y may now be found by:

Obtuse centring rafter back-set (Y) = centring rafter back-set (X) – oblique offset distance

In this case:

$$\text{Centring rafter back-set (Y)} = 3.897 - 2.598$$
$$= 1.299\,\text{m}$$

Check: X + Y = oblique end length

In this case:

$$3.897 + 1.299 = 5.196\text{ m – CORRECT!}$$

Set-out of the top plates may now be completely carried out.

Component lengths and bevels

The first four pillars provided the common rafter lengths and their plumb and level bevels. These may now be set out and cut. The crown end set-out length is the same as for the common rafters; however, as Figure 7.61 shows, the crown end has a different reduction at the ridge. It also has an edge bevel. The concepts used in the set-outs following should by now be familiar and demonstrate both a direct and a geometric development of both the reduction and the edge bevel.

Before we can proceed we must now find the following:

- total rise of the roof
- run, or plan length, of the hips.

From this information we will then go on to develop:

1. hip run/m run of CR (acute and obtuse hips)
2. acute hip length/m run of CR (acute and obtuse hip factors)
3. hip set-out lengths (acute and obtuse hips)
4. hip order lengths (acute and obtuse hips)
5. hip plumb and level bevels (acute and obtuse hips)
6. hip edge bevels (acute and obtuse hips)
7. creeper edge bevels (acute and obtuse hips)

Total rise = half span × CR rise/m run

In this case:

$$\text{Total rise} = 2.250 \times 0.466$$
$$= 1.049\,\text{m}$$

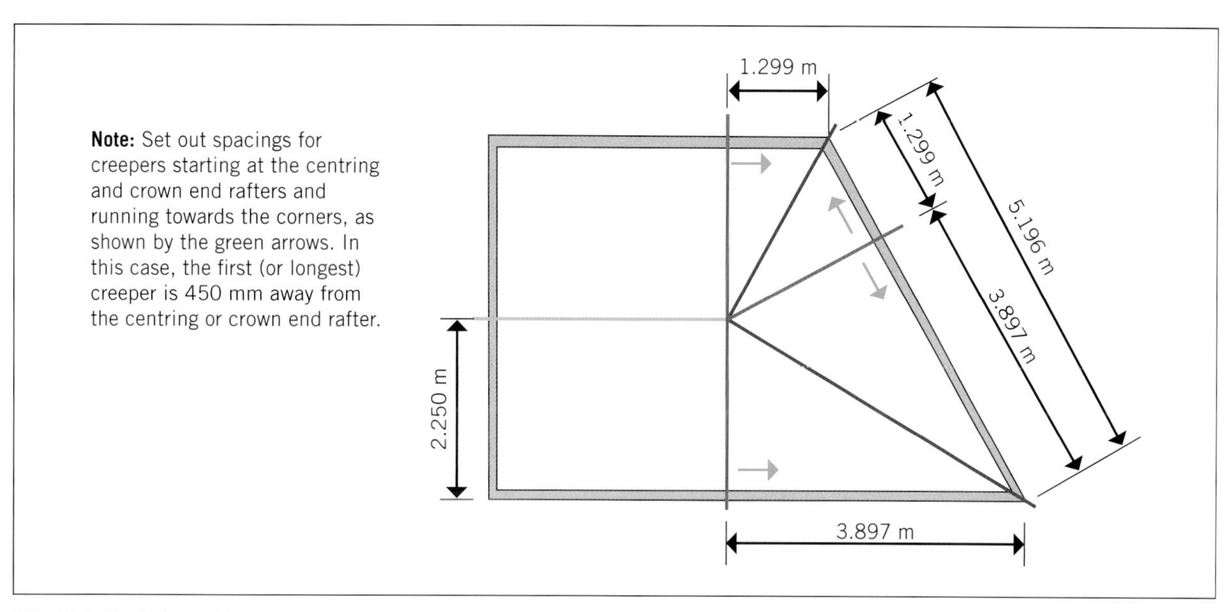

FIGURE 7.60 Oblique hip set-out dimensions

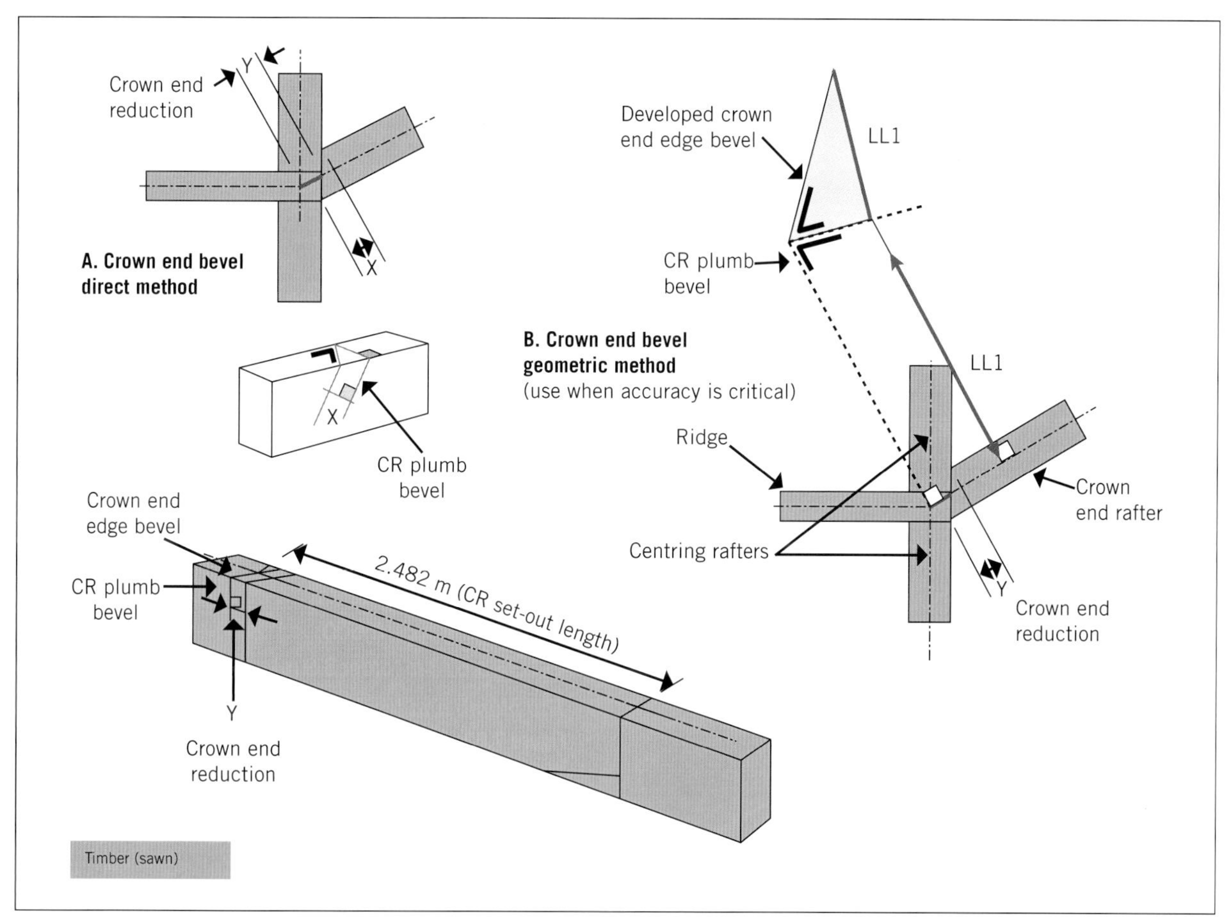

FIGURE 7.61 Set-out of crown end rafter

1 *Hip run/m run of CR (acute and obtuse hips)*
To build on an approach already developed in the skewed gable project, the full run of each hip will be found first. Once again, these may be measured directly off the project (following the string lines) or may be calculated. As we have all the lengths, it is easiest for most students to use Pythagoras' theorem (though trigonometry is just as practicable).

$$\textbf{Run of acute hip} = \sqrt{\text{half span}^2 + \text{acute back-set}^2}$$

In this case:

$$\text{Run of acute hip} = \sqrt{2.250^2 + 3.897^2} = 4.500 \text{ m}$$

$$\textbf{Run of obtuse hip} = \sqrt{\text{half span}^2 + \text{obtuse back-set}^2}$$

In this case:

$$\text{Run of obtuse hip} = \sqrt{2.250^2 + 1.299^2} = 2.598 \text{ m}$$

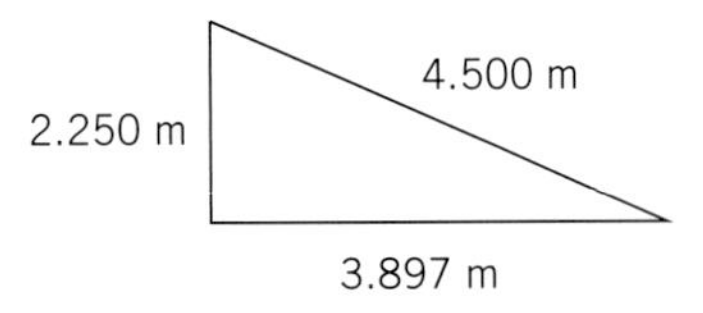

Note: Because this is based upon a 30° oblique end, you will notice a correlation between the span, half span and hip lengths. This will **not** occur when using angles other than 30° (but provides a neat check in this case).

The hip run/m run of CR may now be found for each of the hips in much the same manner as it was for the skewed gable length/m run of CR; that is:

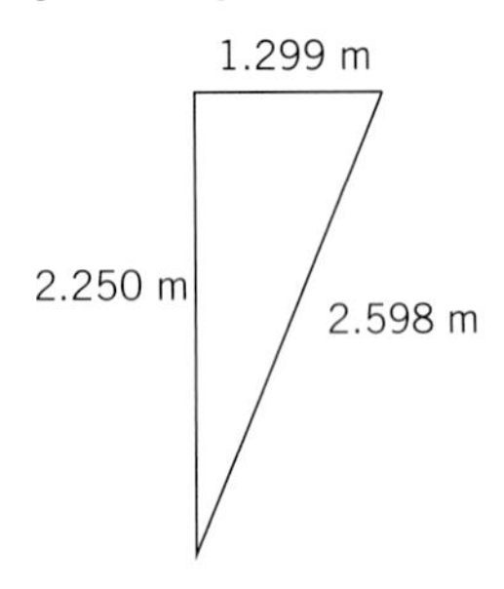

Hip run/m run of CR = hip run ÷ half span

In this case:

$$\text{Acute hip run/m run of CR} = 4.500 \div 2.250 = 2.000 \text{ m}$$

$$\text{Obtuse hip run/m run of CR} = 2.598 \div 2.250 = 1.155 \text{ m}$$

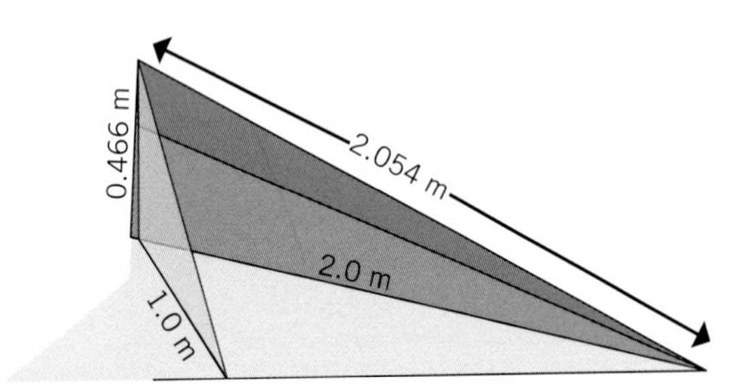

2 *Hip length/m run of CR* (see the earlier section 'The mathematics: The "Seven Pillars" revisited' Pillar 5: hip factor)
As with the skewed gable, this is a derivative of the standard hip factor calculation.

$$\text{Acute hip length/m run of CR} = \sqrt{2.0^2 + 0.466^2}$$
$$\text{Acute hip factor} = 2.054 \text{ m}$$

$$\text{Obtuse hip length/m run of CR} = \sqrt{1.155^2 + 0.466^2}$$
$$\text{Obtuse hip factor} = 1.245 \text{ m}$$

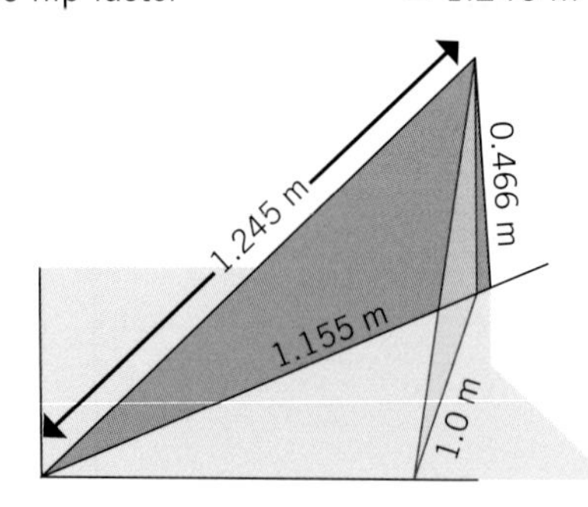

Note: Finding the other lengths follows a similar path as that laid out for the skewed gable rafter.

3 *Hip set-out lengths*
As with a standard hip, the set-out length is:

Acute hip set-out length
$$= \text{half span} \times \text{acute hip factor} = 2.250 \times 2.054 = 4.622 \text{ m}$$

Obtuse hip set-out length
$$= \text{half span} \times \text{obtuse hip factor} = 2.250 \times 1.245 = 2.801 \text{ m}$$

4 *Hip order lengths*
Like a hip, the order length is:

Acute hip order length
$$= [(\text{half span} + \text{eave width}) \times \text{acute hip factor}] + \text{material depth}$$
$$= [(2.250 + 0.600) \times 2.054] + 0.125$$
$$= 5.979 \text{ m (order 6.0 m)}$$

Obtuse hip order length
$$= [(\text{half span} + \text{eave width}) \times \text{obtuse hip factor}] + \text{material depth}$$
$$= [(2.250 + 0.600) \times 1.245] + 0.125$$
$$= 3.673 \text{ m (order 3.9 m)}$$

5 *Hip plumb and level bevels*
Figure 7.62 shows that these bevels are found by making a right angled triangle, matching the rise to the run of the appropriate hip. By scaling these dimensions, the angles can be determined on a steel square or a piece of timber, or a full-sized set-out on a concrete floor. Mathematically they may be determined as they were for the skewed gable.

6 *Hip edge bevels*
Again, these derive from standard hip practice and the use of level lines, which should now be familiar (see Figure 7.63).

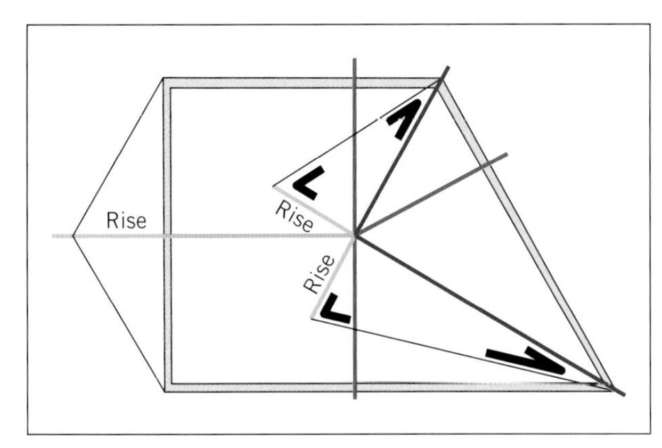

FIGURE 7.62 Finding the hip plumb and level bevels

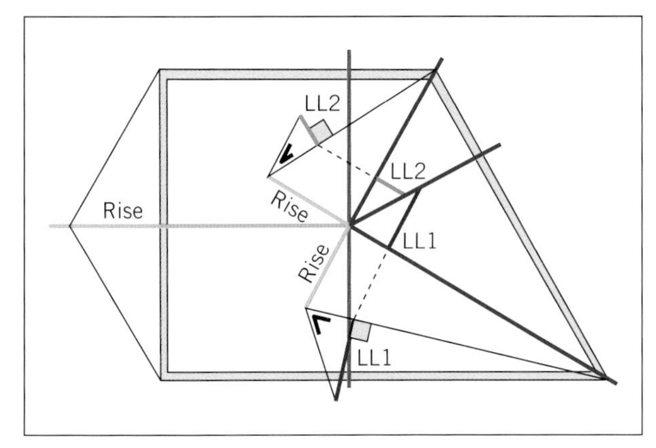

FIGURE 7.63 Using level line principle to find hip edge bevels

7 *Creeper edge bevels*
Once more, this is a standard creeper meeting a hip. Graphically its development is very similar, as Figure 7.64 demonstrates. Alternatively, it may be done on a steel square by matching the true length of the CR to the centring rafter back-set. This latter approach offers you insight into how it is done mathematically should this be required.

Setting out and cutting components

With the top plates set out as described previously (creeper set-out running away from centring rafters), the ridge can be patterned off the plates, common rafters cut, and the central part of the roof erected as per normal practice. The crown end rafter can also be cut and positioned as described earlier. This will help to brace the roof at this point. Set-out of the hips is almost exactly the same as for a normal hipped roof (see Figure 7.33 earlier in the chapter), the only difference being the amount of reduction taken at the apex. For this, a full-sized set-out of the apex is best completed on a small board, as shown in Figure 7.65.

Creeper reduction and creeper lengths

As with the skewed gable roof, the horizontal reduction of the creeper run does not equal the rafter or creeper spacing as the hips are not at 45° to the plates. Again, the horizontal reduction must be found by means of a ratio factor that provides the creeper horizontal run for every metre it moves further away from the corner. As discussed previously, this factor is used to find the run of the longest creeper and the run used to determine the creeper shortening. Once the run of each is found, it is a simple matter of multiplying by the rafter factor (CR length/m run) to find the required lengths.

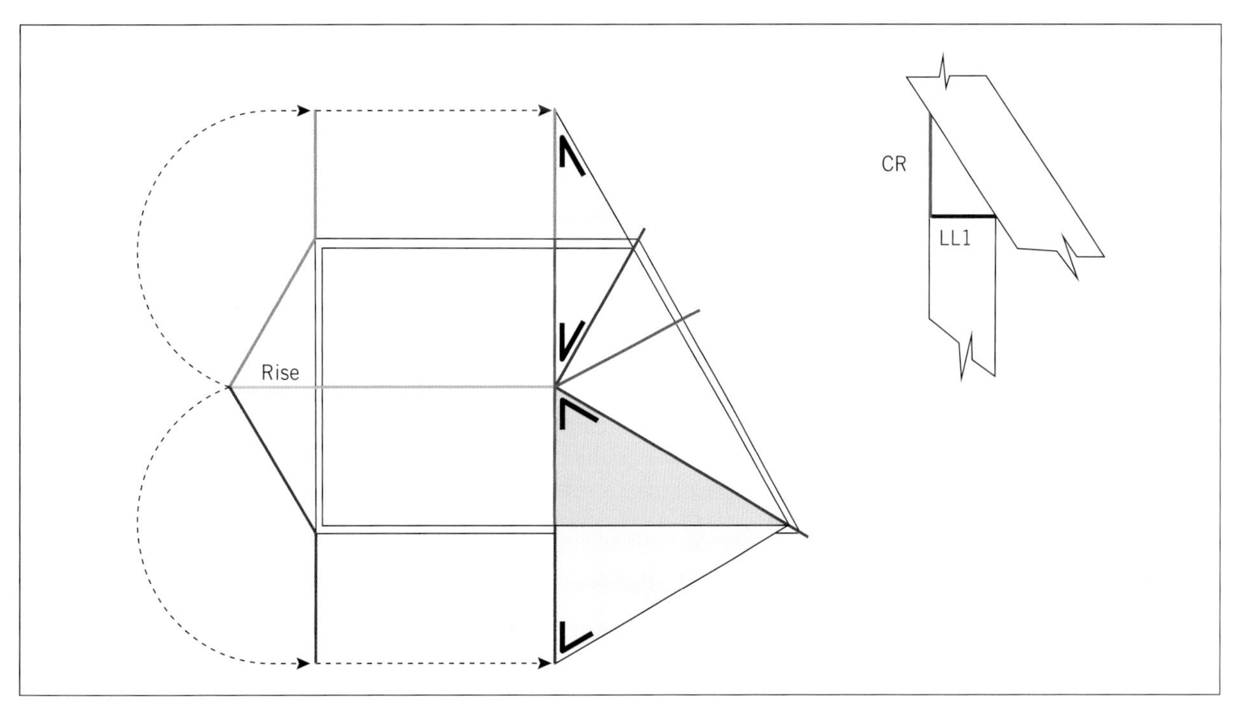

FIGURE 7.64 Finding the creeper edge bevels

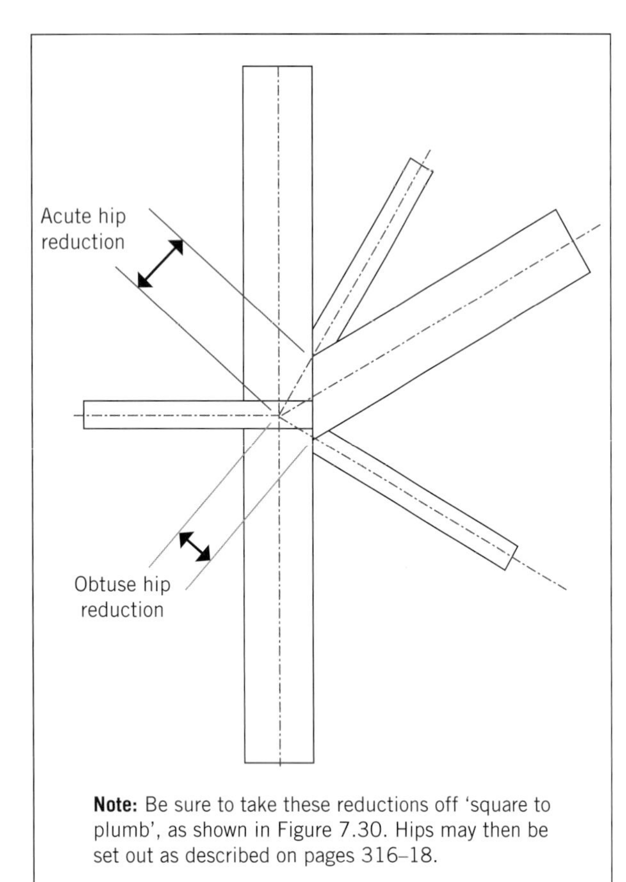

FIGURE 7.65 Hip reduction

First, you need to find the ratio between the half span and the acute centring rafter back-set.

Creeper oblique factor = half span ÷ back-set distance

In this case:

Acute creeper oblique factor = 2.250 ÷ 3.897
= 0.577 m

Acute creeper oblique factor = 2.250 ÷ 1.299
= 1.732 m

To find the horizontal run to the long point of the longest creeper, measure directly off the plate from the face of the hip to the long side of the creeper position (see Figure 7.66). Multiply this by the creeper oblique factor. This is the run of the long side of the first, or longest, creeper.

Run (to long point) of first creeper = creeper oblique factor × Y

Assume:

Measured Y at the acute hip to be 3.440

Measured Y at the obtuse hip to be 0.850

Run (to long point) of first acute creeper
= 0.577 × 3.440
= 1.985 m

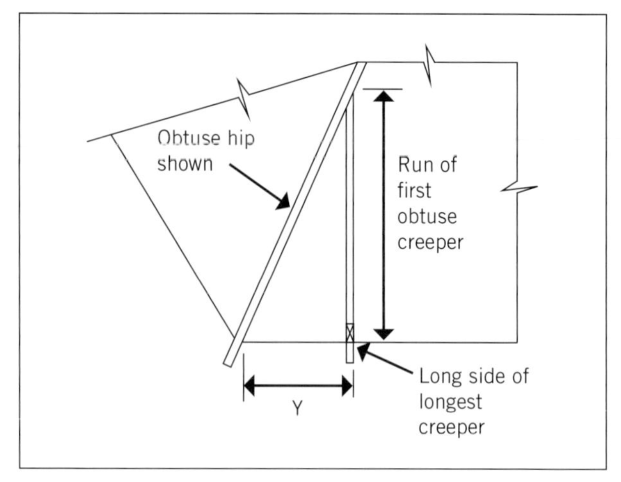

FIGURE 7.66 Measuring length Y, first creeper offset

Run (to long point) of first obtuse creeper
= 1.732 × 0.850
= 1.472 m

Long point length of first creeper is found by:

Creeper long point length
= run of first creeper × rafter factor

In this case:

Acute creeper long point length
= 1.985 × 1.103
= 2.189 m

Obtuse creeper long point length
= 1.472 × 1.103
= 1.624 m

The creeper shortening is now found in a similar manner:

Creeper shortening = rafter spacing × creeper oblique factor × rafter factor

In this case:

Acute creeper shortening
= 0.450 × 0.577 × 1.103
= 0.286 m

Obtuse creeper shortening
= 0.860 × 1.732 × 1.103
= 1.643 m

Use this information to set out the creepers as shown earlier in Figure 7.53. Remember, this time you have to cut the creepers as pairs, one for each side of the hip.

Figure 7.67 shows the completed framing.

FIGURE 7.67 The completed framing

LEARNING TASK 7.5

RESEARCHING OBLIQUE HIP ROOFS

Search for and watch these videos on oblique hip roofs on the Buildsum YouTube channel:

- Oblique bevels
- Oblique end roof – geometrically develop roof member positions
- Oblique hip bevel positions
- Oblique end hip roof bevels, steel square method
- Oblique bevels full size development.

Now, on the sketch below, correctly set the roof frame parts indicated.

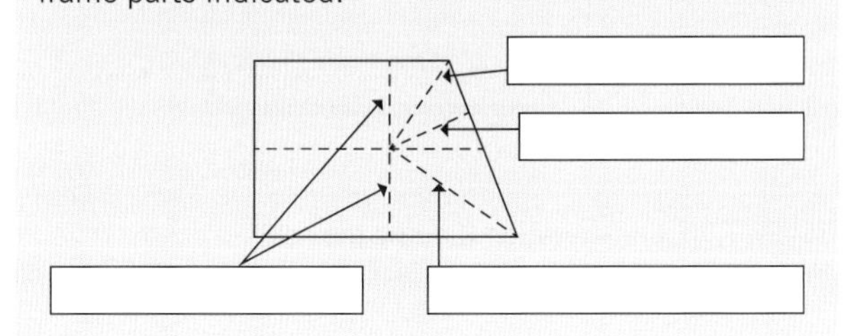

Unequal pitched roofs

Roofs that incorporate two or more roof surfaces running at different pitches or angles to the horizontal are often found in alterations, or where water catchment is required to be directed to one side or another. On other occasions, an unequal pitched roof is built purely for aesthetic reasons. Whatever the reason underlying the design, these roofs offer the builder interesting challenges as the ridge is no longer centred over the span and any hip or valley rafters have an extra edge bevel that must be found (see Figure 7.68). Likewise, there is more than one creeper edge bevel to develop.

Roofs such as this also challenge one of roofing's basic tenets: that hip or valley rafters bisect a corner.

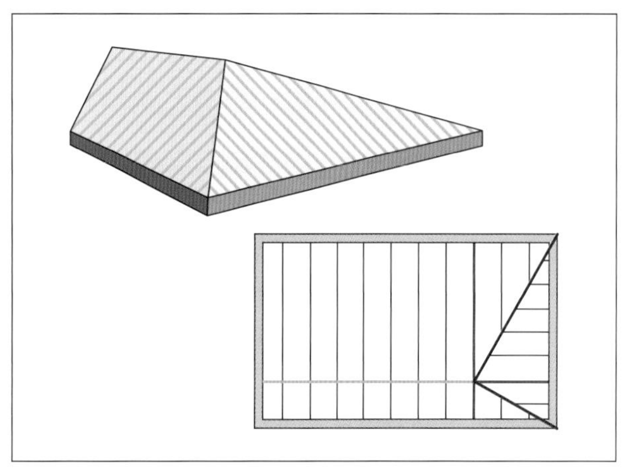

FIGURE 7.68 Unequal pitch

As this discussion will show, the hips of these roofs bisect the corner only if the roof planes are of equal pitch. When this is not the case, the position of the hip must be found in another way.

For this section, the roof characteristic of 25° used to this point will be continued for one roof plane only. For the others, the hip end will be at 40° and the other side of the main roof structure will be at 55°. As the focus will be on the challenges arising from the unequal pitch, a simple rectangular plan will be used (see Figure 7.69).

Set-out

The first issue arising from a roof of this type is that there is no easy point from which to start the set-out. With the hips not bisecting the corners, it is not possible to cast string lines as a practical means of solving the problem (as was demonstrated earlier with the oblique hip). The location of the ridge is also an unknown factor until an end elevation is drawn. This is the clue to the practical or graphical solution. However, usually this can only be done to a reduced scale as there is seldom the space on either the slab or sheet

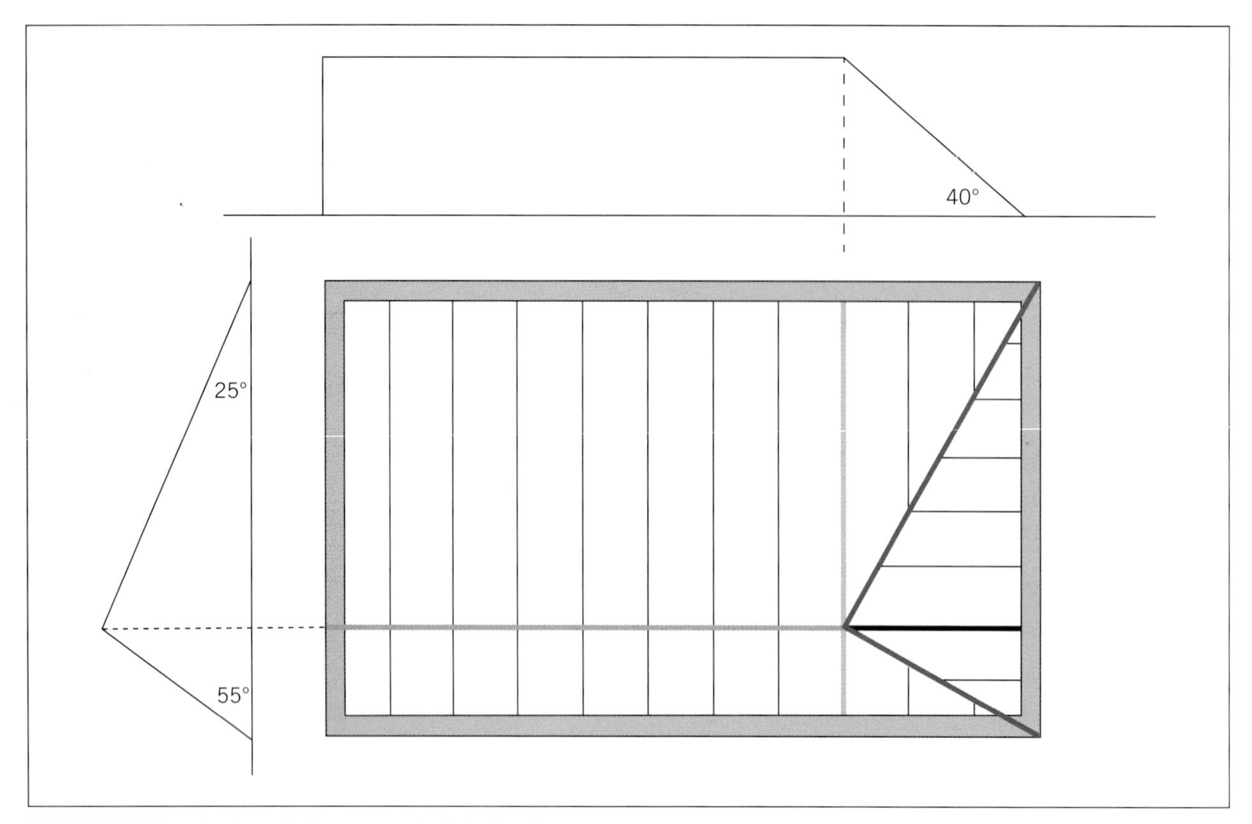

FIGURE 7.69 Framing layout of an unequal pitched roof

floor to draw it full size, unless this is done before the walls go up.

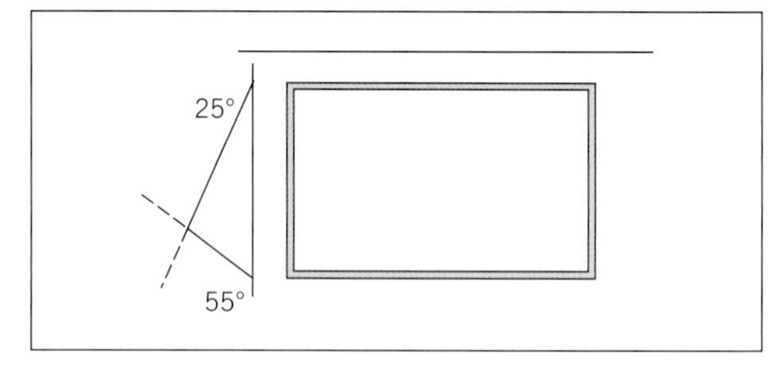

FIGURE 7.70 End elevation (finding the rise)

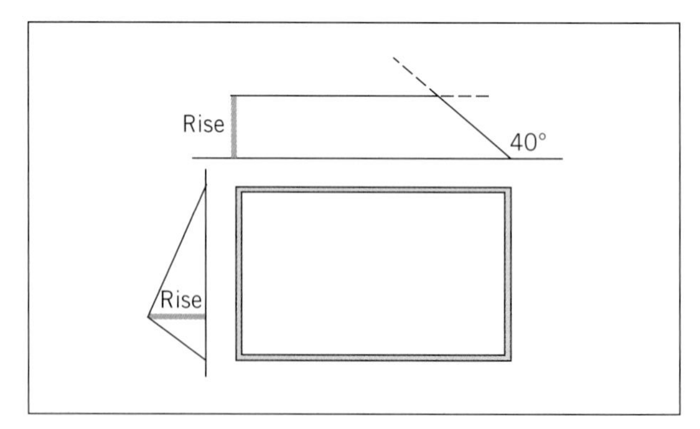

FIGURE 7.71 Front elevation (finding the centring rafter back-set)

The graphical or practical layout

This approach requires that you begin with an end elevation and trace the rafter angles precisely. Where these traces meet provides you with your ridge position and the rise can then be measured. Once the rise is known, you can draw the side elevation, showing the ridge and hip end traces. Lines can then be drawn from the apex of each of these elevations over the plan of the wall plates. Where these cross in the plan is the apex of the roof where the hips will meet. This also provides the location of the centre lines of the centring rafters (centring rafter back-set) and the crown end rafter (see Figures 7.70, 7.71 and 7.72).

The mathematical approach

The mathematical approach is quicker, and more accurate, than the graphical approach providing that you have built the frame correctly. The mathematical approach is also at times the only means of solving the issue prior to commencement of construction. This may be critical in determining material section sizes, location of appropriate footings for load-bearing points, and the like.

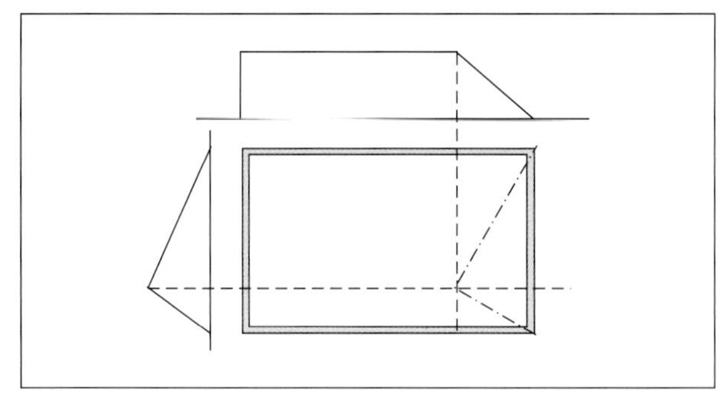

FIGURE 7.72 Plotting ridge and centring rafters and hips to plan

As with the graphical method, first sketch an end elevation with the known information; that is, the roof pitch angles and the span. The 'missing' angle is found by adding the two known angles and subtracting them from 180° (sum of the angles within a triangle).

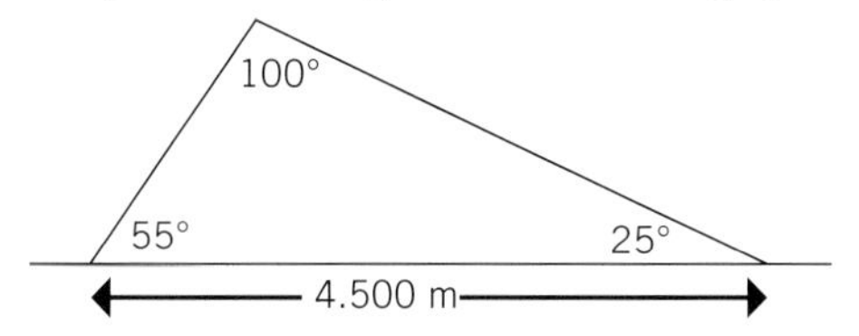

We now must find one of the rafter lengths before we can find the rise. We do this using the sine rule.

This rule states the following:

$$\frac{a}{\sin A} = \frac{b}{\sin B} = \frac{c}{\sin C} \text{ where:}$$

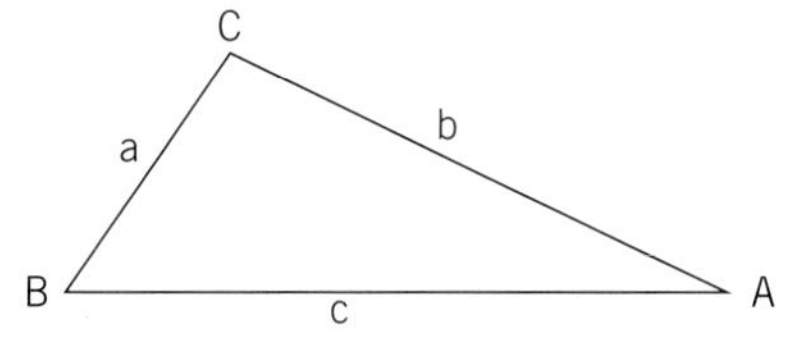

As side C is known, sine C will form one side of the equation; the other side can be either 'A' or 'B'. In this case A will be chosen (it makes no difference). So:

$$\frac{a}{\sin A} = \frac{c}{\sin C}$$

In this case:

$$\frac{a}{\sin 25^\circ} = \frac{4.500}{\sin 100^\circ}$$

$$\Rightarrow \quad a = \frac{4.500}{\sin 100^\circ} \times \sin 25^\circ$$

$$a = 0.985 \times 0.423$$

$$a = 1.931 \text{ m}$$

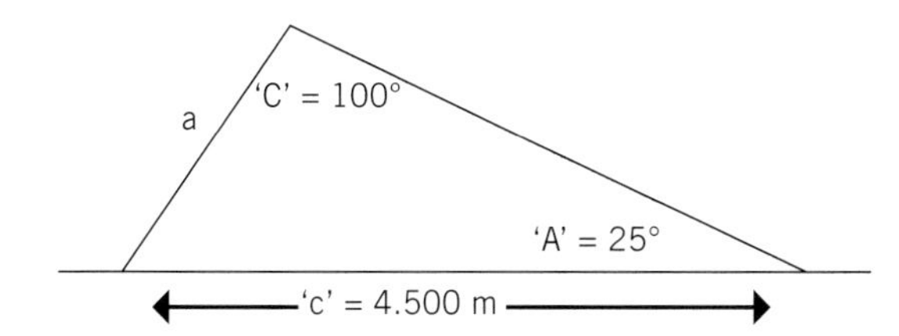

The rise can now be found using:

$$\sin 55^\circ = \frac{\text{rise}}{1.931}$$

$$\text{rise} = \sin 55^\circ \times 1.931$$

$$= 1.582 \text{ m}$$

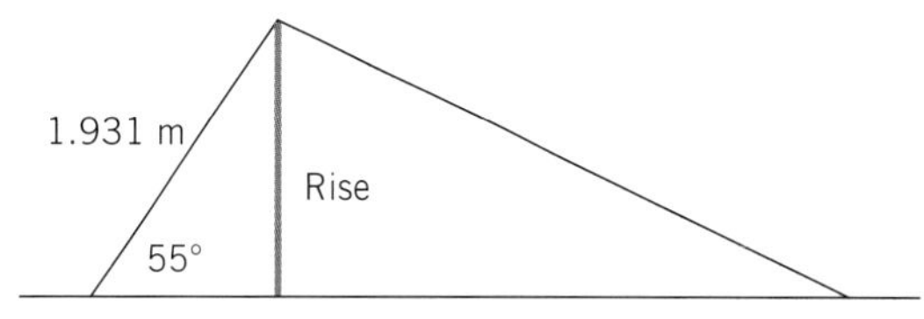

From here we can find the ridge location using either Pythagoras' theorem, or the cos or tan function; that is:

$$\cos 55^\circ = \frac{X}{1.931}$$

$$X = 1.108$$

Note: It is good practice to repeat this calculation for the other side. The sum of the two results should equal the span of the building.

With the rise known, following the graphical approach developed earlier we can now find the run of the crown end rafter. This gives us the centring rafter back-set. This is done using the rafter pitch for the hip end (i.e. 40°); that is:

$$\tan 40^\circ = \frac{1.582}{Y}$$

$$Y = 1.885$$

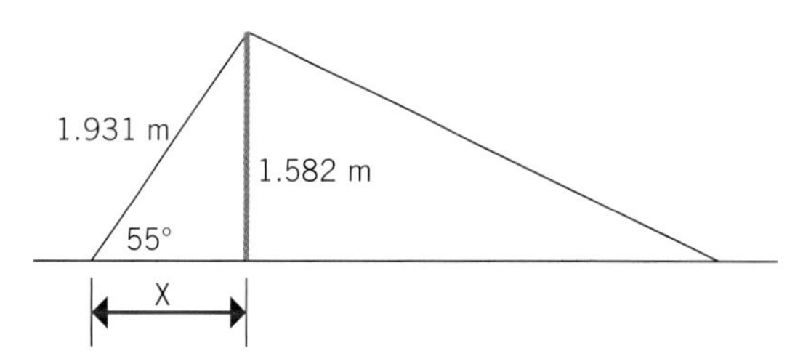

The plates can now be set out using the dimensions found (see Figure 7.73). Creepers are set out to standard

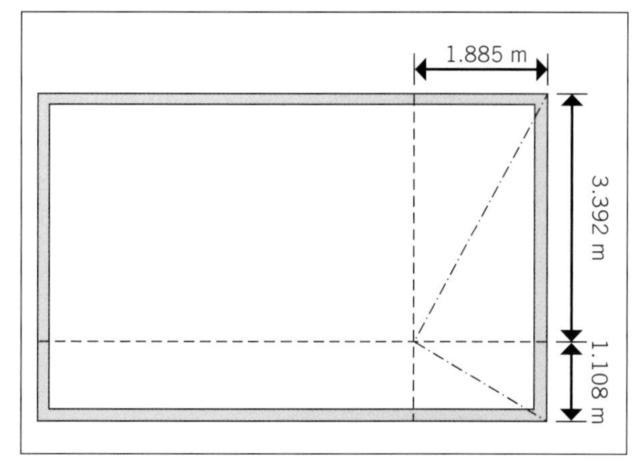

FIGURE 7.73 Mathematically determined lengths applied to plates

common rafter spacings, starting from the centring rafter back-set and the crown end location, as has been described earlier (see Figure 7.53).

Note: In following the format for laying out the creepers, you will find that the tops of the creepers do not align. This is acceptable; however, it does mean more care must be taken to ensure that the hip remains straight. There are alternative approaches, but they increase labour and material costs without providing any benefit other than the creeper tops meeting.

Positioning the ridge and hips

While the phrase 'centre line' has been used above to describe the location of the ridge and hips, these lines actually do not lie on the centres of these members. The true position of these components must now be determined. This is done graphically, usually on a piece of sheet flooring or bracing. (It can be done mathematically, but the approach we will use is more practical.)

The ridge and crown end rafter

Draw a scale end elevation, large enough so that full-sized material sections can be laid upon it. Position the ridge material as shown in Figure 7.74 – 'balanced' so that the top edge of the ridge is touching both rafter traces. This is your ridge position. Use the same approach to locate the crown end (assuming that it is of a differing thickness).

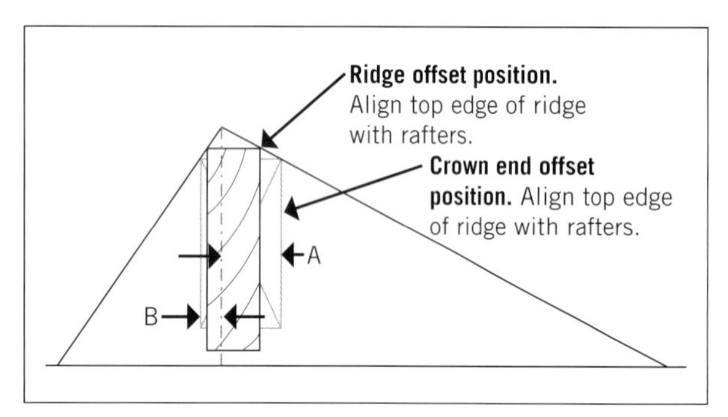

FIGURE 7.74 Offset position of ridge and crown end components

Some important notes for Figure 7.74:

1. Measure distances A and B. These are the common rafter reductions. (Note that in an unequal pitched roof it is no longer simply half the ridge thickness.)
2. Leave room around this set-out to draw the other developments.

The hips

The hips must also be offset from the 'centre line'. This is done in a similar way to the ridge and crown end. Set out the corner large enough so that full-sized hip material can be laid on edge upon it (for clarity, see Figures 7.75 and 7.76). (If dealing with an oblique hip

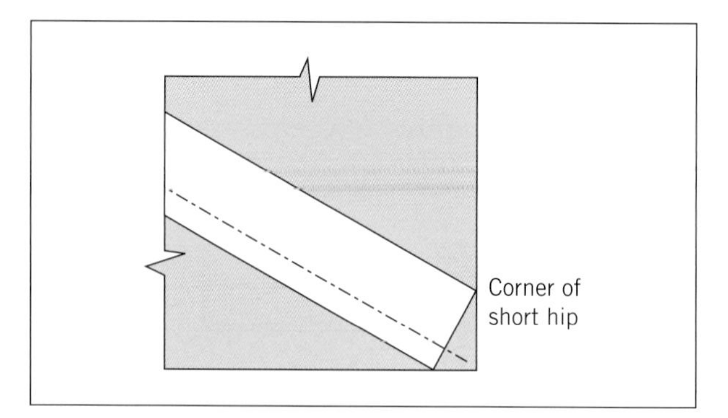

FIGURE 7.75 Short hip centre line offset

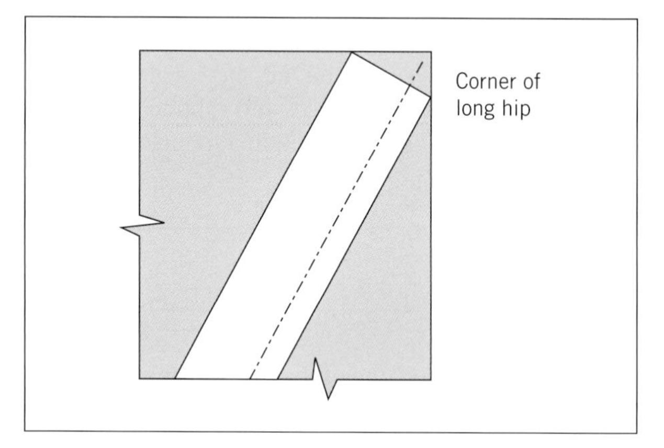

FIGURE 7.76 Long hip centre line offset

with an unequal pitch, the approach is identical.) This offset is important for three reasons: first, it gives the correct position to trim the corner of the plates to accept the hip bird's mouth; second, by placing the hip in this position, each hip edge will be aligned with the roof surface that it intersects; and third, the bird's mouth can be cut to the rafter plumb height, as normal, while again aligning with the roof surfaces.

Having determined the hip, ridge and crown end offsets, a full-sized set-out of the apex, or 'crown end cluster', can be drawn. It is from this that the edge bevels for the hips will be taken, using either the level line system used in previous examples, or its derivative, direct measurement. In addition, this set-out will provide us with the shortenings for the hips.

Now the elevations of the hips and crown end rafter need to be drawn. These are best drawn to a scale that will fit the board (each elevation need not be to the same scale; it is the angles that are critical here). From these you will get the plumb and level bevels for the hips and the crown end rafter; they are also necessary to complete the level line developments (see Figure 7.78). Use Pythagoras' theorem to calculate the run of each hip from the information derived earlier. A typical layout is shown in Figure 7.77.

Develop the bevels as demonstrated in the previous examples. If uncertain, refer to the basic principles at the beginning of the chapter. Note that in this case there

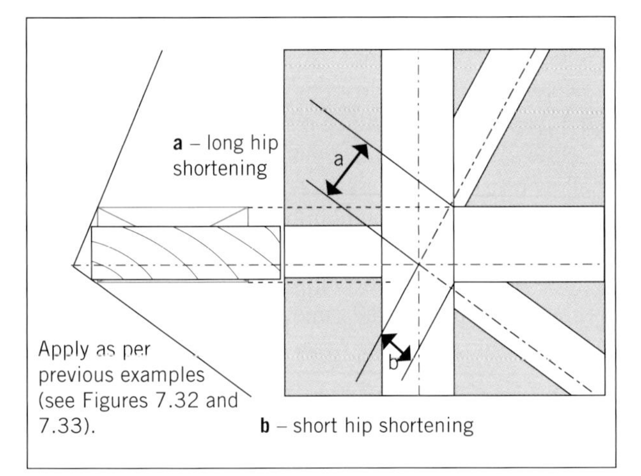

FIGURE 7.77 Hip shortenings

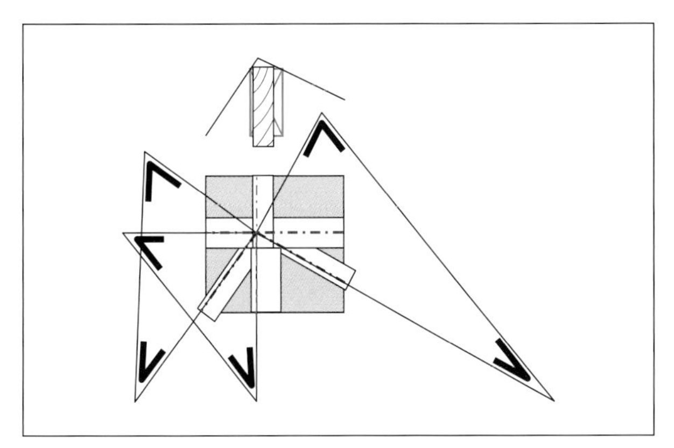

FIGURE 7.78 Hip and crown end edge bevels

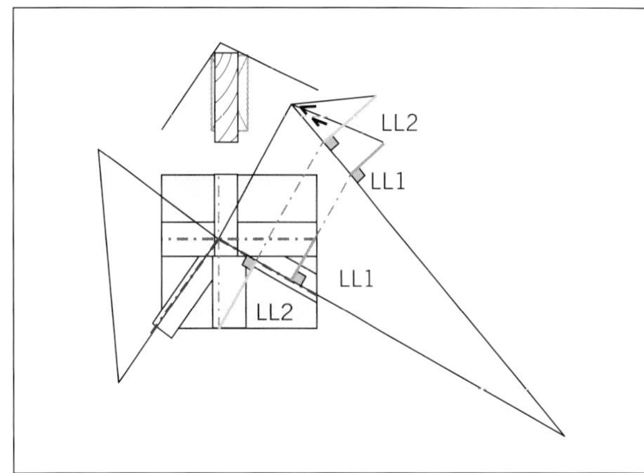

FIGURE 7.79 Edge bevels for long hip

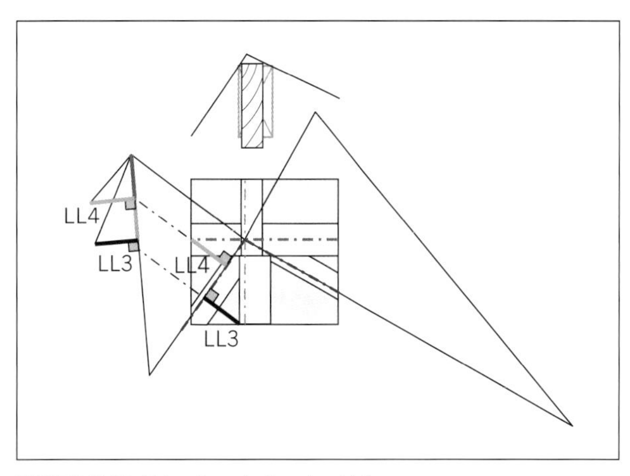

FIGURE 7.80 Edge bevels for short hip

are two edge bevels for each hip. The edge bevels for the long hip are shown in Figure 7.79. Figure 7.80 shows the developments for the short hip. When doing the level line developments, deciding where to position the line is generally based simply on making the largest practicable triangle – this will ensure that the accuracy of the bevel is increased.

Determining the component lengths

The lengths of the hips and crown end can be determined using the same principles as in each of the previous sections of this chapter. You will have noticed that so far we have not followed the 'Seven Pillars' in this section. This is not because they have ceased to have relevance as useful mathematical approaches, but because they must now be used repeatedly to obtain the lengths required. Revision of the previous sections and the following brief outline should allow you to calculate the lengths of the components.

What is known are the:

- rise
- rise/m run of 25° common rafter
- run of both common rafters
- run of crown end rafter
- run of both hips
- set-out length of 55° common rafter
- all required bevels.

What must be found are the:

- rise/m run for 55° and 40° rafters: found by: rise ÷ run of rafter
- CR factors for all rafters (3 off): found by: Pythagoras' theorem applied as shown in 'Basic principles of roofing' at the beginning of the chapter.
- hip factors (2 off): see pp. 326–7 of 'Oblique hips' section
- common rafter set-out length (2 off): see 'Basic principles of roofing'
- common rafter order length (2 off): see 'Basic principles of roofing'
- crown end set-out length: see 'Basic principles of roofing'
- crown end order length: see 'Basic principles of roofing'
- hip set-out length (2 off): see 'Basic principles of roofing'
- hip order length (2 off): see 'Basic principles of roofing'
- creeper shortening (4 off): see pp. 327–8 of 'Oblique hips' section.

Once obtained, the setting out of the components can begin.

Set-out of components

The set-out of the components of unequal pitched roofs differs in one important aspect from all other roofs. The 'centre line' does not run in the centre of the edge of the material on some components – notably, the hips – and, instead, is offset (see Figures 7.77, 7.79 and 7.80). This has an impact on the location of the long point on the hip and the corner of the building around which the bird's mouth must be developed. It is critical that this offset centre line is traced correctly. If it is not, the hip will not site correctly and will be out of plane with the roof surfaces.

The hips

The set-out of the long hip is shown in Figure 7.81. Note the application of the shortening, the location of the 'centre line' (its position is measured off the full-sized set-outs done previously; see Figure 7.83) and the application of the edge bevels (note that they 'swap sides' depending on whether they are being used at the top or the bottom).

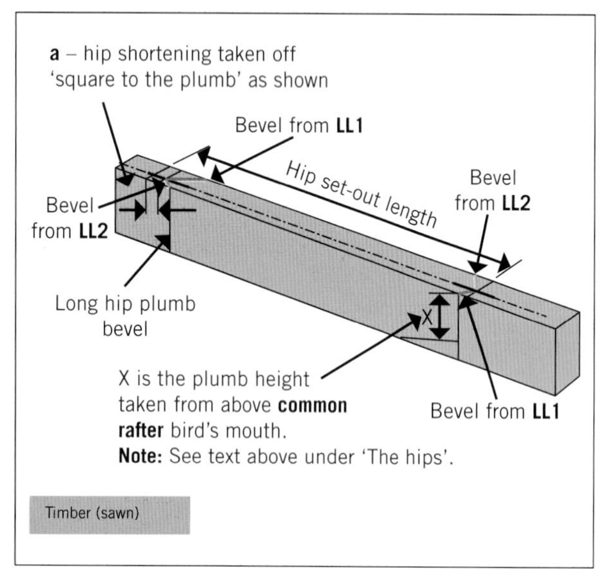

FIGURE 7.81 Set-out of long hip

Note that X, the plumb height above the bird's mouth, is determined by the height that best suits both the steepest and the shallowest roof pitch. As can be seen in Figure 7.82, as the rafter becomes steeper the height X becomes greater if the one-third bird's mouth principle is maintained. Sometimes the solution is simply an average of the two; at other times it requires judgement as to which will give adequate seating for the shallowest rafter, yet not take too much out of the steepest.

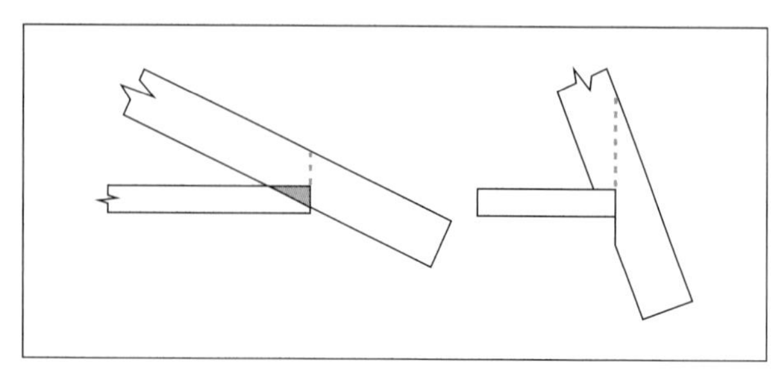

FIGURE 7.82 The implications of retaining the one-third bird's mouth set-out principle on height X as the pitch of the rafter changes

Note: All components, whether rafters (steep or shallow pitched side), crown end rafter, creepers or hips, **must use the same plumb height** (X) above the bird's mouth.

The set-out of the short hip is similar to the set-out for the long hip, with the same plumb height (X) being used on both rafters (see Figure 7.83).

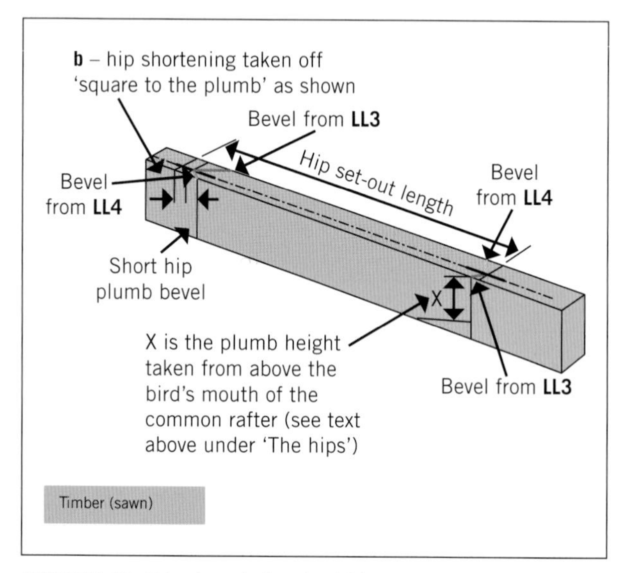

FIGURE 7.83 Edge bevels for short hip

Creepers

Creeper set-out is the same as for the oblique hip except that creepers are not paired. Each set of creepers must be set out individually. Apart from the repetitive nature of this set-out, it is not a difficult task. A review of the 'Oblique hips' section earlier in this chapter should answer any questions on this procedure.

Note: Be sure to use the same plumb height (X) above the bird's mouth on each of the four set-outs.

Purlins and struts

We learnt in Chapter 10 of *Site Establishment, Formwork and Framing* (Laws 2020) that **purlins** are used so that smaller section sizes can be used for rafters, and **struts** are used to support the purlins' off-load bearing walls or strutting beams.

The size of purlins and struts is laid out in Section 7 of AS 1684.2 Residential timber-framed construction – Part 2: Non-cyclonic areas, but you need to be aware there are differences for cyclone-affected zones.

The roof cladding materials will need to be considered when determining the rafter section size; for example, sheet roofing is much lighter than tiles and so a smaller section size may be used for rafters depending on the span and purlin/strut supports used.

Purlins and struts are used in pitched (coupled) roofs and their design specifications will depend on the spans of rafters and the section sizes used as well as the internal layout of the building; for example, where large, open-plan areas are used inside the building it will usually be necessary to incorporate a strutting beam (similar to a hanging beam) to bear the weight of the roof.

There are a number of different combinations for installing strutting beams, and these are detailed in Section 7 of AS 1684.

The sizing of purlins is critical. The purlin size can determine the number and spacing of struts (see the supplementary Span Tables of AS 1684).

Note: Different span tables apply depending on whether the building is in a cyclonic or non-cyclonic area and the species and grade of timber.

If two rows of purlins are needed they should be evenly spaced between wall plates and the ridge.

A typical purlin can be a glue-laminated section or made up from two right-sized sections nail-jointed together to achieve the full dimension. Use the rules and tables in Section 7 of AS 1684 to determine the right size. Purlins in general should be run in single lengths wherever possible. If they have to be end-jointed, they need to be joined over a strut using a halving or lap joint and nailed (see Figure 7.84).

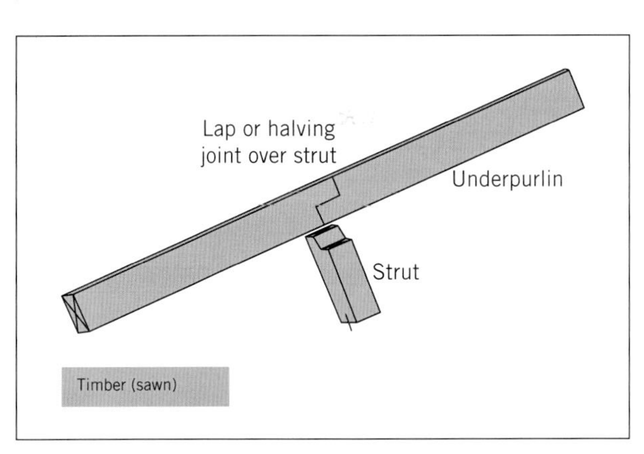

FIGURE 7.84 Underpurlin joint

Purlins can be lapped and nailed over a strut, but must be jointed in a specific way as detailed in the Standards. They can be cantilevered by a maximum of 25 per cent and supported by hip and valley rafters as long as the Standards are followed.

Struts not only support underpurlins but can also support ridges and hip and valley rafters. They can only be supported by load-bearing walls and/or strutting beams or combination beams. They cannot be supported by hanging beams or counter beams.

Struts should be vertical or perpendicular (at 90°) to the purlin and rafters. They should be bird's mouthed to or halved to the underpurlins (see Figure 7.85) and where they are not, metal strapping or framing anchors need to be fixed according to the Australian Standard.

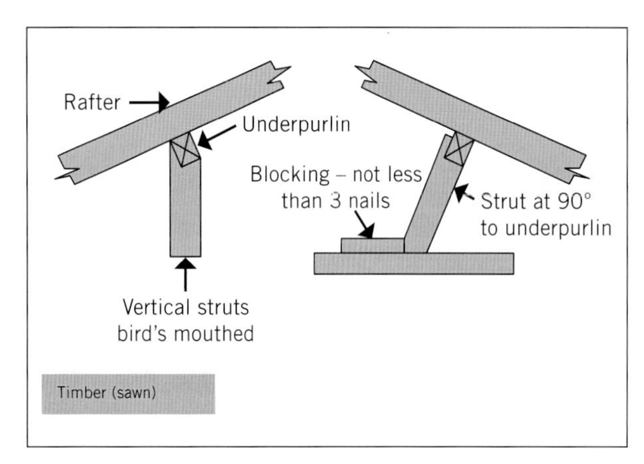

FIGURE 7.85 Simple strutting

Fan struts are pairs of opposing struts fixed to the same point and cannot be set at a greater angle than 45° (see Figure 7.86).

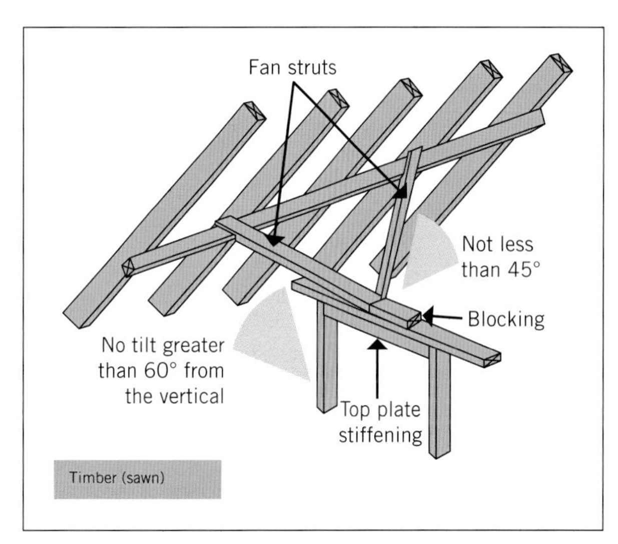

FIGURE 7.86 Fan struts

Strut section sizes depend on the stress grade of the timbers used and the span length of the strut. You need to consult AS 1684, Section 7 and the relevant tables in the Standard to determine the right section size.

Cable struts

Cable struts are a unique method using steel cables (wire ropes) with load-bearing rods acting to support roof members (see Figure 7.87).

Cable strutting can be used on occasions where it may not be practical to use the other forms of strutting. Cable struts also provide a means of strengthening purlins. They are often used in older roofs that have sagged due to poor construction or where the original

FIGURE 7.87 Cable strut

timbers were undersized or insufficient for the spans involved (see Figure 7.88).

Cable strutting provides a faster, lighter solution to supporting roof loads than adding timber struts, strutting beams or metal brackets, resulting in reduced vertical loads on the structural elements of the building.

LEARNING TASK 7.6

RESEARCHING UNEQUAL PITCHED ROOFS

Consider the following question in a teacher-led class discussion: What are the critical problems that need to be solved when building unequal pitched roofs? In the discussion, consider these issues:

- the importance of trigonometry
- why we need to offset the ridge, hips and crown end rafter
- the varying eave height and width.

Next, answer these questions and/or fill in the missing words:

1 The offset of the ridge always moves towards the ____________________ of the roof.
2 ____________ are the most useful trig values in working out the angles and lengths in an unequal pitched roof.

Using the information in the diagram below:

1 Calculate the length of side 'a' and side 'b'.
2 Calculate the perpendicular rise height.

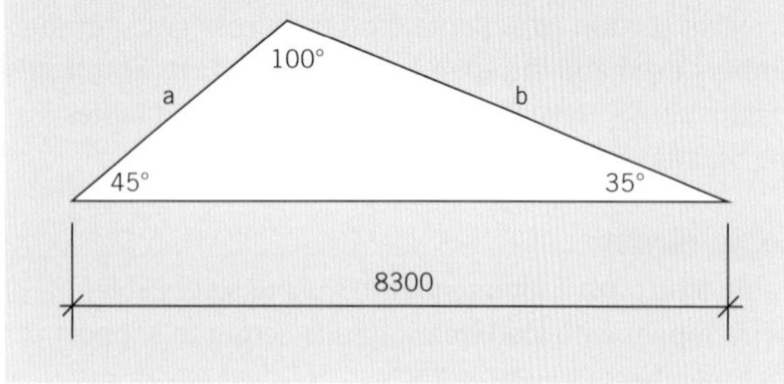

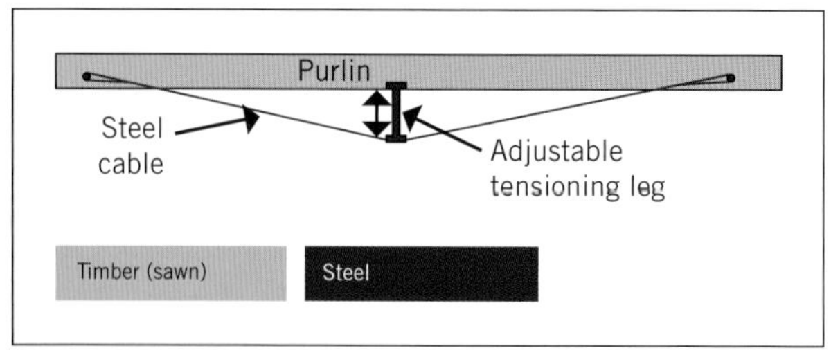

FIGURE 7.88 Cable strutting can be used to strengthen and support underpurlins

Soldier walls for extended eaves (soffits)

The eaves width sometimes varies requiring the eaves framing to be extended. This happens when the building line changes to allow for an entry and the roof line remains the same (see Figure 7.89). When this happens a **soldier wall** will be required and the rafters over that portion will have a different bird's mouth position.

Setting out a common rafter and framing of eaves (or soffits) are explained in Chapters 10 and 12 of *Site Establishment, Formwork and Framing* (Laws 2020). There are two main alternatives to solve this problem:

a Extend the entry frame to the new height to support the rafters and the new position.
b Build a soldier wall to sit on top of the regular house frame.

Calculating height of the soldier wall and position of a new bird's mouth

In reality it is the same as calculating a soldier wall for a jerkin head roof (see 'Calculating the height and width of the soldier wall' earlier in this chapter). To calculate the new bird's mouth position, you use the basic rise/m run calculation; for example, roof pitch is 30° rise/m run = tan 30° = 0.5773/m run. Work back from the common rafter bird's mouth if you know the length of the return wall 'a'.

In this example 'a' = 1200 mm or 1.2 m:

Rise 'b' = 1.2 m × 0.5773 m = 0.693 m = height of the soldier wall

Now apply $\sqrt{a^2 + b^2}$ to find distance 'c'

$$c = \sqrt{1.2 + 0.6932} = \sqrt{1.44 + 0.4802} = \sqrt{1.9202}\ \text{m}$$

'c' = **1385.7 mm**

Refer to *Site Establishment, Formwork and Framing* (Laws 2020) Chapter 10 'Construct pitched roofs' to refresh your understanding of common rafter set-out and Chapter 12 'Construct eaves'.

Depending on the length of the new eaves **trimmers** (soffit bearer), larger timber sections may be required, so consult AS 1684 and span tables to determine the grade and correct sizing of materials if an eaves trimmer support (also known as 'over batten') will be required to keep eaves (soffits) aligned and prevent sagging. Two methods are indicated in Figure 7.90.

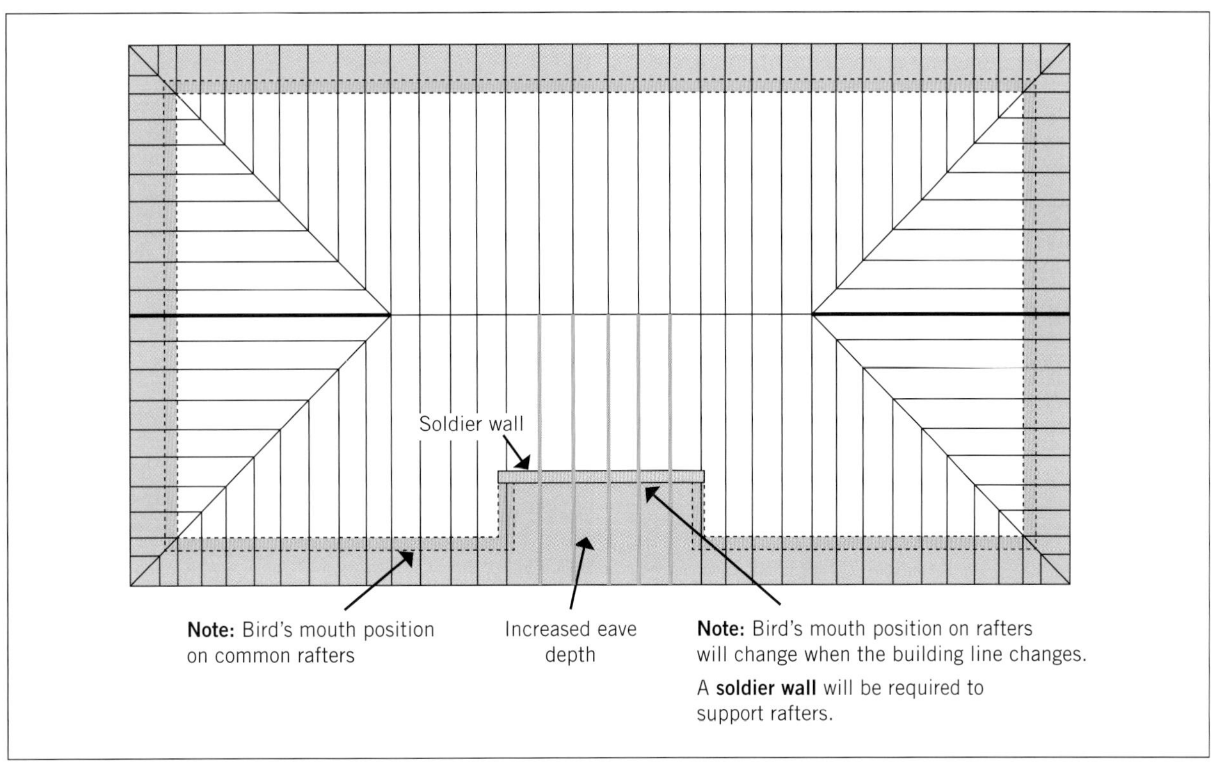

FIGURE 7.89 A changed building line

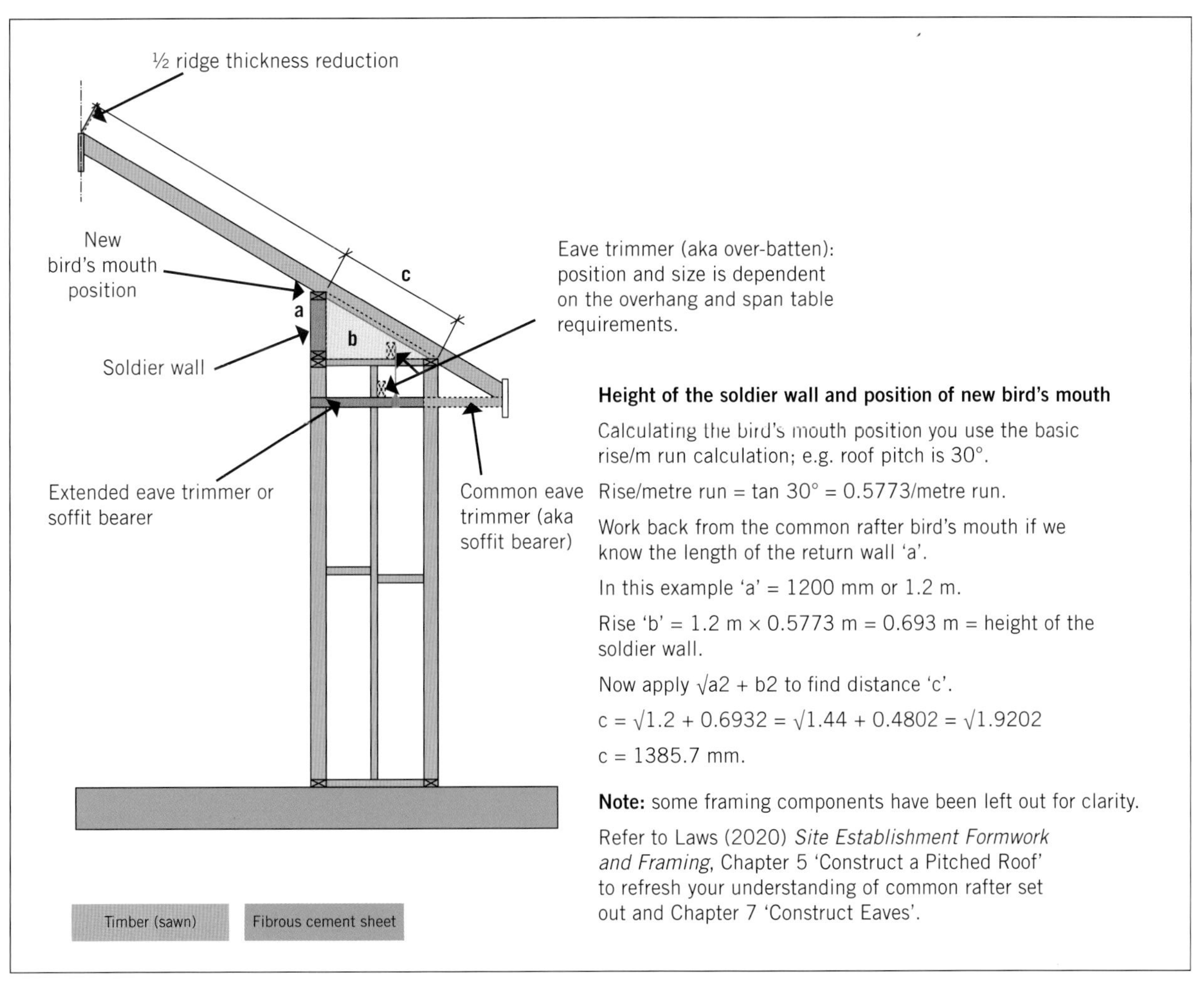

Height of the soldier wall and position of new bird's mouth

Calculating the bird's mouth position you use the basic rise/m run calculation; e.g. roof pitch is 30°.

Rise/metre run = tan 30° = 0.5773/metre run.

Work back from the common rafter bird's mouth if we know the length of the return wall 'a'.

In this example 'a' = 1200 mm or 1.2 m.

Rise 'b' = 1.2 m × 0.5773 m = 0.693 m = height of the soldier wall.

Now apply √a2 + b2 to find distance 'c'.

c = √1.2 + 0.6932 = √1.44 + 0.4802 = √1.9202

c = 1385.7 mm.

Note: some framing components have been left out for clarity.

Refer to Laws (2020) *Site Establishment Formwork and Framing*, Chapter 5 'Construct a Pitched Roof' to refresh your understanding of common rafter set out and Chapter 7 'Construct Eaves'.

FIGURE 7.90 Soldier wall calculations

For lighter construction methods it is common practice to use metal framing systems to support eaves (soffit lining).

Dormers

A dormer is any roof structure projecting out from the plane of a pitched roof. Most often they contain windows to provide light into a room constructed in a roof space (i.e. **dormer windows**). They are most often constructed similarly to a gabled roof at the front with a valley where it intersects the pitched roof (see Figure 7.91).

FIGURE 7.91 Dormer window iron roof

They can also be **skillion** or curved roof shapes (see Figures 7.92 and 7.93).

FIGURE 7.92 Dormer window skillion roof

Dormer roofs (with windows) are best suited to steeply pitched roofs to provide maximum floor space and head height. They can also be incorporated into a roof as a means of venting (see Figure 7.94).

Each type presents its framing challenges. The pitched dormer is set out along centre lines like any other pitched roof. On a new build you may have the luxury of time to set out and construct the framing, but on a renovation you need to consider how long you can reliably have the roof space open to the weather.

FIGURE 7.93 Dormer window with curved roof

FIGURE 7.94 Dormer venting

There is always more than one way to frame the dormer. In this section we focus on a pitched dormer (see Figures 7.91 and 7.95).

A couple of key points to consider:

1. When cutting into an existing roof space, install deep floor joists first (or similar framing) to support flooring materials. New builds will already have these in place.
2. If the pitch on a pitched dormer is different from the main roof structure you will have to make new calculations for plumb level bevels for common rafters; and plumb, level and edge bevels for creepers and valley rafter.
3. Wear your personal protective equipment (PPE), including a safety harness. Use a scaffold to provide fall protection. Other edge protection systems may get in the way of construction.

Steps (new buildings)

These steps assume the main roof structure has been constructed.

1 Set out the position of the structure on the flooring; that is, frames, ridge and common rafters.
2 Using rise and run calculations, calculate the position of the head trimmer and mark it on the adjacent common rafters. Install the head trimmer, noting that the top edge should be splayed to the pitch of the rafters. This will make it easier to align the valley rafter.
3 Plumb up from the set-out, the intersection of wall frames and common rafters for the main roof structures. Set out and build frames as per the relevant AS 1684 framing requirements. Stand and fix and temporarily brace frames in position making sure the front wall is parallel to the main roof fascia and side frames are at right angles. Install the waling piece to the front wall, at the same time making sure the front top edge aligns with the top edge of the common rafters.
4 Mark the position of the common rafters on the top plates and the centre line of the ridge on the top plate and head trimmer.
5 Clamp a temporary post to the front frame to provide a fixing point for the ridge. Measure, cut and fix the ridge in position. Check and make sure it is level. Stick to the centre lines.
6 Mark, cut and fit the common rafters as you would for any pitched roof. See *Site Establishment, Formwork and Framing* (Laws 2020) Chapter 10 'Construct pitched roofs'. Leave gable end rafters to last.
7 Set out the valley rafters. Check to see they are going to fit before cutting. As indicated in Figure 7.95, the top edge of the valley rafter aligns closely with the top edge of the first set of common rafters.

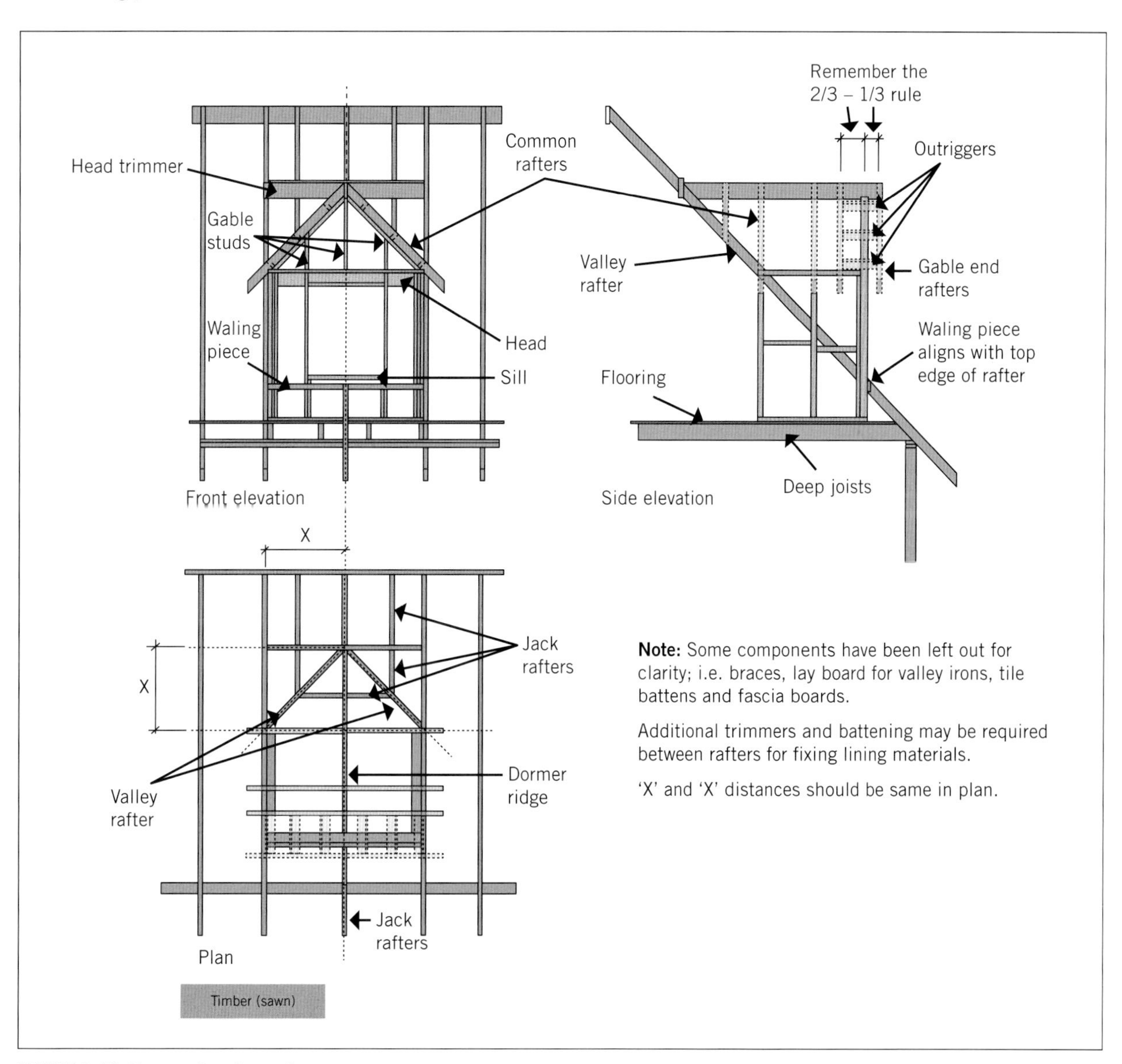

FIGURE 7.95 Dormer framing ortho

8 Install the outriggers that will support the gable end rafters. Apply the two-third to one-third rule to positioning; that is, when installing outriggers, only one-third of its length can extend beyond the building framing (see AS 1684, Section 7). These can be aligned to take tile or roof sheet battens, providing additional support to the outriggers and gable rafters.
9 Mark out, cut and fit the valley jack rafters and jack rafters.
10 To finish off the framing, install gable end rafters, roof bracing, valley boards, gable studs, roof battens and fascias.

Note: Collar ties across rafters may be needed, depending on the span of the dormer.

FIGURE 7.96 Semi-octagonal roof

Octagonal roofs

Octagonal roofs are often used to create gazebo, roof turrets and semi-octagonal roof sections.

In this section we describe two models of semi-octagonal roof structures (see Figures 7.97 and 7.98).

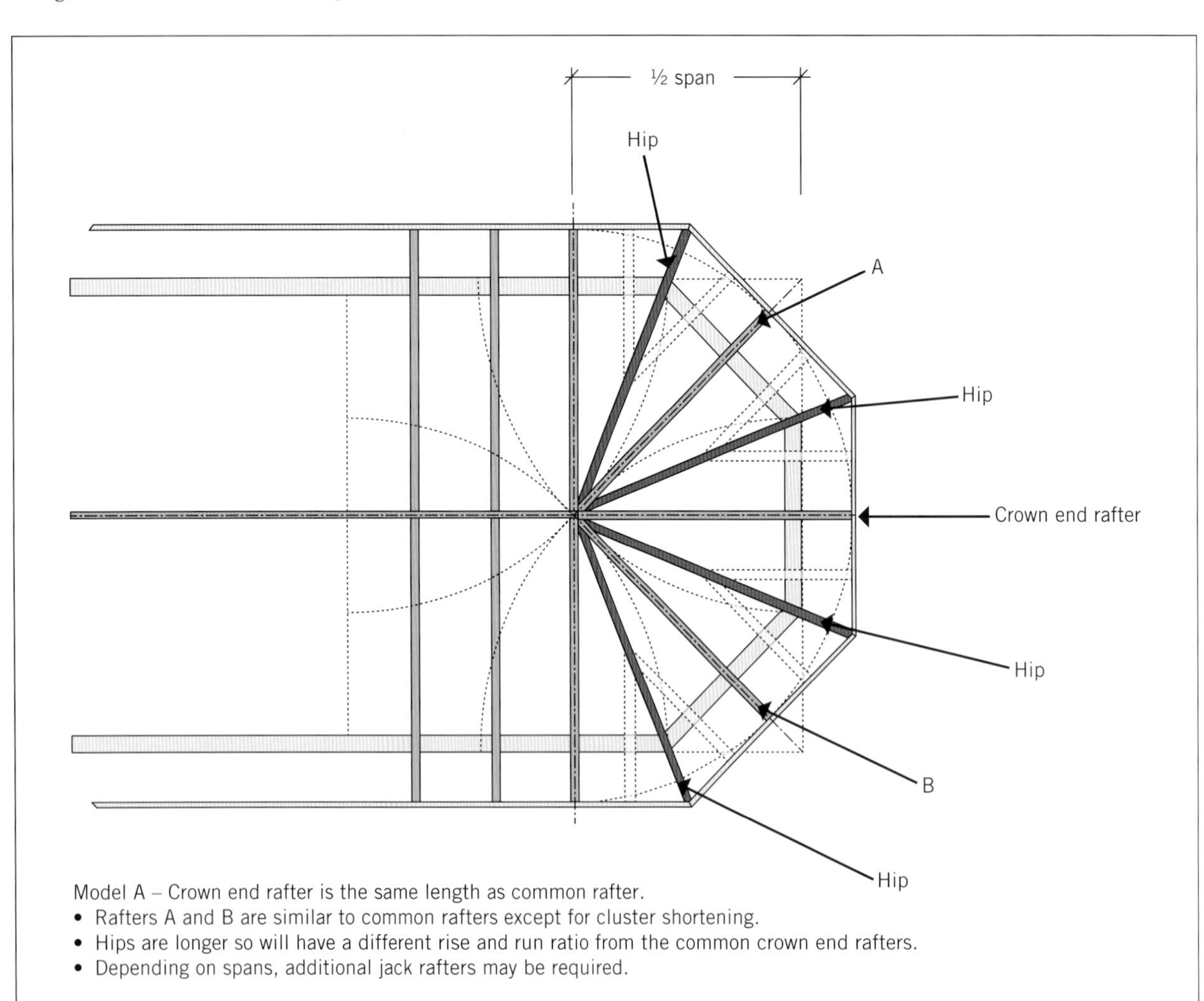

FIGURE 7.97 Semi-octagonal roof – Model A

The 'Basic principles of roofing' described earlier in the chapter applies, and the same basic calculations and geometry are used.

In Model A, the steps involved are similar to those of a regular roof. More information on the set-out can be found on YouTube at Buildsum: 'Semi-octagonal roof geometric bevel development' https://www.youtube.com/watch?v=4PzUlvG02m4.

Step 1

The crown end rafter is installed first (see Figure 7.97). This is set out in the same way as for a regular hip roof. See *Site Establishment, Formwork and Framing* (Laws 2020) Chapter 10 'Constructing pitched roofs' for more detail.

Step 2

Rafters A and B are set out like regular hips using the same plumb and level bevels as a common rafter (see Figure 7.99). The plumb bevel for rafters A and B along with the edge bevel creeper are shown in Figure 7.100.

Remember to make deductions in the rafter length equal to half the thickness of the common rafter and crown end rafter (see Figure 7.101).

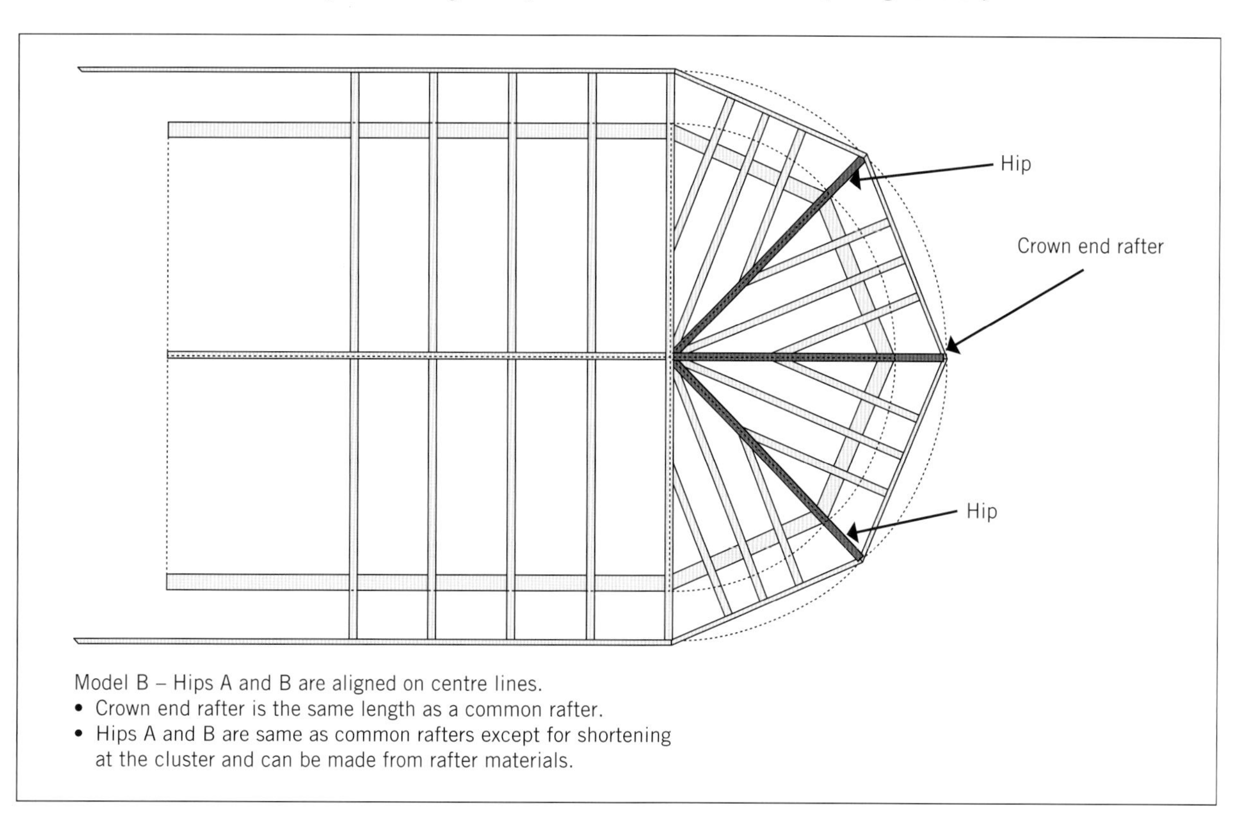

FIGURE 7.98 Semi-octagonal roof – Model B

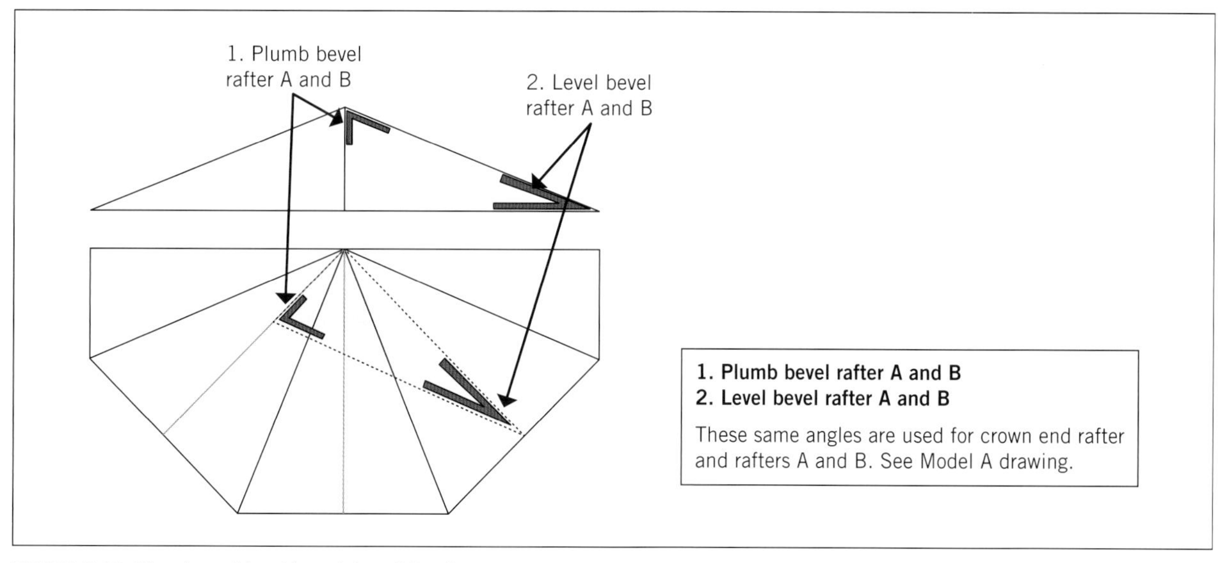

FIGURE 7.99 Plumb and level bevel A and B rafters

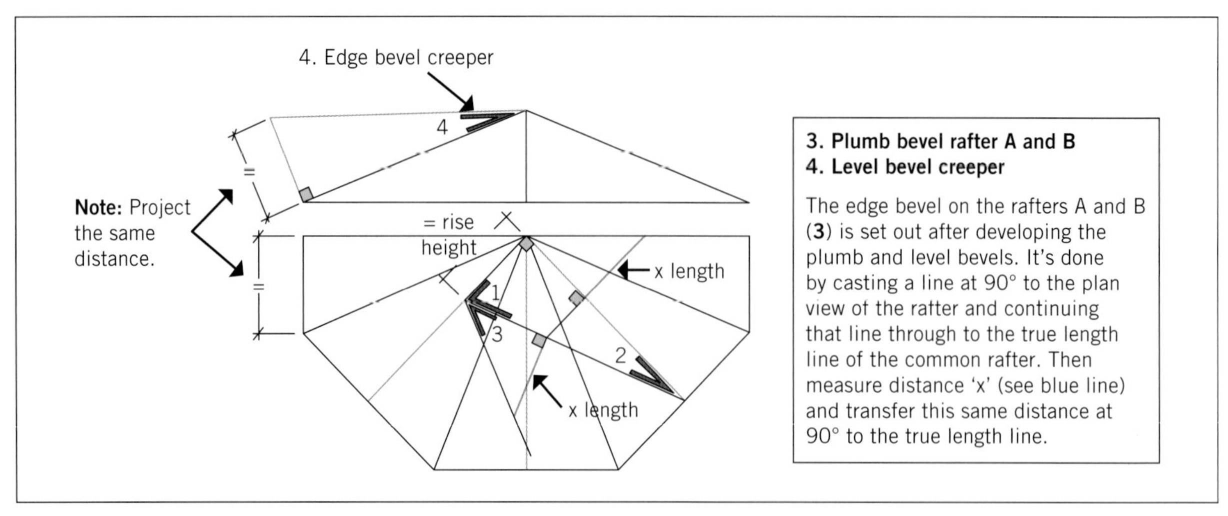

FIGURE 7.100 Edge bevels rafter A and B and edge bevel creeper

Step 3

In Model A, the hips have extended eave overhang to allow for the parallel running of the fascia to the wall structures. This makes them longer and so you need to set them out the same as for a normal hip, except the hip angle is 22.5° rather than the 45° on the plan (see Figure 7.97 and Figure 7.102). This works for all four hip rafters. The geometric set-out for the hip plumb and the level and edge bevels are shown in Figure 7.103.

The geometric set-out for purlin bevels is shown in Figure 7.104.

An alternative to making the calculations on the lengths of hips is to use them in situ. A good example of this can be found on the Buildsum YouTube channel ('Establish hip length – measured in situ': https://www.youtube.com/watch?v=PwU8pWOMcaw).

In Model B the hips are the same as the common rafters except at the top where deductions at the cluster impact the length (see Figure 7.98).

The construction is basically the same, starting with the crown end rafter, followed by the hips and then in-filled with creepers, purlins and struts.

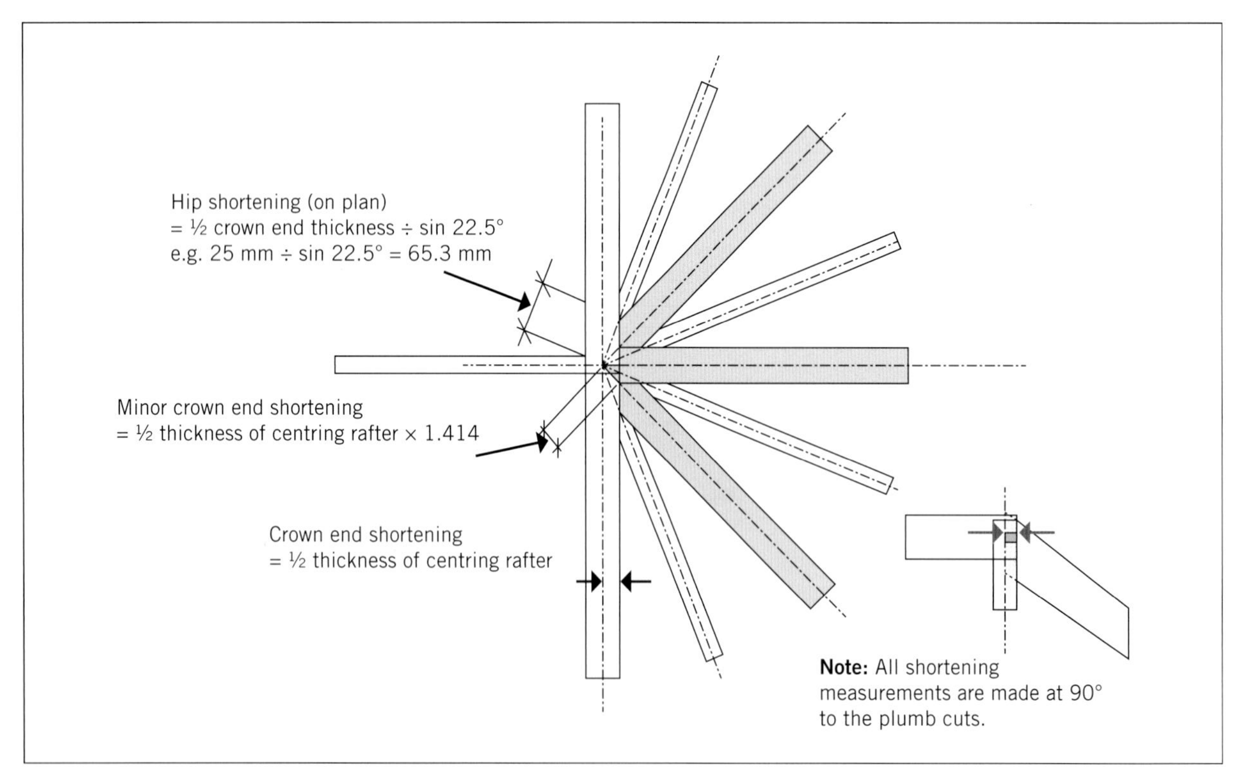

FIGURE 7.101 Cluster deduction Model A

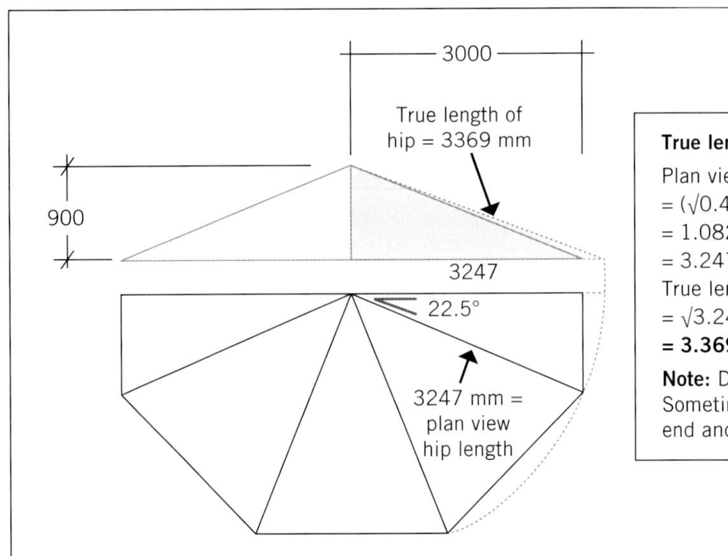

True length of hip calculations:

Plan view hip length = $(\sqrt{\tan 22.5^2 + 1.0^2}) \times \frac{1}{2}$ span

= $(\sqrt{0.4142 + 1.0^2}) \times 3.0$ m

= 1.082 m (hip factor) × 3.0 m (½ span)

= 3.247 m

True length of hip = 3.247 m × rise height

= $\sqrt{3.247^2 \times 900^2}$

= 3.369 m

Note: Deductions for cluster will need to be made. Sometimes best practice is to mark in situ once crown end and other rafters are in place.

FIGURE 7.102 True length of hip calculations

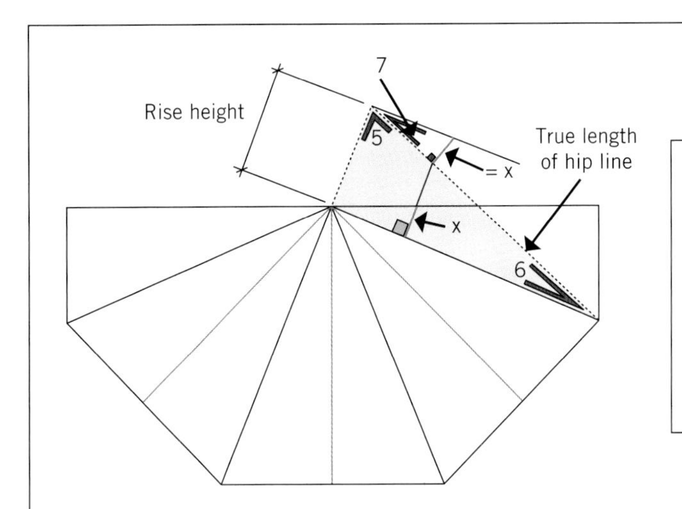

5. Plumb bevel hip

6. Level bevel hip

7. Edge bevel hip

The edge bevel on the hip is set out after developing the plumb and level bevels. It's done by casting line at 90° to the plan view of the hip and continuing that line through to the true length line of the hip. Then measure distance 'x' (see blue line) and transfer this same distance at 90° to the true length line.

FIGURE 7.103 Plumb bevel, level bevel and edge bevel hip

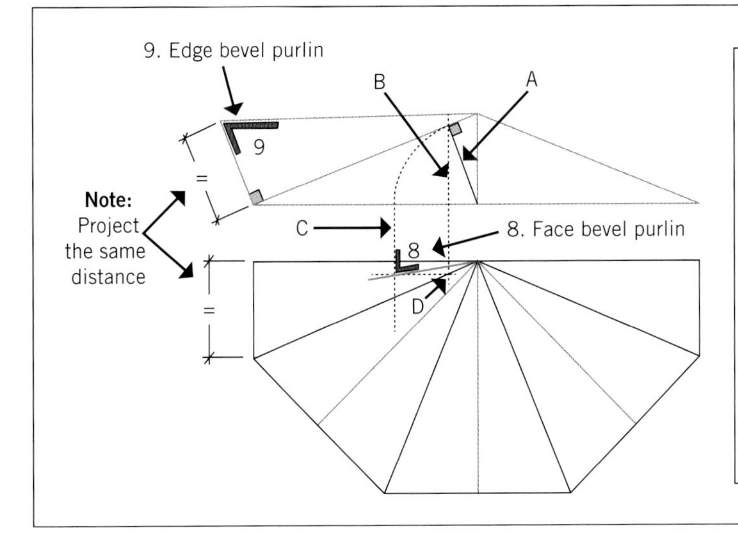

8. Face bevel purlin

9. Edge bevel purlin

The face bevel on the purlin is done by casting line at 90° to the elevation view of the common rafter to the ½ span point on the bottom chord of the triangle (see line A). Project the line down using a compass (or measuring to the horizontal) and project down from the elevation lines B and C to the plan view. Where line B intersects the hip at point D scribe a horizontal line to the intersect line C. Draw a line from the apex of the roof cluster to the intersections of lines C and D to form the face bevel purlin.

Note: The edge bevel purlin is shown as opposite the level bevel creeper (4 in Figure 7.100).

FIGURE 7.104 Face and edge bevels for purlin

Intersecting roofs with different pitches

Intersecting roofs of unequal pitch are similar in construction methods to a Scotch Valley roof. Take an opportunity to review Chapter 10 in *Site Establishment, Formwork and Framing* (Laws 2020), particularly the section on Scotch Valley construction, and the section earlier in this chapter, 'The mathematics: the Seven Pillars revisited'.

Intersecting roofs of unequal pitch present a few challenges where decisions on construction methods impact the setting out and calculations (see Figure 7.105).

When setting out intersecting roofs of unequal pitch you will be required to have different sets of bevels for each roof (see Figure 7.106).

A good idea is to use two pieces of timber to set out all bevels early on. Watch 'Hip roof bevels direct method' from Buildsum's YouTube channel: https://www.youtube.com/watch?v=nhP6lR4u9DY.

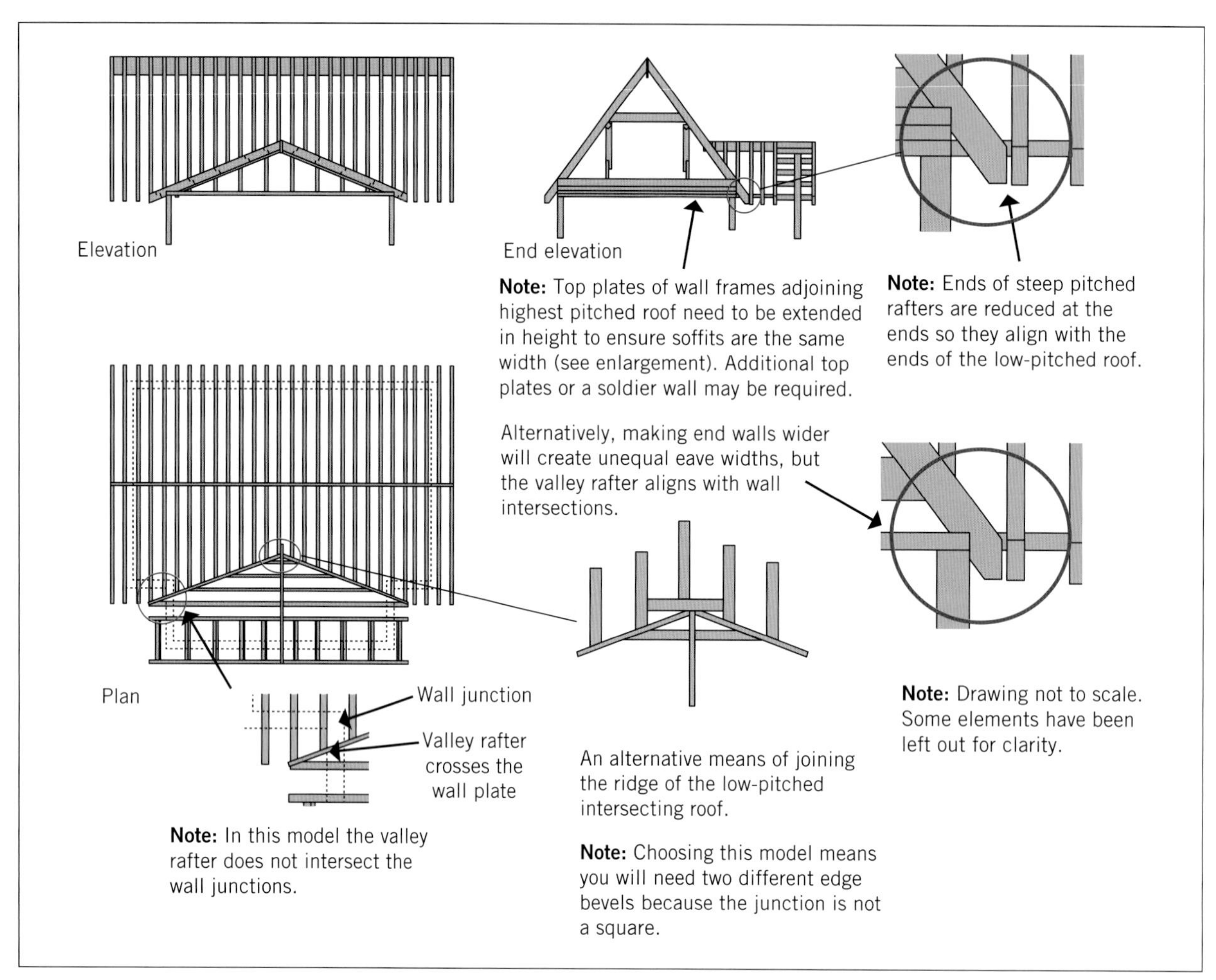

FIGURE 7.105 Unequal pitched roof

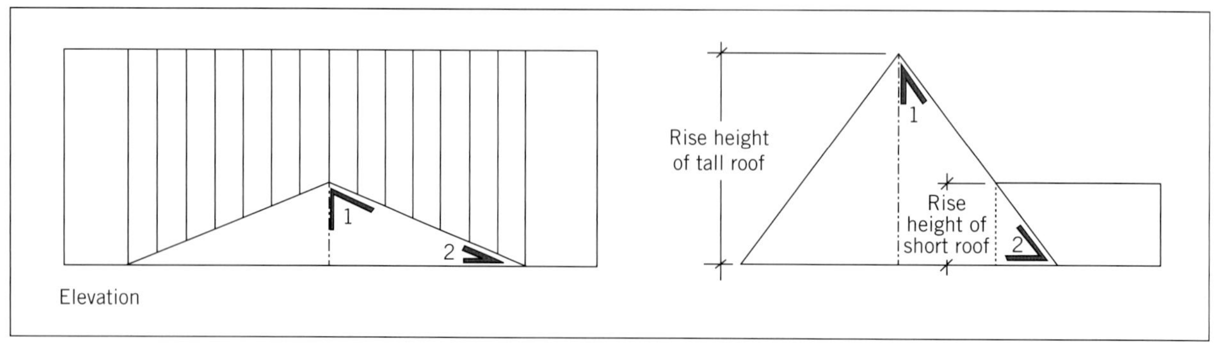

FIGURE 7.106 Plumb and level bevels for tall and low roofs and valley rafters

Challenges

Valley rafter or pitching plates?

A valley rafter is shown in Figure 7.105. An alternative model would be to use a pitching plate (or valley board) laid across the top of extended rafters on the high-pitched roof, adding another layer of framing complexity (see Figure 7.107). If you choose to go with a pitch board, you will need to span the walls across the top plates of the low-pitched roof with either multiple plates or a beam equivalent to the raised height of the tall pitched roof extension. The rafters on the tall pitched roof become common rafters without the extension for eaves. A pitching plate is laid across the rafters instead of a valley rafter. Watch the Buildsum YouTube video 'Broken hip and valley roof Scotch Valley' (https://www.youtube.com/watch?v=wSjet1eh-1Q) on how to install the pitch boards.

Equal or unequal eaves widths

Decide whether to have eave widths the same all around the building or different widths on each roof. To achieve equal width eaves on unequal pitched roofs requires you to raise the wall heights on the walls below a steep pitched roof section. This will affect how you frame out the ceiling. The use of a suspended ceiling system will overcome that internal problem.

If you are having unequal eave widths, all wall frames can remain at the same height.

It doesn't matter if the eaves width is the same or not, the ends of the steep pitched rafters will need to be trimmed off to make the soffits sheets align all round (see Figure 7.105).

Follow the basic steps outlined in Chapter 10 of *Site Establishment, Formwork and Framing* (Laws 2020) 'Scotch Valley construction'.

Locating the low-pitched roof ridge intersection point

Locate the position of the ridge for the low-pitched roof relative to the steep pitched roof. On the plan view of Figure 7.105, you see the ridge is aligned to attach to a rafter. This is OK if the exact position of the intersecting roof is not critical, as may be the case when adding a pergola or carport. But where the ridge line is critical to the alignment of rooms and eaves, a trimmer should be installed between rafters to allow the exact placement of the ridge. See the framing alternative in Figure 7.105.

To calculate the position down along a common rafter you will need to do a few calculations (see Figure 7.108). Practise working these out by completing Learning Task 7.7.

LEARNING TASK 7.7 SHORT ROOF RIDGE INTERSECTION

1. Calculate the intersection position, C, of the tall roof rafters and the ridge for the short roof.
   Note: Fill in the missing dimensions first.

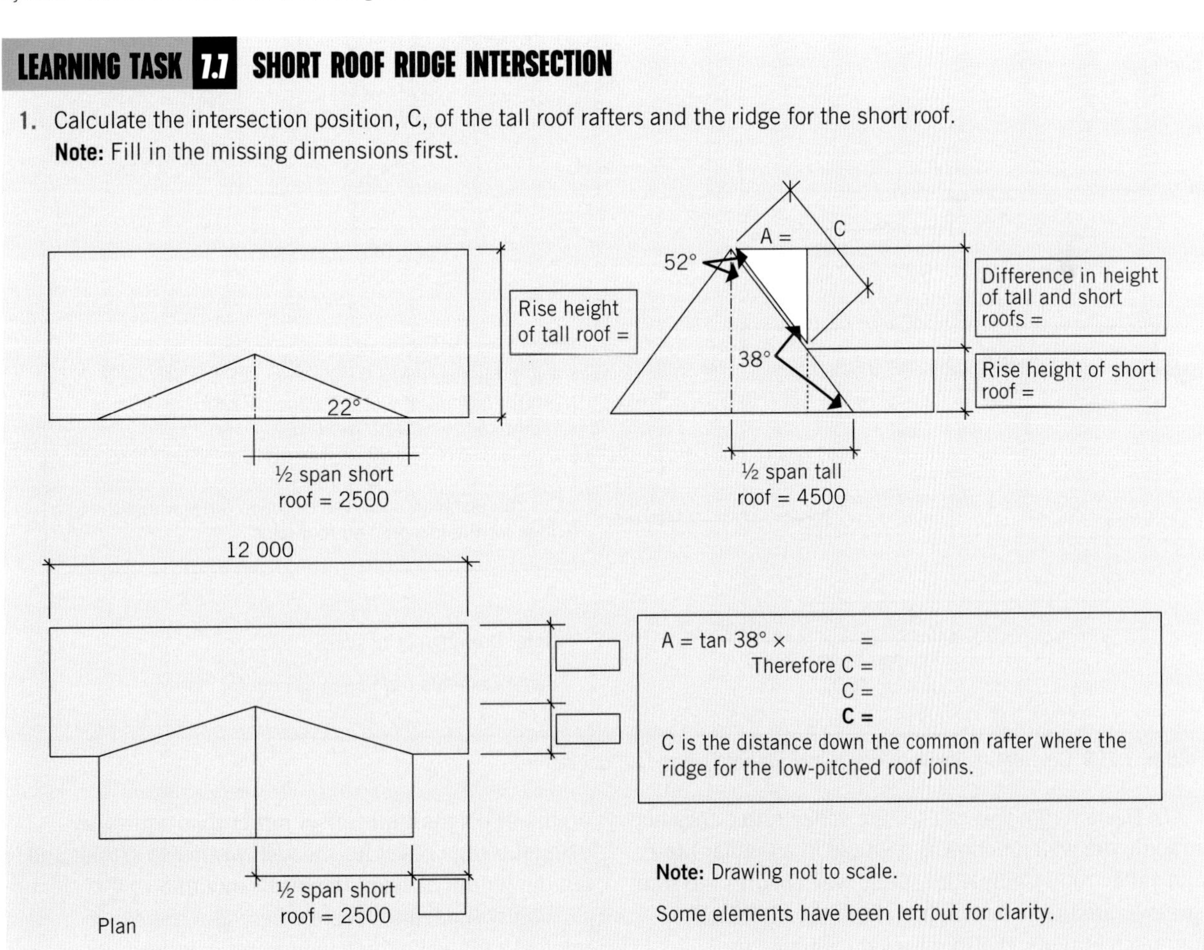

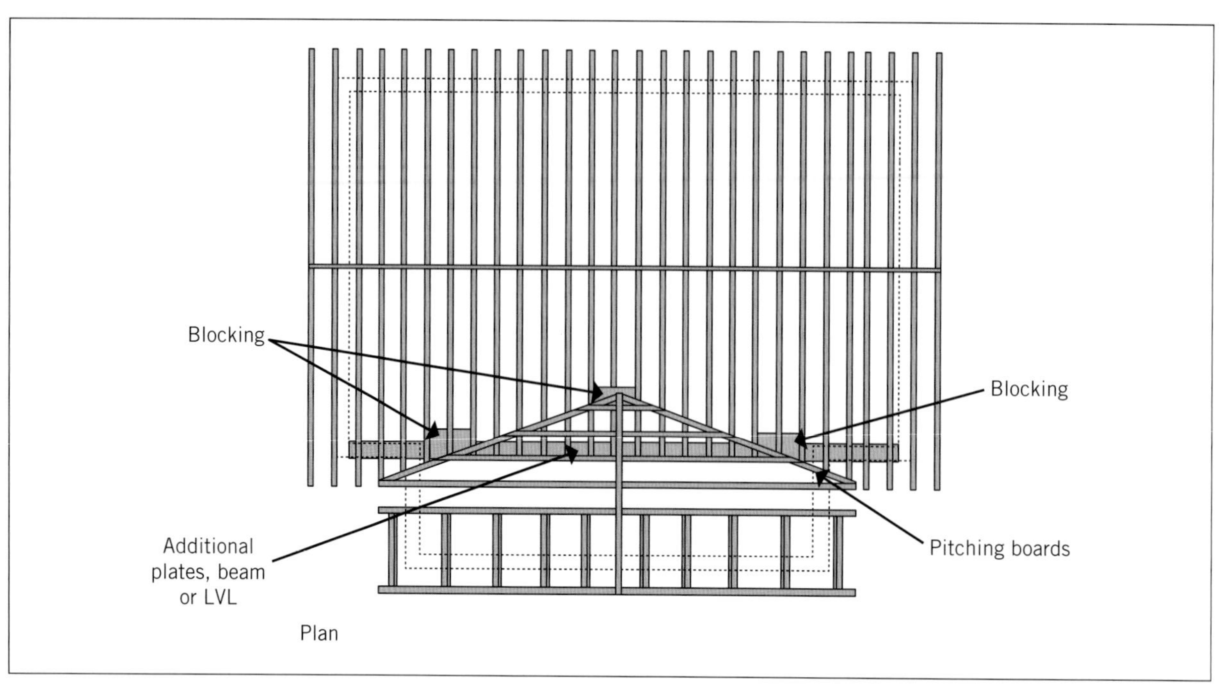

FIGURE 7.107 Pitching plates

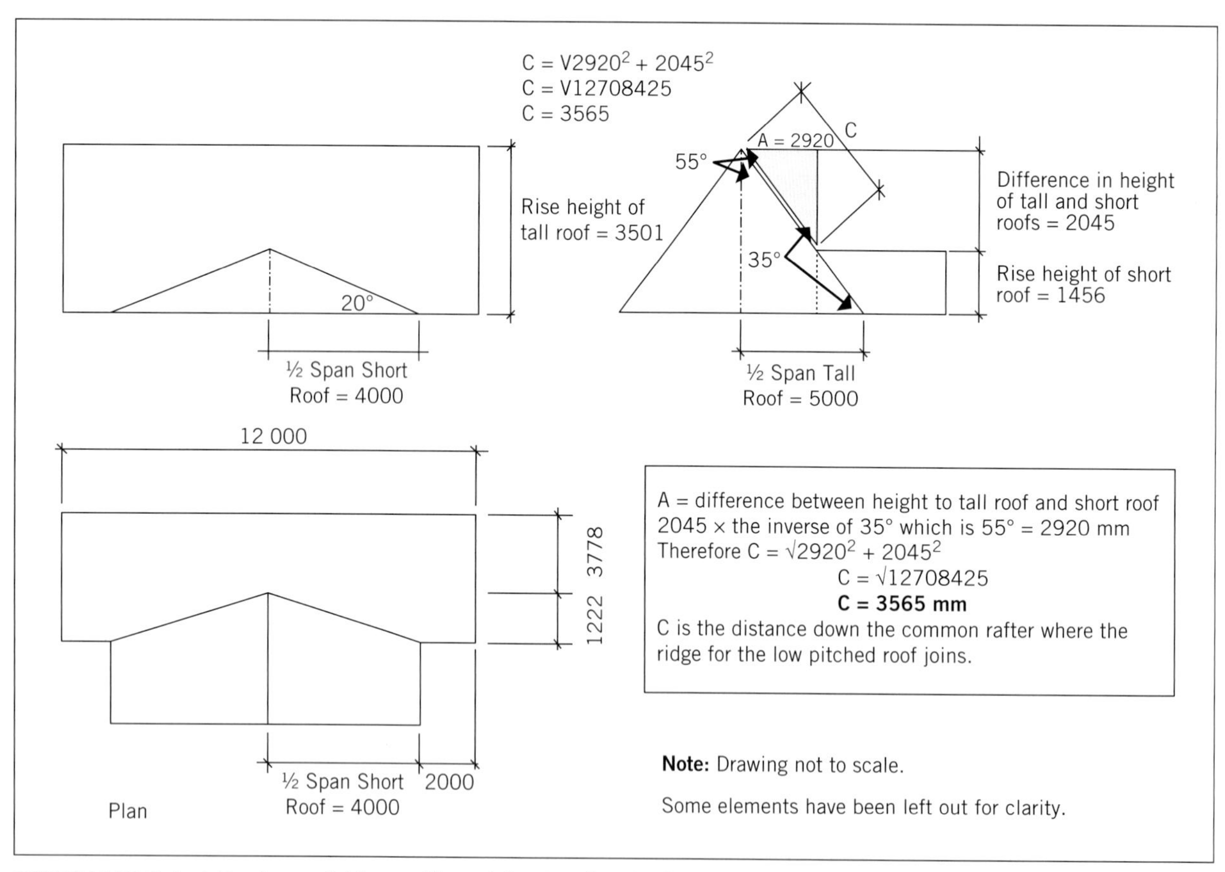

FIGURE 7.108 Calculating low roof ridge position relative to tall roof rafters

In Figure 7.105, you can see the valley rafter does not straddle the wall junction as it would in a regular hip and valley roof. Completing it this way ensures the soffit or eaves are the same width on both the high- and low-pitched roofs. It also means the plumb cut on the bird's mouth is offset rather than square by using the edge bevel for the valley rafter to set out and cut the plumb bevel of the bird's mouth.

You will need to locate where along the low-pitched roof walls the valley rafter crosses over.

LEARNING TASK 7.8

VALLEY RAFTER CALCULATIONS

1. Calculate the true length of the valley rafter C.

C =
1500 mm rise height
A
½ span short roof = 4500
1300 mm

Plan view length A =
A =
A =

The true length of valley rafter C =
C =
C =

You will have already built the wall frames and determined how the intersecting roofs are going to be built (i.e. valley rafter or pitching plate). Work from the centre line of the main roof to mark out where the valley rafter or pitching plate will finish on the low-pitched roof walls. A string line, spirit level and tape can be used to find the point.

Valley rafter length

To calculate the valley rafter length, you apply the same principles as for a hip. The difference here is that you know what the half span of the roof is, but not necessarily the angle of the valley rafter. You can calculate the valley rafter length by scaling the distance out from the ridge/steep pitched roof intersection on the plans if it's not already indicated. You can also scale it from the drawing if it's not already shown (see Figure 7.109).

Another alternative is to measure and mark the piece in situ.

Once you know the true length of the valley rafter, you apply the same principles in setting out the valley

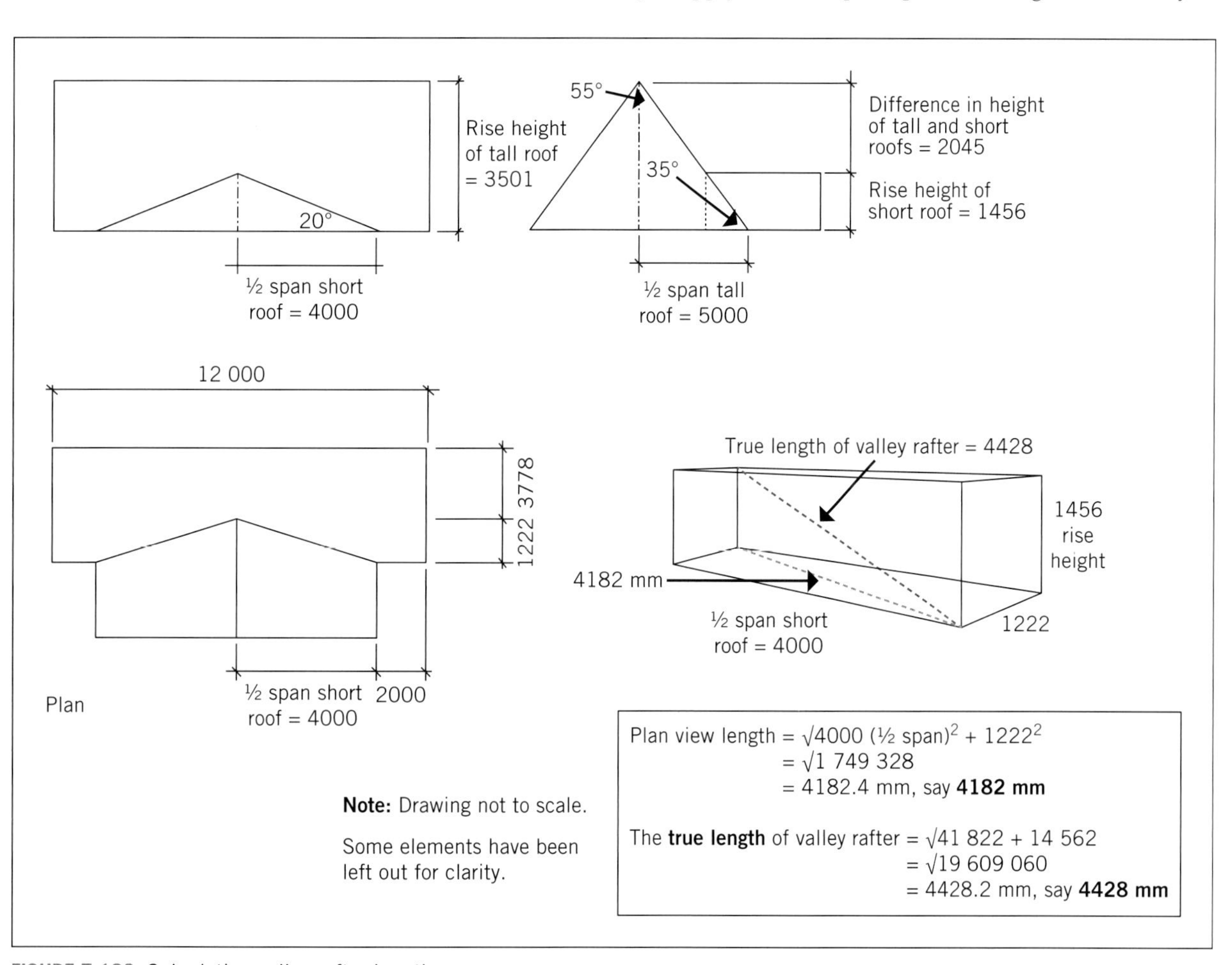

FIGURE 7.109 Calculating valley rafter length

rafter as you would for a regular valley rafter; that is, make deductions for top cluster and add amounts for eaves overhang.

Once the ridge and common rafters for the low-pitched roof are installed, the valley rafter is installed.

Setting out the bevels

The edge bevels for the creepers and cripple creepers on both roofs will be different. Separate set-outs and pattern rafters will be required.

To refresh your memory on setting out pattern rafters watch the Buildsum video, 'Hip roof pattern rafter' (https://www.youtube.com/watch?v=TcN3O7owg9I) and 'Hip roof bevels direct method' (https://www.youtube.com/watch?v=nhP6lR4u9DY).

The main plumb and level bevels for common rafters and the valley rafter set-outs for both roofs are shown in Figure 7.110.

The edge bevel for the valley rafter is set out in Figure 7.111.

The set-outs for the edge bevels on creepers are shown in Figure 7.112.

In summary, joining roofs of unequal pitch is challenging, but with thought and patience it can be done successfully. The key points are:

- you must decide whether or not to keep eaves widths the same
- two different set-outs are required
- the use of pitching plates may make it easier, but may also use more materials and require rafter adjustments and additional plating to wall frames to maintain ceiling heights internally.

The use of roof trusses as an alternative to a pitched roof may make the problem of intersecting roofs of different pitches much simpler.

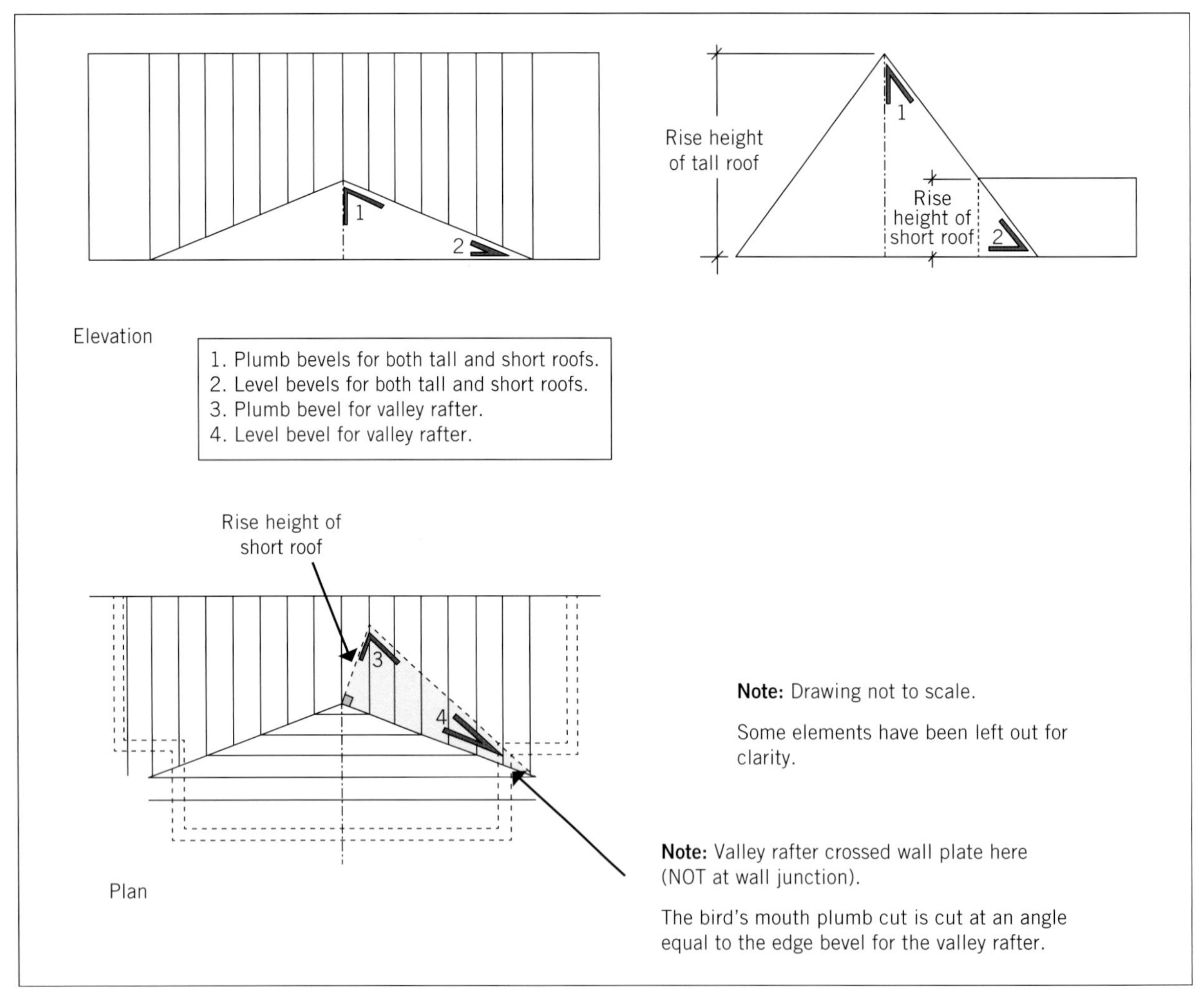

FIGURE 7.110 Plumb and level bevels for tall and low roofs and valley rafters

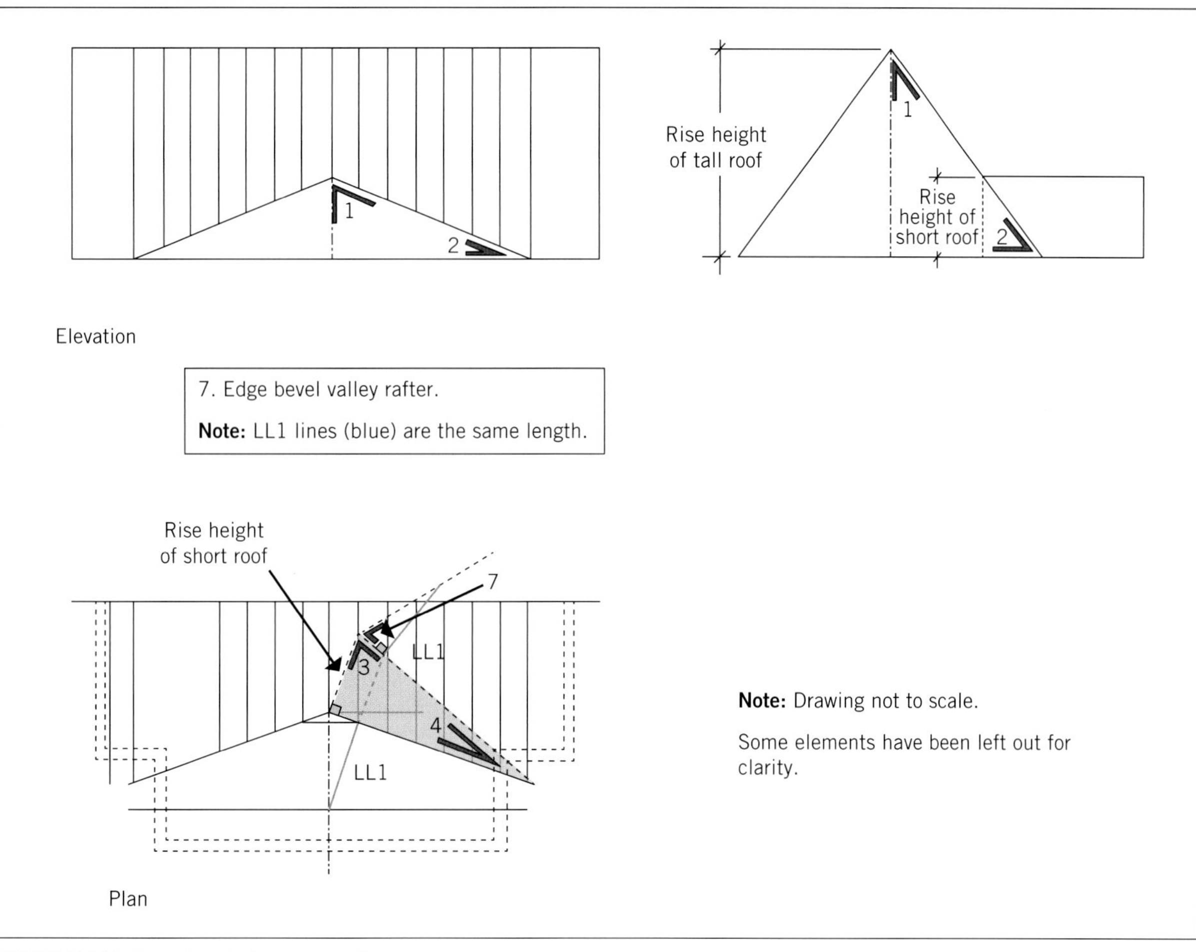

FIGURE 7.111 Edge bevels for valley rafter

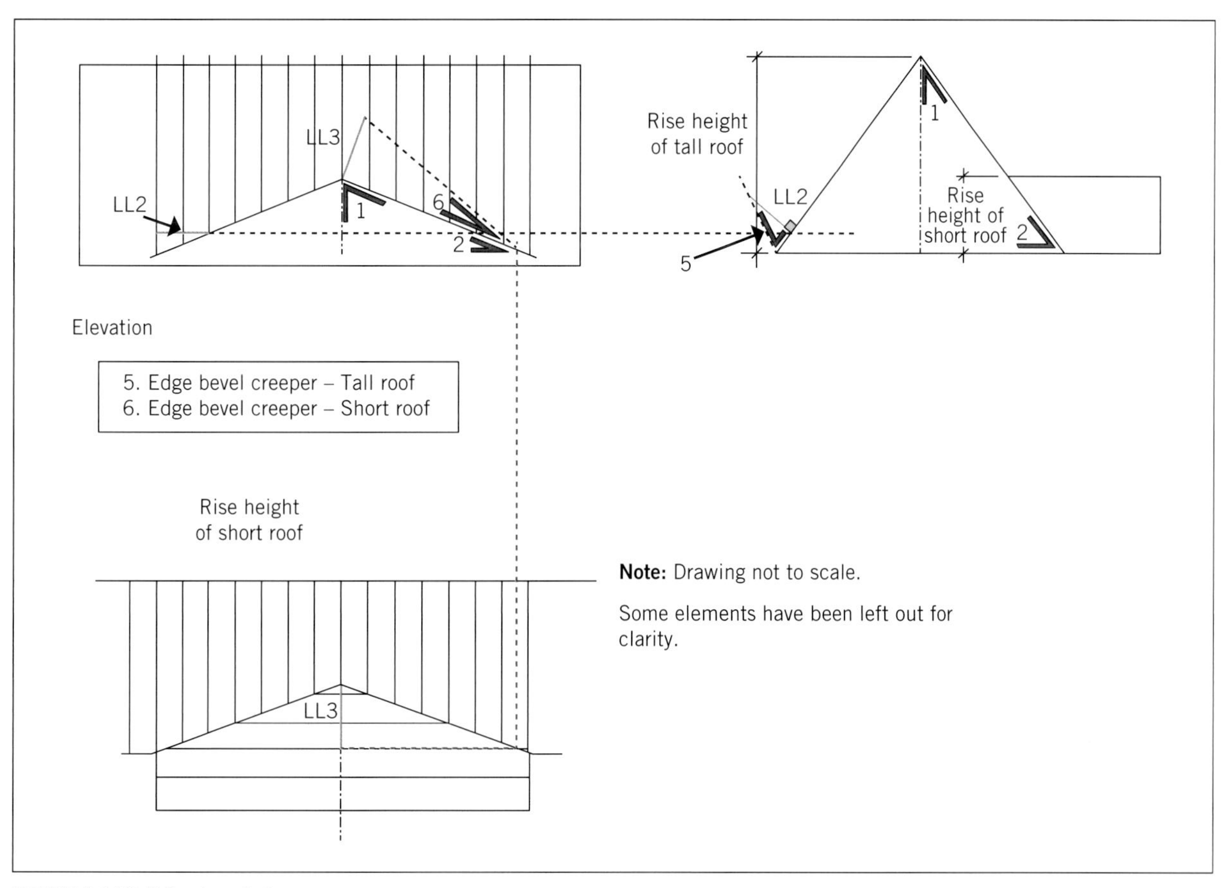

FIGURE 7.112 Edge bevels for creepers

SUMMARY

This chapter addressed the topics outlined below relating to constructing advanced roofs:

- sustainability in roof design, materials and construction methods
- the basic principles of roofing:
  - roofs are made of triangles
  - roofs are set out to centre lines
  - ridges run parallel to wall plates
  - rafters run at 90° to plates
  - hips and valleys bisect intersecting roofs
- the 'Seven Pillars' calculations used in roofing:
  1. common rafter rise per metre run (CR rise/m run) = Tan y
  2. common rafter length per metre run (CR factor)
  3. common rafter set-out length (CR set-out length)
  4. common rafter order length (CR order length)
  5. hip length per metre run of common rafter (hip factor)
  6. hip set-out length
  7. hip order length
- the eight bevels required to set out and cut a roof (see Chapter 10 of *Site Establishment, Formwork and Framing*; Laws 2020):
  1. plumb bevel – common rafter
  2. level bevel – common rafter
  3. plumb bevel – hip
  4. level bevel – hip
  5. edge bevel – hip
  6. edge bevel – creeper
  7. face bevel – purlin
  8. edge bevel – purlin
- gambrel (or Dutch gable) roofs, which are hipped roofs with an extended ridge. The ridge set-out is determined by the position of the last common rafter. This includes the calculations for jack rafters and reductions and adjustments required to align various roof members
- jerkin head roof, which is useful where more roof space is required for a living or storage area. Not often used in contemporary architecture, the roof has an extended ridge but the hips and ridges move with the ridge. This roof form requires a soldier wall. Calculations for jack rafters and reductions and adjustments required to align various roof members are also explained
- skewed gable and oblique hip roofs, which are 'splayed ended roofs' requiring creeper rafters with each length varying according to its position. They are also built to centre lines
- unequal pitched roofs – not built on centre lines. They incorporate two or more roof surfaces using different pitches and are often found in alterations, or where water catchment may be critical to the design. The design of the roof means extra bevel calculations need to be made because of the unequal pitch on each side
- shortening' or 'offsets' of measurements when determining the cut lengths required, which are critical in the development and calculations of all lengths (see Figures 7.6, 7.30, 7.31, 7.32, 7.34, 7.61, 7.65 and 7.77):
  - half ridge thickness for common rafters
  - half rafter thickness for crown end rafters
  - crown end cluster offsets for hip calculations
  - hip reductions for different types of construction.
- purlins and struts, which support rafters and transfer roof loads to load bearing walls. The purlin size can determine the number and spacing of struts (see the supplementary Span Tables of AS 1684). The internal layout of the building may require special strutting (e.g. fan struts) and the use of strutting beams
- roofing principles as they apply to dormer windows
- octagonal or semi-octagonal roofs, which require multiple centring rafters and hips. Calculations for jack rafters, and the reductions and adjustments required to align various roof members, are also explained
- applying the principles to setting out intersecting pitched roofs of unequal pitch. Each roof of unequal pitch must be treated as an individual roof, meaning common rafters and jack rafters will be different on intersecting roof planes, with bevels and plumb cuts different on intersecting pairs of planes.

REFERENCES AND FURTHER READING

Texts

Laws, A. (2020), *Site Establishment, Formwork and Framing* (4th edn), Cengage Learning Australia, Melbourne, VIC.

Staines, A. (2012). *The Roof Building Manual* (5th edn), Pinedale Press, Caloundra, Qld.

Web-based resources

The Buildsum YouTube channel has many useful video resources on roofing that use Australian terminology, metric measurements and the Australian Standards: **https://www.youtube.com/user/Buildsum**.

Safe Work Australia has developed a number of Model Codes of Practice (MCoP) relevant to advanced roof construction. These include:

- MCoP – Construction work
- MCoP – How to manage work health and safety risks
- MCoP – Managing the risk of falls at workplaces.

The MCoP must be approved as a code of practice in that state or territory. Check with the WHS/OHS regulator in your state or territory to see if MCoP has been adopted or an alternative guide is available.

SlidePlayer (**https://slideplayer.com**) has many useful video and PowerPoint presentations on developing roof bevels. You can embed the links into your own presentations. For fast search results use appropriate terms like 'oblique roof'.

Relevant Australian Standards
AS 1684.2 Residential timber-framed construction – Non-cyclonic areas
AS 1684.3 Residential timber-framed construction – Cyclonic areas
Australian Standards can be purchased from the Standards Australia website: **https://www.standards.org.au/**, or see your teacher or librarian for assistance in accessing Australian Standards online.

GET IT RIGHT

CONSTRUCT ADVANCED ROOFS

Questions

Below are two pictures showing examples of varying eave widths along the roof length.

1 What could be a problem with eaves in the first picture that have the gutter inside the eaves and the outer walls?

____

____

2 What could be done to make the roof easier to waterproof?

____

____

3 What could happen if nothing is done?

____

____

HOW TO

USE A ROOFING SQUARE TO SET OUT A ROOF'S ANGLES

To see a demonstration of how the calculations are done, view mastertradeskills.com videos at: https://www.youtube.com/playlist?list=PLG-1AejO-HiMHWaeVyuaKUqb554cutetm.

Watch the 'Hip roof bevels steel square' video for more detail: https://www.youtube.com/watch?v=DJ34JxKZcEQ.

Tools required:

- framing square
- stair gauge (or stair buttons)
- combination square or similar
- sliding bevel
- pencil.

Steps involved		
1	**Plumb and level bevel – common rafter** Calculate and set out the rise/m run to find the common rafter plumb and level bevels cut angles. For example: Common rafter rise/m run Where the run = 1.000 m and roof pitch is 25° Rise/m run tan 25° = 1:000 m = 0:466 rise This can be scaled by dividing by 2; that is: • run = 1 m ÷ 2 = 500 mm on the blade • rise 466 mm ÷ 2 = 233 mm on the tongue. Keep the longest length on the blade and shortest on the tongue. **Note:** Both the **common rafter** and **hip plumb** and **level bevel** are used to mark bird's mouth cuts on each respective piece.	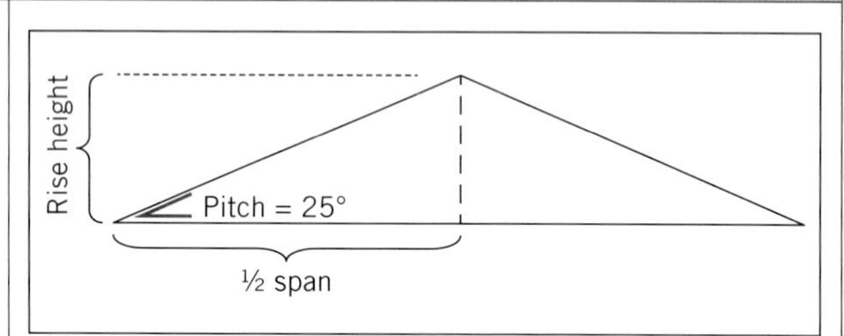  Source: Greg Cheetham 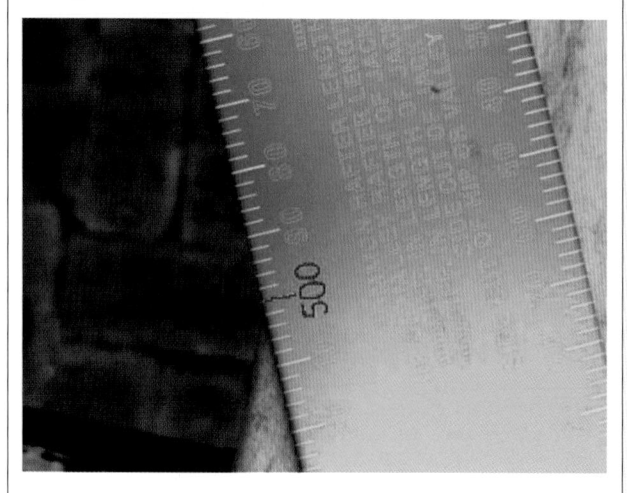 Source: Greg Cheetham

>>

>>

		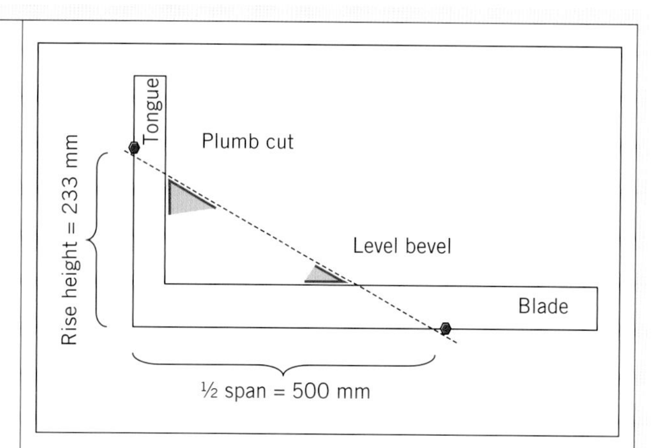
2	**Plumb and level bevels – hip** Similar to common rafter level bevels. Calculate and set out the rise/m run to find the hip – plumb and level bevels cut angles. For example: Hip – rise/m run Use Pythagoras' theorem - ($c = \sqrt{a^2 + b^2}$); that is: $c = \sqrt{1^2 + 1^2} = 1.414$ Where the run = 1.414 m and roof pitch is 25° and the rise remains 0.466 m. You need to divide both rise and run by 3 to fit them on the roofing square. For example: 1.414 ÷ 3 = 0.4713 (say, 471 mm) on blade 0.466 ÷ 3 = 0.1553 (say, 155 mm) on tongue.	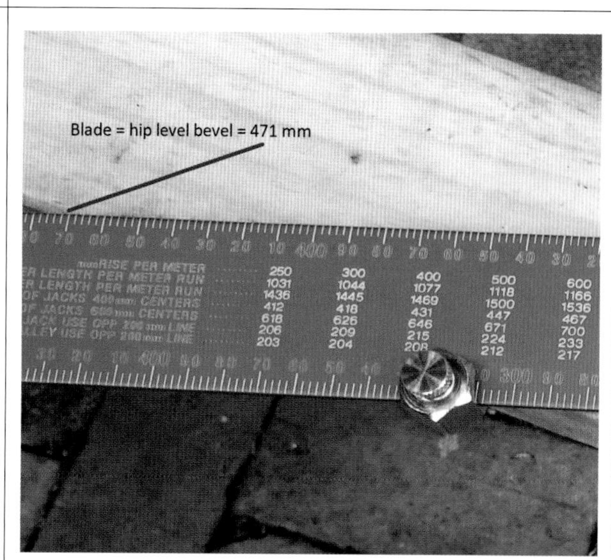 **Source:** Greg Cheetham 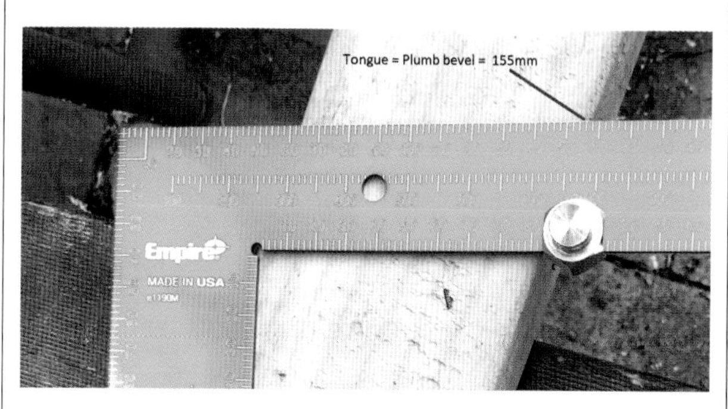 **Source:** Greg Cheetham 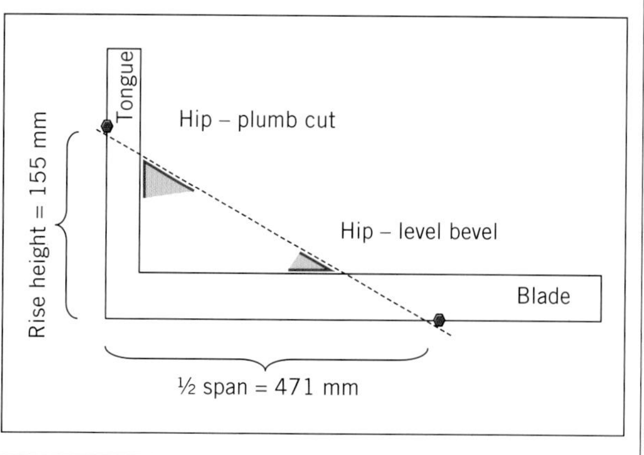

>>

>>

3	**Edge bevel – hip** Having worked out the hip – plumb and level bevels, we use the plumb bevel to develop the edge bevel. Using level bevel hip arm (blade), slide it along until the thickness of the hip material aligns with the top edge of the hip material. In this instance we use a piece of 90 mm × 45 mm timber because it makes the diagrams bigger. The alternative is to hold an off-cut of hip material parallel to the plumb bevel and mark the point at which the material intersects the edge. Square the line across the edge and mark a line between the apex of the plumb and where the squared line meets the opposite edge. This gives the hip edge bevel. Finally, transfer the angle using a sliding bevel to form the 'arrowhead' of the edge bevel hip.	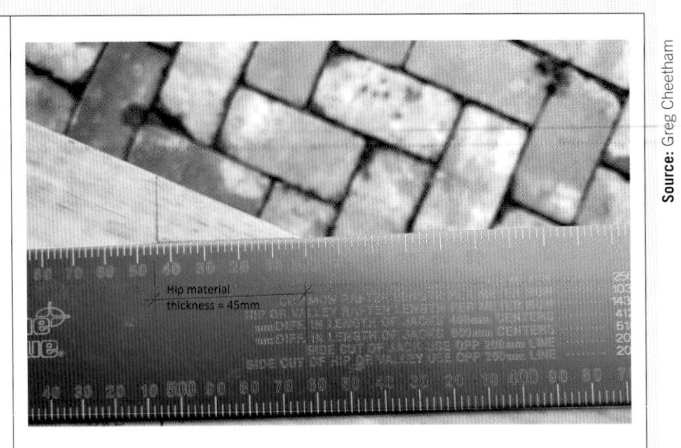 **Source:** Greg Cheetham  **Source:** Greg Cheetham 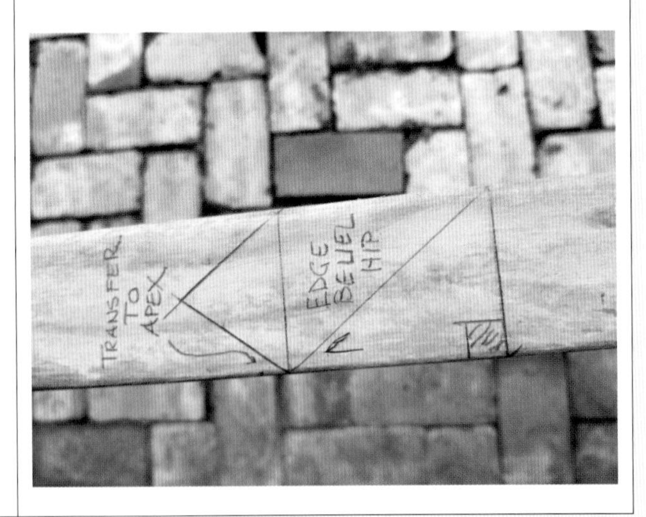 **Source:** Greg Cheetham

>>

>>

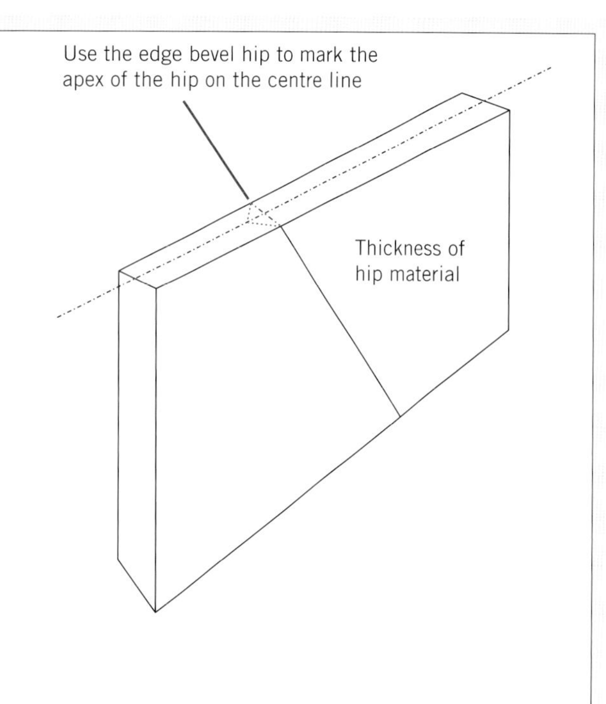

4 Edge bevel – creeper

We use the plumb and level bevel common rafter for this work. The same practicc used in developing the edge bevel – hip (EB Hip) is used to develop the edge bevel – creeper (EB Creeper).

Use the level bevel common rafter and mark the thickness of creeper materials at 90° to the plumb bevel common rafter.

Again, the alternative is to hold an off-cut of creeper material parallel to the plumb bevel and mark the point where the material intersects the edge.

Square the line across the edge and mark a line between the apex of the plumb level and where the squared line meets the opposite edge. This gives the edge bevel creeper angle.

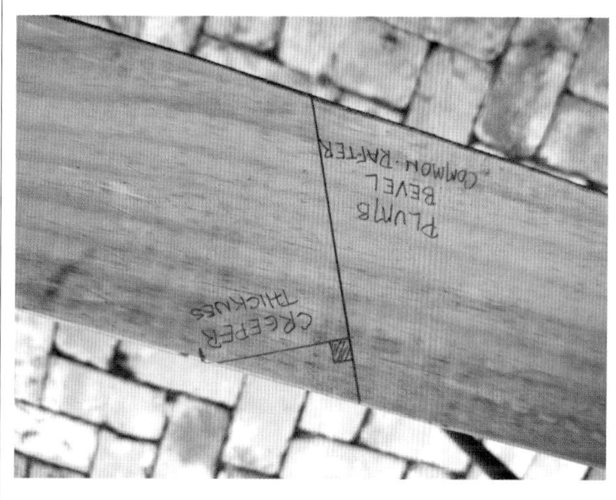

Source: Greg Cheetham

Source: Greg Cheetham

>>

>>

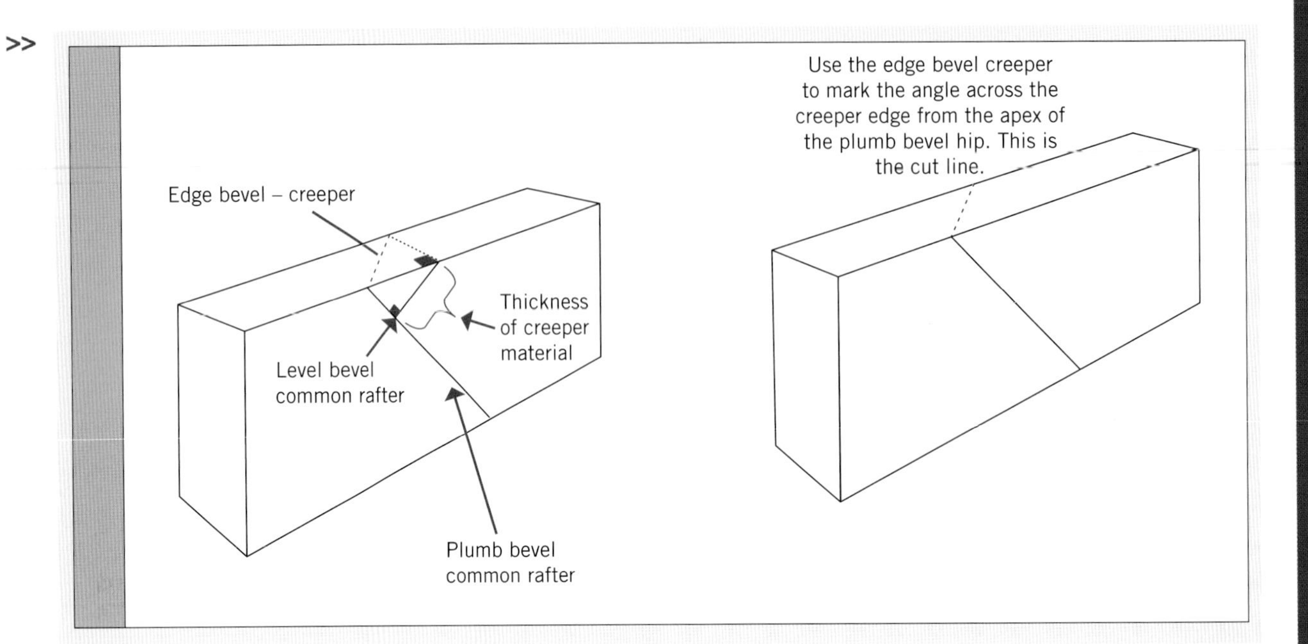
Edge bevel – creeper
Level bevel
common rafter
Thickness
of creeper
material
Plumb bevel
common rafter
Use the edge bevel creeper
to mark the angle across the
creeper edge from the apex of
the plumb bevel hip. This is
the cut line.

WORKSHEET 1

To be completed by teachers	
Student competent	☐
Student not yet competent	☐

Student name: ______________________

Enrolment year: ______________________

Class code: ______________________

Competency name/Number: ______________________

Task

Roof type:

Jerkin head

Characteristics:

Pitch: 40°

Span: 6000 mm

Ridge extension: 1800 mm

Rafter spacing: 600 mm

Rafters: 125 × 35 mm

Hips: 150 × 35 mm

Eave: 450 mm

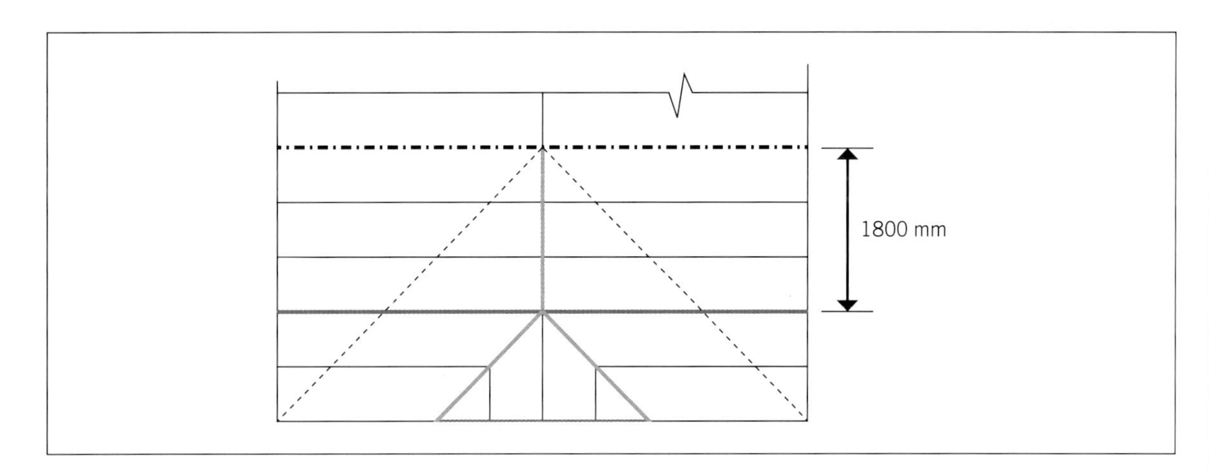

Find:

1 The height of the soldier wall.

2 The set-out length of the first creeper.

WORKSHEET 2

To be completed by teachers	
Student competent	☐
Student not yet competent	☐

Student name: ______________________________

Enrolment year: ______________________________

Class code: ______________________________

Competency name/Number: ______________________________

Task

Roof type:

Skewed gable

Characteristics:

Pitch: 35°

Span: 6800 mm

Rafter spacing: 450 mm

Rafters: 125 × 35 mm

Hips: 150 × 35 mm

Eave: 600 mm

Skew angle: 25°

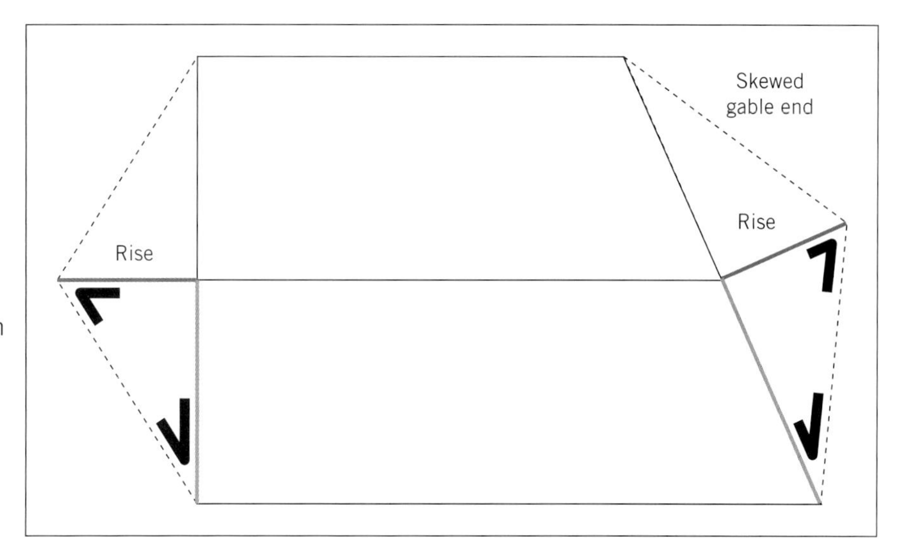

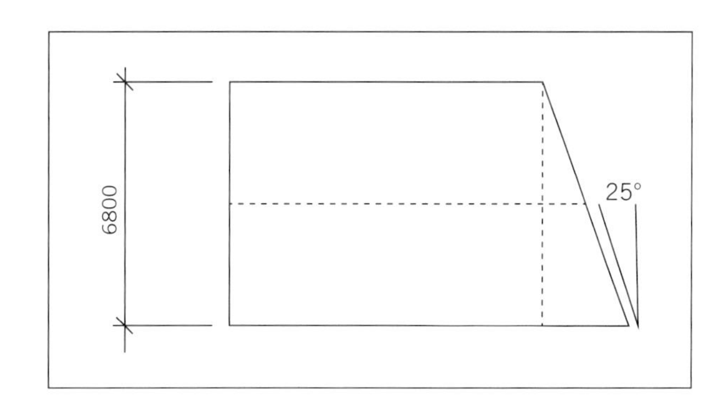

Find:

1 The rise

2 The skew offset distance

3 The run of the skewed gable rafters

4 The creeper shortening

Creeper skew factor =

CR rise/m =

WORKSHEET 3

To be completed by teachers

Student competent ☐

Student not yet competent ☐

Student name: ______________________

Enrolment year: ______________________

Class code: ______________________

Competency name/Number: ______________________

Task

Roof type:

Unequal pitch

Characteristics:

Pitch: Main 30°/65°; Hip end 35°

Span: 5400 mm

Rafter spacing: 450 mm

Rafters: 150 × 45 mm

Hips: 200 × 35 mm

Eave: 600 mm

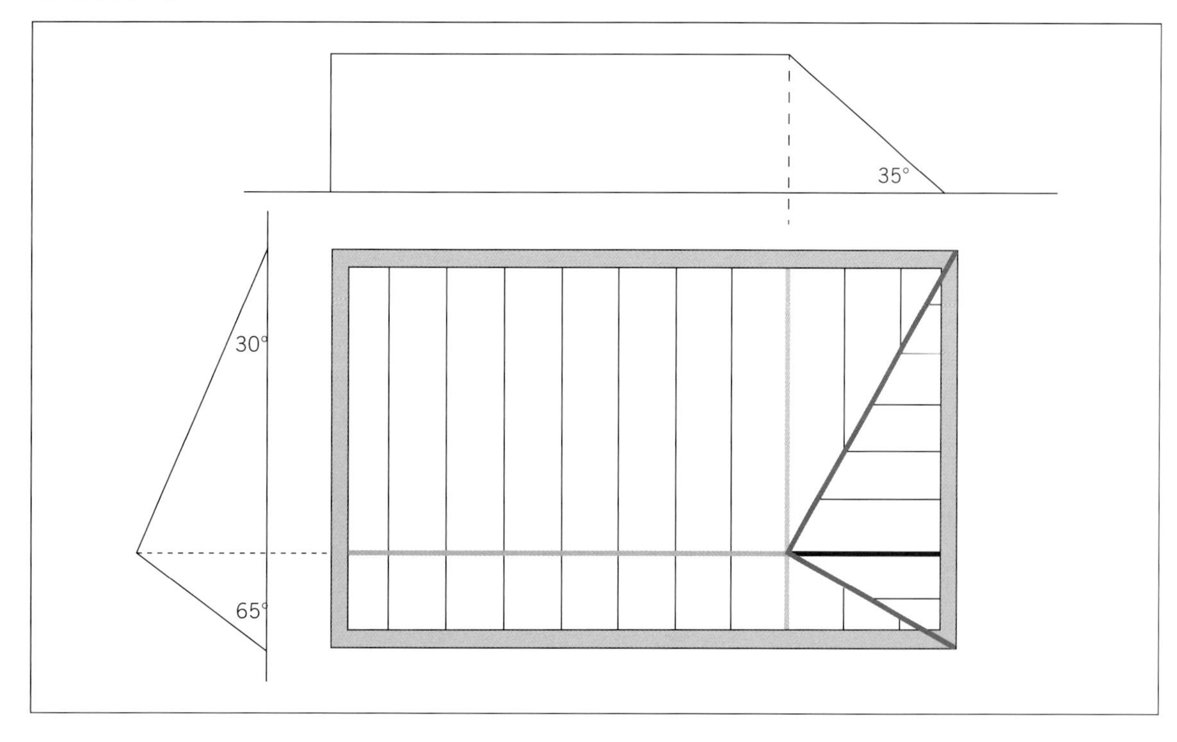

Find:

1 The rise

2 The centring rafter back-set distance

3 The hip factor

4 The creeper shortenings

WORKSHEET 4

To be completed by teachers	
Student competent	☐
Student not yet competent	☐

Student name: ____________________

Enrolment year: ____________________

Class code: ____________________

Competency name/Number: ____________________

Name the parts for this gambrel roof:

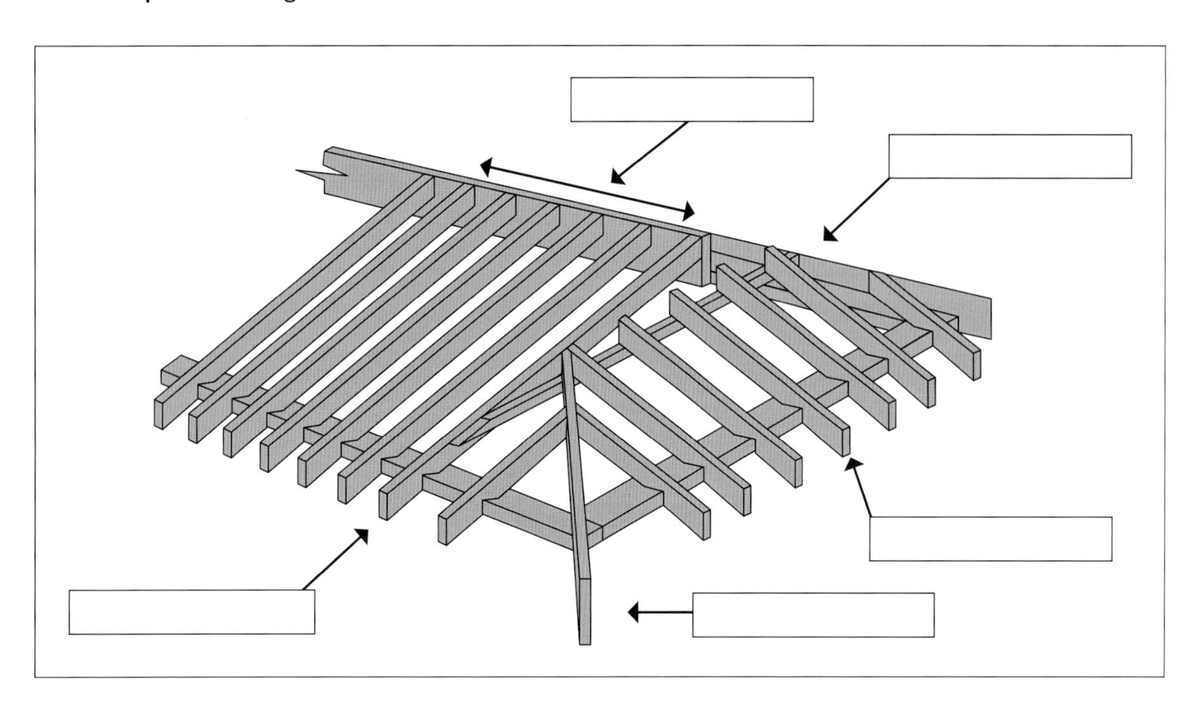

Obtain a copy of Safe Work Australia's *Preventing Falls in Housing Construction Code of Practice* (February 2016) and answer the following questions. (A copy of the Code can be obtained from: (https://www.safeworkaustralia.gov.au/system/files/documents/1705/mcop-preventing-falls-in-housing-construction-v2.pdf).

1 Is construction of 'stick roofs' considered a high-risk construction work activity? Explain why/why not.

____

____

2 Falling objects can pose a significant risk and cause serious injuries to workers at construction workplaces or to members of the public if control measures are not implemented to eliminate or minimise the associated risks.

a What is a PCBU (person conducting a business or undertaking) required to provide and maintain to ensure a safe system of work is carried out?

____

____

b *Preventing Falls in Housing Construction Code of Practice*, Section 9, outlines the processes and procedures to be initiated to prevent falls when working on roofs. How can much of the risk be reduced, given it is difficult to use harness systems when building a 'stick' roof?

c What are the passive and active fire controls relevant to roof construction and where would you find information on them?

3 Obtain a copy of AS 1684.2 Residential timber-framed construction – Non-cyclonic areas and answer the following questions.

a Why are underpurlins used in a pitched roof?

b Joints in underpurlins should be fully supported by what framing member?

c Describe an alternative method for jointing underpurlins.

d A strutting beam must be installed to support a strut when a non-load-bearing wall is accessible. Describe how it must be installed and explain why.

e When are fan struts used?

f Why are collar ties used?

g How many nails are used to fix a strut to an underpurlin?

h When installing fan struts, what are spreader cleats used for?

WORKSHEET 5

To be completed by teachers

Student competent ☐

Student not yet competent ☐

Student name: ______________________

Enrolment year: ______________________

Class code: ______________________

Competency name/Number: ______________________

1 What are three advantages of a pitched roof over a truss roof?

____

____

____

2 What are two disadvantages of a pitched roof?

____

____

3 Fill in the missing dimensions in this oblique roof shape.

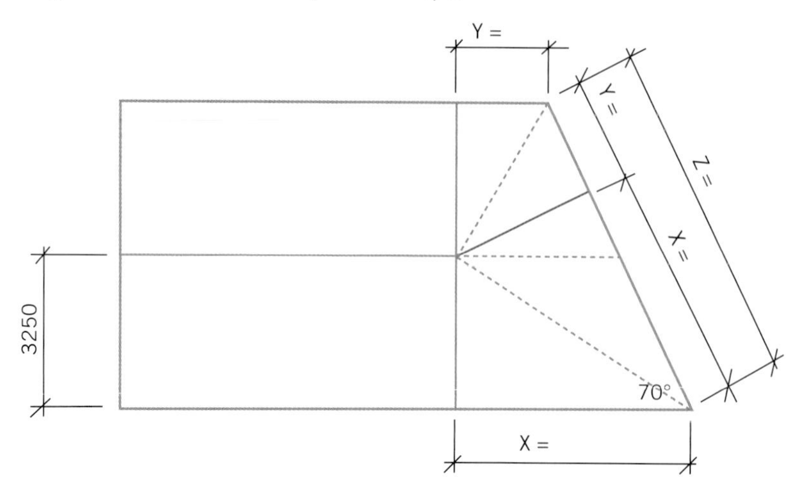

4 Fill in the missing words:

a Roofs are made up of ______________________ and are set out to ______________________.

b Ridges run ______________________ and level to ______________________.

c Regardless of the roof shape, rafters run at ______________________ to the wall plates.

d Hips and valleys ______________________ internal and external corners.

5 As a general rule, how many bevels are required in a pitched roof?

6 Answer True or False:

In a gambrel or Dutch gable roof the ridge is shortened:

True ☐ False ☐

WORKSHEET 6

To be completed by teachers

Student competent ☐

Student not yet competent ☐

Student name: __________________________

Enrolment year: __________________________

Class code: __________________________

Competency name/Number: __________________________

1 From the diagram below, use the direct method to set out the eight bevels required for the roof frame members on a 600 × 70 × 45 mm length of timber.

Rafter material: 70 × 45 mm

Ridge material: 200 × 35 mm LVL

Hip material: 150 × 42 mm LVL

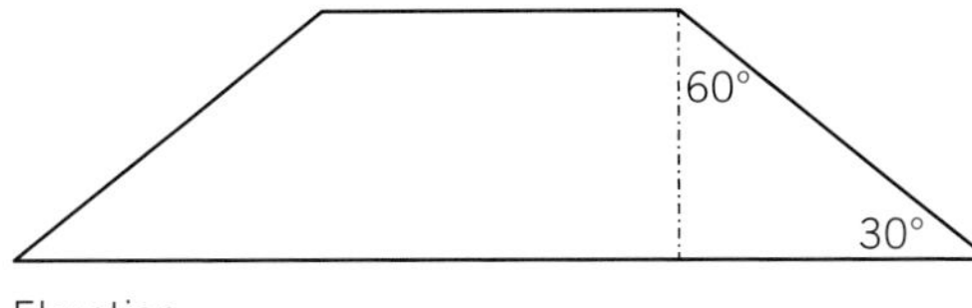

Elevation

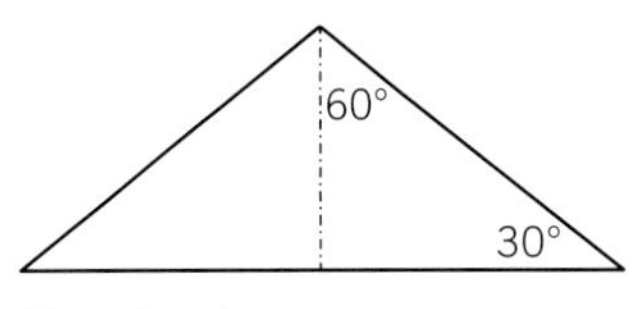

Side elevation

Note: Drawing not to scale.

To review how it's done, watch Buildsum's YouTube video 'Hip roof bevels direct method' (**https://www.youtube.com/watch?v=nhP6IR4u9DY**).

2 From the diagram below, use the direct method to set out the eight bevels required for the roof frame members on a 600 × 70 × 45 mm length of timber.

Rafter material: 70 × 45 mm

Ridge material: 200 × 35 mm LVL

Hip material: 150 × 42 mm LVL

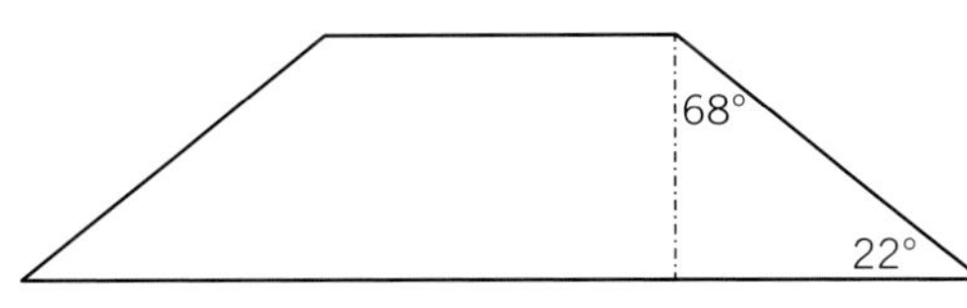

Elevation

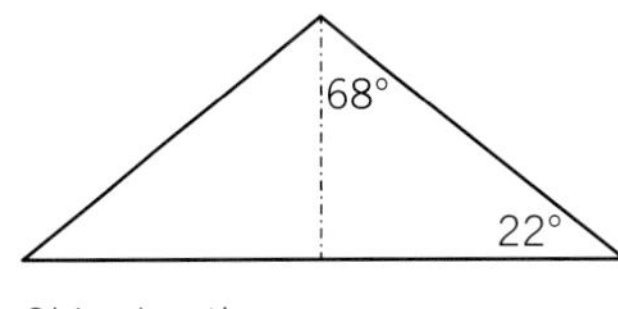

Side elevation

Note: Drawing not to scale.

WORKSHEETS 7

WORKSHEET 7

Student name: ______________________________

Enrolment year: ______________________________

Class code: ______________________________

Competency name/Number: ______________________________

To be completed by teachers	
Student competent	☐
Student not yet competent	☐

1 What is a dormer?

____

2 What do you need to consider when cutting a dormer window into an existing roof?

____

3 Fill in the missing words.

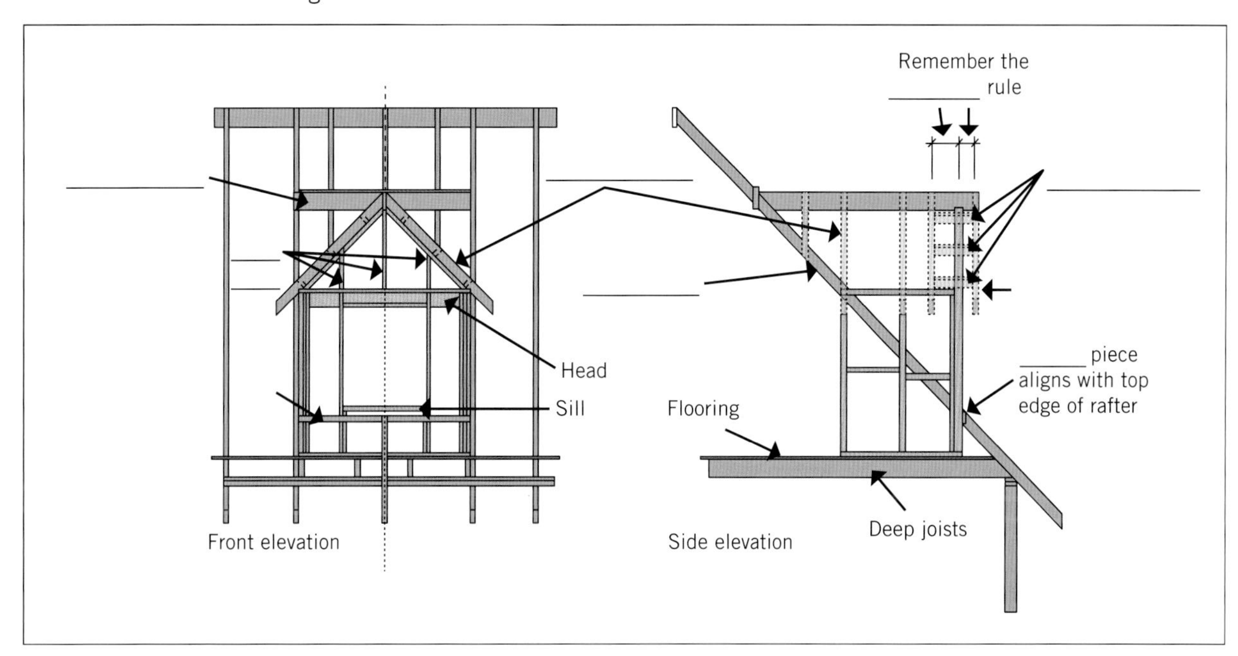

4 When constructing an octagonal roof, once you have installed the ridge and common rafters, what is the next roof component to go in?

____

5 In Model B of the octagonal Roof set-out, the hips are very close in as the crown end rafter. Explain what the difference in length is related to.

____

6 Fill in the missing words/phrases relating to intersecting roofs of unequal pitch.

To achieve equal width eaves on unequal pitched roofs requires you to ______________________ on the walls below a steep pitched roof section. This will affect how you frame out the ceiling. The use of a ______________________ will overcome that internal problem.

7 Answer True or False

Intersecting roofs with different pitches are similar in construction methods to a Scotch Valley roof.

True ☐ False ☐

8 Find the answers to the missing dimensions shown in the diagrams below.

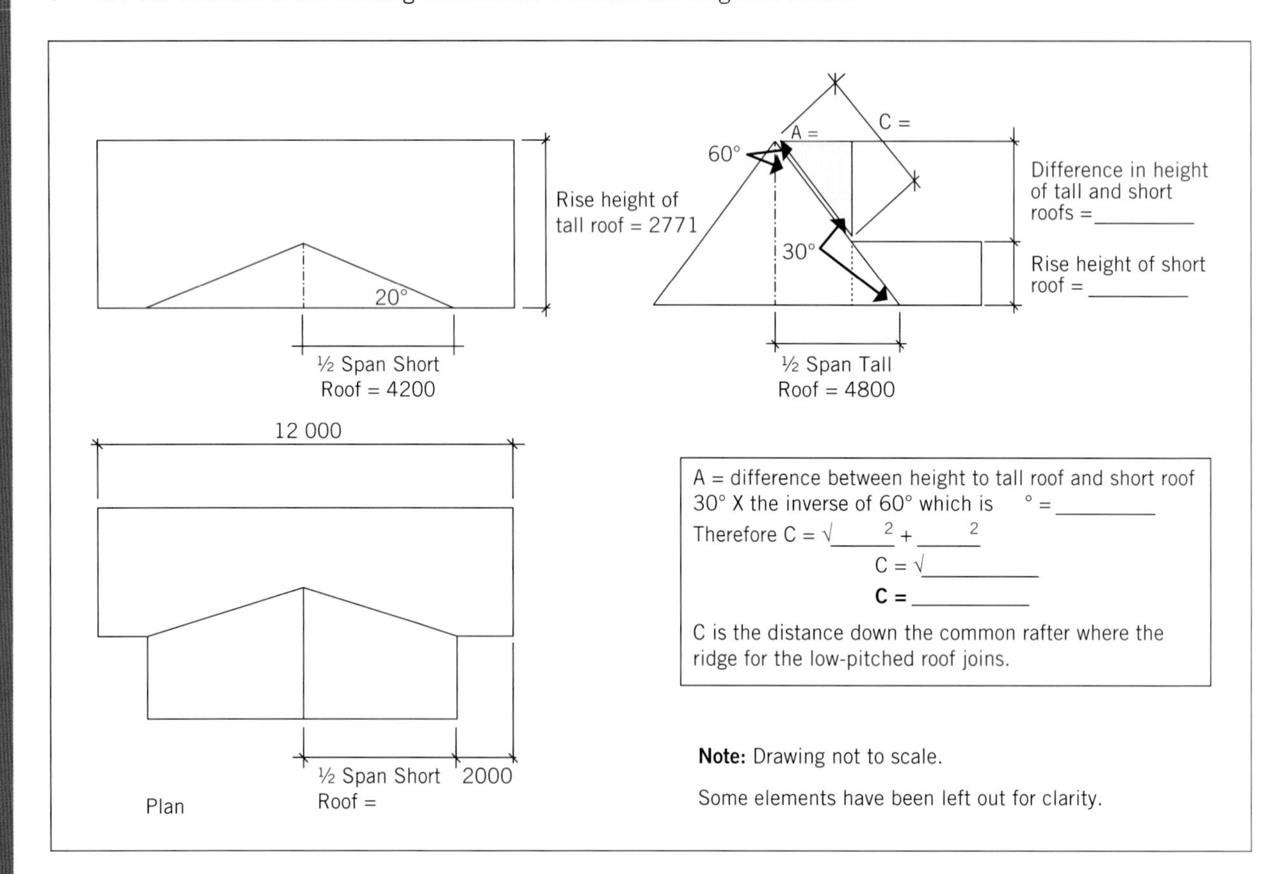

8 INSTALL EXTERNAL CLADDING

This chapter aligns with the unit of competency, 'Install exterior cladding'. The unit outlines the knowledge and skills required to install cladding materials applied to external framed walls. We will discuss and explore making buildings weatherproof. Cladding materials like weatherboards, plastic, metal and fibre-cement sheeting and the knowledge required to install cladding materials are covered.

In particular the chapter addresses the key competencies of:

- locating and applying information on cladding materials and systems
- work health and safety requirements and regulations
- sheet materials used in this process, such as plywood, fibre-cement sheeting, weatherboards, and plastic and new cladding materials
- traditional and alternative methods for preparing the substrate materials and frames to accept cladding materials, including preparation of masonry walls
- fixing and finishing of materials, including the set-out and cutting of weatherboards and other strip cladding materials.

Practical projects as a minimum should involve the cladding of at least two intersecting walls (see Worksheet 1 as an example). The projects should include a range of wall cladding materials, including sheet and board materials fitted around a window or door opening, and the use of set-out rods, weatherproofing and flashings. Consult the performance and knowledge evidence guides in the unit of competency for more detail.

Learning objectives

By the end of this chapter you will be able to:

- describe the purpose of cladding and nominate a number of advantages and disadvantages for different cladding materials used in residential construction, including timber, fibre-cement sheet, hardboard, vinyl and aluminium planking, corrugated iron and structural insulated panelling
- prepare for the safe installation of a range of cladding materials, including straightening walls and studs, installing building wraps (sarking) and installing flashing
- identify the different types and profiles of timber cladding, fibre-cement sheeting, metal cladding and hardboard materials, and understand the basic National Construction Code and Australian Standards requirements for different classifications of materials
- calculate material quantities required using sheet and plank type materials
- install a range of sheet and plank products to a range of building structures, including framed and masonry
- apply quality finishing techniques around doors, window openings, internal and external corners.

About cladding

Cladding can be defined as any product, or products, used to cover the external framework of a residential, commercial or industrial building.

Solid timber was the alternative to masonry construction in the early part of the twentieth century and was used extensively during that time (see Figure 8.1). Solid timber is still used today, but mainly for sympathetic period renovations. Such boards are known as weatherboards and they come in a range of styles. In Australia, it is important to use the National Construction Code – Volume 2 for technical information on fixing cladding.

Source: Greg Cheetham

FIGURE 8.1 Traditional weatherboard house

Weatherboards may be used to clad the entire structure or combined with a range of other materials for effect (see Figure 8.2).

Source: Greg Cheetham

FIGURE 8.2 Fibre-cement sheeting and weatherboards

Cladding materials can be used to enclose bay windows (see Figure 8.3), on walls in first-floor additions, on gable end roofs and in dormer windows (see Figure 8.4).

Source: Greg Cheetham

FIGURE 8.3 Shingles over masonry

Source: Greg Cheetham

FIGURE 8.4 Dormer window

In modern construction, it is not uncommon to see what used to be roof sheeting, or corrugated iron, used as wall cladding, such as in farm sheds. In fact, there is a tradition of this use; for example, the NSW Parliament buildings are corrugated iron clad.

In more modern residential construction a range of composite materials, including aluminium face hardboards and metal sheeting, are used to complement and be sympathetic in style and look to glass monolithic structures used in modern architecture (see Figure 8.5).

Source: Greg Cheetham

FIGURE 8.5 Modern wall cladding – painted hardboard with metal cladding

As a result of the 2014 Lacrosse apartments fire in Melbourne and the 2017 Grenfell apartments fire in London, most states have introduced or are planning the introduction of more stringent legislation to prevent the use of non-conforming building products in buildings and for the rectification of such building materials.

Non-conforming building products are products and materials that claim to be something they are not (e.g. they don't meet Australian Standards) or use deceptive marketing practices. Non-compliant building products are products and materials used in construction that do not comply with the National Construction Code (NCC) where the NCC states they are not suitable, even if they meet an Australian Standard or performance test, or have markings.

Cladding materials that comply with codes and standards and are certified must always be used, or you will have to remediate the job at your own cost. Other penalties also may apply.

Cladding materials have significant advantages over masonry products in that they are:

- relatively cheap
- strong
- easily insulated
- easily cut and fitted
- readily renovated, replaced or restored.

Cladding materials also have disadvantages, namely:

- that materials may be costly to buy
- time taken to install
- that they often require a second treatment such as painting, rendering or other applied surface finishes
- ongoing maintenance costs.

Preparation for safe and quality installation

When preparing to install cladding, make sure you have prepared the site correctly. It is important to complete a risk assessment of the site and determine what hazards may be present. A site induction may also be required to go through specific safety requirements. Always refer to the Safe Work Method Statement (SWMS) and be aware of the Job Safety Analysis (JSA) for each task. Hazards should be identified, and where possible controls and safety signage installed. Safety concerns should be monitored and reviewed during construction, especially if the work is carried out over several days or when other trades are working on site. You need to check the following safety issues:

- All materials must be onsite and stacked in suitable locations so that materials handling can be done safely. If the cladding is being attached to a first-floor frame, consider how the material will be raised to the appropriate height. This can be done manually or with gin wheel, ropes or a block and tackle.
- Appropriate scaffolding or fall arrest systems must be in place.
- Signage and barricades must be erected.
- Power tools and leads need to be checked for serviceability.
- Environmental controls must be in place. This is important if you need to remove old fibre-cement sheeting and there is a concern that it may contain asbestos. Dust control is another issue related to the removal of old cladding, as dust from old painted surfaces containing lead particles may be released from the **roof space** when preparation is being carried out for a roof conversion or the addition of dormer windows.
- Appropriate additional personal protective equipment (PPE), such as serviceable harnesses, needs to be available, particularly when working at heights.
- Installers need to have access to and have read the appropriate safety data sheets (SDS).
- Ensure materials comply and conform to codes, standards and regulations and are 'fit for purpose'.

A safety harness won't stop you falling from heights, but it will prevent serious injury or even death if you do fall – as long as it is attached to an approved anchor point.

It is always a good idea on any job to create a simple quality control checklist. The quality control checklist can provide you with a simple job plan, as well as identify potential problems early. The sorts of things you may need to look for include:

- checking for straight walls
- checking for wall wraps (sarking) and flashings
- checking door and window frame alignment
- checking manufacturers' technical information installation requirements.

Table 8.1 is a simple checklist you may find useful when preparing for and fixing wall cladding. The comments boxes can be used to record faulty work where necessary, identify who performed the task and note any rectification actions required.

TABLE 8.1 Quality control checklist

	Quality control checklist	Check	Responsible person	Comments
1	Check and straighten the wall frame.			
2	Attach/adjust door and window frames and associated flashings to the wall frame.			
3	Fit the plinth board to the wall.			
4	Fit flashing along the top of the plinth board if necessary.			
5	Fit the sarking to the outside of the frame.			
6	Calculate the set-up dimensions and material quantities.			
7	Mark the locations for the bottom edge of the weatherboards on the stops.			
8	Fit the weatherboard stops.			
9	Mark, cut and prepare the individual boards.			
10	Fit the cladding.			
11	Clean up.			

LEARNING TASK 8.1

Students are to investigate how product compliance and conformity is managed in their state or territory.

Useful resources for each state include the following:

- Australian Capital Territory – Planning: https://www.planning.act.gov.au
- New South Wales – NSW Fair Trading: http://www.fairtrading.nsw.gov.au
- Northern Territory – Housing, property and land: https://nt.gov.au/property
- Queensland – Queensland Building and Construction Commission: http://www.qbcc.qld.gov.au/
- South Australia – Building rules, regulations and information: https://www.sa.gov.au/topics/business-and-trade/building-industry/building-rules-regulations-and-information
- Tasmania – Consumer, Building and Occupational Services (CBOS): https://www.cbos.tas.gov.au
- Victoria – Victorian Building Authority: https://www.vba.vic.gov.au
- Western Australia – Building and Energy Division: https://www.commerce.wa.gov.au/building-commission

Use 'non-conforming building products' as search term if necessary.

In a teacher-led discussion you should consider the points below:

1 What may be the impacts on customers when builders don't use conforming or compliant cladding products?
2 Does your builder insurance cover you if you use a product that is later found to be non-conforming? Are you required to replace it?

Note: The Australian Senate has conducted an inquiry into non-conforming building products. The results of this inquiry are driving further reform initiatives on non-conforming building products nationally.

Straightening wall frames

When cladding is fitted to a bowed or sprung wall, the uneven surface will look ugly to the eye, and fixing will be more difficult as the boards or sheets will have to be sprung to fit the irregularities in the frame. If the wall or opening is out of square, trimming will be difficult and compromises will have to be made.

Walls need to be vertical, straight and level to make cladding and lining easy. Distortions in frames can occur after construction and before erection of the cladding for several reasons, including:

- the use of poorly seasoned timbers
- the presence of penetrations or cuts in frames from construction activities
- additional loads being transferred to the frames after construction (e.g. heavy ceiling-mounted water heaters and air-conditioning units or first-floor additions can cause frames to bend, bow or move sideways).

Checking for straight, plumb and level

Before commencing any cladding job, either in new work or a renovation, you need to check the walls for straight, plumb and level, and any distortions should be made good before any cladding is installed.

These are the basic terms used to describe crooked frame members:

- high – when the face of the plate or stud is bowed out
- low – when the face of the stud or plate is bowed in
- sprung – when the entire length of the stud is bent in or out, causing problems inside and out
- warped – when the wall is twisted out of plumb alignment.

Use the string line and block method to check the wall plates for straight (see Figure 8.6). A laser level can be used to check for straight.

A straight edge can also be used to check the height of the frame for straight and plumb at the same time (see Figure 8.7). A builder's level (minimum length 1.8 m) can be used for short sections.

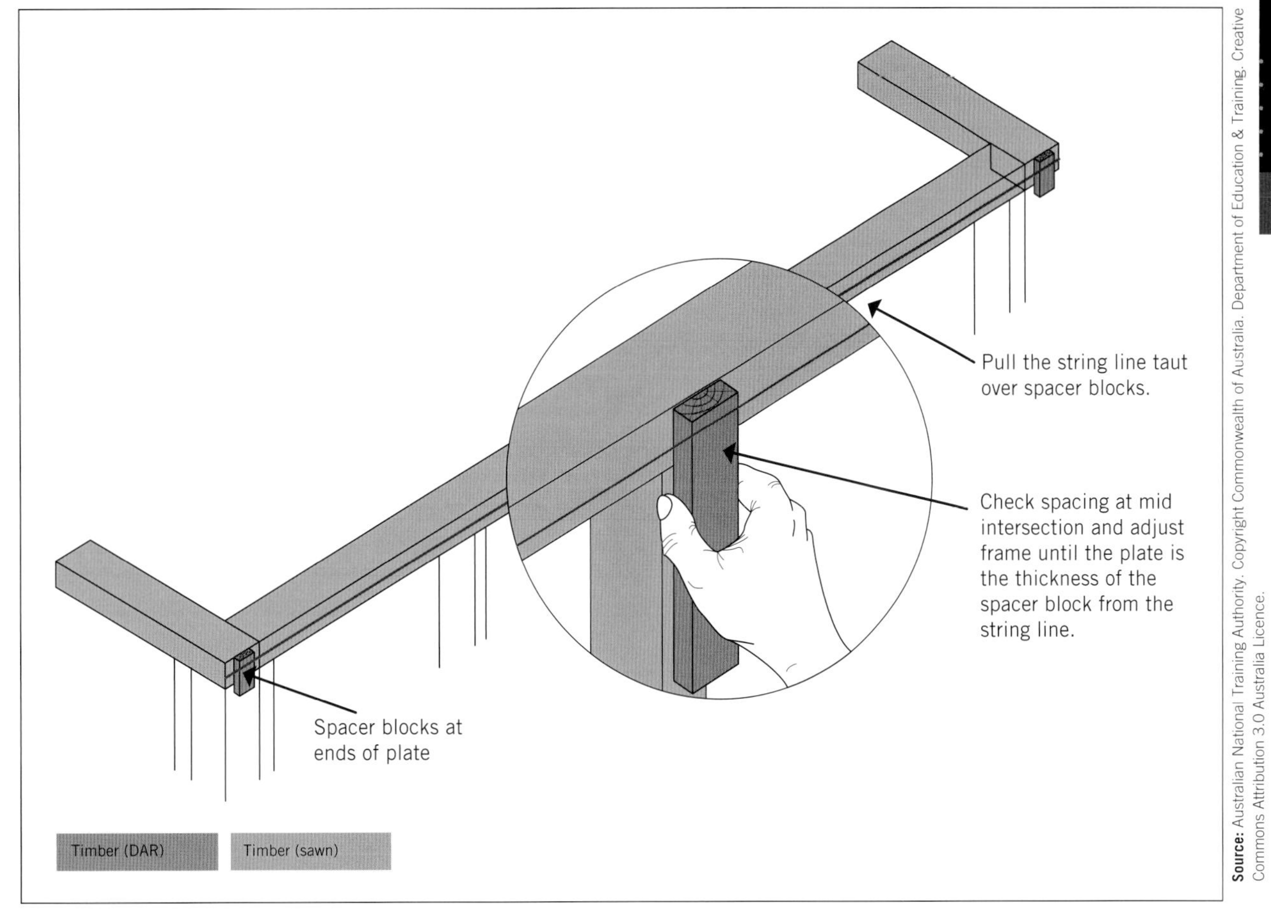

FIGURE 8.6 String line and block method

FIGURE 8.7 Checking for straight

The steps involved are:

1 Straighten the top and bottom plates first.
2 Check and straighten each stud. Usually planing will be required if the frame is bowed out.
3 Check along the length of the wall across all the studs to make sure you have achieved a straight wall.

Straightening bowed studs in a wall

Occasionally bowed studs are inadvertently used when constructing wall frames, or frames suffer from poor material handling and storage. Whatever the cause, these sprung studs require straightening before cladding and linings can be installed (see Figure 8.8). The process is called 'crippling' the studs. This process is frowned upon today because if not done correctly it will weaken the stud, but it is allowed under certain conditions – see AS 1684 Residential timber-framed construction, Section 6 for details.

Best practice is to check frames or timber delivered to the site for defects, and to reject the materials if necessary. That way suppliers know and understand your quality requirements. Creating a good working relationship with your suppliers is a critical quality assurance component of any business.

Note: Up to 20 per cent of studs may be crippled except for studs supporting 'concentrated loads'; for example, studs at the sides of door and window openings must not be crippled.

Where cladding is being retrofitted, the process can only be carried out once the exterior of the frames has been exposed.

Make a saw cut to a maximum half the depth of the stud on the concave edge of the stud. Then cut a wedge from hardwood and drive the wedge into the saw's **kerf**, forcing the stud to straighten. Next, skew a nail through the wedge to fix it in place. Finally, reinforce the cut stud by fastening two cleats on either side of the cut, directly over the cut and wedged area. The cleat should be 600 mm long and fastened evenly with a minimum of 4 × 50 mm nails. See AS 1684, Section 6.2.1.1 Straightening of studs (crippling) for more details.

Planing is an alternative (see Figure 8.9) as long as the AS 1684, Section 6.2.1.1 requirements are followed.

Note: Some states do not allow the crippling of studs at the sides of openings (i.e. windows and doors) or studs supporting concentrated loads (i.e. corner studs). Check your state's regulations or timber framing manuals before deciding on crippling or planing.

Note: If you cannot cripple the stud in the right direction because of internal wall lining, you may need to pack the low face (see Figure 8.9).

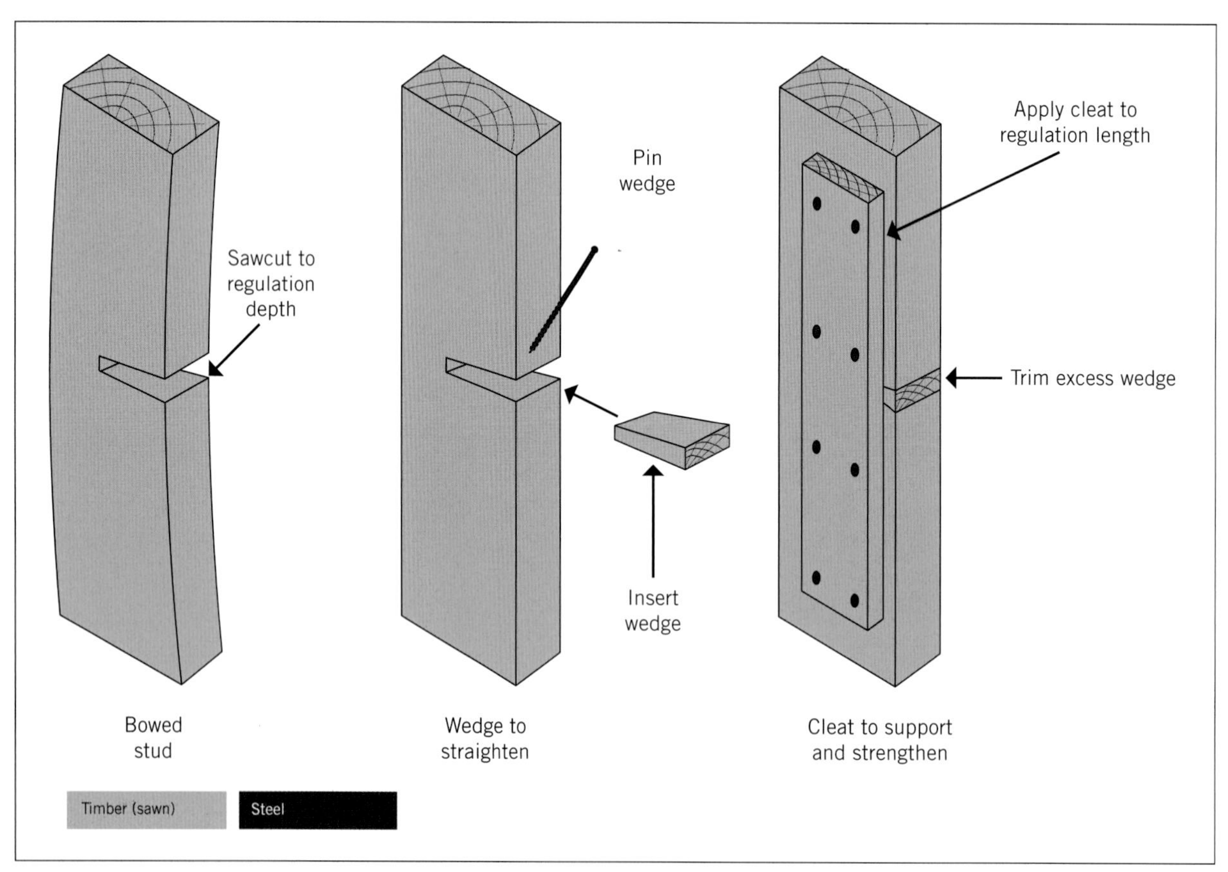

FIGURE 8.8 Straightening studs

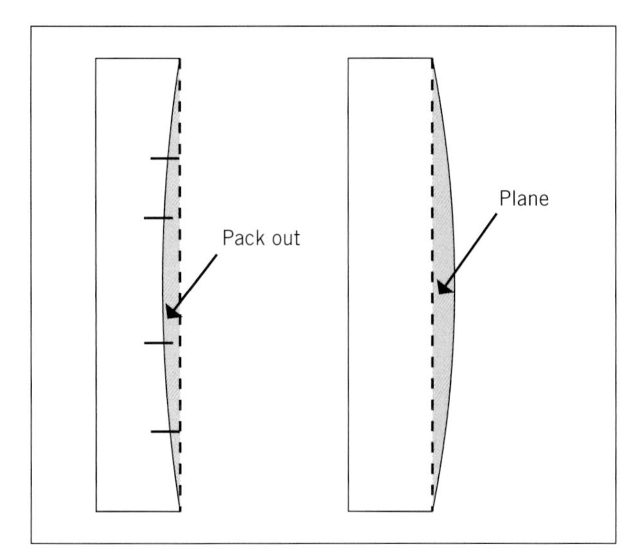

FIGURE 8.9 Pack or plane

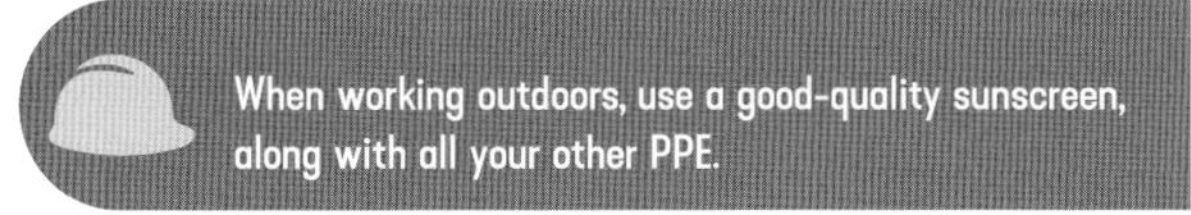

Sarking

Builder's paper or foil is commonly called 'sarking'. Sarking is sold in rolls 1350 mm wide, and has a shiny foil surface on one side and a dull surface on the other. It provides insulation and weatherproofing behind wall and roof cladding. It is available in three different strength grades: lightweight, medium and heavyweight.

Basic sarking is a vapour barrier and is used with insulation to increase the energy efficiency and comfort of a home. But the increased air tightness can increase respiratory problems for some people because of the increased condensation and associated mould, mildew and decay.

To minimise the risks, use vapour-permeable membranes. There are now 'smart membranes' that react differently depending on external climatic conditions.

More information can be found on the federal government's Your Home website under 'Passive design': http://www.yourhome.gov.au/passive-design.

Sarking is fastened to the outside of the wall and is applied before cladding goes up. The shiny surface is fitted against the frame, with the dull side facing outwards to reduce problems with glare when fixing. Sarking should be rolled out horizontally along the wall, with the bottom row being fixed first and the second or top layer overlapping 150 mm on the first run.

Ensure sufficient overlap at window and door openings and corners, so that the material can overlap door and window flashings and sarking on adjoining walls.

Sarking can be easily damaged, so fit it to the building as soon as practicable before the cladding is to be fitted. Use appropriate foil fasteners or shade cloth fasteners because regular nails or screws will penetrate the sarking (see Figure 8.10). Alternatively, you can prepare your own temporary fasteners using 50 mm hardboard or plywood squares and clouts. The hardboard or plywood acts as a washer to prevent the clout penetrating the sarking. Make sure you remove these fasteners as you install the cladding; otherwise the additional thickness of the hardboard or ply will cause the cladding to bulge out.

Source: Ed Hawkins

FIGURE 8.10 Ventilated sarking and fastener

When fixing sarking to steel frames, use hex head screws and washer plates as temporary fasteners, noting they will have to be removed as the cladding is installed. Space the fasteners evenly and take account of typical wind strengths in the area to decide on the number of fasteners required to hold the sarking in place until cladding is complete.

Alternatively, you can fix the wall wrap to the frames using batten screws fixed at each stud. This provides a breathing gap and allows for potentially easier fastening of cladding materials as self tapping screws can be used that do not need to be removed (see Figure 8.11).

Insulated building wraps can also be used as a further means of insulating buildings. Air cell (similar to bubble wrap) is a common insulating building wrap used today. The advantage is that you can apply it using button head screws that do not have to be removed before fixing over with sheeting. Air cell should be fitted loosely between studs to allow for the insertion of spacers to keep a void space between the wall wrap and cladding to facilitate vapour evaporation.

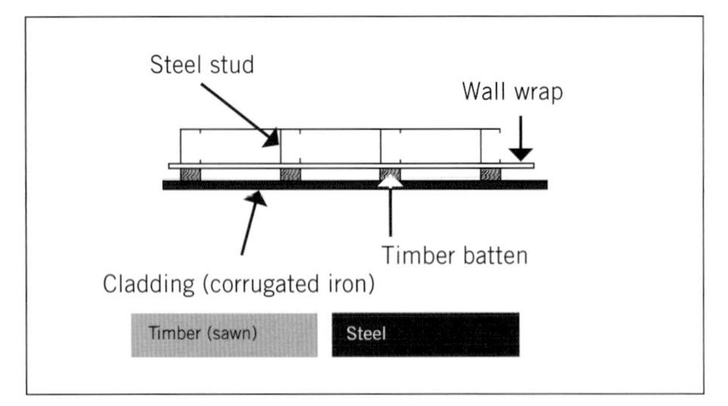

FIGURE 8.11 Timber battening to steel frames

LEARNING TASK 8.2

PRACTICAL ACTIVITY – STRAIGHTENING STUD LENGTH

In the workshop or at your work site find a bowed length 2.4 m / 90 × 45 mm. On the concave edge make the necessary cut and insert a wedge to straighten the stud length.

If students are performing this activity in the workplace, they should provide photo evidence and a supervisor's signature of its successful completion.

Evidence should indicate:

1 the original bowed material (size of bow)
2 the student making the cut and inserting the wedge
3 the straightened stud with cleat supports attached.

Note: In all photo evidence the student should be clearly recognisable.

Flashings

Make sure flashings are in place above and below windows and doors (see Figure 8.12). Bitumen-covered aluminium flashing materials, such as Alcor, are suitable for flashings.

Where metal cladding such as galvanised iron or Zincalume® steel is used, lead or copper flashings should not be used because the galvanic reaction is likely to cause corrosion where the two metals are in direct contact with each other. Flashings made from the same materials or appropriate aluminium, bitumen-coated plastic or rubber-based flashing material will work well with these products (see Figure 8.13).

You also need to ensure that copper penetrations such as water pipes, tap heads and the like do not come into contact with Zincalume® or Colorbond® metal products. For more information about this, visit the BlueScope Steel website listed at the end of the chapter.

Adjusting door and window frames during renovation

When installing cladding over an existing wall finish, door and window frames will need adjusting to accommodate the increased thickness of the walls.

Source: Greg Cheetham

FIGURE 8.12 Flashing above (top) and below (bottom) windows

This is usually done by first removing the internal and external architraves, storm moulds, and other external door and window frame components. The next step is to determine how the frames have been installed and decide on a method for releasing the frames so they can be moved forward to accommodate the new cladding thickness. A useful tool for this kind of work is a sabre saw (see Figure 8.14). This may mean securing new or wider reveal linings to windows and doors and changing flashings, especially if you are installing metal cladding that may be of an incompatible metal. For further information on this process, see Chapter 3.

Source: Greg Cheetham

FIGURE 8.13 Aluminium and rubber-based flashing around an aluminium window

Source: Milwaukee Power Tools Australia

FIGURE 8.14 Sabre saw

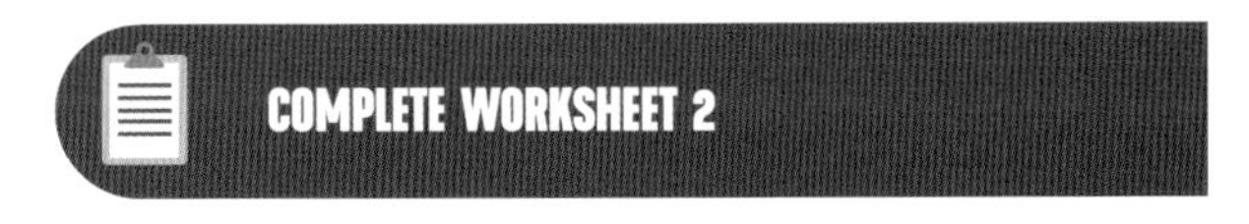

Types of cladding

Solid timber weatherboards

The traditional weatherboard that was used in Australia was rough sawn or split from a log and had a tapered profile usually about 150 mm wide by 25 mm thick. Today weatherboards are available in a large range of milled timber profiles from a range of timber species. They are typically fastened horizontally but can be fixed vertically. See Volume 2, NCC Housing Provisions – Part 7.5 of the National Construction Code for the specific nailing requirements.

Solid timber cladding is machined from a range of local and imported hardwood and softwood timbers. The most common locally grown species are:

- hoop pine – native species, plantation grown (also known as arakaria); not recommended for cladding unless preservative treated and coated on both sides
- radiata pine – imported species, plantation grown
- slash pine – plantation grown (also known as Florida pine or yellow pine); not recommended for cladding unless preservative treated and coated on both sides
- cypress pine – native species, managed forests (also known as white cypress).

Imported species include:

- western red cedar – North American species, plantation grown, mainly available as rough-sawn, splayed cladding materials; favoured for its durability and light weight
- Baltic pine (also known as Norway spruce or whitewood); not recommended for cladding unless preservative treated and coated on both sides.

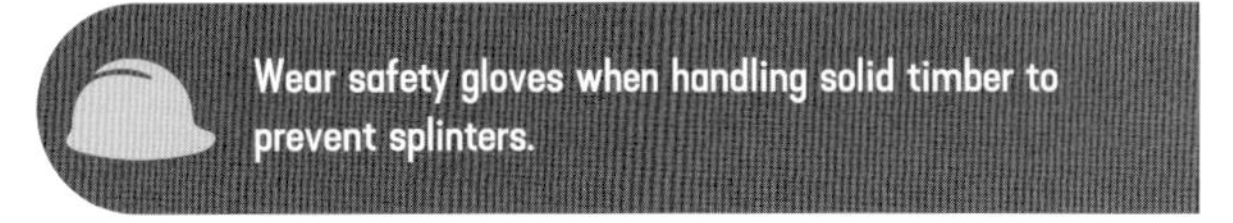

Durability class of timber

Durability Class 1 and 2 timbers are recommended for external use. You can only use durability Class 3 and 4 timbers if they have been suitably preservative treated under pressure. Light Organic Solvent Preservative timbers are recommended for use as cladding. The preservative is applied after they are milled to profile. All external timbers should be treated with some form of surface coating to ensure their durability and extend their service life.

Table 8.2 shows various species of timber in each class.

Note that most of the timbers used for cladding are of Class 4 durability, meaning that most timbers need to be preservative treated when used as cladding.

AS 1604.1 Specification for preservative treatment – Sawn and round timber provides the specifications for preservative-treated timbers in Australia.

AS 2796 Timber – Seasoned hardwood – Milled products specifies the requirements for sawn and milled hardwood products, including the allowable moisture content and guidance on tolerances, profiles and grades. It also specifies the requirements for glue-laminated, finger-jointed and end-matched hardwood materials. Timber machined profiles will vary between manufacturers.

If hardwood cladding is specified, a naturally durable species should be used, particularly in harsh climates such as tropical, alpine and semi-arid regions. Where timber decay, rot or insect infestation are less

TABLE 8.2 Durability classes of timber

Durability class	Suitable species
Class 1: Highly durable	Cypress, tallow wood, turpentine
Class 2: Durable	Western red cedar, spotted gum, river red gum, jarrah, Sydney blue gum, stringy bark (yellow and white)
Class 3: Moderately durable	Brush box, rose/flooded gum, keruing, messmate, karri, silver-topped stringy bark
Class 4: Non-durable	Hoop pine, slash pine, radiata pine, Douglas fir, mountain ash/Tasmanian oak, meranti, Baltic pine

destructive, such as in the southern states of Australia, commonly available timbers of durability Classes 3 and 4, such as alpine ash, mountain ash or messmate, can be used as long as suitable preservative treatments are used to coat all surfaces, including the end grain of cut lengths, before installing the cladding.

There is no Australian Standard that covers imported timbers used in cladding and as such builders have to rely upon specifications and advice from local timber advisory services on suitable materials to use.

LEARNING TASK 8.3

TIMBER PRESERVATION SYSTEMS

Investigate the range of timber preservation systems used and report on the different systems, outlining their advantages and disadvantages. Useful websites include:

- Wood Solutions: https://www.woodsolutions.com.au/wood-species
- Timber Preservers Association of Australia (TPAA): http://www.tpaa.com.au

Also obtain a safety data sheet (SDS) for each product. **Alternative activity: Investigate durability classes in timber**

Obtain a copy of the *Timber service life design – Design guide for durability* (05), published by Wood Solutions and available online at: https://www.woodsolutions.com.au/publications/timber-service-life-design-guide.

Note: You must create a free account before you can access the guide. Use the search term 'Timber service life design' in the search tab if you have trouble locating it.

Now answer the questions below:

1 How should timber-to-timber joints be treated?
2 Why should external timber be protected from weather and how can this be done?

Profiles

As with all cladding, timber weatherboards are designed to keep the weather out. Rebates, grooves and overlaps are the typical methods used to joint boards and provide weatherproofing.

A range of profiles, which have also been copied in alternative materials, have been created over the years to allow the user to create a functional and attractive surface finish.

Figure 8.15 shows some of the profiles available.

Sizes

Timber cladding generally comes in boards from 100 × 25 mm to 200 × 25 mm nominal size. Check for the exact **coverage**. Wider boards are available, but must be milled from seasoned timber. Most boards milled to a profile will have a 19 mm or 18 mm thickness, while rough-sawn splayed boards in western red cedar, for example, will be 19 mm thick, tapering to 4.5 mm along the narrow edge.

For full details on sizes and profiles, conduct an internet search using terms such as: 'weatherboard profiles', 'cladding profiles' and 'timber profiles catalogue'.

Advantages

Advantages of solid timber weatherboards include that they are:

- available in a wide range of shapes/profiles
- cost-competitiveness compared to masonry
- good insulators
- easily workable with hand or power tools
- low embodied energy materials
- a renewable resource.

Disadvantages

Disadvantages of solid timber weatherboards include that they:

- are prone to insect and fungal attack
- are fire prone
- suffer from splits and splinters
- require sealing during and after installing, and need ongoing maintenance.

Shingles and shakes

Mainly cut from western red cedar, these rectangular shaped timber sections are occasionally used as roof coverings and wall claddings.

Shakes are timber panels rough split and are often tapered in thickness, while shingles are generally cut to shape and thickness (see Figure 8.17). Shakes are more often used for roofs, while shingles are used to clad walls.

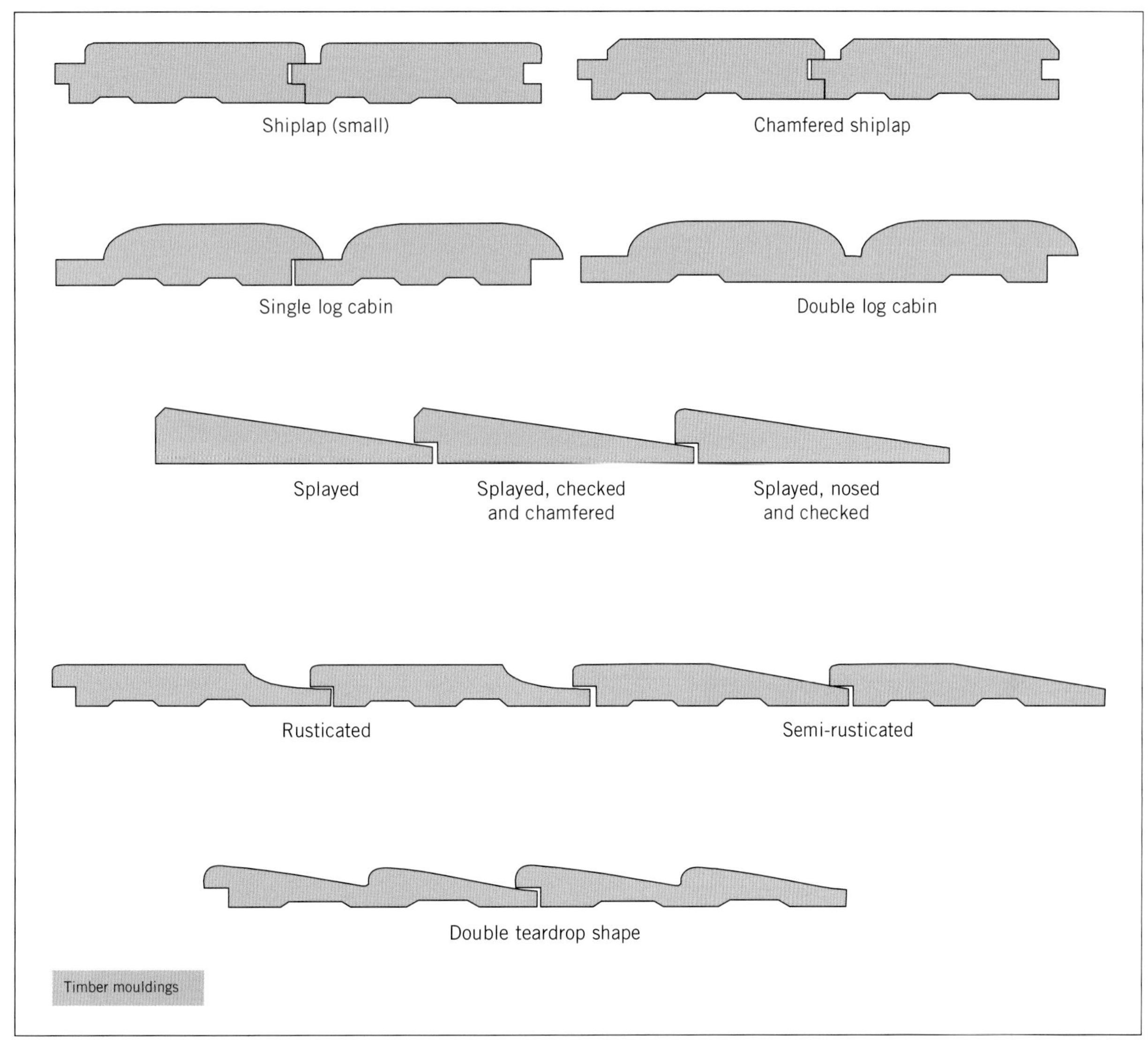

FIGURE 8.15 Timber profiles

Source: Greg Cheetham

FIGURE 8.16 Shingles used to clad gables

Sizes

Shingles and shakes can be bought in strapped packs ready for transport, stacking and storing. They come in lengths of 450 mm to 600 mm and widths of 100 mm to 300 mm.

Because of their size, shingles and shakes need to be installed on a groundwork of timber battens, set out to provide suitable cover and support to the row below (see Figure 8.17). Typically, a minimum of 40 mm cover is left on either side, with at least half of the shingle length being covered over by the next row.

Ventilated sarking paper must be used behind the battens and the shingles or shakes.

Shingles and shakes have similar advantages and disadvantages to those of solid timber.

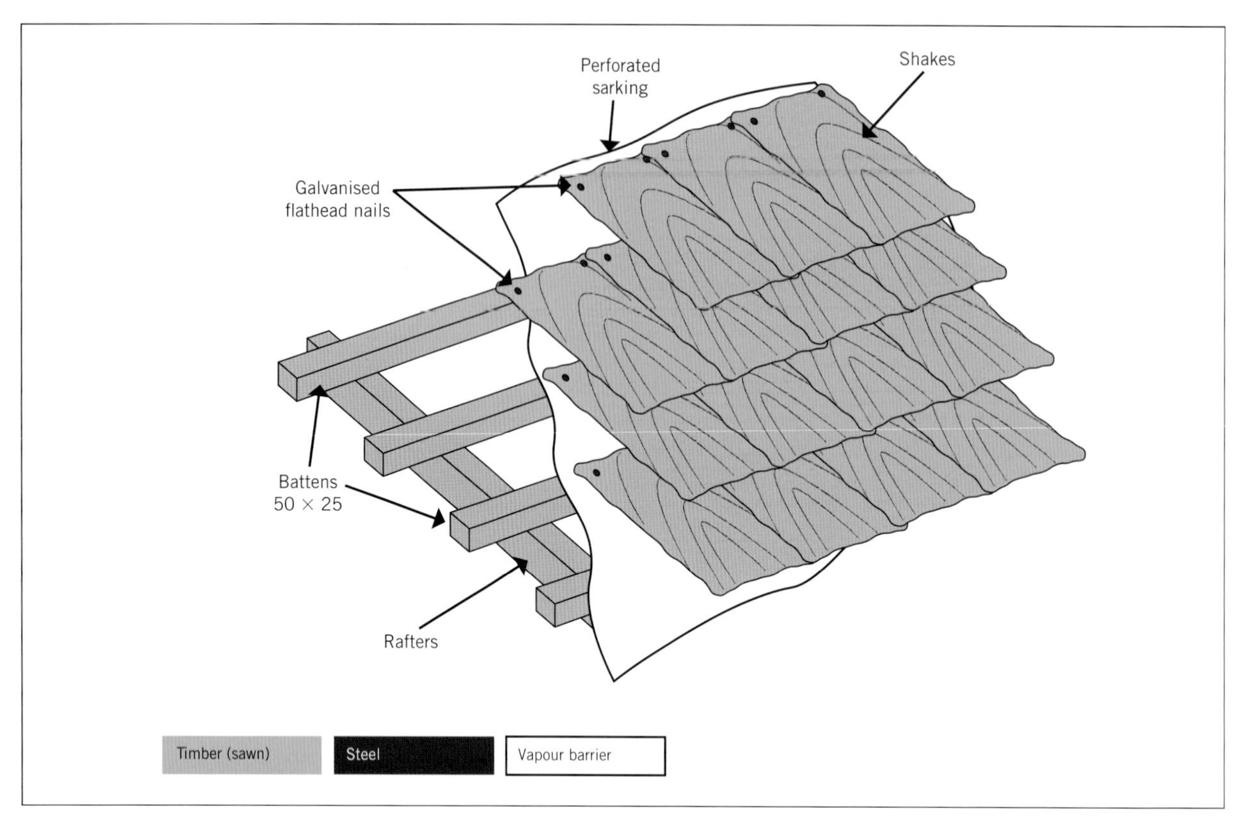

FIGURE 8.17 Installing shingles or shakes

Fibre-cement sheets

Fibre-cement sheets are made from a mixture of cement, fine **silica** and fine cellulose for binding and flexibility (see Figure 8.18). They are manufactured by a number of companies in Australia. James Hardie is the predominant Australian and international manufacturer, while BGC Fibre-cement, based in Western Australia, manufactures and distributes fibre-cement sheets nationally. The two main products available for use in residential and commercial work are James Hardie's Hardiflex™ and BGC's Durasheet™.

Fibre-reinforced cement-based products should conform to AS 2908.2 Cellulose cement products – Part 2: Flat sheets, making them resistant to water penetration and rot.

Note: See the National Construction Code – Volume Two, Housing Provisions – Part 7.5 Fibre-cement sheet wall cladding for technical information on fixing.

The standard sizes for fibre-cement sheets are as follows:

Width (mm)	Lengths (mm)
900	2400, 3600, 4200
1200	1800, 2100, 2400, 2700, 3000, 3600, 4200
1350	2400, 3000, 3600

The typical sheet thickness is 6 mm, with a weight of approximately 9 kg/m^2 at 6 mm. A 4.5 mm sheet thickness is also available, but this should only be used in high areas such as first-floor additions, gable ends and soffit linings.

Source: Greg Cheetham

FIGURE 8.18 Fibre-cement cladding

Both sheets and planks are available in a range of finishes, including flat, stucco (replicating a hand-trowelled finish) and fake rough-sawn timber-grain finish.

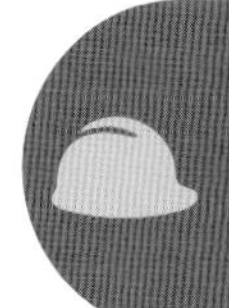

Exposure to crystalline silica dust over time can lead to serious irreversible health conditions such as silicosis, a progressive lung disease that can lead to disability and death. Check your state or territory's Safe Work website for information.

Advantages

Advantages of fibre-cement sheets are that:

- they have low cost-effectiveness compared to masonry and timber cladding
- patterned surfaces are available
- they are rot resistant
- they are easy fixing
- they have low embodied energy relative to brick veneer
- they are low maintenance.

Disadvantages

Disadvantages of fibre-cement sheets are that:

- they are brittle and can suffer impact damage
- special tools are required for working with them
- they are poor insulators
- the cement component is not renewable.

Vinyl and plastic planking

Available mainly as a weatherboard replacement product, vinyl and plastic planking comes with a wood-grained pattern surface in a range of pre-finished colours and patterns (see Figure 8.19). It is a strip planking product requiring cutting and joining over framed or battened groundwork. The material is available with a polystyrene foam insulation backing. It is easily cut and worked.

Source: Ed Hawkins

FIGURE 8.19 Vinyl cladding

Advantages

Advantages of vinyl and plastic cladding are:

- it is lightweight
- it does not require any maintenance
- it is easily worked
- it has cost advantages over other forms of strip planking
- it is recyclable
- it won't dent like metal siding.

Disadvantages

Disadvantages of vinyl and plastic cladding are:

- it has no real structural strength
- it has a limited colour range
- its colours tend to fade
- it is susceptible to mould and fungus build-up.

Installing vinyl cladding

Vinyl cladding is installed similarly to other planking products (metal sidings and weatherboards) in that it can be installed horizontally, vertically and diagonally. The material can be cut using a handsaw, hacksaw, tin snips or utility knife.

Note: Even if a manufacturer recommends using a portable power saw with the blade set in the saw backwards, this action contravenes WHS/OHS regulations and should be avoided at all times. A more suitable power saw solution is to use a sabre saw with a metal cutting blade or very fine-toothed blade inserted. This will give as good a cut and not break any WHS/OHS regulations. If cutting or trimming around penetrations or obstacles using tin snips, avoid closing the blades completely to achieve a neater cut.

Make sure you check that blades and cutters are sharp before using any cutting tool or machine.

A utility knife or scoring tool can be used to score and snap vinyl cladding. Use a straight edge to guide the scoring of the face surface before bending up to snap along the scored line.

Vinyl material can be stapled in position, provided the staples are a Class 3 finish (i.e. galvanised) and the staples are inserted through the perforations in the top edge of the plank (see Figure 8.20). Placing staples through the vinyl may crack it.

As with all strip or plank cladding, walls must be prepared, any doors and windows must be replaced or relocated as necessary, and the starter strips, corner posts, flashings, trims around penetrations, eaves, and trimming material or J channel need to be installed.

Once the starter strip (see Figure 8.21) has been set out and fixed around the perimeter of the building or wall, the first plank can be laid in the starter strip and fastened off using the appropriate fasteners. The next plank is set on top of the first, making sure that it sits well down over, and level to, the original fixed sheet and is fastened off (see Figure 8.21).

Visit the Vinyl Council of Australia website (https://www.vinyl.org.au/in-greenstar) to find out more information on Green Start Ratings for vinyl

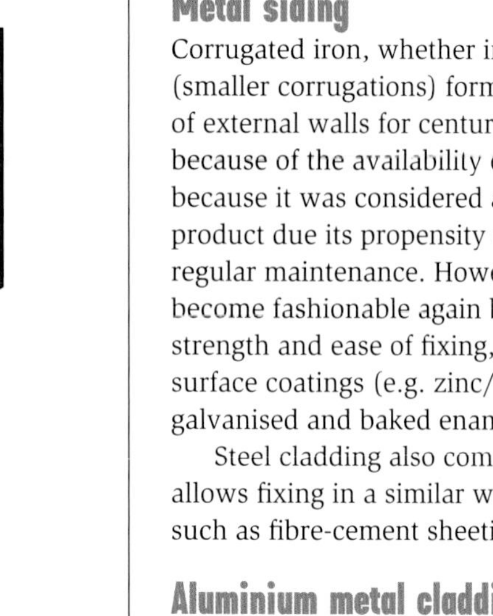

FIGURE 8.20 Stapling planks

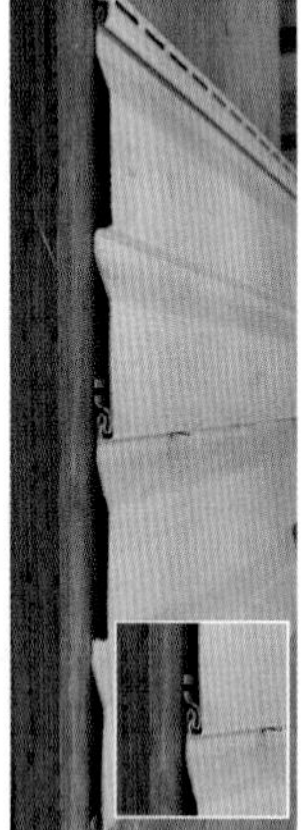

Source: Ed Hawkins

FIGURE 8.21 Starter/finishing capping (left); typical vinyl cladding joint (right)

cladding products. Check with manufacturers and regulators to ensure products meet combustibility standards and are safe to use.

In NSW, regulations now exist defining 'combustible cladding' as cladding comprising materials that are capable of readily burning, including aluminium composite panels, timber, polystyrene and vinyl.

In the regulation, building owners, depending on the type and size of building, must notify the Department of Planning, Industry and Environment with:

- details of the building and the combustible cladding materials
- details on inspection by a properly qualified person of risks and mitigation strategies.

Visit the NSW Department of Planning, Industry and Environment website for details: https://www.claddingregistration.nsw.gov.au/what-combustible-cladding. Other states and territories have similar reporting requirements.

Metal siding

Corrugated iron, whether in roofing form or mini-orb (smaller corrugations) form, has been used for cladding of external walls for centuries. It went out of fashion because of the availability of brick and timber, and because it was considered a cheap and nasty building product due its propensity to rust and its need for regular maintenance. However, metal siding has become fashionable again because of its structural strength and ease of fixing, and improvements in surface coatings (e.g. zinc/aluminium coatings and galvanised and baked enamels).

Steel cladding also comes in planked form, which allows fixing in a similar way to other planked products, such as fibre-cement sheeting.

Aluminium metal cladding

Aluminium metal cladding is a tough material that is impervious to normal environmental conditions such as wind, rain, and insect and fungal attack (see Figure 8.22). Some systems come with a polystyrene foam backing to improve the heat and acoustic insulation properties.

Source: Greg Cheetham

FIGURE 8.22 Industrial cladding showing a range of cladding materials

Aluminium cladding must be stacked level and flat off the ground and kept dry, either by storing it under cover or covering it with plastic. The finishes on aluminium cladding – powder coated, baked enamel or anodised – can be damaged by sand and cement and general building site wastes.

The cladding can be cut using normal woodworking power tools. A jigsaw or hacksaw can be used for making the cut-outs and other small shaped cuts, while a power saw fitted with a 'negative hooked', 60-tooth, tungsten carbide teeth blade is recommended for straight cuts. When cutting with power saws, a soluble white oil or wax-type lubricant should be used to limit the saw blade from jamming when cutting the materials.

Sheet aluminium cladding is usually surface fixed in the valleys or crests, while planked cladding is secret-, or concealed, fixed using patented or special fastener systems. Concealed fixing is carried out similarly to fixing vinyl or other planked cladding materials by first attaching a starter strip level when horizontal fixing. See the manufacturer's specifications and installation instructions before starting.

Advantages

Advantages of aluminium metal cladding are that it:

- is strong, durable and recyclable
- resists corrosion when properly surface finished and correctly installed
- is lightweight and thermally efficient
- is readily available.

Disadvantages

Disadvantages of aluminium metal cladding are that:

- it dents and marks relatively easily, making it hard to repair
- if not installed properly, temperature movement can cause severe expansion/contraction noises
- compatibility of metals can cause corrosion of fixing and fasteners and around penetrations
- it is energy intensive to produce, so environmental sustainability may be an issue
- depending on type, it may not be compliant with regulations and standards.

Metal fasteners for sheet metal cladding

Hex head screws

Hex head screws with neoprene washers are the most common means of fixing metal corrugated sheet cladding (see Figure 8.23). For Colorbond® steel cladding, a washer made from EPDM (ethylene propylene diene M-class rubber) is recommended instead of neoprene.

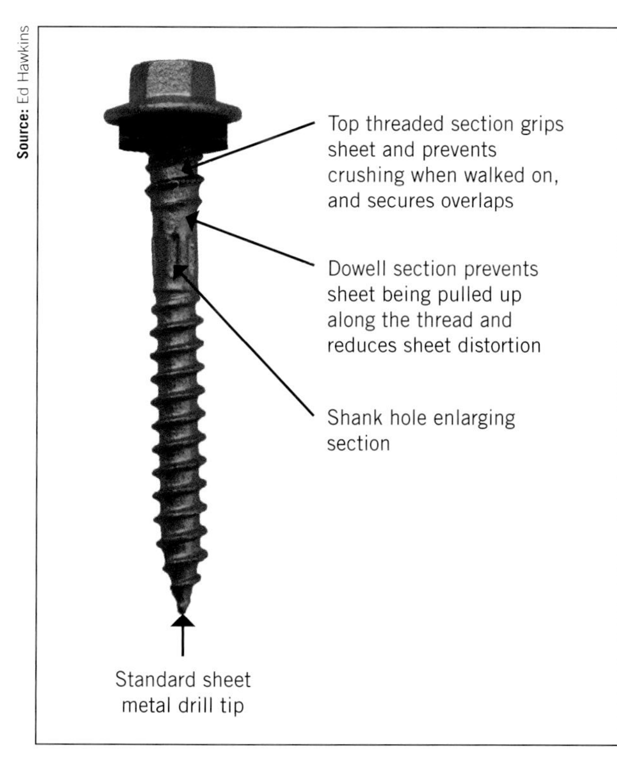

Source: Ed Hawkins

FIGURE 8.23 Hex head screw for sheet fixing

The screw size will depend upon the size of corrugation being used and the material being fixed into. Fasteners should be a minimum of Class 3, as per AS 3566 Screws – Self-drilling – For the building and construction industries.

Fasteners come in a range of colours to match the materials being fastened.

The finish will depend upon the prevailing weather conditions of the area (e.g. for harsh climates, powder-coated finishes may be recommended, while extreme climates or environments may call for stainless steel fasteners). Check with the manufacturer for specific advice.

Hex head screws can be used for fixing into both timber and steel groundwork.

For a neater finish on exterior cladding, smaller screws can be used, as shown in Figure 8.24. The screws shown in this figure were placed in the troughs in a horizontal clad home. If placed on the hills, there is a greater risk of the head eventually pulling through and distorting the sheet.

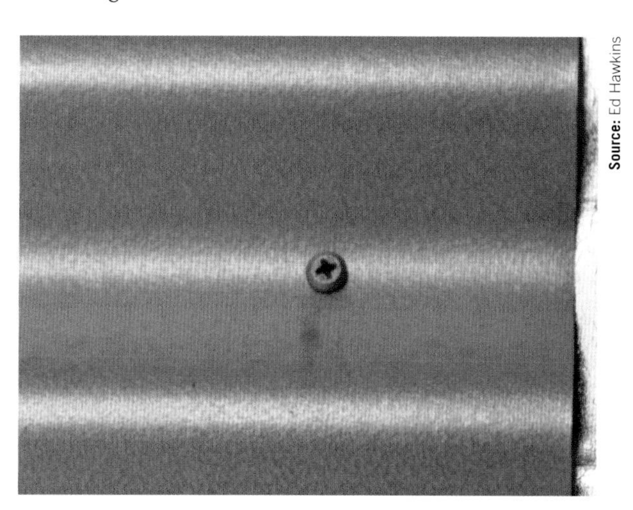

Source: Ed Hawkins

FIGURE 8.24 Stainless steel self-tapping screw in trough

Blind (or pop) rivets

Blind rivets are available in flat head, countersunk head and modified flush head, with standard diameters of 3, 4 and 5 mm (see Figure 8.25). They are made from soft aluminium alloy, steel, copper and Monel metal (a nickel–copper alloy).

Blind rivets are used to attach fittings and secure joints in sheet metal products. They are available in finishes to match the finish on the cladding and compatible metals.

Nails

Nails are only used to fix brackets used in the various cladding systems and must be galvanised. You should consult with the manufacturer or supplier to determine the suitable fastener spacing specifications.

Source: Ed Hawkins

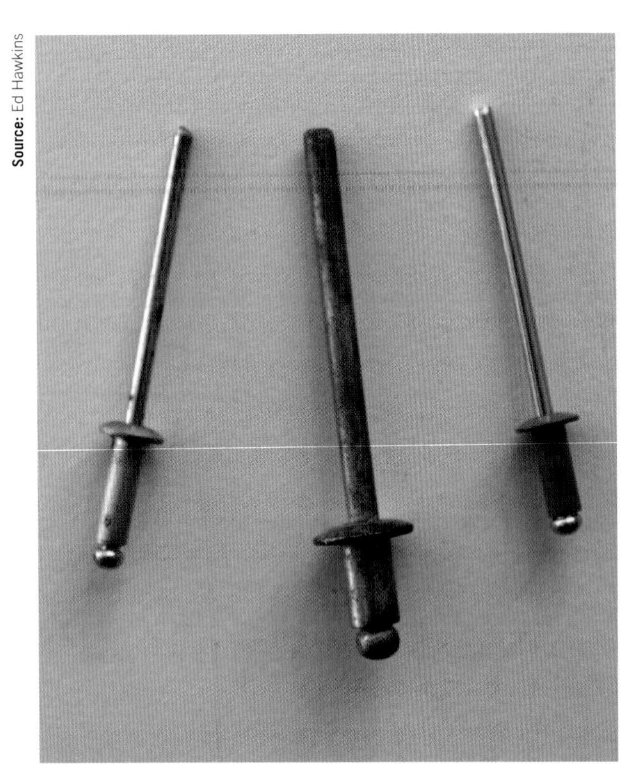

FIGURE 8.25 Blind (or pop) rivet

Cutting metal materials

Sheet product can be cut using a portable power saw fitted with a metal cutting blade. An abrasive disc *should not be used*.

Ensure that the coloured surface on Colorbond® is face down when cutting to reduce the risk of **swarf** (metal filings) being embedded in the exposed cladding surface. These will rust and leave stains on the surface of the cladding.

Electric shears are best used to cut sheets, but all waste should be removed to prevent rust spotting on sheets.

You can obtain more information about installing metal cladding from the websites of BlueScope Steel, Metroll Ltd and Stratco, and about fasteners from the Bremick website. Websites are listed under 'References and further reading' at the end of the chapter.

Plywood

Type A (phenol-formaldehyde bonded) plywood is often used as an external cladding. It has either a stained and varnished or painted finish, and is available in a wide range of sheet sizes and veneer finishes. An A grade face finish is required when a clear or stained finish is needed and a B grade face is needed for a regular paint finish. The back surface can be finished with a downgraded C or D grade finish.

Plywood cladding has three main surface finishes:

- rough sawn surfaces, available in grooved finish
- V-grooved (smooth veneered), usually used internally
- smooth veneer face.

Plywood cladding should be H3 or H3a preservative treated to protect against insects and rot when used externally. H3 means suitable for use in a place where the timber is not in contact with the ground but is exposed to weather. H3a means the same as H3 but must be regularly maintained with a paint coating. For more information on hazard classes in timber and preservative treatments, visit the Timber Preservers Association of Australia (http://www.tpaa.com.au) or Wood Solutions (https://www.woodsolutions.com.au).

Sizes

The most common plywood sheet sizes are 2400 × 1200 mm and 2700 × 1200 mm. Sheets 900 mm wide are also available. Sheet thicknesses vary between manufacturers; however, standard panel thicknesses range from 3 mm up to 25 mm, with 6 mm to 12 mm being the common thicknesses used for external cladding. Check the National Construction Code, Housing Provisions Standards – Part 7.5 and AS/NZS 2269.0 Plywood – Structural – Specifications for full details.

Grades of plywood

There are five appearance grade veneer qualities specified for plywood in AS/NZS 2269, which also specifies the requirements for the manufacture, grading, finishing and branding of structural plywood. These grades are:

- A: high-quality appearance grade veneer suitable for clear finishing of decorative surfaces such as furniture and internal wall panelling
- S: an appearance grade permitting natural timber defects such as knots as a decorative feature
- B: used under high-quality paint finishes
- C: used as flooring to be covered or overlaid (all defects are filled)
- D: used in structural plywood elements and bracing (unfilled defects).

LEARNING TASK 8.4 PLYWOOD RESEARCH

Investigate and report on the Australian Standards that relate to the external use of plywood in buildings.

Resources you can use are:

- Engineered Wood Products Association of Australasia (EWPAA): http://www.ewp.asn.au
- Wood Solutions: https://www.woodsolutions.com.au.

Advantages

Advantages of plywood are its:

- large range of sheet sizes
- structural strength, stability and workability
- ability to be fixed vertically as well as horizontally
- high durability
- low embodied energy
- environmental sustainability if made from forest waste.

Disadvantages

Disadvantages of plywood are that it:

- can degrade if not maintained
- is expensive when compared to fibre-cement sheeting
- is not generally recyclable.

Fixing plywood sheets

Plywood sheets are usually fixed vertically (see Figure 8.26), but can also be fixed horizontally. The critical point for vertically fixed sheets is to make sure you install the sheets plumb. Use a plumb bob suspended from the top plate to ensure the sheet installation is straight to begin with. Continue to check each sheet as it is installed.

Source: Ed Hawkins

FIGURE 8.26 Plywood cladding

When fixing sheets horizontally, use a level string line set to a datum or along the row of noggins to ensure they are level.

Tempered hardboard

Tempered hardboard is a dense manufactured board that is strong, both grain- and defect-free, and easily worked with regular woodworking tools (see Figure 8.27). It is manufactured from hardwood fibres and a small amount of paraffin wax as pressed hardboards, usually 9.5 mm thick for exterior usage.

Source: Greg Cheetham

FIGURE 8.27 Tempered hardboard finish

Most commonly used in strip planking forms such as Weathertex®, it is available in different surface finishes. The face edges and ends are anti-fungal treated and the back is moisture-resistant sealed.

Tempered hardboard is available in 3600 × 200 mm and 300 mm wide strips, as well as sheet form.

When working with hardboard you should minimise dust creation and personal exposure to dust fumes or vapours. The SDS recommends a class P1 or P2 respirator be worn when machining hardboard.

Advantages

Advantages of hardboard are its:

- stability and durability
- ability to be easily worked
- range of finishing methods available
- low embodied energy
- environmental sustainability if made from forest waste.

Disadvantages

Disadvantages of hardboard are that it:

- is expensive when compared to fibre-cement sheeting
- needs maintenance to ensure durability
- is not generally recyclable.

Fixing hardboard sheets

The sheets may be joined vertically or horizontally, with joints treated similarly to fibre-cement sheeting. Horizontal joints require a flashing. Flashing is also required around windows and doors. A vapour-permeable sarking material should be used beneath the hardboard. Vertical sheet edges can be slightly bevelled 2 mm to 3 mm for a neat finish, or PVC (polyvinyl chloride) or aluminium extrusions can be used to join the sheets. (**Note:** Ends should be sealed after cutting.) Alternatively, the sheets may be fixed and timber battens applied over the surface of the joints.

Adhesive fixing

Tempered hardboard can be fixed using construction adhesive, provided the framework has been properly prepared. To attach the sheets, apply adhesive to framing members or existing walls in a continuous bead. Sheets are located correctly as soon as the adhesive has been applied. They should be temporarily fixed with two hardboard nails along the top edge. Press and brace the sheet firmly against the frame for maximum adhesive contact, then temporarily block the sheet out from the wall and allow the construction adhesive to become touch dry before removing the block and bringing the sheet back in touch with the framing members. Use a hammer and block with a protected face to tap the sheet over the framing member to improve the adhesion. Remove any excess adhesive by squeezing it out before it dries.

You can obtain an installation guide for hardboard fixing and installation from the Weathertex website (listed at the end of this chapter) or by contacting the manufacturer.

Expanded polystyrene cladding systems

The most common form of expanded polystyrene (EPS) cladding is fixed directly to the frame and finished with a variety of rendered or site-applied finishes. Because buildings react to the environment, cladding materials expand and contract and can develop cracks. Cracking can be controlled through the use of control joints, but these can impact design aesthetics and weatherproofing.

EPS usually contains high-impact greenhouse gases, which can increase its embodied energy relative to other insulation materials. A small percentage of imported foams also include ozone-depleting substances that are banned in Australia. Polystyrene (or foam) cladding is non-structural and a lightweight building material used as an alternative to rendered brick facade.

EPS panelling (and similar autoclaved aerated concrete products) are not automatically 'deemed to satisfy' the National Construction Code. They must be offered as an 'alternative solution' with evidence to satisfy regulators and compliance monitors such as registered building surveyors. The material and construction systems must meet the performance requirements of the code, especially relating to fire-resistance and combustibility. See the National Construction Code 2022, Building Code of Australia – Volume One – Section C Fire resistance, particularly Specification 6 Structural tests for lightweight construction for details of acceptance of design and construction requirements.

AS 1366.3 Rigid cellular plastics for thermal insulation specifies the requirements for rigid cellular polystyrene in the form of sheets, board, blocks and cut shapes for thermal insulation purposes (see Figure 8.28). You should also be aware of the *Handbook of Australian Fire Standards – Part 4: Building materials, products and construction* (HB 37.4) that deals with issues related to fire and fire tests concerning building materials, products and construction.

The EPS panels come in a range of sheet sizes and thicknesses and can be manufactured to specification by some manufacturers.

A minimum density of 19 kg/m^3 is recommended for wall cladding. Other densities are available to meet specifications.

This kind of wall cladding is most often fixed to timber or metal framed structures. Before installing any system always consult the manufacturer's recommended fixing methods.

Source: Stratco

FIGURE 8.28 Expanded polystyrene wall cladding

Wall panels are fixed to framework using hot dipped galvanised screws. The length and type depend on the frame materials and thickness of panels used (see Figure 8.29).

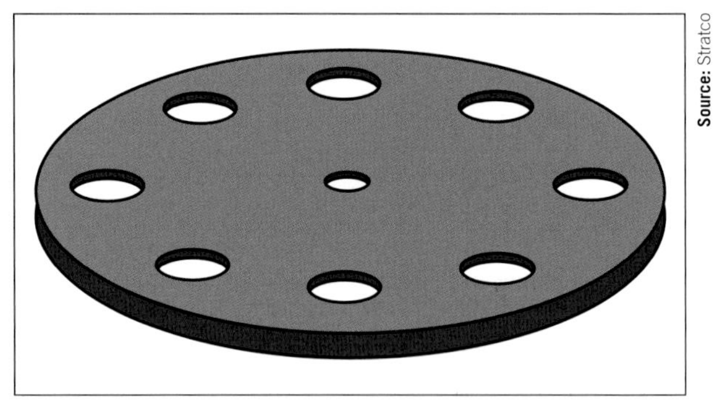

Source: Stratco

FIGURE 8.29 Plastic washers for galvanised fixing screws in polystyrene cladding panels

Follow these general rules when installing on framed walls:

- Make sure the frames are straight and plumb.
- Correctly install all windows, DPC, flashing and termite barriers and external corner angles.
- Use vapour-permeable sarking.
- Use blocking where 'off-stud' jointing of panels occurs and for installing wall mounted accessories; that is, taps, electrical fittings/meter boards and security shutters. Some manufacturers recommend double studding behind vertical panel joints to prevent foam breakout at joints.
- Set the recommended starter bead to frames to ensure panel alignment.
- On rebated slabs and suspended slabs, ensure the rebates are clean and DPC is applied.

- All adjoining edges must be glued with recommended adhesive, including corners.
- When cutting, use a straight edge.
- Leave a gap around door and window frames to allow for sealant (3 mm approximately). It is critical to maintain thermal efficiency that this is done carefully.

Finishing

Once the panels have been installed and all gaps sealed, a base coat (usually acrylic resin-based render) is applied to a specified thickness on a groundwork of alkaline-resistant mesh (usually fibreglass). This is followed by a second coat, usually of a finer grade, followed by the final finish texture coat (see Figure 8.30).

Advantages of EPS cladding systems are that they:

- increase the speed of construction, which reduces building costs
- provide excellent thermal insulation (depending on panel thickness)
- are durable once rendered over
- are termite resistant
- provide a strong water and vapour barrier.

Disadvantages of EPS cladding systems are that:

- specialised construction skills are required; the EPS is quick to install but finishing requires multiple layers that can extend construction times and have varying life cycle environmental impacts
- non-flame-retarded polystyrene foam insulation board will burn toxic fumes
- material strength is poor, so non load bearing
- they are prone to cracking if not properly rendered
- they have quality assurance issues.

Structural insulated panel

Structural insulated panel (SIP) systems are a form of panel integrating different types of foam panelling sandwiched between a number of different layers of metal, plywood, fibre-cement or oriented strand board (OSB). They can be used as structural components such as walls, floors and roofs, and are particularly favoured for temporary buildings such as site sheds, refrigerated rooms and warehouses with metal faces. SIPs have higher insulating values than regular EPS systems because of the sandwich construction methods (see Figure 8.31).

Onsite construction time and costs are reduced because offsite manufacture of the products means fewer tradespeople are required for assembly. They are environmentally sustainable, with the lifecycle cost generally lower than a conventional framed construction depending on the overall building design being integrated with other technology.

An SIP structure must be designed to resist wind loads and other applied forces and meet all

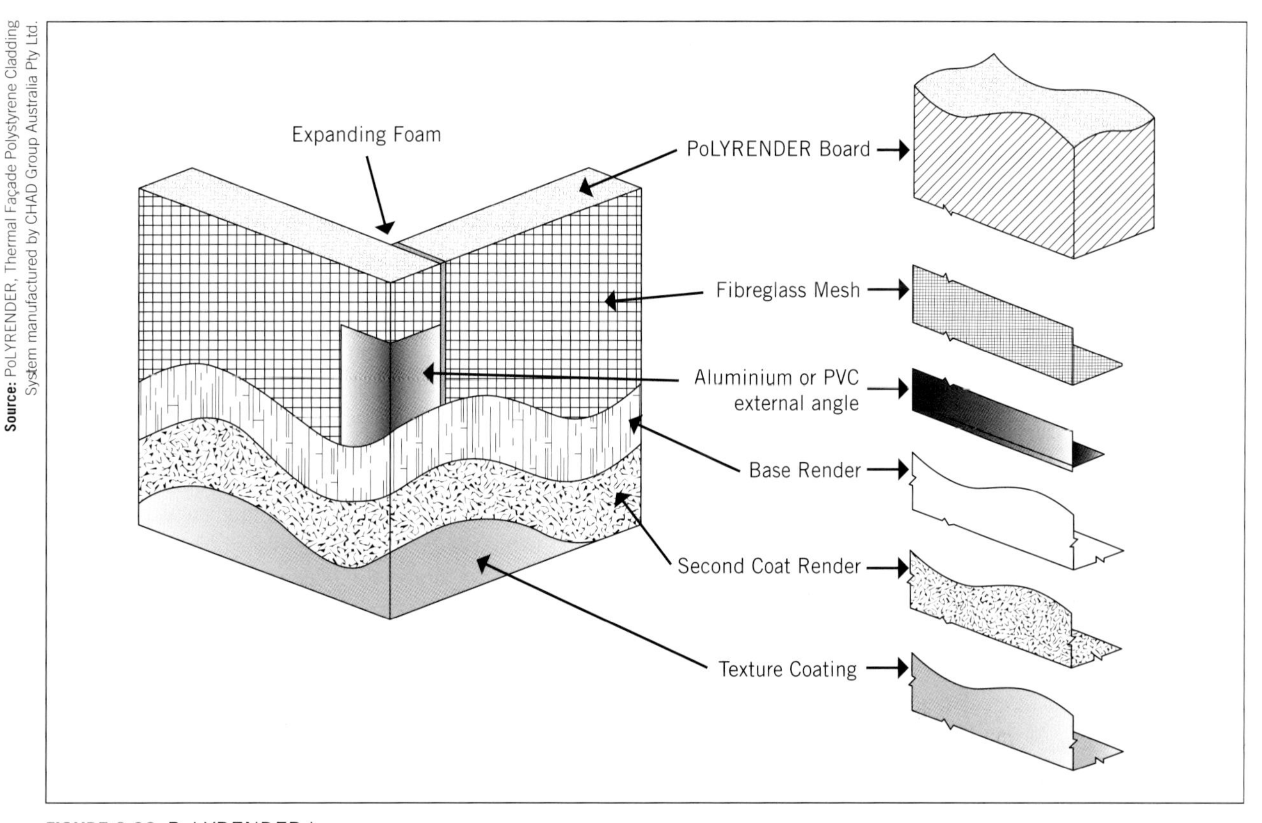

Source: PoLYRENDER, Thermal Façade Polystyrene Cladding System manufactured by CHAD Group Australia Pty Ltd.

FIGURE 8.30 PoLYRENDER layers

Source: SIPs Industries

FIGURE 8.31 OSB faced SIP

Building Code of Australia and Australian Standards requirements. The Wood Solutions website provides an overview of their application and other design and load considerations (see list at the end of the chapter).

Typically, a recess around the perimeter of SIPs is to allow the insertion of structural timber to assist in splicing panels to each other and to form a continuous base plate bolted to the slab. The panel then sits over the plate and is nailed to the side of the base plate. Similar methods are used for other panel intersections (see Figure 8.32).

Note: There are a number of jointing systems used depending on the manufacturer and materials used to sandwich the panels.

Innovative cladding products

Sustainable innovative cladding materials are being developed all the time. Some new products include:

- hemp-based products such as Hempcrete™
- products based on agricultural waste (e.g. straw and husks) such as Durra Panel™.

These products are manufactured to be waterproof and meet Australian Standards and National Construction Codes. They have lower embodied energy and less environmental impact than traditional products. Other properties include higher acoustic ratings and fire resistance. More information is available at https://www.yourhome.gov.au/materials/cladding-systems.

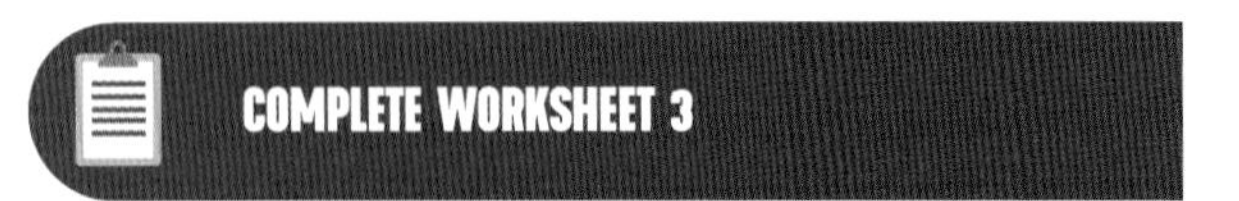

LEARNING TASK 8.5 INSULATED PANELLING

Students are to work in pairs to investigate a minimum of two EPS and two SIP manufacturers/suppliers in their state or territory, identifying:

- manufacturer – provide a web and physical address and contact details
- types of panel systems – including core materials, face finishes and jointing methods (i.e. acrylic render or oriented strand board)
- whether their products are certified under the CodeMark Certification Scheme and/or other certification regimes.

Use search terms such as 'polystyrene wall cladding systems + Australia' and 'structural insulated panels + Australia'; or instead of Australia, insert your state or territory.

Students must investigate the Australian Building Codes Board's (ABCB) revised CodeMark Certification Scheme requirements by explaining the role of an unrestricted building certifier in the certification process. Students must identify if the company has an approved certificate of conformity or other certification for the product.

To assist in research, visit the ABCB product certifications pages: https://codemark.abcb.gov.au

Calculating quantities

The *overall height* is the height of the wall between the eave lining and the top edge of the rebate on the plinth board (see Figure 8.33).

Effective cover is the exposed width of each installed board (see Figure 8.34). Horizontal weatherboards overlap each other when installed, making the actual height of each board the full board width minus the overhang. The overlap is generally recommended by the manufacturer of the boards or determined by the depth of the rebate. Splayed boards with no rebate should be overlapped a minimum of 25 mm or up to 35 mm.

The *actual cover* (also known as weather cover) is the amount of cover that is used so that the wall can be covered using full boards; that is, no ripped boards (see Figure 8.35). This dimension is used to set out the positions of the boards on the wall.

When calculating board quantities, it is not uncommon to find that the final calculation comes up with an answer that is not equal to full boards; for example:

Overall wall height = 2400 mm
Effective cover of boards = 130 mm
So the calculation is 2400 ÷ 130 mm = 18.46 effective board widths

Source: SIPs Industries

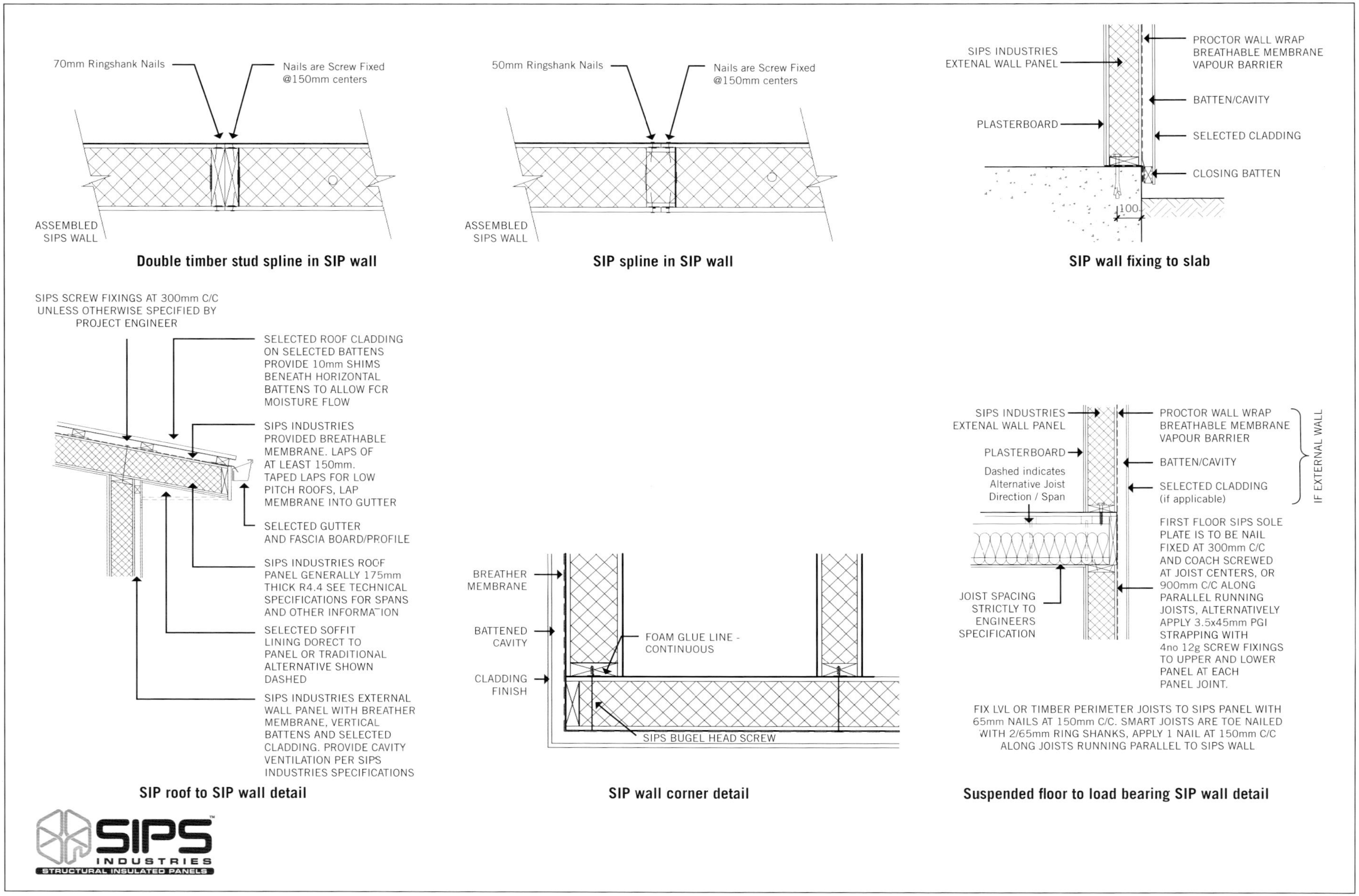

FIGURE 8.32 Typical SIP joints

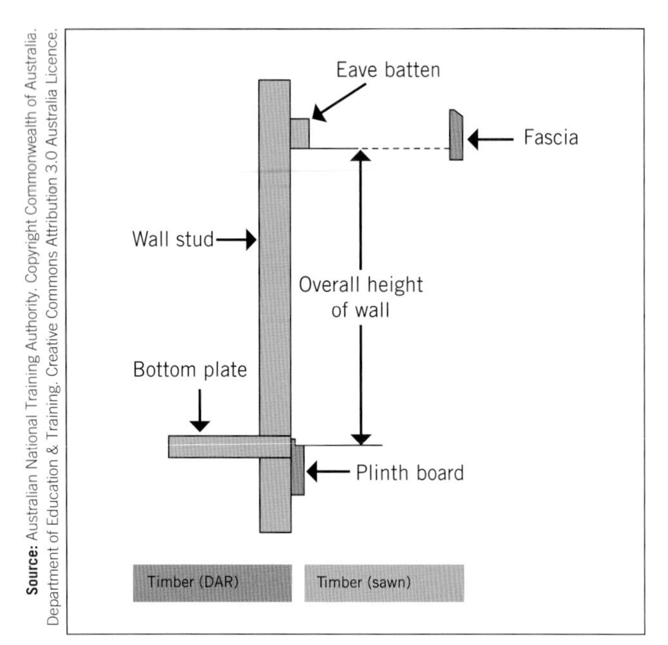

FIGURE 8.33 Overall height

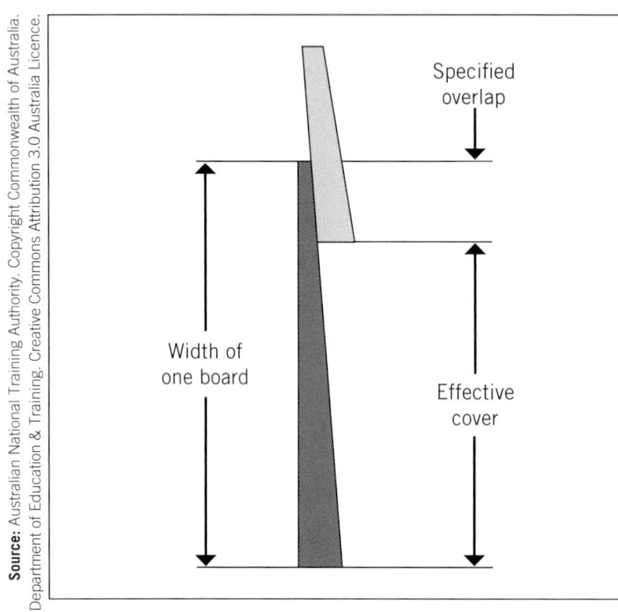

FIGURE 8.34 Effective cover

You would not rip one board down to 0.46 of a board width. Instead, you would increase or decrease the effective cover by dividing the wall height by the number of full boards you choose; for example:

2400 ÷ 19 full boards = 126.31 mm (effective cover)

If you decrease the number of boards:

2400 ÷ 18 full boards = 133.3 mm (effective cover)

As long as you are not reducing the effective cover to less than the manufacturer's minimum overlap, it is practical to go with the round-off figure.

Once the calculations have been made, you can mark off the heights on the corner stops to provide a guide to work to as you progress with the installation.

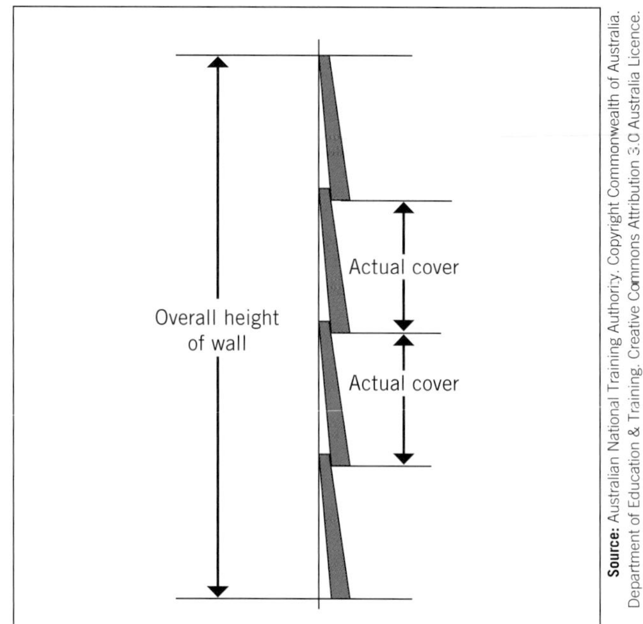

FIGURE 8.35 Calculating actual cover

Note: This is really only practical with splayed lining boards; for rebated and tongue and grooves, effective and actual cover are the same.

Calculating weatherboard and other plank quantities

How many lineal metres are required to cover a wall depends upon the length of the wall and the number and size of the openings in the wall.

Now that we know the effective cover of the boards and how many board widths will be required to cover the wall, we simply multiply the number of boards required by the length of each wall (or the total perimeter distance of the building).

For example, if the total perimeter of the building shown in the plan in Figure 8.36 is 47.56 m, then 47.56 m × 18 boards = 856.08 m.

Now you have to **deduct** the total m^2 of all openings; for example:

Entry door	1 / 2040 × 820 = 1.673 m^2
Rear entry door	1 / 2040 × 820 = 1.673 m^2
1 pair sliding glass doors	1 / 2040 × 2120 = 4.325 m^2
Windows	1 / 1218 × 1200 = 1.462 m^2
	3 / 1215 × 1200 = 4.374 m^2
	2 / 912 × 1000 = 1.824 m^2
	2 / 909 × 1000 = 1.812 m^2
	1 / 606 × 800 = 0.48 m^2
	2 / 1212 × 1000 = 2.424 m^2
	1 × 612 × 800 = 0.5 m^2
	Total m^2 = 20.547 m^2 (say, 20.5)

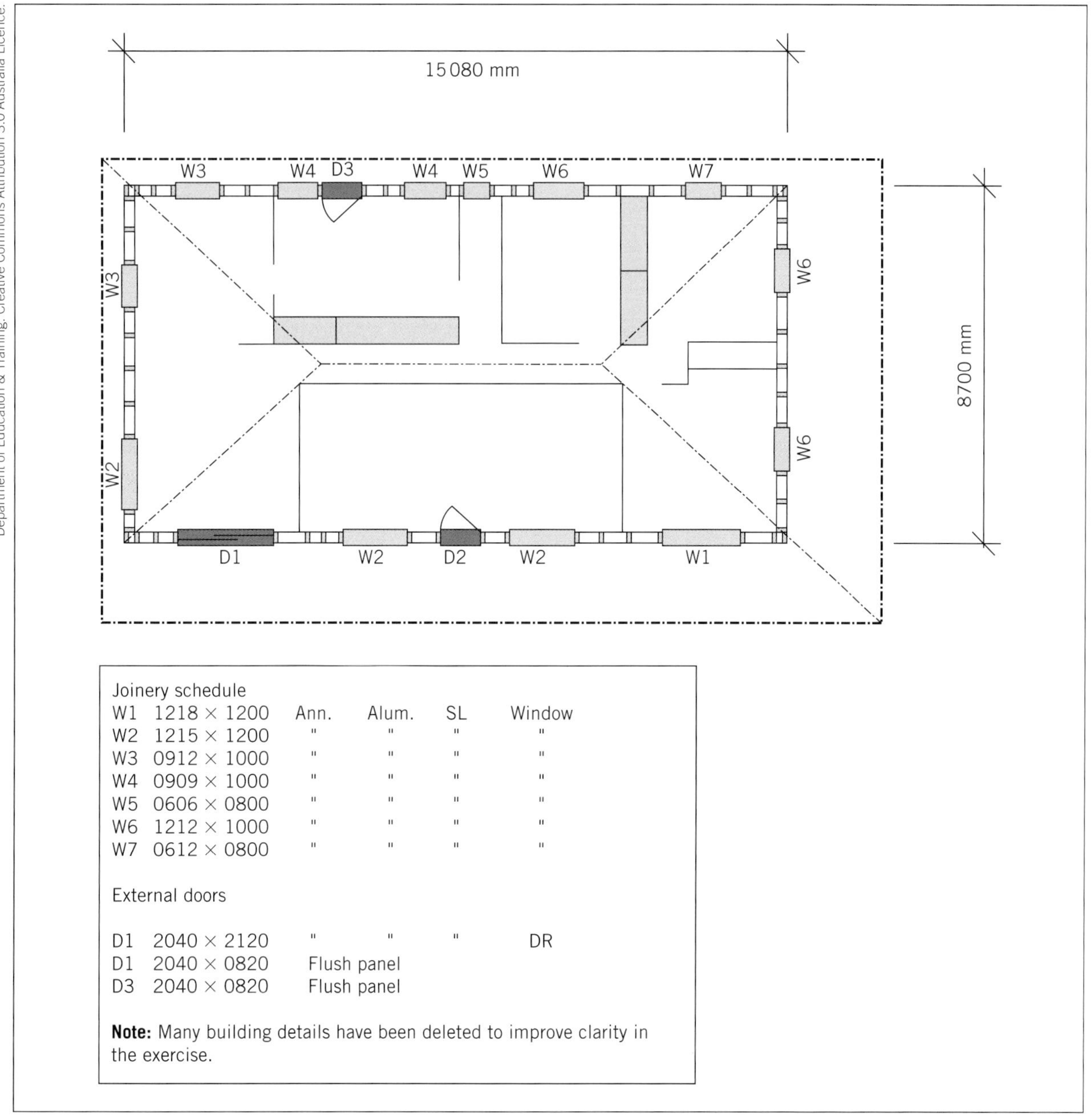

Joinery schedule

W1	1218 × 1200	Ann.	Alum.	SL	Window
W2	1215 × 1200	"	"	"	"
W3	0912 × 1000	"	"	"	"
W4	0909 × 1000	"	"	"	"
W5	0606 × 0800	"	"	"	"
W6	1212 × 1000	"	"	"	"
W7	0612 × 0800	"	"	"	"

External doors

D1	2040 × 2120	"	"	"	DR
D1	2040 × 0820	Flush panel			
D3	2040 × 0820	Flush panel			

Note: Many building details have been deleted to improve clarity in the exercise.

FIGURE 8.36 Actual house plan

Total m^2 of the building = 47.56 m × (18 × 0.1333 m [refer to earlier effective cover of boards calculation])

= 47.56 × 2.4

= 114.144 m^2

Minus the amount for the doors and windows = 114.144 – 20.5 = 93.64 m^2

Lineal metres of board required to cover 93.64 m^2 = 93.64 ÷ 0.1333

= 704.1 m (say, 705 m)

Lengths needed to do the job:

Best stock length = 4.8 m

Divide the total metres by the stock length = 705 m ÷ 4.8 m

Total number of 4.8 m boards = 146.9 boards (say, 147 boards)

Now add 5% for waste and off-cuts = 147 boards × 1.05

Order = 154 boards at 4.8 m or 154/4.8

Calculating sheet materials

For sheet material, all we need to do is divide the surface area of all walls by the surface area of the chosen sheet size.

Using the examples from the building above, the calculations are:

Total surface area to be covered

$= 93.64\ m^2 \div 4.32\ m^2$ (3600 × 1200 mm sheet)

$= 21.675$ sheets

= say, 22 sheets + 5% for waste

$= 22 \times 1.05 = 23.1$ sheets

Order = 23 sheets (rounded down)

Installation

In this section you will learn about the various methods and practices used to install cladding materials.

Always wear a P2 respirator when cutting fibre-cement sheeting. Note that the 'score and snap' cutting method produces less silica dust.

Installing fibre-cement sheets

Fibre-cement sheeting contains silica, which can cause cancer; inhaling excessive amounts of the dust can cause respiratory diseases. When cutting or working with fibre-cement sheeting, make sure the area is well ventilated, and minimise dust by using the score and snap method, fibro cutters or mechanical shears wherever possible. When using other portable power tools or power saws, make sure there is an HEPA (high-efficiency particulate arrestance) grade vacuum cleaner available and signage to warn others in the area, and wear a respirator with a minimum P1 filter. When cleaning up, spray water on the dust and vacuum up the waste.

Do not dry sweep dusty waste. Use a spray bottle or light water spray to dampen down dust then use a vacuum cleaner. Always wear a dust mask.

Cutting fibre-cement sheeting

Cutting fibre-cement sheeting requires special tools and techniques. Corded or uncorded electric shears are the most common tools used to cut single sheets (see Figure 8.37). Fibre-cement cutting shear adaptors are available for power drills too.

FIGURE 8.37 Power shears

Source: Greg Cheetham

Multiple sheets being cut to the same size can be stacked up, with the straight edges clamped over and a portable power saw with a masonry blade used to cut several sheets at a time. This is not recommended because of the amount of silica dust it will generate. However, if you do use this method, make sure you and others nearby are wearing dust masks, and wet down the waste before cleaning up.

A hand guillotine (fibro cutter) is the most common hand-operated tool used for making straight cuts (see Figure 8.38).

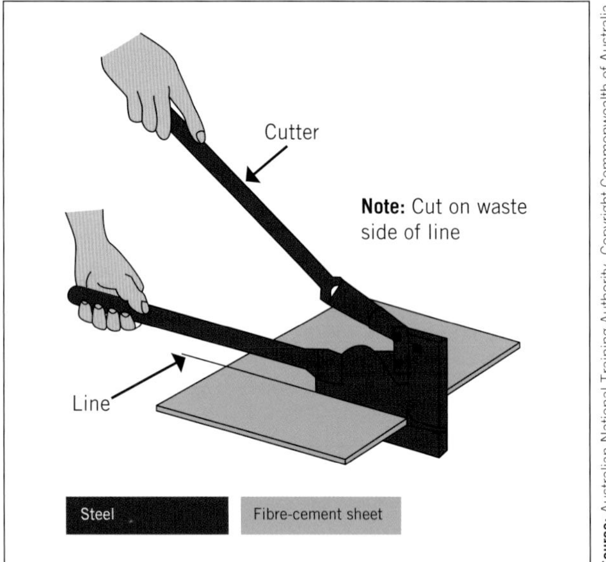

FIGURE 8.38 Hand guillotine

Source: Australian National Training Authority. Copyright Commonwealth of Australia. Department of Education & Training. Creative Commons Attribution 3.0 Australia Licence.

Mark the sheets carefully, including penetrations, and check again before cutting.

Lay each sheet flat on 100 × 50 mm timbers, saw horses or similar supports to allow the guillotine to slide across the sheet without obstruction. Don't forget to cut on the waste side of the line with the seen surface face up.

You can also use a score and snap action with a tungsten carbide-tipped scoring tool combined with a straight edge (see Figure 8.39). Score on the face side of

Timber (DAR) Fibre-cement sheet

Source: Australian National Training Authority. Copyright Commonwealth of Australia. Department of Education & Training. Creative Commons Attribution 3.0 Australia Licence.

FIGURE 8.39 Straight cuts, using score and snap method

the sheet only and bend the sheet upwards to ensure a clean cut.

The favoured power tool for making penetration in fibre-cement sheeting is the reciprocating saw (also known as an oscillating saw or multi tool; see Figure 8.40). This is an excellent tool capable of making straight or curved penetrations and comes with a range of blades and adaptors such as sanding pads, making it a very versatile power tool.

Source: Milwaukee Power Tools Australia

FIGURE 8.40 Reciprocating saw

Penetrations can also be made by drilling small holes around the marked-out penetration and punching out the waste with a hammer from the face side (see Figure 8.41). The edges can be cleaned up with a rasp.

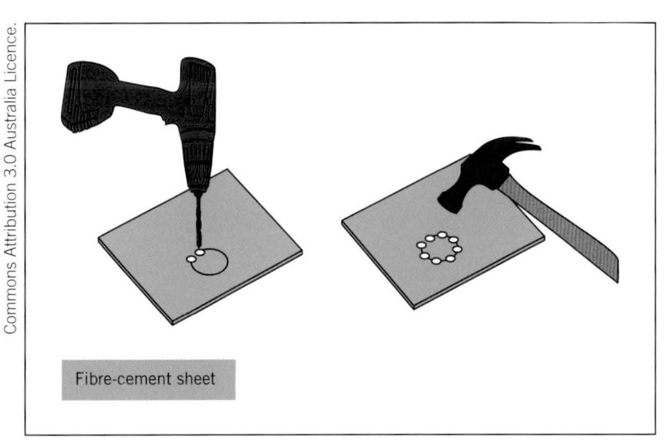

Source: Australian National Training Authority. Copyright Commonwealth of Australia. Department of Education & Training. Creative Commons Attribution 3.0 Australia Licence.

FIGURE 8.41 Penetrations in fibre-cement sheeting

Large penetrations (such as those needed for vents) can be cut out using a score and snap method. Score around the penetration; make a hole in the centre of the penetration; saw cut using a keyhole saw to the edges of the penetration; then snap face side up to release the waste (see Figure 8.42).

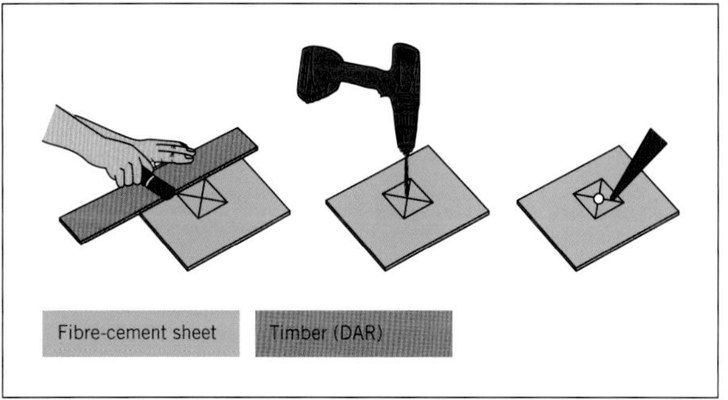

Source: Australian National Training Authority. Copyright Commonwealth of Australia. Department of Education & Training. Creative Commons Attribution 3.0 Australia Licence.

FIGURE 8.42 Score and snap large penetrations

Fasteners for timber-framed fixing

Class 3 external grade fasteners must be used for fixing all cladding materials, whether they be nails or screws; that is, they must be galvanised, zinc coated or stainless steel, and must be from a compatible metal, especially when fixing metal cladding materials.

For fibre-cement sheeting, the regular fastening is a hot dipped galvanised flat head nail 40 × 2.8 mm, with a minimum length of 38 mm (see Figure 8.43). James Hardie's external cladding instruction manual has an excellent table outlining the type and size of nails suitable for fixing fibre-cement sheet using gun nailers (see the James Hardie website listed at the end of this chapter), but consult with your machine's manufacturer to find their equivalent or recommended product.

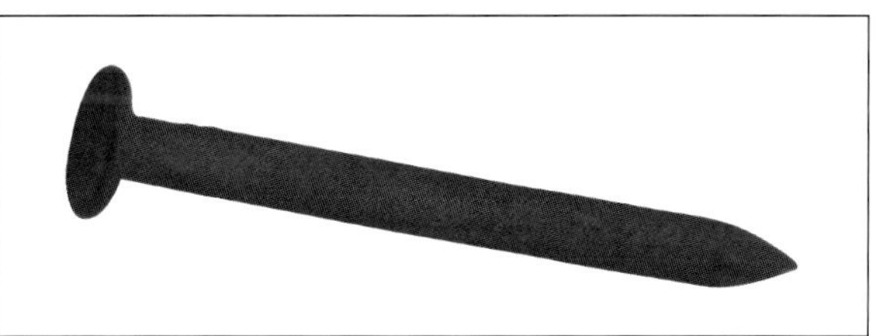

Source: Ed Hawkins

FIGURE 8.43 Fibre-cement sheet nail – hot dipped galvanised

Nail guns must be fitted with a flush drive attachment to prevent the nails being punched through the surface. Some guns have adjustable headpieces to help adjust fastener depth drive. Do not overdrive the nails; if you cannot adjust the depth, leave the nail **proud** of the surface and drive it flush with a hammer. Make sure you push the nailer and the sheet hard against the stud when nailing to prevent the back from blowing out.

Stud sizes

Timber studs should be a minimum of 42 mm thick, and 38 mm for steel stud framing. If the studs are undersized, then a continuous batten or additional stud needs to be inserted.

AS 4055 Wind loads for housing provides guidance on the appropriate stud spacing and fastener spacing to use in different cyclonic and non-cyclonic wind areas. Basically, depending upon the location, stud spacing will usually be 450 mm centres in a load-bearing wall, with fastener spacing between 200 and 300 mm, depending on the wind area rating.

Fasteners for steel-framed fixing

Screws can be used externally on fibre-cement sheeting for both steel and timber framing. If they are used, they need to be a class 3 fastener.

Screws are generally only used to fasten fibre-cement sheeting externally in timber frames when the joints will be covered with strip battens (see Figure 8.44), or set and rendered or finished later.

FIGURE 8.44 Fibre-cement sheet joints covered with battens

Screws are used to fasten all fibre-cement sheets in steel-framed construction.

If the surface is to have an applied finish, such as render or faux render, a recessed edge sheet material is recommended.

The size of screw to be used depends on the thickness of steel being used in the construction. For 0.55 to 0.75 mm thick steel, 20 mm self-drilling screws should be used. For steel framing with a thickness of 0.8 to 1.6 mm, 8 × 32 mm self-drilling screws are necessary.

Note: Self-embedding screws must not be used in 4.5 mm thick sheets because the sheet is too thin compared to the head of the screw and when used with a power driver has a tendency to embed too deeply.

Fixing fibre-cement sheets

Fibre-cement sheets should be fixed at 200 mm centres along edges. The fasteners, whether nails or screws, should not be any closer than 12 mm to the edge, or closer than 50 mm to the corner. Intermediate stud fixing should be at 300 mm centres. Do not use stud adhesive. Sheets are fixed along all sheet edges over studs in wall-cladding applications.

Fibre-cement sheeting may be set vertically or horizontally as cladding. By sheeting vertically, you can usually avoid having to run a central cover strip and PVC or galvanised Z flashing.

Vertical jointing

Timber battens are a common method used to vertically seal fibre-cement sheets (see Figure 8.45), particularly on gable ends in period houses. In modern constructions, PVC straight joiners are the preferred method (see Figure 8.46).

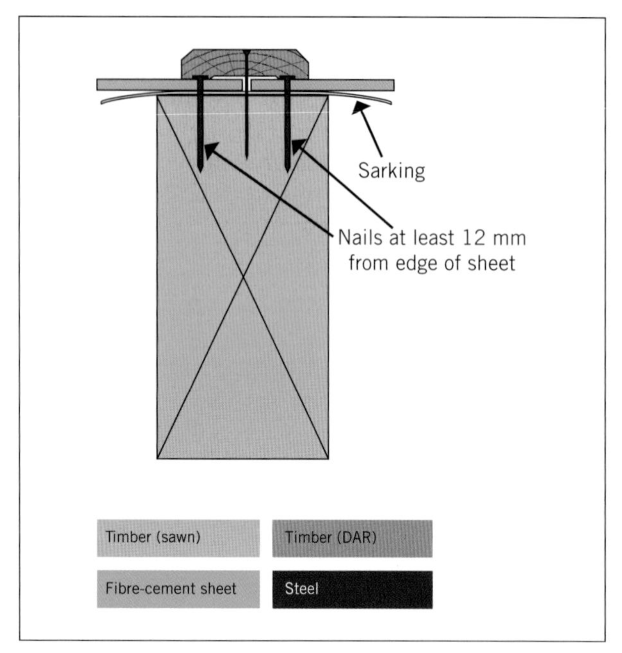

FIGURE 8.45 Vertical fixing with cover batten

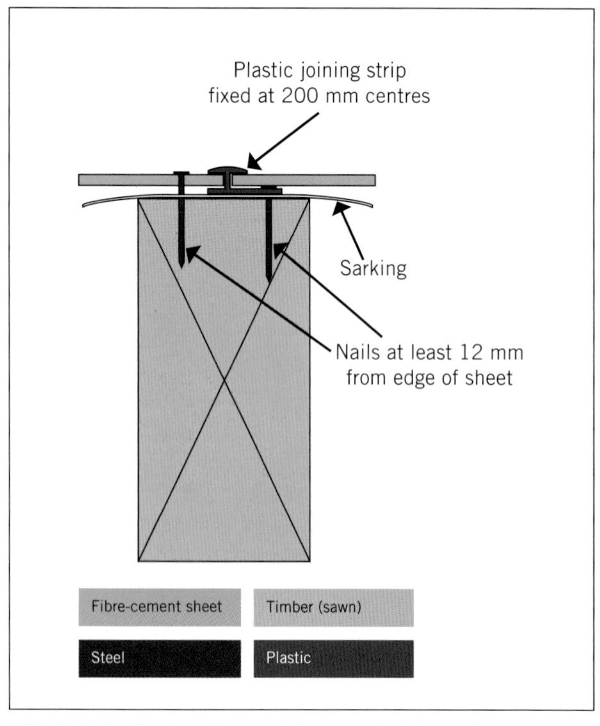

FIGURE 8.46 Vertical fixing with plastic joining strip

If Stucco™ sheets are vertically joined, they are nailed off to studs, leaving a 3 mm gap between the sheets, and the gap is sealed with a flexible joint sealant.

Horizontal jointing

A PVC or galvanised metal Z flashing strip is used to waterproof horizontal joints in 6 mm fibre-cement sheeting (see Figures 8.47 and 8.48). A PVC drip strip, similar to a vertical joiner, is used for 4.5 mm sheets (see Figures 8.49 and 8.50). Sheets are joined on a row of noggins using vertical joining strips, as for vertical fixing.

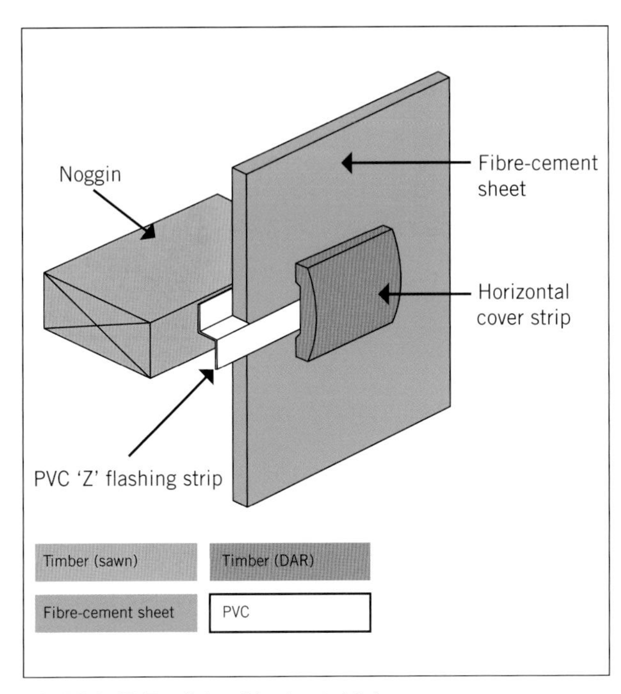

FIGURE 8.47 Traditional horizontal fixing

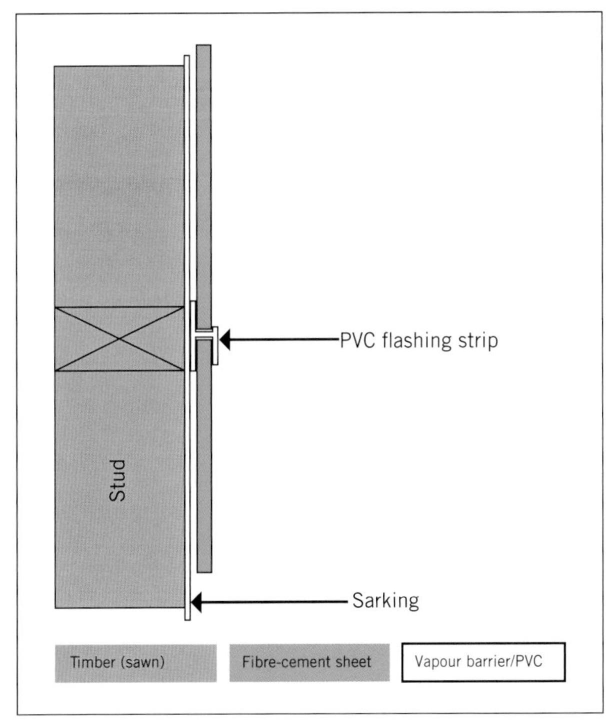

FIGURE 8.48 PVC flashing strip

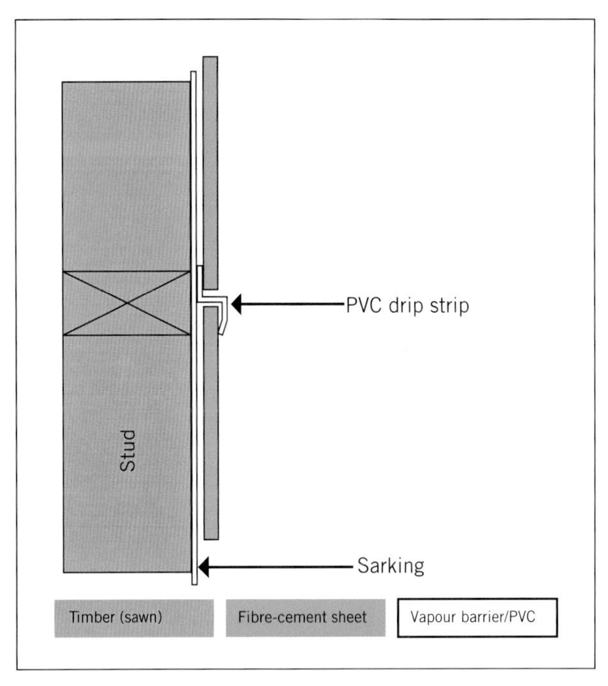

FIGURE 8.49 PVC drip strip

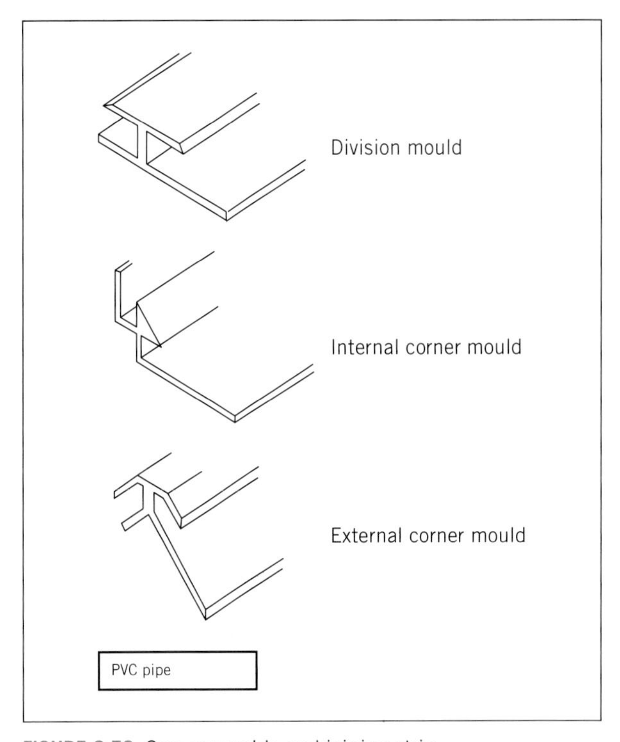

FIGURE 8.50 Corner moulds and joining strip

Internal and external corners

Traditionally, internal and external corners were covered by battens in a similar way to weatherboards (see Figures 8.56 and 8.57) and quad moulds were used to cover the joints at internal corners. Now the most common method is to use plastic extrusions as the appropriate alternative (see Figure 8.51).

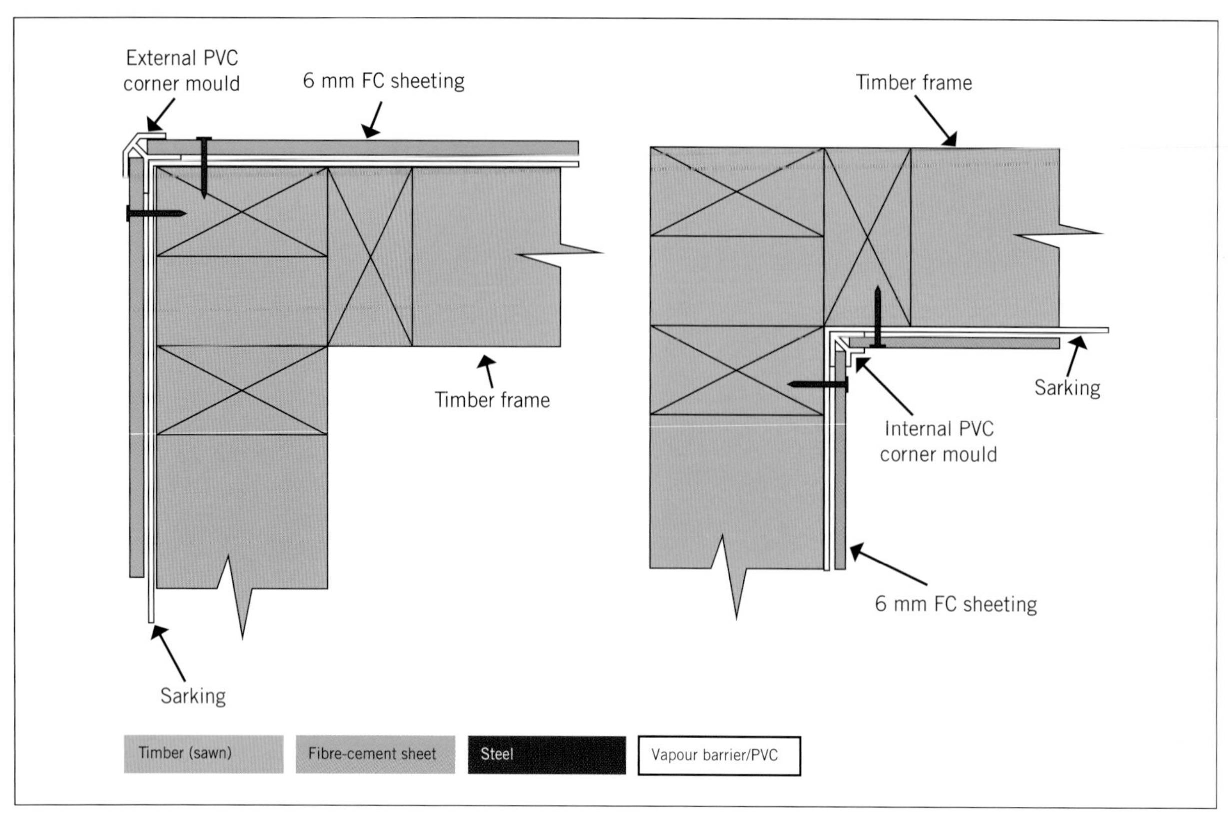

FIGURE 8.51 Internal and external corners

Control joints

When fibre-cement cladding joins up to masonry walls, control joints must be incorporated into the framework, using 9 mm backing rods and appropriate sealants (see Figure 8.52).

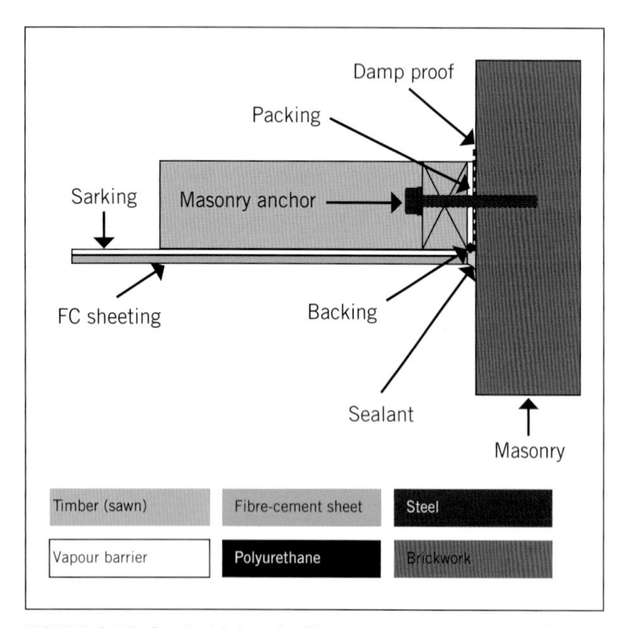

FIGURE 8.52 Control joints in fibre-cement to masonry joints

Installing timber weatherboards

In most cases, the starting point for the installation of planked cladding will be on top of the **ant capping**, which will be in place on top of piers and/or dwarf walls that enclose the floor frame material. Pre-set foot moulds, starter strips or plinth boards are installed using a string line set up level along the bottom edge of the frames (see Figures 8.53, 8.63 and 8.64).

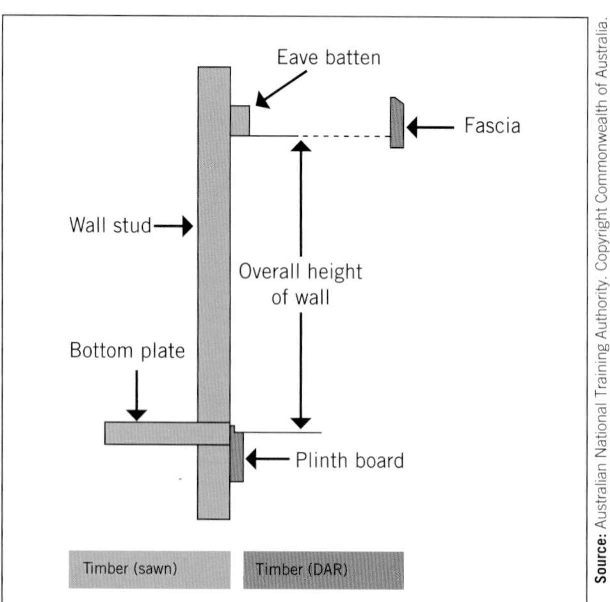

FIGURE 8.53 Measuring the height of walls for cladding

When setting the height of the first board, you need to take into consideration the effective cover of each board and the distance to the eave lining, if eaves are a consideration. It always looks better to finish at the top of the wall as close as possible to a full-width board.

You also need to install internal and external corner battens to which the material is butted or fitted at the corners.

Straight-line joints are cut over a stud to allow for fixing at each board. The joints must be staggered across the row. Mitre or scarf joints are better because when the cladding shrinks, which it inevitably does, there is still some timber behind it to accept a sealant (i.e. sealer, primer and/or paint). Each end of the boards is fastened, using two 50 mm hot dipped galvanised or stainless steel nails. As a general rule, nails should be two to 2½ times as long as the thickness of the material they are intended to hold.

If the weatherboards or other strip forms of cladding are to be installed vertically or at an angle, additional noggins for fixing need to be installed. Alternatively, walls may be battened off with battens at 600 mm centres (see Figure 8.54). Timber battens can be used. Galvanised metal **furring channel** can also be used with some cladding materials; for example, corrugated iron sheeting.

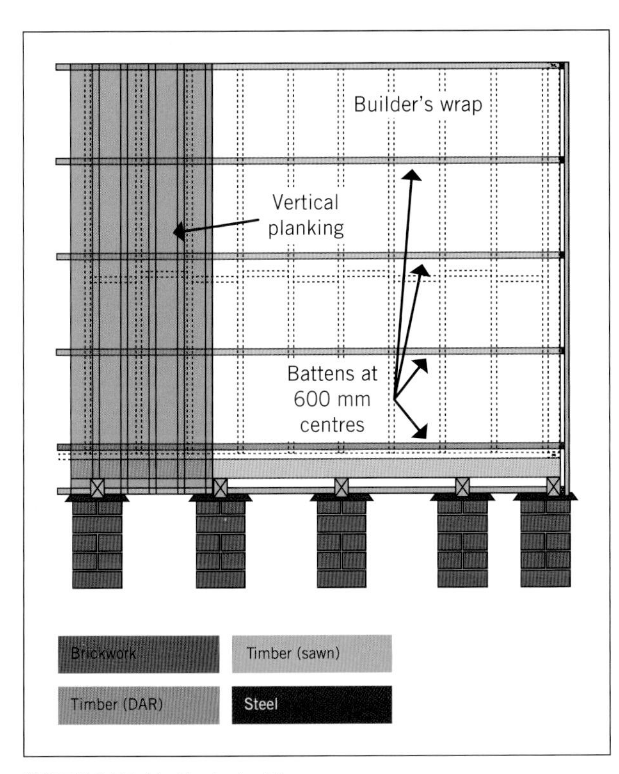

FIGURE 8.54 Vertical planking

Ensure that flashings are in place. Check that the door and window frames have been set in place as this will allow abutment of the lining boards flush with the frames. This is necessary for the finishing of the architraves.

Where splayed boards are being used, a template or jig (lap gauge) should be made to assist in ensuring that overlaps are even all the way around the building (see Figure 8.71).

Straight joints

Weatherboards are designed to be joined on a stud by butt jointing and skew nailing to hold the ends tight.

Follow these steps:

1 Cut the boards slightly over-length (1–2 mm) with ends square and slightly undercut so that the faces will fit tight (see Figure 8.55).
2 Prime the ends with an appropriate primer.
3 Pre-drill the ends of the boards to prevent the ends splitting when nailing.

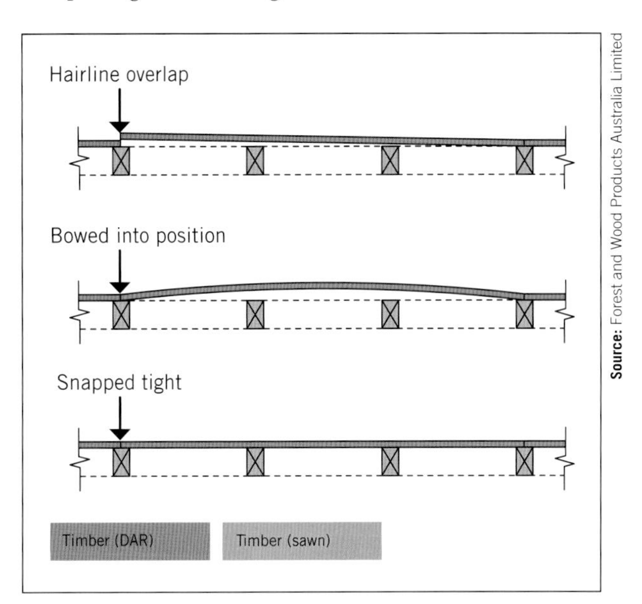

FIGURE 8.55 Straight-line joints

If you are only replacing a couple of boards and no sarking is present, cut flashing material and fix it behind the joint to prevent moisture penetrating to the frame.

Weather stops

Weatherboard stops or stop beads provide a neat appearance at corners and increase weatherproofing; they also help to locate the weatherboards being fitted to the wall (see Figures 8.56 and 8.57).

Timber measuring 35 × 35 mm is commonly used for internal corners (see Figure 8.58), while 70 × 35 mm material is used for external corners – other sizes may be more suitable depending upon the cladding material you are using.

Before installing the stops, remove the **arris** from the corners and prime all surfaces, including ends. Make sure the stops have been marked with cladding heights

Source: Forest and Wood Procucts Australia Limited

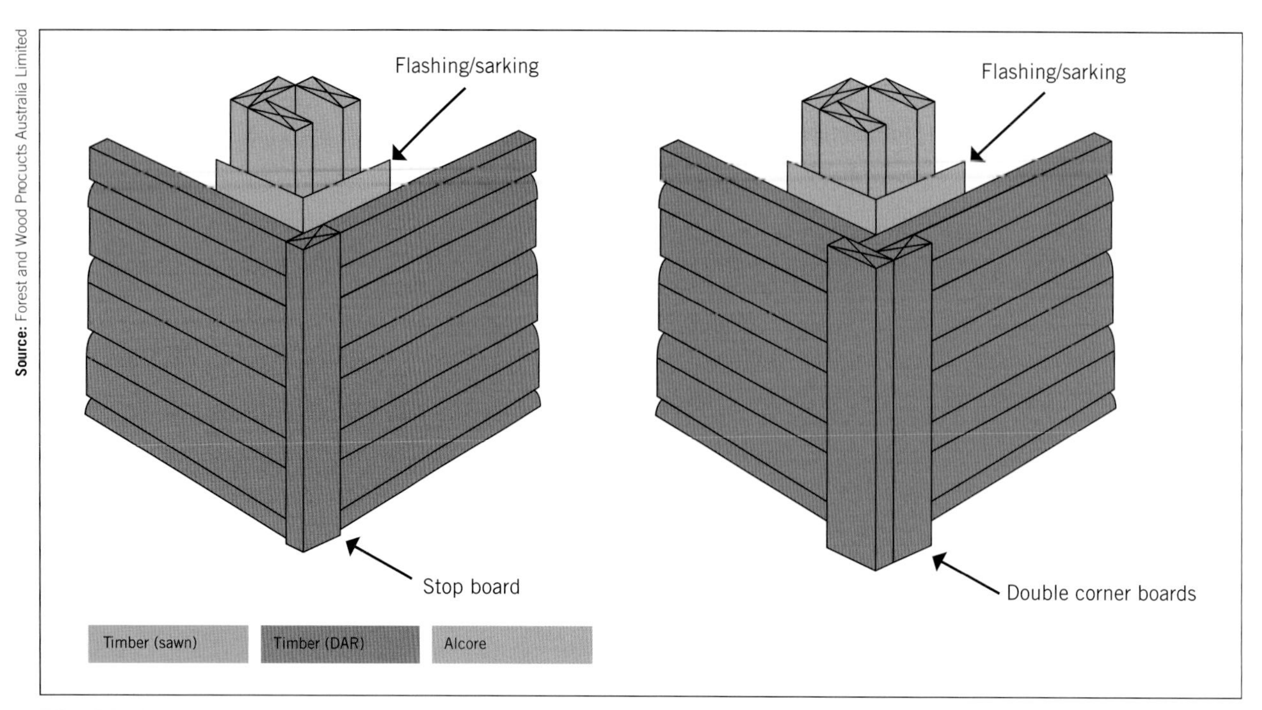

FIGURE 8.56 Weatherboard stops on external corners

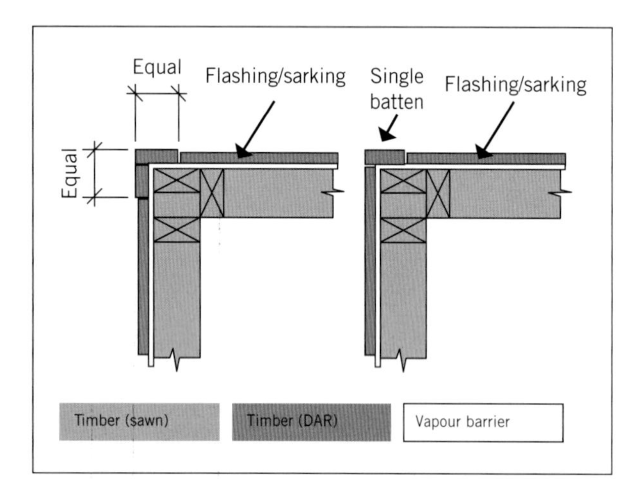

FIGURE 8.57 Weatherboard stops

(actual cover) on the faces before the plinth board is attached (unless the plinth board is to be mitred).

Ensure that the narrow edges of the stops are fixed so that they face the front of the building and extend beyond the face of the weatherboards to ensure full cover and protection.

Note: You can also make a gauge batten that can be used to mark out the position of each row of weatherboards on internal and external stops to ensure that the set-out is equal all round (see Figure 8.59). Simply line up all the weather stops and transfer the marks with a combination or builder's (also known as roofing) square and pencil (see Figure 8.60).

Alternatively, boards can be scribe fitted to each other, but this is time-consuming and requires accurate marking out and cutting (see Figure 8.61). If you choose this method, make a template shape to transfer the

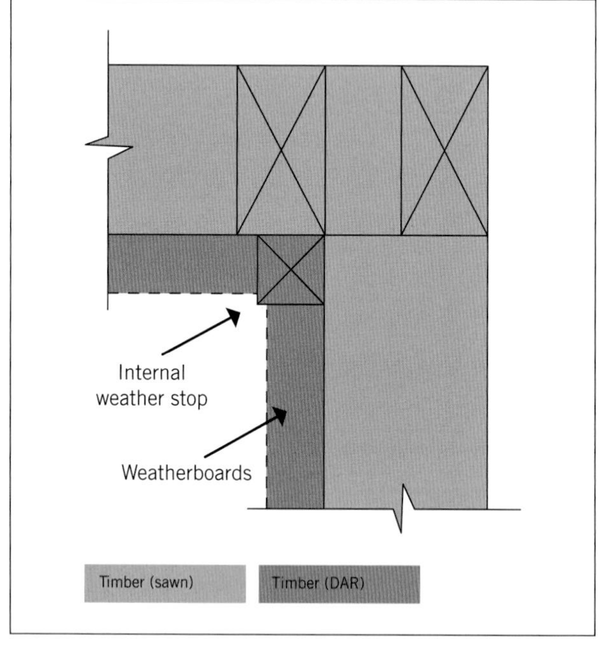

FIGURE 8.58 Internal corner weather stops

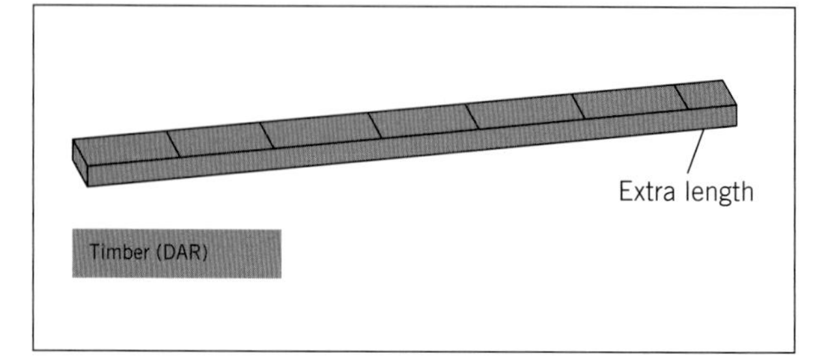

FIGURE 8.59 Simple actual cover gauge

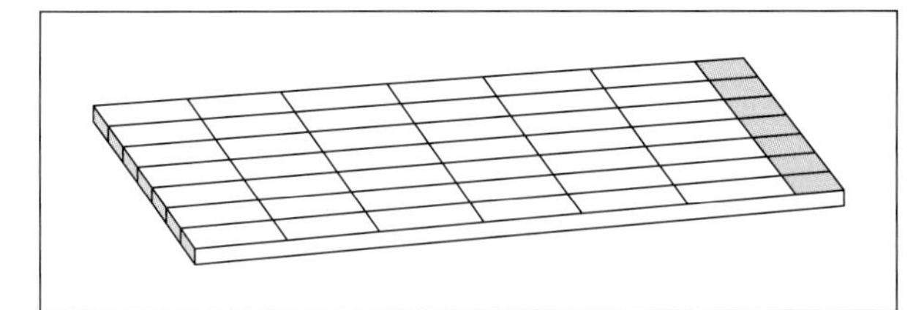

FIGURE 8.60 Marking out multiple weather stops

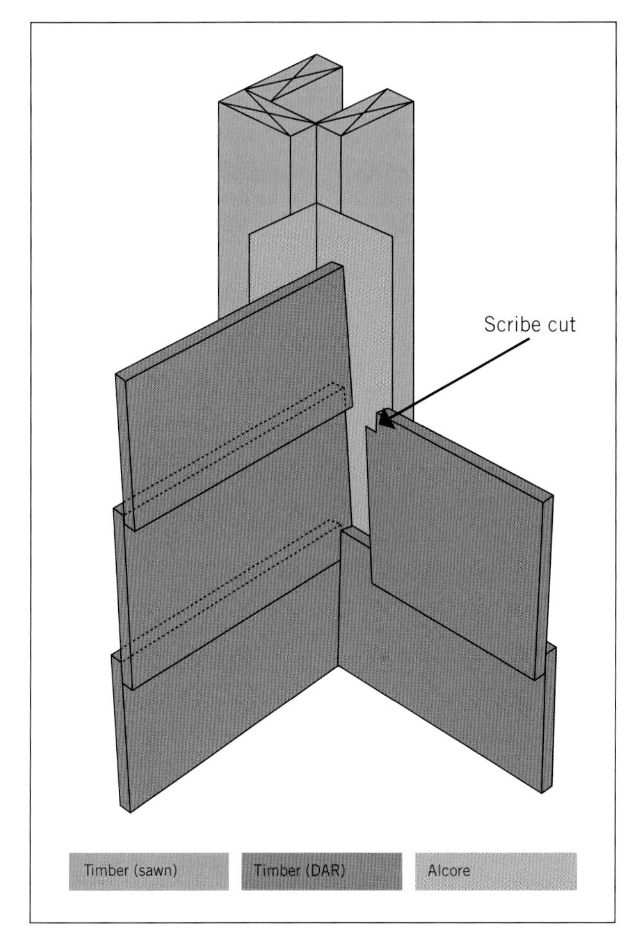

FIGURE 8.61 Scribe-cut wall boards

scribe profile on to the end of the board. Cut the scribed boards over-length and spring the boards into place. Any remaining gaps will have to be filled before painting.

Plinth boards

A plinth provides a tidy finish to the bottom edge of clad walls. It should be used on sheeted walls and strip-planked walls to protect the bottom edge of the sheets. These are particularly prone to damage in low-to-the-ground pier constructions.

Plinth boards are used in levelling and setting up for the first board. They are usually made from timber with a bevelled or rebated edge (see Figure 8.62) to improve weatherproofing and allow for ease of installation of the first boards. Plinth boards are fixed level around the perimeter of the house before the cladding is installed (see Figure 8.63).

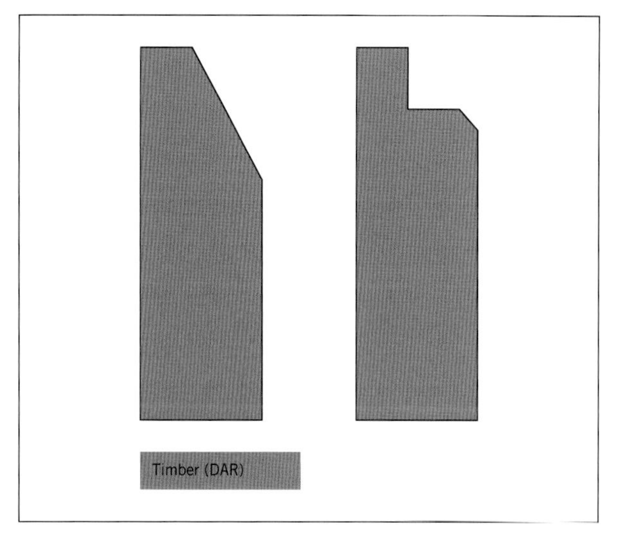

FIGURE 8.62 Typical plinth boards

The first board in splayed cladding or overlapping cladding (but **not** in rebated weatherboards) has to be packed out so that subsequent boards have the same angled face and also to provide a drip line for rain and moisture build-up. This is done in one of two ways. If the top edge of the plinth board is rebated (see Figure 8.64), the first weatherboard is fitted in the rebate. If the plinth board is not rebated, the first board is packed out to establish the angle and provide a drip line (see Figure 8.65).

The bottom edges of the first weatherboards can be bevelled to further improve weatherproofing.

Alternatively, the plinth board can be laid to a level string line stretched from one end of the wall to the other (see Figure 8.66). Take care, as a long length can sag over long distances; a fishing line or a laser level might be a better alternative. If you are not using a plinth board, you can set the first board out to a string line. If the boards are overlapping, remember to use a packer behind them to set the correct angle.

Fastening timber planks

The most common method of fastening timber weatherboards to new frames is to use 65 × 2.8 mm hot dipped galvanised bullet head nails. Set your nail gun to punch the heads below the surface 1–2 mm. If hammering nails, leave them proud so as not to dent the surface of the weatherboards, then punch below the surface 2 mm to allow for filling and finishing (see Figure 8.67).

Note: The nails should not be fastened through the top of the underneath board. If you do this, the boards will usually shrink and large splits can occur (see Figure 8.68).

Weatherboards with rebated bottom edges are fastened at each stud using two 38 × 2.5 mm (minimum) galvanised bullet head nails through the face. The nails are usually placed in the thickest part of the milled face to prevent the timber splitting, but if you are installing a

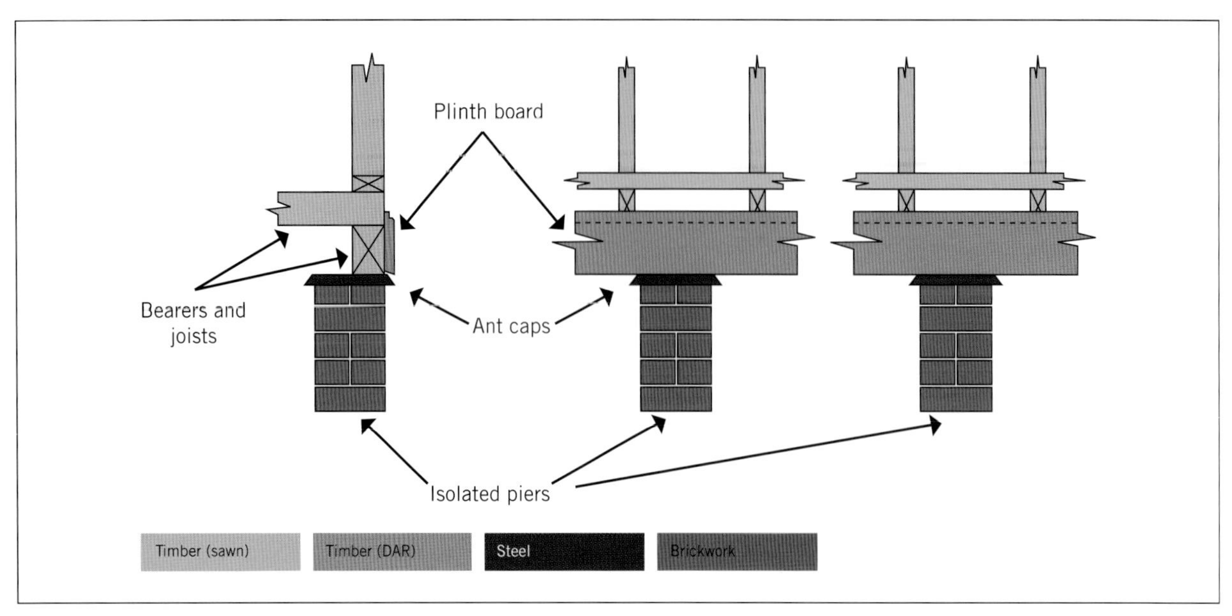

FIGURE 8.63 Position of plinth board

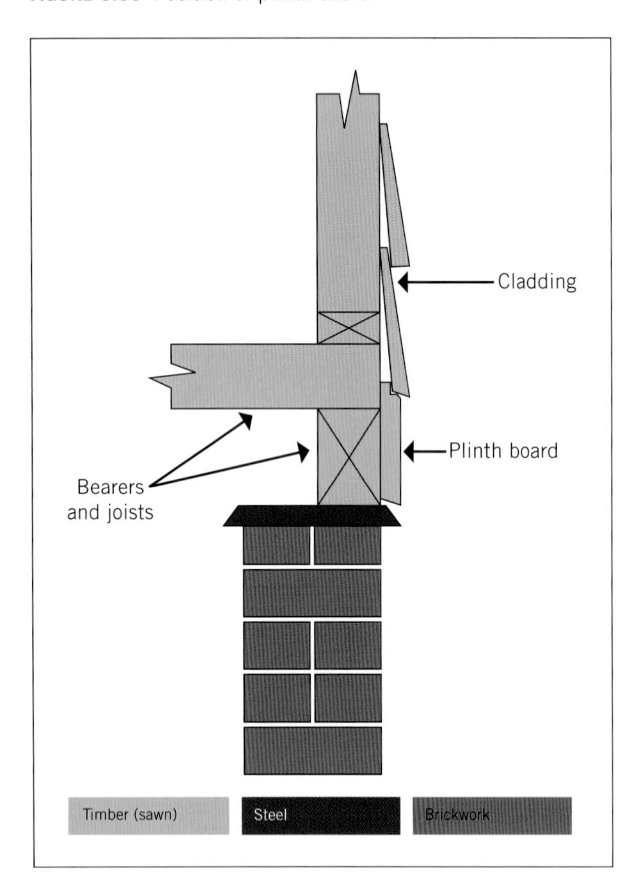

FIGURE 8.64 Rebated plinth board

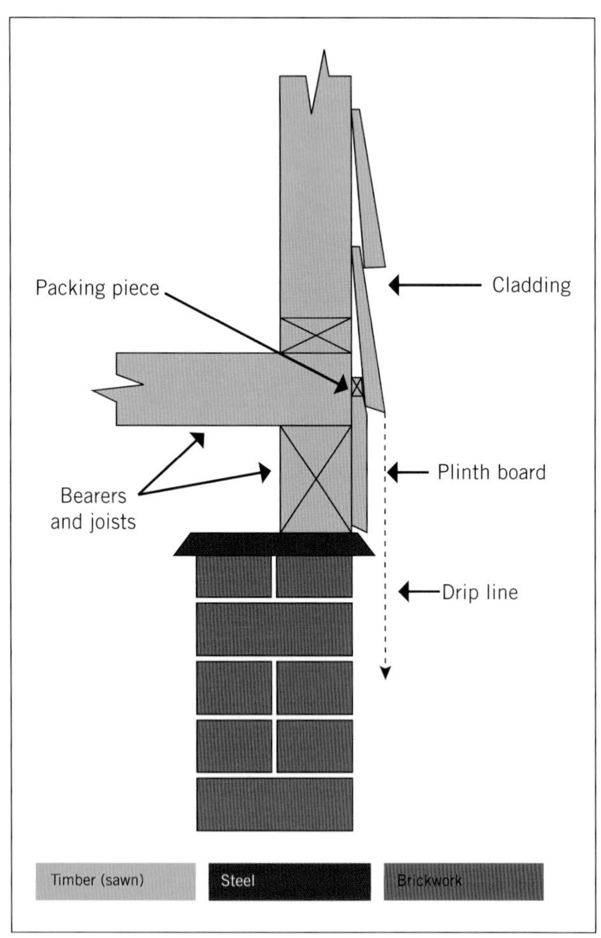

FIGURE 8.65 Installing packing piece in a bevelled plinth board

double teardrop or a double log cabin, one nail can be placed in the recess and the second on the high face of the board to make them less obvious. This may be difficult if using a nail gun to fasten the boards.

When fixing weatherboards, fibre-cement sheeting or fibre-cement planks to steel studs, choose the appropriate self-drilling, self-tapping screw. Allow for the thickness of the overlapping boards and feathered edge and take care not to overtighten the screw and strip the thread by using a drill with an adjustable torque setting. Keep the screws as close as possible to the return face of the stud to prevent the flange from collapsing in (see Figure 8.69).

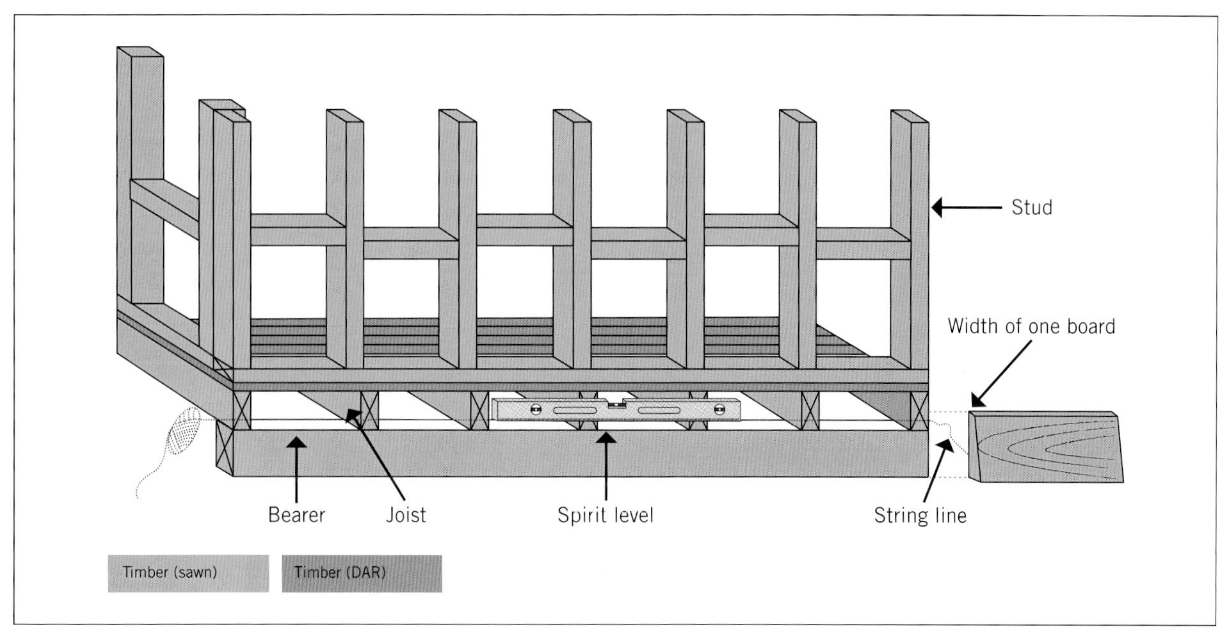

FIGURE 8.66 Setting to a string line

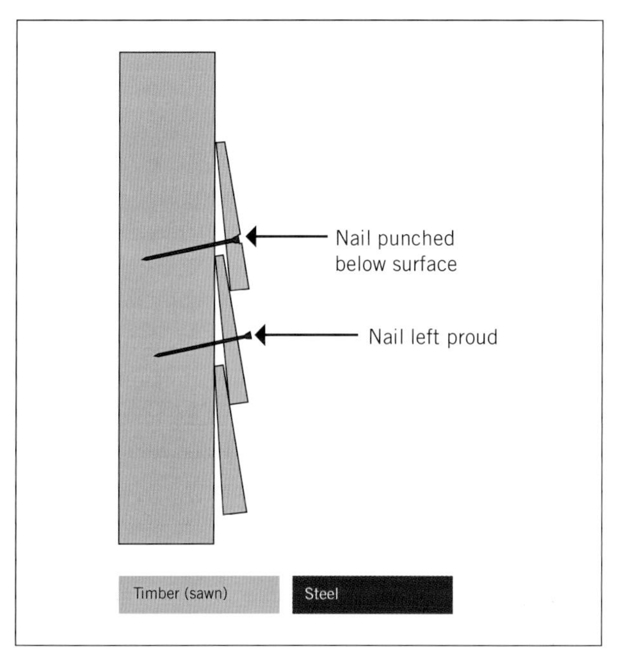

FIGURE 8.67 Nailing off splayed boards

Source: Ed Hawkins

FIGURE 8.68 Split boards

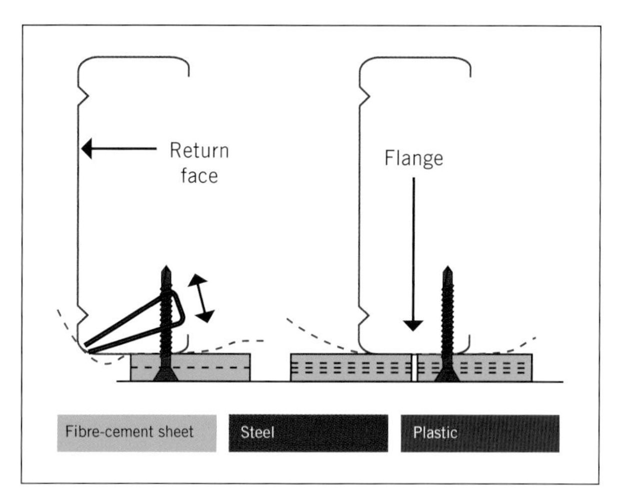

FIGURE 8.69 Fixing and jointing on steel studs

When jointing on steel studs, fasten the weatherboard, sheet or plank to the outer side of the flange first, then fasten the plank closest to the return face to keep the faces even (see Figure 8.69).

Overlapping weatherboards (depending on their thickness) and tempered hardboard planking can also be secret nailed (see Figure 8.70).

Ensuring overlap of boards is the same

A lap gauge is a simple tool used to ensure that the amount of overlap of boards is the same and maintains the actual cover width evenly. It also ensures that the actual cover (weather cover) of the boards is the same. You will need to make a new lap gauge each time you do a cladding job that requires actual cover measurements to be determined, as these will usually be different for each job. For ease of installation, you need three lap gauges cut out to exactly the same dimensions, one at each end and one in the middle.

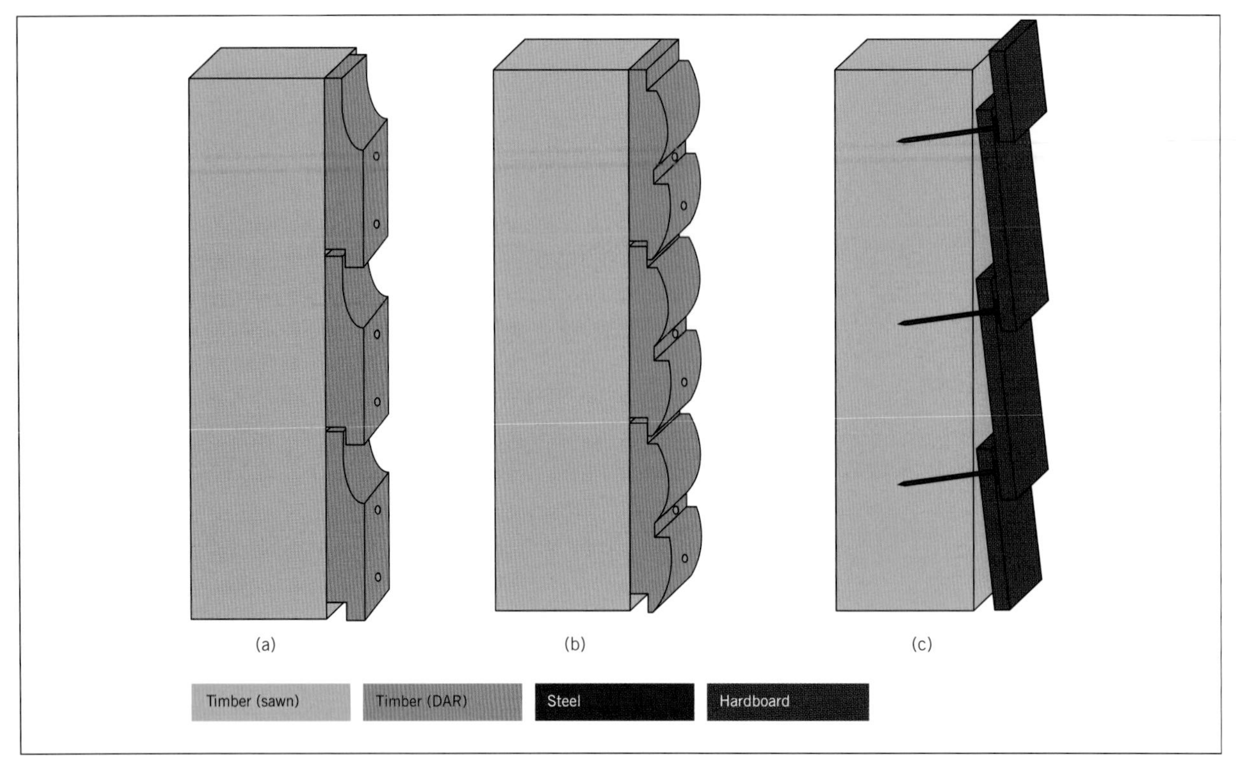

FIGURE 8.70 Nail fixing weatherboards and hardboard planking: (a) rusticated boards, (b) double log cabin, (c) concealed nailing in overlapping planks

Follow these steps:

1. Mark the corner battens and/or studs with the heights for several rows of boards.
2. Work out the actual cover to make the lap gauge (see Figure 8.71).
3. Set the plinth board in position and fix it off, making sure it is level.
4. Using the lap gauge, locate the end of the new plank in position so that it overlaps the top edge of the plinth board. Check that it lines up with your markings on the studs or corner battens.
5. Make sure the step in the gauge is hard up to the bottom edge of the plinth board at each stud and the bottom edge of the new board is sitting on the top

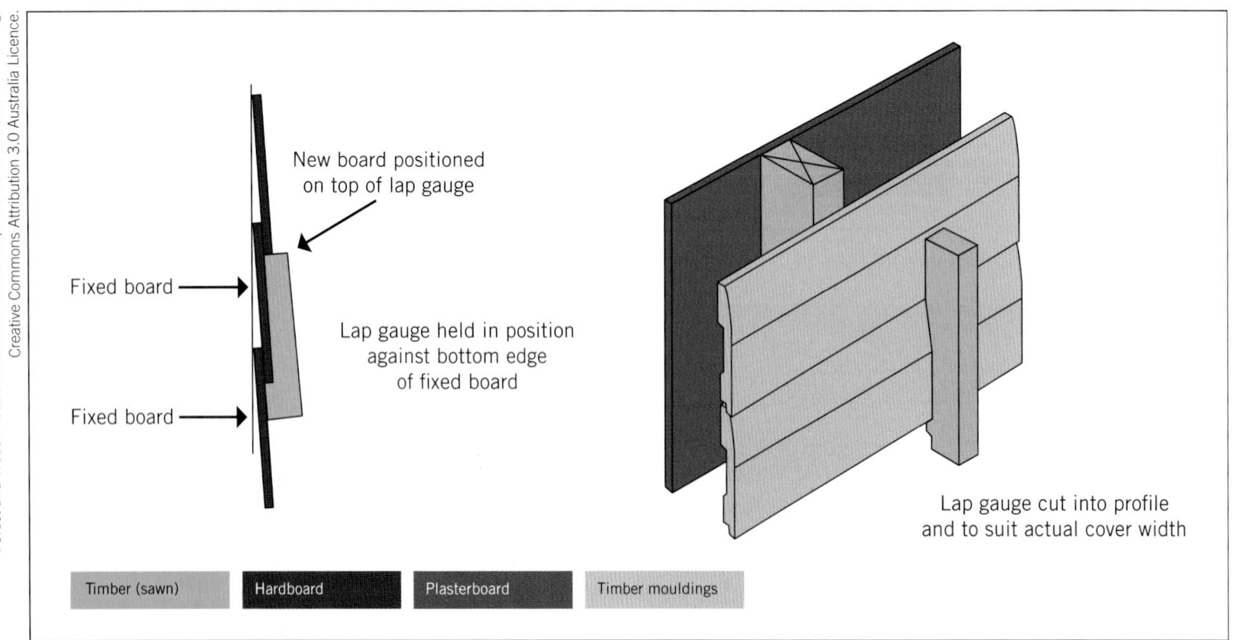

FIGURE 8.71 Using a lap gauge

end of the gauge. Make sure the lap gauge is held vertically. Nail off as you go along.

6 Repeat the process for reach row of boards checking regularly you maintain straight and level planking and lap coverage.

There will be more detail required around openings that will be explained later.

Installing vertical or diagonal cladding

If the weatherboards are being installed vertically or diagonally, additional rows of noggins, spaced 600 mm centre to centre, will have to be installed to allow for fixing of the boards. A 5 mm gap at the foot of the boards, between flashing and board ends, must be left to help prevent rotting.

The boards can be ripped to fit tight against the studs and a stop bead, quad or scotia mould can be cut over the face of the adjoining boards (see Figures 8.72 and 8.73).

Note: Do not use these techniques on horizontally fixed boards. The gaps in milled boards provide ideal nesting spots for spiders and insects.

External corners can be joined similarly to internal corners, with a stop being fixed to a stud and the boards cut up to the stud and sprung into place. Alternatively, they can be mitred, but using battens or beads is the recommended finish.

Fibre-cement planking is fitted in internal and external corners using special metal corner brackets that support the ends of the sheets on hooks and are fastened to the studs using galvanised sheet nails (see Figure 8.74). These make cutting the planks much simpler and allow for a little movement at corner joints.

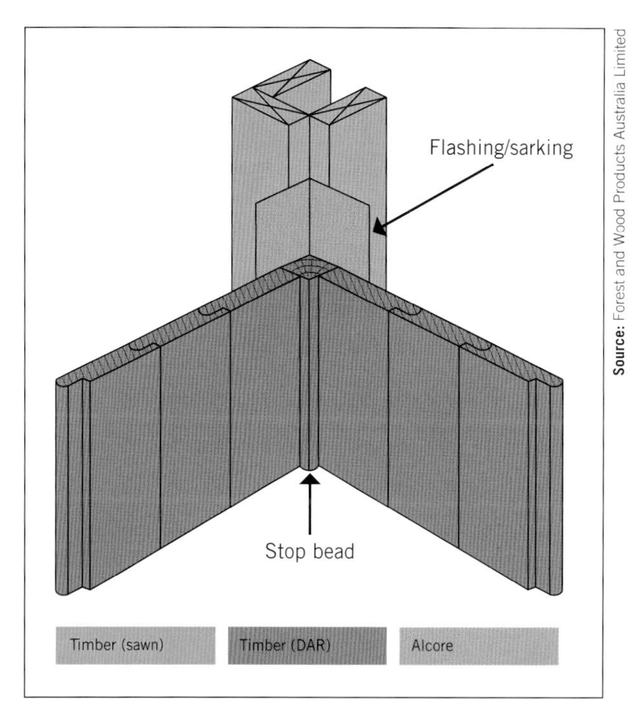

FIGURE 8.72 Vertical cladding on an internal corner

Vertical joints in horizontal fibre-cement planking

When off-stud jointing of fibre-cement planks is needed, special metal brackets are available from the manufacturer (see Figure 8.75). However, these may not be necessary for every kind of planking (see Figure 8.76).

Remember, nails or fasteners cannot penetrate the plank underneath, and end joints in consecutive rows must be staggered. Leave a 3 mm gap between ends to allow for movement and appropriate sealants.

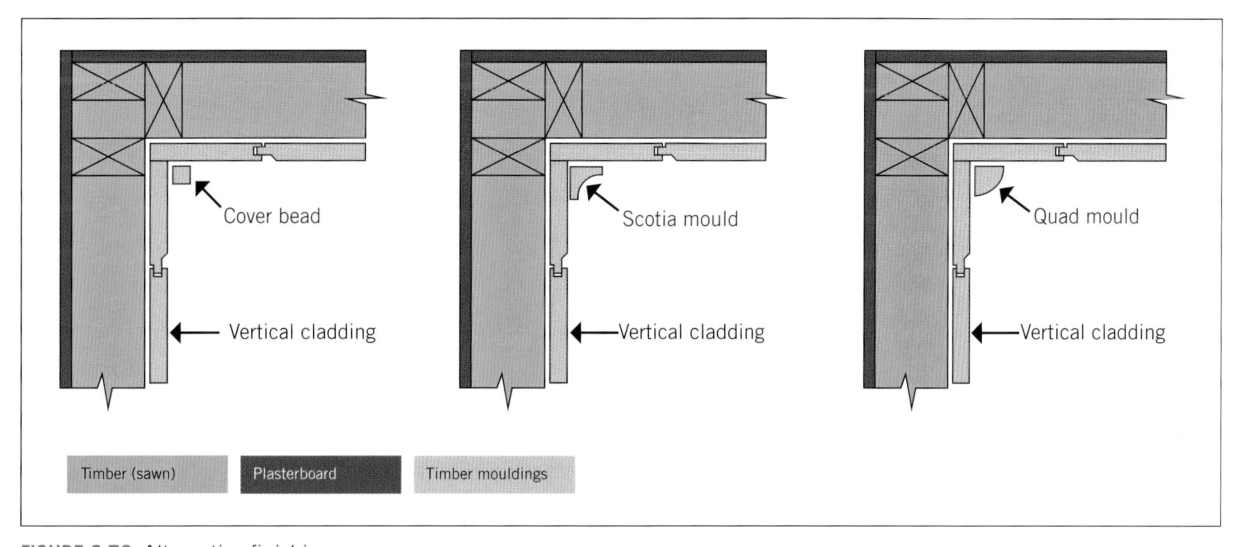

FIGURE 8.73 Alternative finishing corners

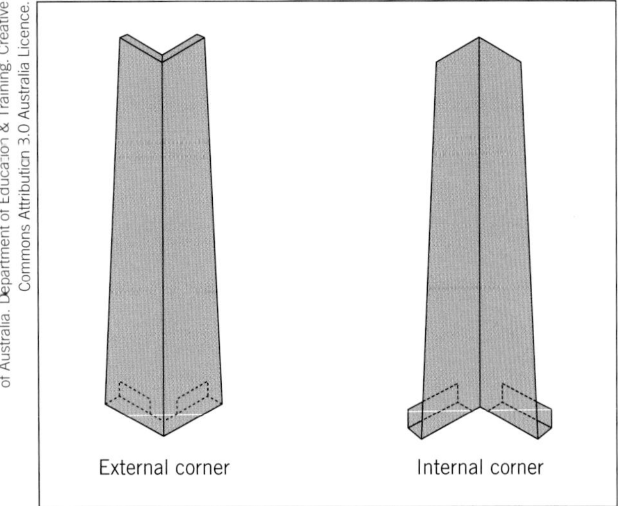

FIGURE 8.74 Alternative corner fittings for fibre-cement sheet

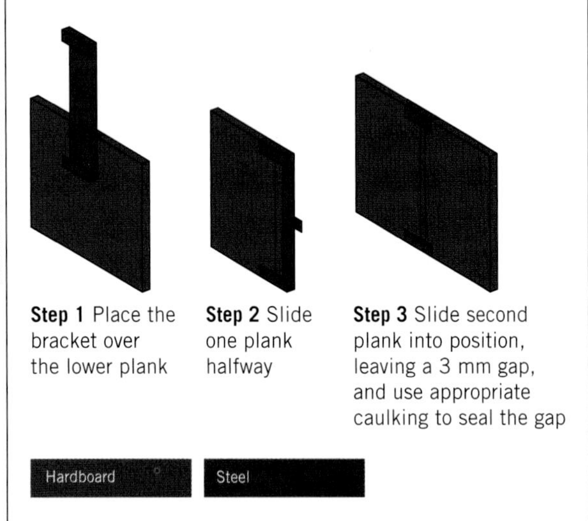

FIGURE 8.75 Metal soaker fixing for off-stud

On-stud jointing requires a gap of 3 mm between the ends of boards to allow for filling with sealant. When applying sealant, use masking tape on both sides of the joint, fill the joint, remove excess with a damp rag, and then remove tape before the sealant sets.

Fixing cladding to masonry materials

Sometimes you may have to fix cladding materials to an existing brick or masonry wall to create a new look or hide cracked brickwork.

Before going ahead with the job, make sure the masonry **substrate** is suitable for fixing the cladding to. Check to see if the wall is straight, clean and free from moisture problems. Rotted groundwork will cause the new cladding to fail and possibly fall off.

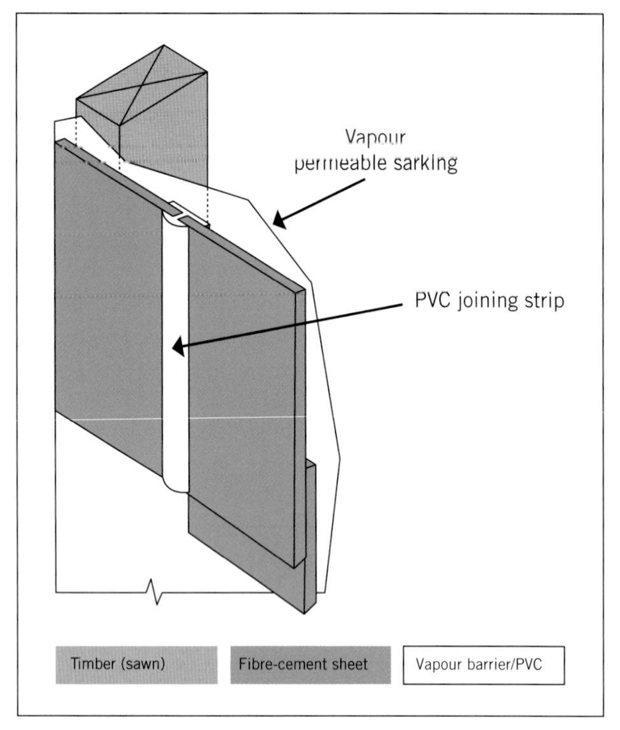

FIGURE 8.76 PVC jointing off-stud

The most common method is to fix timber battens vertically to the wall at 600 mm centres (see **Figure 8.77**), or to suit the cladding material being used.

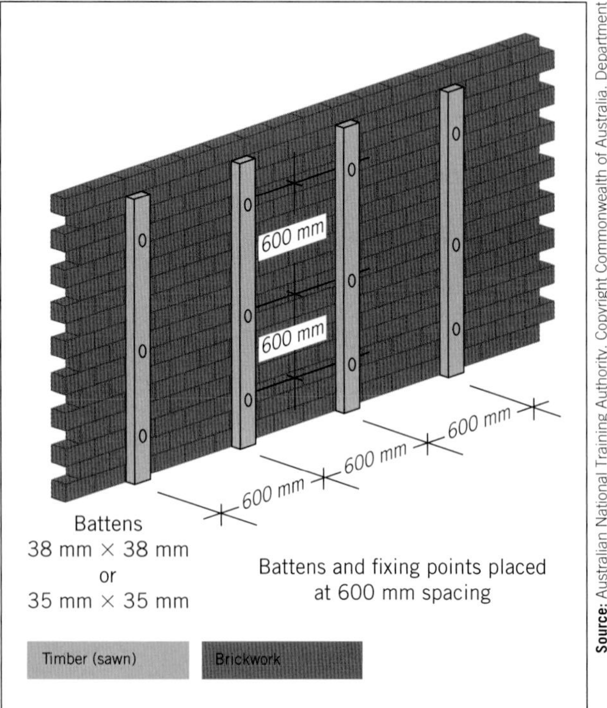

FIGURE 8.77 Batten fixing on masonry walls

As the structural strength of the building remains with the original walls, the battens can be much thinner, but need to be the same width and spacing as the stud framing for a similar load-bearing stud wall.

Fastening battens can be done by using:

- construction adhesive fixing and following the adhesive manufacturer's instructions – make sure the battens are firmly held in position by propping or masonry fasteners until the adhesive is completely set
- an appropriate screw and masonry plug method
- a power-activated or power-driven fastener system.

Advantages

Advantages include:

- better weatherproofing through the addition of a layer of sarking
- simple fixing methods used to fasten cladding
- battens can be packed or planed to create plumb, straight framing for cladding.

Disadvantages

Disadvantages include:

- increased wall thickness
- possible difficulty in trimming around windows and doors.

Finishing off around doors and windows

The process for installing doors and windows into timber frames, complete with flashings above and below, was covered in Chapter 3. In this chapter, we concentrate on finishing off the cladding around doors and windows and eaves lining.

Important points to remember are:

- cladding must not interfere with the operation of doors and windows
- joints between doors and windows need to be weatherproof (hence the flashing).

As we know, weatherboards are fitted to the walls from the bottom up, **not** the top down, so the first obstruction will usually be the door sill or threshold. In new construction, the threshold should already have been cut and installed to allow for the fixing of the cladding material (see Figure 8.78).

To fit the boards around the threshold, follow these basic steps:

1. After cutting the board to the full length, brace or temporarily fix the board in position beneath the sill or threshold and mark the position of the outer sides of the door frame on the board or plank.

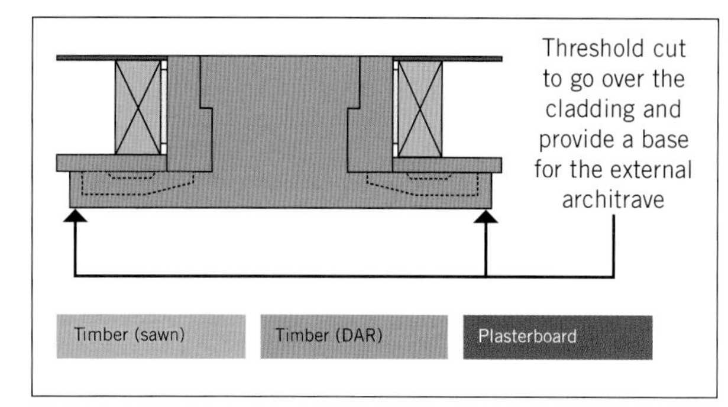

FIGURE 8.78 Thresholds and sills cut around the cladding

2. Remove the board and measure the distance from the bottom of the sill to the bottom of the already fitted board and subtract the actual cover from the distance measured, leaving you the width of the board to remain exposed (see Figure 8.79).

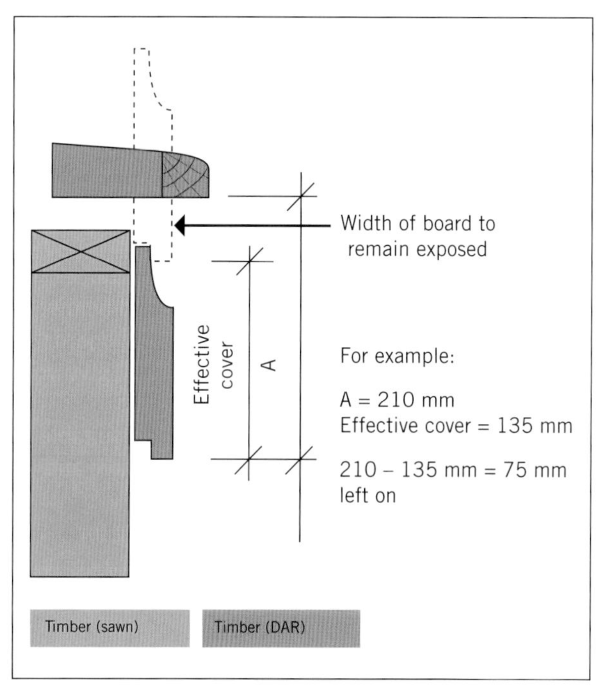

FIGURE 8.79 Calculating the amount of board left on

3. Mark a line along the board between the two lines you marked in Step 1. The line should be the distance you calculated in Step 2 above the bottom of the board. For a board to fit under a sill you should be removing a piece from the top edge of the board (see Figure 8.80).
4. Cut out the marked section of the board to be removed and fasten the board under the sill. Take care when handling the board as you have weakened it by reducing its width.

Finally, cut the cladding to fit snug to the side of the frame. It is very important to keep the boards aligned on either side of the window or door frame.

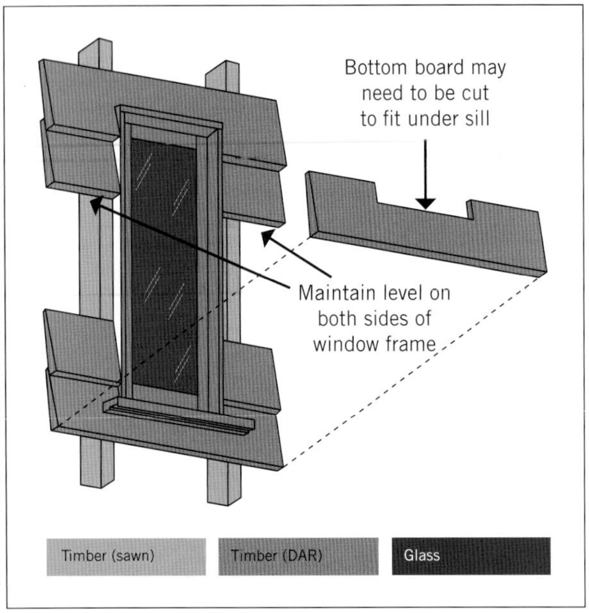

Bottom board may need to be cut to fit under sill

Maintain level on both sides of window frame

Timber (sawn)

Timber (DAR)

Glass

FIGURE 8.80 Cutting boards or planks around a door or window frame

When you reach the last board before the top of the window, it will have to be cut in across the top of the frame. To do this, you need to make similar calculations as you did for the sill boards: measure the distance from the top of the door or window frame head to the bottom of the already fitted board below, then subtract the actual cover of the cladding boards from the distance already measured. This will give you the amount to be removed.

Make sure the flashings are in place and a suitable amount is available for laying over the architrave and trimming off to a neat line of overhang.

The cut-outs for penetrations and trimming for electrical power meters (see **Figure 8.81**), gas meters and other utilities, where applicable, are done as the materials are installed.

If you are installing vinyl, metal or other planking materials that require specialised trim accessories to be installed, it is often recommended that these sections be installed before starting to install the planking. Check with the manufacturer for full installation instructions.

Clean up

When finished, the work the area needs to be cleaned, and all materials sorted for landfill, recycling and reuse

FIGURE 8.81 Power meter trimming

according to the work site requirements and workplace procedures.

Clean up the work area, removing any debris, tools and excess materials. Stack and store excess materials according to workplace procedures.

Complete a visual inspection of all tools and equipment to ensure safe operation. Regular checks and maintenance of tools and equipment should be done each time they are used. If faults are found, follow workplace procedures to record problems, tag the tools or equipment, and relocate them away from other functioning tools.

Restore the surroundings to their original condition, ensuring that any damage caused during installation is repaired.

Obtain the necessary sign-off and approvals from the client or project manager. Ensure that all necessary documentation has been obtained and filed appropriately.

SUMMARY

In this chapter we have looked at sheet materials such as plywood, fibre-cement sheeting and planking, weatherboards, and plastic and metal siding materials and structural insulated panels (SIPs).

Important points include the following:

- Before installing cladding, always check the National Construction Code – Volume One, Parts C and F; Volume Two – Deemed-to-Satisfy Provisions, H1D7 – Roof and wall cladding; Housing Provisions Standards, Part 7.5 – Timber and composite wall cladding; and manufacturers' technical manuals or sheets.
- Cladding materials are strong and relatively cheap, and can be easily cut and fitted, insulated and renovated, replaced or restored.
- When choosing cladding, consideration needs to be given to sustainable building practices and the environment in which the building is to be placed (e.g. fire-prone areas, alpine areas or tropical regions).
- Pre-installation checks should include:
  - ensuring all materials are onsite and safely stacked and stored
  - ensuring scaffolding barricades and signage have been erected
  - checking tools and equipment, including PPE, for serviceability
  - checking environmental and safety controls are in place and that quality control procedures have been implemented.
- Check and adjust walls for straight and plumb, locate any penetrations that may be required and check if there are to be any additional loadings placed on the walls, such as air-conditioning units.
- Sarking, or builder's wrap, is used behind cladding to improve insulation and weatherproofing.
- Flashings are installed at all openings before commencing cladding or sarking. Flashing materials must be compatible with the cladding materials (e.g. don't use lead or copper flashings with Zincalume™).
- When retro-cladding, door and window frame reveals may need to be adjusted to account for the different thickness of the cladding material.

This chapter also described the cladding materials commonly used and the methods used for fixing and installation. The main materials are as follows.

Timber

- Milled products include shiplap, splayed, log cabin, double log cabin and rusticated.
- Timber must be from durability Class 1 or 2 or Class 3 if preservative treated and should comply with AS 2796 Seasoned hardwood – Milled products and AS 4785 Timber – Softwood – Sawn and milled products.
- All timber should have a suitable exterior finish, such as paint or oil.

Fibre-cement sheets and planks

- These products must comply with AS 2908.2 Cellulose – Cement products.
- Sheets and planks come in different finishes: flat, stucco and fake rough-sawn timber grain.
- Cutting fibre-cement sheeting is done with electric shears or a hand guillotine. Score and snap is an alternative method and small penetrations can be made with reciprocating saws.

Vinyl and plastic planking

- This is a strip planking product for use over framed or battened groundwork. It usually has a polystyrene foam insulation backing to improve thermal efficiency.
- It can be installed horizontally, vertically and diagonally.
- It can be cut using a handsaw, hacksaw, tin snips, utility knife, sabre saw or jigsaw.
- It has no structural strength and the colours tend to fade.

Metal siding

- Metal siding includes corrugated iron and plank products. It has some structural strength as a bracing material.
- Surface finishes include galvanised, baked enamel and zinc/aluminium.
- It can be applied vertically and horizontally, and can be painted. Aluminium siding is also available.

Plywood

- Use Type A bonded exterior grade, H3 treated with a surface treatment applied (e.g. oil, paints or stain).
- Use A or S grade face veneers for visual effect with C or D grade backing veneers.

Tempered hardboard

- It is available as a sheet or plank product, with the planked product being the most common.
- Cost can be a disadvantage, and it requires ongoing maintenance, while the dust from working with it poses potential health risks.

Expanded polystyrene (EPS) cladding systems

- EPS cladding is a non-structural and lightweight building material used as an alternative to rendered brick facade. EPS cladding is fixed directly to the frame and finished with a variety of rendered or applied finishes.
- Advantages include that it's:
  - a lightweight building product
  - an excellent thermal insulator
  - easy to work with and fast build.
- Disadvantages include that:
  - it's not 'deemed to satisfy', but rather must be accompanied by suitable performance evidence for regulators and certifiers

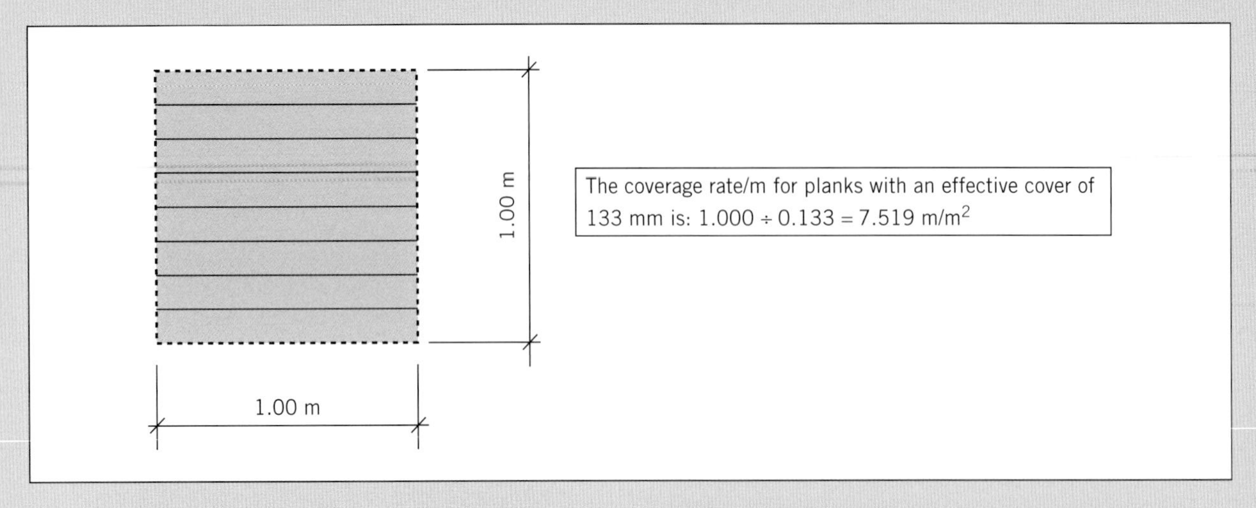

- its surface finishes can be subject to cracking (though cracking can be controlled through the use of control joints)
- it may not comply with fire regulations in some circumstances.

Installing EPS

Make sure the frames are straight and plumb.

Correctly install all windows, DPC, flashing and termite barriers and external corner angles.

Use vapour-permeable sarking.

Use blocking where 'off-stud' jointing of panels occurs and for installing wall-mounted accessories.

Structural Insulated Panel (SIP) systems

- SIPs are foam panels with external layers of metal, plywood, fibre-cement or oriented strand board (OSB) and often used for site sheds and refrigerated cold stores.
- Advantages include that:
  - they can be 'structural' elements
  - they have higher insulating factors than EPS
  - onsite construction means savings in time and labour.
- Disadvantages include that:
  - special structural design requirements are needed for different locations (e.g. cyclonic, non-cyclonic and bushfire-prone areas)
  - their high embodied energy means they're not so sustainable.

Calculating quantities

- When calculating quantities for plank products, the 'effective cover' is critical. Minimum overlap requirements should be obtained from the National Construction Code and manufacturers.
- Many builders calculate quantities using the lineal metres per m^2 method by dividing 1 m^2 by the effective cover of the material.

Installation

Installing fibre-cement sheeting

- Dust from fibre-cement sheeting is a health hazard. Avoid generating dust by using score and snap method, guillotines or shears. Use:
  - an HEPA grade vacuum cleaner and spray water to dampen the area when cleaning up
  - signage to warn others in the area
  - a P1 or P2 respirator when cutting or drilling.
- Fasteners must be galvanised, zinc coated or stainless steel, and compatible with other metals they may contact.
- Nail guns must be fitted with a 'flush drive' to prevent the nails being punched through the surface. Otherwise leave the nail proud of the surface and drive it flush with a hammer.
- Use screws on metal frames.
- Use screws on timber frames when the joints will be covered with strip battens or set and rendered.

Installing planked products

- The starting point for the installation of planked siding is on top of the ant capping.
- Install internal and external corner battens and install additional noggins, battening or metal furring channel for angled planking.
- Join planks on a stud or use specially designed brackets for off-stud jointing.
- Weatherboard stops or stop beads are used at corners.

Fixing cladding to masonry materials

- Ensure the masonry is straight, clean and free from moisture problems and cracking.
- Fix timber battens vertically at 600 mm centres to the wall to suit the cladding material being used.

Finishing off around doors and windows

- Cladding must not interfere with the operation of doors and windows. Fit and cut cladding carefully around doors and windows and ensure they are aligned level each side of the opening.

REFERENCES AND FURTHER READING

National Construction Code

Copies of the National Construction Code can be obtained free of charge from the Australian Building Codes Board website: **http://www.abcb.gov.au**.

National Construction Code 2022, Building Code of Australia – Volume One.

National Construction Code 2022, Building Code of Australia – Volume Two.

Web-based resources

Aluminium cladding

Nu-Wall Aluminium Cladding: **http://www.nuwall.co.nz**

Fibre-cement cladding

BGC Plasterboard Fibre Cement: **https://bgcinnovadesign.com.au**

James Hardie Ltd: **http://jameshardie.com.au** – for contact details in your state or territory.

Metal cladding

BlueScope Steel: **http://www.bluescopesteel.com.au/building-products/walling-and-cladding/walling-and-cladding**

Metroll Ltd: **http://www.metroll.com.au**

Stratco: **http://www.stratco.com.au**

Tempered hardboard

Weathertex Pty Ltd: **http://weathertex.com.au/**

Structural insulated panels

CHAD Group: **https://www.chadgroup.com.au/**

SIPS Industries: **https://www.sipsindustries.com.au/**

Wood Solutions: **https://www.woodsolutions.com.au/applications-products/structural-insulated-panel-systems-sips**

Fasteners

Bremick: **http://www.bremick.com.au**

Bondor Thermal and Architectural Building Systems: **http://www.bondor.com.au**

Bostitch Australia: **https://bostitchtools.com.au**

Paslode Australia: **https://www.paslode.com.au**

Senco: **https://www.senco.com/fasteners**

Other

Wood Solutions: **https://www.woodsolutions.com.au/**

Video resources

James Hardie: **https://www.youtube.com/@JameshardieAu/playlists**

SIPS Industries: **https://www.youtube.com/@sipsindustries1742**

Aluminum Cladding systems: **https://www.youtube.com/results?search_query=nu-wall+aluminium+cladding**

External wall cladding installation: **https://www.youtube.com/results?search_query=external+wall+cladding+installation**

Relevant Australian Standards
AS 1366 Rigid cellular plastics sheets for thermal insulation – Part 1: Rigid cellular polyurethane – depending on type (i.e. EPS moulded, extruded or other polymer)
AS 1562 Design and installation of sheet roof and wall cladding – Part 1: Metal
AS 1562 Design and installation of sheet roof and wall cladding – Part 3: Plastic
AS 1604.1 Specification for preservative treatment – Sawn and round timber
AS 1604.2 Specification for preservative treatment – Reconstituted wood-based products
AS 1684 Residential timber-framed construction (known as the National Timber Framing Code)
AS 2796.2 Timber – Hardwood – Sawn and milled products – Part 2: Grade description
AS 4785.1 Timber – Softwood – Sawn and milled products – Part 1: Product specification
AS 4785.2 Timber – Softwood – Sawn and milled products – Part 2: Grade description
AS 5068 Timber – Finger joints in structural products – Production requirements
AS/NZS 2269.0 Plywood – Structural – Specifications
AS/NZS 4859.1 Materials for the thermal insulation of buildings – General criteria and technical provisions
ISO 22452 Timber structures – Structural insulated panel walls – Test methods
Australian Standards can be purchased from the Standards Australia website: **https://www.standards.org.au/**, or see your teacher or librarian for assistance in accessing Australian Standards online.

GET IT RIGHT

INSTALL EXTERNAL CLADDING

Questions

Below are pictures of two jobs showing the detailing of internal and external cladding corners. Answer the following questions about them:

1 What is wrong with the external corner method?

____

____

2 What is wrong with the internal corner method?

____

____

3 How should they be done correctly?

____

____

4 What should be done to rectify these corners?

____

____

HOW TO

CRIPPLE A STUD

Note: Up to 20 per cent of common studs, including those in bracing walls, may be crippled. Studs at the sides of openings and studs supporting concentration of load must not be crippled.

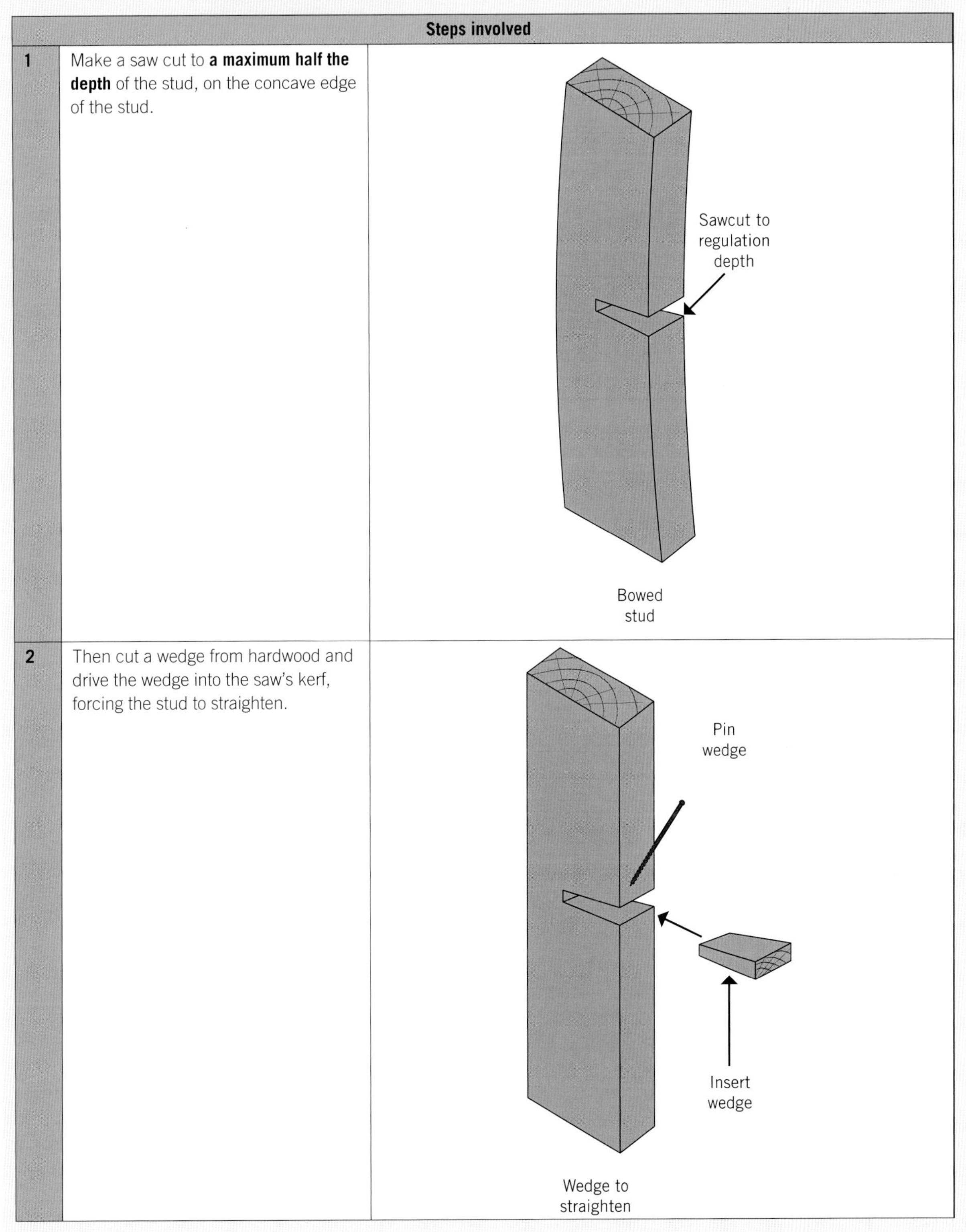

Steps involved	
1	Make a saw cut to **a maximum half the depth** of the stud, on the concave edge of the stud.
2	Then cut a wedge from hardwood and drive the wedge into the saw's kerf, forcing the stud to straighten.

>>

>>

3	Skew a nail through the wedge to fix it in place. Clean off the waste wedge with a chisel or plane, making it flush with the stud.	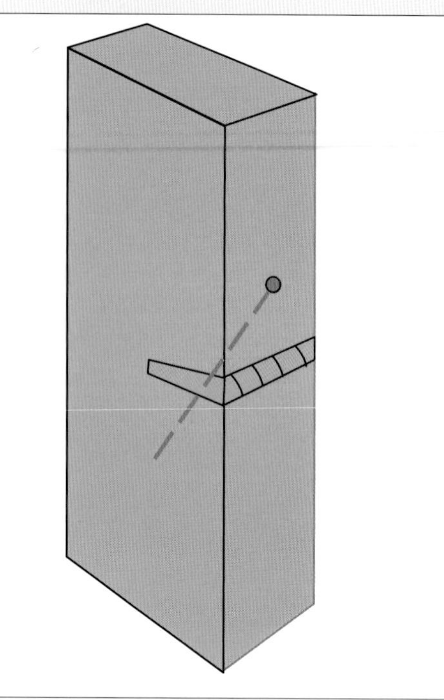
4	Reinforce the cut stud by fastening two cleats on either side of the cut, directly over the cut and wedged area. The cleat should be 600 mm long and fastened evenly with a minimum of four 50 mm nails.	

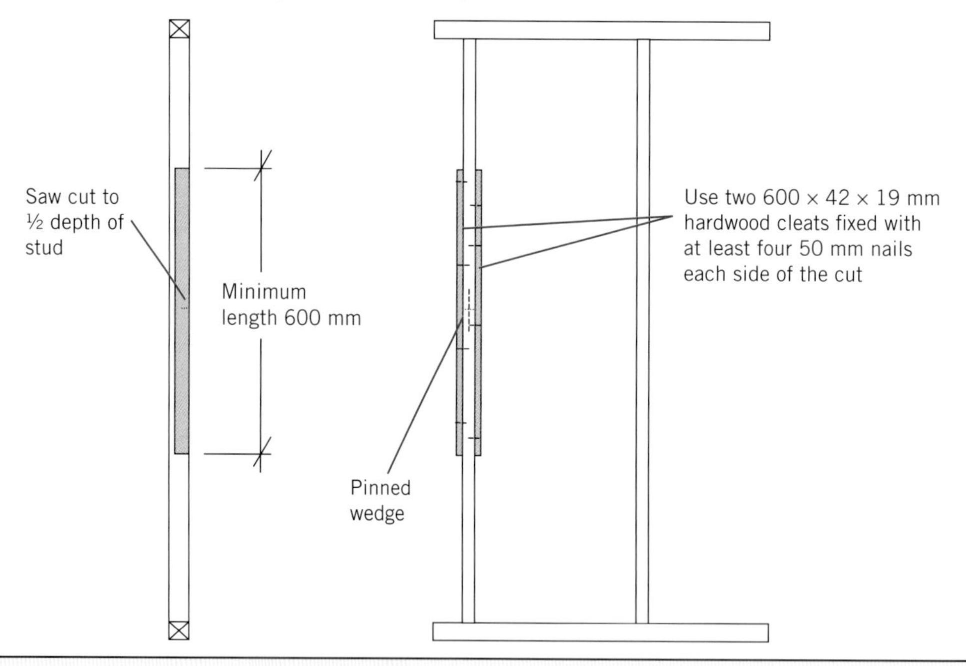

WORKSHEET 1

To be completed by teachers	
Student competent	☐
Student not yet competent	☐

Student name: ______________________

Enrolment year: ______________________

Class code: ______________________

Competency name/Number: ______________________

Specifications:
Framing timber: 90 × 35 mm MGP 12 Pine (or F8)
Fastening: nails 75 and 90 mm bullet head
Screw: 90 mm
Bracing: Speed brace or metal angle fixed with clouts
Regular stud spacing: 450 mm centres

Wall cladding exercise

Working independently, students are to:

a Work in pairs to construct one wall frame then work in a team to join frames together to make a square room structure. Ensure safe work practices. 'Tom' and brace frames during construction.

b Once frames are stood, students are to work in pairs to clad the exterior as indicated:

- Horizontal plank clad – timber, fibre-cement or hardboard planking
- Vertical planking – 10 mm rebated gap
- Diagonal planking

Note: Corner boards and stop boards must be used on external corners and at junctions. Walls should be sarked and flashing installed at windows and door openings.

c Internally, the walls should be treated as external walls and lined with fibre-cement sheeting. Appropriate joining strips and corner moulds should be used.

Practical activity – installing planked cladding

Note to teachers: Prepare skills bays where students can practise installing planked cladding materials.

Students: You are required to horizontally fasten the hardboard cladding material to the prepared timber frames. You will need to complete one internal corner and one external corner and work in with adjoining trainees to complete the exercise.

Before starting, ensure you have your personal protective equipment (PPE), and all the necessary tools and safety equipment, including barricades and signage where necessary.

Note: All your work must be carried out according to the following task specifications. After each task, your assessor **must** check your progress and provide feedback.

External frames – Horizontal planked cladding

Task A

Prepare a safe work method statement (SWMS) for the task ahead after (or as part of) a toolbox meeting with your supervisor, colleagues and/or teacher.

Task B

1 Calculate the appropriate actual and effective cover measurements for the wall frames you are to clad.

2 Check and correct wall frames for straight, plumb and other frame defects that may affect cladding installation.

3 Install flashings and appropriate sarking to frames.

4 Mark out and install plinth boards and corner battens/stop beads as required.

Task C

Install cladding materials level, including:

1 at least one 'on the stud' joint and one 'off the stud' joint

2 concealed fixing method and visible fixing method

3 fixing architrave or similar materials around door and/or window frame.

Task D

1 Remove architrave, cladding, sarking and flashing materials and make good frames in preparation for the next cladding exercise.

2 Stack and store reusable materials in an appropriate site storage facility. Recyclable materials are to be placed in appropriate recycling facilities.

Task E

Calculate the number of boards and total lineal metres of boards to be used for the horizontal cladding of the entire framed structure.

Sheet cladding

Task A

Prepare a safe work method statement (SWMS) for the task after (or as part of) a toolbox meeting with your supervisor, colleagues and/or teacher.

Task B

1 Calculate the number of sheets required for cladding the entire framed structure (internal faces only).

2 Adjust wall frames as necessary, including installation of additional noggin to take tap head penetration.

3 Install flashings and appropriate sarking to frames.

4 Mark out and install plinth boards and corner battens/stop beads as required.

Task C

Install sheet cladding materials horizontally:

1 Install flashing at horizontal joint.

2 Fix batten over horizontal joint.

3 Finish vertical joints using a PVC joining strip.

4 Fix architrave or similar materials around door and/or window frame, with flashings to be made and trimmed to neat finish.

Task D

1 Remove architrave, cladding, sarking and flashing materials and make good frames.

2 Stack and store reusable materials in an appropriate site storage facility. Recyclable materials are to be placed in appropriate recycling facilities.

WORKSHEET 2

To be completed by teachers	
Student competent	☐
Student not yet competent	☐

Student name: ______________________

Enrolment year: ______________________

Class code: ______________________

Competency name/Number: ______________________

Preparing frames for cladding

Answer the following questions.

1 Answer True or False.

a Wall cladding provides the building with a weatherproof cover.

True ☐ False ☐

b Wall cladding provides an ecologically sustainable element of the building.

True ☐ False ☐

c Sarking provides good insulation for a wall.

True ☐ False ☐

2 Fill in the missing words:

Non-conforming building products (NCBP) are products and materials that claim to ______________________, don't meet Australian ________________ or use ________________ marketing practices.

3 Fill in the missing words:

Non-compliant building products are products and materials used in construction that do not comply with the ______________________.

4 If timber-framed walls are not straightened before cladding, what effect can this have on the cladding process?

____

____

____

____

5 Describe the three steps used to check and straighten a timber wall frame.

a ______________________________

____

b ______________________________

____

c ______________________________

____

6 Briefly describe a 'low' stud face and the process of straightening it.

____

____

____

7 Briefly describe a 'high' stud face and the process of straightening it.

____

____

____

8 Describe, in steps, the process of straightening a sprung timber frame member using the 'crippling' technique.

____

____

____

____

____

9 When is the 'crippling' method of straightening a sprung frame stud not allowed to be used?

____

____

____

10 Answer True or False:

Studs to the side of door or window openings and studs supporting concentrated loads (roof trusses) cannot be crippled.

True ☐ False ☐

11 What two fastening methods are used to fix sarking to timber-framed walls?

a ______________________________

____

b ______________________________

____

12 Select the best correct answer.

Which of the statements below is *not* a relevant environmental or regulatory requirement when cladding or re-cladding a building?

a Non-friable asbestos sheeting can be removed by a non-licensed tradesperson. ☐

b The selection of vapour barriers will depend on the cladding material being used. ☐

c Durability Class 3 and 4 timber must be preservative treated if used for cladding. ☐

d Lead in dust from old paint and in roof spaces is a health hazard and control measures should be in place to manage the dust. ☐

13 Select the best correct answer.

Which of the tools listed below is the best for cutting fibre-cement sheeting?

a Angle grinder ☐

b Hand guillotine ☐

c Portable power saws with masonry blade ☐

d Corded or uncorded electric shears ☐

14 Fill in the missing words:

Durability ______________ timbers are recommended for ____________ use.

15 Answer True or False:

Durability Class 3 and 4 timbers are recommended for external use. You can only use durability Class 1 and 2 timbers if they have been suitably preservative treated.

True ☐ False ☐

WORKSHEET 3

To be completed by teachers	
Student competent	☐
Student not yet competent	☐

Student name: ____________________

Enrolment year: ____________________

Class code: ____________________

Competency name/Number: ____________________

Fixings and fastenings

Answer the following questions.

1 Fill in the missing words

Weatherboards are designed to be joined on a stud by butt jointing and ____________ nailing.

The ends of the boards are cut ____________ (1 mm over-length) and slightly undercut to produce a tight surface ____________ the ends of the boards and ____________ the ends of the boards to prevent the ends ____________ when nailing.

2 From the range of fasteners listed below, select the correct fastener to use when fixing a timber weatherboard to a timber frame.

a 45 mm wide crown staple

b 65 × 2.8 hot dipped galvanised clout

c 50 × 2.7 mm hot dipped galvanised nail

d 75 × 2.5 mm ringed flat head nail

3 Answer True or False:

LOSPs are favoured preservative treatments for timber because of environmental concerns and issues surrounding the disposal of CCA treated timbers and offcuts.

True ☐ False ☐

4 Describe the three main characteristics of sheet fixing screws used for steel frame fixing.

a ____________________

b ____________________

c ____________________

WORKSHEETS 8

5 Describe the four main factors to consider when choosing a fixing method for cladding a wall frame.

a ______

b ______

c ______

d ______

6 When screw fixing cladding to a steel frame, why must you position the screws as close as possible to the corners (flange) of the steel studs?

______

______

______

______

7 Explain why you would use a hammer and a punch to insert a nail when fixing a timber weatherboard to a timber frame.

______

8 Outline the three main methods of fixing battens to masonry walls.

a ______

b ______

c ______

9 Briefly outline the four main things to look for when inspecting a masonry wall before cladding.

a ______

b ______

c ______

d ______

10 Select the best correct answer.

When fixing sheet cladding stud spacing and fastener fixings, spacing is important and will depend on:

a The climatic zone and wind area rating ☐

b The sheet materials and thickness ☐

c The type of fastener being used ☐

d The relevant Australian Standard ☐

11 Match the steps by drawing a line to the correct step sequences:

Step 1	Work out the actual cover to make the lap gauge.
Step 2	Set the plinth board in position and fix it off making sure it is level.
Step 3	Using the lap gauge, locate the end of the new plank in position so that it overlaps the top edge of the plinth board.
Step 4	Mark the corner battens and/or studs with the heights for several rows of boards.
Step 5	Repeat the process for reach row of boards, checking regularly you maintain straight and level planking and lap coverage.
Step 6	Make sure the step in the gauge is hard up to the bottom edge of the plinth board at each stud and the bottom edge of the new board is sitting on the top end of the gauge.

WORKSHEET 4

To be completed by teachers	
Student competent	☐
Student not yet competent	☐

Student name: ______________________

Enrolment year: ______________________

Class code: ______________________

Competency name/Number: ______________________

Answer the following questions.

1 Calculate the number of boards required and the *actual cover* of the boards for a wall 1300 mm high, which is to be covered using timber weatherboards 190 mm wide, with a nominal overlap of 22 mm.

____

____

____

____

____

2 Calculate the number of boards, in stock 4.5 m lengths, required to cover a rectangular house that is 11.5 m long and 10.5 m wide. The walls are 3200 mm high. The boards to be used have a width of 180 mm, and a lap of 40 mm is intended. Use a wastage allowance of 5 per cent.

____

____

____

____

____

3 Answer True or False:

Compared to masonry walls, fibre-cement sheeted walls are a good thermal insulator.

True ☐ False ☐

4 Nominate five possible environmental and safety issues associated with cladding a building:

a ______________________

b ______________________

c ______________________

d ______________________

e ______________________

WORKSHEET 5

To be completed by teachers	
Student competent	☐
Student not yet competent	☐

Student name: ______________________

Enrolment year: ______________________

Class code: ______________________

Competency name/Number: ______________________

Answer the following questions.

Horizontal cladding

1 Describe the four features of cladding and cladding systems that determine how weatherproof the cladding is.

a ______________________

b ______________________

c ______________________

d ______________________

2 What are the two different methods for locating the position of boards as they are fitted to a wall?

a ______________________

b ______________________

3 Describe the three functions of weatherboard stops.

a ______________________

b ______________________

c ______________________

4 What is a plinth board and what are two functions of the board?

Description: ______________________

____

Function 1: ______________________

Function 2: ______________________

5 What is the purpose of a scribing block?

______

______

Fibre-cement cladding

6 What is a building plank?

______

______

7 List three advantages of using fibre-cement building planks.

a ______

b ______

c ______

8 Describe three effective methods used to cut fibre-cement building planks.

a ______

b ______

c ______

9 Define and describe a lap gauge.

______

______

Vertical cladding

10 What is vertical cladding?

______

______

11 In vertical cladding, additional rows of noggins are installed. What is their maximum spacing vertically, centre to centre?

______

______

12 What is an alternative to noggins when fixing vertical cladding?

____

____

13 List the five main considerations when installing vertical cladding boards.

a ____

b ____

c ____

d ____

e ____

14 What sort of joint should be used to join plinth boards at internal and external corners of walls?

____

15 Calculate the number of 2725 × 900 × 7.5 mm fibre-cement sheets that will be required to cover the wall shown in the drawing below.

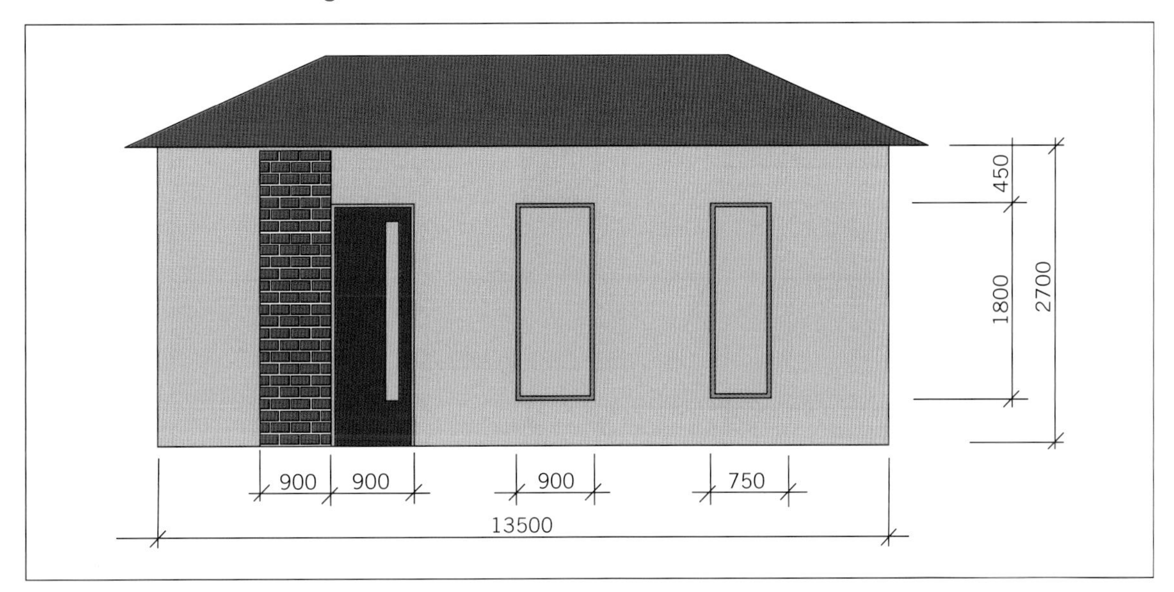

16 Calculate the total lineal metres of lining boards and number of fibre-cement sheets required to cover the front wall of this building (including return wall).

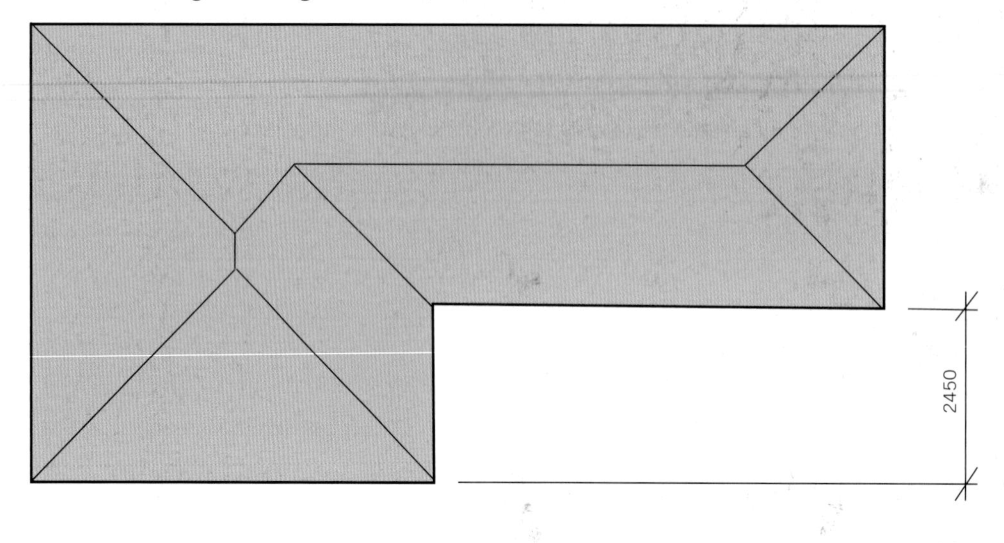

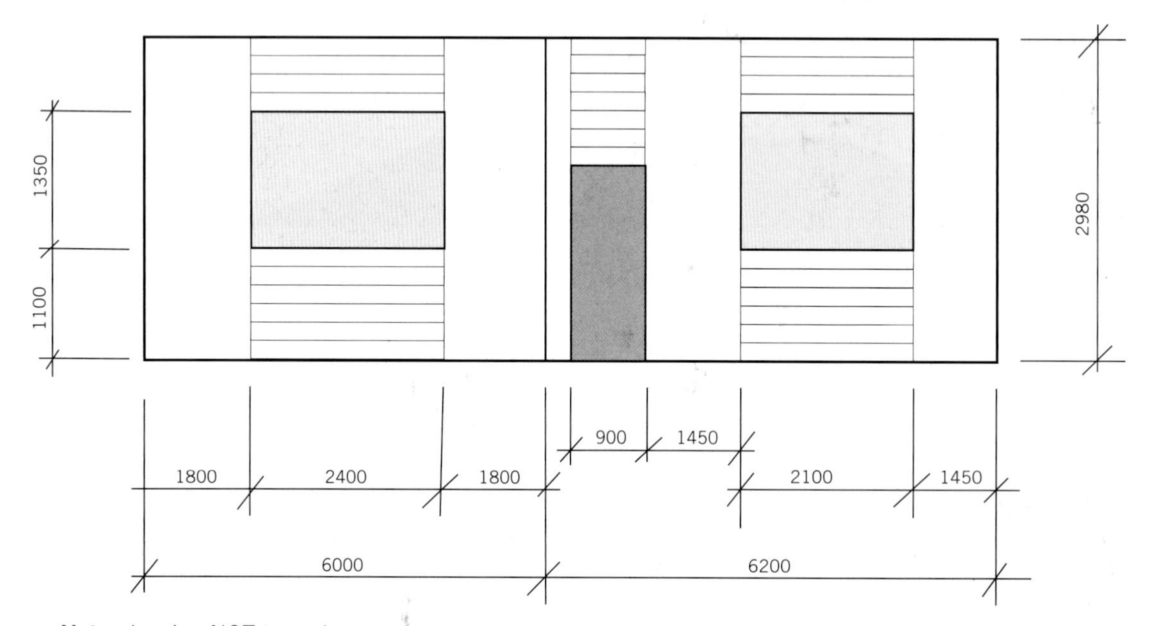

Note: drawing NOT to scale.
Lining boards' effective cover is 130 mm.
Fibre-cement sheet size is 3000 × 900 × 8.5 mm

PART 4

MACHINING AND COMPONENT MANUFACTURE FIELD OF WORK – JOINERY

Part 4 of this textbook deals with manufacturing components for windows, doors and built-in cabinetry. It builds on the knowledge learnt about windows and doors in Chapter 3, with additional content covering components used in built-in cabinetry.

It is based on the unit of competency and the content described in Chapter 9: Manufacture and assemble joinery components.

The key elements in Part 4 are:

- Planning, preparing and accurate setting out of components. This includes setting up operations and machines to operate safely.
- Dressing and shaping of materials to meet design specifications and tolerances.
- Producing and assembling of tight-fitting joints.
- Finishing and basic assembly techniques of windows and doors.

In Part 4, the skills and knowledge also reflect some key elements associated with other units of competency, such as CPCCJN3100 – Process materials to produce components using static machines.

The safe use of the common machines found in a typical modern joinery shop is critical in the application of the skills and knowledge presented in Part 4.

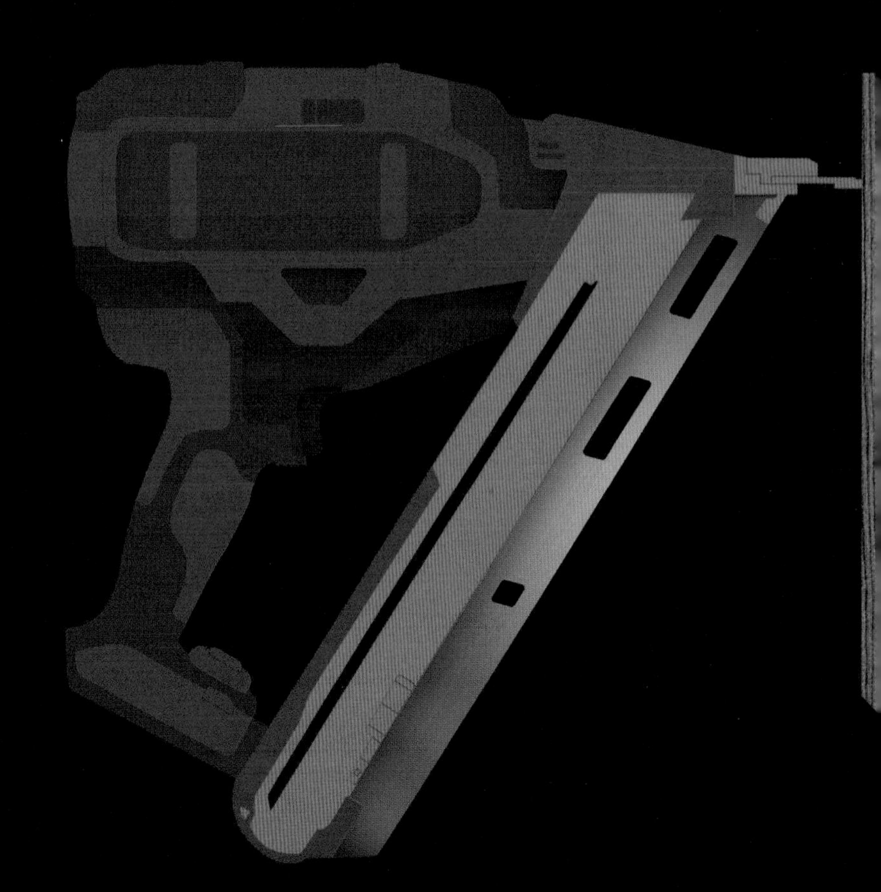

MANUFACTURE AND ASSEMBLE JOINERY COMPONENTS

9

This chapter aligns with the unit of competency 'Manufacture and assemble joinery components' and but has application across a wide range of joinery products. The unit outlines the skills and knowledge required to set up machinery and manufacture a range of joinery components for timber-based doors and window units. It also covers machining and manufacturing processes for window frames and sashes, doors and door frames.

In particular the chapter addresses the key competencies of:

- planning and preparation. This includes sourcing and interpreting work instructions, and selecting tools, equipment and the right materials. Each step in the process requires planning the methods, processes, safety requirements and quality assurance measures to be used
- identifying a range of different types of doors and windows, including typical jointing techniques for frames and panels, sash joints and mouldings and curved work
- preparing components and sub-assemblies for packaging
- setting up machines and equipment for manufacture and assembly, which includes a range of static machines and their safe operating procedures. This chapter also provides a basic understanding of computer numerical control (CNC) machine principles.

Mass-production systems require significantly different equipment and machinery and use processes beyond the scope of the typical joinery workshop, so in this chapter we will also look at modern manufacturing methods of machining and assembly for both small- and large-quantity production systems.

Learning objectives

This chapter focuses heavily on timber joinery for doors and windows, but knowledge and skills of kitchen and bathroom cabinet construction methods is also considered.

By the end of this chapter, you will be able to:

- plan and prepare to manufacture joinery components for doors, windows and joinery cabinetry
- identify door, window and cabinet components, creating cutting lists, and interpret plans and specifications
- identify and mitigate safety risks and hazards in the manufacturing process
- set up and use common static machinery to accurately make joinery components for doors, windows and basic cabinets safely
- prepare joinery components, including windowsills, thresholds, and door and window frames and cabinets
- use finishing tools, machinery and assembly techniques to prepare components for final assembly, including testing component fit and dry assembly
- prepare components and sub-assemblies for packaging.

Introduction

In ancient times a person was considered wealthy if they could afford a door for their house (see Figure 9.1), and grand public buildings often had doors that framed the entrance to the building.

Source: Greg Cheetham

FIGURE 9.1 Medici family palace, Florence, Italy

Today, doors are an essential item in the building envelope, not only providing a means of access and exit, security and protection from the elements (wind, rain, heat and noise), but also serving as architectural design features, as they can be made from a range of different materials (see Figure 9.2).

Windows provide security and protection from the elements, but are not generally a means of access from one space to another. They provide a means of lighting a space and can provide ventilation, insulation and privacy. In years to come, windows or glass panelling are expected to be a means of power generation.

Before the invention of glass, walls had openings for light and ventilation that were secured by shutters (see Figure 9.3) or covered with cloth (canvas or hessian) to provide privacy and keep out the elements or to keep in the warmth. Today, shutters have made a comeback and are used internally on windows as an alternative to curtains and blinds.

There is now a range of windows available that can provide privacy, security and protection from the elements, as well as let in light and breezes (see Figure 9.4), keep out heat and eliminate noise transmission. They can be self-cleaning (e.g. Viridian Renew™ Self-Cleaning Glass), and in the future, through the use of nanotechnology, may provide a source of electricity similar to photovoltaic solar panels simply through sun shining on the glass.

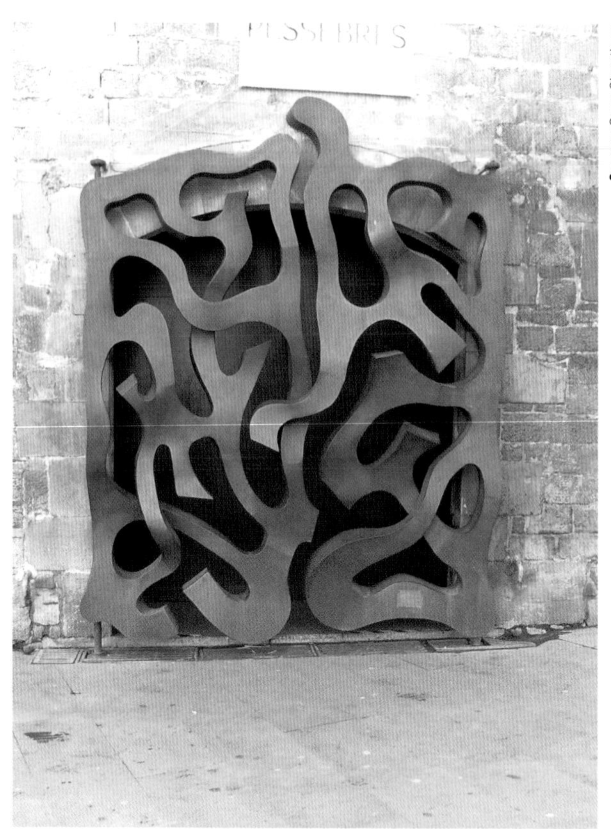

Source: Greg Cheetham

FIGURE 9.2 Gaudí doors in Barcelona, Spain

Source: Shutterstock.com/Oleg Zaslavsky

FIGURE 9.3 Shutters

Source: iStock.com/Tammy616

FIGURE 9.4 Modern louvre window

Planning and preparation

When planning to manufacture doors and windows, knowledge and understanding of door and window types is essential, as is the ability to produce plans, working drawings or computer-aided design (CAD) set-outs. In addition, detailed knowledge and understanding of the requirements of the relevant Australian Standards (AS) for their manufacture (including timber selection, design criteria and construction methods) is critical.

Australian Standards

There are many regulations relevant to the manufacture of doors and windows in buildings. The main Australian Standards covering the design, construction and manufacture of windows are:

- AS 1288 Glass in buildings – Selection and installation, which covers other items including skylights, recycled products and heritage items
- AS 2047 Windows and external glazed doors in buildings, which covers most glazed windows and external doors, louvres and shopfronts
- AS 4055 Wind loads for housing
- AS 4420.0 Windows – Methods of test – General introduction and list of methods.

There are other Australian Standards that deal with timber, metal and PVC (polyvinyl chloride) extrusion quality, and hardware used in the construction of doors and windows.

The National Construction Code specifies that window manufacturers must certify windows and glazed external doors to meet mandatory minimum specifications under AS 2047 and AS 1288. These windows are subject to strict performance testing as specified in other Australian Standards that include testing for wind deflection, operating under extreme conditions, water penetration, and air and acoustic efficiency. All residential windows and doors are marked with a Performance Label or sticker confirming compliance with AS 2047.

In general, timber doors no longer have a specific Australian Standard dictating how they are to be made and assembled. Rather, the Standards deal with the materials from which such doors are manufactured. For example, AS 2754.2 Adhesives for timber and timber products – Polymer emulsion adhesives provides the standards for adhesives to be used for timber, while AS 2796.3 and AS 4785.3, repectively, provide the standards for hardwood and softwood timbers used in 'furniture products' (see 'References and further reading' at the end of this chapter for full details of these Standards).

Environmental considerations

Buildings today are rated for their energy efficiency. There are a number of house energy rating systems in Australia. The Nationwide House Energy Rating Scheme (NatHERS), Basix (Building Sustainability Index) and Window Energy Rating Scheme (WERS), along with the state and territory star rating systems, assist designers and builders to achieve these efficiencies. Doors and windows are critical to ensuring the energy efficiency and acoustic insulation of any building, as these are the most vulnerable points in a building envelope.

GREEN TIP

There is a need to use residential housing to help reduce greenhouse gas emissions and use energy more efficiently, and to make living more comfortable, economical and sustainable.

The risk of bushfire is a very significant environmental consideration. The Bushfire Attack Level, or BAL, will determine the materials and construction methods able to be used in certain locations; for example, double glazing. A written report may be required from an accredited Bushfire Consultant. Building envelope materials and methods should be designed in compliance with AS 3959 – Construction of buildings in bushfire-prone areas.

The National Construction Code – Volume 2, Part H6 and Housing provisions require the edges of all operable doors and windows to have a seal to 'restrict air infiltration' around the perimeter. The seals may be a draught protection device fitted to the bottom edge of a door (see Figure 9.5) or a compressible strip fitted to the rebates or edges. Some fittings will require grooves or rebates to be run so they can be fitted. This needs to be accounted for in the planning stage, especially when the building is air-conditioned.

Traditionally, glass in doors and windows was often not much more than 3 mm thick, but today laminated glass is used for security and double glazing (and even triple glazing) in doors and windows. This means that rebates in doors and timber windows need to be set at 12 to 18 mm deep as a minimum, depending on the glass selected.

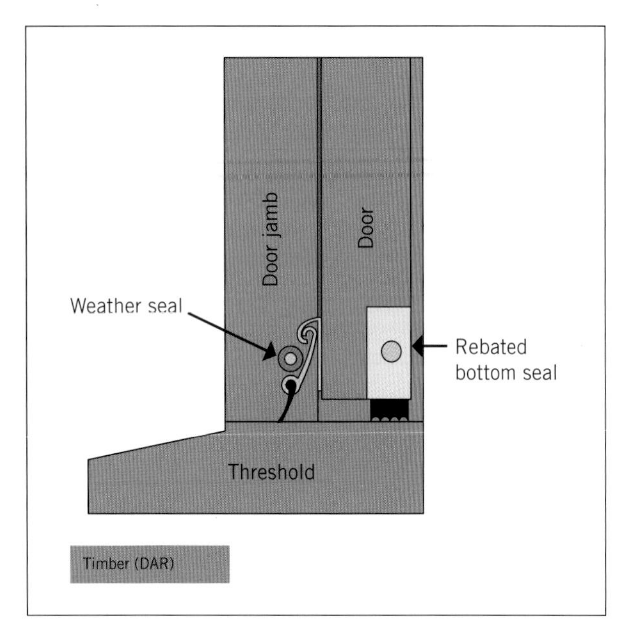

FIGURE 9.5 Seals are a form of draught protection

Setting up the joinery and processes

When setting up to manufacture doors and windows, the layout of the workshop is critical to the efficiency of the process. For example, you need to consider:

- the size of the factory required in relation to the size and number of machines
- the number of personnel to be employed
- the amount of materials storage and racking required for pre-production stock; that is, solid timber, glass and other materials
- the flow of work through the production process; that is, materials storage → machine shop → assembly → finishing/glazing → despatch
- space requirements for materials movement.

It is equally important when planning the work processes to:

- identify quality assurance issues; for example, selecting the correct materials (timber adhesives, hardware and finishing medium) and understand Australian Standards and National Construction Code quality assurance requirements
- identify work health and safety/occupational health and safety (WHS/OHS) issues; for example, preparing work method statements and dust extraction, apply risk control measures for machinery and use signage and barricades where necessary
- ensure correct PPE is used and maintained for each task. Modify PPE when changing work activities if required.
- check machines for safe operation
- ensure you have the appropriate tools for the various tasks
- prepare a checklist of procedures and quality assurance checks to accompany the jobs through the manufacturing process.

Pre-production checks may incorporate:

- sketches to support plans and specifications
- set-out rods showing the full size of components and joint details
- a cutting list outlining component quantities and sizes and other details.

With the advent of computer numerical control (CNC) machinery and computer-integrated design technology, the modern joinery shop manufacturing doors and windows has changed significantly by using a systems design approach, especially for mass production techniques. The bespoke door and window industry also uses modern machinery, but the types and configurations of machinery within the joinery shop still require the same careful planning to achieve efficient productivity. (See 'Corporate – Machinery Manufacturers' at the end of this chapter for industry links.)

Material selection for doors

In general, the timber should be durable, well seasoned to between 9 and 14 per cent moisture content, straight and free of defects. For the best results, use timber that is quarter sawn; back-sawn boards tend to cup significantly and shrink in width.

Hardwoods should be structural appearance grade or select grade. For doors made with a range of materials, consult AS 2688 – Timber and composite doors.

Softwood should be clear grade or select dressing grade, as outlined in a number of Australian Standards (see 'References and further reading' at the end of this chapter).

Any sapwood material must be immunised. This is particularly relevant when using imported timbers such as Pacific maple and meranti.

Material selection for windows

Western red cedar is commonly used for window frame and sash construction in metropolitan areas (but not for sills – see Chapter 3 for more details on sill materials) because of its durability, ease of working and finishing characteristics. However, it is not suitable for use in rural areas where birds such as the cockatoo can wreak havoc on such softwoods. Victorian ash (also known as mountain ash, Tasmanian oak and alpine ash), coachwood and kwila (also known as merbau) are becoming increasingly popular in rural areas.

Window sills should be made from durability Class 1 or 2 timbers such as tallowwood, turpentine forest red gum, spotted gum, river red gum, jarrah or Sydney blue gum, depending on the location of the window in the building (e.g. well protected under a verandah from the weather). See also Appendix G in AS 2047 Windows and external glazed doors in buildings.

Other timbers can be used as long as they comply with the requirements of the Standards.

Sash, stile and rail material needs to be straight and free from defects. The moisture content will depend upon the location, but is generally between 9 per cent and 14 per cent.

Sustainability

Sustainability is a critical issue in the timber door and window manufacturing industries today. To ensure the ongoing supply and quality of materials, a number of certification programs have been started and are often written into specifications for building projects. The primary programs include forest certification and chain-of-custody certification, where timber is tracked through the supply chain from tree to manufacturers and retailers.

The two dominant international certification schemes are the Programme for the Endorsement of Forest Certification (PEFC) and the Forest Stewardship Council (FSC) schemes. Both schemes operate in Australia. PEFC has endorsed the Australian Forest Certification Scheme (AFCS) and the FSC operates in Australia under interim standards from internationally accredited FSC certifying bodies.

GREEN TIP

Current information on the certification of sustainable forest and production companies and updates on the development of standards is available from the Responsible Wood website at https://www.responsiblewood.org.au/, and from the FSC at https://anz.fsc.org/.

Technology

Access to suitable raw materials is becoming more difficult, so technology and manufacturing processes need to adapt to these changes. Laminating and timber jointing technologies have seen increased use of lesser grades of timber that would once have been waste timbers. The use of timber in combination with plastic sections has improved the durability of these materials, thus extending the useful life of these resources. In future years we will need to continue to develop new and better ways to sustain the industry if we want to continue to use quality timbers in joinery works.

Laminating timber

Laminating timber is the process of gluing timber sections together to create larger sections. It allows for use of lower quality timbers in the core while more expensive species can be glued on the outside. This is common in the production of solid core doors.

Finger jointing

Finger jointing timber is the process of gluing shorter lengths of timber end to end to create longer sections. The finger joints create a larger gluing surface and combined with adhesives can create a very strong joint line (see Figure 9.6).

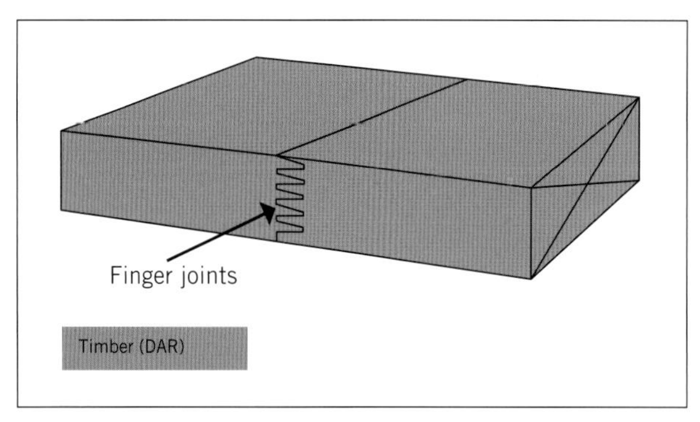

FIGURE 9.6 The finger joint creates a larger gluing surface

LEARNING TASK 9.1 PLANNING AND PREPARING

Register yourself on the Supply Chain Sustainability School website http://www.supplychainschool.org.au and navigate to the e-learning module.

This is a one-hour module that takes you through the environmental and social impacts that materials can present to the construction and joinery industry.

Click on 'Learning Resources' then search for 'materials' to find the 'Materials e-learning module'.

Your task is to work through the module and complete the self-assessment quiz (15 questions) correctly. On successful completion of the quiz, print and submit the certificate along with your answers to the following two questions:

1. Using an example from your workplace, explain how your company demonstrates its sustainability in terms of environmental and social practices.
2. When making decisions on manufacturing doors and window components for buildings, how would you improve your product purchasing processes?

These waste management case studies may assist in answering the questions:

- Net zero waste homes: https://www.sustainability.vic.gov.au/energy-efficiency-and-reducing-emissions/in-a-business/by-sector/zero-net-carbon-homes
- AKD Softwoods: https://www.akd.com.au/sustainability

Other helpful links:

- Australian Forestry Standard: https://www.responsiblewood.org.au/standards/australian/
- The Australian Standards for:
  AS 4708 Sustainable forest management: https://www.standards.org.au/standards-catalogue/standard-details?designation=as-nzs-4708-2021
  AS 4707 Chain of custody for forest products: https://www.standards.org.au/standards-catalogue/others/afsl/as--4707-colon-2021

Types of doors

Flush doors

Flush doors comprise both hollow core and solid core doors (see Figures 9.7 and 9.8).

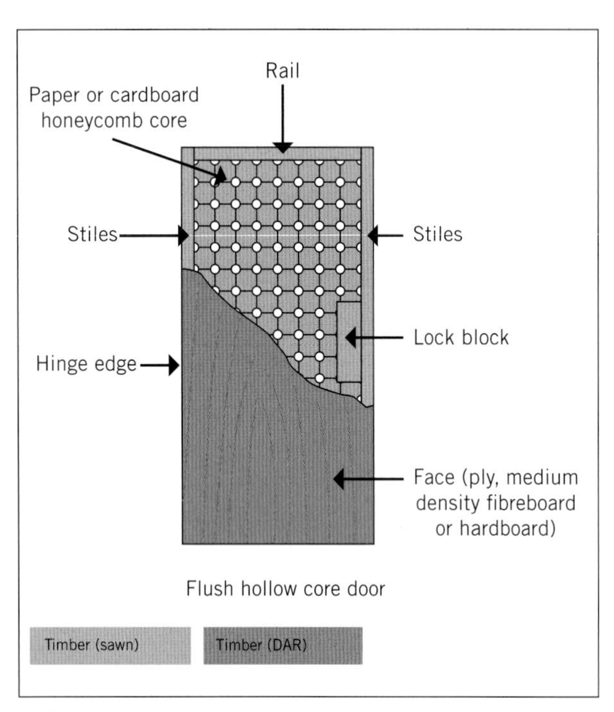

FIGURE 9.7 Hollow core door

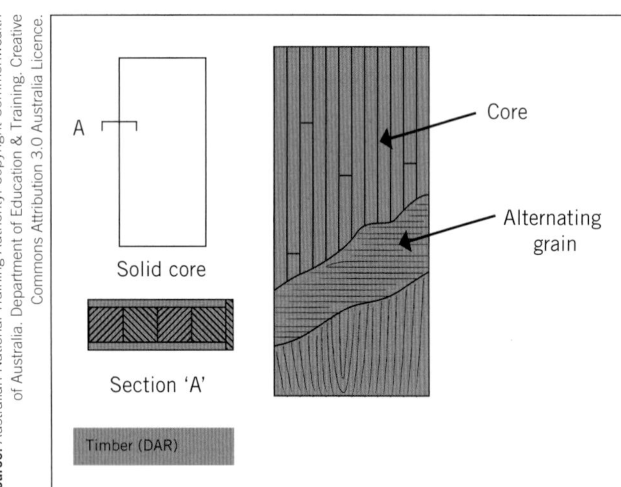

FIGURE 9.8 Solid core door

Internal flush doors

Hollow core doors are for internal use, and solid core doors are used externally.

The frames of **hollow core doors** are made from seasoned, square dressed timber. In mass-produced hollow core doors, a paper or cardboard core is laid inside the frame to keep the faces evenly separated. When this type of core is unavailable, a mid-rail and blocks the same thickness as the frame should be incorporated to keep the face materials evenly spread. Alternatively, light ply slats can be inserted to keep the faces apart.

If nothing is done to keep the faces evenly spread, temperature changes in a building can cause the face materials to cup inwards and create a sunken face surface on the door that looks unsightly. The core material also improves the acoustic properties of these doors.

These doors are also being produced with moulded **medium density fibreboard (MDF)** skins to resemble framed and panelled doors. Modern construction methods use particleboard or MDF as a core material, faced with a veneer or plywood on both sides. These doors should only be used internally. A 10–15 mm edge strip is applied to the doors for fixing hinges and locks and to provide the appearance of a solid door.

The use of MDF as a core material has allowed manufacturers to router designs into the outer faces of the doors and router out for glazing. This routing and carving is carried out using CNC machinery in mass-production factories. To see a demonstration of this type of technology, conduct an internet search using the following terms: CNC router carving, CNC router, or Biesse or Weinig.

External flush doors

To ensure their durability, these doors need to be made from a suitable grade of timber. The minimum quality for softwood used in doors and door frames should be select grade, as defined in AS 4785 Timber – Softwood – Sawn and milled products – Parts 1 to 3, and the moisture content should be between 9 per cent and 14 per cent depending upon the specific locality. Hardwoods should be Select Grade or Medium Feature Grade, as defined by AS 2796.2. (See 'References and further reading' at the end of this chapter for details of these Standards.)

Traditionally, solid core doors have a core of seasoned straight-grained timbers glued together with high water-resistant adhesives such as urea-formaldehyde. The core is faced with a cross band veneer at 90° to the core and then overlaid again at 90° using a decorative face veneer (see Figure 9.8). Plywood may also be used to face the core material.

Both hollow core doors and solid core doors require a surface finish (usually painted) to be applied to **all** faces and edges.

Ledged and ledged and braced doors

Ledged doors are used mainly as shed, gate and garage doors, but are not commonly used today except as gates. However, high-quality ledged and braced types are occasionally seen as entry doors on high-end value homes, and religious and commercial buildings.

Ledged doors perhaps were the earliest form of door: timber slabs held together by cross battens, which we now call ledges, to counteract the effects of twisting, bowing and **cupping**. They consist of two or three ledges fastened across the boards using nails or screws (see Figures 9.9 and 9.10). Today, these doors are constructed using tongue and groove lining boards laid vertically, or occasionally diagonally.

The boards are nailed through from the face side, with the nails being long enough to pass through the ledge (approximately 6–10 mm) and be clinched across the face of the ledge and punched below the surface of the ledge (see Figure 9.10).

The top edges of the ledges should be bevelled slightly to allow moisture to run off if the door is to be used in an outdoor location. This helps prevent rot in the ledges.

Ledged and braced doors are constructed using the same techniques, except that a brace is inserted diagonally across the reverse face of the door. The brace runs down from the closing edge towards the hinge edge.

The brace is often cut into the ledges to provide greater strength and resist sagging. The end of the brace is cut square and marked to one-fifth of the width of the brace material (see Figure 9.11), and then tapered back to the width of the brace material.

Framed ledged and braced doors

Here, a framework is prepared using mortise and tenon joints. This door has a framework consisting of **stiles** and **rails**, which are **mortised** and **tenoned** together, with diagonal braces fitted between the stiles and ledges, or rails. The face of the door is covered with tongue and groove boards, which may have beaded or V-joint edges. To allow for the sheeting, the bottom and middle rails are reduced in thickness to allow the surface of the panelling to finish flush with the outside face of the outer frame. The top rail can be rebated or grooved to provide weather resistance (see Figure 9.12). The lining boards are nailed (or secret nailed) or screwed through the face to the rails and stiles.

Framed ledged and braced doors have been developed in a range of styles. While they are not a popular choice for modern construction, they can still be encountered during renovations and heritage restoration, and their size to weight ratios makes them particularly useful for large doors or gates.

Figures 9.13 and 9.14 show some examples of framed ledged and braced doors.

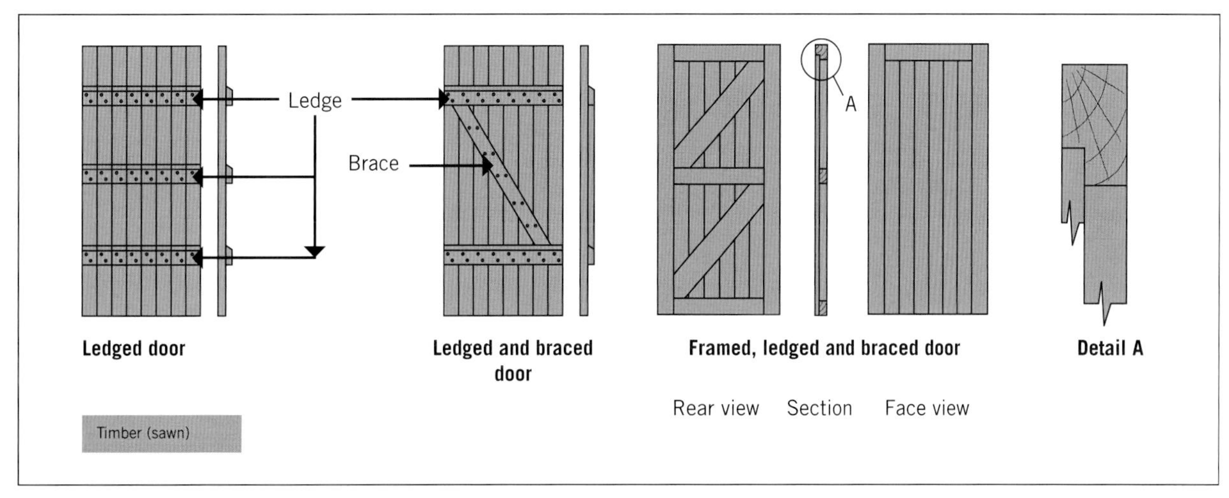

FIGURE 9.9 Ledged door and ledged and braced door

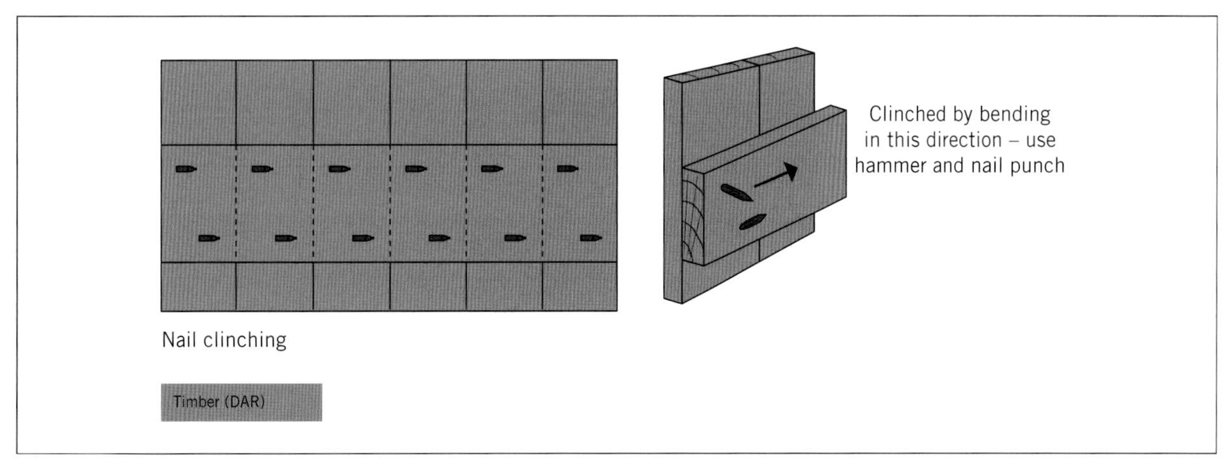

FIGURE 9.10 Nail clinching

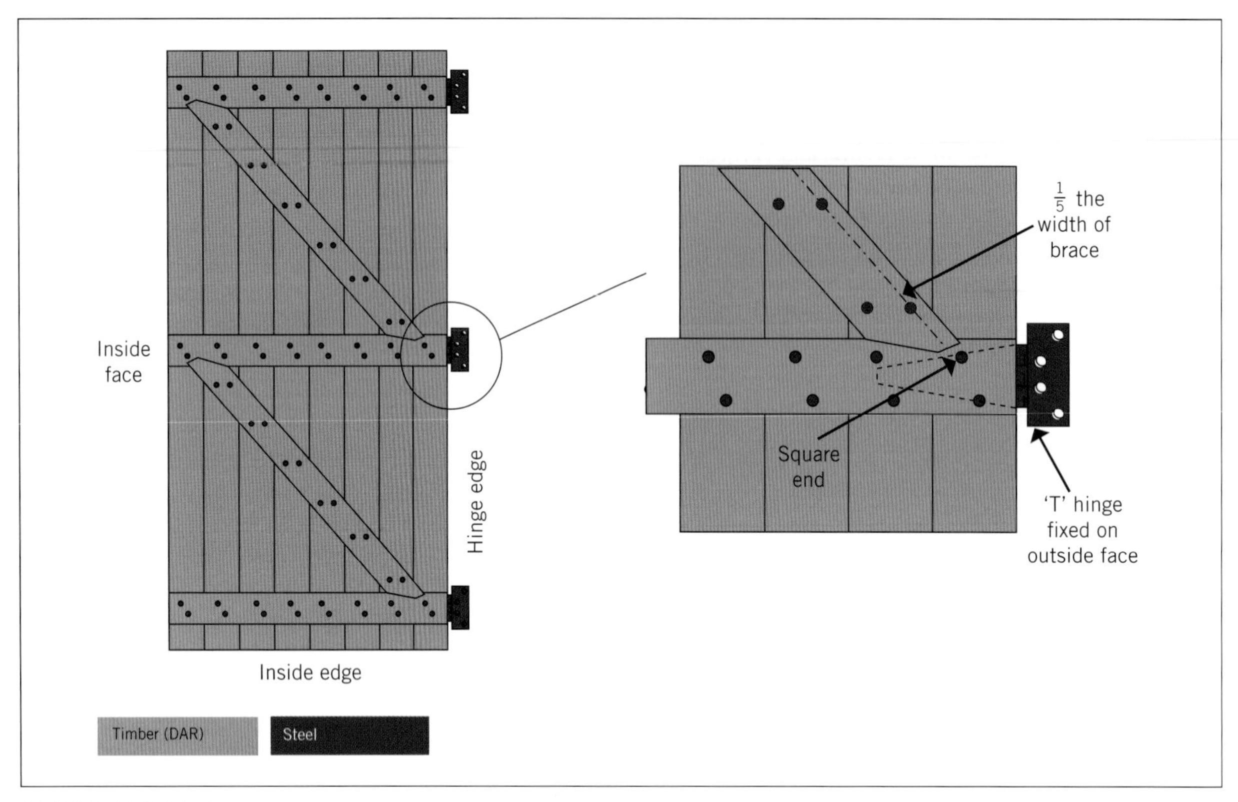

FIGURE 9.11 Cut-in brace

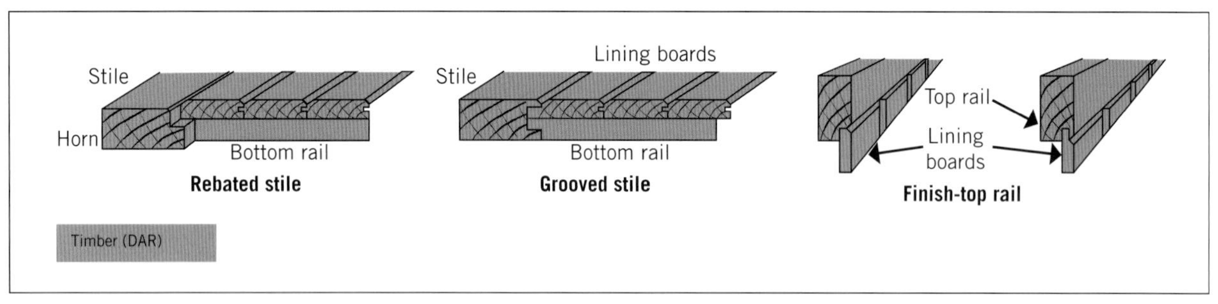

FIGURE 9.12 Inserting lining board variations

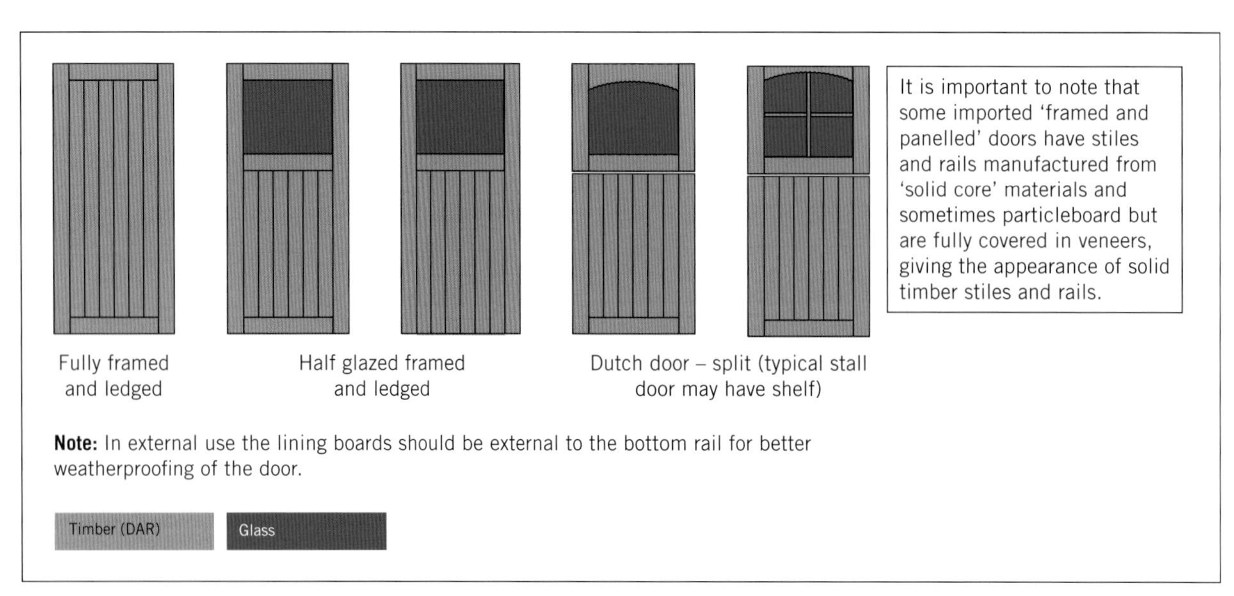

FIGURE 9.13 Framed, ledged and braced styles

Source: Agefoto Stock/Eric Hernandez

FIGURE 9.14 Dutch door

Framed and panelled doors

Framed and panelled doors, for the purposes of this book, consist of solid timber frames, forming rectangular or other shapes, into which a range of panels and moulded materials may be fitted (see Figure 9.17).

Other components of doors

While frames for doors are made up of the stiles and rails, there are other components you need to be familiar with:

- **Glazing bar:** Used to break up a glazed section into smaller panes (see Figure 9.15). Traditionally used in earlier times because glass was not available in large panes.

Source: Ed Hawkins

FIGURE 9.15 Glazing bar joint

- **Glazing bead:** Used to hold glass in a rebated frame and glazing bar as an alternative to glazing putty (see Figure 9.16).

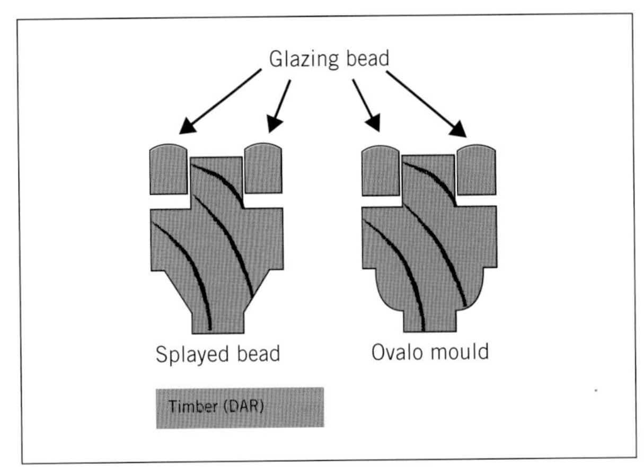

FIGURE 9.16 Glazing bar profiles

- **Gunstock stiles:** Used to reduce the width of the stiles in the glazed area to increase the size of the glazed area and improve the light and vision through the door (see Figure 9.17).
- **Mid-rail (or lock rail):** Provides fixing points for mortise locks and doorknobs and knockers; generally as wide as the bottom rail. Double tenons are used at each end. Angled shoulders are required to match gunstock styles when used (see Figure 9.17).
- **Muntin (or mullion):** Intermediate vertical member used to divide panels (see Figure 9.17). There are also frieze muntins.
- **Frieze panel:** Topmost panel or panels in a multi-panelled door used to provide light or special decoration (see Figure 9.17).
- **Frieze rail:** Supports the frieze panel (see Figure 9.17).
- **Recessed panel:** A panel whose face finishes below the surface of the framing timbers (see Figure 9.17).
- **Raised panel:** A panel that varies in thickness from its edges to the centre and may be splayed or fielded (see Figures 9.16, 9.19 and 9.20). It does not always finish flush with the surface of the surrounding framework.
- **Bolection mould:** Moulding fitted over a panel and frame (see Figure 9.21).
- **Inlay mould:** Mould that is applied to the rebate formed by the frame and panel (see Figure 9.21). The inlay mould can also be used as a panel mould.
- **Panel mould:** Moulding that is planted onto a flush panel door (solid or hollow; see Figure 9.22). These mouldings can also be used to enhance the appearance of plain framed and panelled doors.

There are no typical door sections. However, in selecting the right-sized material, you need to consider the following questions:

- Is the door internal or external?
- Is the door to be a feature (e.g. primary entrance door) or simply a barrier (e.g. ledged door to a shed)?

Top rail
Mortise and tenon joint
Stub tenon
Glazing bar
Glass panel
Mid-rail or lock rail
Gunstock or reduced stile
Muntin
Raised panel
Double tenon joint
Bottom rail
Frieze panel
Frieze rail
Recessed panel
Raised panel
Timber (DAR)
Glass
Timber mouldings

FIGURE 9.17 Parts of framed and panelled doors

Source: Shutterstock.com/Nicolesa; Shutterstock.com/Palo_ok; iStockphoto/billyfoto; Dreamstime.com/Malgorzata Kistryn

FIGURE 9.18 A range of framed and panelled doors: simple to intricate

- Is the door to be painted, stained or clear finished?
- What hardware will need to be fitted and what is the amount of material required to support the hardware?

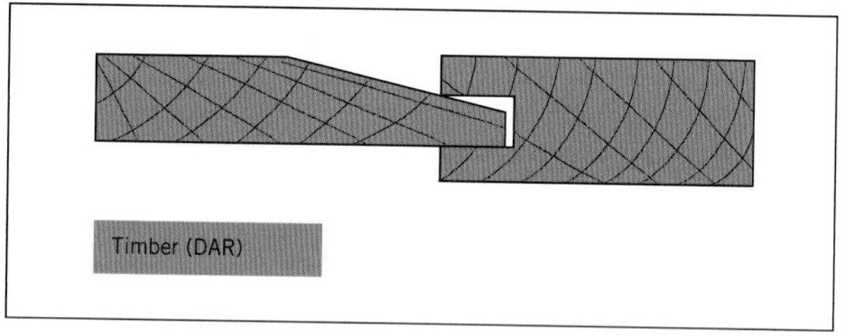

FIGURE 9.19 Splayed panel

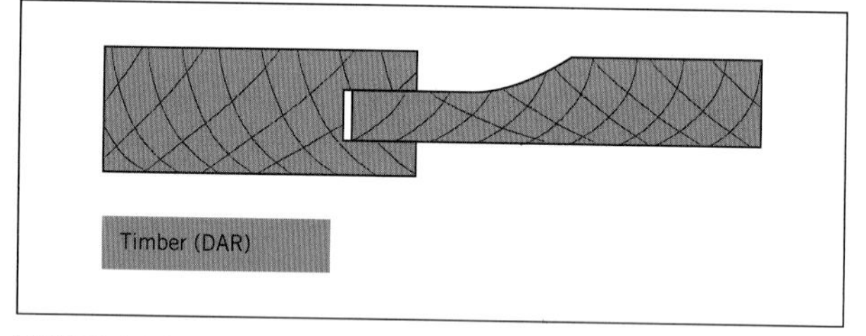

FIGURE 9.20 Fielded panel

For framed and panelled entrance doors, typical sizes are:

- 100 × 50 mm (38 mm for internal doors) – DAR (dressed all round) timber is typical for top rails and stiles.
- 150–200 × 50 mm (38 mm for internal door) – DAR timber is typical for bottom rails.

However, what is required often depends on the design of the door. In some circumstances, boards may need to be edge jointed to form wider boards suitable for components such as panels and centre rails. This is discussed later in the chapter.

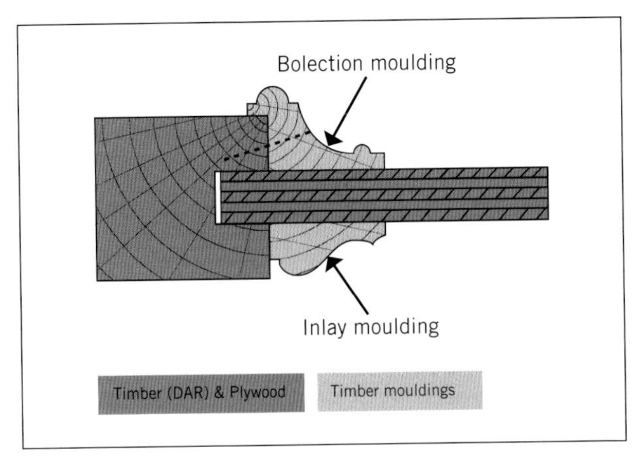

FIGURE 9.21 Bolection and inlay moulds

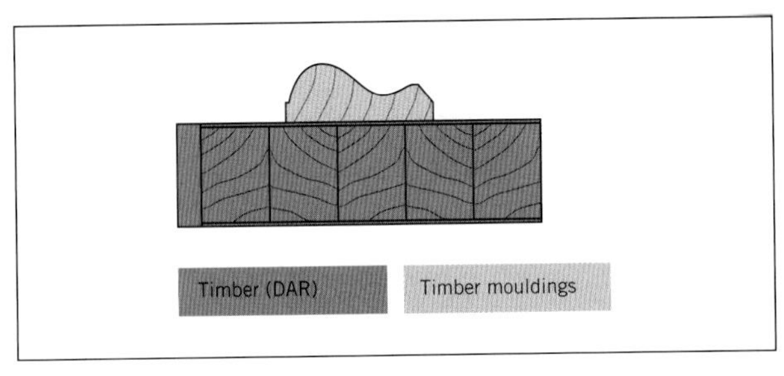

FIGURE 9.22 Panel mould

Making a framed panel door

Framed panel doors are made from rails and stiles and infill panels, traditionally using mortise and tenon joints to form the frame joints, as shown earlier in Figure 9.17.

Alternatively, stiles and rails can be moulded and grooved to accept stubbed tenons with scribed shoulders. Some manufacturers incorporate dowels into the frame construction to increase the strength of the joints.

Frames are also manufactured using rebated stiles and rails with long and short shoulder joints, or by grooving the frame stock to accept the panels.

Proportions of mortise and tenon joints

Mortise and tenon joints are the traditional joints used in the manufacture of timber frames, whether those frames are used in the construction of windows or doors.

The proportions for each part of the joint have been worked out over centuries and provide the right balance of strength and stability for the joint. The same basic proportions are used in the construction of bridle joints.

The primary proportion is one-third, this being the thickness of the tenon in relation to the thickness of the stock being used (see Figure 9.23). Hence, an internal framed door that is 35 mm thick will generally have a tenon 12 mm thick (approximately one-third the thickness of the framed door and equal to the nearest chisel, machine cutter or drill bit). This gives support to the cheeks of the joint and leaves sufficient material on

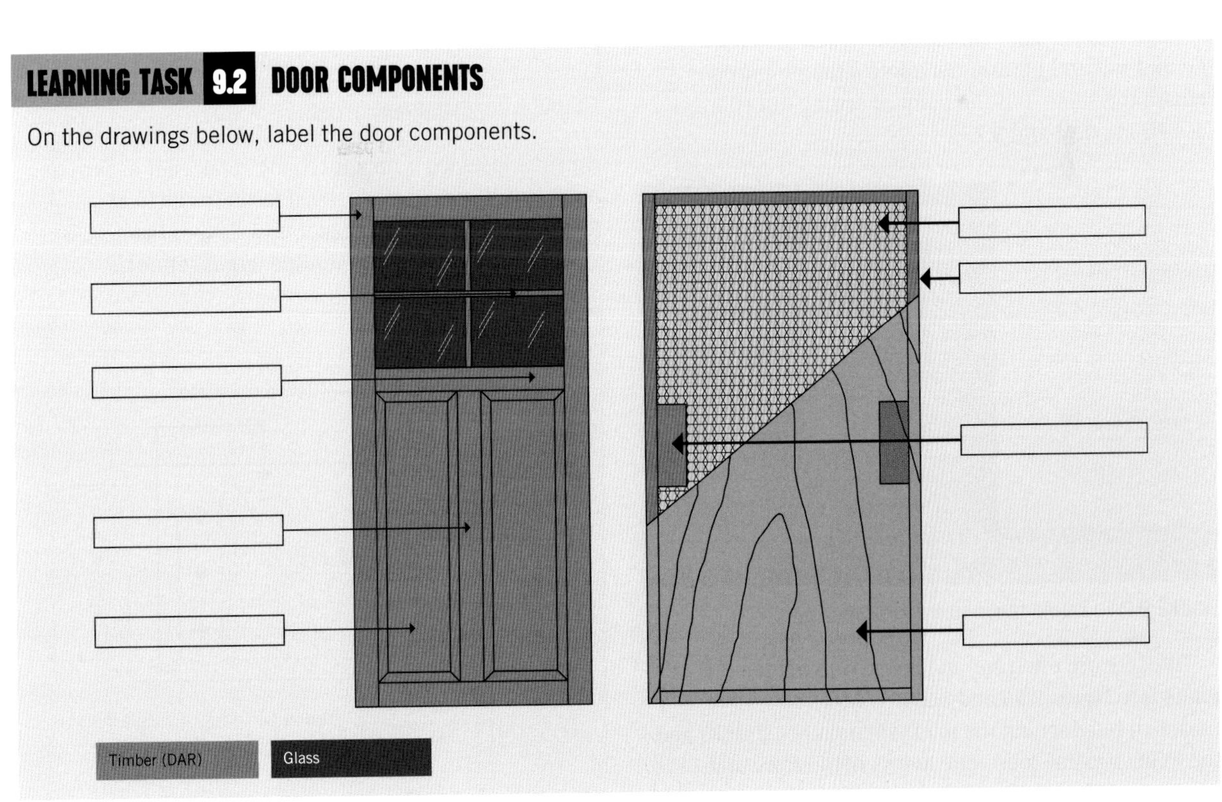

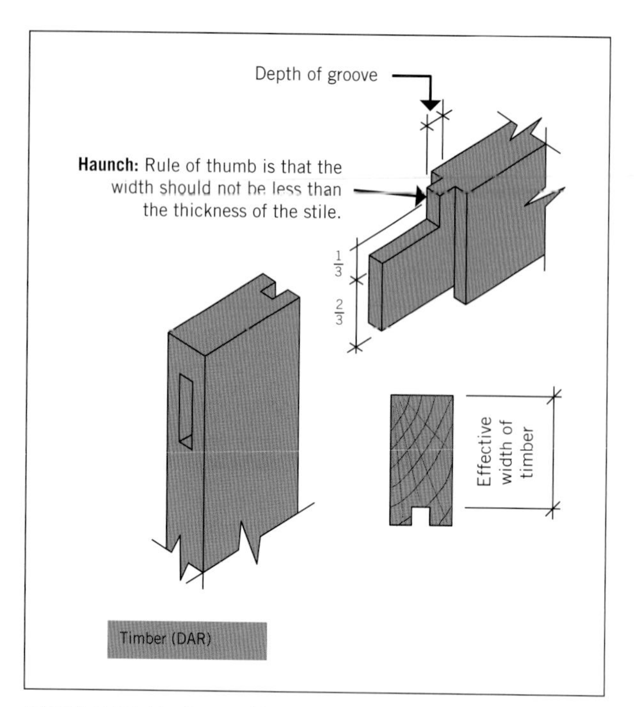

FIGURE 9.23 Mortise and tenon proportions

the tenon to provide strength and resistance to twisting or snapping.

The width of the tenon is dependent upon the width of the timber stock being used and the *effective width* of the timber (see Figures 9.23 and 9.27). For example, when used in a narrow timber frame, such as that used in a window sash, a single tenon is adequate, while in a door frame often a double tenon is used in the mid-rail and bottom rail joints – the proportions remain the same, but the point of setting out changes (see Figures 9.24 and 9.27).

FIGURE 9.24 Double tenon in a door frame

Wedges are often cut and inserted to tighten the joints (see Figures 9.25 and 9.27). Wedges also serve as a means of retightening the joint when timber shrinks and the joints become loose due to loss of moisture and exposure to the elements. Today, modern production techniques and modern adhesives have reduced the need to wedge tenon joints. Joints that can easily come apart

FIGURE 9.25 Wedges in mortise and tenon joints

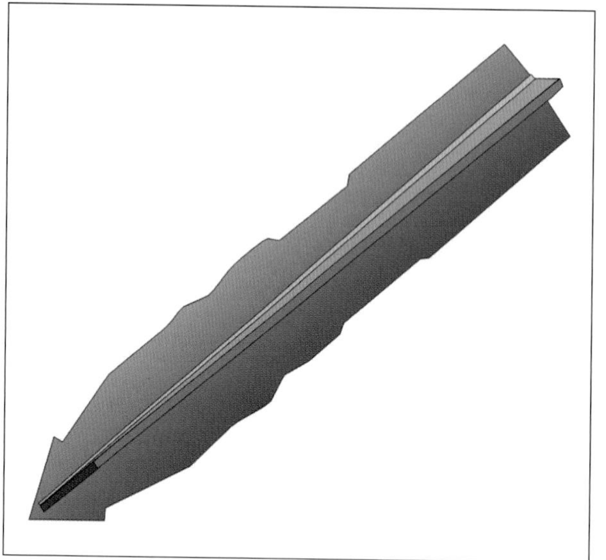

FIGURE 9.26 Star dowel

must be fastened with a mechanical fastener as well as adhesive. Types of mechanical fasteners include stainless steel brads and nails, and star dowels (see Figure 9.26).

Figure 9.27 shows the double tenon set-out.

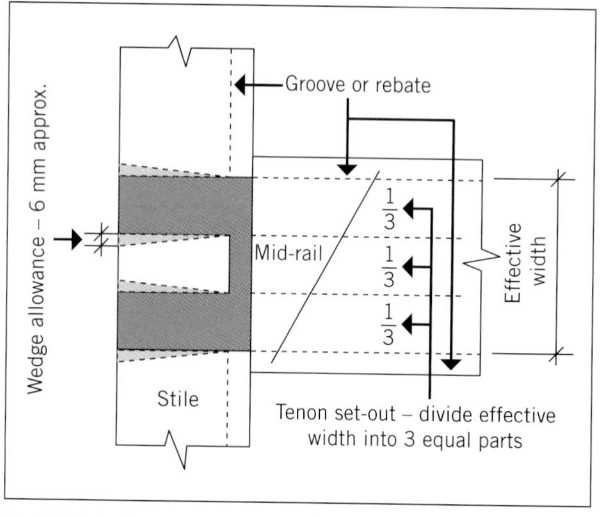

FIGURE 9.27 Double tenon set-out

Traditionally, mortise and tenon were held together with pins, or 'pegs' as they used to be known. This is still an effective method for guaranteeing that the joint won't come apart. The pin is usually inserted after the joint has been glued and clamped (see Figures 9.28 and 9.29).

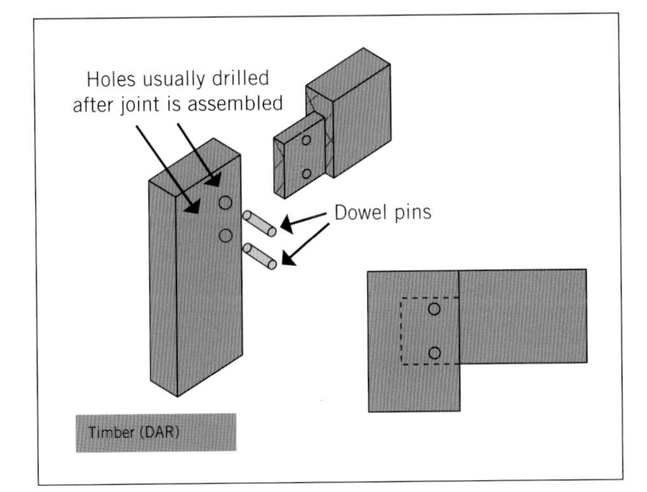

FIGURE 9.28 Pegged mortise and tenon joint

Source: Ed Hawkins

FIGURE 9.29 Gates joined with peg joints

Grooves and rebates in stiles and rails

Several static machines and portable power tools can be used to work grooves and rebates into stiles and rails. The choice of machine depends on the production capacity required. For large quantities of material a multi-header would be appropriate, while in a small shop that is only occasionally producing rebates and grooved products, a table saw might be used.

Static machines include:

- pin router – overhead and inverted types, using wing cutters
- spindle moulder
- multi-header
- jointer
- table saw.

Ensure there is appropriate guarding of the machines and cutters and, wherever possible, use fences, stops and **jigs** to keep workers' hands away from cutters.

Never wear loose clothing when using machinery, and keep sleeves and shirt tails tucked in. It is too easy for them to get caught in revolving cutters.

Portable power tools include:

- router – grooving bits (wing cutters) for small grooves; rebating or straight cutters for larger work
- laminate trimmer – for small rebates and grooves only.

When grooving long edges on rails and stiles with a hand-operated router or trimmer, it is good practice to increase the width of the stile by clamping an extra piece along the edge to improve its stability (see Figure 9.30).

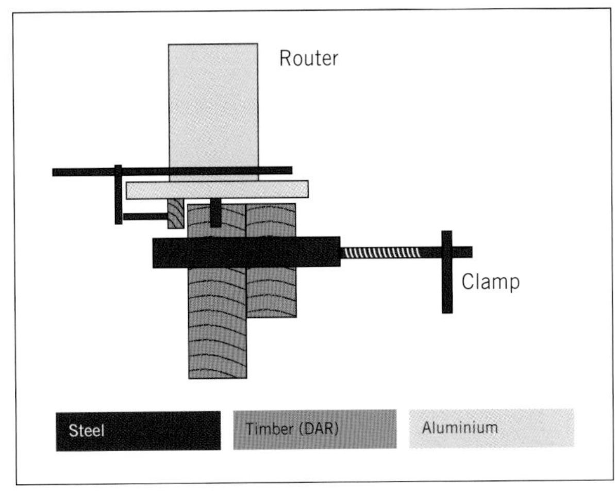

FIGURE 9.30 Grooving a stile or rail with a portable router

Jigs enable multiple identical pieces to be produced (see Figure 9.31). Jigs should be made of materials that are robust and will not deteriorate rapidly with multiple uses (often specialist tradesmen produce them). They should be properly labelled and stored to avoid damage.

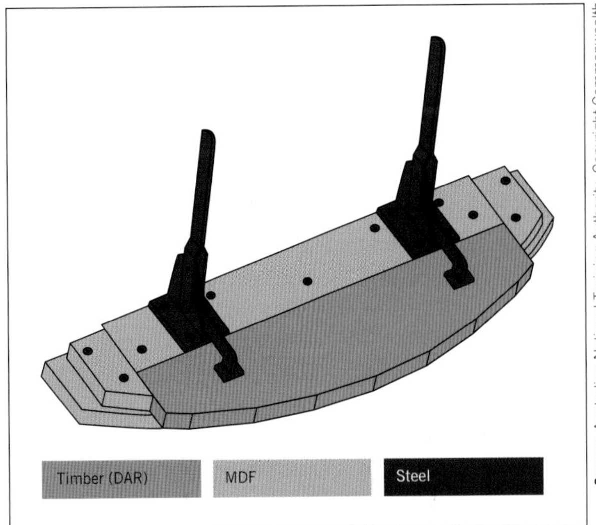

Source: Australian National Training Authority. Copyright Commonwealth of Australia. Department of Education & Training. Creative Commons Attribution 3.0 Australia Licence.

FIGURE 9.31 Simple jig using toggle clamps

Scribed joints

Stiles and rails can be moulded and grooved to accept stubbed tenons with **scribed** shoulders (see Figure 9.32). Some manufacturers incorporate dowels into the frame construction to increase the strength of the joints (see Figure 9.33). The frames are also manufactured using rebated stiles and rails with long and short shoulder joints, or by grooving the frame stock to accept the panels.

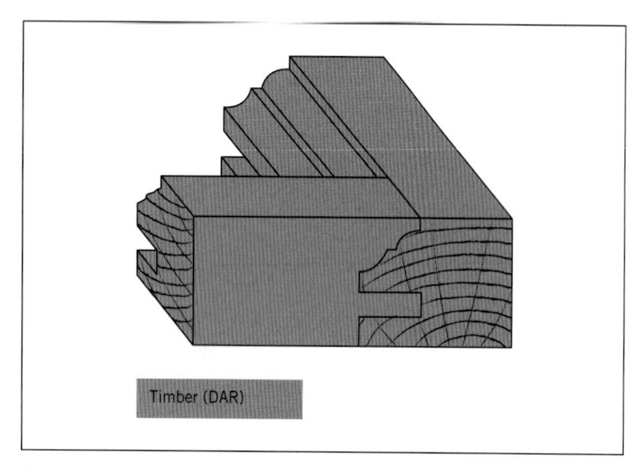

FIGURE 9.32 Scribed joint

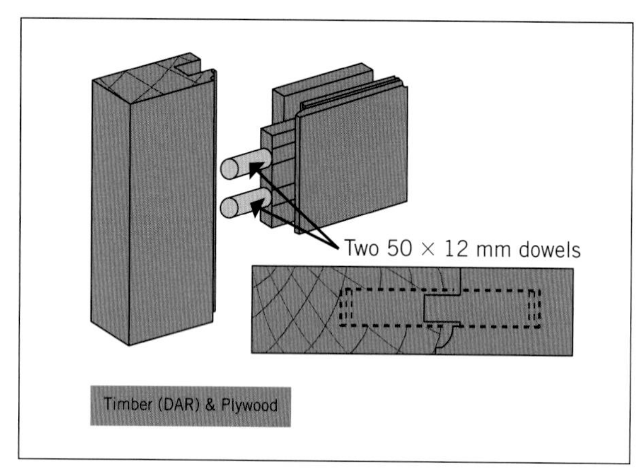

FIGURE 9.33 Scribed joint with dowels

Long and short shoulder mortise and tenon joints are used in the construction of framed and panelled doors, and framed and ledged and braced doors (see Figure 9.34).

Haunched joints are used at the top and bottom rails to reduce twisting, and bare-faced joints and tenon joints are for middle and bottom rails in ledged and braced doors (see Figure 9.34). The stiles and rails can be rebated to take lining boards or panelling (as is the case in framed and panelled doors). Inlay or **bolection** mould can be used to hold the panels in place in the rebate (see Figure 9.35).

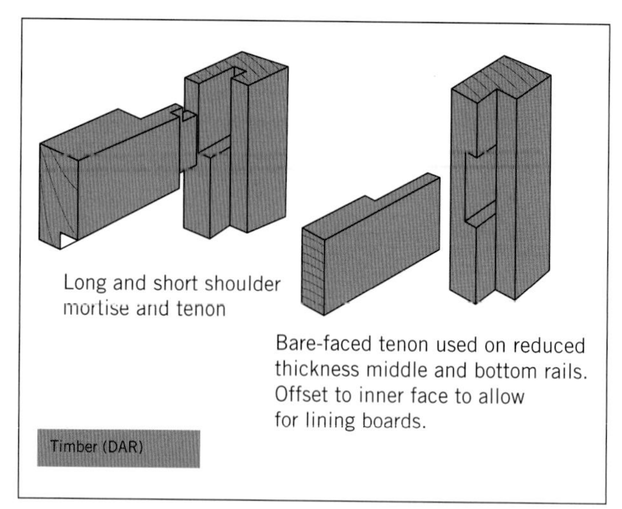

FIGURE 9.34 Long and short shoulder mortise and tenons

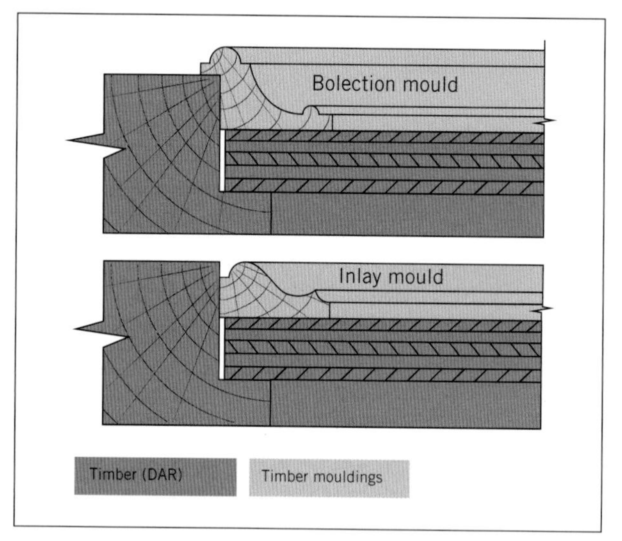

FIGURE 9.35 Bolection and inlay moulds

Panel joints for doors

When selecting timber for panels, be sure to choose timber that is straight and free of defects. For the best results, use timber that is quarter sawn. Back-sawn boards tend to cup significantly and shrink in width. Remember, always wear the appropriate personal protection equipment (PPE) when selecting and machining timber.

Most solid timber panels require the edges of the boards to be clamped together to achieve a wide panel. Figure 9.36 shows typical joints used in this process.

Panel sub-assemblies

Panel sub-assemblies from solid timber have to be made, machined and glued up, and left overnight to dry, which can affect production scheduling.

When machining material for panels, try to match the grain of the timbers and dress the face sides and edges only, as after gluing they can be put through a thicknesser to reduce them to the right thickness and keep the surfaces even. Use face marks to keep the faces right way up and an arrowhead to keep the boards set out as a panel (see Figure 9.37). If you are making more than one panel, the boards can become mixed up, and the arrowhead will tell you which boards go next to each other. Have clamps on the bottom and over the top to prevent the panels springing up (see Figure 9.38).

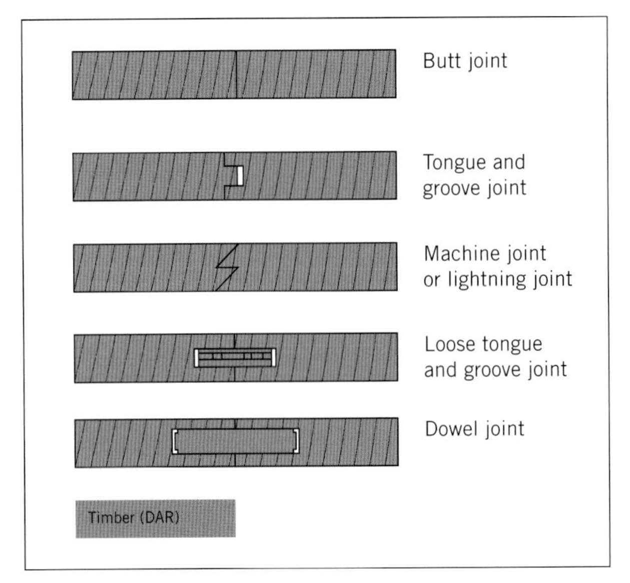

FIGURE 9.36 Widening joints

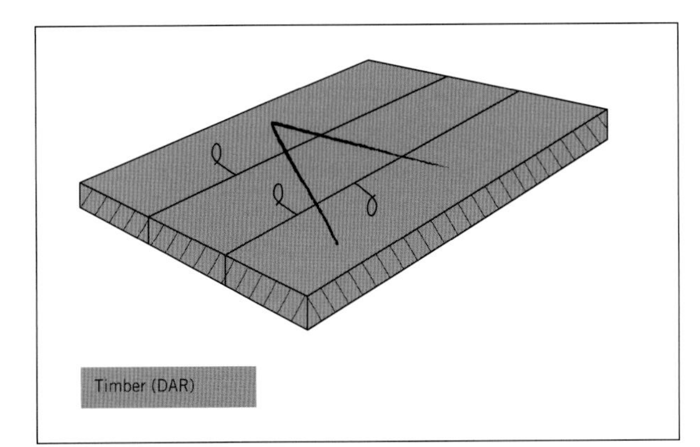

FIGURE 9.37 Face marks and arrowhead

Source: Greg Cheetham

FIGURE 9.38 Clamping panels

Clamp cleats across the face to keep the faces aligned if the boards are butt jointed.

In a mass production joinery shop a multi-staged pneumatic or hydraulic panel press is used (see Figure 9.39).

Door or window frame stile to head joints

The simplest method of door jamb or window frame construction is where the rebated section of the head is

Source: Weldline Engineering Pty Ltd

FIGURE 9.39 Weldline clamping system

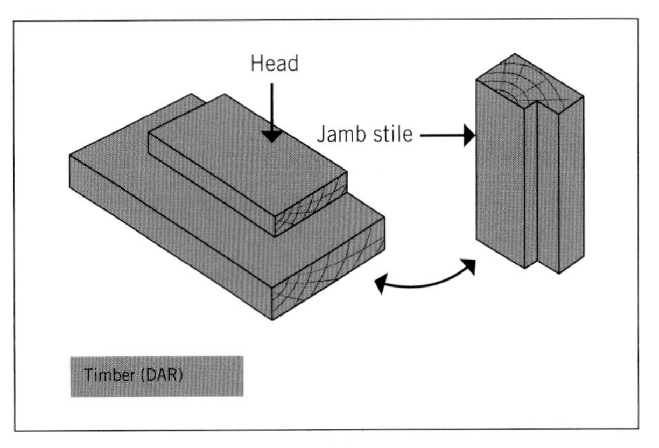

FIGURE 9.40 Door head to stile joint

flushed off with the face of the rebate and the joint glued, screwed or nail fixed (see Figure 9.40).

Onsite, the waste material can be cleaned out using a compound mitre saw with a depth stop. By hand, use a tenon saw or handsaw, or carefully use a portable power saw, for cutting the shoulders, and use a 25 mm chisel and mallet for the waste.

If being done in a factory where stock is being produced, there are several machine options, depending on the machines available:

- radial arm saw with a trenching head
- double-end tenoner (for larger production runs)
- spindle moulder
- CNC timber milling centre.

For stronger joints, traditional joints such as the bridle joint and mortise and tenon joint are used (see Figures 9.41 and 9.42). Proportions for these joints are basically the same – the minimum thickness of the tenon will equal the minimum thickness of the material; that is, 18 mm.

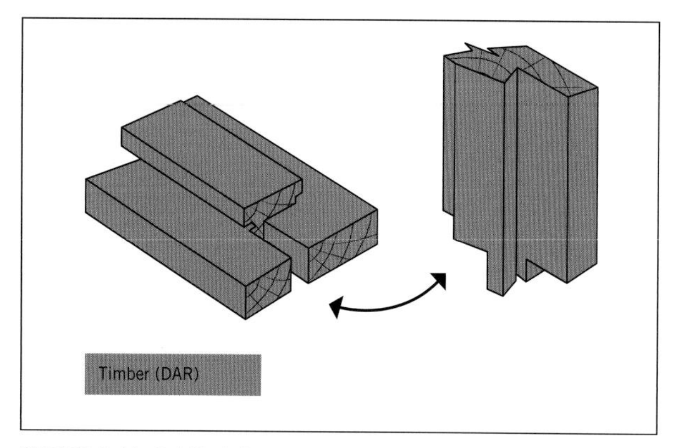

FIGURE 9.41 Bridle joint (double rebated)

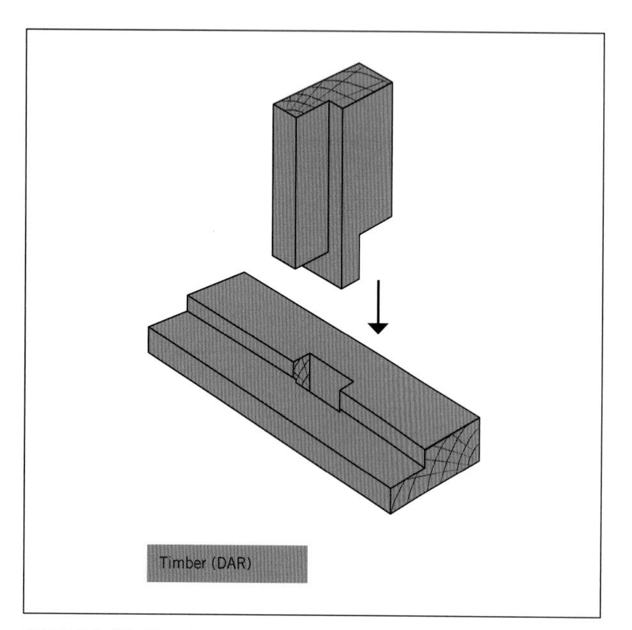

FIGURE 9.42 Mortise and tenon joint (single rebated)

For small production runs, the shoulders of the tenons can be cut using a radial arm saw or sliding table saw and the cheeks can be removed with a bandsaw.

For larger production runs, trenching heads can be set up on the radial saw; a double-end tenoner is a practical alternative.

Sill and threshold joints

Materials for door thresholds are covered in Chapter 3.

Thresholds should be manufactured according to AS 2047 Windows and external glazed doors in buildings.

Thresholds, like window sills, should have a casting groove cut into the bottom of the sill. The **casting groove** prevents the threshold from twisting by reducing the internal stresses in the timber. More than one groove can be made.

A drip groove of no less than 3 mm radius, and set back from the front edge of the material 10 mm, is needed to prevent rainwater entering a building interior underneath the threshold (see Figure 9.43).

It is practical to run a groove in the back edge of the sill to accept the nosing. The nosing is bare-faced tongued into the sill (see Figure 9.43). This allows the nosing to swell and shrink with the weather without causing an unsightly gap that will need filling by the painter. Tenons are cut on the frame stile, and mortise or housings are cut into the sill to accept the tenons. The joint can be screwed or glued and screwed together.

LEARNING TASK 9.3

SAFE WORK METHOD STATEMENTS

Using the search term 'safe work method statement template' on the internet, find and download a template for a safe work method statement (SWMS) or job safety analysis (JSA). There are lots available, but try your state or territory's Work Safe authority first.

Complete the SWMS to construct and assemble the parts required to manufacture the door in Worksheet 1, question 8. In filling out your SWMS, consider these key points:

- handling timber
- joint construction
- setting up and checking machinery
- assembly techniques
- cleaning up.

Types of windows

Windows are any glazed opening in a building that transfers light from outside into a room or space. Technically, skylights, integrated door and window modules, and glass blocks can be included in this definition.

Window sashes and frames are made from a range of materials, including solid timber, aluminium and PVC (polyvinyl chloride). We will not address the construction and manufacture of PVC or aluminium windows here because these types of windows are best suited to specialist window manufacturers with significantly different machinery from that used in regular joinery workshops.

The different types of windows are discussed in Chapter 3. However, briefly, the most common forms of timber windows are casement sashes, including:

- double-hung
- awning: hinged or opening from the top edge – opening outwards
- hopper: hinged or opening from the bottom edge – opening inwards
- casement: hinged on the sash stiles (left or right hand)
- sliding sash
- louvres.

Note that all windows can be manufactured in frames accommodating both windows and doors.

AS 2047 Windows and external glazed doors in buildings

AS 2047 Windows and external glazed doors in buildings is specifically designed to provide minimum design and manufacturing standards to companies mass-producing windows with the following objective:

> … to provide window designers and manufacturers with a generic window code, setting out the performance requirements and specifications in the design and manufacture of all windows regardless of materials.

AS 2047 does not specifically cover architecturally designed windows, windows constructed onsite, skylights and roof lights (except those vertically mounted) or heritage windows. There are other limitations to the Standard as well, so refer to the Standard whenever in doubt.

As a general rule, however, it is good trade practice to base construction methods on the minimum standards outlined in AS 2047 to ensure that performance and durability of the product is maintained.

As we are specifically focusing on the manufacture of timber windows, section manufacturers and small joinery companies should use Appendix G of AS 2047 as a guide to the minimum standards for timber profiles, sealant devices and hardware requirements in designing and manufacturing windows.

Heritage windows

The Australian Standards allow for the construction of heritage windows that exactly match existing products as long as glazing is carried out according to AS 1288 Glass in buildings – Selection and installation. These windows can then be marked as 'Heritage'. Otherwise, the windows can still be produced but they cannot be marked as 'Heritage' or 'meeting Australian Standards'. This is often the case with heritage products.

Sill and threshold sections

Windows are slightly different from doors in that there are standard profiles used in industry for all window products. Many timber suppliers carry stock size profiles, which are available to smaller joinery shops to allow construction of regular timber windows. For details on profiles, refer to AS 2047 Windows and external glazed doors in buildings.

See Figures 9.43 to 9.48 for a range of the stock components used in window construction. Note: Some details are not shown and you will need to consult

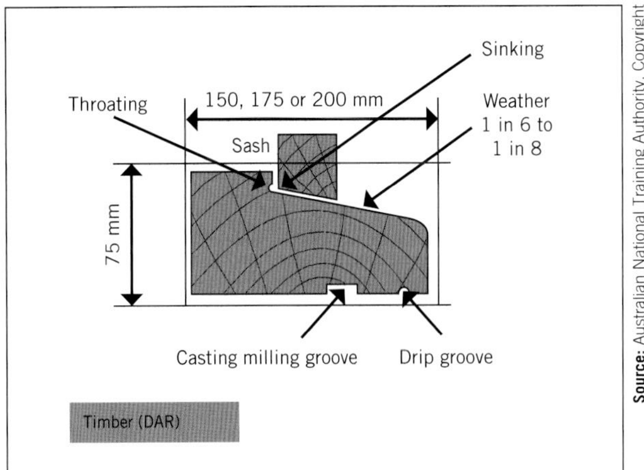

FIGURE 9.43 Casement window sill section

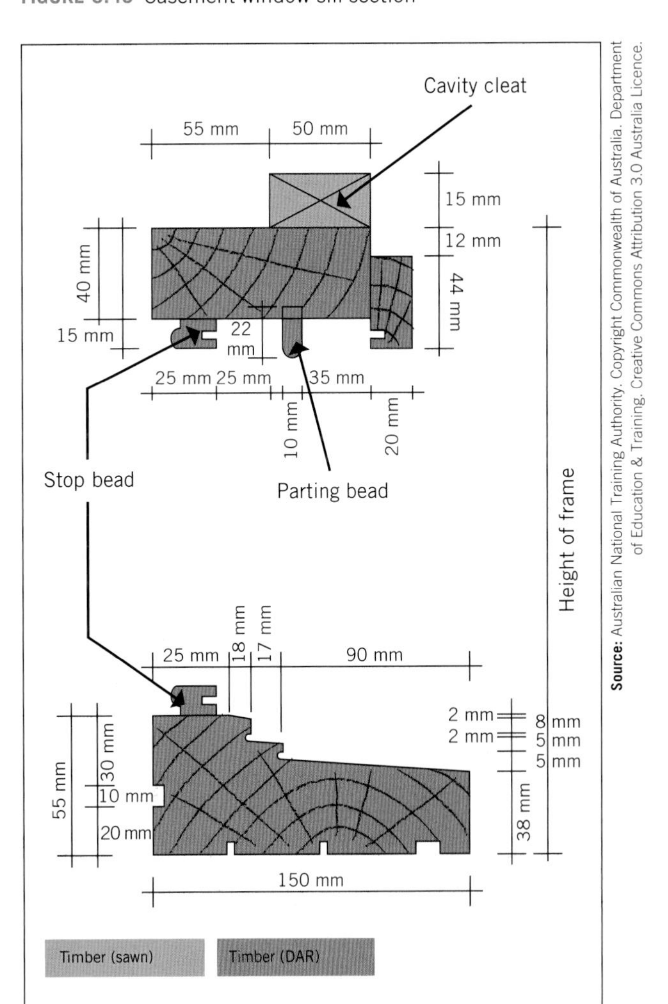

FIGURE 9.44 Double-hung window sections – spiral spring balance

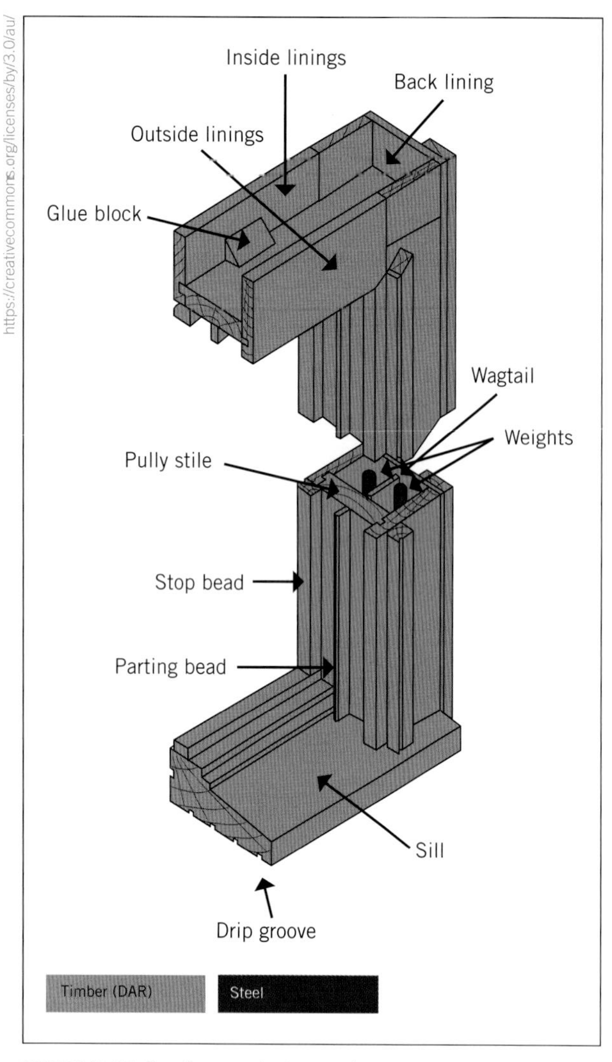

FIGURE 9.45 Box frame window parts

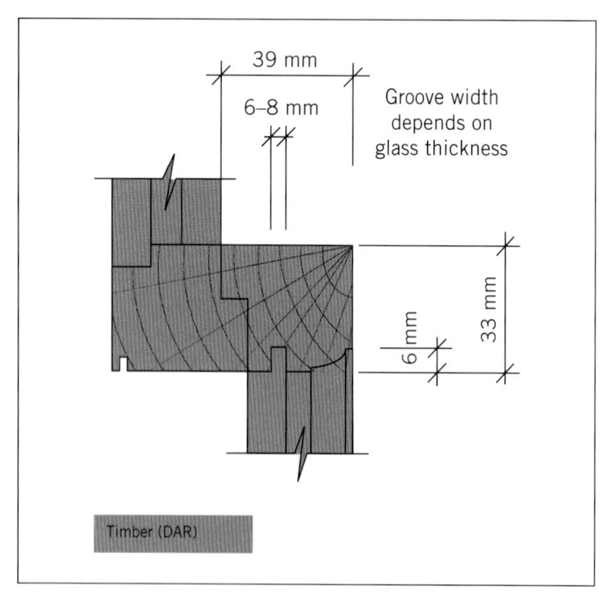

FIGURE 9.46 Meeting rail details

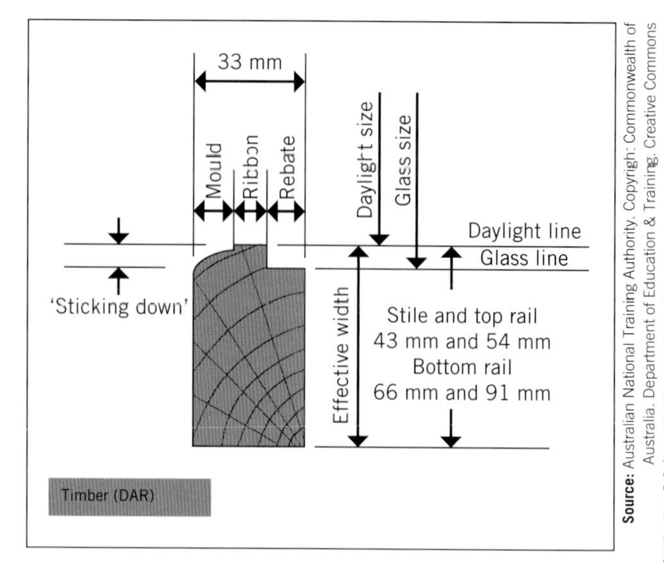

FIGURE 9.47 Sash stock sizes

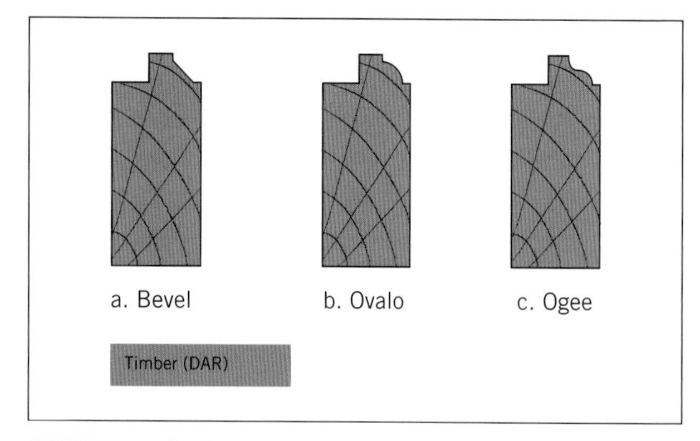

FIGURE 9.48 Sash stock mouldings

AS 2047 for further specific machining details in regards to weather strips and sealant devices.

Appendix G in AS 2047 Windows and external glazed doors in buildings provides more guidance on window sections and sizes.

Special note: You may have to complete some sections with grooves and other mouldings for special fittings such as spiral balances and weather seals. Australian Standards and manufacturers' brochures need to be consulted in order to meet these requirements.

Following is some special sill terminology that is useful to know:

- **Throating:** The undercutting in a rebate on a sill, which is done to help resist water penetration (see Figure 9.43).
- **Sinking:** The angle referred to on the bottom of the sash that matches the weathered area.
- **Weathered:** The term used to describe the sloping surface of the sill.
- **Transom:** The intermediate horizontal frame member.
- **Casting groove, or milling groove:** Single or multiple grooves that run in the bottom of the sill to prevent warping or twisting.

Sash frame materials

Typical sash stock sizes are given in Table 9.1 (see also Figures 9.46 and 9.47).

Traditional mouldings

Figure 9.48 shows typical sash stock mouldings.
Following is some special sash terminology:

- **Parting bead:** Timber bead run in stiles and mullions, used to separate running channels for top and bottom sashes.
- **Staff bead (or stop bead, in some states):** Used on the inside of a double-hung window frame to form a rebate for the bottom sash to run in; a doorstop bead is also referred to as a 'plant on stop'.
- **Wagtail (or feather):** Ply or solid timber strip used in box frame construction to separate channels for weights; prevents weights from 'clanging' together.

Some manufacturers also use plastic-coated and composite timber materials for which a lesser grade of timber or laminated timber can be used. These types of material can have greater structural strength and dimensional stability compared to other materials. Typically, the core timber material in plastic-coated windows is Douglas fir or Oregon.

PVC windows

PVC windows are not common in Australia; however, they do have some advantages over timber windows, particularly because double- and even triple-glazed units can be inserted in the frame. These units are filled with an alternative gas that reduces the amount of UV radiation entering a building, and therefore makes the windows significantly more energy efficient. For more information about PVC windows, visit some of the websites listed at the end of this chapter.

LEARNING TASK 9.4 SASH OPERATING SYSTEMS

Conduct an internet search for videos about modern joinery window construction and identify the range of sash operating systems available for the following window types:

- double-hung sash
- spiral balance
- casement sash (side hung)
- hopper and awning sash (top or bottom hung)
- sliding sash
- sash cords and weights
- propitiatory window operating systems.

Preparing window components

Sash joints

Sash joints will either be mortise and tenon or bridle joints. If bridle joints are used, they should be secured with a fastener such as a stainless steel nail to prevent the sash from 'collapsing' (see AS 2047 Windows and external glazed doors in buildings – Selection and installation for a definition of 'collapsing') should the adhesive fail. Mortise and tenon joints should not collapse even if the adhesive fails.

Specialised mortise and tenon joints are used in the construction of sash joints. The joint, instead of having a haunched mortise and tenon (see Figure 9.49), has what is termed a 'franked' mortise and tenon. This is where the ribbon section of the stile is squared off to form a spur and the end of the tenon shoulder is cut away to form a '**franking**' (see Figure 9.50). Making the joint in this way prevents the rails from twisting.

Source: Ed Hawkins

FIGURE 9.49 Haunched mortise and tenon

The shoulders of the tenon may be fitted to the stile in several different ways, depending upon the production system. For example:

- The shoulder may be scribed to fit neatly to the moulded edge of the stile (see Figure 9.50).
- The moulded edge of the stile can be flattened off to align with the glazing rebate. A mason's mitre is then created on the end of the moulded section on both the stile and the rail. The disadvantage of this method is that as the timber shrinks the mitre opens up. Alternatively, the moulded section can be cut to match the moulded angle, or shape, on the rail.

Waste material from the haunch area can be used to cut the wedges as the thickness will match the tenon.

TABLE 9.1 Typical sash stock sizes

	Height	Thickness	Rebate or glazing groove
Bottom rail	66 and 91 mm	33 mm	12 × 6 mm
Stiles/top rails	43 and 54 mm	33 mm	12 × 6 mm
Meeting rail	32 mm	39 mm	Groove depth 6 mm. Width depends upon glass thickness (6 or 8 mm)

Source: Ed Hawkins

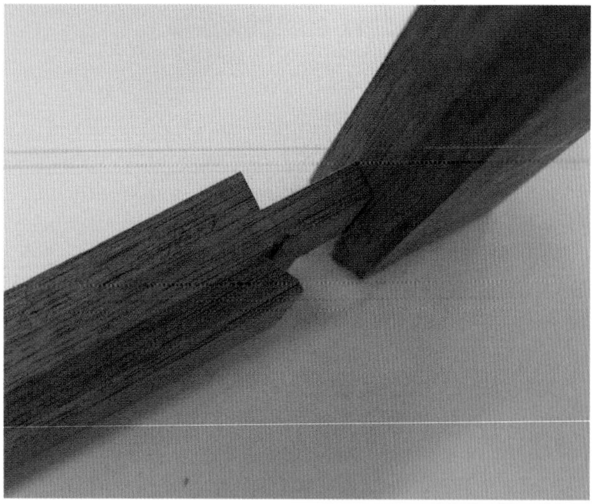

FIGURE 9.50 Scribed shoulder mortise and tenon

Meeting rail joints

Meeting rails are used in double-hung sash window construction. These components allow the top and bottom sashes to close neatly and fill the gap between the two sashes created by the parting bead.

A range of joint configurations has been developed over the years; however, the splayed joint is the most enduring and provides a deterrent to break-in by having the angle away from the outside, making it difficult to access the window lock mechanism via the joint (see Figure 9.51). Additional rebates can be added (see the section 'Sill and threshold joints' earlier in this chapter).

Source: Australian National Training Authority. Copyright Commonwealth of Australia. Department of Education & Training. Creative Commons Attribution 3.0 Australia Licence.

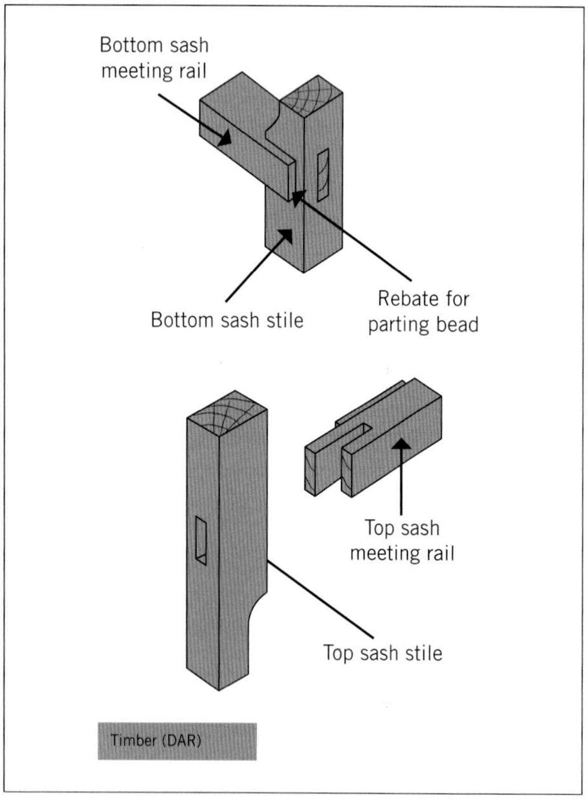

FIGURE 9.51 Meeting rail joints

Care needs to be taken when preparing these components to ensure the external section will not be broken off during manufacture and assembly of the sash.

Joints in curved head frames

The joints used in curved head frames can cause the corner material to break out if it is not made correctly. A mason's mitre shoulder should be used to reduce the risk of break-out due to the short grain of the timber (see Figure 9.52).

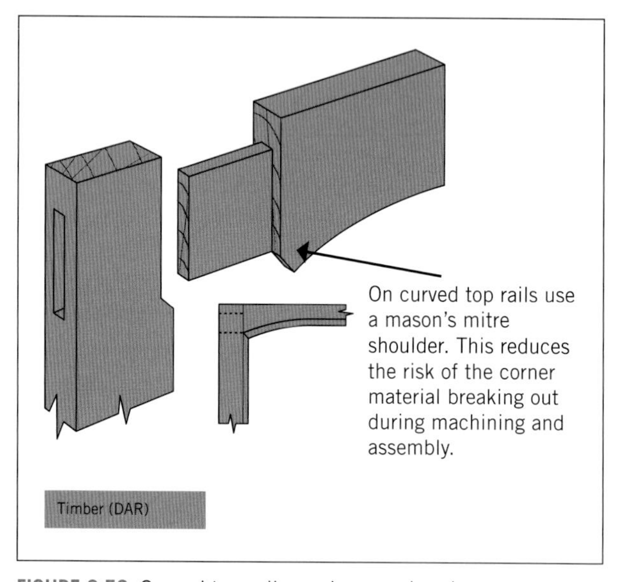

FIGURE 9.52 Curved top rails on doors and sashes

Laminated heads for door and window frames

Curved heads for door and window frames are usually laminated to achieve the right curve. This requires a jig to be made to the correct radius or curve if an elliptical shape is required.

The thickness of the material required to bend is determined by the size of the smallest radius (see Figure 9.53). For a radius less than 200 mm, segmental build-up is the preferred method of producing the curved shape. It is possible to combine the two methods.

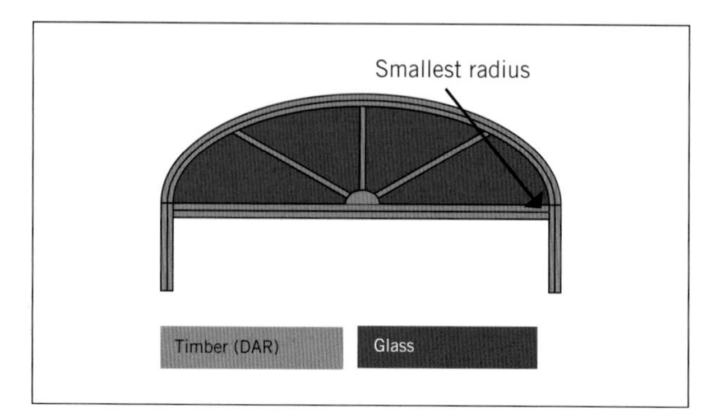

FIGURE 9.53 Smallest radius

Ellipse 2 Ellipse 1

Ellipse 2 = Ellipse 1 + thickness of laminated materials

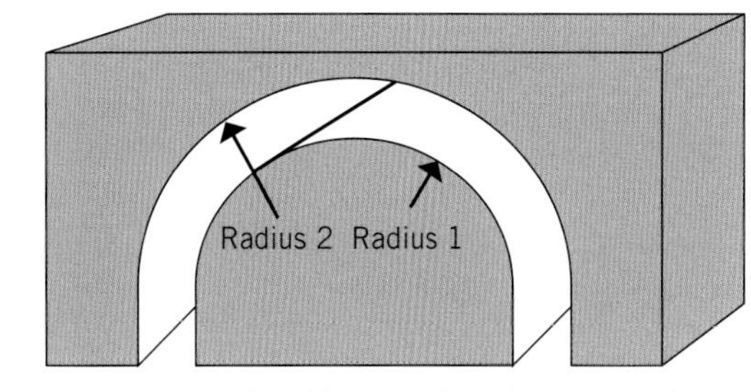

Veneered MDF

FIGURE 9.54 Elliptical and round formers

Source: Shutterstock.com/Bobkeenan Photography; Shutterstock.com/Pta Pta; iStockphoto/ChristinLola

FIGURE 9.55 Glazing bar joints

Laminating timber is sometimes called 'cold bending', and a simple jig can be used for this process. The process can be tricky and slippery, particularly when the layers of timber have been glued up. It is a good idea to make the lengths overhang the jig, and to tie the ends together with cable ties, so the sections stay together but can still slide easily as they are bent around the jig. Not all timbers are suitable for bending.

When constructing **cauls**, the radius used on the inside curve is smaller than the radius used for the external caul by the total thickness of the material being laminated (see Figure 9.54). If the inner and outer cauls are not matched accurately, the laminated shape will come out with gaps between the layers.

A simpler jig involves setting out and machining the internal radius or elliptical shape on a suitable thickness material (several layers may be required). Use a hole saw to make holes big enough to accept the G clamps. Keep the edge of the holes about 25 mm back from the edge of the jig.

Make sure you use packing blocks between the workpiece and the clamps to prevent bruising.

Laminated curved heads are glued and screwed to frames. The curved head is made longer than necessary and the stile is extended beyond the **transom**, and a lap joint is created with the laminated section held between (see Figure 9.56). This counteracts any spring in the laminated head that may be encountered as a result of spring-back after the laminating process.

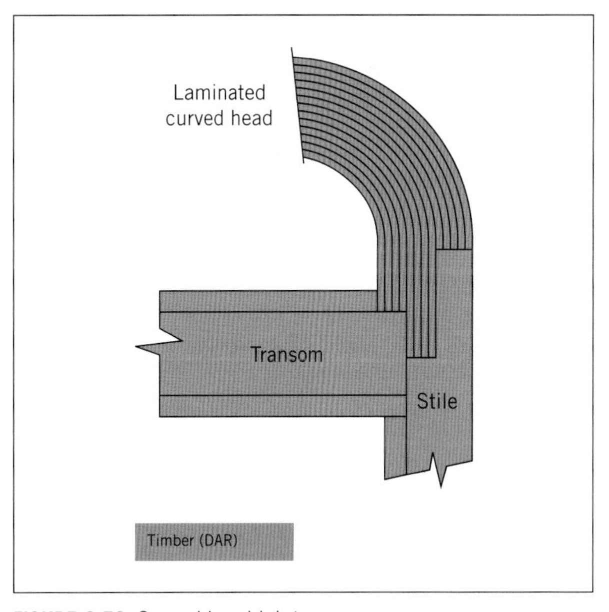

FIGURE 9.56 Curved head joints

For more information on making curved components for windows and doors, obtain a copy of one of the books listed under 'References and further reading' or via the electronic links listed at the end of this chapter.

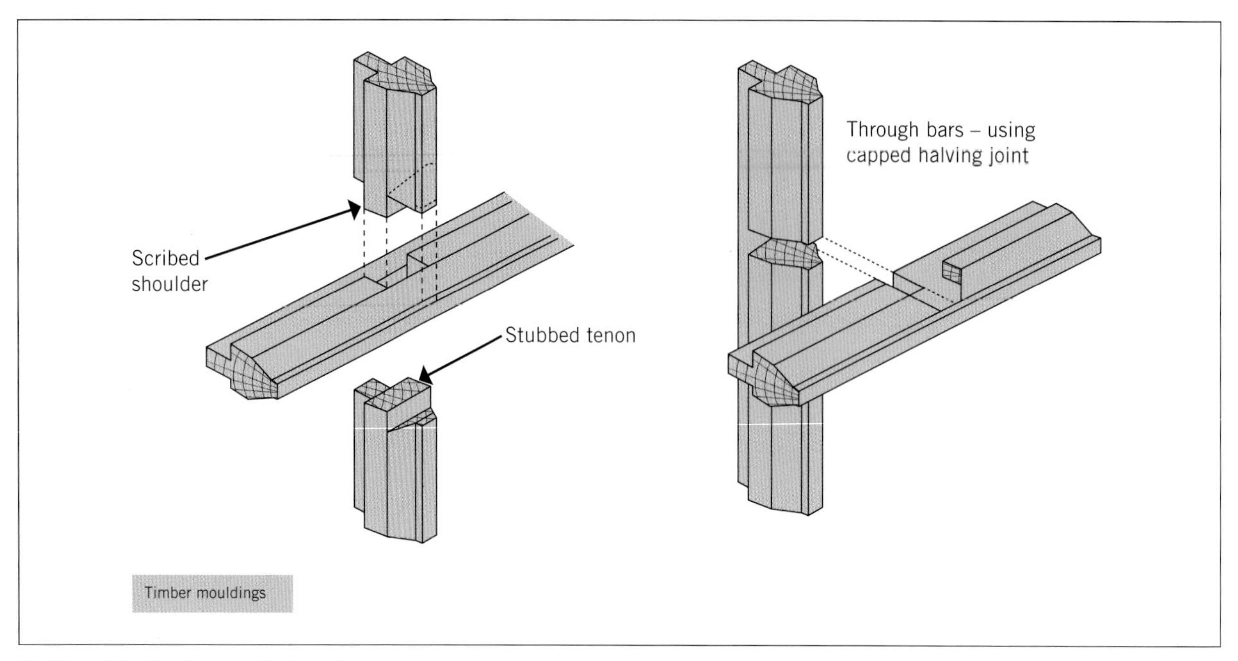

FIGURE 9.57 Glazing bar joints – diagram

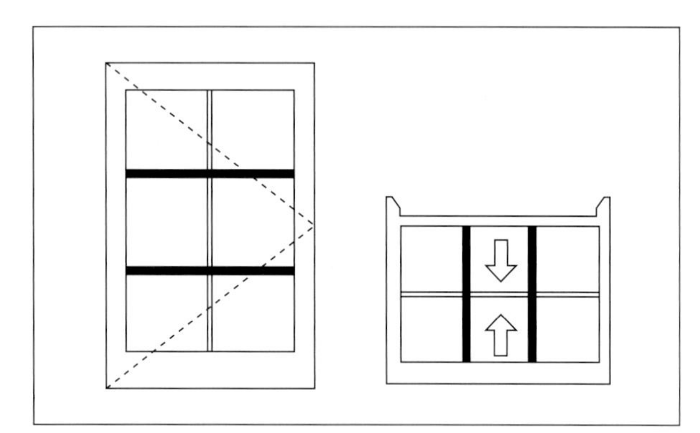

FIGURE 9.58 Glazing bar direction

Glazing bars

Glazing bars are used in both door and window construction. There are a number of methods used to cut and assemble glazing bars.

These bars are joined together to give the appearance that they are all separate short pieces coming together at a common point (see Figure 9.55). In fact, they may have through bars for strength, with short-cut bars mortise and tenoned into them, or they may have through bars in both directions halved over one another (see Figure 9.57).

When using through bars with cut bars, the through bars support the sash in its weakest direction (see Figure 9.58), either going from top to bottom or side to side.

When assembling windows, the glazing bars can be pre-assembled and then the main sash frame can be assembled around the glazing bar. Alternatively, the long bars can be inserted into the stiles or rails, the cross glazing bars then added and the stiles or rails assembled afterwards.

Setting out bespoke doors and windows

Specifications and drawings

The overall design of the window or door you need to manufacture and assemble will be detailed in drawings and specifications. The documents will tell you:

- the dimensions of the component parts
- how they are fixed together/types of joints
- the overall size of product
- the types of materials required
- the position of components
- the finish required
- special conditions.

It is important to follow the details exactly to make sure you produce the right size and quality product for your client; otherwise you risk potential warranty problems.

The critical measurements required when setting out doors are:

- door frame height
- door frame width
- door height (with clearance allowances for floor finish)
- door width.

For windows, the critical measurements are:

- overall window opening size
- overall window frame size
- sash sizes. (**Note:** Sash stiles and heads are from the same material, while it is common for a wider material to be used for the bottom sash rail; see Figure 9.59.)

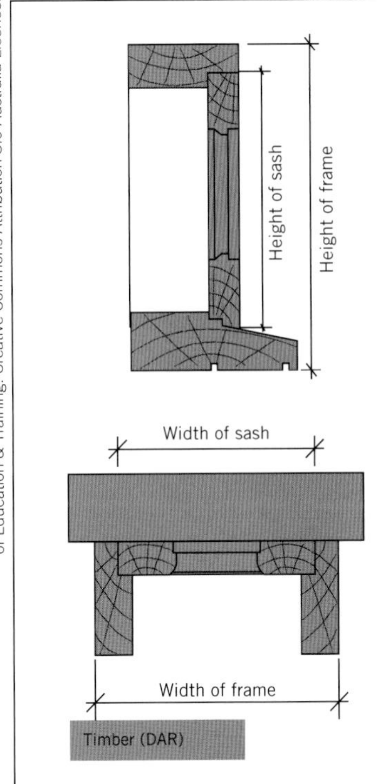

ITEM	DESCRIPTION AND QUANTITY	SKETCH AND CALC.
Frame stiles	105 × 40 Dry dressed (timber type) 2/1.2	
Frame head	105 × 40 Dry dressed (timber type) 1/0.9	
Sill	150 × 60 Dry dressed (timber type) 1/1.2 (rebated and sunk to standard casement detail)	
Sash stiles	Standard ovalo (timber type) 2/1.2	
Top rail	Standard ovalo (timber type) 1/0.9	
Bottom rail	Standard ovalo (timber type) 1/0.9	

FIGURE 9.59 Window sizes and specifications

Special order doors and windows are set out on a rod or set-out board using the plans and specifications. The process is similar for both doors and windows. The tools and materials you need are:

- a set-out board long enough to take the longest measurements plus 200 mm on which to run the combination square
- a combination square and suitable sharp pencil
- a 5 mm tape measure or folding rule
- plans and specifications.

You need to have:

- an understanding of the relevant Australian Standards (see list in the 'References and further reading' section at the end of this chapter)
- an understanding of machinery and machine operations
- materials knowledge
- knowledge of door and window construction methods, joints and tolerances.

Before setting out and making any doors or windows, it is a good idea to discuss with the builder the finished sizes for the windows and agreed allowances so that everything fits properly when the windows and doors are delivered to the site and installed.

Your teacher will work with you in developing appropriate set-out rods as part of the practical exercises associated with this chapter.

More information on technical drawing and shop drawings is available from Wikipedia at: http://en.wikipedia.org/wiki/Technical_drawing.

Built-in cabinet components

AS/NZS 4386.1 – Cabinetry in the built-in environment – Commercial and domestic provides detail on requirements for domestic and commercial kitchen, bathroom and other built-in cabinets. The four main points are:

1. The melamine faced materials should be used for internal basic **carcass** construction. This doesn't include doors and benchtops of plinth boards.
2. Solid timber or veneered board products should have a finish applied that conforms to AFRDI 108.94. There are other AFDR certifications that may apply. These are listed in the 'References and further reading' section at the end of this chapter.
3. All exposed edges (after assembly) must have edge tape applied or be sealed with lacquer or paint.
4. The construction materials must be moisture resistant; for example, HMR (high moisture resistance) particleboard or MR (moisture resistant) MDF (medium density fibreboard).

Making and installing cabinets that do not meet minimum construction standards requirements could seriously damage your reputation and cost you time and money to repair or replace.

The main materials used to fabricate domestic kitchen cabinets are 16 or 18 mm melamine faced or laminated HMR (high moisture resistance) particleboard or medium density fibreboard MR. Alternatives include the following:

- plywood – 'A' bond for wet areas or 'C' bond for other areas
- solid timber – vinyl-wrapped timber is sometimes used for top rails and fixing rails
- hardboard products – for specialised applications only.

In the past, cabinets were made with plywood or particleboard ends, with front and back rails, additional fixing rails and face frames. Plinths were integrated into the carcass (see Figure 9.60). Face frames are used today to give that traditional feel. Bottom rails are often used in wall cabinets to provide fixing top and bottom.

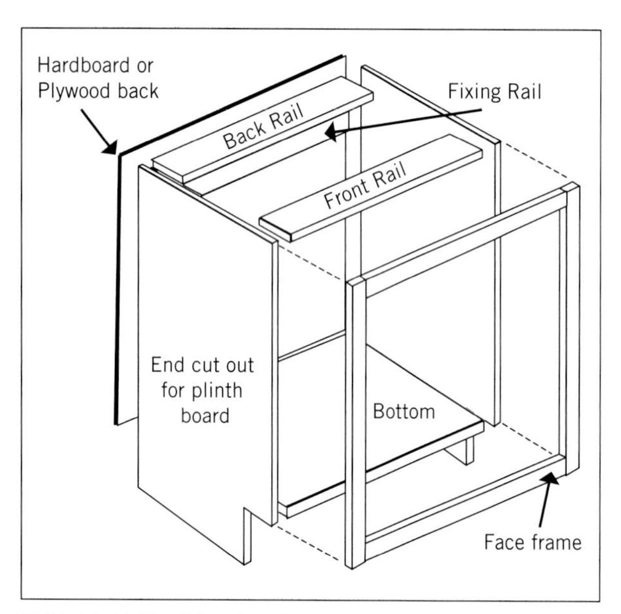

FIGURE 9.60 Traditional cabinet

Machining these components takes a little more time, but some material savings can be made to offset the cost additional machining. Modern construction uses solid panel construction for most carcass members.

The basic components of a cabinet are shown in Figure 9.61.

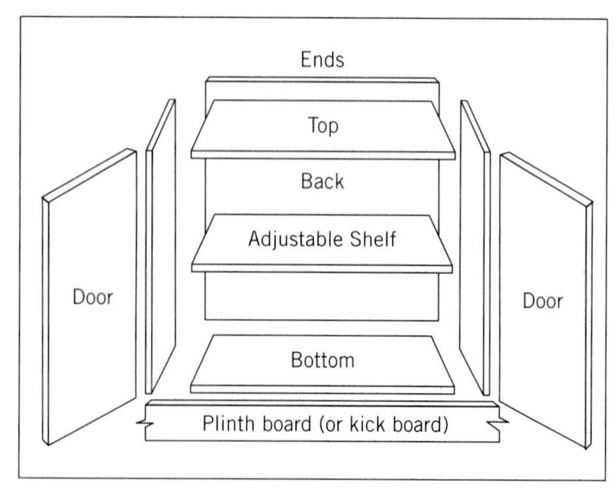

FIGURE 9.61 Basic cabinet components

The carcass is generally made from all the same material, and less edging is required. Plastic height-adjustable legs make installation easier, especially on uneven floors (see Figure 9.62). The legs can be disconnected from the base plate for transport. The base plates protect the bottom of the cabinets in the factory and act as a skid plate.

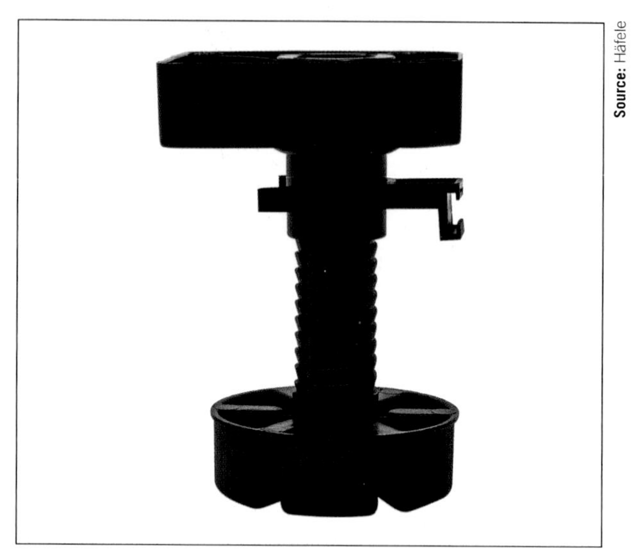

Source: Häfele

FIGURE 9.62 Häfele height-adjustable leg

Solid backs mean fixing positions are not limited and provide better fixing for **GPOs** (general power outlets) inside cabinets.

Square and rectangular flat panel components can be cut with a number of machines.

Small shops may use a sliding table saw (see Figure 9.88 later in the chapter).

Flatbed CNC machines (see Figure 9.101 later in the chapter) are also used today. They have an advantage of using spiral cut router bits (see Figure 9.63) to cut materials, leaving a crisp clean edge. As these machines are programmed and can change their own cutters, it is important to make sure the right cutter selection is made for the job.

Source: Shutterstock.com/BartlomiejMagierowski

FIGURE 9.63 Spiral router cutter

There are many other types of cutters available to complete a range of specialised wood panel-based milling operations. See the list of manufacturers at the end of this chapter.

Panel saws (see Figure 9.91 later in the chapter) have two blades. These must be sharp and perfectly aligned to produce an edge suitable to take edge treatments. (Watch the following video on the correct adjustment of scoring saw blades: https://www.youtube.com/watch?v=BBPnUObFPus.)

When producing flat panel components, it is important to check:

1. dimensions for parallel
2. diagonal dimensions for square
3. edges for chips and misaligned scribe cuts
4. special machining details such as grooves and rebates for fit.

Joining systems used in carcass construction may include:

1. butt joint and screw fixing – the use of pilot holes and clearance holes is recommended, especially when using CNC machining; this makes joint alignment more accurate and reduces splitting, especially in the edge of flat panel products
2. 'biscuit' joints – made using a portable 'biscuit' jointer power tools and require glue and clamping
3. dowelled joints – made with common machinery and require glue and clamping
4. knock-down fittings (cam and dowelled joint) – can be used in combination with dowels or biscuits. They provide a jointing mechanism that is quick to assemble and acts as a clamp. Unglued or dry joints can be disassembled and reassembled multiple times. There are many different types – see hardware provider catalogues for a range of 'KD' fittings.

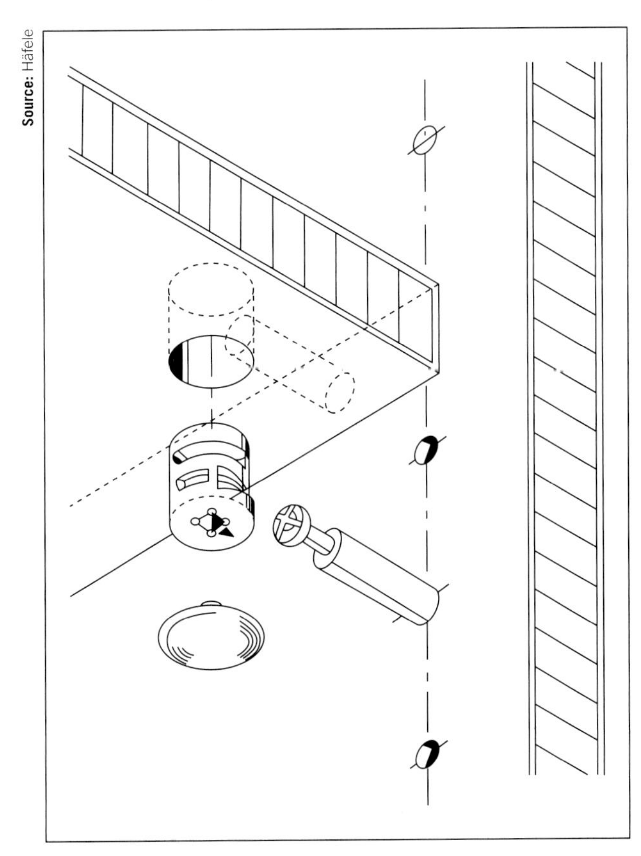

Source: Häfele

FIGURE 9.64 Typical knock-down fitting

System 32 cabinet hardware compatibility

The System 32 component system is a cabinetmaking model developed for hardware, machine and cabinet manufacturers to standardise both component dimensions and production processes. Key design principles are based on 32 mm row hole boring, typically using 5 mm and 8 mm holes. For the fixing standard cup type hinges plates and common drawer systems, the front row of holes are set 37 mm back from the front edge of the panel.

Advantages include the following:

1. The setting up of single row drilling machines is simplified.
2. If the distance of the rear row of holes to the rear edge is 37 mm, no retooling of the machine is required (panels can be both left and right handed).
3. If the rear row of holes is 37 mm from the rear edge on wall cabinets, special fixing hardware can be accommodated.
4. The distances between holes are in multiples of 32 mm, which facilitates drawer systems installation.

Some hardware manufacturers sell handheld jigs and multi-spindle boring machines suitable for smaller joinery shops where expensive CNC machinery is not available (see the link for the BlueMax Mini Modular Plus in the 'Video resources' section at the end of the chapter).

Edge treatments for flat panel components

Flat panel manufactured boards require edge treatment to:

1. conceal the unfinished edges
2. provide a decorative edge
3. protect the edge from damage or moisture ingress
4. strengthen the edge to take edge-fixed hardware.

Pre-glued and unglued edge tapes are melamine-, PVC- or ABS-based, and are applied using either automatic or hand-fed banding machines.

Pre-glued tapes have a hot-melt adhesive already applied, usually by a hand-fed edge bander, while unglued tapes have hot-melt adhesive applied by an automatic edge bander, via heated tanks. (See the 'Edge banders' section later in this chapter for more on these machines.)

Common edging materials include the following:

- ABS (acrylinitil butedien styrol) is the most common edging material used today. It is available in a range of widths (with 21 mm the most common,

up to 38 mm) and between 1 and 2 mm thick. ABS is available in a wide range of solid colours, woodgrains and patterns. Its thickness provides good protection against bumps and knocks.
- Melamine edge is 0.4 to 0.6 mm thick. Mainly produced to accurately match melamine faced pre-finished panels, it is brittle and can chip easily.
- PVC edge is between 0.5 and 1.0 mm thick and considered more durable than melamine edge. It is mostly used for carcasses as it's only available in a limited range of colours.
- Plastic laminate edge can be applied by hand or edge bander. It provides a durable edge, but shows a black line. It also may be post-formed to provide a curved edge.
- Timber veneered edge is usually matched to veneered boards. It's often thin (between 0.4 mm and 0.7 mm) so is not resistant to hard knocks. Pre-glued timber veneered edging can be applied to shapes with a hot iron or via an edge banding machine. It can be fleece backed for the postforming of rounded edges.
- Solid timber edge can be applied by hand or appropriate edge bander. It provides good protection and can be applied by butt joint or a variety of other methods, depending on how it is to be used.
- Aluminium edge is available in a range of profiles. It can be used for edge treatments and handles, and provide a hard edge when needed.

In calculating component sizes you will need to adjust to account for the edging materials you will be using.

Figure 9.65 shows different methods of applying edge treatment to manufactured boards.

Setting up machinery

Setting up static machines requires a high degree of accuracy and care in order to produce high-quality components. This chapter is general in nature and does not provide sufficient detailed information on the set-up and operation of individual machines. Significant experience needs to be gained under the direct supervision of a trained and experienced machine operator.

The important aspects to setting up a machine include:
- selecting the right machine
- checking that cutters are sharp and in good condition
- ensuring the cutters are correctly installed in the machine
- checking that all guards are in place to prevent injury
- ensuring that operators are fully trained in the machine's techniques.

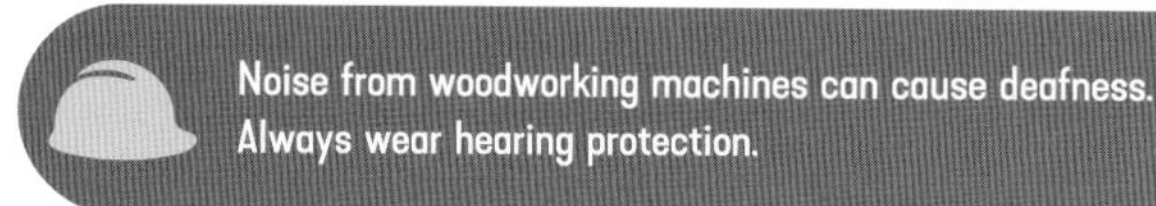

The materials for any product move in planned stages from a raw state to a finished product state. Likewise, components for doors and windows are manufactured in a sequence of planned stages.

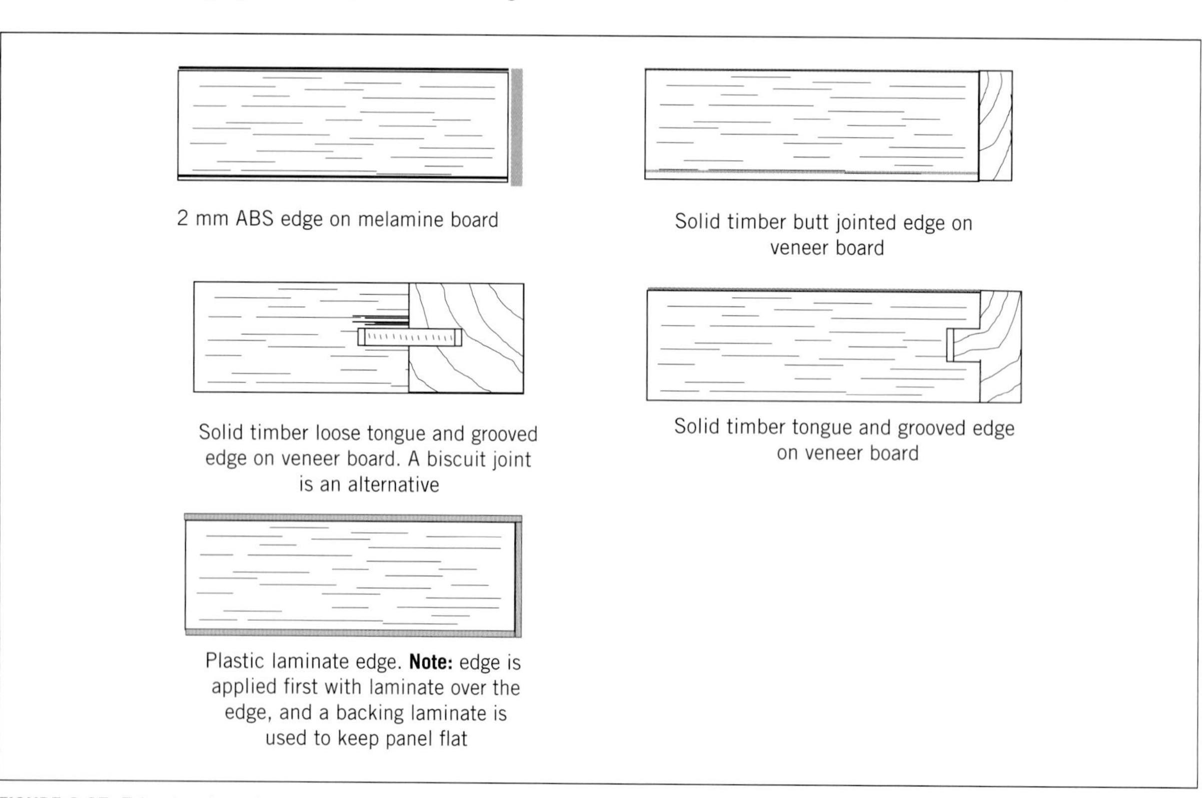

FIGURE 9.65 Edge treatments

The planned stages are:

1 **Rough sizing:** Material is rough-cut close to finished size using a ripsaw (table saw) and docking saw (radial arm or compound mitre saw). Short lengths may be accumulated in one length to save on manual handling and reduce machining problems associated with short stock.
2 **Dressing of raw stock:** Timber is passed over a jointer to dress a face side and edge (see Figure 9.66), ensuring the faces are true straight and square before adding the face marks and finally passing through a thicknesser to bring the material to a finished size.

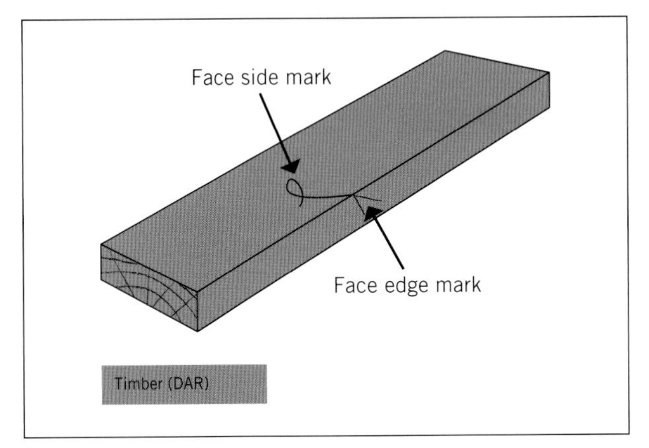

FIGURE 9.66 Face side and face edge marks

3 **Final finishing into usable components:** After thicknessing to DAR state, the component parts are finally docked to length or machined to shape; that is, moulded or curved and joints set out and machined.

Special components

Doors and windows often have curved components (see Figures 9.67 and 9.68). Curves can be cut from solid timber, laminated from strips or built up like brickwork. It depends upon the components as to the most suitable method used. For example, the bull's eye vent in Figure 9.69 has a sash frame cut from several sections of solid timber, while the frame is built up to its full width using segmental build-up (see Figures 9.70 and 9.71) or brickwork.

The rough shapes of each block are cut from a solid using a bandsaw, and sub-assemblies are made before final finishing in spindle moulders and/or drum and bobbin sanders. A template of the shape required is traced on to the rough stock and then cut on the bandsaw on the waste side of the line (see Figure 9.72).

Source: Shutterstock.com/littleny

FIGURE 9.67 Curved laminated head

Typically the finger joint is used to join the timbers together on end (see Figure 9.70). These joints are prepared using a special jig on a spindle moulder or, in a modern factory, a CNC machine.

Machining the end grain poses significant issues of timber chipping out and also safety issues, so the jig should incorporate toggle clamps and handles to keep the operator's fingers away from cutters, and backing blocks behind the end of the timber being machined to prevent timber chip-out.

As components are cut, mark them with a part number related to the cutting list number and job number for easy identification. Use chalk to write the IDs on the timber – it is easy to wipe off with a damp cloth and is clearly visible.

Carefully stack the components on a pallet or trolley so that they cannot fall. Use spacers or alternate the stacking direction for stability (see Figure 9.73). When moving the material around the various stations, make sure the cutting lists and job plans, specifications and workshop drawings remain with the materials.

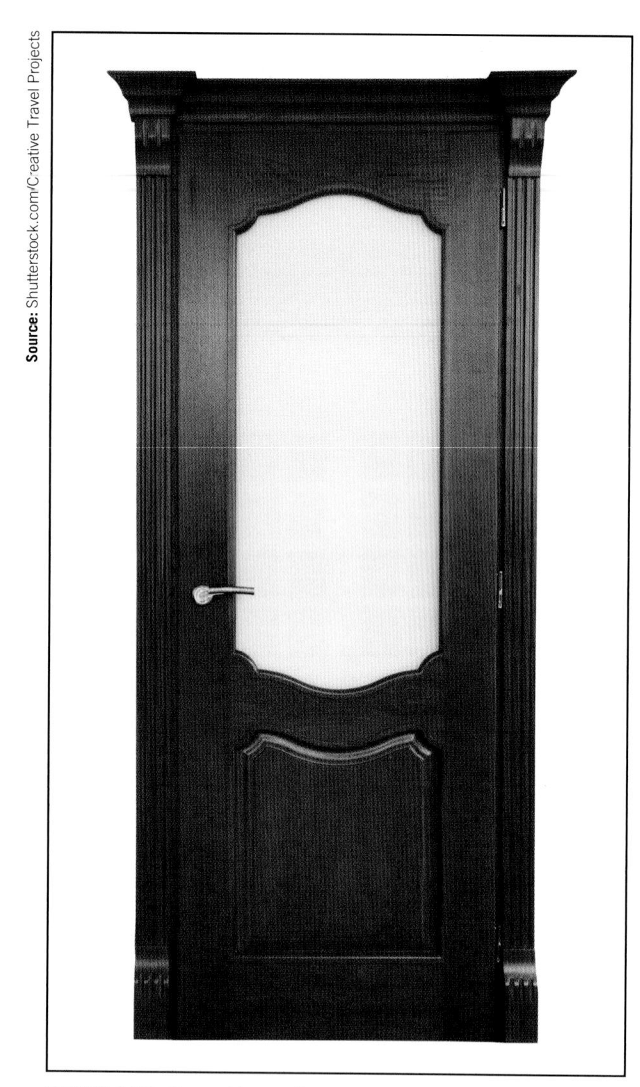
Source: Shutterstock.com/Creative Travel Projects

FIGURE 9.68 Curved top rail

Source: iStockphoto/IDFK303

FIGURE 9.69 Bull's eye vent

Source: Ed Hawkins

FIGURE 9.70 Segmental build-up in frames

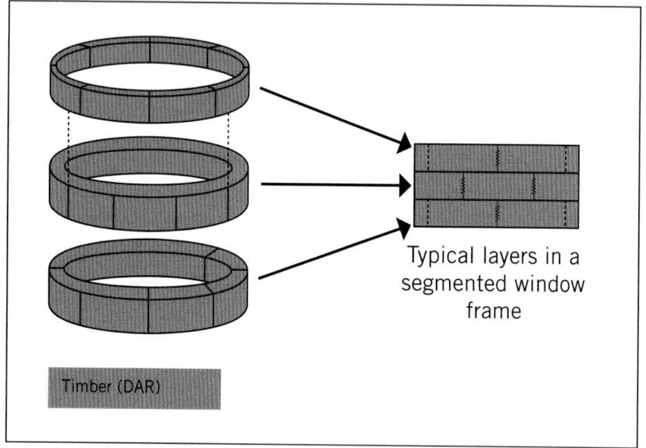

FIGURE 9.71 Segmental build-up

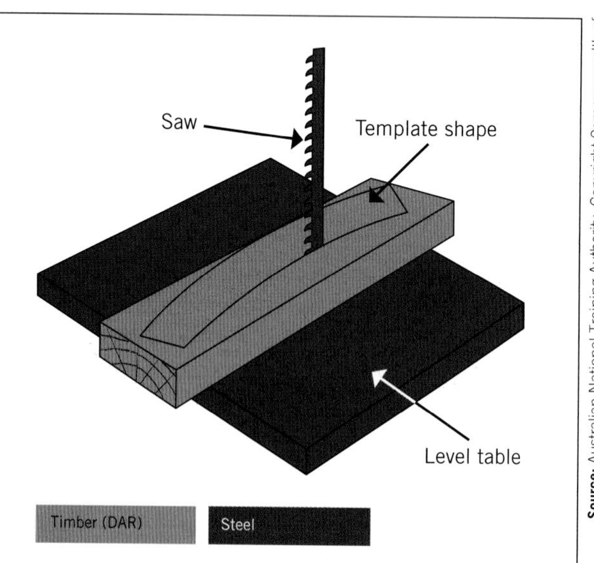

FIGURE 9.72 Rough cutting on the bandsaw

Source: Australian National Training Authority. Copyright Commonwealth of Australia. Department of Education & Training. Creative Commons Attribution 3.0 Australia Licence.

FIGURE 9.73 Timber stacked ready for machining

LEARNING TASK 9.5 MACHINE SEQUENCING

Correctly write out a sequence of machining for manufacturing the window sash bottom rail shown in the diagram below.

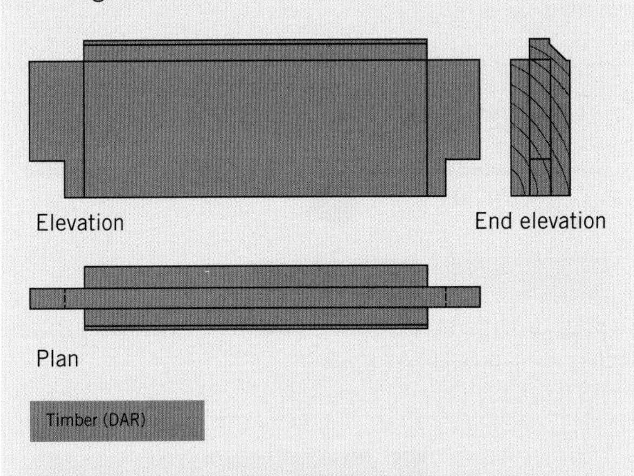

Machines to choose from include the bandsaw, multi-header, spindle moulder, planer (thicknesser), jointer (buzzer), router, double end tenoner and router.

You need to explain which machine you would choose to do each process and why you chose that machine.

Note: There could be more than one correct sequence depending on which machine you choose.

Machines

Machine safety

The same machines are often used to manufacture doors and windows. The safe use of machines is critical to the production of good-quality doors and windows. All machines must have appropriate safety signage located at the machine, including pictorial signs and safe operating procedure (SOP) manuals. A maintenance manual should also be located close by. A machine maintenance plan should be kept for each machine being used. Regular maintenance reduces disruption due to unexpected breakdowns, saving both time and money. At a minimum, a basic maintenance plan or record should contain the following features:

- a description of the regular task to be carried out
- how often the work is to be done; for example, monthly, fortnightly or annually
- who is responsible for performing the maintenance with contact details; for example, in-house or contractor
- the type of maintenance work; for example, grease and oil bearings, check and tension chains, belts and pulleys, air regulators and oilers
- estimated time to perform the tasks
- special tools, materials and equipment required to perform the tasks.

In Australia, the guarding of machinery is critical and is covered by both state and national WHS/OHS legislation.

Australian (and New Zealand) Standards relate directly to the design, manufacture, supply, employers and users of machinery to minimise the risks of working with, or near, machinery.

The machines you use should be installed and set up in such a way as to minimise the risk of harm to you or the materials. This includes the following:

- design, installation and use of proper and effective guards and interlocking devices including acoustic shields and covers
- stop/start and emergency stop switches
- location of the machine to ensure safe and adequate access to perform tasks
- installation of adequate and effective dust extraction (and exhaust extraction where required)
- properly signed to ensure safe operation (see Figure 9.74) and access to SOPs and maintenance instructions and register.

General safety guidelines include:

- Avoid loose clothes (long sleeves, ties, etc.).
- Wear safety footwear (non-slip soles).
- Keep work areas clean.
- Use correct tools for making adjustments.
- Check that all adjustments are locked tight.
- Do not remove guards to operate any machine.
- Use sharp cutters and blades at all times.
- Disconnect power when changing cutters or blades.
- Never leave a machine partly set up unless you leave a notice stating the condition of the machine.
- Ensure adequate lighting.
- Ensure stop/start switches are easily located.
- Never reach for switches when timber becomes jammed.

- Avoid slip, trip and fall hazards.
- Use bins or boxes for off-cuts.
- Leave passageways clear of obstacles.
- Always do a test cut.
- Remember your duty of care to others.
- Ensure SOPs, posters and manufacturers' manuals are available at all machines.

All machines must be operated by trained, qualified operators. The following sections describe the machines commonly used in the building and joinery trade.

You only get one pair of eyes in your life so protect them. Always wear safety glasses.

Jointer (or buzzer)

A jointer is used to dress a straight face side and edge before bringing it to the nominal size via a thicknesser or planer. Jointers can be set up to make rebates in the edges of timber and to taper stiles for gunstock stile doors.

Safety precautions

- Check cutters for sharpness and gaps and chips.
- Check guards for operation.
- Check that the depth of cut does not exceed 2 mm.
- Check in-feed and out-feed tables are aligned correctly (i.e. parallel).
- Do not use material too small to handle; that is:
  - less than 300 mm long (in any section size)
  - less than 20 mm square
  - when planing the face of thin material (say, less than 12 mm thick).
- Do not attempt to plane off too much with one cut – plane no more than 3 mm at a time depending on the timber density.
- Never pass your hands directly over the cutter block (see Figure 9.74).
- Use push sticks or push blocks.
- Dress the face side first, then the edge – bowed face down (see Figure 9.75) and sprung edge down.
- Use fences and stops for tapers; that is, as in gunstock door stiles.
- Use an additional in-feed and out-feed table or roller to support long lengths of stock.

Planer (or thicknesser)

A planer (thicknesser) is used to dress the timber to width and thickness. Three-bladed heads provide a better finish than two-bladed heads.

The machine shown in Figure 9.76 is an 'under over' machine. Under over machines can be used as jointers and then converted to thicknessers. This saves on vital factory floor space and machine costs.

Note: Only material that has had a face side dressed true and straight should be put through a thicknesser. The dressed face of stock goes *face side down* on the bed of the machine and the cutters remove waste from the top to achieve a flat finish.

FIGURE 9.74 No hand zone

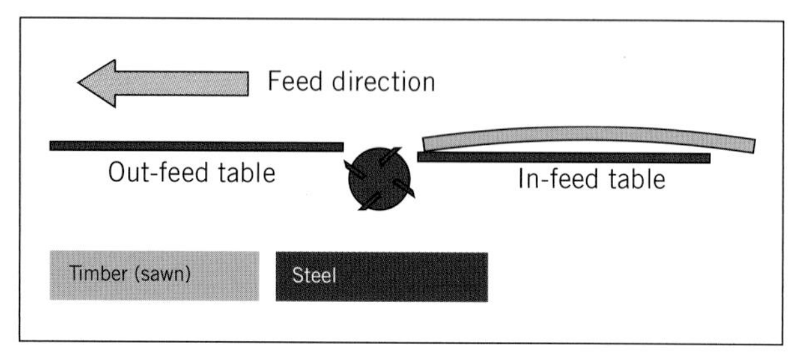

FIGURE 9.75 Bowed face down

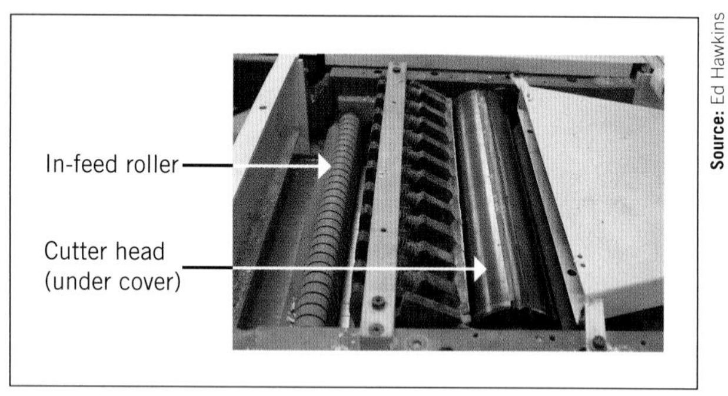

FIGURE 9.76 Cutter head and feed rollers

Safety precautions

- Check guards are in place before switching on the machine.
- Check the calibration of the rise and fall table using scrap material before running stock.
- Use an additional in-feed and out-feed table or roller to support long lengths of stock (see Figure 9.77).
- Check in-feed pressure rollers for excessive gum build-up.
- Never try to remove too much material in one pass.
- For best results, check the grain direction of the timber before feeding it into the machine.
- Ensure dust extraction is fitted and working properly before starting the machine.
- Check all timber is free of nails and screws.
- Do not attempt to thickness timber that is too short.

Source: Ed Hawkins

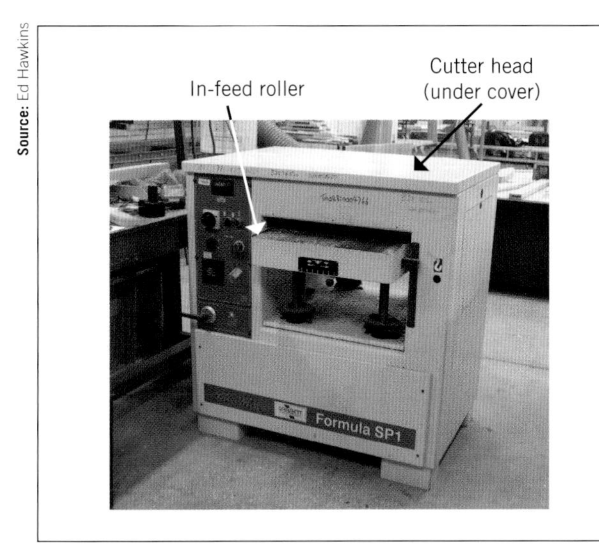

FIGURE 9.77 Planer (thicknesser)

Multi-headed planers and moulders are used to dress timber and lengths of other materials. The lengths are profiled and in some circumstances sanded in one pass through the machine. The materials are fed into the machine via a series of feed rollers. A large range of cutter heads and saw blades can be fitted to the machine.

Drum sander

A drum sander consists of at least two wide drums (600 mm minimum) wrapped in abrasive paper (with different grits on each drum; see Figure 9.78). Components or completed frames are passed through the machine and the frames and/or parts are sanded smooth and joints flushed. The width of the frame or panel that can be sanded is governed by the width of the drums.

A wide belt sander is an alternative machine that can be used.

A linisher, also known as a belt sander or belt grinder, is a power tool used for sanding, grinding and polishing various materials, typically metal or wood. It consists of an abrasive belt (usually made of sandpaper or abrasive cloth) that rotates on two rollers and is driven by an electric motor.

Source: Greg Cheetham

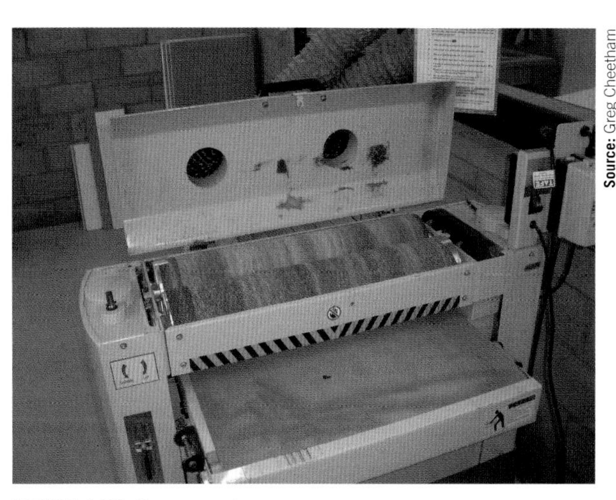

FIGURE 9.78 Drum sander

Safety precautions

- Isolate power before changing belts or opening the cabinet.
- Use only recommended abrasives.
- Clean abrasive material belts regularly.
- Allow belts to reach maximum speed before feeding in stock (these are not true surface finishing machines).
- On belt machines, check the tracking and tension of belts before operating.

A disk sander can be used to efficiently smooth and shape timber surfaces. It can be mounted on a bench or used as a handheld device. The grit disks can be changed depending on the requirements. Coarser grits (e.g., 80 or 120) are used for initial material removal, while finer grits (e.g., 220 or 320) are used for finishing.

Spindle moulder

The spindle moulder is used to mould edges and run rebates simultaneously for medium volumes of stock materials. When large volumes of material require dressing and profiles added, a heading machine is required.

Start and finish your work with 10 fingers. Protect them from cutters and blades by using push sticks, fences, guards and isolation systems.

The machine in Figure 9.79 has an automatic feed unit attached. This keeps the material pushed against the fence and keeps the machine operator's hands out of the way. This machine can be used to end-dress lengths of material ready for joining, using lighting or finger joints, as well as for machining tenon cheeks. It should only be used by trained wood machining operators.

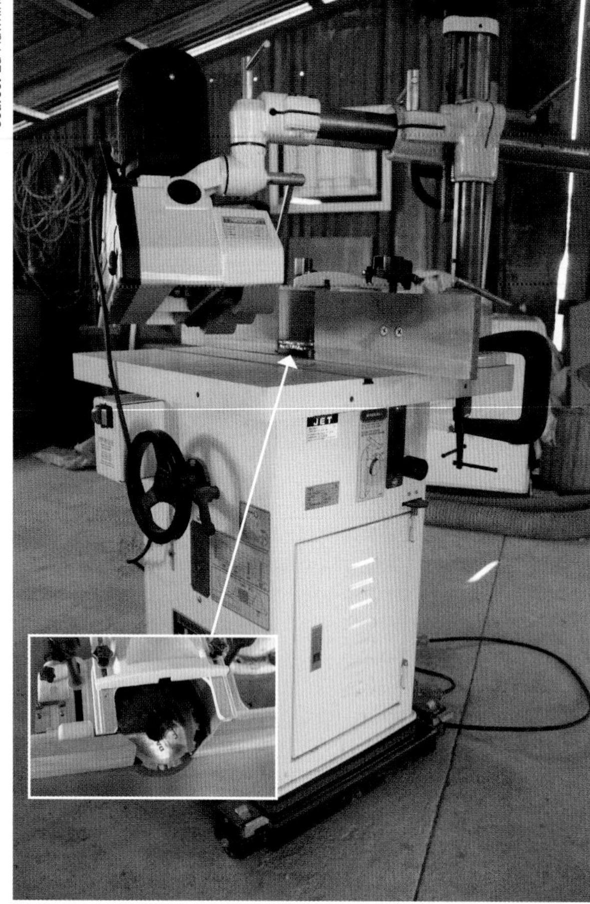

FIGURE 9.79 Spindle moulder with feed rollers

Safety precautions

- Isolate the power before changing cutters or adjusting fences and stops.
- Keep the work area clean.
- If using cutters in slotted collars, the cutters must be embedded at least two-thirds the width of the cutter (see Figure 9.80). Solid heads and cutters must be balanced and sharp, and unused sections of the cutters must be set below the spindle table.
  Note: Plain slotted collars are illegal in Australia.

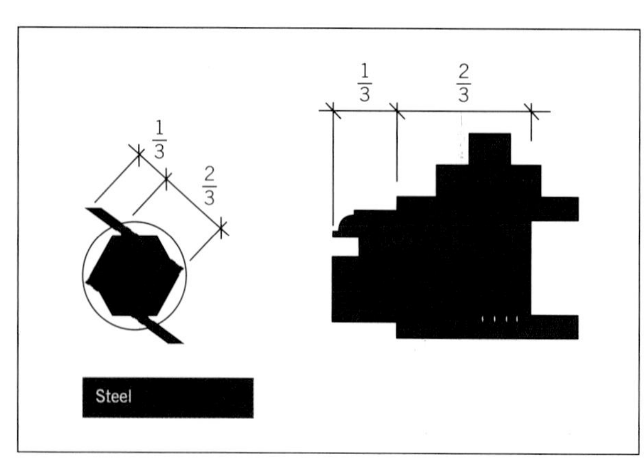

FIGURE 9.80 Spindle cutters

- Check the locking nut is tight before turning on the machine.
- Do not slow down spindles with your hand after turning off the machine.
- Feed material into the cutters against the rotation of the spin (see Figure 9.81).

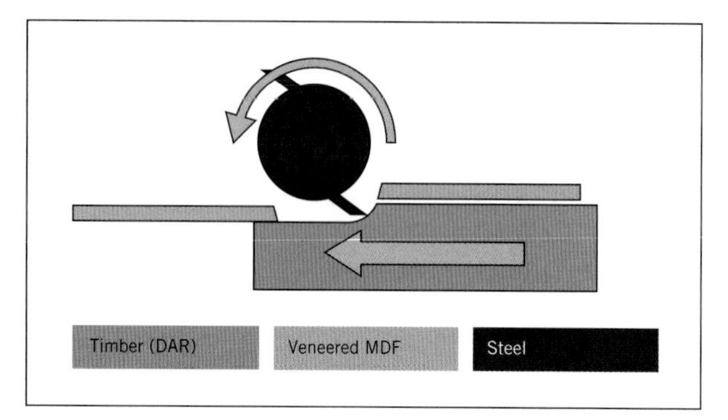

FIGURE 9.81 Feed stock into cutters

- When moulding end grain, use a backing block behind the cut to reduce break-out, or construct a jig with a backing block to prevent break-out.
- Where possible, use an electric drive feed unit to push straight material into the cutters.
- Locate guards correctly. Ring guards are used to guard cutters during spindle operations on curved work.
- Use jigs whenever practical.

Figures 9.82, 9.83 and 9.84 show an interchangeable blade head, interchangeable blades and a pin lock cutter, respectively.

Source: Ed Hawkins

FIGURE 9.82 Interchangeable blade head

Extreme caution: Never place your hands on the out-feed side of the stock being machined on a spindle moulder (see Figure 9.85). If the machine catches the timber, your hands can be pulled back into the cutters.

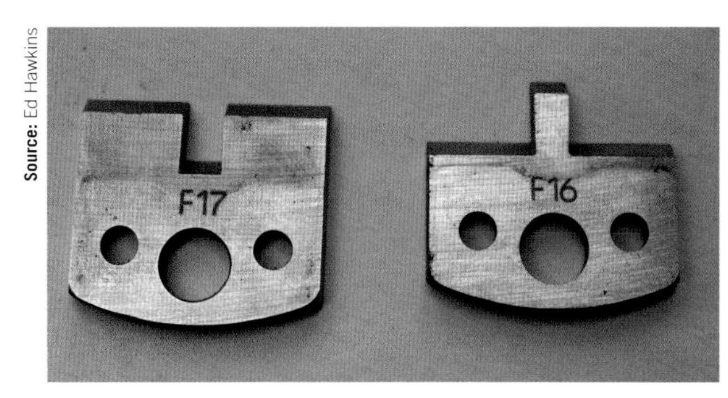

Source: Ed Hawkins

FIGURE 9.83 Interchangeable blades

Source: Ed Hawkins

FIGURE 9.84 Pin lock cutter

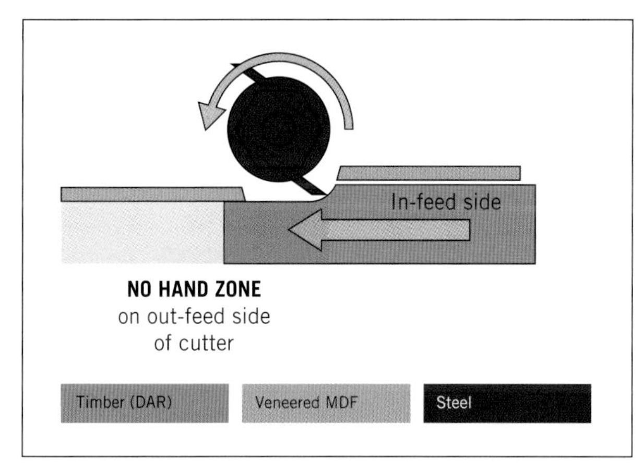

FIGURE 9.85 Spindle moulder no hand zone

Radial arm saw

Radial arm saws are used to break material down from long lengths into usable lengths (see Figure 9.86). They can also be set up with a trenching head to cut tenon cheeks. The head can be angled to cut angled shoulders for the mid-rails used with gunstock stiles.

Source: Greg Cheetham

FIGURE 9.86 Radial arm saw

Safety precautions

- Before operating a radial arm saw, check the saw return: pull the saw forward, then let go to see if the saw returns to position behind the fence.
- Make sure adjustable front guards are as close as possible to the workpieces.
- Use fences and stops for accuracy and safety.
- Use a holding stick instead of your hands to clear waste from around the blade.
- When cutting bowed or sprung material, keep the bowed face down and the sprung edge against the fence (see Figure 9.87).
- Use a backing block behind the material to ensure a clean cut.

Caution: If it is not controlled properly, the radial arm saw can 'climb over' the material being cut and 'run' directly at the operator. Keep a firm grip on the controls. Docking saws can be automated with electrically operated digital read-out stops, air clamps and hands-free operation.

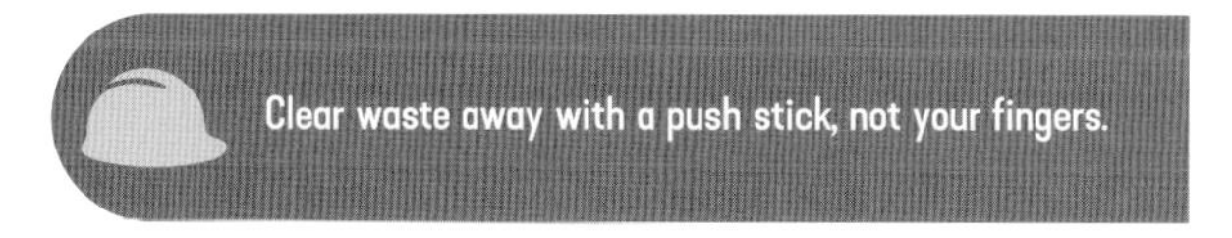

Clear waste away with a push stick, not your fingers.

Sliding table saw

The sliding table saw can be used to dock materials to length, cut the shoulders on tenons, trim and cut panels to size and, depending upon the riving knife and hood guard configuration, cut grooves and rebates in rails (see Figure 9.88).

Safety precautions

Inexperienced table saw operators tend to concentrate their attention on the feed side of the saw, as this is the area where material is being cut. Injuries occur at this end of the blade from the operator coming into direct contact with the cutting edge; such injuries

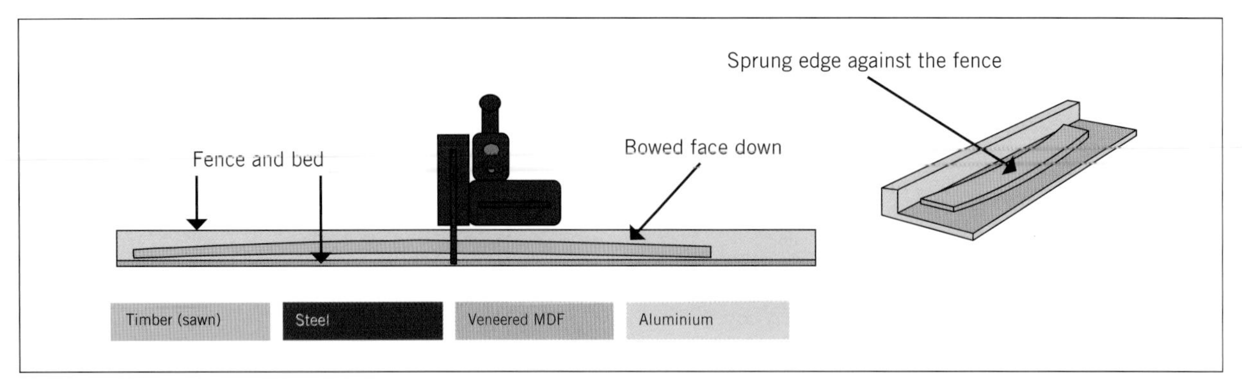

FIGURE 9.87 Cutting bowed or sprung material

Source: Ed Hawkins

FIGURE 9.88 Sliding table saw

are usually caused by carelessness. However, injuries are more often caused by the rotation at the back of the blade picking up the swarf and flicking it into the face of the operator or the person tailing out (i.e. the person working at the back of the saw taking the cut sections), or by having hands or clothing forced against the ascending rear edge of the blade and being pulled on to the blade. Correct adjustment of the riving knife helps to prevent these accidents (see Figure 9.89).

Factory-sited saws with a diameter of 450 mm or more must have at least 1200 mm of fixed take-off table or support frame behind the blade, primarily for the protection of secondary personnel tailing out.

The following are more general safety precautions:

- Never remove crown guards or the riving knife.
- Check the alignment of rip fences to ensure the timber won't jam and be flicked back (see Figure 9.90).
- Check the cross-cut fence for square.
- Check the calibration of fences to ensure that digital and marking read-outs correspond with measurements.
- Check dust extraction is on before starting to cut.
- Use a backing block to prevent break-out of timber on the back of the cut.
- Check and adjust, if necessary, the alignment of the scribing blade if fitted (see Figure 9.91). If no scribing blade is fitted, use a backing board on the table to ensure a clean cut on the underside of the material.

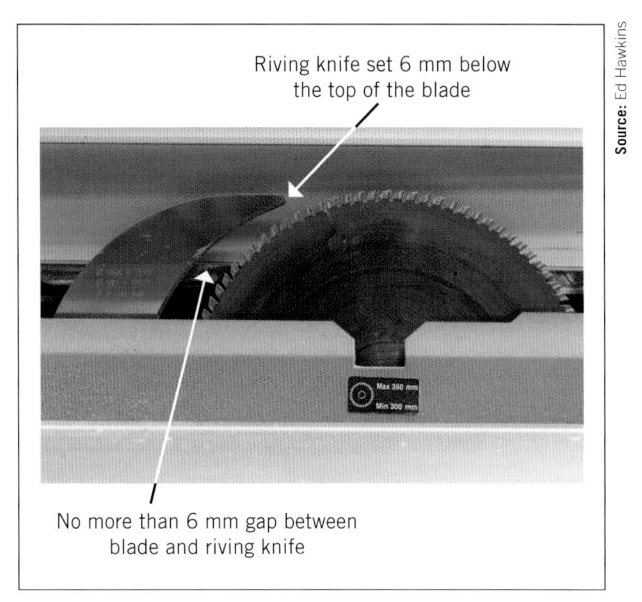

Source: Ed Hawkins

FIGURE 9.89 Riving knife adjustment

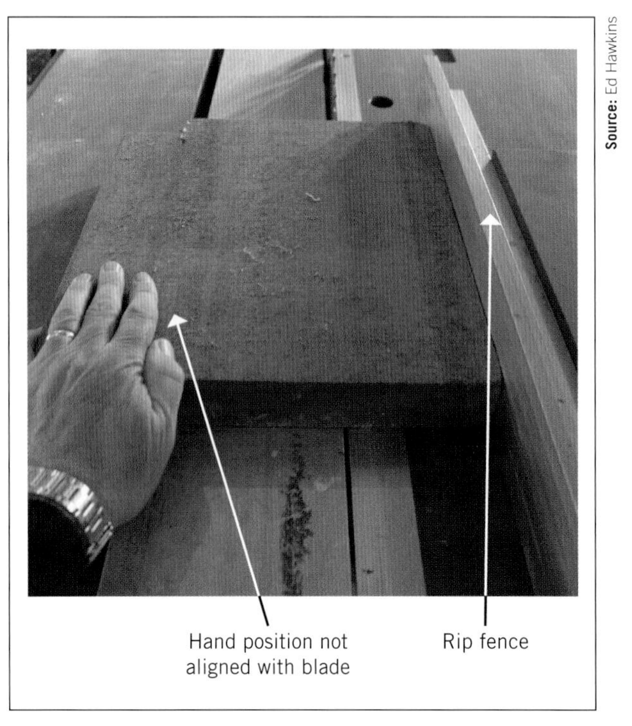

Source: Ed Hawkins

FIGURE 9.90 Rip fence and hand position

Riving knife
Main blade
Scribe blade – spins in opposite direction to main blade. It scores the underside of panel products.

Source: Ed Hawkins

FIGURE 9.91 Panel saw – blade set-up

- Check that the sliding table operates smoothly.
- Set the depth and height of the blade before starting the machine.
- Adjust the crown guard to be as close as practical to the material (see Figure 9.92).
- Clear the table of all off-cuts and excess dust.

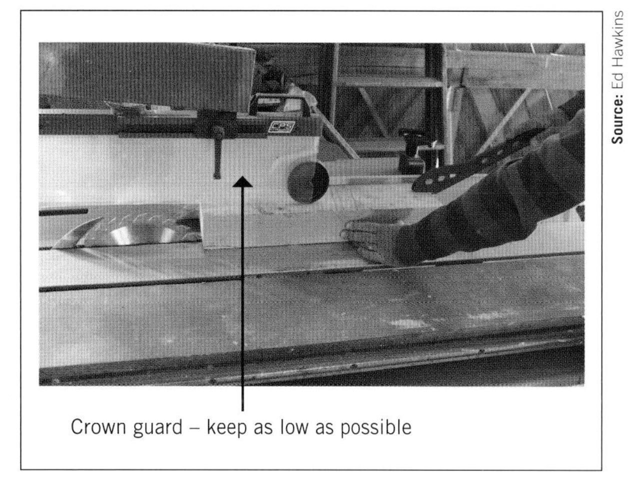

Source: Ed Hawkins

FIGURE 9.92 Push stick and crown guard

Make sure the riving knife is properly adjusted to prevent timber jamming on the blade and timber kickback; use a feather board.

Tenoner

The single-end tenoning machine produces a tenon on one end of the stock, while a double-end tenoner can produce a tenon on each end of the stock simultaneously (see Figure 9.93). It can produce:

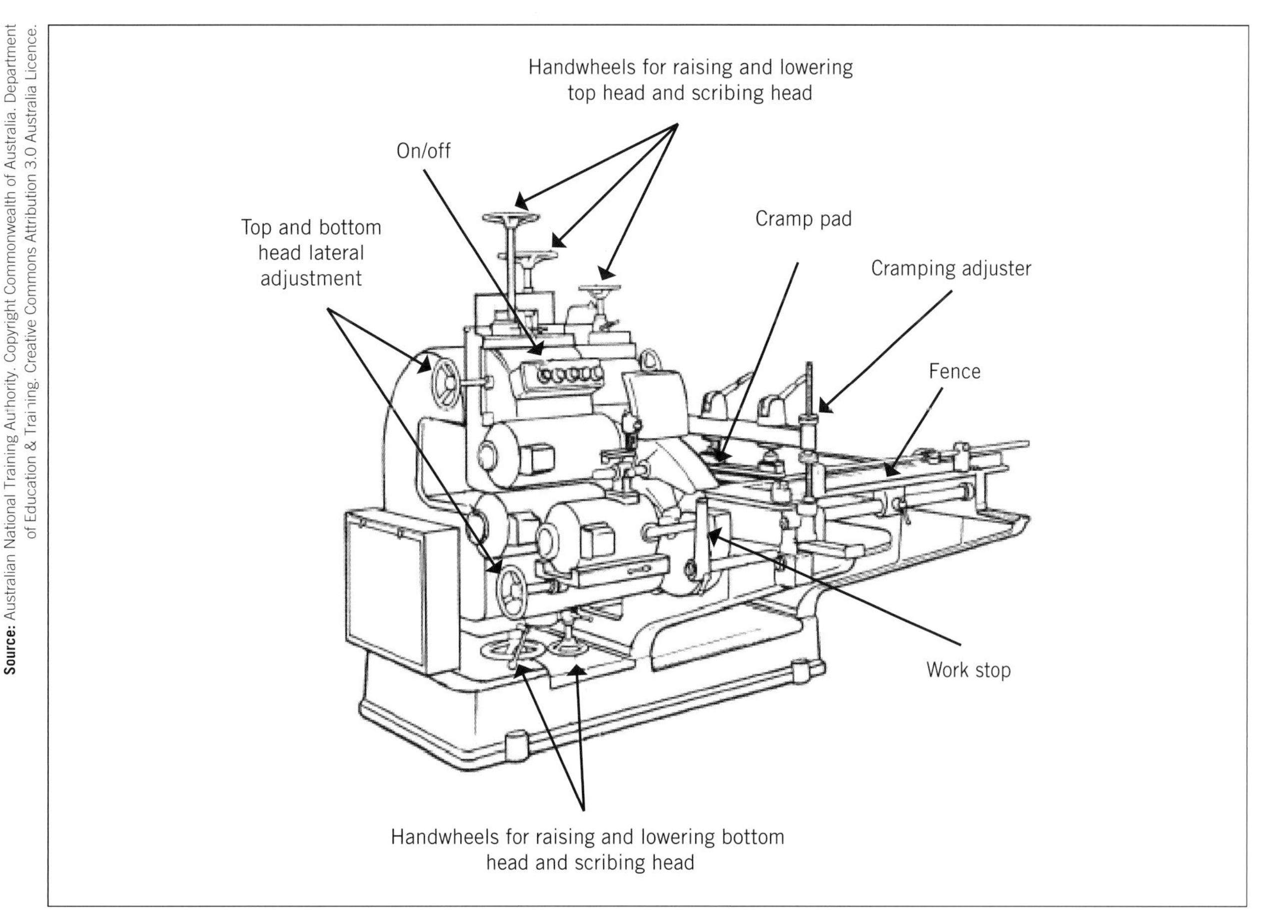

FIGURE 9.93 Tenoner

- through tenons
- bare-faced tenons
- haunched tenons.

This is a specialist piece of machinery and must only be operated by trained and experienced wood machinists.

Automatic tenoners are capable of machining a tenon on the ends of multiple pieces of timber simultaneously.

Bandsaw

The bandsaw is possibly the easiest and most docile of all woodworking machines (see Figure 9.94). The blade cuts in a downward motion and does not kick back like a circular saw does. Adequate lighting must be provided over the work area and a waste box must be located nearby.

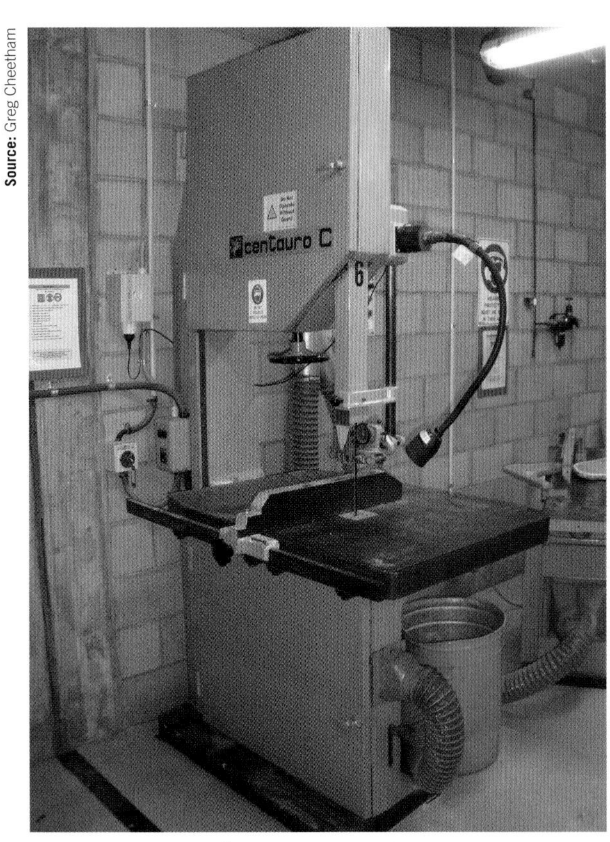

Source: Greg Cheetham

FIGURE 9.94 Bandsaw

When cutting, all areas of the blade must be covered except for the part that is cutting. Some machines have a foot or hand brake that should be used to stop the blade rotation before the operator leaves the machine.

The size of a bandsaw is determined by the diameter of its wheels, which range from 300 to 700 mm for normal workshops. The throat size is the distance between the table and the base of the column. Some bandsaws have three wheels, providing this type of saw with a larger throat than those with two wheels.

Bandsaws can be used for cutting the cheeks on tenons and curved sections in door and window components. The table can be tilted for bevel and compound cuts. Wheels are usually rubber-covered to keep noise down and protect the teeth of the wheel. These should be checked regularly for wear and damage.

Safety precautions

- Ensure that all guards are in position before cutting.
- Check the blade for sharpness and tension before starting to cut. A blunt blade will *not* cut straight.
- Check guide wheel adjustment before starting the machine (see Figure 9.95).
- Check that the friction wheel is properly aligned (see Figure 9.95).
- Use a push stick (not your hands) to remove waste material from the table to a bin.
- Avoid tight curves by using relief cuts.
- Never force the cut, or the blade will snap.
- Keep your hands at least 50 mm either side of the blade alignment.

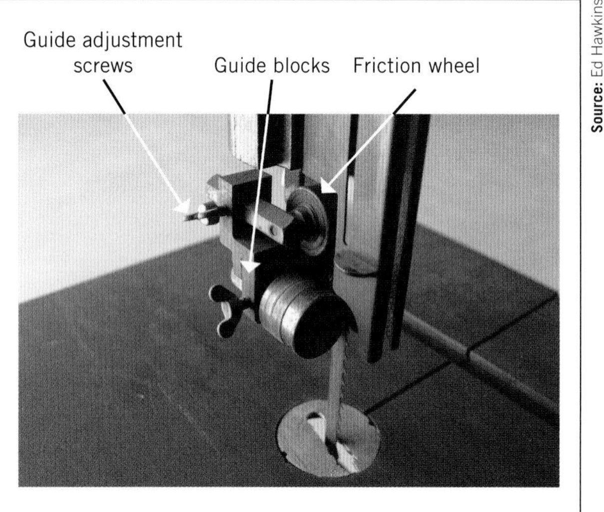

Source: Ed Hawkins

FIGURE 9.95 Bandsaw guide block adjustment

Mortising machine

There are a few different varieties of mortising machines, including the chisel, chain (or oscillating bit router; see Figure 9.96) and square chisel mortisers. The chain mortiser is perhaps the most common machine found in the industry at the moment.

The typical procedure for setting up a chain mortiser is as follows:

- Isolate the machine from the power source.
- Ensure the chain is sharp and that moving parts and slides are lubricated.
- Cramp the workpiece in position and make sure it lines up with the set-out positions of the mortise and haunch. For through tenons, make sure a scrap piece of material is set below the mortise to prevent break-out when the chain pierces the stock; this also prevents damage to the work table.
- Ensure the face side of the material is set against the fence.

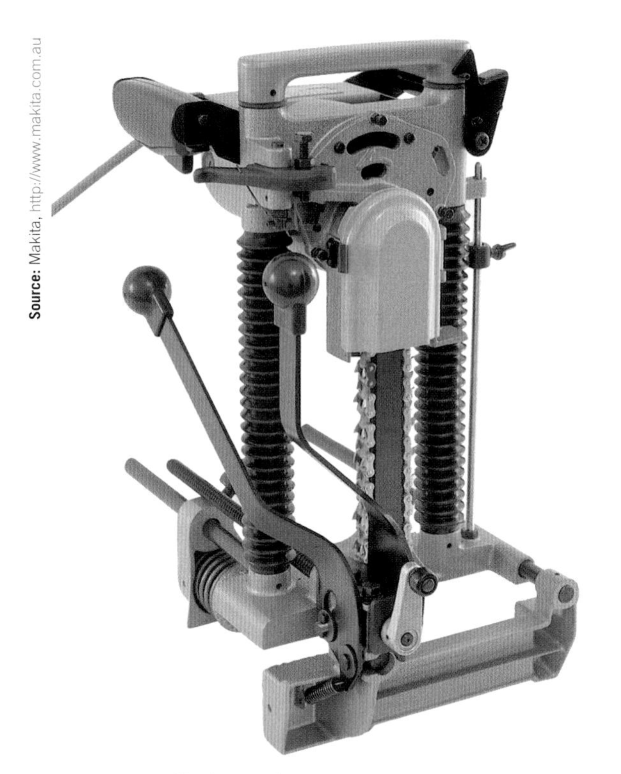
Source: Makita, http://www.makita.com.au

FIGURE 9.96 Chain mortiser

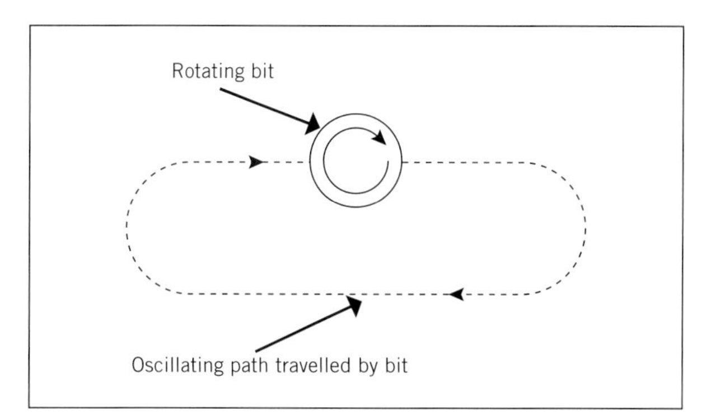

FIGURE 9.97 Oscillating bit mortiser action

- Adjust the position of the stock relative to the chain with the levers (or hand wheels, depending on the type of mortising machine).
- Adjust the depth stops – one for the mortise, and one for the haunch if necessary. If the mortise is stubbed tenon, the component will need to be shifted to one side to allow the chain carriage to be brought down, and the depth stop will need to be adjusted accurately to suit the stubbed tenon. You can use a tape or rule to measure from the work table to the tip of the chain to assess the correct depth.
- Replace the real component with a test piece.
- Check that all levers are in their correct starting position before switching on the machine.
- Allow the chain to reach maximum speed, then bring it into contact with the test piece. Take the chain to its full depth and then remove it. Do not force the pace of the cut. Wait for the chain to stop before checking set-out and make any adjustments as necessary.
- Replace the test piece with the real component and repeat the process, using the wheels or levers to move the work-table carriage along to produce the right width of mortise.

The same basic operational procedures are used for a hollow chisel mortiser, except that a square hollow chisel is used to produce the mortise (see Figures 9.98 and 9.99). The hollow chisel size should match as closely as possible the ribbon size on the sash stock. The hollow chisel mortiser requires several plunges of the combined drill bit and a square chisel to remove the waste. Careful alignment of the chisel to the fence is essential to achieve a quality cut.

A horizontal borer is typically used in machining and manufacturing processes to create holes or bores. It is specifically designed for horizontal drilling or boring operations, where the cutting tool moves parallel to the workpiece's surface. The spindle on a horizontal borer can be equipped with various types of cutting

Source: Greg Cheetham

FIGURE 9.98 Hollow chisel mortiser

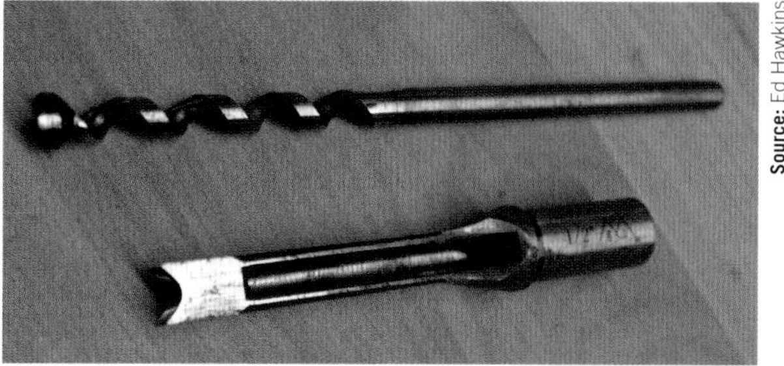
Source: Ed Hawkins

FIGURE 9.99 Hollow chisel and bit

tools, such as drills, boring bars, or end mills, depending on the specific machining requirements.

A number of machine manufacturers are listed at the end of the chapter. Many of these specialist machine companies produce machinery dedicated to the window manufacturing industry and are capable of turning out components for 50 or more windows per day.

Edge banders

There are many different types of edge banders available:

- handheld machines – good for shaped work
- hand-fed static machines – can trim and end cut
- automatic self-feeding straight line edgers
- soft-form edgers – for edge strips to curved and shaped edges.

They all apply either pre-glued or unglued tapes (see the earlier section 'Edge treatments for flat panel components').

Handheld and hand-fed machines are useful for small workshops and can be used by most carpenters/joiners with a little training. The larger machines need a well-trained person to operate, adjust and perform routine maintenance on the machine.

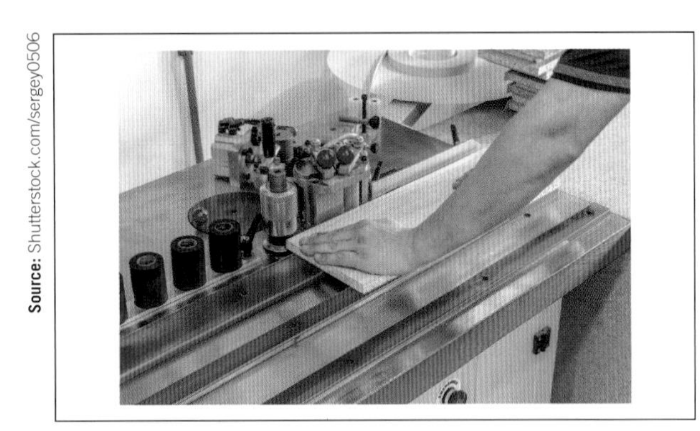

Source: Shutterstock.com/sergey0506

FIGURE 9.100 Hand-fed edge bander

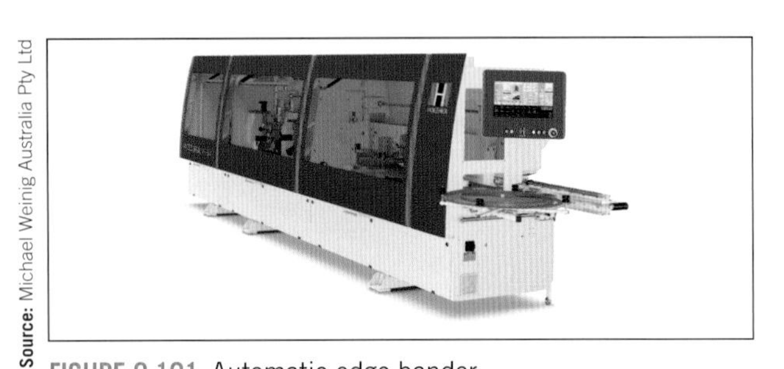

Source: Michael Weinig Australia Pty Ltd

FIGURE 9.101 Automatic edge bander

Edge banding machines can be temperamental beasts if not set up and operated properly. To achieve a quality finish using both pre-glued and unglued tapes it's important to ensure:

- the edges being banded are not chipped
- no scoring blade misalignment is present.

For pre-glued edge tapes it's also important to ensure:

- temperatures are controlled and set accurately – too hot and tapes can burn; too cold and they won't stick
- the feed speed is controlled.

Edge bander machines have many moving parts, some of which operate at high temperatures. Only trained operators should use them. Always ensure all guards are in place and that appropriate PPE is worn.

CNC machine centres for joinery

Computer numerical control (CNC) machinery has its functions and motions controlled by a prepared computer program containing coded alphanumeric data. The program can control the motions of either the workpiece or the tool, depending on what is required; it is also capable of processes such as the depth of cut, feed speed, and other functions such as turning the cutters on or off, and turning coolant on or off.

The original CNC machines started with basic panel saws and drilling machines before the development of three-axis machining whereby the cutter head travelled in three directions: up and down, side to side and backwards and forwards.

Modern machines can do five-axis machining, making production faster and providing the ability to machine more complex shapes without changing machines (see Figure 9.102).

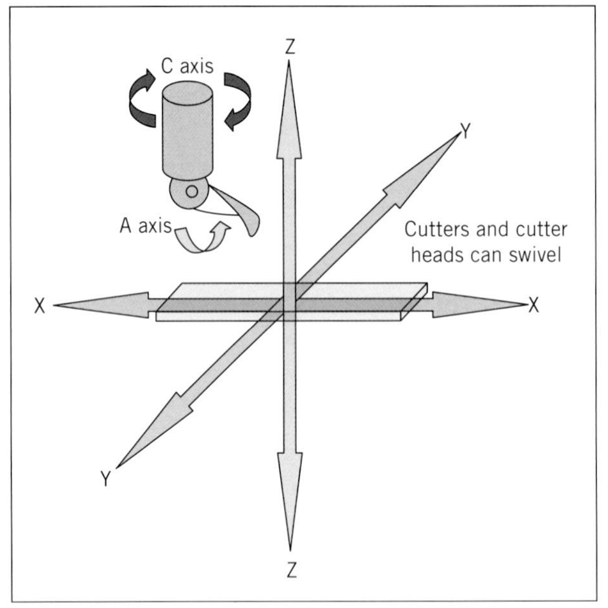

FIGURE 9.102 Five-axis machining

Modern joinery shops are embracing these new technologies at a rapid rate (see Figure 9.103). With the cost of technology coming down and the cost of labour rising it makes good economic sense to invest in CNC machine centres.

The choice of machine centre and machinery will depend on a number of factors, including:

- the materials being used in production
- the quantity of doors or windows to be produced
- the size of the manufacturing facility
- cost factors, such as machinery costs and lease or buy decisions.

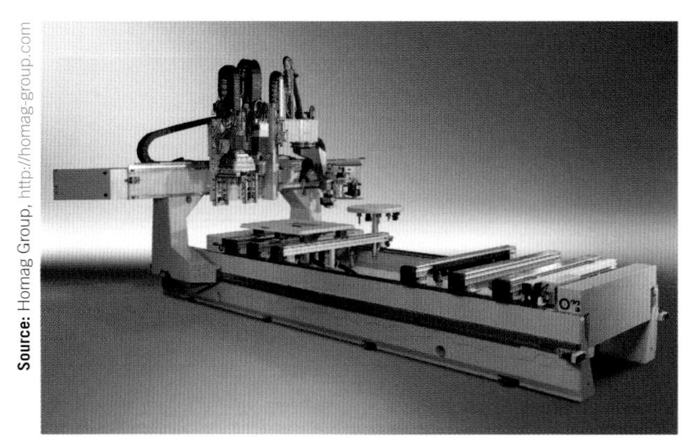
Source: Homag Group, http://homag-group.com

FIGURE 9.103 CNC machinery

The advantages of CNC machinery include:
1 high accuracy in manufacturing
2 short production time
3 greater manufacturing flexibility
4 sophisticated shaping using up to five-axis machining
5 reduced human error.

The disadvantages of CNC machinery include:
1 high start-up costs
2 cost of maintenance
3 the need for a CNC-trained operator.

Machine manufacturers today provide complete machine shop solutions, and are capable of producing all the parts of a door or window using one-off processes or by batch and mass production, with systems geared towards 'just-in-time' manufacturing processes.

Hardware machining requirements can also impact on the type and size of machinery used.

There are a number of safety checks to note when using CNC machinery.

Before you start:
- Take care when handling sharp cutters and tooling.
- Ensure all tools will clear other machine parts when mounted in the magazine.
- Ensure the correct materials specifications and sizes are adhered to.
- Check pre-matching of datums – this is critical.
- Check machine lubrication and cooling systems are correct and functioning.
- Check cutter head numbers match the program numbers in the tool changer.

After each program run:
- Ensure the machine returns to the 'safe' position.
- Ensure the operating program returns to the beginning of the production sequence.

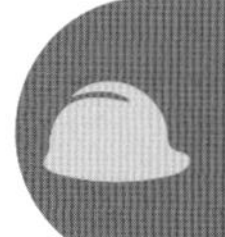

Never open or enter a safety barrier area until the machine parts have come to a complete stop, and always follow the protocols for shutting down a CNC.

Flat panel components use beam saws for high-volume production runs, where hundreds of the same component are cut on a beam saw, cutting up to 125 mm or several sheets at the same time. Sophisticated machines can be programmed to load, cut, turn and transfer sheets and components to other machines.

Software for operating CNC machinery

CNC machines require software to carry out their automated processes. Computer-aided drafting software enables the user to create flat panel products using digital templates from libraries. The software translates the template information into CNC machine processes, such as cutting shapes and drilling holes. Many machines come with their own operational software, but some generic after-market software packages are compatible with a range of machines, meaning you may not need to purchase new software when you buy a different brand of machine.

Optimising software is used to create cutting patterns for sliding table saws, beam saws and nested machine centres. Optimisers work out the best cutting patterns for the board size, which minimises waste and increases productivity.

Do your research before deciding which software package will suit your current and future needs.

LEARNING TASK 9.6

DOOR AND WINDOW MACHINERY

Search for and watch some online videos about modern door and window manufacturing processes by using the following search terms:
- timber window door CNC production
- window manufacture processing centre + video
- timber window machinery
- windows: frame assembly press – fully automated.

Then discuss in class the various systems identified with a focus on benefits and limitations, safe operating procedures and WHS/OHS and machine guarding, and the future for manufacturing doors and windows.

Alternative

In a teacher-led discussion consider the future for the manufacture and assembly of joinery components in a diminishing resource environment. Key points include:
- reduce, recycle, reuse
- timber: from the forest to finished product – how we can produce more timber
- the potential to recycle products – what types of products can be recycled
- using lesser quality timbers – why, when, how
- using waste materials – what are waste materials; what are the sources of waste materials
- technology to assist in reducing, recycling and reusing.

Assembling the frames

Try a dry assembly of the frames before final assembly. Check to make sure that:

- the size is accurate to plans and specifications
- the joints are closed within the specified tolerances
- the assembly is square (see Figure 9.104)
- the assembly is in wind (see Figures 9.105 and 9.106).

If the frame in the clamps is twisted or out of square, it may be possible to adjust the clamps to pull or return the frame to true straight and square. You will need to check the components and joints to find the cause of the problem and rectify it before final assembly.

Once the components for the doors or windows have been machined and checked, they are ready for final assembly. This may involve gluing up, sanding, glazing and fitting up of the hardware to the windows or doors.

Modern frame assembly techniques make use of compressed air rams and robot-like technology to automate the assembly process.

There are many other aspects related to the assembly of components and sub-assemblies that we have not covered in this chapter. Additional information on these topics can be found by visiting some of the websites at the end of this chapter or your local TAFE library.

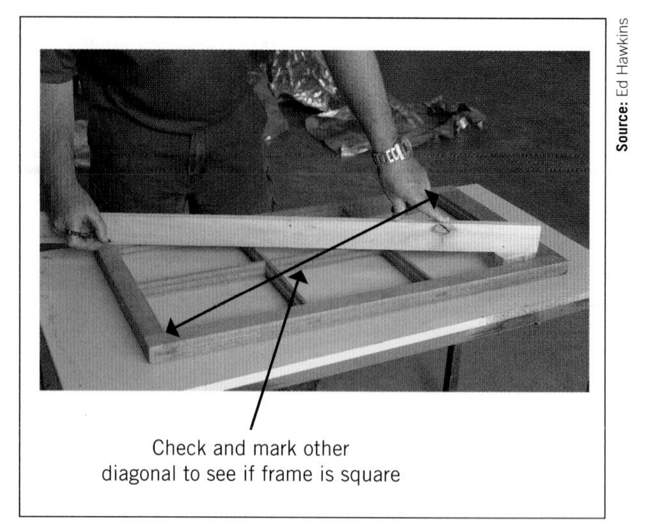

FIGURE 9.104 Checking for square

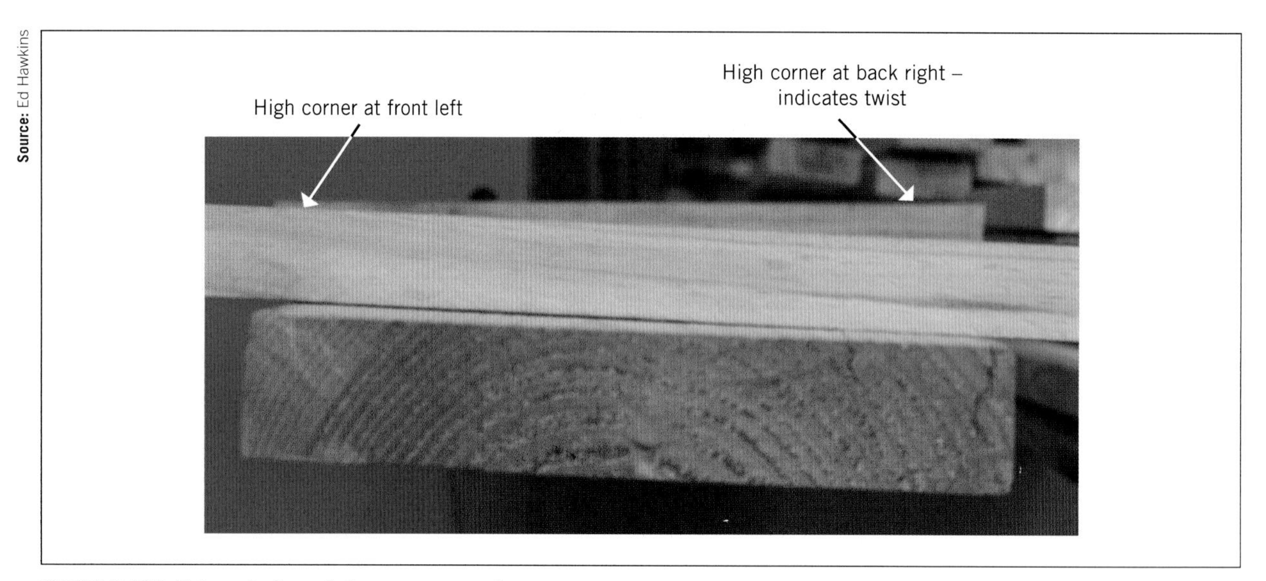

FIGURE 9.105 Using winding sticks on components

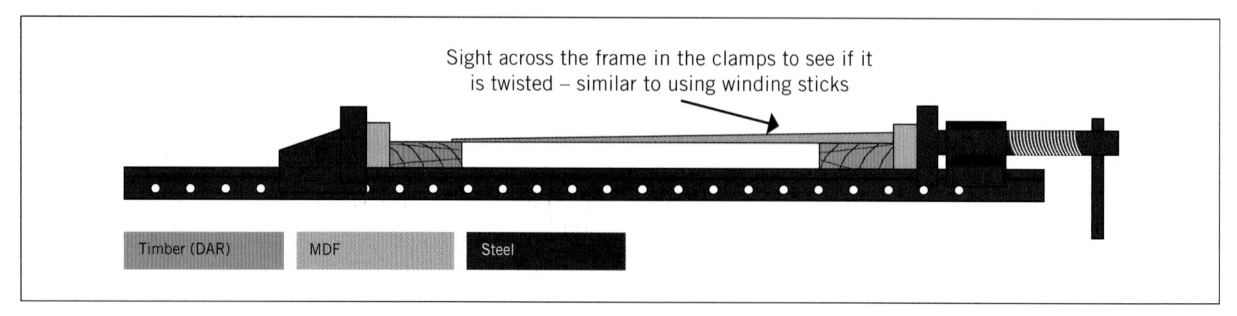

FIGURE 9.106 Checking frames for wind

Clean up

Clean up all work areas when work activities are completed. Wear a respirator if cleaning fine dust particles. Stack and store unused materials in a dry place.

Reuse excess materials or off-cuts and recycle where possible. Follow workplace procedures for waste management adhering to environmental plans and laws.

Shutdown machinery as per the manufacturer's guidelines. Clean and check machinery according to operational guidelines. This may include replacing or sharpening blades, oiling components or replacing belts or cutting equipment. Report faults as per workplace procedures.

SUMMARY

This chapter has addressed the skills and knowledge required to set out and manufacture a range of joinery components, including timber doors, window units and basic cabinetry. It has covered:

- the planning and preparation to manufacture joinery components for doors, windows and cabinetry
- the importance of a range of Australian Standards for such things as glass in buildings and wind loads, along with environment considerations, energy ratings schemes and Bushfire Attack Levels (BALs)
- setting up of workshops and machine shop layouts
- material selection for windows and doors, and the impact of international certification schemes of material availability and selection processes, including timber jointing technologies that reduce wastage
- the various types of doors (flush, hollow core, ledged and braced, and variations on these), plus frame and panel construction, the frame joints used, selection of timber for doors and windows, and choosing material that is durable, well-seasoned, straight and free from defects (quarter sawn timber is best if available)
- the different components and joints used to make a framed panel door including:
  - mortise and tenon joint and bridle joint proportions:
    - thickness being one-third of material
    - width being the 'effective' width
    - haunch being no more than one-third the 'effective' width of timber
  - types of mouldings and glazing bars
  - how to run grooves and rebates
  - other joints, such as scribed (or matched) joints, barefaced joints, and stubbed and dowelled joints
  - panel sub-assembly methods and procedures
- types of windows and typical box frame components
- joints used in timber window manufacture, including meeting rail joints
- curved head construction, including laminating, segmental brickwork and glazing bars
- basic construction methods used for built-in cabinets:
  - key points from the Australian Standards, including edge treatments
  - box construction using flat panel methods and face frame construction models
  - advantages of using adjustable legs and solid backs with cabinets
  - key points when using CNC machine centres and panel saws:
    - selecting cutters
    - adjusting scoring blades
    - basic checks of components
  - types of carcass joints:
    - butt and screwed
    - biscuit or dowel joints
    - knock-down fittings
  - key design principles of the System32 cabinet component model
  - edge treatments and materials for flat panel construction including pre-glued and unglued hot melt edge bandings and solid timber methods
  - types of edge banding machines – handheld, hand-fed, automatic and soft-form
- setting up, safe operation and maintenance of various machines used in manufacture of joinery products, including:
  - tooling – drill bits, blades and knives
  - testing – before operating, and using backing blocks to prevent break-outs
  - using push sticks and feeder mechanisms to keep hands safe
  - doing a test run first to ensure the program operates correctly
- using CNC machine centres – computer-controlled machines operating in up to five axes. Choosing a CNC machine will depend on the product, materials, quantities and cost of machines. Advantages of CNC machines include accuracy, flexibility and sophisticated shaping. Disadvantages include the high cost of machines and maintenance and availability of trained operators
- when assembling frames, the need to check:
  - the size is accurate to plans and specifications
  - the joints are closed within the specified tolerances
  - the assembly is square and in wind.

REFERENCES AND FURTHER READING

National Construction Code

Copies of the National Construction Code can be obtained free of charge from the Australian Building Codes Board website: **http://www.abcb.gov.au**.

Texts

SafeWork NSW, *Kitchen Manufacturing and Joinery*: **https://www.safework.nsw.gov.au/__data/assets/pdf_file/0017/50138/safety-guide-joinery-kitchen-manufacturing-8501.pdf**.

South Western Sydney Institute of TAFE NSW (2020), *Basic Building and Construction Skills* (6th edn), Cengage Learning Australia, Melbourne.

There is also a large range of books available detailing all aspects of timber joinery and manufacture of various timber products that students should investigate. Some will be available in your local council and TAFE libraries.

Web-based resources

Industry and government associations

Australian Fenestration Rating Council (AFRC): **http://www.afrc.org.au**

Australian Glass and Window Association: **https://www.agwa.com.au/**

BASIX Certificate Centre: **http://www.basixcertificatecentre. com.au**

Nationwide House Energy Rating Scheme (NatHERS): **http://www.nathers.gov.au/**

Window Energy Rating System (WERS): **http://www.wers.net**

Corporate – Cutters and blades

Carbatec: **https://www.carbatec.com.au**

Dimar Australia: **https://dimar.com.au**

Leuco Australia: **https://www.leuco.com/EN/AU/web/solutions/leuco-p-system**

Corporate – Door manufacturers

Corinthian Doors: **https://www.corinthian.com.au/**

Jeld-Wen Windows and Doors: **http://www.jeld-wen.com.au**

Trend Windows and Doors: **http://trendwindows.com.au**

William Russell Doors: **http://www.wrusselldoors.com.au**

Corporate – Glass manufacturers

Australian Glass Group: **http://www.australianglassgroup.com.au**

Viridian: **http://viridianglass.com**

Corporate – Windows: Timber, aluminium and UPVC

Enviro Vision Windows and Doors: **http://www.envirovision.com.au**

PVC Windows Australia: **http://www.pvcwindows.com.au**

Rehau: **http://www.rehau.com/au-en**

Wood Solution Technical Design Guide 10 – Doors and Windows (you will need to register to download this free guide): **https://www.woodsolutions.com.au/blog/free-technical-design-guide-doors-windows**

Corporate – Machinery manufacturers

Altendorf Australia: **https://altendorf.com.au**

Biesse Australia: **http://www.biesse.com/au/**

HOLZ-HER: **http://www.holzher.com**

SCM Group: **http://www.scmgroup.com/en**

Wadkin Moulder Pty Ltd: **http://www.wadkin.com**

Weinig Group: **http://www.weinig.com**

Learning resources

CNC machining (general information): **http://www.technologystudent.com**

Video resources

Altendorf WA6 scoring blade set-up: **https://www.youtube.com/watch?v=oOnNfQO6spO**

BlueMax Mini Modular Plus – Automatic drilling and insertion machine: **https://www.youtube.com/watch?v=3Q4ARwUHmWQ**

Correct adjustment of scoring saw blades: **https://www.youtube.com/watch?v=BBPnUObFPus**

Leuco 'P' Systems: **https://www.leuco.com/EN/AU/web/solutions/leuco-p-system**

Safety Video: Table Saw 1 (Altendorf F45): **https://www.youtube.com/watch?v=wIT2Uzhs-Zo**

Workshop panel saw Altendorf F45 set-up, walk-through, features, performance: **https://www.youtube.com/watch?v=kGDG88_ooA4**

Cabinet hardware catalogues

Blum Australia: **https://www.blum.com/au/en/company/news-blum-australia/blum-catalogue-2018-2019**

Häfele Australia: **https://www.hafele.com.au/en**

Hettich Australia: **https://web.hettich.com/en-au/home**

Nover and Co: **https://www.nover.com.au**

Relevant Australian Standards

Door and window construction and installation	AS 1473 Guarding and safe use of woodworking machinery. Note these Standards were withdrawn in April 2017 but still provide significant advice on safe operation of machines. Check with your local WHS/OHS regulator for current requirements.
	AS 2047 Windows and external glazed doors in buildings. Covering most glazed windows and external doors, louvres and shopfronts.
	AS 4024.1 Safety of machinery. The series provides guidelines for machine designers, manufacturers, suppliers, employers and users to help reduce the risks of working with, or near, machinery.

>>

>>

Door and window construction and installation	AS 1288 Glass in buildings – Selection and installation covering skylights, recycled products and heritage items
	AS 2047 Windows and external glazed doors in buildings
	AS 2082 Timber – Hardwood – Visually stress-graded for structural purposes
	AS 2796 Timber – Hardwood – Sawn and milled products – Part 1 to 3: Timber for furniture components
	AS 4785 Timber – Softwood – Sawn and milled products – Parts 1 to 3: Covering product specs, grade descriptions and timber for furniture components
	AS/NZS 4420 Windows, external glazed, timber and composite doors – Methods of test. Note this revised standard amalgamates all previous Parts 1–8
	Australian Standards can be purchased from the Standards Australia website: **https://www.standards.org.au/**, or see your teacher or librarian for assistance in accessing Australian Standards online
Australasian Furnishing Research and Development Institute Certifications	AFRDI 104.94 Performance of furniture and furniture components in various climates
	AFRDI 108.94 Requirements – surface finishes used on timber furniture
	AFRDI 891002 Quality of furniture and joinery. Guide to the measurement of dimensions.

GET IT RIGHT

MANUFACTURE AND ASSEMBLE JOINERY COMPONENTS

Questions

Below are two scenarios: one of joinery components that have been assembled (completed), and one of components ready for assembly (incomplete).

1 What is wrong with each piece of work?

____

____

2 How should it be done correctly?

____

____

3 What should be done to rectify these joints?

____

____

Completed joints

Incomplete joints

WORKSHEET 1

To be completed by teachers

Student competent ☐

Student not yet competent ☐

Student name: ____________________

Enrolment year: ____________________

Class code: ____________________

Competency name/Number: ____________________

Answer the following questions.

Most answers will come straight from the chapter. However, you may need to consult the relevant Australian Standards for further information.

1 Name two frame joints used in the construction of framed doors.

a ____________________

b ____________________

2 Answer True or False:

The common proportional thickness of a tenon joint is two-thirds of the thickness of the materials.

True ☐ False ☐

3 List four items that should be checked in reference to quality control on a finished window leaving the workshop.

a ____________________

b ____________________

c ____________________

d ____________________

4 Name six different types of windows.

a ____________________

b ____________________

c ____________________

d ____________________

e ____________________

f ____________________

5 The National Construction Code 2022 – Volume Two Housing Provisions, Part 4 Health and Amenity outlines the performance criteria for satisfying ventilation requirements. Describe two ways you can satisfy the criteria, as outlined in Part 10.6 Ventilation.

6 Name three common sash joints.

a

b

c

7 Briefly describe the following terms:

a Parting bead

b Fanlight

c Stop bead

d Mullion

e Horn

f Transom

8 Your task is to construct and dry assemble the panel door shown in the drawing below.

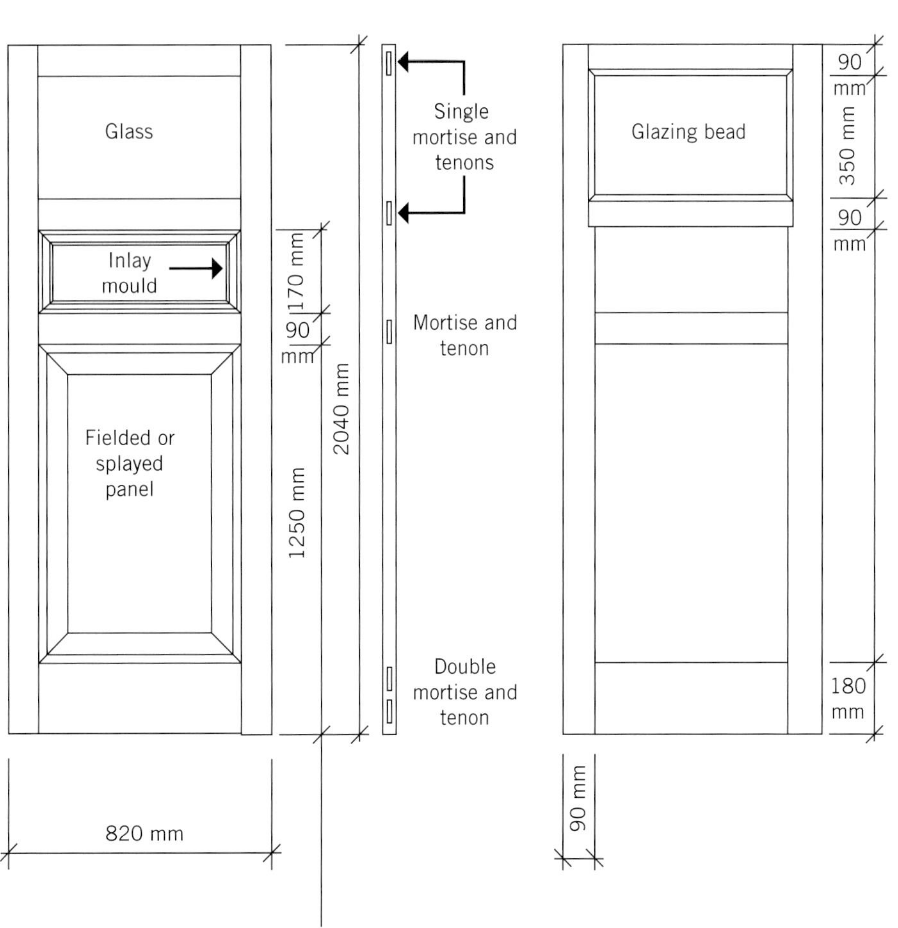

SPECIFICATION

Make the components for a **half** scale model (1:2) of the door shown above.

- Joints as indicated:
  - top rail – long and short shoulder haunched mortise and tenon
  - mid-rails – 1 off long and short shoulder through mortise and tenon
    1 off through mortise and tenon
  - bottom rail – double haunched mortise and tenon
- Inlay mould to suit cutters
- Fielded or splayed panel can be used; requires sub-assembly
- Glass: minimum 3 mm clear float

This practical project provides **partial** evidence of competence for the following unit of competency:

- CPCCJN3003A Manufacture components for door and window frames and doors

This practical project may have relevance to other units of competence. A thorough evaluation via a validation process is required to establish the validity for any other units of competency.

WORKSHEET 2

To be completed by teachers

Student competent ☐

Student not yet competent ☐

Student name: ______________________

Enrolment year: ______________________

Class code: ______________________

Competency name/Number: ______________________

Answer the following questions.

Most answers will come straight from the chapter. However, you may need to consult Australian Standards for further information.

1 Name two sash joints used in the construction of windows and doors.

a ______________________

b ______________________

2 List six basic safety precautions applicable to working with *all* machines.

a ______________________

b ______________________

c ______________________

d ______________________

e ______________________

f ______________________

3 What are the three basic steps in production?

a ______________________

b ______________________

c ______________________

4 Name and describe two methods of producing curved shapes for curved head frames.

a ____

____

b ____

____

5 Your task is to construct and dry assemble the window shown in the drawing below.

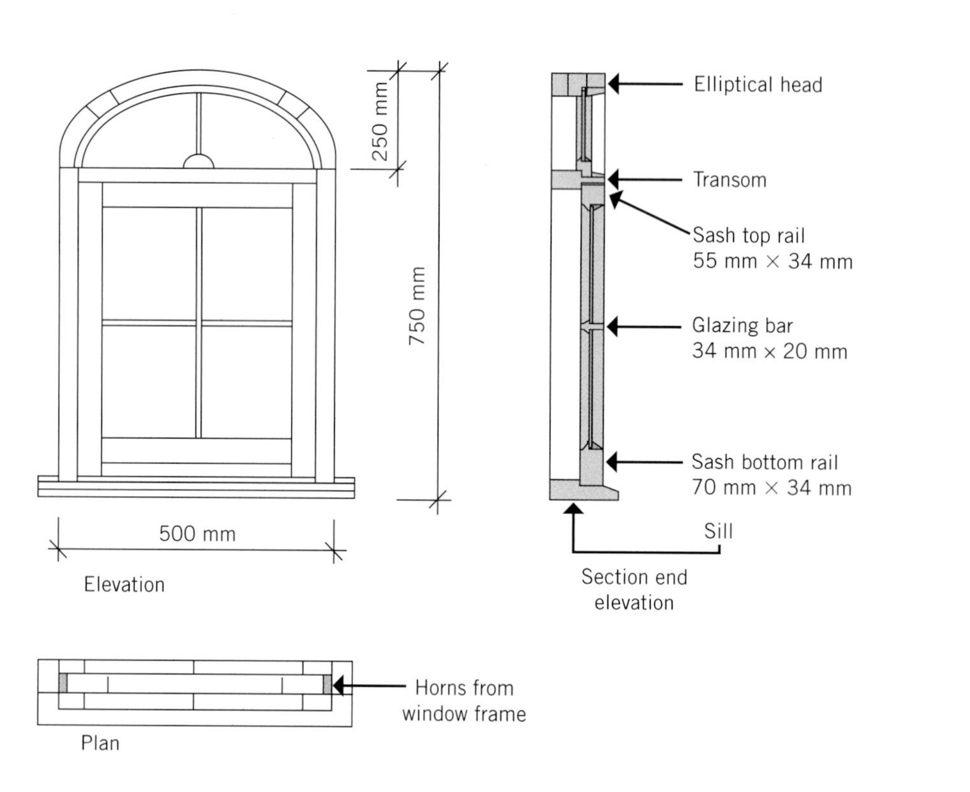

SPECIFICATION

Make the components for casement sash shown above.

- Joints as indicated:
  - sash top rail – haunched mortise and tenon
  - sash bottom rail – haunched mortise and tenon
- Glass: minimum 3 mm clear float
- Glass in sash linseed oil putty faced off
- Glass in elliptical head – timber bead and silicone finish

This practical project provides **partial** evidence of competence for the following unit of competency:

- CPCCJN3003A Manufacture components for door and window frames and doors.

This practical project may have relevance to other units of competence. A thorough evaluation via a validation process is required to establish the validity for any other units of competency.

WORKSHEET 3

To be completed by teachers	
Student competent	☐
Student not yet competent	☐

Student name: ______________________

Enrolment year: ______________________

Class code: ______________________

Competency name/Number: ______________________

Answer the following questions.

1 Complete the safe operating procedure for a circular cut-off saw.

Hazard identification	
Identify personal hazards and protective equipment to be used	• High speed ____________ ____________ and saw blades. • Cuts to ____________ and ____________. • ____________ and ____________ injuries from ____________ and ____________ safety ____________ protection. • ____________ clothing – no ____________ cuffs collars shirts or singlets. • ____________ hair. • Injuries to ____________ wear ____________ ____________. • Inhaling ____________ ensure ____________ is fitted and____________.
Safe operating procedure	
Machinery/equipment checks before starting the operation	• Access to emergency ____________ ____________. • Ensure the machine and area is ____________ and ____________ ____________. • All guards are____________ and ____________. • Blade or cutters are sharp, ____________ and ____________ correctly. • Ensure adequate lighting. • Ensure that the ____________ switch is working correctly.
Never	• Attempt an operation if you are ____________ of what you are doing. • Place or move hands ____________ of the cut or ____________. • Stand directly ____________ material being fed into a machine.

During the operation	• Examine the timber, _____________ any _____________ material. • Position _____________ correctly. • Use a _____________ to move _____________ away from blade or cutters. • Disconnect power when changing _____________.
When you finish	• Turn _____________ and isolate machine. • Ensure area is left _____________ and all _____________ are removed or stored correctly.

2 What would a machine maintenance schedule contain?

____

____

____

3 What are the advantages and disadvantages of CNC machines?

Advantages:

____

____

____

____

____

Disadvantages:

____

____

____

____

____

WORKSHEET 4

To be completed by teachers	
Student competent	☐
Student not yet competent	☐

Student name: ______________________

Enrolment year: ______________________

Class code: ______________________

Competency name/Number: ______________________

Answer the following questions.

1 What are two common timber jointing methods used to reduce waste and save timber?

a ______________________

b ______________________

2 Internal flush doors (hollow core) have what material used internally? Explain why this is the case.

____

____

____

3 Window sills should be made from what class of timber? Name two suitable timber species.

a ______________________

b ______________________

4 Select the recommended moisture content range for timbers used in doors and windows from the possibilities below:

a 5–25 per cent

b 9–14 per cent

c more than 18 per cent

5 Select the best correct answer.

When setting up to manufacture doors and windows, which of the following is *not* critical to the layout of the workshop in terms of efficiency?

a The size of car parking required

b The amount of materials storage and racking required

c The amount of space required for dust extraction equipment

d Space required to move material

6 List in the correct order the three main planned stages for making components for windows and doors.

a ______________________________

b ______________________________

c ______________________________

7 What must you never do when using a jointer/buzzer?

____

8 Answer True or False.

Dust extraction must be fitted to all woodworking static machines.

True ☐ False ☐

9 Tick Yes or No.

Is it OK to use 'plain' slotted collars on spindle moulders in Australia?

Yes ☐ No ☐

10 When moulding the end grain of short grain in timbers, how can you prevent break-out or chipping?

____

____

____

WORKSHEET 5

To be completed by teachers

Student competent ☐

Student not yet competent ☐

Student name: ______________________

Enrolment year: ______________________

Class code: ______________________

Competency name/Number: ______________________

1 Using System 32 mm row hole boring principles, what would be the height of a standard kitchen cabinet end and free-standing base (approximately 150 mm high, as shown in the drawing below) using 16 mm white melamine particleboard? Provide the exact height of cabinet end and base height.

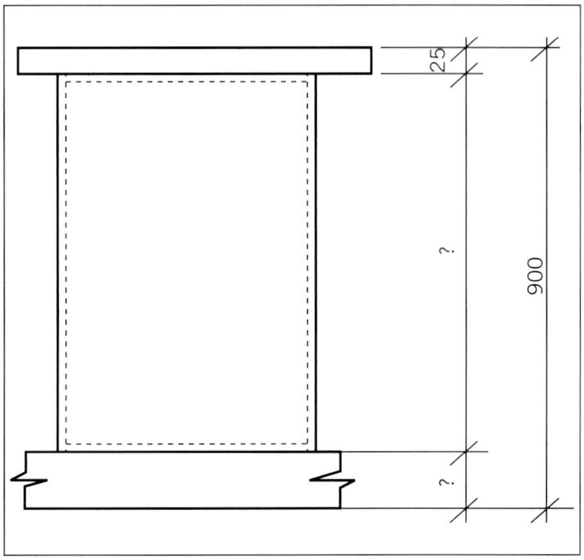

25
?
900
?

____

____

2 Name the Australian Standard that applies to cabinets in the built environment.

____

3 What type of straight cutting router bit provides the best finish on the edge of manufactured boards?

____

4 Which of the following machines is best for high-volume flat-panel cutting?

a Sliding table panel saw

b Flat bed CNC router

c Beam saw

____

5 What are the four key checks when producing flat panel components?

6 List two advantages of using knock-down (KD) fittings to assemble cabinets.

7 Name the three main types of edge tape used on cabinet components.

8 Describe two checks that must be made when using an edge banding machine.

GLOSSARY

A

acute An angle less than 90°.

aggregate A mass of fragments or particles loosely compacted together and used in concrete.

anchor A mechanical device used to fix materials to a masonry surface.

ant cap A galvanised sheet metal plate fixed on top of piers and perimeter walls to help identify termite tracks.

apron piece A piece of timber fixed below the nosing of a window.

architraves Boards or moulding fitted around windows, doors and other wall openings to hide the gap between the walls and frames.

arris A sharp edge formed by the meeting of two surfaces.

B

backing rod A flexible cylindrical foam rod or strip used to fill gaps and provide a backstop for caulking in wet areas; also known as a bond breaker.

BAL Bushfire attack level; a measure of the severity of a building's potential exposure to ember attack, radiant heat and direct flame contact.

balusters The vertical members between the string or landing and the handrail. The handrail and the balusters together are referred to as the balustrade.

barricade A barrier to prevent people's movement through an area.

bead A small section of timber (less than 19 × 19 mm) used to fill gaps or to enhance the appearance of finishes.

bearer (ledger) A horizontal member supported on shores, piers or hangers and carrying joists.

bird's mouth Angled cut in rafters, hips or stair soffits or joists.

bisect To cut exactly in half; for example, to bisect a 90° angle means to create two 45° angles.

block-out A recess used in formwork to prevent concrete filling a small void, hole or penetration in a concrete slab or wall.

bolection A moulding projecting beyond the face of a panel or frame.

brace A horizontal or raking member resisting a compressive or tensile stress in a bracing system.

C

cant An oblique or angled line of a surface.

carcass The cabinet body onto which doors are hung and drawers are fitted.

carriage piece Sometimes called a hanging beam or strongback. Raking members, frequently 100 × 75 mm, placed on top of riser boards and used in conjunction with cleats to prevent the riser boards bending or bulging along their lengths. In stair construction, the carriage piece sits directly below treads and risers in wide stair cases.

casement Any glazed, hinged or movable sash.

casting groove A wide groove on the underside of wide skirting boards, architraves and window sills that helps to prevent cupping and twisting in timber and makes fitting to uneven surfaces easier.

caul A removable form used in shaped formwork.

chamfer A 45° angle, usually on the edge of timber or concrete columns or beams.

chasing The term used for cutting a channel in masonry work to lay cables or plumbing.

cladding The exterior facing of a framed-up building.

cleats Timber members nailed across a number of boards to hold them together; can be a block fixed to a main member to provide a bearing or resist a thrust.

closed strings Have the treads and risers housed into the strings and the top edge parallel to the bottom edge.

clout A type of flat-head nail with a broader head than a regular flat-head nail.

coach bolts (or carriage bolts) Large bolts with round heads, used for fixing wooden panels to masonry or to one another.

competent person Someone who has acquired through training, qualification or experience the knowledge and skills to conduct the task.

control joint An expansion joint or movement joint designed to cope with expansion and contraction of construction materials.

cornice Shaped material (usually plaster) used to hide the gap between ceilings and walls; also may be used in joinery.

coverage Effective cover of boards, by deducting the overlap of boards from the finished width of the board.

crown moulding a plaster or timber mould used as a cornice to hide gaps and decorate joints between walls and ceilings, or to cap moulds on cabinetry.

cupping Timber shrinkage term describing back-sawn distortion.

cut strings Have the top edge cut to the lines of the treads and risers.

D

dado The lower part of a wall, below the dado rail and above the skirting board.

datum In construction, a datum point is a known point of reference from which further measurements can be made. It can be referring to finished floor level, an existing building or another benchmark.

dead loadings Weight of concrete and reinforcing a formwork/falsework.

decking The sheeting to a soffit form.

dormer window A window that is fitted into the roof slope to allow outside access for air and light into a roof space or habitable room. Dormer windows do not usually extend to the fascia line.

F

falsework Props and supports for formwork.

fibre-cement lining Also known as 'fibro'; wall and floor lining material made from wood pulp, cement and other materials. Is a replacement material for asbestos cement sheeting.

fillets Used to evenly space out balusters in handrails and strings.

finger jointing The process of gluing pieces of timber end to end to extend the length of timber sections. Can also be used for corner joints. Also known as a comb or box jointing.

flashing Usually a strip of metal used to stop water penetration where roof surfaces or other roof penetrations such as chimneys exist. In bathrooms, flashings can be PVC, fibreglass or rubber-based materials laid in vertical and horizontal corners to make the area waterproof. Also applied above doors, and above and below windows.

flitch A section of timber (approximately 300 × 150 mm or larger), usually squared up or natural edged, that is used for matching veneers in a sequence.

flush door Flat-faced door.

foot The bottom edge of a string sitting on the floor.

formwork The shutters, sheaths and decks that provide the concrete with its shape.

four light Refers to the number of panes of glass in a window or door. It may be: two light, three light, four light and so on.

French doors Usually a pair of light-framed doors with full-length glass panes.

furring channel Metal furring channel sometimes referred to as 'hat or top hat section' that is used to level off or raise surfaces of another material; for example, roof trusses and walls.

G

going (tread) The horizontal distance measured from the front to the back of the tread (going) minus any overhang from the next tread above. All treads must have a non-slip finish or skid strip close to the front of the nosing. Treads can be wider than the going because of added nosings.

gooseneck A form of handrail joint allowing handrails and landings to connect where the landing's balustrade is set at a different height from the stair balustrade.

GPO General Purpose Outlet or power point; they can be single or double point.

H

handrail The rail fixed parallel above the string, to provide support when climbing or descending the stairs. It may also provide security and support around a landing.

haunched joints The haunch is part of a sash corner joint that prevents the tenon twisting, and fills a groove.

hip The sloping edge formed when the outer ends of two roof planes meet. Opposite of valley.

hob To close off the area between the bath and the walls or to create a wall to support plumbing or to separate the bath from the shower recess.

hollow core door A flush door with plywood, medium density fibreboard (MDF) or hardboard faces on framework.

J

jack rafters Shortened rafters used in gambrel (Dutch gable) and some other roof forms.

jig A device that holds a piece of work and guides the tool operating on it.

joists Parallel lengths of timber (or steel) supporting other building elements such as wall frames. In floors they lay at right angle to bearers and support flooring materials. Ceiling joists support ceiling linings.

K

kerf The width of a saw cut or width of a material that is removed by other cutting processes.

L

landing The area at the top or bottom of stairs or in between two flights.

lateral forces Horizontal forces, such as wind or concrete moving sideways as it is poured.

ledged door A wood door without stiles that is constructed of vertical boards held together by horizontal boards (or ledges).

lintel A structural horizontal member that spans the space or opening between two vertical supports.

live loads Includes the weight of people, tools, machinery and equipment on formwork and scaffolding.

louvre A glass blade opened and closed using a mechanical device.

M

margin line Sets the rise and tread intersection distance from the top edge of the string.

medium density fibreboard (MDF) An engineered wood product made by breaking down hardwood or softwood forest thinning into wood fibres, often mixed with wax and adhesives to form panels.

membrane A thin sheet of material, usually plastic, forming a barrier or lining.

mortise Recess made on a stile to take a fitted tenon.

N

newels Vertical posts, generally at the end of a flight of stairs, to which the string and handrail are connected.

no. off A common term to describe the quantity of items taken off (observed) on working drawings when developing a door or window schedule, hardware and/or materials list.

nosing Overhang of tread on riser.

O

oblique An end not parallel or at right angles to an adjoining frame.

obtuse An angle greater than 90° but less than 180°.

P

pilaster Vertical members used to break up the run of panelling.

pinch bar A bar used to lever materials with an end adapted for pulling nails and spikes.

pitch board Shows the margin, the rise height and the tread width.

plywood The layering of timber veneers (in odd numbers of layers) each at 90° to the other to form flat panels used in construction and furniture production.

post-tensioned concrete Where steel strands or tendons are inserted into ducts or conduits as reinforcement in concrete elements. Once the concrete hardens, tension is applied to the ends of the tendons and locked off.

pre-tensioned concrete Where wire strands or tendons are carefully laid out and tensioned to a predetermined plan and design load in a concrete slab or beam. The ends of steel tendons are released, and the stress is transferred to the concrete element. This is predominantly used in off-site precast elements such a bridge decks, girders and beams.

proud To sit above the surface.

purlin A horizontal, structural timber roof member inserted to support rafters; it is supported by struts back to a load-bearing wall.

Pythagoras' theorem $A^2 + B^2 = C^2$ formula for working out the length of sides in a right angle triangle.

Q

Quirk An angle or space dividing a moulding or architrave around a frame.

R

rails Horizontal members in a framed and panelled door.

raked Angled plane or inclined surface.

reduced levels A surveying term; is a calculated elevation in relation to a particular datum.

reshoring The temporary propping installed below a freshly stripped concrete slab after the slab has deflected under its own weight, transferring the load to the structural columns and walls of the building.

reveal A space between window and door frame and the internal lining material. It may be covered using timber or, if the construction is double-brick, cement render. Often windows come with timber reveals linings attached.

ridge Highest member of a roof. Supports upper ends of rafters and runs parallel to wall plates.

riser Distance from the top of one tread to the top of the next tread.

riser boards The boards that form the vertical face of a step.

roof space Distance between the ceiling and the roof covering.

runners Timbers (75 × 25 mm) used to support the bottom edge of string shutters.

S

sash Moving frame within a window.

scabbling A mechanical process of removing a thin layer of concrete from a structure, using compressed air.

scarf joint Used to join lengths of timbers or to join special shapes. It may be a simple mitre joint or more complex keyed joint.

schedule (door and window) A table outlining all doors and windows required on a job, indicating sizes, types, locations and special requirements.

score To mark or cut a line. The type of tool required is dependent on the material being cut; for example, a utility knife is used to score plasterboard.

scribed A joint where one moulding is cut to the profile of the second.

services The range of non-building materials (such as electrical, plumbing, drainage, air-conditioning ducting and telecoms) required for a building to achieve its function.

sheathing The plywood or flat face of the formwork.

silica A natural mineral found in construction materials such as concrete, bricks, tiles, mortar and engineered stone.

sill The lowest horizontal member in a window frame or door. An entry door sill is often referred to as a threshold.

shutter The sheath plus supporting framework.

skewed gable Roof pitched on a building with an angled end other than 90°.

skillion A single pitch flat sloping roof style.

skirting The boards or mouldings that run around the perimeter of a room at floor level, hiding gaps between the floor and wall.

soffit Underside of an architectural structure such as a balcony, eave or staircase.

soldier wall A short wall placed atop another wall to extend the height of framing.

sole plates Horizontal pieces of timber or other rigid material beneath a short prop or jack to distribute the load from the member above.

spa A bath containing hot aerated water.

spandrels In this book a spandrel is panelling enclosing the space beneath a staircase. Spandrels are also used in structural glazing applications.

sparging The process of cementing up using a 'dry' mortar mix.

spiral stair A flight of stairs, circular in plan, winding around a central post, with steps that radiate from a common centre.

sprung studs Studs that curve edgewise.

staging Temporary platforms arranged as a support for different levels of scaffolding.

starter bars Pieces of steel reinforcing that are set into footings, slabs or walls in order to provide a tying point for other reinforcement bars. They have to overlap if constructing a joining wall, slab, column or beam.

stiles Vertical members in a framed and panelled door.

stirrup head A means of adjustment that can be used at either the top or bottom of a scaffold support structure.

stop cocks An externally operated valve regulating the flow of water or gas in a pipe.

storm mould Fitted to the junction between masonry walls and window frames to cover gaps and prevent moisture penetration.

string A length of timber (or sometimes steel) that runs diagonally between floors to support stair treads and risers.

strongback A length of timber fixed with cleats above the riser board to prevent bowing of the riser board in flights wider than 1 m. Strongbacks are sometimes referred to as hanging beams.

strut A timber roof framing member that transfers roof loads through rafters and purlins down to load-bearing walls.

substrate A term for any materials used to support a surface finish.

swarf The fine chips produced by a machining process when working with timber, timber-based products, metals or other materials.

T

tendon A steel cable used for pre- and post-tensioning in concrete.

tenon Projection cut on the end of a rail to fit in a mortise.

threshold A strip of wood or stone forming the bottom of a doorway and crossed when entering a house or room.

throat The distance in millimetres from the rise and going intersection to the soffit, measured at 90°.

tom A temporary prop usually made from timber. The sizing depends on the load being supported.

tongue A projecting piece machined on the edge of a piece of timber to fit in a groove, ensuring faces are aligned and joints are strengthened.

transom A horizontal door or window frame member separating doors from windows above, or separating glazed sections in shopfronts or glass curtain walls.

trimmer A horizontal member used in roof framing between rafters and barge rafters and boards. Trimmers also can be used to provide fixing points for plumbing and electrical fitments.

V

valley The sloping edge formed when the inner ends of two roof planes meet. Opposite of hip.

vanity A cabinet that sits below a basin in bathrooms and hides plumbing and drainage works.

W

walers Horizontal member used in shutters to keep concrete forms from bulging (also called wales).

waling piece Horizontal member spanning rafters in gambrel or Dutch gable roofs to form a ledge to attach jack (shortened) rafters.

wall strings Located on the wall side of the flight and are usually closed.

wedges Used in pairs to tighten, loosen, raise or lower props or other members.

weep holes A small opening that allows water to drain from within a masonry wall structure or cavity.

well string Located on the open side of the flight and may be closed or cut strings.

winder Treads within a straight flight, used to change direction of the stairs around corners while continuing the slope.

wreath A twisted and curved handrail section that allows the handrail to maintain the same profile as it rises and changes direction.

Y

yoke A collar used in drainage works to provide a lead for waste grates and drains. The yoke can provide a leak-control waste system to enable moisture to drain if it gets below the floor tiles.

INDEX

Note: Page numbers in **bold** represent defined terms.

C

T

U

V